Nanostructured Materials Preparation via Condensation Ways

Anatolii D. Pomogailo
Gulzhian I. Dzhardimalieva

Nanostructured Materials Preparation via Condensation Ways

Springer

Anatolii D. Pomogailo
Gulzhian I. Dzhardimalieva
Russian Academy of Sciences Inst.
Problems of Chemical Physics
Chernogolovka, Moscow region, Russia

ISBN 978-90-481-2564-7 ISBN 978-90-481-2567-8 (eBook)
DOI 10.1007/978-90-481-2567-8
Springer Dordrecht Heidelberg New York London

Library of Congress Control Number: 2014947802

Printed on acid-free paper

Springer is part of Springer Science+Business Media (www.springer.com)

Dzhardimalieva Gulzhian dedicates her work to Prof. Anatolii D. Pomogailo on the occasion of his 75th anniversary.

Preface

We are very pleased to present this book to the English speaking readers. Any author is contented to have his book published by the world-known *Springer* publishing house, which has published most fundamental scholar works. It is also important for us because we have already published our two previous monographs in Springer Series in Materials Science: A.D. Pomogailo and V.N. Kestelman, *Metallopolymer Nanocomposites*, (Springer, Berlin Heidelberg: New York, 2005) and A.D. Pomogailo, G.I. Dzhardimalieva, and V.N. Kestelman, *Macromolecular Metal Carboxylates and Their Nanocomposites*, (Springer, Berlin Heidelberg, 2010).

Publishing of this book was initiated by Sonia Ojo, Senior Editor of Springer UK publishers, who proposed to us writing this book in April 2008 at the Congress of American Chemical Society in New Orleans, and we are endlessly thankful to her for tolerance to delays which were inevitable under conditions of permanent time pressure during writing of this book.

A credit for making this book cannot be given just to the authors. We are grateful to our colleagues for daily help. Our colleague, Dr. of Chemical Science Aleksander Rozenberg (1942–2009) was at the upstream of these studies. We only had time to discuss a plan of this work with him.

We hope that this book will be useful for novices in science, because it has relatively simply written introductions to each chapter, and also for specialists experienced in nanoparticle and nanocomposite development, since it contains a great deal of references to original works in highly rated journals. In this connection we apologize to the authors whose works are not referred to in this book, because their remarkable works are inadvertently missed in this publication for the sake of avalanche of publishing activity (predominantly since 2000).

Chernogolovka, Moscow Region
Russian Federation
May 2014

Anatolii D. Pomogailo
Gulzhian I. Dzhardimalieva

References

A.D. Pomogailo, V.N. Kestelman, *Metallopolymer Nanocomposites*. (Springer, Berlin/ Heidelberg/New York, 2005)

A.D. Pomogailo, G.I. Dzhardimalieva, V.N. Kestelman, *Macromolecular Metal Carboxylates and Their Nanocomposites*. (Springer, Berlin/Heidelberg, 2010)

Contents

1 **General Introduction** 1
References 11

2 **Reduction of Metal Ions in Polymer Matrices as a Condensation Method of Nanocomposite Synthesis** 13
2.1 Formation of Metal Nanoparticles During Chemical Reduction . . . 13
2.1.1 Nucleation 14
2.1.2 Growth 17
2.2 Chemical Methods of Reduction 23
2.2.1 General Characteristics of Low Molecular Reducing Agents 24
2.2.2 The Main Factors Controlling the Reduction Process 28
2.2.3 Reduction in Statu Nascendi in Polymer Matrix 40
2.2.4 Polymers as Reducing Agents and Templates 62
2.2.5 Co-reduction (Bi-metal Nanocomposites) 71
2.3 Electrochemical Reduction 75
2.3.1 Electrochemical Deposition 75
2.3.2 Direct Electric Reduction of Metal Ions in Electrolyte Solution 78
2.3.3 The Method of Anode Oxidizing Electrolysis 78
References 80

3 **Physical-Chemical Methods of Nanocomposite Synthesis** 91
3.1 Synthesis of Nanocomposites by Microwave Irradiation 92
3.1.1 Basic Principals 92
3.1.2 Microwave Heating in Synthesis of Colloid Nanocrystals . . . 94
3.2 Photo Chemical Reduction 97
3.2.1 The Main Approaches to the Synthesis of Matrix-Stabilized Nanoparticles by Photochemical Irradiation 97

3.2.2 Photochemical Reduction in Solutions 102
3.2.3 Photo Reduction on Solid Substrates 108
3.2.4 Photo Induced Processes in Nanoparticles 111
3.3 Radiation-Chemical Reduction 113
3.3.1 Kinetics and Mechanism of Radiation-Induced Synthesis of Nanocomposites 115
3.3.2 Fabrication of Polymer Nanocomposites *In Situ* 117
3.3.3 γ – Radiolysis in Synthesis of Semiconductor Nanomaterials 118
3.3.4 Synthesis of Bimetallic Nanocomposites 120
3.3.5 In Situ Radiolysis 121
3.4 Sonochemical Reduction 122
3.5 Physical and Chemical Deposition from the Vapor Phase 127
3.5.1 Plasmochemical Synthesis 128
3.5.2 Evaporation and Spraying Methods 129
3.5.3 Laser Ablation 132
References 133

4 Physics and Chemistry of Sol-Gel Nanocomposites Formation 141
4.1 General Characteristics of Sol-Gel Reactions 142
4.2 Combination of Reactions of Polymerization and Sol-Gel Synthesis of Nanocomposites *in-situ* 149
4.3 Formation of Inorganic Precursor in Presence of Organic Polymers: Polymer Sol-Gel Synthesis 161
4.4 Hybrid Nanocomposites Based on Heteroelemental Oxide/Oxide Ceramics 177
4.5 Morphology and Fractal Models of Hybrid Nanocomposites 190
References 195

5 Physical Chemistry of Intercalated System 205
5.1 Composition, Structure, and Intercalation Properties of Layered Materials 207
5.2 General Characteristics of Intercalated Systems 221
5.3 Special Features of Polymerization of Monomers in Interlayer Space 225
5.3.1 Emulsion Polymerization in Interlayer Space 226
5.3.2 Interlayer Blocked Radical Polymerization and Copolymerization 226
5.3.3 Interlamellar Catalytic Polymerization 229
5.3.4 Intercalation Assembling 231
5.3.5 Polymerization with a Ring Opening 233
5.3.6 Redox-Intercalation Polymerization 233
5.4 Hybrid Nanocomposites Produced by Direct Intercalation of Macromolecules in *Host* Layers 241
5.4.1 Intercalation of Polymers from Their Solutions 242
5.4.2 Intercalation Composites Produced from Polymer Melts 245

5.4.3 Nanocomposites Based on Non-polar Thermoplastics Produced by Reactor Mixing 247
5.4.4 Composites Formed in Melt of Polar Matrices 252
5.4.5 Melt Intercalation of PVC 256
5.4.6 Nanocomposites PEO-Silicate Obtained by Melt Technique 257
5.4.7 Intercalated Nanocomposites Formed in Polyamide Melts 259
5.4.8 Intercalated Polymer Mixtures 260
5.4.9 Intercalated Network Composites 261
5.5 Supramolecular Assembling in Nanolayered Materials 267
5.6 Structure of Inclusion Nanocomposites Metal Chalcogenide-Polymer 271
References 276

6 Thermolysis of Metallopolymers and Their Precursors as a Way for Synthesis of Nanocomposites 287
6.1 Kinetic Approaches to the Investigation of Thermolysis of (Metallo)polymer Systems 289
6.2 General Characterization of the Principal Methods of Thermolysis of (Metallo)polymers and Their Precursors 294
6.2.1 Thermogravimetric Methods 296
6.2.2 Linear Pyrolysis Method 298
6.2.3 Volumetric Methods 300
6.2.4 Thermolysis Induced by High-Energy Radiation 302
6.2.5 Spray Pyrolysis 305
6.2.6 Calorimetric Bomb as a Method to Study the Enthalpy of Combustion (the Enthalpy of Formation) of Nanoparticles 308
6.3 Thermolysis of Metalloprecursors 308
6.3.1 Metal Nitrates 309
6.3.2 Thermolysis of Metal Carboxylates 312
6.3.3 Thermolysis of Metal Complexes 322
6.4 The Conjugate Thermolysis – Nanoparticle and Polymer Matrix Formation Process 324
6.4.1 Thermal Transformations of Salts of Unsaturated Carboxylic Acids 325
6.4.2 Co-condensation of Metal and Monomer Vapours 333
6.4.3 The Methods of Formation of Metal – Carbon Nanocomposites 334
6.5 Thermolysis of Metal-Containing Monomers in the Frontal Mode 338

6.6 Thermolysis of Precursors in the Polymer Matrices 342
6.6.1 Effect of Metal-Containing Components on the Destruction of the Polymer Matrix 344
6.6.2 Metal Salt – Polymer Matrix System 345
6.6.3 Thermolysis of Complex Compounds in Polymer Matrices 347
6.6.4 Thermolysis of Organometallic Compound – Polymer Matrix Systems 352
6.6.5 Metal Thiolates and Sulfides in the Polymer Matrix 354
6.6.6 Polymer-Mediated Thermal Synthesis of Mono- and Multimetallic Alloys and Ceramics 358
6.6.7 Thermolysis of Metal Carbonyls in Polymer Matrices 364
6.6.8 Comparison of Different Methods of Solid-State Thermolysis of Metallopolymers 367
6.7 Computer Modelling of the Kinetics of Nanoparticle Formation During Solid-State Thermolysis 370
References 375

7 Bionanocomposites Assembled by "From Bottom to Top" Method 389
7.1 Bioreducing Agents in the Synthesis of Nanocomposites 390
7.1.1 Vegetable Biomasses and Extracts as Reduction Agents of Metal Ions 391
7.1.2 Microbiological Synthesis of Metal Nanoparticles 395
7.1.3 Fermentative Synthesis in the Formation of Nanoparticles 402
7.2 Sol-Gel Process as a Way of Production of Template-Synthesized Bionanocomposites 403
7.2.1 Biomedical Nanocomposites 406
7.2.2 Biocomposites as Means for Drug Transfer 411
7.2.3 Biomineralization and Bioconcentration 414
7.3 Intercalation Processes in Development of "Green" Nanobiocomposites 417
7.3.1 Biodegrading "Green" Bionanocomposites 421
7.3.2 Nanocomposites Based on Polysaccharides 425
7.3.3 Biomedical Application of Hybrid Nanocomposites 430
References 436

Conclusions and Future Prospects 449

Index 451

Author Bios

Prof. Anatolii D. Pomogailo, Dr. Sci. (Chem.) is the Head of Laboratory, Institute of Problems of Chemical Physics Russian Academy of Sciences and Prof. of Department of Applied Mechanics, Moscow Aviation Institute (National Research University). He was born in 1939. He graduated from Odessa State University, Department of Physics and Chemistry of Polymers. He received his Ph.D. in 1970 from the Institute of Chemical Sciences (Kazakhstan Academy of Sciences). Since 1972, Dr. Pomogailo is working in Institute of Problems of Chemical Physics, Russian Academy of Sciences. He received his Doctorate in Chemistry in 1981 from the Institute of Chemical Physics, Russian Academy of Sciences in Moscow. He received the title of Professor in 1991. Since 1982 he has been the head of a group involved with metal-containing polymers that was transformed in the laboratory in 1993. The scope of his scientific interest is: physics and chemistry of ultradispersed state, metallopolymer nanocomposites, immobilized-metal complexes, catalysis by macromolecular complexes, (co)polymerization of alfa-olefines, metal-containing monomers and polymers on their base. Professor A. Pomogailo is a well-known specialist in the field of the chemistry of high molecular compounds, including organometallic compounds and metallocomplex catalysts and materials science. Dr. Pomogailo is author of over 450 articles, 14 monographs including ones published by Springer (A.D. Pomogailo, V.N. Kestelman. Metallopolymer Nanocomposites. Springer. Berlin Heidelberg, 2005; Anatolii D. Pomogailo, Gulzhian I. Dzhardimalieva. Macromolecular metal carboxylates and their nanocomposites, Springer, Heidelberg, 2010) and about 100 patents. Prof. Pomogailo has been as invited professor and researcher of Lawrence Livermore National Laboratory (1989), Auburn University (2001),

Warsaw Technology University (2002–2008), Bremen University (2004, 2007), Turin University, Italy (2008), Institute of composite and biomedical materials CNR, Neapol, Italy (2011, 2013).

Gulzhian I. Dzhardimalieva, Dr. Sci. (Chem.), Ph.D. is the Head of Laboratory, Institute of Problems of Chemical Physics Russian Academy of Sciences and Professor of Department of Applied Mechanics, Moscow Aviation Institute (National Research University). She received her Ph.D. in 1987 from the Institute of Chemical Physics, RAS in Moscow, and her Doctorate in Chemistry in 2010 from Institute of Problems of Chemical Physics RAS in Chernogolovka. The scope of her scientific interest is: metal-containing monomers and polymers on their base, macromolecular metal carboxylates, coordination polymers, and metallopolymer nanocomposites. Dr. Dzhardimalieva is author of about 100 articles and 2 monographs. She has been as invited researcher of Auburn University, USA (2001), Warsaw Technology University, Poland (2002, 2011), Bremen University, Germany (2007), Turin University, Italy (2008), and Institute of composite and biomedical materials CNR, Neapol, Italy (2011, 2013).

Abbreviations

ABS	Acrylonitrile-butadiene-styrene copolymer
ABS	Aminobenzenesulfonate
AcAc	Acetylacetone
AFM	Atom force microscopy
AIBN	N,N′-azo-bis-isobutyronitrile
APS	Ammonium peroxydisulfate
AN	Acrylonitrile
An	Aniline
AOT	bis(2-Ethylhexyl) sulfosuccinic acid
APN	Allylnovolak resin
ATRP	Atom transfer radical polymerization
bipy	Bipyridine
BMA	n-Butyl methacrylate
BPO	Benzoyl peroxide
BTDA	3,3′,4,4′-Benzophenone dianhydride tetracarboxylic acid
CDMI	Caprolactam-blocked methylene diisocyanite
CEPD	Copolymer of ethylene-propylene-diene
CMC	Carboxymethylcellulose
CMC	Critical concentration of micelle formation
CMPT	Combined polymerization of metal-containing monomer with the thermolysis
Cp	Cyclopentadienyl ligand
ChT	Chymotrypsin
CVD	Chemical vapour deposition
DABSA	2,5-Diaminobenzene sulfonic acid
DCC	Dicyclohexyl peroxydicarbonate
DDS	3,3′ Diaminodisulphonephenyl
DMAAm	N, N-dimethylacrylamide
DMAC	N,N(dimethylacetate amide)
DMDA	N,N-(dimethyl)dodecyl ammonium

DMSO	Dimethylsulfoxide
DNA	Deoxyribonucleic acid
dppp	1.3 Bis(diphenylphosphino)propane
DSC	Differential scanning calorimetry
DTA	Differential thermal analysis
DTG	Differential thermogravimetry
ELS	Epoxy resin layered silicate
EPy	2-Ethylpyridine
EVA	Ethylene-vinyl acetate
EVOH, poly (E-*co*-VA)	Copolymer of ethylene with vinyl alcohol
FC	Field cooling magnetic susceptibilities
FCC	Face-centered cubic lattice
FLG	Few layer graphenes
FMR	Ferromagnetic resonance
FP	Frontal polymerization
FTIR	Fourier transformation infrared spectroscopy
FWHM	Full width at half maximum
HDPE	High density polyethylene
HEMA	2-Hydroxyethyl methacrylate
HEPES	4-(2-Hydroxoethyl)-1-piperazineethanesulfonic acid)
HEX	Hexagonally packed cylinders
HF	High-frequency
HMDSO	Hexamethyldisiloxane
HMEM	2-[*p*-(2-Hydroxy-2-methylpropiophenone)]-ethyleneglycol methacrylate
HTSC	High-temperature superconducting ceramics
IPC	Isophthaloyl chloride
IPN	Interpenetrating polymer network
Irgacure-2959, I-29591	[4-(2-Hydroxyethoxy)phenyl]-2-hydroxy-2-methyl-1-propan-1-on
ITO	Indium-tin oxide
KZTPP1	(Cyclo[dineopentyl(diallyl)] pyrophosphate dineopentyl (diallyl) zirconate)
LAM	Lamellae
LCST	Lower critical solution temperature
LDHs	Layered double hydroxides
LDPE	Low density polyethylene
LGC	Layered graphite compound
LP	Linear pyrolysis
MA	Maleic anhydride
MAA	Methacrylic acid
$MAcr_n$	Metal acrylate
MCM	Molecular sieves

MD	Molecular dynamics
$MFum_n$	Metal fumarates
MHEC	Hydroxyethylcellulose
$MItac_n$	Metal itaconates
MLAM	Modulated lamellae
MMA	Methylmethacrylate
$MMal_n$	Metal maleinate
MMT	Montmorillonite
MRI	Nuclear magnetic resonance imaging
MOCVD	Metal-organic deposition from gas phase
MVD	Multifunctional vectors of drugs
NADPH	Nicotinamide adenine dinucleotide phosphate
NIPA	N-Isopropylacrylamide
NLS	Novolak resin layered silicate
NMP	N-methyl-2-pyrrolidone
ODA	4,4(Oxydianiline)
OMMT	Organically modified MMT
ORMOCER	Organically modified ceramics
ORMOSIL	Organically modified silicates
OTs	*n*-Toluene sulfonate
P4VPy	Poly-4-vinylpyridine
PA	Polyamide
PA-6	Poly-ε-caprolactam
PA	Pyromellitic anhydride
PAA	Poly(acrylic acid)
PAMAM	Polyamidoamines
PAAm	Poly(acrylamide)
PAI	Polyamidoimide
PAmA	Polyamic acid
PAN	Polyacrylonitrile
PAn	Polyaniline
PB-PEO	Poly(butadiene-b-ethylene oxide)
PBT	Polybutylene terephthalate
PCL	Poly-ε-caprolactone
PDMA	Poly(2,5-dimetoxyaniline)
PDMAC	Poly(diallyldimethylammonium chloride)
PE	Polyethylene
PE-b-PMAA	Polyethylene-block-polymethacrylic acid
PEG	Polyethylene glycol
PE-*gr*-MA	Polyethylene-grafted-maleic anhydride
PEI	Polyethylene imine
PEO	Polyethylene oxide
PETP	Polyethylene terephthalate
PHEMA	Polyhydroxyethylmethacrylate

PHAs	Polyhydroxyalkanoates
PHPS	Perhydropolysilazane
PI	Polyimide
PMMA	Poly(methyl methacrylate)
PMPS	Poly(methylphenylsilane)
PLA	Poly(lactic) acid
PLAM	Perforated lamellae
PMA	Polymuconic acid
PMMA	Poly(methyl methacrylate)
PMVK	Poly(methyl vinyl ketone)
PNIPAAm	Poly(N-iso-propyl-acrylamide)
PNIPAAm/MACACS	Poly[N-iso-propyl-acrylamide-(maleate carboxymethyl chitosan)]
PP-gr-MA	Polypropylene-grafted-maleic anhydride
POM	Polyoxometalates
POSS	Polyhedral oligosesquioxane
POSS-MA	Methacryloyl-functioned POSS
PP	Polypropylene
PPN	Penetrating polymer network
PPO	Polypropylene oxide
PPy	Polypyrrole
PS	Polystyrene
PS-g-(PMA-b-PMMA)	Polystyrene-grafted-(polymethacrylic acid-block-poly(methyl methacrylate)
PS-PMAc	Poly(styrene-block-methacrylic acid)
PS-P4VP	Poly(styrene-block-4-vinylpyridine)
PSS	Poly(sodium 4-styrenesulfonite)
PSS	Poly(styrene sulfonic acid)
PTFE	Polytetrafluoroethylene
PU	Polyurethane
PVA	Polyvinyl acetate
PVA	Polyvinyl alcohol
PVB	Polyvinyl butyral
PVC	Polyvinylcarbazole
PVC	Polyvinyl chloride
PVdF	Polyvinylidene fluoride
PVI	Polyvinylimidazole
PVM-MA	Poly(methyl vinyl ether-co-maleic anhydride)
PVP	Poly(vinylpyrrolidone)
PVPy	Polyvinylpyridine
RAFT	Reversible addition fragmentation chain transfer
RAPET	Reaction under autogenic pressure at elevated temperature
RNA	Ribonucleic acid

SAN, poly (St-*co*-AN)	Copolymer of styrene with acrylonitrile
SAXS	Small angle X-ray scattering
SD	Spinodal decomposition
SGA	Self-generated atmosphere
SPR	Surface plasmon resonance
SPU	Segmented polyurethane
SPIOs	Superparamagnetic iron oxides
TA	Thermal analysis
TAF, Teflon-AF	Copolymers of tetrafluoroethylene and 2,2-bis(trifluoromethyl)-4,5-difluoro-1,3-dioxole
TEGM	Triethylene glycol methacrylate
TEOS	Triethoxysilane
TESPMA	N-Triethoxysilylpropylmaleamic acid
TG	Thermogravimetric analysis
TGDMA	Triethylene glycol dimethacrylate
THF	Tetrahydrofuran
TMOS	Tetramethoxysilane
TSI-PN	Prepolymer with triethoxysilylpropylic groups
TSPU	Thermosetting polyurethane
UCST	Upper critical solution temperature
USPIO	Ultrafine superparamagnetic iron oxides
VBS	Vinylbenzenesulfonate sodium salt
WAXS	Wide-angle X-ray scattering
XRD	X-ray powder diffractometry
YSZ	Yttrium stabilized zirconia
Z	Aggregation number
ZFC	Zero field cooling magnetic susceptibilities

Chapter 1
General Introduction

The nanocomposite science concerning the class of composite materials, whose typical feature is nanometer size of their structural elements (metal particles, their oxides, chalcogenides, carbides, nitrides, etc.) has appeared in the last decades of twentieth century at the junction of different fields of knowledge: physical, organic, colloid, polymer chemistry, biology, and materials science. One of the driving forces of this is that extensive ways of materials science development have undergone essential depletion, giving way to searching for methods of improvement and modifying of materials properties, miniaturization of their structural elements. Along with continuous perfection of the known composites, which provide technical and economic effect due to unique combination of properties, some tends appear in development of new materials actively interacting with environment. These materials are called "intellectual", "smart", "wise", "integral", etc. due to their ability to "sense" their physical state, environmental impacts, and react in special way to these "sensations", i.e. self-diagnose nucleation and growth of a defect, make its correction, and stabilize a material state in critical zones. Due to variety of properties of the "intellectual" materials they can be used in different structural elements, for example in air-space technique (cases, fairings, compartments, frictional units, etc.), and they will allow one to control and predict a state of different articles and structures in a required time and even in difficultly accessed parts, to improve significantly life time of the systems, as well as reliability [1].

Successes in chemistry and technology of nanometer materials science have brought to outbreaking development in many fields in science and industry. General laws of formation of nanoparticles, methods of their optimization and manipulation, special methods of fabrication of articles based on these materials, main fields of their applications have become clear. It follows from expertise of specialists that in the following 20 years 90 % of contemporary materials used in industry will be replaced by new ones, in particular, "intellectual", which will allow development of structures and devices, and will provide technical progress in twenty-first century. We shall highlight three main advantages of the nanocomposites as compared to conventional composites: they have lighter weight due to lower content of a filler; lower cost due to

A.D. Pomogailo, G.I. Dzhardimalieva, *Nanostructured Materials Preparation via Condensation Ways*, DOI 10.1007/978-90-481-2567-8_1

less amount of a filler, improved properties (including mechanical, thermal, optical, electric, barrier, etc.) [2]. While having higher operating characteristics, they envisage weight reduction of articles by 30–50 %. Under given operating conditions and increase in loading an element based with the same weight on these materials provides increment in rigidity by 60–80 %, increment in reliability by 20–25 %, and increase in guarantee term by 60–70 % as compared to conventional composites. The key advantage of nanocomposites is that a possibility of improvement of their properties without change in technological conditions, and procedures of surface modifying can be quite easily integrated into the technological process. Providing that, the surface treatment methods are preferred, because often particularly the surface characteristics determine quality of a whole article [3]. Besides, with account for economic aspects, modifying of a surface is more attractive as compared to the ways of change in bulk characteristics of materials and articles. This approach makes it possible to use most efficiently combination of properties of a basic material and the modified surface layer or modification of articles.

Though definitions "nanocomposites", "nanomaterials", "nanofillers" used for description of materials are introduced relatively recently, these materials exist for a long time. For example, soot is used as reinforcing filler for rubber since 1904. Roman masterpiece of fourth century known as Lykurgus cup is made of glass with suspended nanometer (~70 nm) gold and silver particles in the ratio 3:7. The cup shows unusual properties of dichroism, which appears in green in reflected rays and in red in transmitted light [4]. It has recently been shown that Maya blue dye used from ancient times is nanostructural hybrid material: needle-shaped palygorskite crystals $H_{10}Mg_2Al_2Si_7O_{24}.4H_2O$ form superlattice with the period 1.4 nm, and Mg crystal nanoparticles are impregnated in this matrix. Beautiful blue light appears only in presence of Mg nanoparticles, indigo molecules, and palygorskite superlattice. Sizes of the nanoparticles are also important, in many cases it can be found by a change in color of solution. For example, in the systems with Faraday red gold sol stabilized with formaldehyde color of the solution changes from red to blue due to aggregation of the nanoparticles [5]: red (Au particle size is 23.7 ± 0.9 nm), violet (24.3 ± 0.7 nm), lilac (32.5 ± 1.4 nm), and blue (33.2 ± 2.0 nm). Since 1940 SiO_2 nanoparticles are widely used in many articles, and since 1960 metal nanopowders have been developed [6]. In 1989 [7] it was shown that compacts of fine crystallites (~10 nm) have unique physical properties, which can be used in engineering. The term "polymer nanocomposite" has a clear definition. Nanocomposite is a product of combination of at least two different materials with a distinct interface boundary; at least one of them should have nanometer sizes (1–100 nm) no less than in one direction. This also concerns distances between layers and nets formed by polymer and inorganic ingredients [8]. In other words, polymer nanocomposite is a two-phase material in which organic and inorganic phases are mutually distributed at the nanolevel. Although dimensional range of structural elements of the particles is to some degree conditional, it is comparable with correlation radius of one or other chemical or physical phenomenon (for example, free path of electrons, phonons, coherence length in a superconductor, sizes of magnetic domain, or nuclei of a new phase, etc.), in these systems so called dimensional effects manifest themselves. Nowadays many materials can be obtained at the nano-level with usage of high-energy impacts, electric deposition, spray-pyrolysis, gas phase, vacuum and sol-gel technique, hydrothermal and aerosol processes, intercalation, and micellar templates, chemical reducing agents, etc.

Hybrid organic-inorganic nanostructural materials occupy more and more free space remained between inorganic and polymer chemistry and biology, progress in this area depends significantly on competence of materials scientists for development of contemporary nanomaterials, the task is not only in creation of nanostructures, and in reaching the best knowledge of properties of these ensembles, but in processing and controlling of their assembling in any form, in knowledge of its structure at different spatial levels. Thus, the main key to development of new hierarchically organized hybrid materials is not only in control of the origin, but also of sizes of the nanoparticles, availability of interior interfaces of a nanocomposite. Despite variety of methods of production of nanocomposites it has been found that all versatility of synthetic approaches to preparation of nanometer particles and their self-assembling can be reduced to two principally different ways: "top-down" (descending way) and "bottom-up" (ascending way). The first of them appears in various ways of grinding of coarse particles to nanometer, the essence of the "bottom-up" way is in assembling of nanoparticles from different atoms (or ions with the following reduction) to nanoparticles of a given size in presence of polymer matrix (or its precursor). The latter method, which is considered in this monograph, is more widely used due to its variability and potential abilities than the dispersion method. A researcher can predict in advance properties of obtained nanoparticles: he can choose a content and properties of initial components, stabilizing agents, can estimate their role, predict conditions of nucleation and growth of the nanoparticles at each stage, thus constructing designed materials. Properties, which can be reached in these materials, depend strictly in synergetics between organic and inorganic origin of components and are, undoubtedly, defined by nano/microstructure and degree of its ordering.

In turn, formation of metal-polymer nanocomposites can be implemented by two main methods: "wet" and "dry". The first of them is characterized by one component (or at least at one stage of the multistage process) being used as a solution or dispersion in a solution. The dry method is characterized by entire elimination of a solvent from the technological process. Due to poor solubility of the components, and in some cases chemical instability of them, and due to difficulties of the following removal of some solvents, including ecological aspects, the "dry" method of producing metal polymer nanocomposites is preferred.

Most scientific works are focused on synthesis of polymer nanocomposites and studying of their physical and mechanical characteristics required by operating conditions. Nanocomposites based on thermoplasts or reactoplasts reinforced by nanometer particles or intercalated layers, are actively studied. Nanofillers can present in significant amounts in the interface zone providing strong interaction due to interatomic binding forces: Van-der-Waals in chemical (metal) bond, while in the case of micron particles the fillers concentration in a transition layer is considerably lower. Realization of chemical bond in a contact is a thermomechanical activation process, which goes due to fluctuations of groups or individual atoms in a solid phase topochemical reaction. A structure of semi-crystalline polymer matrix depends not only on prehistory of the treatment, but on presence of nanoparticles having effect on kinetics of crystallization and final morphology. Understanding of the composite reinforced by nanoparticles phenomenon requires profound studies, and will advantage reinforcement of more complicated anisotropic structures in polymer composite systems.

Though many fields of their application are based on properties of individual nanoparticles (sensors, medical diagnostics, homogeneous catalysis, etc.), there are many important fields in which self-organized nanoparticles are needed (nanoelectronics, optoelectronics, heterogeneous catalysis, etc.). The same also concerns bio-inspired strategies of imitation ("simulation") of growth in biomineralization processes, development of innovated multi-scaled structural hybrids (from nano- to multimeter scales) hierarchically organized at the level of structures and functions.

Historically reduction of metal ions to nanoparticles is related to Faraday, who discovered chemical reduction of gold ions in presence of gelatin in 1857, which has still a great value, another important stage in development of this approach is connected with Turkevich name [9], who for the first time realized reproducible synthesis of Au nanoparticles (20 nm) via reaction of cytrate reduction of $[AuCl_4]^-$, etc. [5]. Among the most widely used contemporary methods there is "diborane" method of production of $Au_{55}(PPh_3)_{12}Cl_6$ (1.4 nm) nanoclusters stabilized with phosphine ligands, the method of reduction by alcohols (so called polyolic process) with usage of polyvinyl pyrrolidone, polyvinyl alcohol, etc. as stabilizers.

On the whole, synthesis of nanoparticles of a given diameter with almost monodisperse size distribution and low degree of agglomeration is limited by two boundary conditions: precise control over nucleation and growth of a nucleus and efficient suppression of agglomeration. The most important task of increasing resistance to aggregation of highly reactive nanoparticles with highly developed surface is their surface passivation by different stabilizers. At early stage of the studies low molecular organic compounds (carbonic acids, alcohols, amides, etc.) and natural polymers (gelatin, cellulose, starch, gummiarabic, agar-agar, etc.) were used. A new stage in materials science deals with usage of synthetic polymers: a possibility of steric stabilization is caused by the fact that spatial dimensions of even relatively low-molecular compounds are comparable with the radius of action of London attraction forces or even greater. On the one hand, regulation of stability of nanoparticles by polymer surfactants is the most important part of the polymer colloids modern science which studies laws of formation of disperse systems with highly developed interface surface, various surface phenomena on interface boundaries and adsorption of macromolecules from liquid media on solid surfaces [10]. As a result of steric stabilization the nanoparticles will be surrounded by a protective barrier, a continuous layer of solvated polymer links, and a colloid system becomes unlimitedly stable as long as the protective layer remains undamaged. In the systems with high concentration of disperse phase thermodynamically stable spatial structures form spontaneously as a result of particles linking in dispersion medium: liophobic nanometer particles are allegedly arranged in a web of protective polymer. On the other hand, at this stage already the main demands to structuring and optimizing of metal polymer nanocomposites can be taken into account. Chemical reduction of metal ions in water and non-water solutions in presence of polymers playing a role of stabilizing matrix and/or reducing agents is the most widely spread method in strategy of condensation synthesis of nanocomposites. Very attractive are the methods simultaneously using a stabilizing polymer ligand also as a reducing agent. Advantages of this approach are obvious. It is replacement of some toxic reducing agents, lower cost, higher efficiency, etc. Most often for these purposes are used polymers whose molecules contain a great deal of different functional groups (polysaccharides [11], humic substances [12], peptides, etc.). Promising systems for these purposes are block-co-polymers, polymer linked hydro-gels working as nanoreactors in synthesis of colloid metal particles.

One of universal condensation methods of fabrication of nanoparticles stabilized by oxide inorganic or polymer matrix is sol-gel method based on hydrolysis and condensation of metal alkoxides including different modifications of the method. A special interest is in materials, which are obtained by combination of sol-gel chemical methods with aerosol and spray processes, combination of reactions of polymerization and sol-gel synthesis in which inorganic phase, monomers and polymer links are captured inside oxide net. These self-organizing systems are characterized by strong interface interactions between inorganic and polymer components, they are convenient methods for preparation of organic-inorganic nanocomposites of different types. Polymer-inorganic materials in which nanoparticles can reach 10 nm have high mechanical strength and thermal stability. They have already found their place in various contact lenses, optical wave-guides, printing lithography, heat-resistant coatings, memory and printing units, chemical filters, solid electrolytes, biosensors, semiconductors, catalysts of different reactions, etc. The sol-gel process is very flexible and versatile by types of possible reaction products, motivation of this direction is structuring at the stage of fabrication of materials with given properties, in the recent years special attention has been focused on composites based on heteroelemental ceramics.

The methods of production of hybrid nanocomposites from polymer solutions and *in situ* polymerization in combination with delamination and exfoliation, special type of topo-chemical reactions were discovered in the mid of the last century. In early 1990s a convenient method of their fabrication was obtained on example of PEO (polyethylene oxide) and MMT (montmorillonite), which advantageously demonstrated intercalation process in a polymer melt and potential applications of the formed products in solid phase electrolytes of recharging lithium accumulators.

However, a promising commercial strategy in this problem formed after a team of researchers from Toyota had found unprecedented strengthening of mechanical properties of polymer layered nanocomposites based on nylon, which had extremely big surface contacts of ingredients and high aspect ratio reached in intercalation/exfoliation process, and high homogeneous dispersion of silicate plates in polymer matrix. These functional materials relate to nanocomposites, on the one hand, because of nanometer (1–5 nm) galleries of layered silicates, distances between nets and layers formed by polymer and inorganic ingredients, and, on the other hand, because of sizes of exfoliated fragments. A polymer is integrated into interlayer space of a smectite by penetration from a solution/melt with the following delamination and exfoliation with *in situ* formation of mineral nanoparticles, which are individually distributed in the polymer. Exfoliation is accompanied by formation of monolayers of nanometer thickness with high shape anisotropy. Thus, in this process formation of three main types of hybrid phase composites is possible: microcomposite with separated phases (traditional material), intercalated (including a flocculated one), and exfoliated nanocomposites. Intercalation of organic ingredient in inorganic materials with limited interlayer space, which preserves a layered structure, is, on one hand, a perfect way of structuring of organic-inorganic ensembles, and, on the other hand, proposes additional possibilities for studying of physical-chemical properties of these systems. Nanocomposites of intercalation type based on polyamides are of great interest due to their special properties and they are used as structural, flame-resistant materials; their enhanced barrier characteristics, chemical resistance, low penetrability for solvents, flame-resistance attract interest not only from scientific but from industrial point of view. The driving force of intercalation process is a decrease in free energy of a system: enthalpy changes due to intermolecular interactions, while entropy changes due to configuration interactions, and enthalpy is a dominant factor. Special place have *host* lattices with electron conductivity (semiconductors based on layered metal oxides, their chalcogenides, etc.), which undergo redox reactions with electron or ion transport causing substantial change in physical properties during intercalation.

Typical methods of production of intercalated composites, which are analyzed in this book, can be reduced to *in situ* polymerization (solution mixing); mixing in suspension, latex compounding, and direct intercalation in a polymer melt (reactor mixing). The simplest methods of intercalation of polymers into inorganic structures are one-staged emulsion and suspension polymerization, interlayer blocked radical poly- and co-polymerization of traditional monomers in presence of different organophilic minerals. A special place in this problem have orientation control of a guest-molecule in intercalation system of a host, inter-lamellar catalytic, oxidizing-reduction polymerization, intercalation assembling, reactor mixing, etc. Architecture and morphology of intercalated nanocomposites are studied in detail. Many original approaches are developed for obtaining of nanocomposites based on epoxy resins and layered silicates, a balance between intercalation and solidification rates is optimized. However, many problems of intercalation chemistry of reactoplasts are yet to be solved. Hybrid self-assembling nanocomposites formed as multilayers are developed substantially; very different approaches to layered assembling are realized. This strategy allows fabrication of very different combinations of materials of semiconductor metal-insulator structures with nanometer sites and unusual properties, which is favored by a close contact between components and strict molecular ordering. And finally, a special attention is attracted to hybrid nanocomposites with periodic nanostructure of semiconductors based on dichalcogenides of metals of MS_2 type. They have special optical, electric, and other properties making them promising for applications in microelectronics.

Intercalation chemistry has a great potential for structuring of hybrid polymer-inorganic nanocomposites, which is far from its realization, and this is a reason for intense development of research work in this field.

Many-sided, convenient and well-reproduced method of formation of nanocomposites is thermolysis of suitable precursors. Under certain conditions this process is ecological and easily controlled. One of practical directions of pyrolysis is polymer waste utilization. Its most interesting application is production of carbon nanomaterials, various ceramics, nanocomposites with "core-shell" structure. On the other hand, thermolysis is a complicated process in which all spectrum of chemical transformations of polymers takes place: destruction, linkage of chains, transformation of functional groups and intermolecular regrouping. At the stage of thermolysis it is convenient to integrate different dopants in polymers, including metal containing, which change mechanism of thermolysis, and bring to obtaining of a range of interesting products. The same concerns metal-forming precursors (metal containing monomers), including the ones chemically bound with polymers containing "active" functional fragments (carboxyl, amino-, amido-, imino-, nitrilo-, oxy-, and other groups). Thermal transformations of these macromolecular metal complexes have begun being studied just recently, as does thermolysis in conjugated processes, on examples of self-regulated stabilization of highly dispersion metal particles in the *metal containing precursor—monomer* system with the following polymerization and thermolytic transformations. These systems are characterized by general picture of transformations consisting of a consequence of three main different by temperature macrostages: dehydration (desolvation) of initial metal-monomers (403–473 K); the stage of solid phase homo- and co-polymerization of dehydrated metal-monomer (monomers) (473–573 K); decarboxylation of formed metal-polymer to metal containing phase and oxygen-free polymer matrix at the temperatures >523 K.

Kinetics of thermal transformations of metal polymer systems is described by macro kinetic equations with account for mechanisms of reactions (decomposition model) in low and high temperature areas (including reactions accompanying burning of polymers, components of solid propellants, polymer bindings for thermal protection of spacecrafts, etc.). These processes go with external or internal heating, either under isothermal or non-isothermal conditions in closed or open system. Depending on the solving problems, different methods of control over a degree of transformation are possible, which predetermine a choice of instrumentation for experimental study of thermolysis: thermogravimetry, linear pyrolysis, volumometry, impact of high energy irradiations, spray (spattering pyrolysis), the processes are studied in precursors and in polymer matrices, in which they are integrated. A significant interest is in mono-disperse magnetic heterostructures like MFe_3O_4 ($M = Ag$, Au, Pt, Pd), polymer-mediated thermosynthesis of multimetal alloys and ceramics, and computer modeling of kinetics of nanoparticle formation during solid state thermolysis for obtaining nanocomposites of a given content.

Gas phase methods of synthesis are based on homogeneous nucleation of supersaturated vapor and the following growth incited by encountering and condensation. Supersaturated vapor can be generated in different ways depending on chemical origin of a precursor, but most often this is realized by heating of a solid matter. Depending on a character of heating and cooling, the methods of flame pyrolysis, heating and evaporation in continuous reactors (Ag, Ga, Al, PbS, Pb, Si, Ge), methods of pulsed laser evaporation of metals in inert gas atmosphere (He or Ar) and reacting gases (O_2, N_2, NH_3, CH_4) with forming of nanocrystalline oxides, carbides or nitrides, etc. Content and size of the nanoparticles in them can be controlled by change in pressure, content of atmosphere, power of a laser pulse, temperature gradient between evaporated target and a surface, on which the vapor is condensed. Due to high temperatures of synthesis strong aggregates often form, which are not easily separated to initial particles. There are also limitations to formation of multicomponent phases because of significant fraction of segregation processes at high temperatures.

Plasmochemical methods are widely used. High rates of formation and condensation of a compound and quite high efficiency are provided by plasmochemical synthesis. The main drawbacks of this method are wide spread size distribution and high content of impurities in a powder. Very fine powders of titanium, zirconium, hafnium, vanadium, niobium, tantalum, boron, aluminum and silicon, titanium nitrides; titanium, niobium, tantalum, tungsten, boron, and silicon carbides; magnesium, yttrium, and aluminum oxides are produced.

Production of metal nanoparticles and nanostructured materials under impact of high energies on a chemical system is related to generation of highly active reducing agents such as electrons, radicals, and excited particles. The main advantage of photo and radiation-chemical reduction is a possibility of synthesis of nanoparticles in different media, including solid ones (for example, polymer matrices, films), and at low temperatures with high reproducibility. Methods of production of nanocomposites using the effect of microwave or sonic-chemical impact give rather short reaction times, provide fine particles with narrow size distribution. For example, rapid and uniform heating under microwave irradiation causes efficient reduction of Au and Pd nanoparticles on the surface of oxides in presence of poly(ethylene glycol) and poly(vinyl-2-pyrrolidone) with homogeneous dispersion [13].

Therefore, the "bottom-up" synthetic strategy provides production of hybrid nanocomposite materials including a way from molecular precursors or nanostructuring blocks of a given structure to final products in shape of fibers, films, or monolith. Design of these materials can be supplied by formation of complicated hierarchical structures in combination with usage of a template technology (micropattering) and "bottom-up" methods (Fig. 1.1). Combination of different approaches provides unlimited possibilities for design of hybrid architectures and new materials for various application fields, smart membranes, new catalysts and sensors, new generations of photovoltaic and solar batteries, micro-optical and optoelectronics components and systems, drug delivery systems, etc. [14].

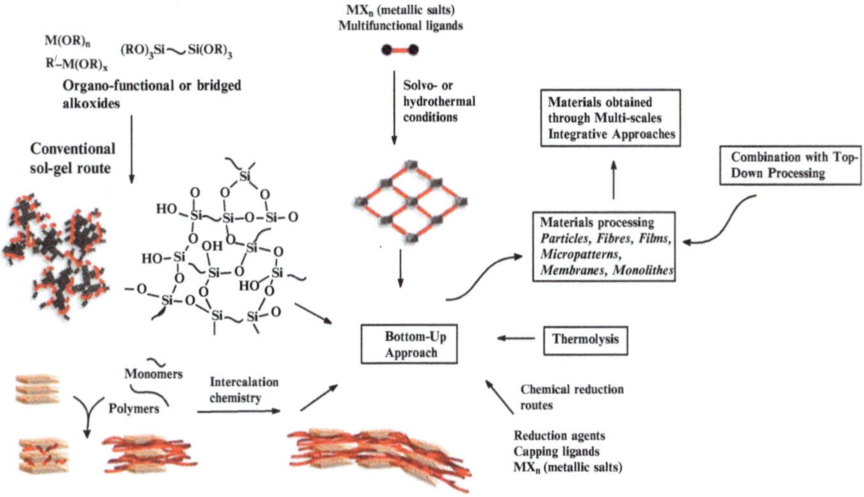

Fig. 1.1 The main synthetic strategies for production of nanocomposite materials

In the recent years particles, which take place in different biologic processes, play a special role in nanometer composites. Interaction of nanoparticles with biopolymers is very important in fermentative catalysis, biosorption, biohydrometallurgy, geobiotechnology, etc. For this purpose are used: peptides and natural or synthetic proteins obtained by chemical methods or genetic modification of microorganisms or plants, nucleic acids (including synthetic ones), biodecomposed complex polyethers, such as poly lactic acid and its derivatives, oligo-hydroxyalkanoates, hydroxyapatites, synthetic and natural (vegetable or animal) polysaccharides, cellulose and its derivatives, alginates, dextranes, gummiarabic, etc.

Almost all synthetic and nano-materials science approaches considered in this book are more or less typical of production of biocomposites. Especially important their application is in medicine, this is a basis for progress in diagnostics and therapy at cellular and genetic levels. Thus, vegetable biomasses and extracts are widely used in the methods of metal ion reduction in mass and extra-cellular production of nanobiocomposites. At that molecular components of biomasses stabilize simultaneously the formed nanoparticles. And, though rates of biogenic syntheses are, as a rule, lower than in reactions of chemical reduction, their advantage is simplicity and minimization of a number of technological cycles, realizing of green chemistry principles at all stages including formation of biocompatible non-toxic products applicable in biomedicine. Soft conditions for synthesis, aqua media, and quite high product yield make plant-mediated processes attractive for scaling, including disposal of toxic metals and radionuclides from wastes. The processes with bacteria participation, oppositely to those involving vegetable objects, demand special stage of cellular cultivation and they are characterized by high selectivity and molecular control over metabolic processes providing reproducible synthesis of nanoparticles. The same relates to highly specific structures such as enzymes and proteins incorporated in bacterial membranes. Processes of biosorption of metal ions by microorganisms are characterized by rather high rates, and bioreduction goes far slower. However, methods of molecular engineering make it possible to design templates which have high affinity to some type of metal ions, for example, to Au(III). Development of integrated materials based on biomolecules and inorganic nanoobjects and assembling of these systems into functional devices makes a basis for nanobiotechnology.

There is an interesting usage of sol-gel technologies providing production of biotechnologically important ferments, biosensor constructions, and ferment electrodes, components of bioactive optical detectors, lipid bi-layer vesicles, encapsulating agents for drug delivery and adsorbents for pharmaceutics and cosmetics, etc. Sol-gel method is realized for immobilizing of ferments acting as bioreactors. Organic polymers are widely used materials for replacement of soft tissues, though they do not have biological activity, they are most often bio-tolerant. At that, intermediate temperatures and soft conditions of hydrolysis and condensation – polymerization of monomer metal alkoxides allow capturing of proteins at the stage of matrix formation without their denaturation. Coating of implants surface made of metal alloys advances their integration in bone tissue and binding with it. Nanocomposites form bioactive ceramics and resolved polymer implant are successful in regeneration of bone tissue, in development of frames in bone tissue engineering.

Bioactive materials based on calcium phosphates (hydroxyapatite and tri-calcium phosphate) and glass/glass-ceramics are widely clinically used in dentistry and orthopaedics. Reconstruction of a damaged tissue, artificial substrate for cell growth, locally active drug delivery system with regulated loading and releasing kinetics are the most widely used application fields of bionanocomposites.

Sol-gel processes are the most widely spread ways of formation of structural bionanocomposite materials in living nature, preferably similar to polymer sol-gel synthesis. Moreover, many of them are borrowed from living nature and are used by it for many different bioprocesses, biosorption, biomineralization, construction of natural composite materials having hybrid structure, for example, bone, borosilicate glasses, etc. During bio-concentration—biomineralization organic matrix (template) performs control over nucleation, growth processes and formation of inorganic materials with perfect morphology; this is used for creation of complicated hierarchic structure of composites with unusual chemical and physical properties.

Intercalation of biostructures, including commercially available ones, in layered materials is accompanied by specific traits. Integration of nanoparticles and biomolecules with unique properties of each object, on the one hand, and present at same nanometer length scale, on the other hand, (enzymes, antigens, and antibodies have typical sizes of nanoparticles 2–20 nm), i.e. both classes of these materials are structurally compatible. One of the interesting directions of this approach is obtaining of biodegrading "green" nanobiocomposites in a variant of reinforcing of polymer matrix for preparation of a framework for development of a tissue, therapy of peridontal bone defects, drug delivery, etc. For example, in montmorillonite (MMT) only some polymers can penetrate into interplanar distance (linear polymer amylose, oppositely to gigantic globes of branched amylopectin), gallery space gives an opportunity for big starch molecules to diffuse between layers and thus promotes interphase interactions, which causes more intense strengthening effect. At that tearing load and edge wetting angle increments are higher for exfoliated than for intercalated nanocomposite, which is related to higher degree of interaction between biomacromolecules and MMT particles in it. These nanocomposites are important in medicine (drugs), genetic engineering (DNA, RNA), biotechnology (proteins, individual cells, enzymes) and in food industry. Antimicrobial properties of modified clays destroy bacterial cellular membranes, suppressing their metabolic activity, and with time cause destruction (lysis) of bacterial cells. Synthetic nanometer apatite crystals manifest increased physiological affinity to host tissues or biological activity, they advance increase in compatibility of materials due to their chemical and structural similarity with natural apatite crystals. A special place has development of magnetically controlled drugs of chemotherapeutic, diagnostic, and hyperthermal activities. Hybrid superparamagnetic nanocomposites are used in MRI for localization of brain and heart infarctions, liver flaws or tumors, where nanoparticles have a tendency to accumulate in high concentration due to a difference in content of a tissues and/or endocytosis channel processes, magnetic hyperthermia. Usage of hybrid nanoparticles and nanocomposites in biocatalytic processes is one of mainstreams of development of bionanocomposites.

We have focused on outline of achievements and problems in the nanocomposite science, which would be filled periodically with new experimental and theoretical data thus forming a complete pattern.

References

1. J. Njuguna, K. Pielichowski, Adv. Eng. Mater. **5**, 769 (2003); Adv. Eng. Mater. **6**, 193, (2004); Adv. Eng. Mater. **6**, 204 (2004)
2. J.B. Donnet, Compos. Sci. Technol. **63**, 1085 (2003)
3. X. Zhang, Y. Huang, T. Wang, L. Liu, Compos. Appl. Sci. Manuf. **38**, 936 (2007)
4. J.B. Lambert, *Traces of the Past: Unravelling the Secrets of Archaeology Through Chemistry* (Addison-Wexcley, Helixbooks, 1997), p. 319
5. R. Zigmondi, Z. Anorg, Chemistry **99**, 105 (1917)
6. R. Kubo, J. Phys. Soc. Jpn. **17**, 975 (1962)
7. H. Gleiter, Prog. Mater. Sci. **33**, 223 (1989)
8. A.D. Pomogailo, A.S. Rozenberg, I.E. Uflyand, *Metal Nanoparticles in Polymers* (Khimiya, Moscow, 2000)
9. J. Turkevich, P.C. Stevenson, J. Hiller, Discuss. Faraday Soc. **11**, 55 (1951)
10. B.V. Deryagin, *Theory of Stability of Colloids and Thin Films. Surface Forces* (Nauka, Moscow, 1986)
11. C.J. Murphy, J. Mater. Chem. **18**, 2173 (2008)
12. S. Zhorobekova, *Macroligand Properties of Humic Acids* (Ilim, Frunze, 1987)
13. G. Glaspell, L. Fuoco, M.S. El-Shall, J. Phys. Chem. B **109**, 17350 (2005)
14. C. Sanchez, P. Belleville, M. Popall, L. Nicole, Chem. Soc. Rev. **40**, 696 (2011)

an opportunity for the starch molecules to diffuse between layers, and thus promote [illegible]

Chapter 2
Reduction of Metal Ions in Polymer Matrices as a Condensation Method of Nanocomposite Synthesis

It would be not an exaggeration to say that development of the methods of synthesis of nanoparticles governs the mainstream of the science concerning nanomaterials [1]. This is, first of all, discovery of carbon nanotubes [2], synthesis of highly organized quantum dots [3], controlled morphology of CdSe nanocrystals [4], etc. New fields of application and unique properties of materials are related, more to a possibility of control of size and shape of particles in nanometer scale than to a change in their sizes [5]. Presently it is expressed in existence of a huge library of nanoparticles including a wide range of compositions, structures, sizes, and this is also true for synthetic methods used for development of nanomaterials. Among successfully developed approaches there are vapor-gas syntheses for production of one-dimensional structures [6, 7]. There are extensive solution methods including such approaches as co-deposition, sol-gel synthesis, microemulsion method, hydrothermal and solvothermal processes, template synthesis, etc. [8–12].

It is known that liquid-phase methods of production of nanoparticles make it possible to control effectively sizes and morphology of particles [13–19]. Many processes of metal ion reduction widely used in synthesis of metal colloids relate to liquid phase reactions [20–22].

2.1 Formation of Metal Nanoparticles During Chemical Reduction

Contemporary ideas about formation of nanoparticles are based on the main principles proposed by J. Turkevich, which include stepwise mechanism of nucleation, growth and agglomeration [23].

On the whole, synthesis of nanoparticles with a certain diameter having almost monodisperse distribution in size and low degree of agglomeration is limited by two

A.D. Pomogailo, G.I. Dzhardimalieva, *Nanostructured Materials Preparation via Condensation Ways*, DOI 10.1007/978-90-481-2567-8_2

principal boundary conditions: precision control over the nucleation stage and seeds growth; efficient impending of the agglomeration process [13, 24–26].

According to LaMer model [27] nucleation is endothermic process. Break of bonds in the initial compound, removal of a solvate shell is energy consuming. At the same time, the processes of seeds growth and agglomeration accompanied by a decrease in enthalpy of the system owing to the bond energy of a lattice are exothermic, i.e. formation of a blocked solid is always energetically beneficial as compared to nanoparticle formation with a typically extensive surface, unsaturated bonds, non-occupied coordination sites. We shall consider each stage separately.

2.1.1 Nucleation

In a typical chemical synthesis of metal nanoparticles a compound metal-precursor reduces with formation of zero-valent metal atoms, building blocks of metal nanoparticles. During fast reduction concentration of atoms reaches the supersaturation point, and from this moment spontaneous homogenous nucleation begins, which is characterized by high-energy barrier (Fig. 2.1). As supersaturation decreases, nucleation ceases, and the formed seeds go on growing. Due to short time of the nucleation the obtained particles have narrow size distribution. At high supersaturation an additional mechanism of a decrease in dispersion of particles is possible [28, 29]. The essence of the conception of "explosive nucleation" [27] is in inducing of an individual nucleation process and excluding of additional nucleation from the further growth process. This method as a method of synthetic strategy is often called "separation of nucleation and growth" [29, 30].

Thermodynamic model of homogenous nucleation is considered in detail in [11].

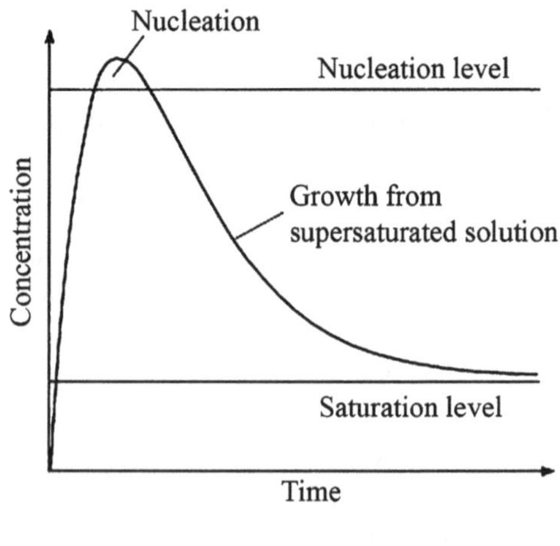

Fig. 2.1 Diagram of the change of precursor concentration under nanoparticle growth [13]

Gibbs free energy of spherical crystals with r radius formed from solution with S supersaturation is defined by the equation:

$$\Delta G = 4\pi r^2\gamma + \frac{4}{3}\pi r^3 \Delta G_v \qquad (2.1)$$

in which γ is the surface free energy per unit area and $\Delta G_\nu = -RT\ln S/V_m$ (V_m is molar volume of the blocked crystal) is a change in free energy between a monomer in a solution and unit volume of the blocked crystal. The r, at which ΔG is maximum, is called critical radius r_c. This is a minimal radius of a seed, which can grow spontaneously in supersaturated solution.

When $d\Delta G/dr = 0$,

$$r_c = \frac{-2\gamma}{\Delta G_2} = \frac{2\gamma V_m}{RT\ln S} \qquad (2.2)$$

One of the necessary conditions for supersaturation during homogenous nucleation follows from Eq. 2.2. The S value should be quite high to provide the critical radius r_c smaller than the germ size of a crystal from which a seed forms during homogenous nucleation. Size of these germs should be no less than 1 nm, which is compatible with sizes of inorganic molecular crystals. One can get the equation for critical Gibbs free energy ΔG_c via combination of Eqs. 2.1, 2.2 necessary for formation of stable seeds:

$$\Delta G_c = \frac{16\pi\gamma^3}{3(\Delta G_v)^3} = \frac{16\pi\gamma^3 V_m^2}{3(RT\ln S)^2} \qquad (2.3)$$

For the reduction reactions an important moment is a problem concerning particles, which aggregate in seeds, and then grow to a nanoparticle: whether they are reduced or unreduced. An example of formation of Pt^{II}–Pt^{I} и Pt^{I}–Pt^{I} dimers formed from the products of hydrolysis of $[PtCl_4]^{2+}$, which is a typical precursor of synthesis of Pt nanocrystals shows that the precursor compound can transform directly in the seeds form or associate with a growing nanocrystal without reducing to zero-valent state [31, 32]. According to the numerical analysis performed by molecular dynamic method, dimers and trimers have higher electron affinity than the initial precursor. It is assumed that reduction will go predominantly by transition of electron from a reducing agent to these dimer and trimer intermediates on the way to clusters and seeds [33].

At that, association of complexes to a cluster or a respective removal of a ligand from the cluster can significantly accelerate growth of metal nanocrystal. Such acceleration is usually regarded as autocatalytic process which is found in many systems (Fig. 2.2) [24–26, 34–36].

It should be noted that the considered mechanism of reduction is preferable under special experimental conditions, for example, when intermediately strong reducing agents are used and/or concentration of a precursor is high, and in this case reduction of the latter to atomic state is not critical. The surface of a cluster or

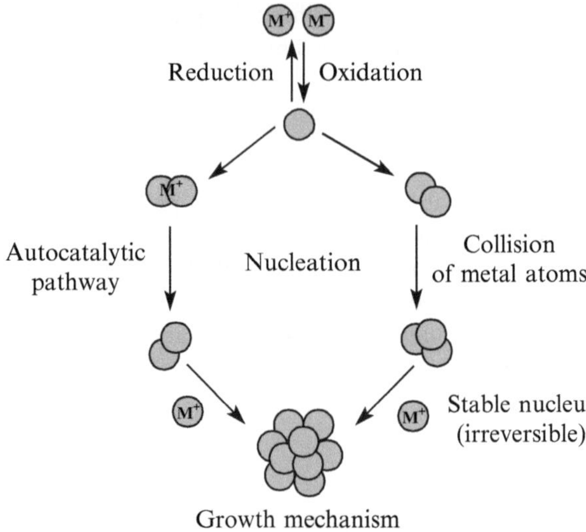

Fig. 2.2 Mechanism of nucleation and growth of nanoparticles [36]

nanocluster is surrounded by positively charged metal ions coordinated with molecules of a ligand or solvent.

Reduction also may be preceded by formation of a complex with a reducing agent. For example, characteristic band in UV-vis spectrum of two-valence copper in Cu^{2+}- polyvinylpyrrolidone [37] disappears at addition of hydrazine-hydrate, and instead of it absorption band appears in the region of 700 nm, which points to formation of $Cu[(N_2H_4)_2]^{2+}$ complex (Fig. 2.3) [38].

Thus, unstable small clusters of atoms or ions formed at the initial stage of reduction of metal ions are the sources of formation of metal nanoparticles. For their studies most often methods of pulse radiolysis [39], electrospray mass spectrometry [40, 41], electrospray photoelectron spectroscopy [42], absorption and emission spectroscopic methods are used.

As many research teams have shown, Ag clusters demonstrate different optical spectra depending on a number of Ag atoms contained in a cluster [39, 43–45].

Atoms of Ag_0 produced during reduction undergo subsequently some transformations with cluster formation [46–50]:

$$Ag_2{}^{+} \rightarrow Ag_3{}^{2+} \rightarrow Ag_4{}^{2+} \rightarrow Ag_8{}^{2+}$$

or

$$\begin{aligned} &Ag^{+} + Red \leftrightarrow Ag_1 + Red^{+};\\ &Ag_1 + Ag^{+} \leftrightarrow Ag_2^{+};\\ &Ag_2^{+} + Red \leftrightarrow Ag_2 + Red^{+}\\ &\dots\dots\dots\dots\dots\dots\\ &Ag_n + Ag_m = Ag_{m+n} \end{aligned}$$

Red is the reduction agent

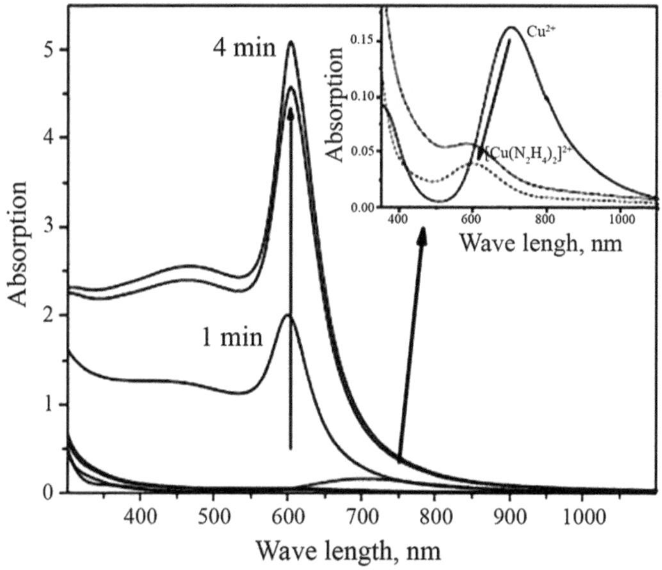

Fig. 2.3 The change of UV-vis-near IR spectra during Cu(II) reduction in DMF at 333 K. In the insert: evolution of the Cu^{2+} band absorption under addition of hydrazine hydrate at 333 K [38]

Particles of Ag_1 and clusters consisting of a small number of atoms are unstable (redox-potential of atomic silver $E^o_{Ag_1/Ag^+} = -1.8V$ (is referred to by [51]) instead of $E^o_{Ag/Ag^+} = +0.799V$ for metal silver). As metal clusters grow, their stability increases. For example, Ag_8^{2+} cluster is quite stable, it lifetime is measured by 10 min. Optical band corresponding to it in the absorption spectrum differs from the bands typical for Ag (360 nm) and Ag_2^+ (310 nm) dimers and disappears as Ag_8^{2+} cluster transforms in silver sol (Fig. 2.4) [39].

Theoretical calculations and experiment [52] have shown that critical size for gold clusters is 2 nm (theoretical calculation gives 1.7 nm). For these particles there is no mode of collective plasmon resonance (Fig. 2.5), i.e. Au_{25}(glutathione)$_{18}$ clusters show one-electron absorption peaks, while 4-nm nanoparticles are plasmon, and transition from cluster to fcc crystal state takes place just about 2-nm size.

Probably, it can be assumed that the analogous types of clusters can be included as intermediates during formation of a metal phase seeds, whose main characteristic features are presence of physical interphase and ability to reduction of absorbed metal ion, i.e. ability of the seeds to growth.

2.1.2 *Growth*

As long as a seed forms, it begins to grow due to adding of atoms. Theoretical consideration of the “diffusion growth” model (referred to by [11]) has shown that growth rate of particles is inverse to their radius, because a number of atoms

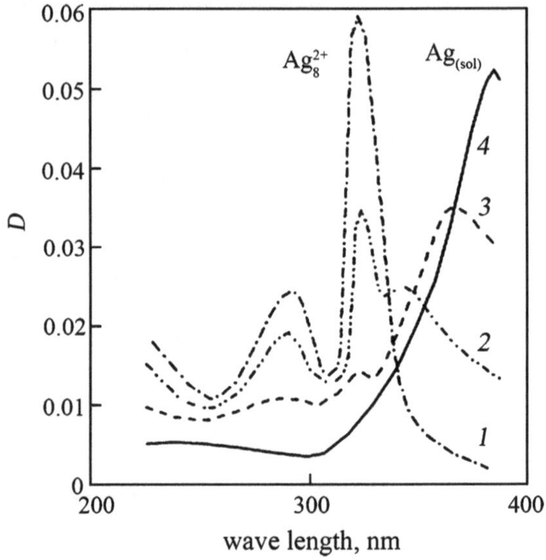

Fig. 2.4 Absorption spectra of the silver salt solution after an electron beam exposure (Ag_8^{2+} cluster transformations in a silver sol). The solution of $1 \times 10.4\,M$ $AgClO_4 + 0{,}1\,M$ 2-propanol. Irradiation duration: 6 s (*1*), 1 min (*2*), 5 min (*3*), 15 min (*4*) [39]

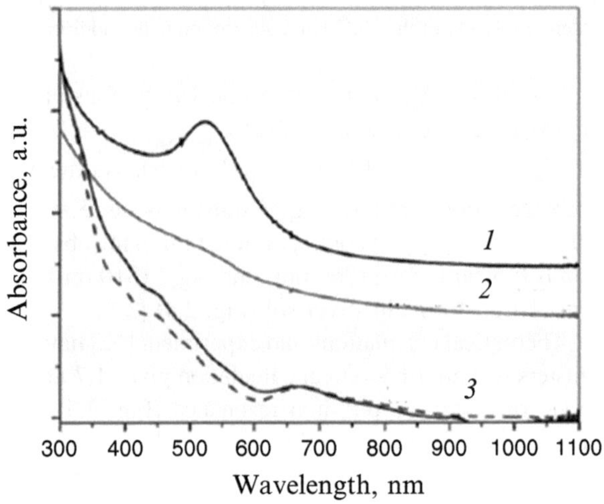

Fig. 2.5 UV spectra of the different sized Au nanoparticles, (*1*) first precipitate (ca. 4 nm) redissolved in water; (*2*) second precipitate (ca. 2 nm) redissolved in water, and (*3*) the *left* supernatant (for comparison, the spectrum of pure $Au_{25}(SG)_{18}$ (*dashed line*) is also shown) [52]

diffusing to the surface of a particle increases in proportion to its squared radius, and the particle volume increases as cube of its radius. Taking this into account, it is shown that for the ensemble of spherical particles a change in radius distribution σ^2 decreases during the growth.

$$\frac{d(\sigma)^2}{dt} = 2V_m D(C_{bulk} - C_s)\left[1 - \bar{r}\,\overline{\left(\frac{1}{r}\right)}\right] \tag{2.4}$$

where $\bar{r}$ and $(\overline{1/r})$ are average values of r and $1/r$, respectively, C_{bulk}, C_{s} are concentrations in solution and on the particle surface, respectively.

In other words, particle size distribution narrows, independently on the initial distribution while all particles grow and no additional nucleation appears. This self-regulating mechanism of size distribution is often called "focusing effect" [29].

On the whole, particle growth is controlled by competition of two processes: a decrease in bulk energy, which advances growth (absorption) and an increase in the surface energy associated with particle dissolution. With account for kinetics of the growth process and its dependence on a particle size, opposite to the "focusing" effect can be obtained: the smaller particle radius, the more difficult its growth, but easier dissolution for higher chemical potential. Two mechanisms of control of particle size distribution are formulated from the results of theoretical studies and numerical simulation [53, 54]: (i) "focusing effect" which is kinetically governed process, when growth is diffusion-controlled and supersaturation is quite high; (ii) Ostwald ripening, which appears at low supersaturation, i.e. under conditions when not all particles remain able to grow and those, whose radii are below critical r_{c}, dissolve, concentration of the particles decreases slowly. Atoms formed during dissolution diffuse to coarser particles and absorb on their surface. On the whole, Ostwald ripening broadens standard deviation to the particle size distribution, i.e. it causes an increase in polydispersion, and the average size of nanoparticle grows.

On the contrary, under conditions of excess stabilizing ligands, so called digestive ripening, growth of nanoparticles is accompanied by narrowing of the size distribution. Single crystalline Ag particles were subjected to such ripening in presence of dodecylamine, the particle size decreased from 12.4 ± 5.1 to 6.1 ± 0.5 nm [55]. Depending on the origin of ligands, particle sizes may not only decrease, but also increase, and the final size, as has been mentioned above, is determined by dynamic ratio of the dissolution rate of metal atoms from particles of some size to the re-absorption rate on particles of other sizes [56, 57].

There are some synthetic methods for separation of the nucleation and growth processes. In so-called "seed-mediated growth" nucleation is physically differs from growth via using of preliminary obtained nanocrystals for a nucleus seed. Actually, in this case heterogeneous nucleation impedes formation of additional nuclei, which would be the case at homogenous nucleation. Pre-seeds are introduced in a reaction matter, and as a precursor is added, the formed atoms absorb on the surface of the formed nucleus. At that, concentration of the precursor remains rather low to prevent homogenous nucleation. This strategy can be used to obtain homogenous particles [58–60] and heterogeneous structures of core-shell type [61, 62]. The considered method makes it possible to finely regulate particle size; however, a necessary condition is monodispersion of the initial seed particles.

The "hot injection" method [3, 63–66] realizes homogeneous nucleation. At fast addition of excess precursor in hot solution of surfactant "explosion" nucleation proceeds. During the nucleation concentration of atoms in the solution decreases dramatically and the nucleation rate decreases.

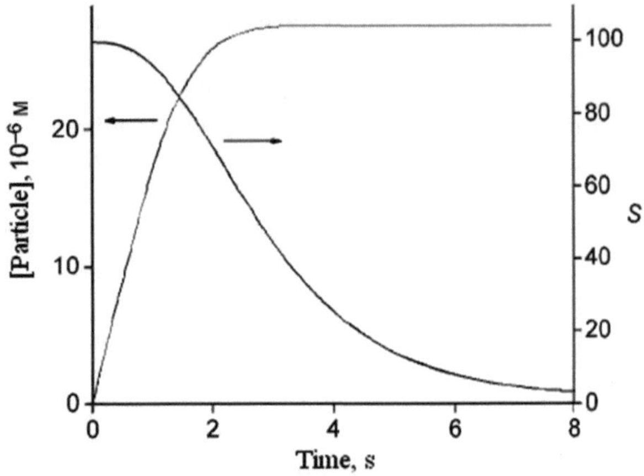

Fig. 2.6 The results from numerical simulations of nucleation and growth of nanocrystals. The particle concentration and supersaturation as a function of time. Values of simulation parameters: S = 100, T = 523 K, $\gamma = 0.11\ \mathrm{J\,m^{-2}}$, $A = 10^7\ \mathrm{s^{-1}}$, $D = 10^{-15}\ \mathrm{m^2\,s^{-1}}$, and $\Delta t = 10^{-2}$ s. [11]

Numerical simulation of homogeneous nucleation [11] during "hot injection" with account for a number of nuclei formed at each time step provides:

$$\frac{dN}{dt} = A\exp\left[-\frac{\Delta G_c}{kT}\right] = A\exp\left[\frac{16\pi\gamma^3 V_m^2}{3k^3T^3N_A^2(\ln S)^2}\right] \tag{2.5}$$

(here a rate of increment in number of particles *N* is defined as the nucleation rate), it shows that inducing of high supersaturation causes fast consumption of a monomer which, respectively, stops nucleation, thus, the nucleation process can be separated from growth (Fig. 2.6). The results of this modeling are consistent with experimental data of a change in concentration of CdSe nanocrystals as a time function during their synthesis [67].

Finally, standard methods of control over direction of reactions and equilibriums in colloid chemistry are electrostatic and steric stabilization of the particle surface just after nucleation stage [20, 68, 69] (Fig. 2.7).

Electrostatic stabilization can be reached by specific ion sorption (H^+, OH^-, SO_4^{2-}, NO_3^-, $RCOO^-$, RSO_3^-, R_4N^+, etc.) by the surface of nanoparticles. Such agents as citrate ions make a double ionic layer on the surface of a nanoparticle which generates repulsive Coulomb forces (Fig. 2.8).

On the contrary, addition of salts containing weakly coordinated organic or inorganic anions such as sodium dodecyl sulfate or sodium p-toluene sulfonate, lithium trifluoromethanesulfonate, or lithium perchlorate causes a decrease in thickness of the double electric layer and thus favors coagulation of nanoparticles [70].

Steric stabilization is realized by absorption of long-chained organic molecules (for example, oleyl amines, oleic acid, trioctylphosphine oxide), polymers and biomolecules. Growth and agglomeration of nanoparticles can also be controlled by separation of reaction spaces, for example, via considerable dilution, and continuous removal of formed nanoparticles in microfluid flow-through setups [71]. The principle of limiting of cluster growth to coarse particles using the effect of their low solubility in organic solvents is shown for Au_n(glutathione)$_m$- [72] and Ag_7- clusters [73, 74].

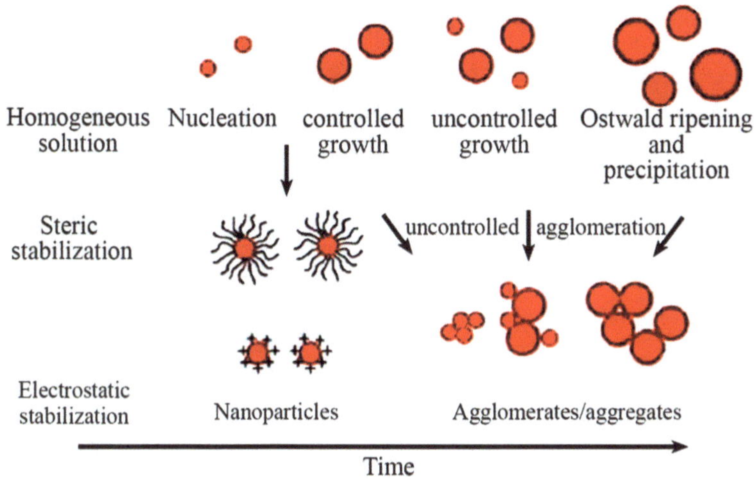

Fig. 2.7 Growth and stabilization of nanoparticles [13]

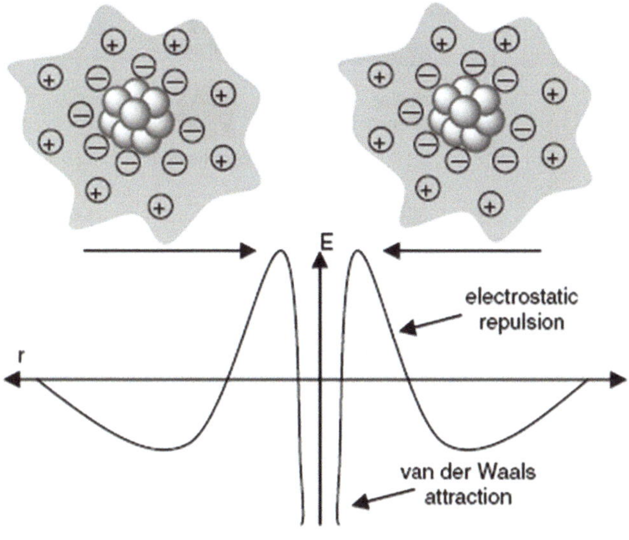

Fig. 2.8 Electrostatic stabilization of nanoparticles

There is one more important problem, whose solution is inextricably associated with understanding of the mechanism of nucleation and growth of nanoparticles, this is control over shape of the particles. Synthetic strategy of production of nanoparticles with a specific morphology, various mechanistic aspects of formation of 1D, 2D, and 3D structures of nanocomposite materials attract keen attention of researchers and are widely discussed in literature [75]. In the recent review [76] concerning detailed analysis of reaction ways of fabrication nanoparticles of

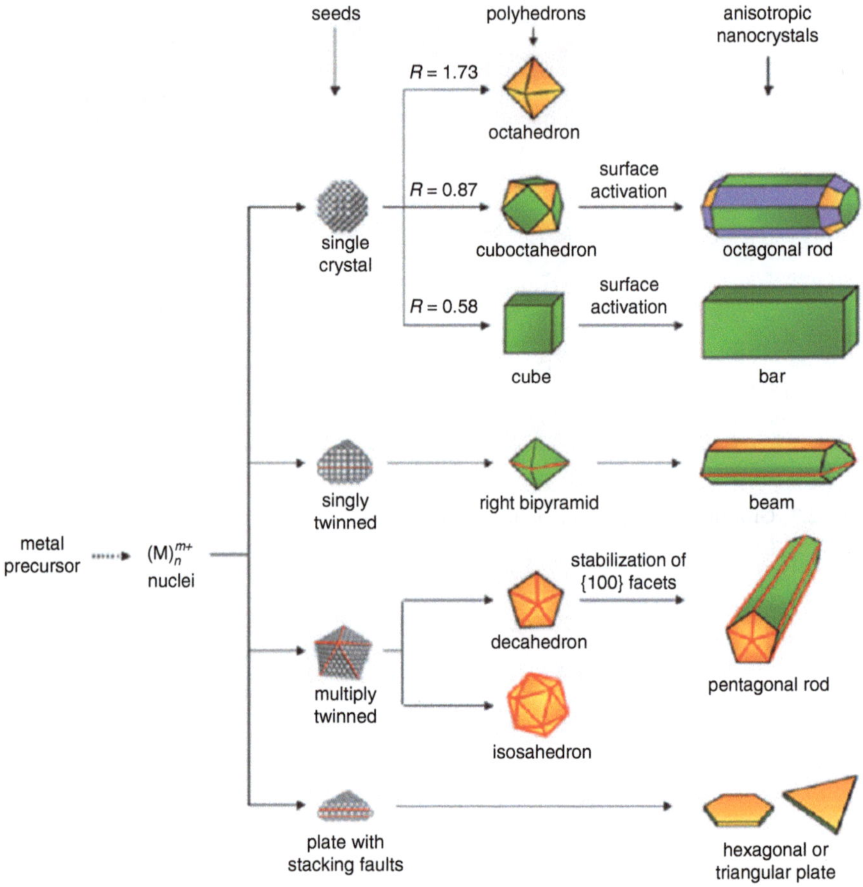

Fig. 2.9 Reaction pathways that lead to fcc metal nanocrystals having different shapes. The parameter R is defined as the ratio between the growth rates along the {100} and {111} directions [76]

different shapes it has been shown that the key role in this would play internal structure of the formed seeds and binding affinity of a stabilizing agent. Structure of the seeds can be ether single or multiple twinned crystals (Fig. 2.9).

Actually, these structures can co-exist during synthesis of one or other nanoparticle. The essence of the problem is fine regulation of growth of seed populations for producing metal nanoparticles with desired shape and morphology. The possibilities for its realization are provided by thermodynamic factors and by kinetic conditions of reactions. This can be achieved by significant decrease in the reduction rate of a precursor [77, 78], action of weak reducing agents [79–81], cooperation of the reduction reaction with oxidizing etching [82], Ostwald ripening [83, 84], etc.

It is known, for example, that Ag_3^+ and Ag_3 clusters in the basic state have a triangle structure with D_{3h} symmetry [85, 86]. Both these clusters have higher electron affinity than Ag^+, which make them prevailing sites for further growth to a nanoparticle. It has been shown that an average edge length of the Ag nanoplate increases as the initial concentration of trimer silver clusters decreases during $AgNO_3$ reduction by polyvinylpyrrolidone [87].

The effective way for shape control of the particles is usage of selective chemically absorbing (capping) agents. This can be ligands, formed as side products during decomposition or reduction of a precursor, or various compounds (low-molecular ions, surfactants, polymers, biomolecules, etc.) specially added to a reaction solution and simultaneously playing a stabilizing role [7, 88–90]. Via chemical binding with the surface of a metal particle a capping agent can change free energy of various planes and therefore their relative growth velocities. For example, polyvinylpyrrolidone as a capping agent is bound most strongly via oxygen atoms with {100} face of Ag and Pd atoms [91], thus providing metal atom bonding to surfaces of other faces. The similar behavior shows citrate ion [92–94]. Moreover, reduction of $HAuCl_4$ using ammonium citrate in presence of Fe^{3+}, Ni^{2+}, Cu^{2+}, Zn^{2+} и Al^{3+} ions causes formation of polygonal gold nanoparticles by ion-inducing mechanism [95]. It is reported that Ag^+ ions can have the same effect [96, 97]. It is suggested that metal ions present in the reacting media are bound predominantly with {111} face of metal atoms, as with the most stable, and prevent growth in this direction.

2.2 Chemical Methods of Reduction

The method of liquid phase chemical reduction discovered by Faraday [98] more than 150 years ago is most widely used technique for production of metal nanoparticles. Another substantial step in development of this approach is connected with J. Turkevich, who for the first time implemented reproducible synthesis of Au (20 nm) nanoparticles by the reaction of citrate reduction of $[AuCl_4]^-$ [23, 99, 100]:

$$xM^{n+} + nxe^- + \text{stabilizer} \rightarrow M^0{}_n(\text{cluster}) \tag{2.6}$$

Among the known reduction methods there are: the "diborane" method of production of $Au_{55}(PPh_3)_{12}Cl_6$ (1.4 nm) nanoclusters stabilized with phosphine ligands; nanoclusters of transition metals stabilized by polyoxoanion and tetrabutyl ammonium; alcohol reduction, so called polyol process [101–104], with the polyvinylpyrrolidone [105, 106], polyvinyl alcohol, etc. used as stabilizers, and reduction of metal salts with formic acid, CO, sodium formate, formaldehyde, and benzaldehyde [107]. In metal nanocolloids silanes, tetrakis hydroxymethyl phosphonium chloride (THPC), hydrazine, hydroxylamine are used as reducing agents. The method of borate reduction is based on the sodium/lithiumborohydrate or tetraalkylammonium hydrotriorganoborates systems, which provide extensive possibilities for production of nanocrystals by chemical reduction of transition metal salts [108, 109].

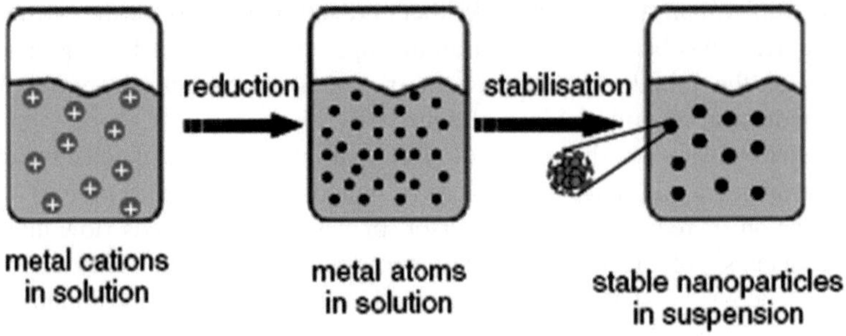

Fig. 2.10 Formation of nanoparticles in the chemical reduction

The chemical reduction method is based on mixing of a reducing agent with a metal-precursor salt in presence of stabilizing agents (ligands, polymers, surfactants) (Fig. 2.10).

Role of the latter is confined to preventing of undesirable particle agglomeration and formation of metal powder. Formation of stable and re-dispersing nanoparticles depends on properties of a stabilizing agent [107].

On the whole, the chemical reduction methods provide a possibility of precision control over growth of nanoparticles, efficient passivation of surface states of nanoparticles, quite narrow distribution of obtained nanoparticles by sizes and control over their morphology [110, 111].

2.2.1 *General Characteristics of Low Molecular Reducing Agents*

Reducing agents are used to reduce Fe from ores (C, H_2, aqua and natural gases, CO, propane, butane), during production of nonferrous and rare metals in metal heat treatment (C, Si, Al, Na, Ca, Mg, La), during precipitation (cementation) of nonferrous metals from aqua solutions and their salts (Fe, Zn); during production of metals and their lower oxides and halogenides, and during chemical deposition of metals, nitrides and carbides from gas phase (H_2, NH_3, CH_4, etc.); during decomposition of chemical processes in solutions ($SnCl_2$, $FeSO_4$, H_2SO_3, N_2H_4, NH_2OH, HCOOH, H_2S, etc.); in organic synthesis (H_2, Na, Zn, $LiAlH_4$, $NaBH_4$, B_2H_6, etc.). These substances can also comprise the main group of reducing agents during formation of nanoparticles. The most often used ones are described below.

2.2.1.1 Sodium Borohydride ($NaBH_4$)

It is widely used in organic synthesis as a reducing agent. It is used to reduce carbonyl compounds to alcohols, imines to secondary amines, nitro compounds, etc. In aqua solutions it is hydrolyzed to Na_3BO_3 and H_2, hydrolysis is accelerated in acid and slows down in alkaline solutions, hydrolysis of sodium borohydrite is catalyzed by salts of transition metals. Oxidation–reduction potential of sodium borohydride in acid and alkaline media is 0.43 and -1.37 V, respectively. One of the schemes of reduction reactions of metal ions M_{2+} can be expressed by equation $2\,M^{2+} + BH^{4-} + 4H_2O \rightarrow 2\,M^0 + B(OH)^{4-} + 2H_2 + 4H^+$, and for alkaline medium by the expression $2\,M^{2+} + BH^{4-} + 4OH^- \rightarrow 2\,M^0 + B(OH)^{4-} + 2H_2$.

2.2.1.2 Lithium Aluminum Hydride ($LiAlH_4$)

It is a strong reducing agent used inn organic synthesis. It is more powerful than other often used agents, for example, sodium borohydride due to weaker Al-H bonds as compared to B-H bonds. It reduces complex ethers, carboxylic acids, and ketone to alcohols, nitro compounds to amines. The temperature of thermal decomposition is -150 °C.

2.2.1.3 Hydrazine Hydrate ($N_2H_4{\cdot}H_2O$)

It is a colorless, slow, steaming in air liquid with a specific faint odor. Its density is 1.03 g/cm^3. Melting point is 118.5 °C at 740 mm of mercury and 47 °C at 2 mm of mercury. It solidifies below -40 °C. The reaction agent is mixed with water and ethyl alcohol, does not solve in diethyl ether, chloroform and benzene. Hydrazine hydrate has a pronounced alkali reaction; its dissociation constant (as a base) is $K_1 = 8{,}5 \cdot 10^{-7}$, $K_2 = 8{,}4 \cdot 10^{-16}$ at 25 °C. Hydrazine and hydrazine-hydrate, hydrazine-sulfate, hydrazine-chloride are widely used as reducing agents of gold, silver, platinum metals from diluted solution and their salts. In organic synthesis hydrazine is used for reduction of a carbonyl group of aldehydes and ketones.

2.2.1.4 Dodecylamine

Dodecylamine. (1-dodecanamine, laurylamine) $C_{12}H_{25}NH_2$, molar weight is 185,34; colorless liquid; melting point is 28,35 °C, boiling point is 247–249 °C, 134–135 °C at 15 mm of mercury; is dissolved in organic solutions, low dissolved in water forming hydrates and liquid crystalline phases. It has low basicity (pK_a 2,68, 25 °C).

2.2.1.5 Sodium Citrate

In the citrate method of formation of nanoparticles citrate ion, which is obtained during water solution of three-substituted sodium salt of citric acid, serves both as a reducing agent and a stabilizer:

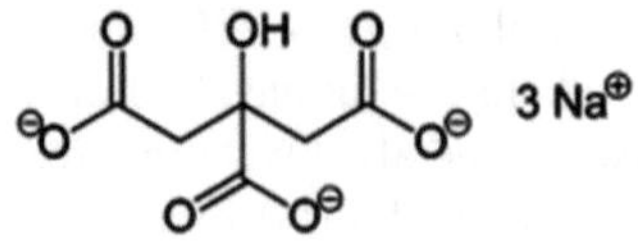

At heating of the solution and oxidizing of citrate ion acetonedicarboxylic and itaconic acids form, they absorb on the particle surfaces and control the particle growth.

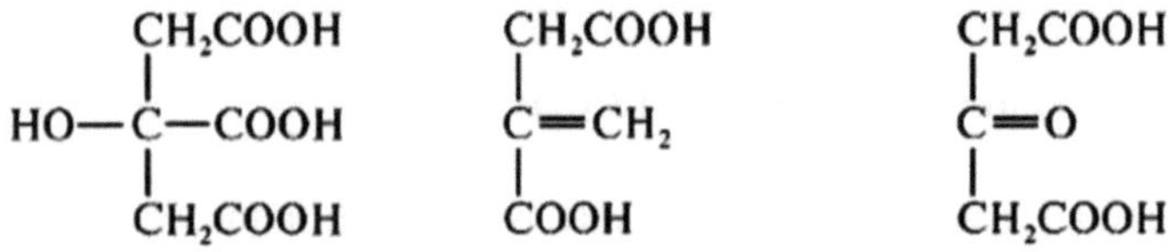

Citric acid Itaconic acid Acetonedicarboxylic acid

In water solutions at $t < 60\ °C$ citric acid and citrates oxidize with formation of 3-ketaglutaric acid as intermediate product, and then followed by formation of formaldehyde and formic acid, which are also reducing agents:

$$C_3H_4(OH)(COOH)_3 \xrightarrow{-CO_2} C_2H_4(CO)(COOH)_2 \rightarrow$$
$$\xrightarrow{-3,5CO_2} HCOH + {}^1/_2HCOOH.$$

Citrate ions absorbed on the surface of colloid particles give them negative charge and thus prevent their aggregation.

2.2.1.6 Ethylene Glycol (HO-CH2-CH2-OH)

Using of ethylene glycol and other alcohols as reducing agents are the basis for polyolic processes of fabrication of nanoparticles. It is assumed [112] that formation of metal nanoparticles in ethylene glycol follows the scheme:

$$HOCH_2CH_2OH \rightarrow CH_3CHO + H_2O$$
$$2CH_3CHO + M(OH)_2 \rightarrow CH_3 - CO - CO - CH_3 + 2H_2O + M$$

The formed nanoparticles are stable in alkali solutions, which supposed their stabilizing by absorption of ethylene glycol and OH^- -type anions [101]. According to the refined mechanism [113], hydroxide ions can take part in acetate ions formation, which are responsible for stabilization of nanoparticles in ethylene glycol, as it has been shown for organic sols of ruthenium:

$$HOCH_2CH_2OH \rightarrow CH_3CHO + H_2O$$
$$3CH_3CHO + 2Ru^{3+} + 9OH^- \rightarrow 3CH_3COO^- + 2Ru + 2H_2O$$

or platinum:

$$HOCH_2CH_2OH \rightarrow CH_3CHO + H_2O$$

$$2CH_3CHO + (PtCl_6)^{2-} + 6OH^- \rightarrow 2CH_3COO^- + Pt + 6Cl^- + 4H_2O$$

2.2.1.7 Ascorbic Acid

Ascorbic acid is a strong reducing agent: enol groups easily oxidize to keto groups. In aqua solution there is a possibility of ascorbic acid self-oxidizing.

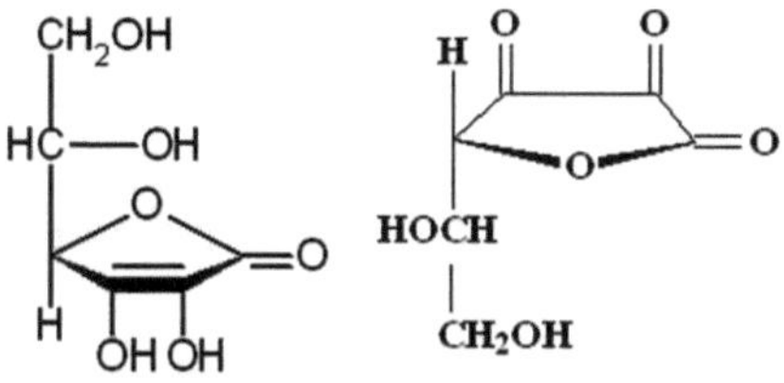

I II

2.2.1.8 L-Ascorbic Acid

L-Ascorbic acid (γ-lacton 2,3-diketo-L-gulonic acid; formula I) has molecular weight 176.12; consists of colorless crystals, the melting point is 190–192 °C (with account for decomposition). Specific optical rotation for D-line of sodium at the temperature 20 °C is $[\alpha]_D{}^{20} + 23°$ (for the concentration 1.6 g per 100 mL of water); high dissolubility in water (22.4 %), lower dissolubility in alcohol (4.6 %), low dissolubility in glycerol and acetone; *pKa* 4.17 and 11.57; λ_{max} 245 (pH <7) и 265 nm (pH >7).

2.2.1.9 Dehydroascorbic Acid

Dehydroascorbic acid (γ-lacton 2,3-diketo-L-gulonic acid; formula II) has molecular weight 174.12; consists of colorless crystals, the melting point is 237–240 °C (with account for decomposition). Specific optical rotation for D-line of sodium at the temperature 20 °C is $[\alpha]_D{}^{20} + 55$, water dissoluble.

In aqua solutions, especially in alkali media, it is reversibly oxidized fast to dehydroascorbic acid (oxidation-reduction potential is +0.058 V), and then irreversibly to 2,3 diketo-gulonic, and then to oxalic acid.

2.2.1.10 Aminoboranes

Aminoboranes are classical examples of adducts of acids and Lewis bases. Most aliphatic aminoboranes are solid crystal matters stable in air and resistant to moisture.

Primary and secondary aminoboranes are thermally stable up to 80 °C. They are soluble in protonic and aprotonic media. On the whole, reduction ability depends on a nature of a complex-forming amine. In the case of alkyl-aminoboranes strength of a reducing agent decreases as a portion of alkyl substituent in the nitrogen compound increases: $H_3N \cdot BH_3 > H_2MeN \cdot BH_3 > HMe_2N \cdot BH_3 > Me_3N \cdot BH_3$ [114]. The reducing ability of aryl-aminoboranes depends on basicity of amines: the lower pK_a of amine, the stronger its reducing properties. In the recent years aminoboranes have being widely used in synthesis of metal nanoparticles, including the processes in aqua and non-aqua media, and in solid state. The main achievements in synthesis of nanocomposites with participation of aminoboranes, their efficiency and advantages as compared with classic reducing agents are analyzed in a recent review [115] and summarized in the Table 2.1.

2.2.2 The Main Factors Controlling the Reduction Process

2.2.2.1 Reducing Ability

A relative reducing ability of two or more substances is found by comparison of changes in Gibbs energy ΔG_p^0 in reactions of these substances with the same oxidizing agent, and in the case of reactions with participation of elementary substances, by energy of formation ΔG_{form}^0 of the oxidation products of the elementary substance. The higher ΔG_p^0, or the modulus value ΔG_{form}^0, active reducing agent is a substance. Often, to compare reducing ability of substances, standard electrode potential E^0 is used. The higher E^0 modulus of half-reaction of reduction with some substance, the stronger reducing properties of this substance (Table 2.2).

Some typical reducing agents have more negative potentials, than the redox-potential of $2H^+/H_2$ pair. As a rule, the most active among them capable of reducing not only noble, but also base metals, have $E \sim -$ 0.5—1.3 V. These reducing agent include hypophosphite, hydrazine, formaldehyde, borohydride, alkyl substituted aminoboranes, etc. If there is a great difference between redox-potential of a reducing agent and a reduced metal ΔE, finely dispersed product of reduction forms fast. The oxidation potential of reducing agents depends on pH: the higher pH, the greater shift of the potential to negative values. In order to obtain wide range of potentials, in which reduction processes can take place, solutions with high pH, containing ligands forming quite strong complexes with ions of a reduced metal, are usually used.

Apart from being a sign of probability of a reaction, a velocity of chemical reaction can be predicted by ΔE^0. A possibility of controlling redox-potential of the reacting components and the reaction on the whole is important for control of properties of the formed metal particles. Electrochemical potential of the majority of metal systems can be changed by their complexation or deposition; the value of the effect will depend on the constant of stability or solubility of final products. For example, at formation of complexes with increasing constants of stability (K_f), the

Table 2.1 Synthesis of nanoparticles in organic and aqueous medium using amine-borane complexes [115]

Nanoparticles	Metal-containing precursor	Stabilizing agent	Reduction agent	Particle size, nm
In organic medium				
Au	$[Au(PPh_3)Cl]$	Dodecanthiol	$tBuH_2N \cdot BH_3$	6
Au	$HAuCl_4 \cdot 3H_2O$	Oleylamine	$tBuH_2N \cdot BH_3$	1–10
PtPb nanorods	$Pt(OAc)_2$, $Pb(OAc)_2$	Hexadecanthiol, Hexadecylamine, Adamantic acid	$tBuH_2N \cdot BH_3$	length $= 45 \pm 12$, width $= 5.9 \pm 2.8$, aspect ratio $= 8.2 \pm 2.8$
Pd	$Pd(OAc)_2$	Oleylamine	$nBu_3N \cdot BH_3$	4
Ni	$Ni(OAc)_2$	Oleylamine Oleyic acid	$nBu_3N \cdot BH_3$	3
Fe@Rh	$Fe[N(SiMe_3)_2]_2$, $[Rh(allyl)_3]$	Tetramethyl-piperidine	$iPr_2HN \cdot BH_3$	1.7 ± 0.8
In aqueous media				
Fe	FeSO4	–	$H_3N \cdot BH_3$	~60 (aggregates)
Cu, Ag, Au	$CuSO_4 \cdot 3H_2O$, $AgNO_3$, $HAuCl_4 \cdot 3H_2O$	Poly(vinylpyrrolidone)	$H_3N \cdot BH_3$	Cu0: 4.3, Ag0: 4.7, Au0: 5.1
Co–Co_2B, Ni–Ni_3B,Co–Ni–B	$CoCl_2$, $NiCl_2$, or their mixture	–	$H_3N \cdot BH_3$	4–8
Ni	$NiCl_2$	Starch	$H_3N \cdot BH_3$/ $NaBH_4$	10
Pd	$Pd(OAc)_2$	Poly(vinylpyrrolidone)	$H_3N \cdot BH_3$	3.2 ± 0.5
Ru, Pd	$RuCl_3$, K_2PdCl_4	Poly(4-styrenesulfonic acid co-maleic acid) sodium salt	$H_3N \cdot BH_3$	Ru0: 1.9 ± 0.7 Pd0: 3.2 ± 1.5
Rh	$RhCl_3$	Sodium laurate	$Me_2HN \cdot BH_3$	5.2 ± 2.7
Ru	$RuCl_3$	Sodium laurate	$Me_2HN \cdot BH_3$	2.6 ± 1.2
Rh	$RhCl_3$	Sodium zeolite Y	$H_3N \cdot BH_3$	–

(continued)

Table 2.1 (continued)

Nanoparticles	Metal-containing precursor	Stabilizing agent	Reduction agent	Particle size, nm
Ni	[Ni(4,4′-bipy)(1,3,5-benzenetricarboxylate)]	MOFs (metal-organic frameworks)	$H_3N \cdot BH_3$	~100
Ni	$NiCl_2$	Poly(vinylpyrrolidone)	$H_3N \cdot BH_3$/ $NaBH_4$	–
Au–Ni@SiO2	[Au(этилендиамин)$_3$Cl], [$Ni(NH_3)_6Cl_2$], $Si(OC_2H_5)_4$	–	[$H_3N \cdot BH_3$/ $NaBH_4$]	3–4
Au@Co	$HAuCl_4$, $CoCl_2$	Poly(vinylpyrrolidone)	$H_3N \cdot BH_3$	~7
Amorphous Co	$CoCl_2$	–	$H_3N \cdot BH_3$/ $NaBH_4$	–

Table 2.2 Redox potentials of reduction agents in aqueous solutions (Referred by [51])

Electrode reactions	E, V
$H_2PO_2^- + 3OH^- = HPO_3^{2-} + 2H_2O + 2e$	E = −0.31–0.09 pH
$HPO_3^{2-} + 3OH^- = PO_4^{3-} + 2H_2O + 2e$	E = +0.14–0.09 pH
$H_2PO_2^- + H_2O = H_2PO_3^- + 2H^+ + 2e$	E = −0.504–0.069 pH
$H_3PO_2 + H_2O = H_2PO_3^- + 3H^+ + 2e$	E = −0.446–0.09 pH
$H_2PO_2^- + H_2O = HPO_3^{2-} + 3H^+ + 2e$	E = −0.323–0.09 pH
$H_3PO_3 + H_2O = H_3PO_4 + 2H^+ + 2e$	E = −0.276–0.06 pH
$H_2PO_2^- + 2H^+ + e = 2H_2O + P$	E = −0.248–0.12 pH
$HCHO + 3OH^- = HCOO^- + 2H_2O + 2e$	E = +0.19–0.09 pH
$HCHO + H_2O = HCOOH + 2H^+ + 2e$	E = +0.056–0.06 pH
$2HCHO + 4OH^- = 2HCOO^- + H_2 + 2e$	E = +0.25–0.09 pH
$BH_4 + 8OH^- = BO_2^- + 6H_2O + 8e$	E = −0.45–0.06 pH
$BH_4 + 8OH^- = H_2BO_3^- + 5H_2O + 8e$	E = −0.40–0.06 pH
$N_2H_4 + 4OH^- = N_2 + 4H_2O + 4e$	E = −0.31–0.06 pH
$N_2H_4 + 2OH^- = 2NH_2OH + 2e$	E = +1.57–0.06 pH
$N_2H_5^+ = N_2 + 25H^+ + 4e$	E = −0.23–0.075 pH
$NH_2OH + 7OH^- = NO_3^- + 5H_2O + 6e$	E = +0.683–0.07 pH
$2NH_2OH + 4OH^- = N_2O + 5H_2O + 64e$	E = −0.21–0.06 pH
$H_2 = 2H^+ + 2e$	E = 0.000–0.06 pH
$S_2O_4^{2-} + 4OH^- = 2SO_3^{2-} + 2H_2O + 2e$	E = +0.56–0.12 pH
$SO_3^{2-} + 2OH^- = SO_4^{2-} + H_2O + 2e$	E = −0.09–0.06 pH
$S_2O_3^{2-} + 6OH^- = 2SO_3^{2-} + 2H_2O + 4e$	E = +0.69–0.09 pH
$HSnO_2^- + H_2O + 3OH^- = Sn(OH)_6^{2-}$	E = +0.33–0.09 pH
$Ti^{3+} = Ti^{4+} + e$	E = −0.04
$Sn^{2+} = Sn^{4+} + 2e$	E = +0.151
$Cr^{2+} = Cr^{3+} + e$	E = +0.41
$Fe^{2+} = Fe^{3+} + e$	E = +0.771
$V^{2+} = V^{3+} + e$	E = −0.25
$R_2NHBH_3 + 3H_2O = R_2H_2N^+ + H_3BO_3 + 5H^+ + 6e$	$E = E_0$–0.05 pH

value of standard redox-potential E^0 of the Ag^+/Ag^0 system can decrease from +0.799 V to +0.23 V, when relatively unstable complex $Ag(SO_3)_2^{3-}$ ($pK_f = 8.68$) forms, to −0.290 V in the case of rather stable $Ag(CN)_2^-$ species ($pK_f = 19.85$) [116]. This tendency is also typical of the case when Ag^+ ions form deposits with decreasing solubility of the products. Thus, in AgCl ($K_{sp} = 1.82 \cdot 10^{-10}$) redox-potential Ag^+/Ag^0 decreases to +0.222 V, whereas deposition of far less soluble Ag_2S ($K_{sp} = 6.3 \cdot 10^{-50}$) causes a decrease in E^0 to −0.69 V.

Another characteristic, which allows efficient control of ΔE^0 is pH of a medium. An effect of pH of the medium can be illustrated using for example a redox system containing hydrazine-hydrate [116]:

$$N_2H_4 \cdot H_2O + 4OH^- \rightleftharpoons N_2 + 5H_2O + 4e^- \quad (2.7)$$

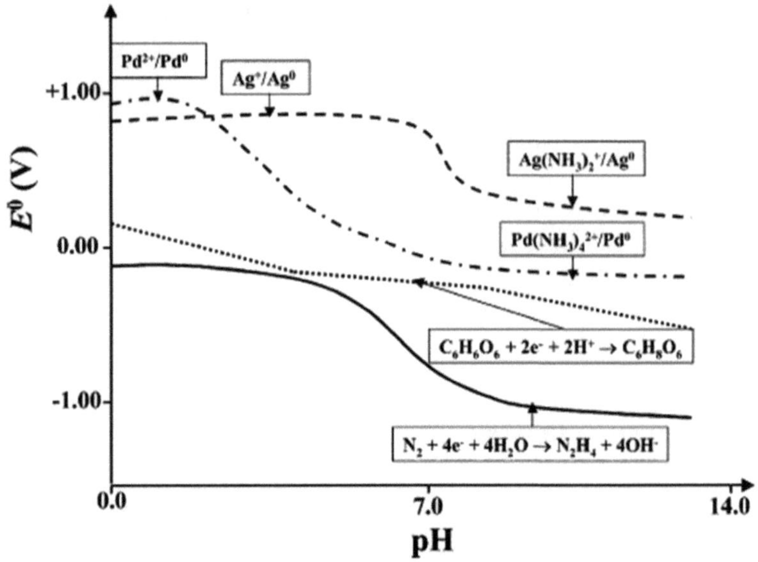

Fig. 2.11 The changes in the standard redox potential of Pd^{2+}, Ag^+, ascorbic acid, and hydrazine species with pH [116]

The value E^0 of this process decreases non-linearly from 21.16 in alkali medium to 20.21 V in acid medium, which proves that reducing ability of hydrazine hydrate decreases as OH^- concentration in a solution decreases. In the case when in reaction system a complex with a metal ion forms, the effect of general redox-reaction on E^0 becomes more complicated. It is known, for example, that ammonia forms a great number of complexes with ions of many transition metals of various stabilities. Figure 2.11 shows E^0/pH diagram, which reflects a complicated dependence of redox-potential on pH of a media in a system containing different alike complexes.

For reducing $Pd(NO_3)_2$ the different tetralkyl-ammonium carboxylates with general formula $(n\text{-}C_8H_{17})_4N^+(R'CO_2^-)$ (where $R' = CH_3$, Cl_2CH, $CH_3CH(OH)$, $HOCH_2$, $(CH_3)_3C$) can be used. A choice of the respective reducing agent allows control over size of forming Pd particles in the range 2.2–5.4 nm (Table 2.3) [117].

Carboxylates with electron-donor substituents (as in the case of pivalate) advance formation of nanoparticles of smaller size, while for the reducing agents with electron-acceptor groups (as in the case of trichloroacetate) the opposite effect appears which correlates with values of their oxidation potentials. The lower is $E_{p\,(Ox)}$, the stronger is reducing agent, the finer are particles. For example, $NaBH_4$ is a stronger reducing agent than hydrogen, and its using brings to formation of Pt particles of smaller sizes [118]. The same was observed during deposition of platinum on the surface of carbon nanotubes under action of $NaBH_4$ or ethylene glycol [70].

Similar laws are typical of many colloid systems.

Monodisperse Ag nanoparticles were used as templates for producing bimetal Ag-Au alloys in the substitution reaction between silver particles and $HAuCl_4$ [55]. Taking into account that a standard reduction potential of $AuCl_4^-/Au$ (1.0 V

Table 2.3 The influence of the redox potential of a reduction agent on the size of palladium nanoparticles [117]

Reduction agent $(n\text{-}C_8H_{17})_4N^+(R'CO_2^-)$	$E_{p(Ox)}$, V	pK_a	d, nm
$R' = Cl_2CH$, dichloroacetate	1.44	1.5	5.4
$R' = CH_3CH(OH)$, lactate	1.16	3.1	2.8
$R' = HOCH_2$, glycolate	1.17	3.8	2.9
$R' = CH_3$, acetate	1.07	4.8	2.0
$R' = (CH_3)_3C$, pivalate	1.08	5.0	2.2

vs. standard hydrogen electrode, SHE) is higher than for Ag^+/Ag (0.80 V), atoms of a silver nanoparticle oxidize to Ag^+ during mixing with $AuCl_4^-$ in the solution. The substitution reduction reaction can be written as:

$$3Ag\,(s) + AuCl_4^- \rightarrow Au\,(s) + Ag^+ + 4Cl^- \tag{2.8}$$

The formed gold atoms are absorbed by the surface of nanoparticles.

Significant interest for the reduction reactions is in polyoxymetalates (POMs), anion oxide structures (Keggin and Dawson structures) with their unique ability to demonstrate widely varied and finely controlled redox-potentials depending on structure and a degree of reduction. Some these POMs, including one-electron reduced Keggin POMs $SiW_{12}O_{40}^{4-}$ b $H_2W_{12}O_{40}^{6-}$ ($SiW_{12}O_{40}^{5-}$ and $H_2W_{12}O_{40}^{7-}$), univalent and bivalent Dawson phosphotungstates $P_2W_{18}O_{62}^{6-}$ ($P_2W_{18}O_{62}^{7-}$ and $P_2W_{18}O_{62}^{8-}$) and bi and tetravalent reduced Dawson phosphomolybdates $P_2Mo_{18}O_{62}^{6-}$ ($P_2Mo_{18}O_{62}^{8-}$ и $P_2Mo_{18}O_{62}^{10-}$) were tested in the reaction of reduction of Ag^+ to Ag^0 nanoparticles [119]:

$$POM(e^-) + Ag^+ \rightarrow POM + Ag^0{}_{colloid} \tag{2.9}$$

As is seen from Fig. 2.12, the value log*k*[1] for the reaction (2.9) is linearly dependent on the redox-potential corresponding to respective POMs, and the more electro-negative redox-potentials, the higher electron transport of electron to Ag^+.

2.2.2.2 The Effect of the Origin of a Reducing Agent

As it was discussed above, origin of a reduction agent can have a significant effect on anisotropy of formed nanoparticles. Often for producing anisotropic structures reducing agents of intermediate force are used. As $AgNO_3$ is reduced in presence of $NaBH_4$, nanoparticles with truncate octahedral structures form, in the case of 1, 2-hexadecandiol nanoparticles of icosahedron shape form (Fig. 2.13) [55].

[1] *k* is constant of the reaction velocity in per min, because this reaction is of the first order by POM and zero order by Ag+, *k* has been obtained from the ratio of the initial rate of re-oxidation of POM measured by a decrease in characteristic absorption at 730 nm to the initial POM concentration.

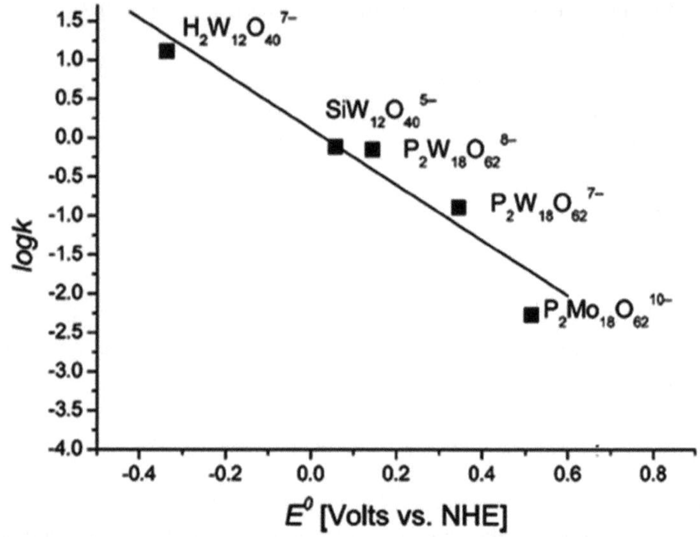

Fig. 2.12 Linear function of log*k* vs. the reduction potential of various reduced POMs for the reduction of Ag^+. [POMs] 1.0×10^{-4} M, [Ag^+] 0.5×10^{-4} M [119]

Moreover, formation of unusual structures, like Au truncate decahedrons favors formation of AgCl-oleyl-amine due to C=C bonds of NH_2 group [120]. Decahedrons had two types of truncation: parallel and perpendicular to the fivefold symmetry axis (Fig. 2.14).

There is an interesting approach called "in situ seeding method" based on introduction into a medium already containing intermediately strong reducing agent a small amount of a strong reducing agent. The aim of the latter is to initiate homogenous nucleation, and the further growth goes due to a weak reducing agent. This method is used to obtain golden nanoparticles shaped as rods/wires, rectangular, cube, and tetrapodlike, and also silver plates and discs [121]. Similarly, the mixture $NaBH_4$-ascorbic acid allows efficient control over nucleation and growth processes and fabrication of various anisotropic particles to grams weight [65, 122]. At higher concentration of borohydride faster nucleation is induced and the formed particles have small size (<5 nm) and spherical shape. If content of borohydride is low, nanoparticles have wide distribution of sizes and shapes; for intermediate nucleation rate nanoparticles become anisotropic and narrow-dispersed. It is interesting to notice that under conditions of seed-mediated growth the process of nanoparticles formation can be realized under medium conditions. It has been shown, for example, that Au seeds enhance reduction effect of ascorbic acid, which, as is known, is unable to reduce $HAuCl_4$ at room temperature [123].

2.2.2.3 The Effect of the Reducing Agent Concentration

A great excess of the reducing agent allows control over sizes of Cu nanoparticles [124, 125] and their shape [126–128].

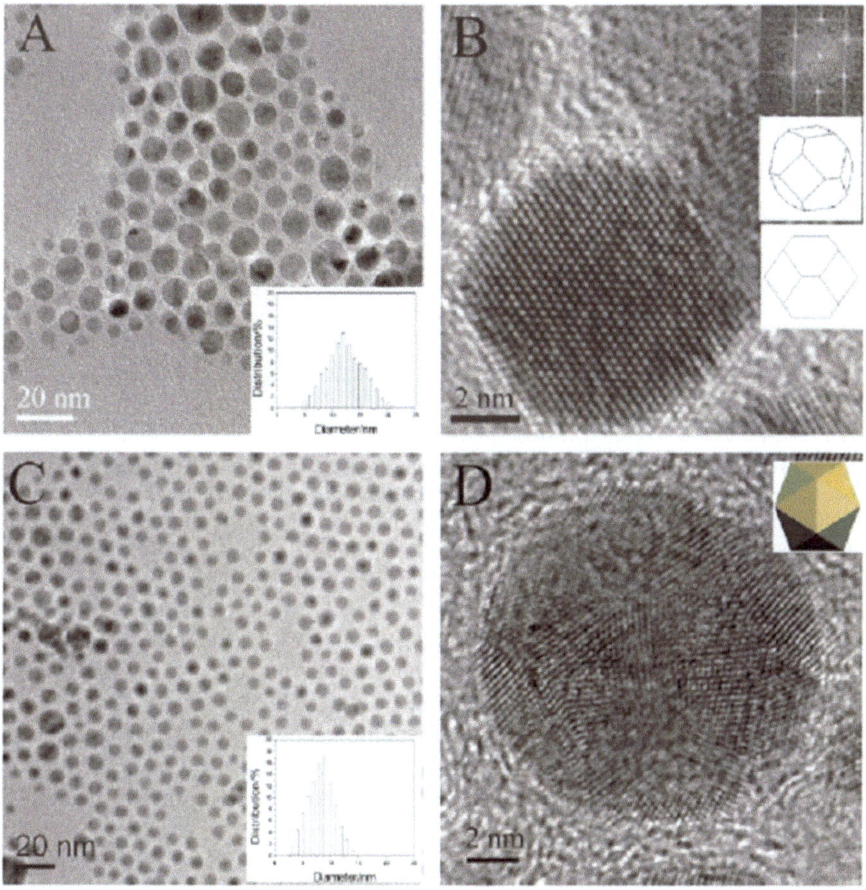

Fig. 2.13 TEM (**a**) and HRTEM (**b**) images and histogram of the Ag nanoparticles prepared by $NaBH_4$ reduction; TEM (**c**) and HRTEM (**d**) images and histogram of the Ag nanoparticles prepared by polyol synthesis; the *inset* shows the schematic illustration of an icosahedron [55]

Variation of the ratio $[N_2H_4]/[AOT]$ (AOT – di(2-ethyl-hexyl)sulfosuccinate Na) from 4.7 to 11.8 leads to increase in sizes of silver nanodiscs and to increase in their polydispersion [129]. Hydrazine hydrate molecules, probably absorb on crystallographic faces, like AOT do, and thus they carry on kinetic control over growth of nanoparticles [88, 130, 131].

The effect of citrate [132] or ferric ammonium citrate (FAC) [95] concentration on golden particle sizes is stronger than on their morphology. And, as is shown for the latter case, average edge length (158 ± 49 nm) of a polygonal nanoparticle at the molar ratio $[FAC]/[HAuCl_4] = 1.0$ far exceeds those found for the molar ratios of the reducing agent and $HAuCl_4$ 0.5 (100 ± 22 nm) and 1.5 (124 ± 24 nm), respectively.

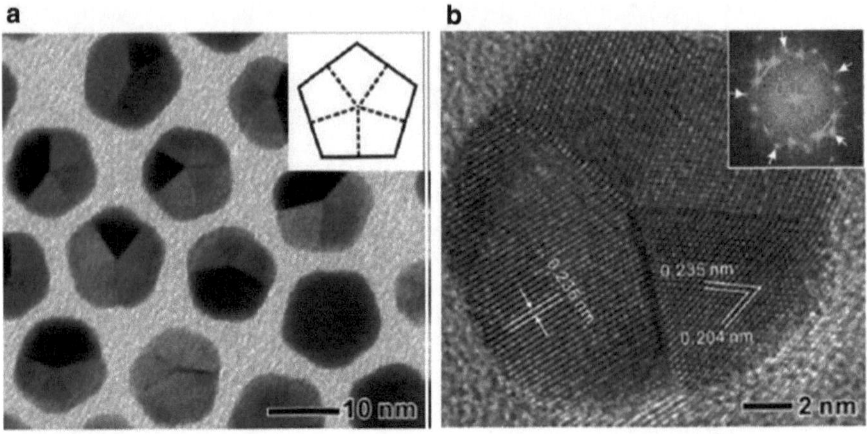

Fig. 2.14 (**a**) TEM image of truncated Au decahedrons. The *inset* shows a two-dimensional schematic drawing of the truncated decahedron. The *dashed lines* indicate the twin boundaries. (**b**) High resolution TEM image taken from an individual truncated Au decahedron recorded along the {011} zone axis and the corresponding FFT pattern (*inset*). In the pattern, the *spots* indicated by *arrows* can be indexed as the {200} reflections from the five single-crystal domains [120]

A change in concentration of the reducing agent $SiW_{12}O_{40}^{5-}$ from 1.0×10^{-4} M to 8.3×10^{-4} M causes an increase in the re-oxidation rate by Ag^{+} ions from $SiW_{12}O_{40}^{5-}$ to $SiW_{12}O_{40}^{4-}$(Fig. 2.15) [119].

An increase in concentration of the reducing agent is reflected in consequent formation of nanoparticles of smaller sizes (Table 2.4). At the lowest concentration $[SiW_{12}O_{40}^{5-}]_0$, 1.0×10^{-4} M the coarsest nanoparticles (>55 nm with wide distribution have been obtained).

2.2.2.4 The Effect of Temperature

At room temperature such reducing agents as citric acid, easily oxidized solvents (alcohols, DMF) and polymers (for example, polyvinyl alcohol) show weak reducing properties with respect to many metal-containing precursors. However, they are able to reduce most metal ions at high temperatures [133–136]. Depending on redox-potentials of a metal compound and a reducing agent, the required temperature of reaction can vary significantly.

In heterophase reactions the reducing regimes have a significant effect both on a degree of reduction and on dispersion of formed nanopowders. Cobalt nanopowders were synthesized by low temperature hydrogen reduction of cobalt hydroxide obtained via interaction of solid salts of cobalt with alkali solution [137]. Increase in temperature from 150 to 320 °C causes increase in a degree of reduction of cobalt hydroxide from 5 to 100 vol.% (Table 2.5).

However, specific surface of the powders decreases from 35.1 to 4.96 m^2/g. During reduction of the metal to 60 %, abrupt decrease in dispersion of the reducing products appears due to sintering of metal particles (Fig. 2.16).

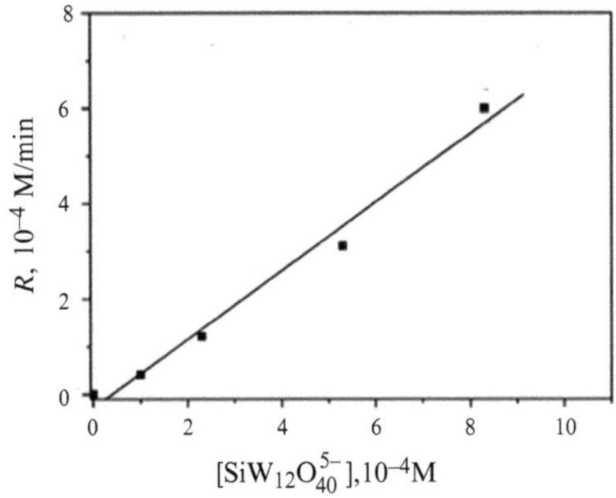

Fig. 2.15 Effect of the initial concentration of $SiW_{12}O_{40}^{5-}$ on the initial rate of reoxidation of $SiW_{12}O_{40}^{5-}$ by Ag^{+}; reduced POM [obtained by illumination of a deaerated solution of $SiW_{12}O_{40}^{4-}$ (1.0 mM), and propan-2-ol (1.0 M)], Ag^{+} (0.84×10^{-4} M) [119]

Table 2.4 Average size of silver nanoparticles obtained upon reduction of Ag^{+} (0.84×10^{-4} M) by various concentrations of $SiW_{12}O_{40}^{5-}$ [119]

$[SiW_{12}O_{40}^{5-}]_0$, M	Diameter, nm	r.s.d.,%
8.3×10^{-4}	28.2	±19.0
5.3×10^{-4}	38.0	± 21.4
2.3×10^{-4}	44.0	±10.6
1.0×10^{-4}	–	–

Note. *r.s.d.* relative standard deviation

Table 2.5 Characterization of Co nanopowders [137]

Reaction temperature, °C	Phase composition, vol.%	Reduction degree, %	Specific surface, m^2/g
150	$Co(OH)_2$-88.5; Co_3O_4-8.85; Co – 2.65; CoOOH-3.39; CoO-0.85; Co-4.24	4	35.1 ± 0.06
200	Co_3O_4-25.87; CoO-12.26; CoOOH-3.42; Co-59.25	59	61.04 ± 0.81
250	Co_3O_4-21.12; CoO-21.12; Co-57.76	58	6.06 ± 0.02
320	Co	100	4.96 ± 0.03
380	Co	100	4.38 ± 0.02
420	Co	100	3.39 ± 0.02

An increase in sizes of Pd nanoparticles (from 1.24 to 3.94) in the temperature interval from 15 to 60 °C during electrochemical reduction of Pd(II) nitrate is associated with some events, including high velocities of a stabilizer desorption from surface of nanoparticles at high temperatures [138]. It is interesting that the temperature-dependent change in conformation of poly(N-isopropyl acryl amide) has an effect on shape of platinum nanoparticles formed during reduction of K_2PtCl_4 by hydrogen [139].

On the whole, temperature is important parameter for regulation of the processes of heterogenous/homogenous nucleation and nucleation/growth. An illustrative

Fig. 2.16 The dependence of the specific surface of Co nanopowder on the degree of conversion during the reduction of cobalt (II) hydroxide by H_2 (the rate of flow 40 L/h) [137]

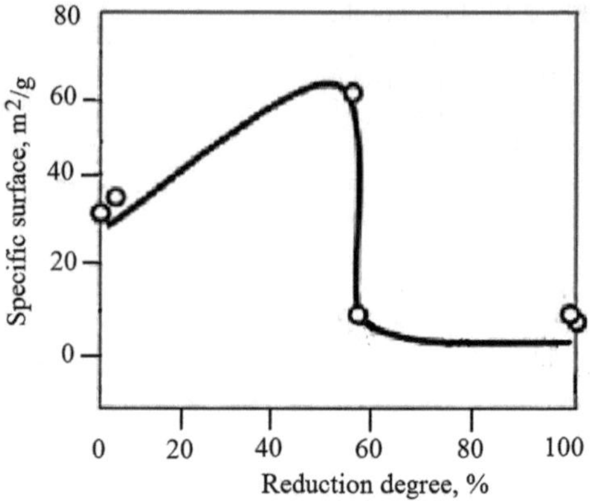

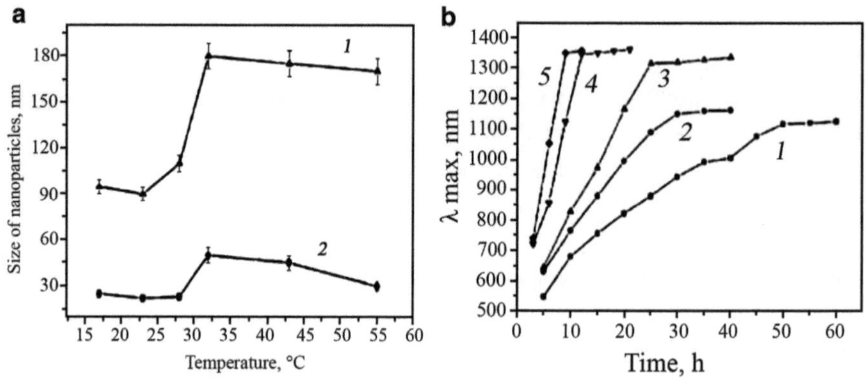

Fig. 2.17 (**a**) The average size of silver nanoplates (*1*) and nanospheres (*2*) obtained at different temperatures and (**b**) the growth rate of nanoplates *vis*. time on the λ_{max} data at temperatures 17 (*1*); 23 (*2*); 28 (*3*); 43 (*4*) and 55 °C (*5*) [140]

example is the systematic studies of chemical reduction of Ag ions in the temperature range from 0 to 55 °C [140]. The low temperatures dramatically reduce the formation and growth of nanoparticles and tens of hours are needed to complete the reaction. The shape and size of nanoparticles formed are rather not uniform. In the range temperature from 17 to 55 °C the rate of reaction as well as the size of the nanoparticles is increased (Fig. 2.17).

It is attracted attention that the increase of the size of particles occurred at 32 °C manifests an elongation of the face of nanoplates from 90 to 180 nm as well as an increase of the diameter of spherical particles from 25 to 48 nm. It was supposed that the growth of particles on this stage can follows to the mechanism of a combined fusion which differs from the Ostwald ripening at room temperature. The analogous peculiarities are characteristic for other noble metals. For example, selective deposition of platinum nanoparticles on carbon nanotubes was in the temperature range from 138 to 160 °C, while as temperature increases to 180 °C, there was aggregation of Pt nanoparticles in solution [70].

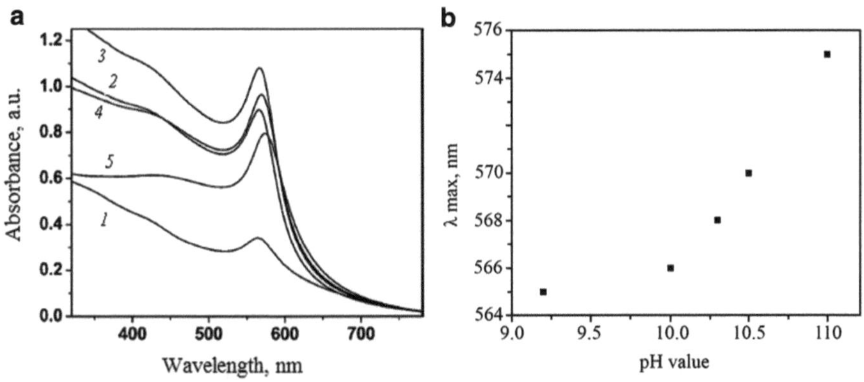

Fig. 2.18 (**a**) UV/Vis absorption spectra of copper nanoparticles synthesized in 1.0 mM aqueous PAA solution at different pH:: 9.2 (*1*), 10.0 (*2*), 10.3 (*3*), 10.5 (*4*) and 11.0 (*5*). (**b**) Absorption maximum of copper nanoparticles versus the pH value [141]

2.2.2.5 The Effect of pH

An important parameter in chemical reduction reactions is pH of a medium, which has an effect not only on sizes of formed nanoparticles, but on their shape and composition. Copper nanoparticles participate if, for example, pH of a solution is in the range 9.2–10.5. According to UV-vis spectra, intensity and position of the maximum change depending on pH (Fig. 2.18) [141].

At lower pH (9.2–10.3) peak of plasmon resonance at 565 nm is typical for copper nanoparticle [142], its intensity increases in these limits, which is the evidence of yield of nanoparticles. As pH to 10.5 and 11, the absorption peak shifts to $\lambda = 570$ nm и 575 nm, which can point to formation of metal copper nanoparticles in copper oxide shell [38, 143].

2.2.2.6 Kinetic Laws

The special features of nanoparticles is presence of so-called surface plasmon resonance (SPR), i.e. dramatic increase in intensity of absorption and spattering at some wavelength of incident light, which is in resonance with eigenfrequency of electron gas vibrations on a particle surface. Parameters of plasmon resonance are: value, spectral position, and FWHM serve for characteristics of nanoparticles and processes of their formation, which allows successful usage of the most informative method of electron spectroscopy for studies. For example, silver nanoparticles absorb light intensely with the maximum of plasmon resonance line in violet range of optical spectrum (390–450 nm). Solutions of gold nanoparticles have also maximum absorption in visible spectrum from 510 to 540 nm, copper nanoparticles have maximum at 550 nm [144].

Kinetics of formation of gold nanoparticles according to the data of UV-vis-near-IR spectroscopy at reduction of $HAuCl_4$ is shown in Fig. 2.19a [145].

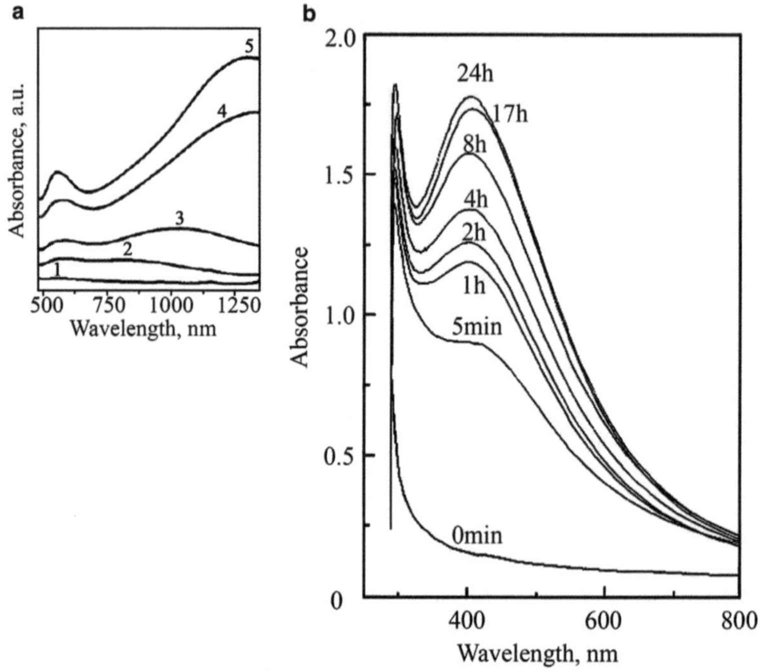

Fig. 2.19 (**a**) UV-vis-NIR absorption spectra of gold nanoparticles measured during the reduction of 10^{-3} M $HAuCl_4$ with *Aloe Vera* extract after 5, 7.5, 8, 9, and 25 h of reaction (curves 1–5, respectively). (**b**) UV/vis spectra of composite of polymer brush and Ag nanoparticles varying the reduction time [145]

The peak of SPR at 560 nm increases monotonously during the reaction, after 7.5 h a new band appears, localized at 817 nm, as the reaction is finished, long-wave absorption undergoes additional red shift. This evolution of the spectrum can be associated with aggregated spherical nanoparticles or with anisotropic structures whose sizes change in time. For example, appearance of new plasmon peak in the long-wave range of the absorption spectrum during irradiation of $AgNO_3$-ethanol-polyvinylpyrrolidone with argon-ion laser ($\lambda = 514.5$ nm) is associated with formation of coarse quasi-ellipsoid nanoparticles and with aggregates of Ag nanoparticles with fractal dimensionality (Fig. 2.20) [146].

Amplitude of plasmon resonance at 405 nm in the polystyrene-gr-poly (methacrylic acid-block-polymethacrylate)-silver nitrate increases during reduction and reaches maximum for more than 24 h, which allowed efficient control over growth of nanoparticles (Fig. 2.19b) [147].

2.2.3 *Reduction in Statu Nascendi in Polymer Matrix*

Taking into account instability of nanoparticles, most preparation techniques include usage of stabilizing agents, which absorb on surface of particles. During

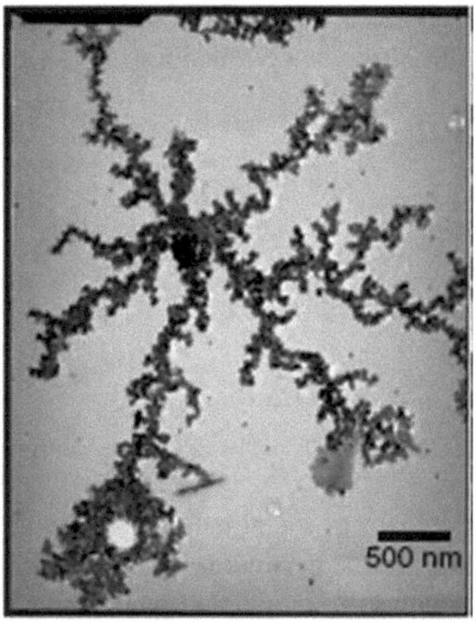

Fig. 2.20 TEM image of the aggregate formed in $AgNO_3$/ethanol/PVP-produced silver colloid [146]

steric stabilization, as was mentioned above, aggregation of the particles is prevented by absorption of large molecules (polymers or surfactants). According to some estimations, efficiency of stabilization with polymers can be expressed by their protective value [148], which for known polymers such as poly(N-vinyl-2-pyrrolidone), poly(vinylalcohol), poly(acrylamide), poly(acrylic acid), and poly (ethylene imine) is 50.0, 5.0, 1.3, 0.07, and 0.04, respectively.

Apart from steric limitations [149]

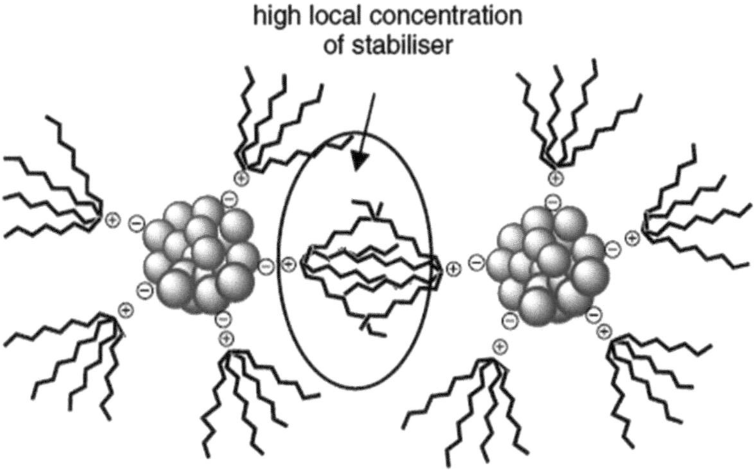

polymer molecule can weakly bound with a surface of a particle via heteroatoms acting as ligands [150].

As a rule, reduction in the medium of polymer template is preceded by formation of macromolecule complex of metal precursor with polymer molecule. The key stage in synthesis of dendrimer—template/Ag—nanocomposite is interaction of

Ag^+ ions with terminal OH-groups of the dendrimer on the basis of 3.5-dihydroxybenzyl alcohol [151] (Scheme 2.1):

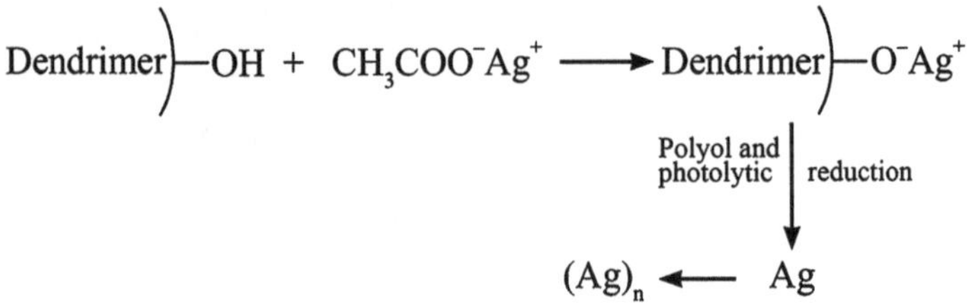

Scheme 2.1 A pathway for the dendrimer/silver nanocomposites formation

Sizes of metal nanoparticles and their size distribution increase as generation of the dendrimer increases, which correlates with increase in the dendrimer size with increase in number of generations:

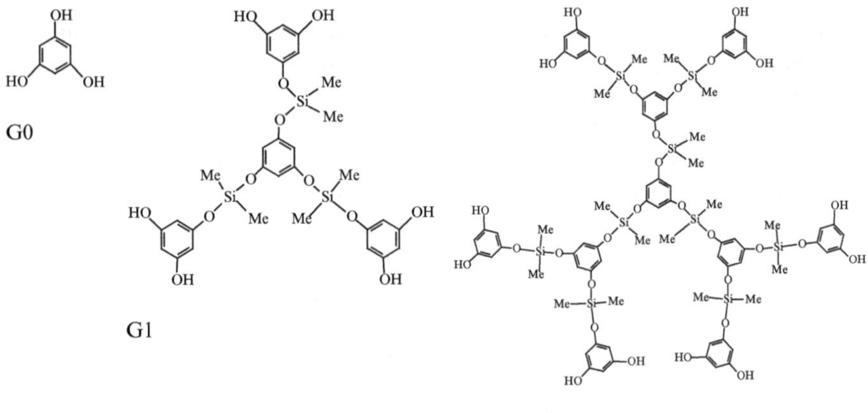

G2

It is known that as number of generations of dendrimers increases, they aggregate [152, 153], which forms template places of various sizes, in which metal nanoparticles grow.

In the case of poly(vinylpyrrolidone) most widely used stabilizing and reducing agents in chemical reactions of producing nanocomposites [154–156], the preserving effect is associated with presence of a free pair of electrons of a nitrogen and a oxygen atoms in polar PVP polar groups, which can be donated in two sp hybrid orbitals of metal ions with formation of complex ions [157] (Scheme 2.2):

Two formed types of complexes can induce various nucleation processes [158]. Moreover, it is assumed that a bond between a PVP and a metal remains after reduction of the metal [159]. The PVP molecule can prevailingly absorb on some crystal faces of a seed, inhibiting or accelerating their growth and, consequently, it advances anisotropic growth of metal nanoparticles [91, 104].

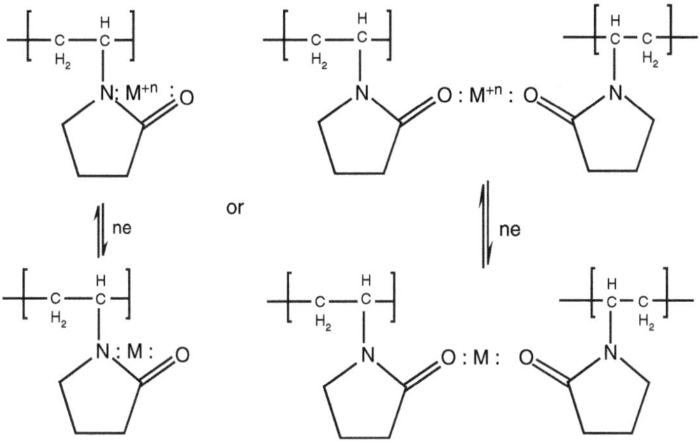

Scheme 2.2 Complexation of metal ions with poly(vinylpyrrolidone) chains

This method can be applied to fabrication of bi- and multicrystalline nanoparticles. For example, scheme of Bi_2Te_3 synthesis includes coordination of Bi(III) and Te(III) ions with polyvinylpyrrolidone, formation of a macrocomplex, and the following reduction in presence of $NaBH_4$ [160] (Scheme 2.3):

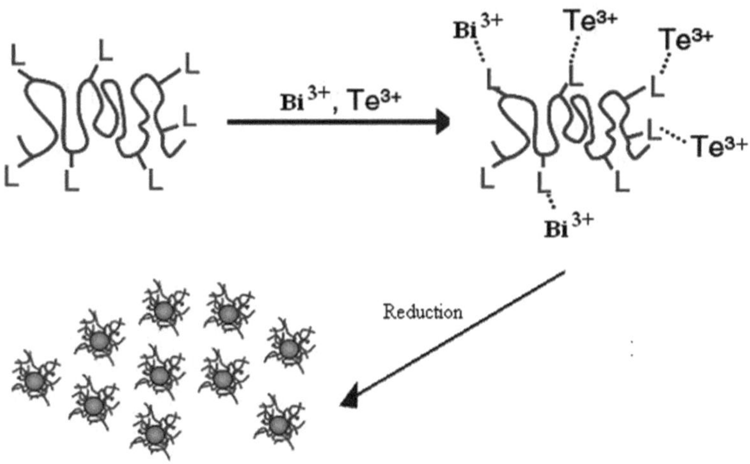

Scheme 2.3 Synthesis of bimetallic nanocomposites

Depending on a way of synthesis, a nature of the used precursors and reagents, particles with various properties, sizes and forms can be obtained. A simple example of synthesis of Ag nanoparticles during reduction of silver nitrate conducted in various media and in presence of various reducing agents [161] can demonstrate importance of all parameters and conditions of the reaction. Colloid dispersions of silver particles from 1 to 600 nm in size are obtained in water

Table 2.6 Synthesis and characterization of Ag nanocomposites (0.2 M $AgNO_3$, 0.1 M aqueous solution of stabilizing agent, ascorbic acid is a reduction agent) [161]

Stabilizing agent	The shape of particle	The size of particle
–	Rice balls consisting of rod like particles	2–4 μm, 300–600 nm
Pluronic PE PEO–PPO–PEO-block-copolymer non-ionic surfactant	Rice balls consisting of rod like particles	2–4 μm, 300–600 nm
Plurafac LF non-ionic surfactant	Rice balls consisting of rod like particles	2–4 μm, 300–600 nm
Gummi arabicum polysaccharide	Rice balls consisting of rod like particles	2–4 μm, 300–600 nm
LutensitA-ES sodium alkylphenol ether sulfate anionic surfactant	Thin large plates	–
Tamol VS vinylsulfonate, Na salt	Crystallites, leafs	From 5–6 to 20 μm
Emulphor FAS 30 alkylphenolate-free ether, Na, salt	Crystallites, leafs, hexagons, plates, spheres	From 2–4 to 3 μm
Polyvinylpyrrolidone	Hexagons, plates, spheres	From 1–3 μm to 300–500 nm

solution under action of ascorbic acid, ferric ammonium citrate, sodium borohydrate and hydroquinone. Polymers of different origins were used as stabilizing media, their variations can be efficiently used to control size and shape of colloid particles (Table 2.6).

Thus, apart from origin and structure of a polymer template, a crucial effect on characteristics of obtained nanocomposites has many parameters of the reduction reaction in polymer matrix in nascendi (concentration of a metal salt, stabilizing and reducing agents, temperature, crystal structure, and sizes of preliminary prepared or in situ grown seeds, etc.). Some of these features are considered below.

2.2.3.1 Reducing Agent

Using of strong chemical agents allows conduction of a reaction under soft conditions (at room temperature and atmospheric pressure). For example, silver nanoparticles were obtained in PMMA solution in 1,2-dichloroethane, in which ammonia solution of silver oxide was introduced (with concentration 10^{-2}–10^{-4} M) in presence of hydrate hydrazine ($2 \cdot 10^{-2}$ M) [162]. Dodecylamine as a reducing agent requires more severe conditions. CdSe nanoparticles stabilized by carbazole containing poly (benzyl ether) dendrons were obtained via reduction of cadmium oxide by dodecylamine initially during heating in vacuum (120 °C) with the following heat treatment at 270 °C in nitrogen atmosphere and fast injection of selenium and trioctylphosphine [163].

2.2.3.2 The Effect of Concentration of Polymer Ligand

It has been shown that sizes and shapes of copper nanoparticles are substantially dependent on concentration of a stabilizing ligand, polyacrylic acid (PAA) [141]. At low concentration of PAA (1.0 mM) nanoparticles were predominantly large (~80 nm) and spherical. If concentration of PAA is 1.33 mM, crystalline faced (cube-octahedron) particles formed with the size ~50 nm. At concentration of PAA 2 mM average size of nanoparticles decreased to 30 nm, and their polydispersion increased. It interesting to point out that under similar conditions, but with usage of polyvinylpyrrolidone, it is impossible to obtain metal nanoparticles of copper, their synthesis is accompanied by formation of Cu_2O nanoparticles in wide range of pH [164]. At high concentration of PVP or short chains (M_W 10000) silver ions reduce on the surface of seed particles without additional heating or other actions [165]. Under these synthetic conditions homogenous star-like, multipod nanoparticles with high yield (100 %) form (Fig. 2.21).

Using of polycrystalline golden nanoparticles (15 nm in diameter) for seeds can produce spheroid or armed Au nanoparticles at simple variation of experimental conditions of the reaction. The key feature is, probably, reduction ability of poly (vinylpyrrolidone) molecules, which has an effect on the reduction kinetics of gold and, consequently, on the deposition rate of Au atoms on the surface of Au seeds. Consequent interactions of PVP via absorption/desorption processes and high ratios

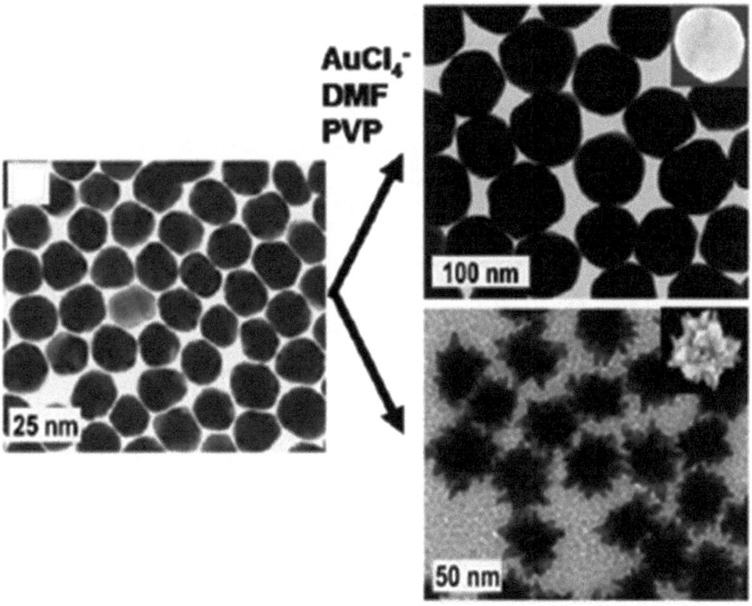

Fig. 2.21 Catalytic growth of the seeds into Au spheres or Au nanostars at high concentration of poly(vinylpyrrolidone) [165]

of poly(vinylpyrrolidone) links to Au atoms cause fast, kinetically controlled and preferable growth along certain crystallographic faces, which finally is ended by formation of star-like nanoparticles.

2.2.3.3 The Effect of Molecular Weight of a Polymer Ligand

A possibility of steric stabilization is defined by the fact that spatial dimensions of even low molecular compounds are compatible with the radius of London attracting forces or even more. If it is assumed that "diameter" of a macromolecule of a linear polymer coincides with the root-mean-square distance between its ends, then the relationship between the average geometrical size of the particle $<r^2>^{0.5}$ and molecular mass of the polymer M can be expressed as [166]

$$< r^2 >^{0.5} \approx 0.06M^{0.5} \tag{2.10}$$

In the case of the polymers with $M = 10^4 < r^2 >^{0.5} = 6$ nm, and at $M = 10^5$ this value is already 20 nm. Consequently, the molecules of the polymers with $M > 10^4$ have the same sizes, which are necessary for stabilization of colloid particles. And a polymer with higher molecular weight has more efficient ability to limitation of growth and agglomeration, as it was found in the system Ni nanoparticles- poly (vinylpyrrolidone) (Fig. 2.22) [167].

Molecular weight of a polymer plays the crucial role in geometry of particles. Poly(vinylpyrrolidone) with molecular weight M_w 1300000 was used to synthesize silver decahedron with high yield [157]. As molecular weight of PVP decreases to M_w 40000, polydispersion particles formed, only 40 % of which had the shape of decahedron, and in the case of even lower molecular weights always spherical nanoparticles formed.

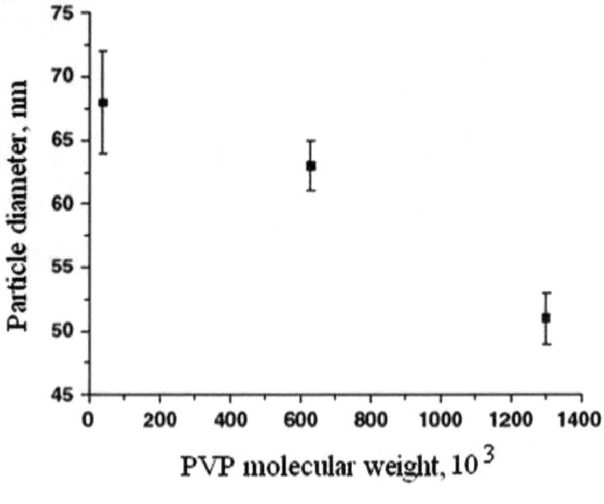

Fig. 2.22 Relationship of poly(vinylpyrrolidone) (PVP) molecular weight and synthesized Ni nanoparticles [167]

2.2.3.4 Supercritical Media

The prospective is the approach based on reduction of metal containing precursors in polymer matrices in supercritical media [168–170]. The main mechanism is diffusion of a solution of organic metal complex solved, for example, in supercritical CO_2 in polymer matrix, not soluble, but swelled. The last fact makes it possible to widely vary origin of a polymer, for example, use different dendrimers [171], natural polymers, such as chitosan [172], etc. In these systems usually H_2 is used as a reducing agent, which is well compatible with supercritical CO_2. Changing density of media can additionally help to regulate size of nanoparticles, as it was shown for Au nanoparticles synthesized in situ in pores of MCM-48 [173]. Size of formed nanoparticles was from 2 to 25 nm under pressure of CO_2 17 and 7 MPa, respectively.

In general case the stabilizing action of highly molecular compound is determined by some additional parameters, in particular, by degree of polymerization, type and number of functional groups, character of their distribution in chain, etc. Interaction of a protective polymer with a forming nanoparticle can be realized via absorption of macromolecules on a surface of the particles or their chemosorption from the solution. Physical absorption is governed by Van der Waals forces, dipole interactions, or hydrogen bonds. Non-covalent interactions of nanoparticles with a macromolecule are very weak, about 10^{-4} J/m^2. There are interactions of nanoparticles with PE, PVA, PEO, PVP, polyethylene glycol macromolecules, etc. [174]. Polymer molecules can be easily displaced from the zone of interaction during Brownian motion of particles [175]. In particular, PEO weakly interact with metal surfaces, therefore, it is a bad stabilizer of colloid gold [176]. For enhancing of stabilizing properties of polymers binary protective systems are sometimes used, as, for example, PVP dodecylamine in polyol synthesis of Ni nanoparticles [167].

Polymer chains during chemosorption can form covalent, ionic or coordination bonds with atoms of surface layers of a metal. Introduction of various functional groups in a polymer causes an increase in the role of acid-base interactions in a polymer-metal system and, respectively, to an increase in adhesion ability of a polymer. Below some polymer systems are considered, which are used as stabilizing media in chemical reactions of production of nanocomposites.

2.2.3.5 Block-Copolymers as Nanoreactors for Synthesis of Nanocomposites

Block-copolymers form micelles in diluted solutions of polar solvents, such as water, or in non-polar organic medium. Micelles of block-copolymers have a definite core-shell structure, consisting of a core of insoluble block and a shell or crown of a soluble fragment of a copolymer. The methods of obtaining of block-copolymers, characteristics of their compositions and properties are widely presented in literature, there are many monographs and reviews considering fundamental principles of their self-organization, structure, and application in synthesis of nanomaterials [177–180].

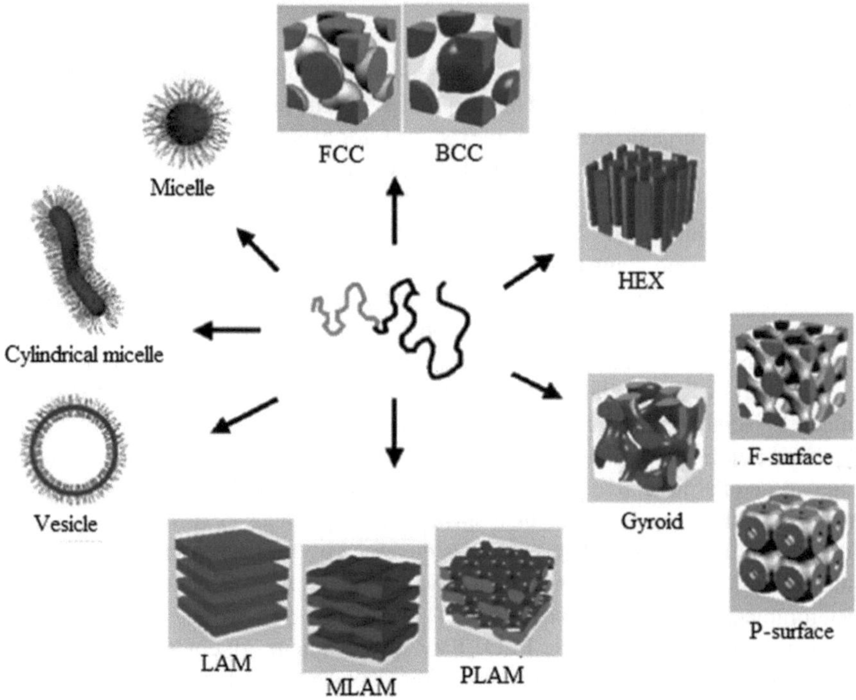

Fig. 2.23 Self-organization structures of block copolymers and surfactants: spherical micelles, cylindrical micelles, vesicles, fcc- and bcc-packed spheres (FCC, BCC), hexagonally packed cylinders (HEX), various minimal surfaces (gyroid, F-surface, P-surface), simple lamellae (LAM), as well as modulated and perforated lamellae (MLAM, PLAM) [179]

Block-copolymers in a solution and in block state (as films) have microphase separation with formation of various types of morphological structures (spheres, rods, vesicles, crew-cut structures, lamellas, cylinders, etc.) (Fig. 2.23) [179, 181, 182].

Shape of micelles can be controlled by various parameters, such as a relative block length, content of ion, composition of solvents, etc. Mainly, morphology depends on the ratio of block length *f* and the segregation parameter χN, where N is the total degree of polymerization $N = N_A + N_B$, χ is Huggins parameter, which characterizes the repulsive force between the polymer blocks. The value χ decreases with increase in temperature according to the dependence $\chi \sim 1/T$. Changing a degree of polymerization provides control over the superlattice unit cell parameter in wide from some nanometers to hundreds nanometers. Figure 2.24 shows this dependence for poly(butadiene-b-ethylene oxide) (PB-PEO) [179].

The initial part of the curve corresponds to short chained non-ionogenic surfactants, which confirms universality of their behavior and block-co-polymers, and, with respect with theoretical predictions, can be described by the relationship: $d \sim d_0 N^{2/3}$ ($d_0 \sim 0.9$ nm).

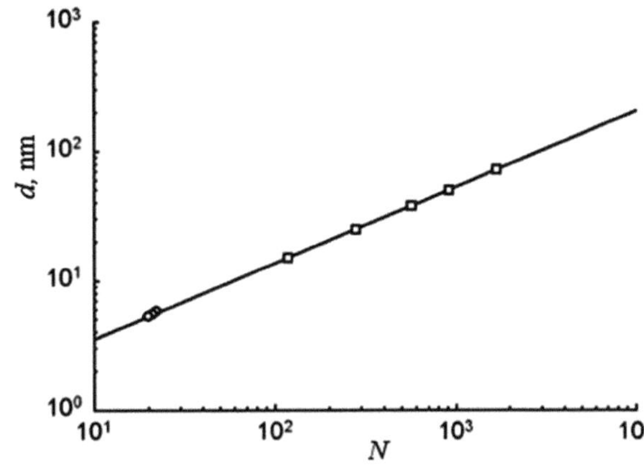

Fig. 2.24 Long period *d* as a function of the degree of polymerization (*N*) of PB-PEO (□) and non-ionogenic surfactants (O)

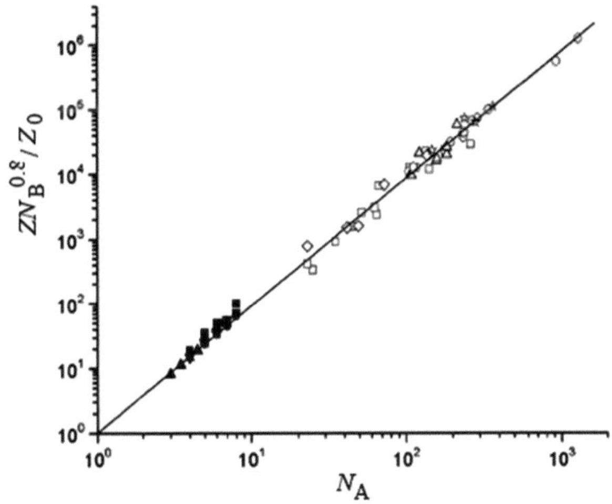

Fig. 2.25 Aggregation numbers Z as functions of the degree of polymerization N_A of the insoluble block. o, □: diblock copolymers (PS-P4VPy, PS-PMAc), Δ: triblock copolymers (PMAc-PS-PMAc), ∇: graft copolymers (PSMSA-g-PEO), *: PS-P2VPy heteroarm star polymers, ◊: PS-PAAc heteroarm star polymers, •: nonionic surfactants, ■: cationic surfactants ($RNMe_3Br$) ▲,▼: anionic surfactants ($ROSO_3Na$, RSO_3Na) [179]

The dependence of a number of aggregates *Z*, i.e. a number of block-polymers in a micelle on a degree of polymerization of a soluble block N_A (Fig. 2.25) can be described by equation (referred to by [179]):

$$Z = Z_0 N^{\alpha}{}_A N^{-\beta}{}_B \tag{2.11}$$

(where $\alpha = 2, \beta \sim 0.8, Z_0$ depends, mainly, on the mixing enthalpy of insoluble polymer block A and solvent), which can be applied to many bi- and tri-block

Table 2.7 Morphology of polymer particles and composites

n/n	Morphology	$\rho = R_g/R_h$
1.	Homogeneous sphere, random coil, monodisperse	0.788
	Θ-conditions	1.50
	Good solvent	1.78
2.	Random coil, polydisperse z = 1	
	Θ- conditions	1.73
	Good solvent	2.05
3.	Regular stars	
	Θ- conditions, f = 4	1.33
	Good solvent, f >> 1	1.079
4.	Rigid rod	
	Monodisperse	>2.0
	Polydisperse	>2.0

grafted and star-like co-polymers and for low molecular cation and anion surfactants, which confirms similarity of their self-assembling mechanism, and diameter of the micelles can be directly determined by a degree of polymerization of blocks, which is important for the controlled synthesis of nanocolloids.

It is known that for rigid spheres $R_g = 0.778R_h$ (R_g is gyration radius, R_h is hydrodynamic radius), and the parameter R_g/R_h changes dependently on architecture of particles and their geometry. In the Table 2.7 are shown values of parameter $\rho = R_g/R_h$ for macromolecules of various architectures [183].

On the whole, the coarser is a particle and higher its anisotropy, the higher is ρ. For example, in the PS(300)-*b*-P[2VP-Au0.5(300)] system gold nanoparticles are agglomerated with increase in R_g and parameter ρ [183], i.e. it can be assumed that the concentration at which R_g and R_h begin to increase corresponds to the critical concentration of micelle formation (CMC) of a block copolymer. Moreover, the longer a block forming a core, the lower CMC of the block-copolymer providing that length of polymer chains of the crown is approximately the same.

The theory predicts that insertion of a particle or a molecular solute in a surfactant micelle leads to perturbation in the initial structure due to the core swelling [184, 185]. However, if concentration of a solved substance is low, the structure of a surfactant micelle is predominantly determined by molecular properties of a polymer. On the contrary, arrangement of a polymer surfactant on macroscopic surface in selective solvents depends on many parameters, including relative affinity of each block of a copolymer to the surface and to the solvent, length of blocks, interphase boundary, and concentration of the polymer. These parameters have a substantial effect on thickness and regularity of the surfactant layer. A significant assumption of many theories of absorption is that curvature of a surface is small as compared to length of a polymer and thickness of a film. And, as a result, it is unclear which type of a monolayer structure is formed by a block-polymer on a strongly curved surface of a nanoparticle: absorbed layer, micelle, etc. It has been shown that curvature of a surface has effect on thickness of the absorbed layer when the curvature radius reaches gyration radius of a polymer ($\rho/R_g \approx 1$) [186]. It is assumed that there is a range of particle sizes, beyond which polymer surfactants combine surface absorption and micelle formation [187]. In other words, block-polymers on the surface of a nanoparticle can form as a grafted layer if it is a coarse nanoparticle or a micelle, if a particle is small. For Au nanoparticles, whose size is more than 10 nm, a core-shell structure forms (Au@PS-*b*-PAA) which is well described by the general model of polymer absorption [188]. Thickness of a layer depends not on absolute concentration of a polymer or Au nanoparticles, but on their relative amount. If Au sizes are small, they dissolve as molecular factions in the core of a micelle of polystyrene-block-poly (acrylic acid) (Scheme 2.4).

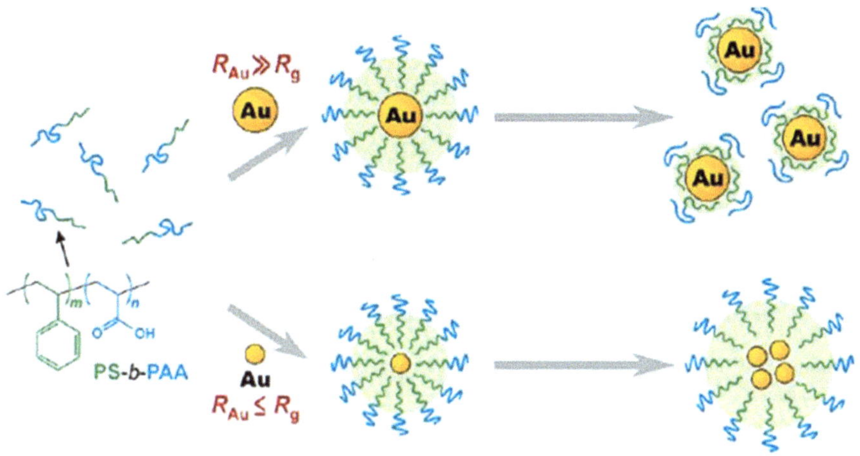

Scheme 2.4 Formation of absorption layers and micelle structures in block-copolymer nanocomposites

Relative and absolute length of the polymer blocks also has an effect on structure of absorbed polymer layer. Moderately asymmetric and symmetric copolymers (PS_{159}-*b*-PAA_{62}, PS_{49}-*b*-PAA_{54}) do not absorb on the surface of nanoparticles, while strongly asymmetric polymers (PS_{250}-*b*-PAA_{13}, PS_{160}-*b*-PAA_{13}, PS_{100}-*b*-PAA_{13}, $PMMA_{240}$-*b*-PAA_{13}) form quite well thick surfactant shell.

Various copolymers used as polymer templates are widely applied in colloid synthesis of metals: in particular, block-copolymers of styrene with poly(acrylic) acid [181], 2- and 4- vinylpyridine [183, 189–191], polyethylene oxide with 4-vinylpyridine [192], and poly(3-caprolactone) [193], etc.

Many representatives of star-like polymers $(AB)_n$, where n is a number of bi-blocked branching, which can form inverse micelles with complicated structures, relate to the block-polymers of the considered type [191, 194–196]. Micelles of this type have monomolecular structure. Thus, (PS-b-P4VPy)4 micelles with petal structure form if toluene is added to (PS-b-PVPy)4 in CH_2Cl_2 solution [194] (Scheme 2.5):

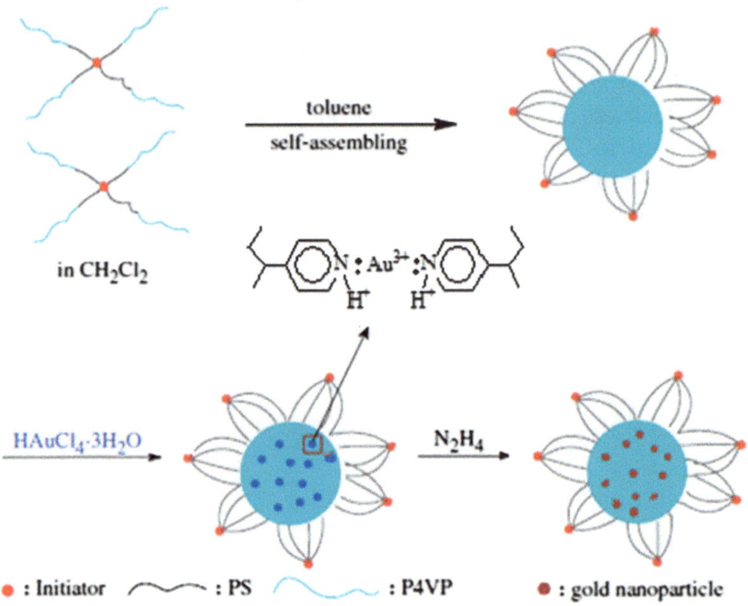

Scheme 2.5 Schematic representation for the preparation of reversed micelles and gold nanoparticles

In this case the exterior blocks (P4VPy chains) form a core and PS chains (interior blocks) serve a petal-like shell. It is interesting to note that independently on length of P4VPy chains micelles preserve spherical shape ($R_g/R_h = 0.82$, where R_g is gyration radius, R_h is hydrodynamic radius, see Table 2.8), though it is known that at high ratio R_g/R_h PS-b-P4VPy has cylindrical morphology [197].

Dense structure of petal-like formation prevents interaction of P4VPy links between two neighbor micelles, and thus prevents their aggregation. Nanoparticles formed in these templates have narrow and homogeneous size distribution in the matrix space (2.5 ± 0.2 nm, 2.4 ± 0.3 nm and 2.0 ± 0.2 nm).

The considered method of synthesis of nanocomposites is mainly based on usage for templates of already formed and well-organized block-copolymer micelles. However, there are systems in which formation of micelles proceeds also in solutions which are "good" for block polymers owing to interaction of inorganic ion with a copolymer or so called ion induced micelle formation [198–201]. Problems of influence of metal ions and nanoparticles on self-assembling processes stay in many ways unclear.

Table 2.8 The parameters of micelles of block-copolymers [194]

Block copolymers	D_h, nm	R_g, nm	R_g/R_h	D_h^{com}, nm	R_g^{com}, nm	R_g/R_h^{com}
$(PS_{25}\text{-b-}P4VP_{12})_4$	26.4	8.4	0.64	39.0	20.7	1.06
$(PS_{25}\text{-b-}P4VP_{32})_4$	44.3	13.6	0.61	51.5	21.3	0.83
$(PS_{25}\text{-b-}P4VP_{64})_4$	74.0	30.4	0.82	59.0	24.0	0.81

D_h, R_g, and R_g/R_h are data of the reverse micelles and, D_h^{com}, R_g^{com}, and R_g/R_h^{com} are data of complex micelles

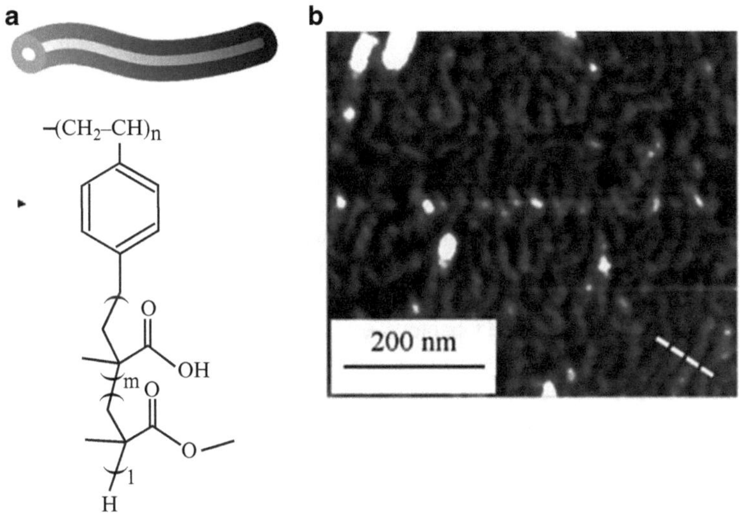

Fig. 2.26 (**a**) Schematic structure and (**b**) AFM photograph of amphiphilic copolymer PS-gr-(PMA-b-PMMA) of double-cylinder-type [147]

Soft conditions of synthesis make it possible to obtain silver nanocomposites based on polystyrene-g-(polymethacrylic acid-block-poly(methyl methacrylate) (PS-g-(PMA-b-PMMA) dissolved in organic acids [147]. Specific structure of polymer template in the shape of two-layered cylinders advances solubility (Fig. 2.26). However, formed nanoparticles have quite wide size distribution spread (7.5 ± 2.3 nm).

However, on the other hand, these systems provide a good possibility for control of particle size and their morphology during change in degree of microphase distribution via variation of a solvent type or molar ratio of hydrophilic and hydrophobic components. In the case of poly(urethane acrylate –styrene) copolymer containing (poly)propylene and poly(ethylene oxide) fragments

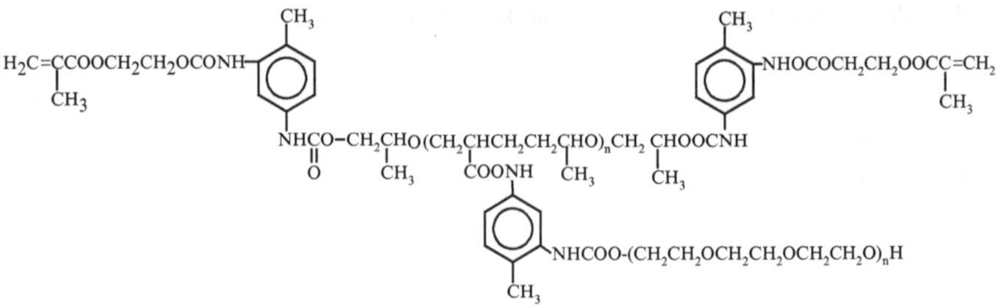

complex formation of cadmium acetate with hydrophilic poly(ethylene oxide) chain favors formation of nanodomain structure with its further fixation during copolymerization [202] on the whole, increases microphase separation.

2.2.3.6 Hydrogels

Polymer cross-linked hydrogels working as nanoreactors in synthesis of colloid metal nanoparticles attract a considerable attention due to their potential application in catalysis [203, 204], photonics [205], drug delivery [206] and biomedicine [207, 208]. In the latter case a decisive role has biocompatibility of these polymers with biomolecules and tissues (see Chapter 7).

Really, substantial free spaces in the linked polymer network in the swelled state work as nanoreactors during nucleation and growth of nanoparticles. It has been shown that size and morphology of nanoparticles can be controlled by concentration of a linking agent and functionality of a reticulate gel [209–211].

Poly(acrylamide) (PAAm) gels are very efficient for synthesis of nanoparticles [212–216]. Usually, PAAm gels are produced by polymerization of AAm in presence of bi-functional cross-linking agent, for example, *N,N′*-methylene *bis*-acrylamide:

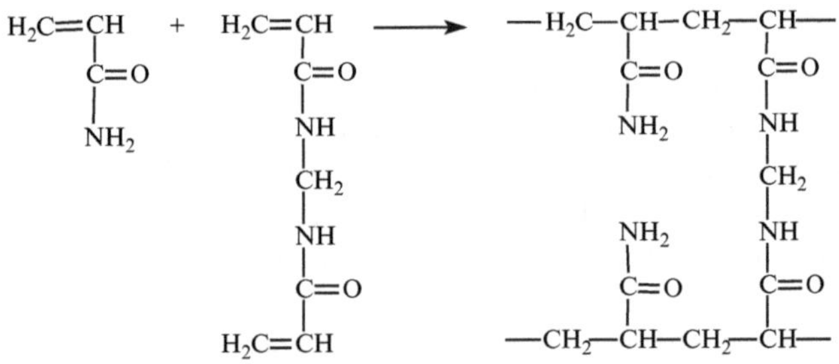

Metal ions are bound with functional groups of a cross linked polymer making monomolecular complexes, their concentration can be varied in quite wide range. In some cases metal ions are introduced already at the stage of polymerization forming in situ polymer complex, which is then subjected to reduction [217, 218].

A special interest in development of nanocomposite materials is drawn to microgels with a response to internal actions (of pH, temperature, ionic force, etc.) [219–221]. Most often used temperature sensitive component in metal/microgel

Table 2.9 Polymers with a response to temperature obtained by RAFT polymerization [222]

Polymer	The response	T °C
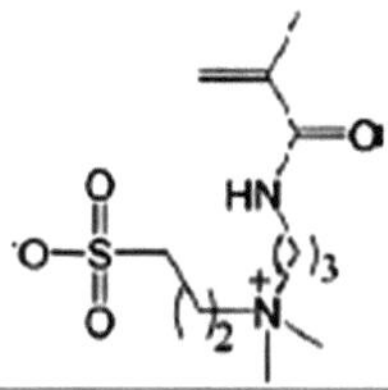	UCST	12
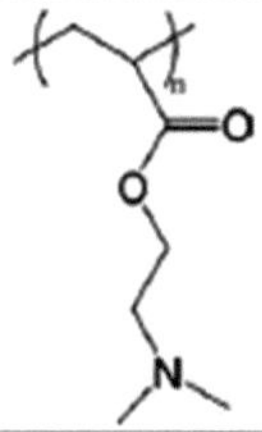	LCST	50
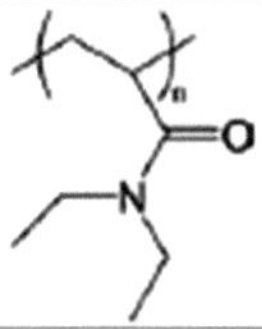	LCST	26–35
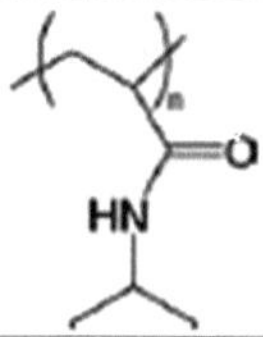	LCST	32
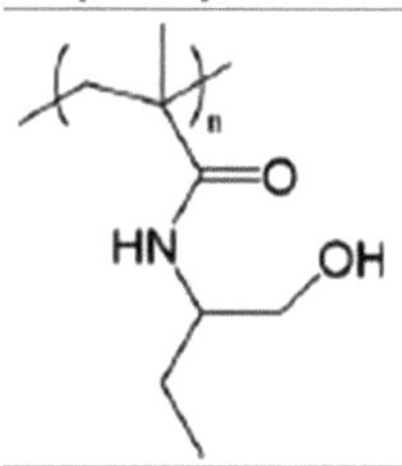	LCST	30

Note. *RAFT* is the reversible addition fragmentation chain transfer, *UCST* is upper critical solution temperature, *LCST* is low critical solution temperature

nanocomposites is poly(N-iso-propyl-acrylamide) (PNIPAAm). The temperature of phase transition, so called (the lower critical solution temperature) (LCST) of PNIPAAm is 32–34 °C, and it can be finely regulated during co-polymerization with other monomers (Table 2.9) [222]. In these microgels such monomers as acrylic acid [223], methacrylic acid [224], acrylic acid-2-hydroxy ethyl acrylate [225], 2-amino ethyl acrylate [226], etc. are used as active centers to bind metal ions. Hybrid Ag/poly[N-iso-propyl-acrylamide-(maleate carboxymethyl chitosan)] nanocomposite was obtained by reduction of silver salt with $NaBH_4$ in the microgel medium (Fig. 2.27) [227]

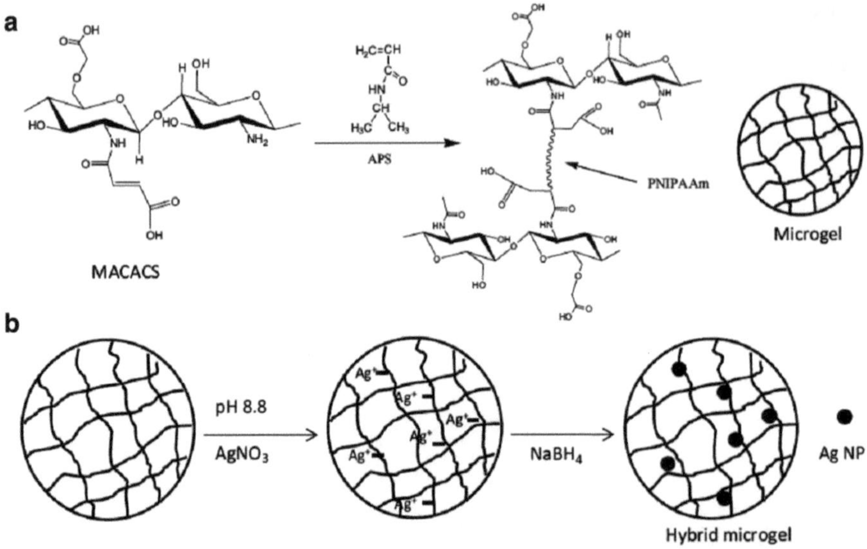

Fig. 2.27 Schematic representation of the synthesis of PNIPAAm/MACACS microgels (**a**), which were used as microreactors for the in situ preparation of Ag nanoparticles (**b**) [227]

The obtained particles are stable in wide range of pH (2–10) and temperature (15–50 °C). Hybrid nanocomposites with silver nanoparticles have the same LCST as the initial polymer microgel (Fig. 2.28).

The temperatures of phase transition of a hybrid nanocomposite were 32 and 35 °C at pH 3 и 8, respectively. For the Pt/poly[(vinylamine)-*co*-(*N*-vinylisobutyramide)] the temperature of flaking was by 0.2–1.6 °C lower than LSTC of the initial copolymer and the difference grows with increase in content of vinylamine comonomer [228]:

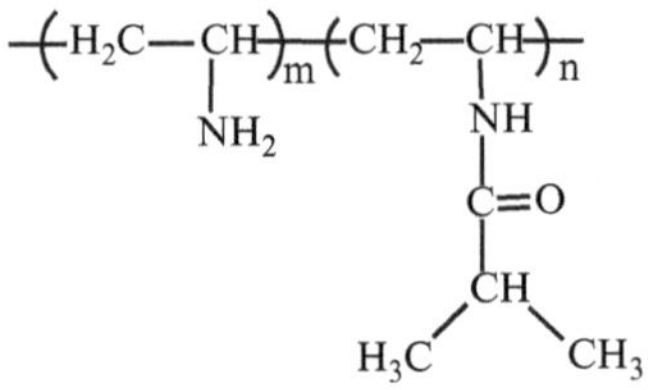

m = 4.1, 8.3, and 19.8 mol %

It should be noted that due to formation of a strong complex of Pt with the polymer chain, strong reducing agent, sodium borohydride, not alcohol, is used for its reduction:

$$H_2PtCl_6 + 6NaBH_4 + 18H_2O \rightarrow Pt_0 + 6NaCl + 6H_3BO_3 + 22H_2 \qquad (2.12)$$

Particularly, formation of the complex, not of colloid platinum was observed in the poly(ethyleneimine) and H_2PtCl_6 system in the ethanol/water mixture [229].

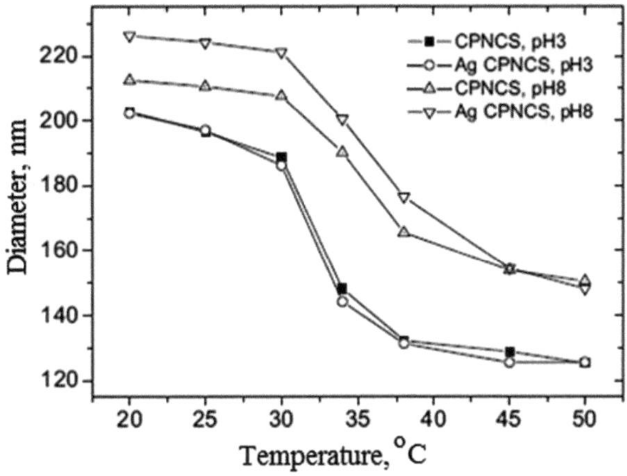

Fig. 2.28 Temperature dependence of the hydrodynamic diameters of the poly[*N*-isopropylacrylamide-(maleated carboxymethylchitosan)] microgels with and without Ag nanoparticles at pH = 3 and 8 [227]

The promising materials for template matrices used for synthesis of nanoparticles, with widespread abilities for regulation of functionality, porosity, ion-exchanging capacity, are ion-exchange resins [230, 231]. A considerable importance has their commercial availability. Usually they are produced as two physical structures: gels and microporous structures. The gels are homogeneous, have no discrete pores, whose function play channels; the microporous resins are characterized by presence of fixed pores. Strong acid resins (gels) are functioned by sulfonic acid groups, while weak acid resins (macroporous) contain carboxylic groups:

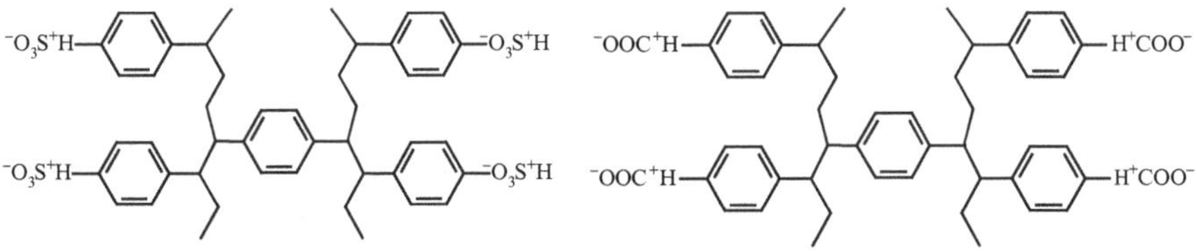

Presence of functional groups in a polymer matrix makes it possible to bind them with metal cations by ion-exchange mechanism and then reduce as, for example, during production of Ni(0)-containing nanocomposites [232] by the following scheme (Scheme 2.6):

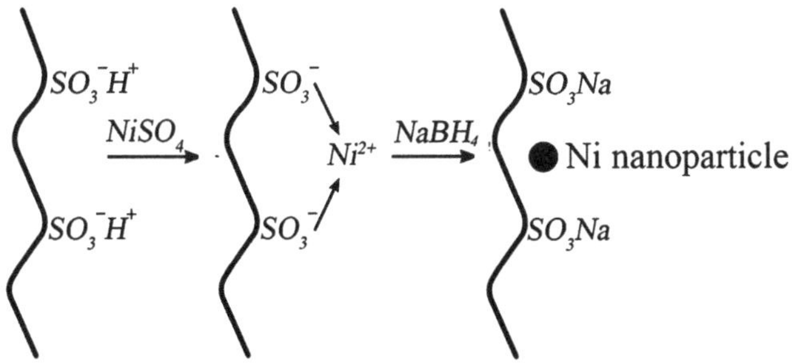

Scheme 2.6 Ion-exchange mechanism of the Ni(0)-nanoparticles formation

Table 2.10 Pd content and loading efficiency for the different prepared nanocomposites [233]

Matrix	Pd loads	Pd content, meq/g	Loading efficiency, %
Sulphonated polyethersulphone with Cardo group	1	0.31 ± 0.02	70.5 ± 0.1
Blend membranes	1	0.08 ± 0.01	30.8 ± 0.1
Blend membranes	2	0.16 ± 0.01	61.5 ± 0.1
Nafion	1	0.53 ± 0.04	48.2 ± 0.1

Redox-process can be described by the following reactions:

$$2Ni^{2+} + 4BH^{4-} + 9\,H_2O \rightarrow Ni_2B + 12\,H_2 + 3B\,(OH)_3 \tag{2.13}$$

$$4Ni_2B + 3O_2 \rightarrow 8Ni + 2B_2O_3 \tag{2.14}$$

It should be noted that average sizes (19 nm) of nickel nanoparticles in strong acid gels are smaller than in the respective macroporous resins (about 40 nm), probably, it is caused by agglomeration of nanoparticles in pores, whose sizes vary in the range 60–70 nm. Repeated loading and reduction cycles provide achievement of content of magnetic active phase in a nanocomposite up to 21 % (gel) and 16 % (microporous resin). The fact that the origin of a polymer membrane plays a crucial role in the loading-reduction cycles is confirmed by the data on concentration of Pd nanoparticles in catalytic membranes (Table 2.10) [233].

2.2.3.7 Polyelectrolites (Polymeric "Brushes")

In the recent years a great interest is attracted to the polymer systems called polymeric brushes [234]. The polymeric brush is a monolayer of polymeric chains bound to some surface by end groups. There are several ways of fabrication of polymeric brushes: chemical, when the end groups are grafted to a surface by a chemical bond, another one is based on ability of polymer molecules to self-organization due to in- and intermolecular interactions [235]. Covalent bonding can be realized in two ways: "grafting to" and "grafting from" technique. In the first case the end functional group of a polymer chain under some conditions settles on the grafted surface. However, this way has some limitations caused by low density of grafting and thickness of the formed polymeric film, through which the polymer molecules should diffuse to reaction places of setting. Therefore the "grafting from" way is preferable. This approach is based on development of a surface-immobilized layer of initializing agent and the following in situ polymerization with generation of the polymeric brush. Thick polymeric films form which are bound by covalent bonds to a substrate with high density of grafting. In this method a convenient way of polymerization is atom-transfer radical polymerization (ATRP) [236–240]. The main advance of the controlled radical polymerization for creation of polymeric brushes is a possibility of a brush thickness control, narrow molecular –weight distribution, and a possibility of synthesis of block-copolymers during re-inducing of the end sleeping chains, etc.

Two-component copolymers of this structure can serve as polymeric carriers for metal nanoparticles. For example, core-shell polymeric brushes of cylindrical or spherical structure are used in synthesis of Ag [147], Au [241, 242], CdS [243], Pd and Pt [242], Au and Pt [244] nanoparticles. The general scheme of production of nanoparticles looks as follows (Scheme 2.7):

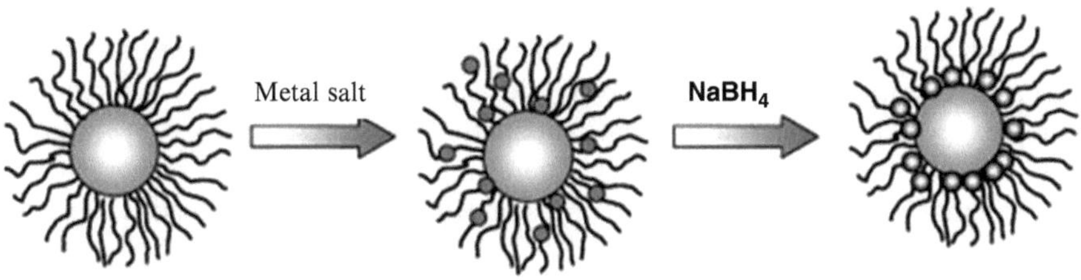

Scheme 2.7 A general scheme of template synthesis of nanoparticles in polymer brushes

As a rule, polymer "brushes" form close-packed grafted chains on a surface of a polymer core, so that the total contour length of the grafted chain far exceeds average distance between them on the surface, as it is, for example, typical of a polyelectrolytic copolymer poly(styrene)-g- poly(2-methylpropenoxyethyl trimethyl ammonium chloride) [242]:

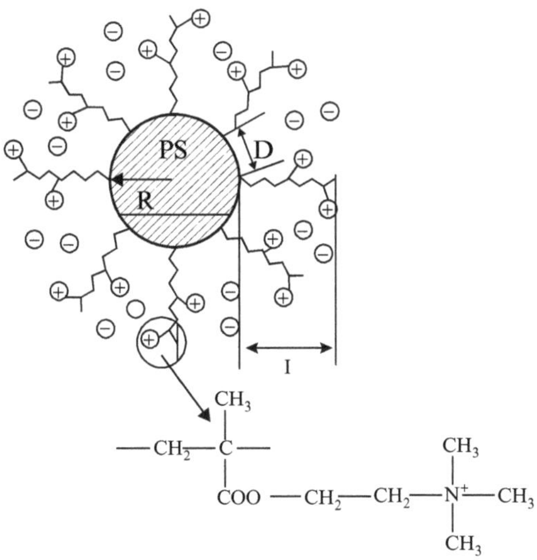

Advantage of this approach is that reduction of a precursor and formation of nanoparticles take place in the polymeric layer.

Multilayer thin films of polyelectrolytes are widely used as nanoreactors in synthesis of metal (Ag, Ni, Pd, Cu) and semiconductor (PbS) nanoparticles [225, 245–248]. A special interest is in polyelectrolytic polymeric brushes shaped as films with response to external actions, so called “smart surface coatings” [249]. Hybrid inorganic/polyelectrolytic brush complexes Si/SiO_2//PS-*b*-PAA-(Ag^+) и Si/SiO2//PS-*b*-PAA(Pd^{2+}) are obtained by the Scheme 2.8 with fixation of the initializing agent on a silicon surface at the first stage with the following synthesis of Si/SiO_2/polystyrene of the polymeric brush with settling by ATRP method of the second bi-block component, *tert*-butyl acrylate, which is subjected to acid hydrolysis with transformation into acrylic acid functional group. Interaction of the obtained polyelectrolytic brush with a metal salt causes condensation of the latter in the layer of blocked polyacrylic acid and formation of the required complex, which is reduced to metal nanoparticles in presence of H_2 (2 atm) at 120 °C during 48 h [250]. It is interesting that content of silver in the nanocomposite was 4.9 %, which corresponded to 87 % of the loaded block of polyacrylic acid (PAA).

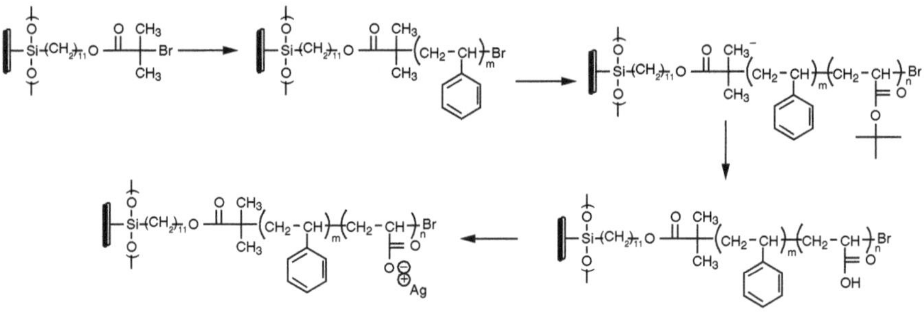

Scheme 2.8 Synthesis of the Si/SiO_2//PS-*b*-PAA(Ag+) brush on a silicon substrate [250]

Polymeric brushes of the considered type response to stimulating external action, i.e. pH of a matter and low molecular electrolyte (NaCl, $CaCl_2$) (Fig. 2.29).

As pH increases from 2 to 10, thickness of a polymeric brush increases from 16 to 26 nm. Increase in size of the polymeric brush at addition of a base is appearance of well-known polyelectrolyte effect. As far as the acid groups along the polymeric chain become deprotonated like carriers causing Coulomb repulsion and, therefore, compelling the polymer to take extended conformation, which causes increase in the brush thickness. On the contrary, with addition of electrolyte, as was expected, thickness of the polymeric layer decreases from 22 to 17 nm with increase in [NaCl] concentration, and increases from 24 to 20 nm with increasing of [$CaCl_2$] concentration.

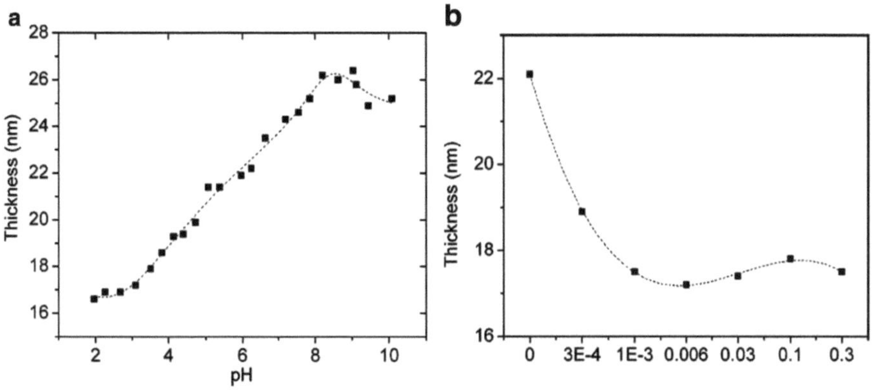

Fig. 2.29 The dependence of the polymer brush thickness of a Si/SiO2//poly(acrylic acid) brush on the solution pH (**a**) and on the concentration of NaCl (**b**) [250]

2.2.3.8 Polyethylene Glycols

Polyethylene glycols are widely used in synthesis of nanocomposites, especially for obtaining of water-solved and biocompatible systems due to complex of valuable properties, such as resistance to oxidation, effect of acids, and bases, moderate thermal stability, etc. Polyethylene glycols have good solvating characteristics and complex forming with respect to many metals [251]. In the literature also special term "PEGylation" is used, which means absorption or coating of the surface of nanoparticle with polyethylene glycol molecules via surface absorption, covalent bond, capture, etc. [252, 253]. It is important preserved by PEG or its copolymers gain not only water-soluble or biocompatible properties, they also become "invisible" for protective systems of organism, thus widening capacities of biomedical application of the nanocomposites [253–255].

Polyethylene glycols are widely used as structurally-oriented templates in synthesis of inorganic nanoparticles with various morphology and sizes [256, 257]. In the PEG media nanoparticles of gold, silver [258] and antimony [259] are obtained. Functional possibilities of polyethylene glycols can be considerably broadened in the case of grafted copolymers based on PEG, such as polyethylene imines [260], PMMA [261], polyamidoamines (PAMAM) [262]. PEGylated nanostructures and their biomedical application (see Chap. 7) are described in detail in the recent review [252].

2.2.3.9 Organo-element Polymers

Element-organic polymers are rarely used as stabilizing ligands in synthesis of nanoparticles. It has been shown that though poly(vinylene arsine) and the polymers obtained via radical co-polymerization of phenyl acetylene with hexaphenyl-cyclohexa-arsine by the Scheme 2.9:

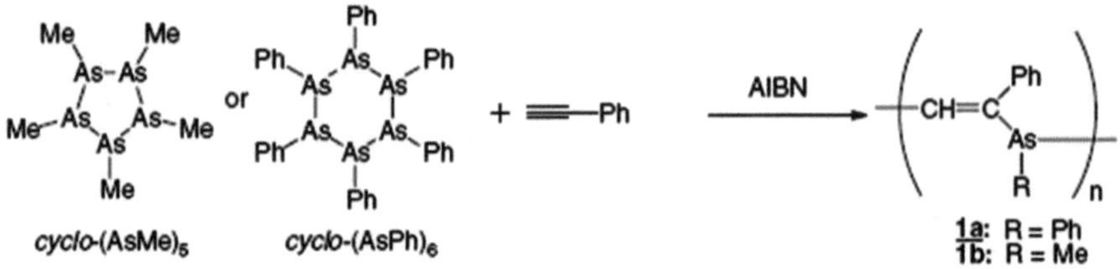

Scheme 2.9 Synthesis of organo-element polymers based on arsenium

favors formation of colloid silver particles, but to the end of several hours there is increase in particle size and its aggregation, the weaker stabilizing agent is phenyl derivative, probably, due to spatial limitations [263].

Combination of simultaneous reduction of a metal ion and formation of a cross-linked polymer in situ under UV irradiation and the following heat treatment at 120 °C of epoxy resin containing silver tri-fluorine acetate provides production of one-, two- and three-dimensional nanostructures [264].

2.2.3.10 Polymeric Microspheres

One of the interests in polymer microspheres is caused by a possibility of metal nanoparticles settling on their surface, which provides substantial accessibility of active centers in the nanoparticles to reagents, for example, in catalytic reactions and, thus, it enhances efficiency of catalysts based on them. Nanoparticles, as a rule, are strongly bound to a microsphere surface and cannot be removed either during severe mixing, or during ultrasonic treatment. Various types of polymers are used as such carriers, for example, microspheres of amino-functioned polystyrene [265], polystyrene-co-poly(N-iso-propylacrylamide) [266], poly(styrene-co-4- vinylpyridine) [267], microspheres with a grafted layer of polyelectrolyte [220, 268, 269]. To immobilize nanoparticles also polymeric ion liquids are applied, first of all, due to their high ability to ion exchange [270, 271] (Fig. 2.30).

Formation of highly dense and well-dispersed Pt nanoparticles [272] (diameter about 2 nm) is defined by low values of interphase tension of imidazole groups on the surface of a microsphere, which causes efficient nucleation in parallel with low rates of Ostwald ripening [273].

2.2.4 Polymers as Reducing Agents and Templates

Very attractive are the methods when a stabilizing polymeric ligand is simultaneously a reducing agent. The advantages of this method are obvious. This is displacement of

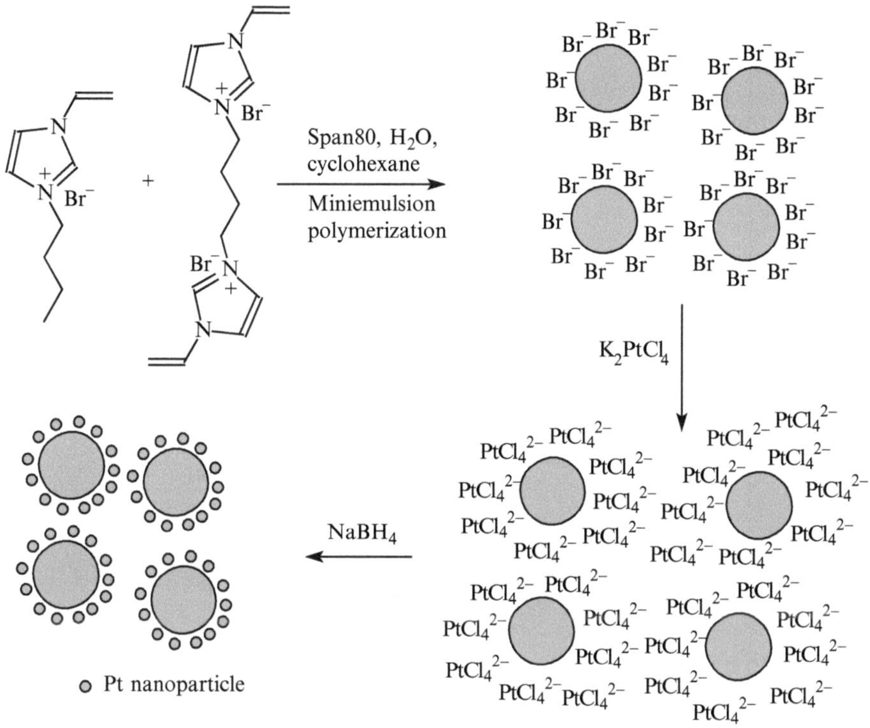

Fig. 2.30 Reaction scheme for the preparation of Pt/poly(1-butyl-3-vinylimidazolium bromide) microspheres [271]

toxic reducing agents, lower expenses, higher efficiency, etc. Most often for these purposes are used polymers whose molecules contain a great number of different functional groups (polysaccharides [274], humic substances, peptides, etc.).

2.2.4.1 Oxygen-Containing Reducing Centers

Representative in this view are cellulose derivatives [275, 276], methyl hydroxyethylcellulose (MHEC), hydroxymethylpropylcellulose, carboxymethylcellulose (CMC). Studies of the mechanism of these process in the system gold-chloride-hydrogen acid- MHEC or its anion derivative CMC [277] point to participation of primary hydroxyl groups in reduction of gold ions, and the formed carboxylic groups interact in some way with the surface of formed microparticles, thus stabilizing them. Dual origin of CMC is demonstrated also for in situ production of Au nanoribbons [278]. It should be noted that formation of intermediate alcohol molecules is associated with self-regulated reduction properties of long chained acryl sulfates [278]:

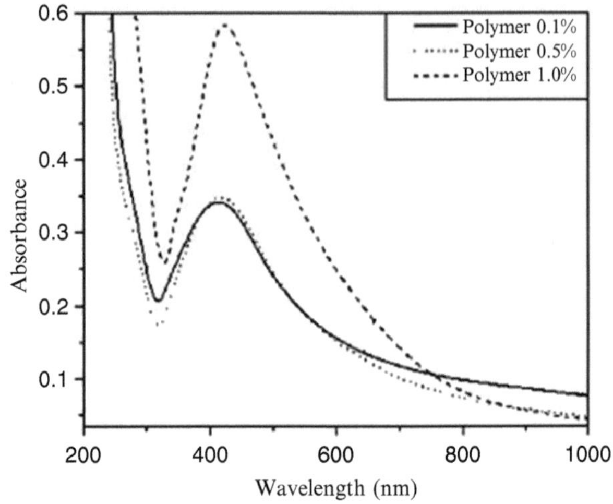

Fig. 2.31 UV-vis spectra of colloidal Ag nanoparticles at a varying concentration of PVM/MA. $[AgNO_3] = 10^{-4}$ M [279]

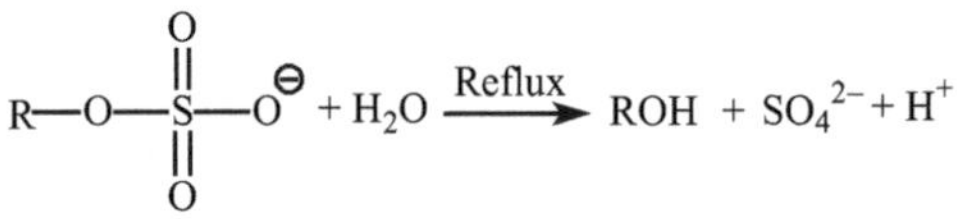

$$R_nOH + M^{2+} + H_2O \rightarrow R_{n-1}COOH + M^0 + 2H^+$$

Reduction ability in the system changes in the range C8 > C10 > C12 > C14 and diameter of the particles decreases as the length of alkyl chain increases.

Reducing character of action of the polymer in the reaction mixture poly(methyl vinylic ether-co-maleic anhydride) (PVM-MA)-$AgNO_3$ points to accumulation of Ag nanoparticles with increase in the copolymer concentration (Fig. 2.31), whereas average diameter stays unchanged [279].

Role of the reducing agent in these systems plays -OCH_3 groups of PVM/MA [280] (Scheme 2.10):

Amphiphilic polyethers are products of polycondensation of sebacic acid and polyethyleneglycol (Mw = 9,200 g/mol and 22,600 g/mol) consisting of alternative hydrophilic and hydrophobic links:

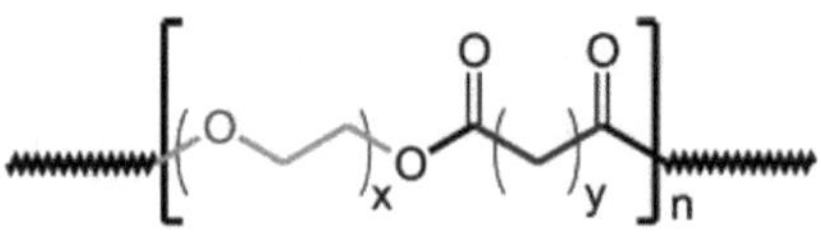

were used as nanoreactors and reducing agents for synthesis of Ag, Au, Pd nanoparticles in benzene or water media [281, 282]. It is known that pseudo-crown-etheric structures PEG can bind and reduce metal ions [258, 283]. In principle, formation of amphiphilic copolymer and metal nanoparticles can be performed simultaneously. Thus, reaction of co-polymerization of urethane

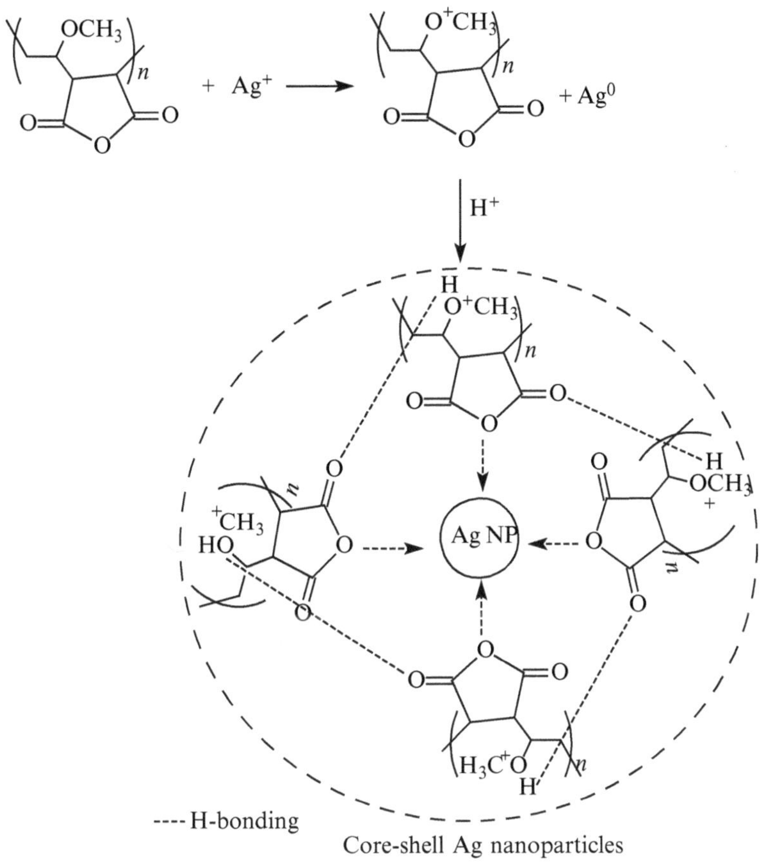

Scheme 2.10 Reduction of Ag^+ by CH_3O-groups of copolymer chain

acrylate oligomer based on poly(propylenoxide)triole, 2,4-toluene-di-iso-cianite, 2-hydroxyethylmethacrylate and polyethylene glycol with styrene in presence of silver nitrate causes formation of polymer nanocomposite film, containing silver nanoparticles [284]. Reduction of metal ion goes with participation of primary or growing radicals in polymerizing system by the scheme:

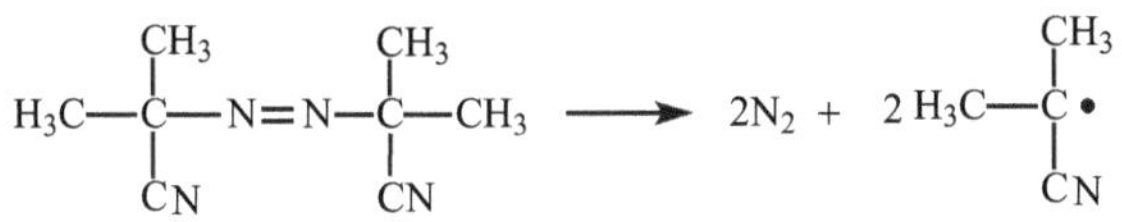
(2.15)

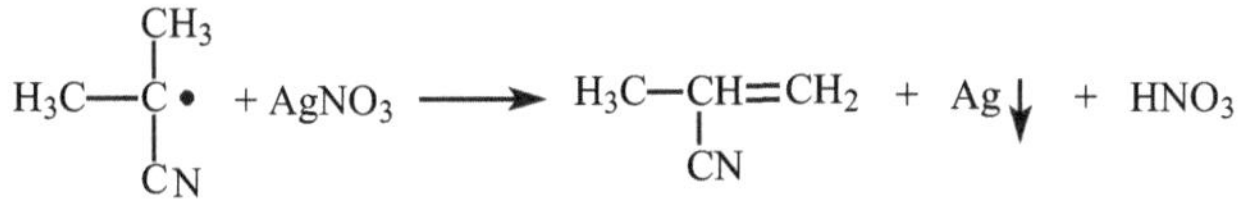
(2.16)

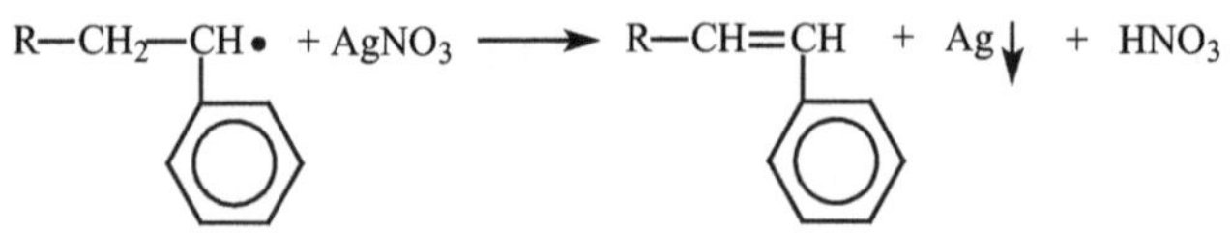

(2.17)

It should noted that similar reactions take place in oxidizing-reducing initializing or radical reduction of metal ions [285, 286], including reactions of intermolecular chain breakage [287, 288]:

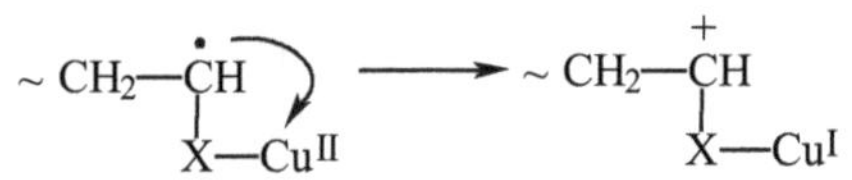

$(E_{oCu(II)\rightarrow Cu(I)}=0.15\ V)$

(2.18)

Mechanism of one-electron reduction is realized in many systems [289, 290]. In general form it can include (i) interaction of cations of metals with labile hydrogen atom with formation of a macroradical; (ii) generation of proton; (iii) reduction of metal ions. In this case protonation of acrylamide with $HAuCl_4$ causes formation of a complex and facilitates electron transport to the amino-group and reduction of the metal ion in the polymerizing acrylamide and N, N-methylene-bis-acryl-amide system [291]. Usually hydroxyl or carbonyl groups form as a result of one-electron oxidization of a polymer chain.

A polymer cross linked hydro gel poly(acrylamide)/poly(N-hydroxymethyl) acrylamide (PAAm-PHMAAm) contains hydroxymethyl fragment $–CH_2–OH$ with labile hydrogen atom, which can be easily removed with formation of macroradicals and the following reduction of a metal ion in the PAAm-PHMAAm/$[Ag(NH_3)_2]NO_3$ [292]:

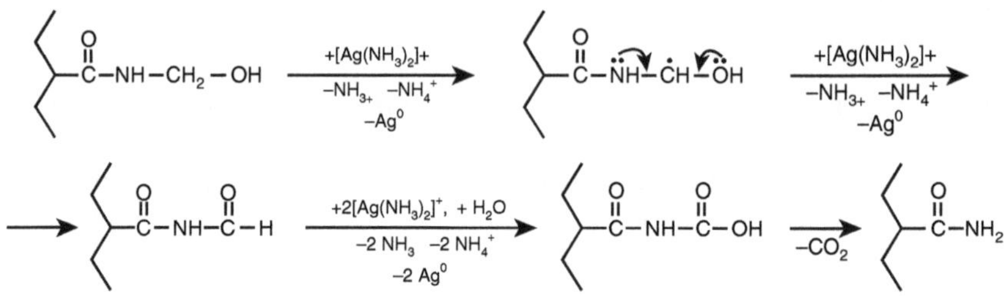

(2.19)

Due to ammonia molecules the reaction medium remains alkaline, Ag^+ cations gain electron from the macroradical and reduces to silver atom. The macroradical generates H^+ proton and two oxygen electrons form C=O double bond (aldehyde group), then aldehyde groups oxidize to carboxyl group.

2.2.4.2 Nitrogen-Containing Reduction Centers

Double function of the reducing agent and stabilizing ligand is typical of many amino compounds. Simple amines, amino acids, polymers with functional amino groups are widely used, for example, in production of Au nanoparticles [293–297], the oleylamine-paraffin hydrocarbon is efficient in synthesis of monodisperse silver particles [298].

Usage of dendrimer polymers with functional amino groups allows avoiding side products. Their structure and chemical properties can be controlled by modification of a core, type and a number of branching, and by terminal functional groups (Scheme 2.11) [149, 299, 300]:

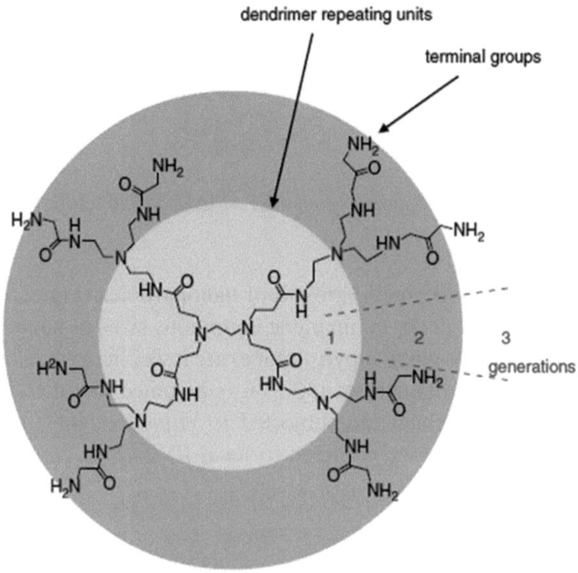

Scheme 2.11 Representation of generational structure PAMAM dendrimer

It is assumed that under conditions of UV radiation during reduction of $AgNO_3$ solution with poly(amidoamine) (PAMAM) electrons transfer from the primary amino groups to metal ions encapsulated in the dendrimer cavity [301]. At the same time mechanistic studies of similar systems confirm that, first of all, tertiary amines of dendrimers, rather than primary or secondary aminogroups play the role of the reducing agents, despite their lower basicity (pKa of the primary and tertiary amines is 9.23 and 6.30, respectively, for the PAMAM dendrimers [302]). A tendency of dendrimer molecules to association is very important [303–306]. Owing to ability of terminal PAMAM amino groups to formation of great aggregates as compared to dendrimer molecules with end COOH-groups, in the first case a polymer molecule contains inner homogeneous in shape cavities, which appears in spherical or nanoprismatic or nanohexagonal silver nanoparticles, respectively [307] (Scheme 2.12):

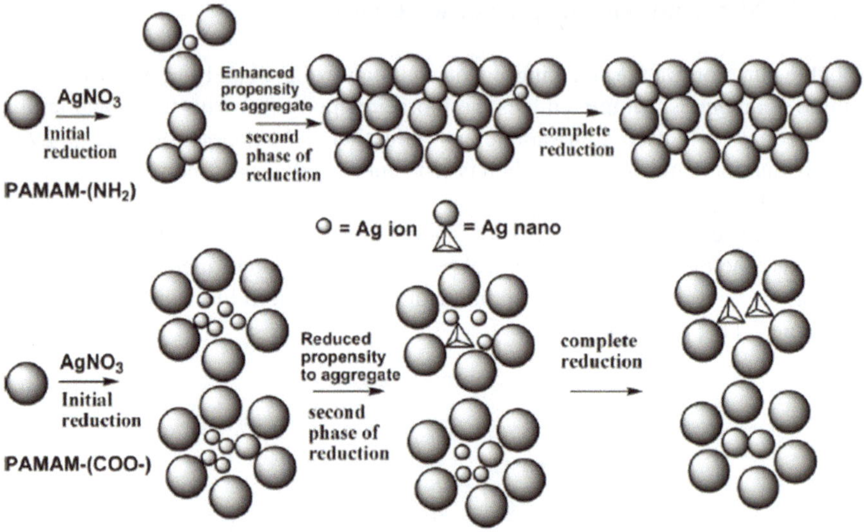

Scheme 2.12 The effect of the peripheral groups in PAMAM dendrimers on the size and shape of the metal nanoparticle

This behavior reflects in kinetic growth of nanoparticles (Fig. 2.32).

It should be noted that under comparable conditions sizes of nanoparticles and Pt clusters reduced by poly(N-vinyl-2-pyrrolidone) are lower than when using additional reducing agent HCHO formaldehyde [308]. It is assumed that metal ions bound in complex with a polymer chain are subjected to reduction, in this case size of a nanoparticle is limited by space confined to local PVP chain containing metal ion (Scheme 2.13):

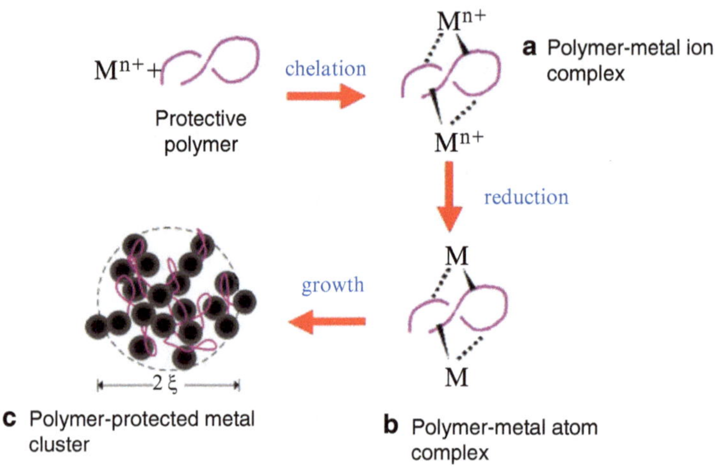

Scheme 2.13 A route for the formation of the polymer–metal complex and nanocomposite

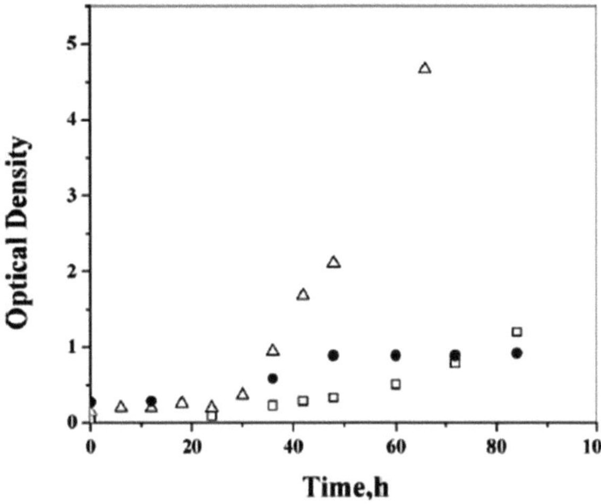

Fig. 2.32 Optical density versus time plot for the formation of silver nanoparticles in amine (Δ)-, hydroxy (•)-, and carboxylate (□) terminated PAMAM dendrimers. ($AgNO_3$ $15 \cdot 10^{-3}$ M, PAMAM $0.20 \cdot 10^{-3}$ M) [307]

It is assumed that particularly electrons of pyrrolidone nitrogen are responsible for reduction of metal ion, which points to appearance of a new C-N band near 1,667 cm^{-1} in FTIR spectrum of Ag^0-poly(N-vinyl-pyrrolidone) [309]. The examples of usage of poly(N-vinyl-pyrrolidone) simultaneously as reducing and stabilizing agent are not uncommon [79, 310–312].

An interesting is the approach in which low molecular reducing agent is chemically bound with polymer via polymer-analogous reaction. This technique is used for metallization of granules of the ternary copolymer of glycidyl methacrylate, methyl methacrylate, and divinylbenzene [313]. At the first stage polymer-fixed hydrazine with concentration of the functional group 2.3 mmol/g is obtained, at the second stage metal ions are reduced from their salt solutions:

(2.20)

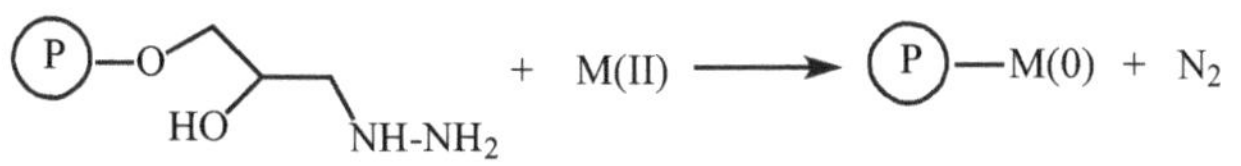

(2.21)

M = Ni, Cu, Ag

These examples can be shown also for inorganic polymer matrices. For example, surface of silica gel can be especially chemically modified by functional groups having reduction properties. Thus, silica gel with grafted silicon-hydride groups occurred very efficient at production of Au- and Ag- [314] or Pd- [315] containing nanocomposites:

$$\equiv SiH + Pd^{2+} + H_2O \rightarrow \equiv SiOH + Pd^0 + 2H^+ \quad (2.22)$$

It may be assumed that many conducting polymers can serve as efficient reducing agents, due to presence in their molecules of potentially easily oxidized centers [316, 317]. Low oxidizing potential of poly(2,5-dimethoxyaniline) (PDMA) [318, 319]:

$$\begin{array}{l}\mathrm{PDMA} \rightleftharpoons \mathrm{PDMA}^{n+} + ne^-, E_{\mathrm{PDMA}} < 0.4\,\mathrm{V} \\ \mathrm{Ag}^+ + \mathrm{e}^- \rightarrow \mathrm{Ag}, E_0 = 0.791\,\mathrm{V}\,vs\,\mathrm{NHE}\end{array} \quad (2.23)$$

favors reduction of silver ions in PDMA matrix forming in situ in presence of poly (styrene sulfonic acid) [320] by the scheme (Scheme 2.14):

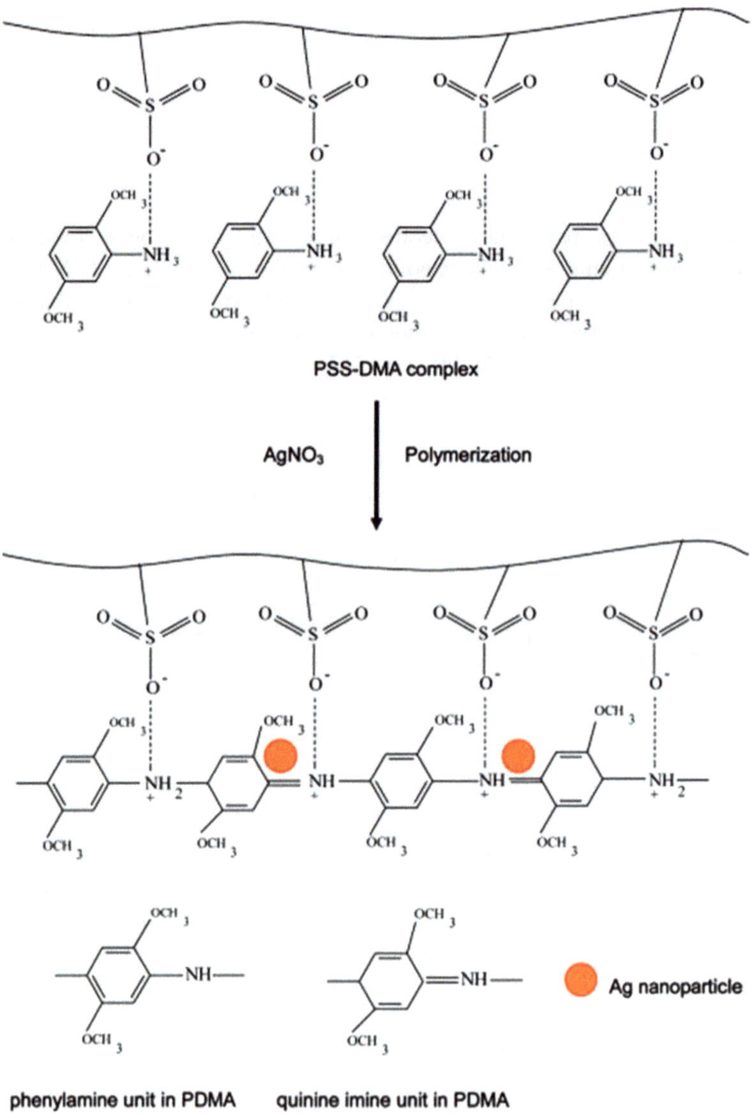

Scheme 2.14 Reduction of metal ions in a matrix of conducting polymer

Ag^+ ions modulate oxidization of DMA with formation of cation-radicals and, respectively, polymerization of molecules of a monomer bound to PSS. The formed PDMA in combination with PSS molecules form a nanoreactor cavity, in which Ag^0 particles localize. Probably, appearance of imine cross-links in the polymer chain is caused by participance of PDMA groups in reduction.

2.2.5 Co-reduction (Bi-metal Nanocomposites)

As shows analysis of soluble synthetic methods, in the case of monometallic systems the process of nucleation and growth can be easily controlled by the reaction conditions: concentration of reagents, molar ratios of precursors and surfactants, temperature, etc. However, in the case of bimetal nanoparticles this becomes a complicated problem, since their kinetic and thermodynamic parameters differ under the same conditions of reactions. In order to avoid separation of the nucleation processes for two metals, it is necessary to choose a suitable reducing agent and reagents.

2.2.5.1 Reduction Agents

Though usage of strong reduction agents, as was repeatedly mentioned, causes an increase in nucleation rates, on the other hand, because of the same reasons, it is more difficult to separate nucleation and growth processes in the case of bimetal substances. For example, co-reduction of $Cu(CH_3COO_2)_2$ и $HAuCl_4$ in presence of $NaBH_4$ at the initial stage causes just formation of Au/Cu aggregates of nanoparticles, and only the following annealing can bring to production of AuCu and $AuCu_3$ intermetallic nanoparticles [321]. Therefore, in these systems also surfactants and polymeric ligands play an important role [322–324]. Poly(vinylpyrrolidone) and poly(2-ethyl-2-oxazoline) are used in these purposes for production of some intermetallic nanocrystals ($FeSn_2$, $CoSn_3$, $NiSn_3$, PdSn) in presence of $NaBH_4$ [325], and for gold intermetallic compounds (Au_3Fe_{1x}, Au_3Co_{1x}, Au_3Ni_{1x}) there is another strong compound *n*-BuLi [326].

However, relatively often to obtain bimetal nanoparticles, weaker reduction agents are used, which provides more efficient control over the nucleation and growth processes [322–324, 327–330]. In some cases a mixture of reduction agents is used, for example, borane tert-butylamine and hexadecanediol in synthesis of Pt_3N nanoparticles [331].

A character of reduction makes it possible to kinetically control the reaction of co-reduction. For example, due to weak reducing ability of poly(vinylpyrrolidone) (PVP), generation and absorption of metal atoms during synthesis of Pd-Pt bimetal nanoparticles is slow, and the formed clusters coalesce easily, which can cause generation of star-like decahedrons and triangle nanoplates [327]. When ethylene glycol is used instead of PVP, the reduction rate increases significantly, fast reduction causes abrupt increase in supersaturation degree and, consequently, fast growth of seed with formation of a single crystal with distorted octahedral shape. Examples of these reactions can be found in the recent review [332].

2.2.5.2 Redox-Potentials of Reagents

It is known that metals with higher redox-potentials are reduced in the first turn. Just chemical behavior of the second metal after completion of its reduction defines a final structure of an article. If the second metal deposits homogeneously on a surface of a preliminary formed seed, the core-shell structure forms; if a metal deposits on some specific face, heterostructures can be obtained. In the case when atom of the second metal diffuses in the lattice of the first metal, with formation of a metal-metal bond, intermetallic compounds or alloys form. Depending on conditions, dynamically more stable state is preferable. If, for example, two different metals A and B are reduced, and redox-potential of A is higher than that of B, the core-shell structure (A)-(B) forms. However, in presence of a surfactant or a stabilizing ligand, which can bind strongly with A, the inverse type of particles with B core and A shell are more stable.

The effective way of approaching of a relative content of reduced metals in alloy and the respective ions in solution is choice of ligands, providing the closest approach of redox-potentials of complexes containing ions of co-deposited metals. However, equality of redox-potentials of the complexes of various metals does not mean that their ions will be reduced at equal rate. A significant role play kinetic factors, which determine a rate of discharge of different ions under specific conditions, providing co-deposition of metals [51].

An interesting synthetic strategy of production of hollow nanostructures is based on interaction of solution of some metal salt with a solid template of more electrochemically active metal. Typical examples include salts of Au^{3+}, Pt^{2+} и Pd^{2+} and silver nanoparticles or nanowires [333]. The schematic illustration of the experimental procedure of this process is shown in Fig. 2.33.

Due to higher redox-potential of the pair $AuCl_4^-/Au$ (0.99 V, vs SHE) as compared to the pair Ag^+/Ag (0.80 V, vs SHE) silver nanostructures suspended in the solution can be oxidized by $HAuCl_4$ according to the following substitution reaction:

$$3Ag_{(s)} + AuCl_4^-{}_{(aq)} \rightarrow Au_{(s)} + 3Ag^+{}_{(aq)} + 4Cl^-{}_{(aq)} \quad (2.24)$$

The elemental gold formed in this reaction is localized in immediate vicinity of the template surface. Here its nucleation and growth to small particles goes, and in the end, a thin shell forms around the Ag template. At the initial stage the shell has open structure, because $HAuCl_4$ and AgCl diffuse continuously through a layer until the silver template is consumed completely. At the temperature of reaction 100 °C the Au shell develops crystal structure via the processes of Ostwald ripening. This

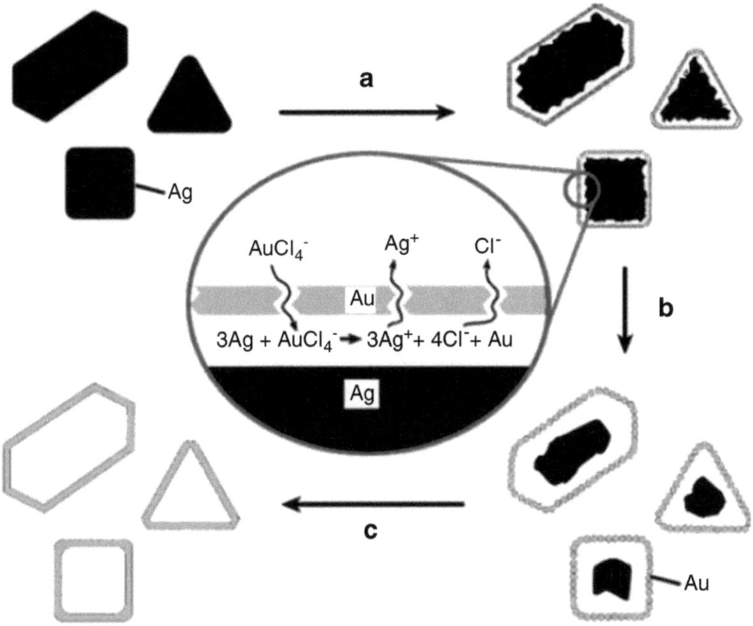

Fig. 2.33 Schematic illustration of production of hollow nanostructures. (**a**, **b**) Addition of $HAuCl_4$ to a dispersion of silver nanoparticles and the replacement reaction; (**c**) Total depletion of a template metal and annealing of the resultant shells [333]

method is used to obtain hollow structures with definite cavity sizes and homogenous high-crystal walls. The main characteristics of these nanostructures are determined by parameters of a template particle. Higher redox-potentials of Pd^{2+}/Pd (0.83 V, vs SHE) и Pt^{2+}/Pt (1.2 V, vs. SHE) pairs make it possible to generate also Pd and Pt hollow nanostructures via interaction of their salts with Ag template:

$$\mathrm{Pd\,(NO_3)_{2(aq)}} + \mathrm{2Ag_{(s)}} \rightarrow \mathrm{Pd_{(s)}} + \mathrm{2AgNO_{3(aq)}} \tag{2.25}$$

$$\mathrm{Pt\,(CH_3COO)_{2(aq)}} + \mathrm{2Ag_{(s)}} \rightarrow \mathrm{Pt_{(s)}} + \mathrm{2Ag\,(CH_3COO)_{(aq)}} \tag{2.26}$$

2.2.5.3 Molar Relationship of Metal Precursors

For bimetal nanostructures the couple: content-controlled synthesis is especially important for the dependence of their physical and chemical properties on content. Thus, the surface plasmon resonance of AuAg alloy is finely regulated by molar ratio Au:Ag in the block-copolymer based on 4-VPy and styrene (PS-PVPy) (Fig. 2.34) [334].

As Au content decreases, SPR peak is blue-shifted to the SPR of Ag nanoparticles. Due to the lower reduction rate of Ag^+ as compared to Au^+, fraction of silver can be increased by increase in the reaction time or in concentration of a precursor. $Au_{0.52}Ag_{0.48}$ и $Au_{0.39}Ag_{0.61}$ nanoparticles of the alloys were synthesized

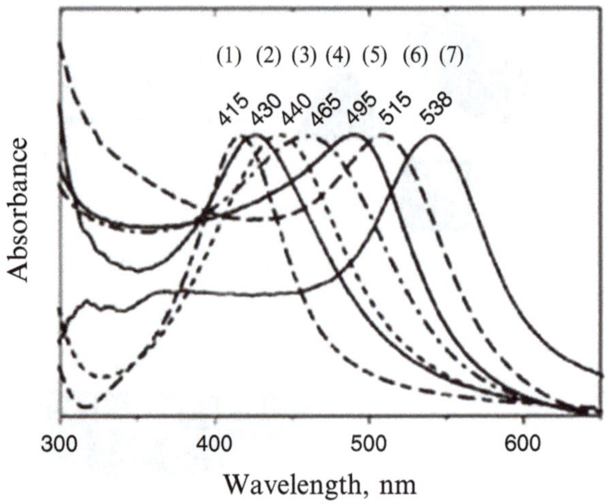

Fig. 2.34 UV-vis absorption spectra of nanocomposites PS-b-PVPy/Ag (1), PS-b-PVPy/AuAg in molar ratio Au:Ag: 1:3(2), 1:2 (3), 1:1(4), 2:1 (5), 3:1(6) and PS-b-PVPy/Au (7) [334]

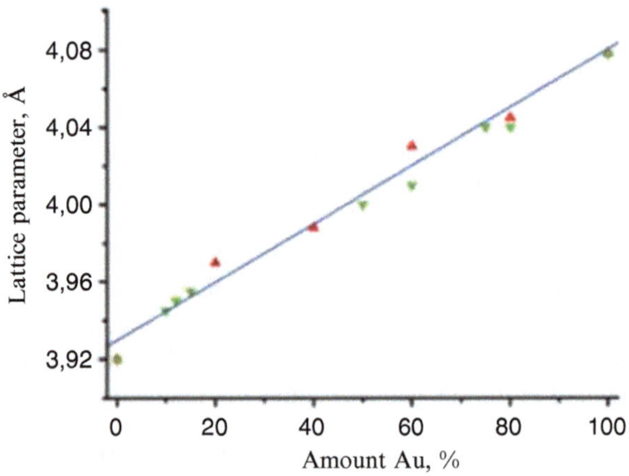

Fig. 2.35 Plot of the lattice parameter dependence on the composition of the alloy particle [244]

during co-reduction of $AgNO_3$ и $HAuCl_4$ (molar ratio 20:1) at 120 °C during 1 and 2 h, respectively [335]. Formation of homogeneous solid solutions was found for the binary system of Au-Pt nanoparticles in the surface polyelectrolyte layer of the polymeric brushes consisting of poly(2-aminoethylmethacrylate hydrochloride) chain [244], which was confirmed by linear dependence of the unit cell parameter of the crystal lattice on Au concentration in the nanocomposite (Fig. 2.35).

Thus, the chemical reduction of metal ions in aqueous and non-aqueous medium in the presence of polymers which can serve both as a stabilizing matrix and/or a reducing agent is one of the preferred synthetic condensation strategies for preparing a wide variety of nanocomposites. According to the classical scheme, reduction of metal-containing precursors (metal salts, monomer and polymer complexes, organometallic compounds, etc.) includes a number of successive transformations: metal atoms → cluster → colloid particle. Usage of strong reduction agents such as borohydrides, aminoboranes, etc. results in the formation of fine, and to a certain degree, monodisperse particles. Meanwhile, more weak reduction agents (citrates, ascorbic acid, alcohols, etc.) are responsible to the lower reduction rates and more narrow particle size distribution. Namely, solution methods of chemical reduction allow using the multiple parameters to control the composition and size of nanoparticles, their shape and dispersity as well as their stability. A crucial point is to reveal the interconnection between main physicochemical parameters: size/shape-function. Systematic studies in this direction are very wide [336–340]. Thus, monodisperse Rh nanocrystals with different shapes were obtained by one-step polyol synthesis owing to adjusting the reaction conditions as temperature, $Rh(acac)_3$ precursor concentration, inert or oxidizing atmosphere, and reaction time (Scheme 2.15) [341]:

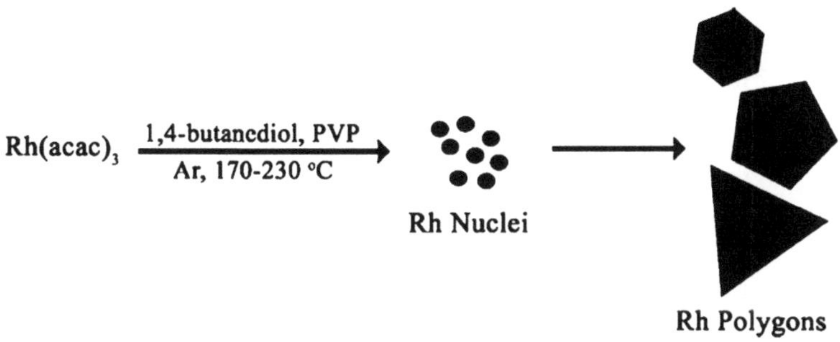

Scheme 2.15 Schematic illustration of one-step polyol synthesis of monodisperse well-shaped Rh nanocrystals

Two-dimensional projects of the nanocrystals are polygons, dominated by hexagons, pentagons (>45 %), and triangles with catalytically active (111) surfaces.

Some characteristics of reaction conditions, composition and sizes of dispersive phases of nanocomposites synthesized by chemical reduction are given in Table 2.11.

2.3 Electrochemical Reduction

Advantages of the electrochemical method of production of nanostructural materials are relative simplicity of the method, high reaction yields, as a rule, absence of side products of reaction. Several approaches to production of these materials are known.

2.3.1 Electrochemical Deposition

This method is based on the reactions of electrochemical deposition of nanoparticles on various surfaces, including polymer templates [342], graphite or carbon substrates, [343, 344], etc. Nanoporous materials used for templates are etched polymers, anodized aluminum oxide, etc. [345–347]. Electrochemical

Table 2.11 Synthesis and characterization of nanocomposites obtained by chemical reduction in polymer matrices

Nanoparticles	Diameter, nm	Reduction agent	Stabilizing agent	The reaction conditions	References
Reduction in polymer matrix in statu nascendi					
Cu, Cu@Cu_2O	48.2 ± 7.8	Hydrazine hydrate (0.1 M)	PVPdn (MW 360 000)	60 °C, DMF, $Cu(OAc)_2{\cdot}H_2O$	[38]
Cu	30–50	$N_2H_4 \cdot H_2O$ solution; 78–82 %	(PAA; Mw = 1,800)	60 °C, water, pH = 9.2–11, $CuSO_4$	[141]
Ag	20–25	NaOH, 5 wt.%	Poly(AAm-*co*-N,N-methylene-bis-acrylamide)	Water, room temperature	[217]
Pd	2.4 ± 0.7–5.2 ± 1.1	H_2 (0.4 MPa)	Dendrite polymers with perfluorine-, perfluorineoligoether-, polysiloxane-, alkyl- and oligoethylene glycole groups	Pd acetylacetonate, CO_2 100 °C, 90 min	[171]
Au	1.25	$NaBH_4$	Polystyrene-gr-poly(2-methylpropeony-loxyethyl)trimethylammonium chloride	$AuCl_4{}^{2-}$	[242]
Pt	2.1	The same	The same	$PtCl_6{}^{2-}$	[242]
Pd	2.4	The same	The same	$PdCl_4{}^{2-}$	[242]
Ag	7.5	$NaBH_4$	PS- N,N-diethyldithiocarbamate-gr-(polymethacrylic acid-b-PMMA)	([$AgNO_3$]/[COO] - 1.2 mol/mol), ([$NaBH_4$]/[Ag]- 2.0 mol/mol), THF, 258 °C	[147]
Pt	2.1 ± 0.2	$NaBH_4$	Poly(1-butyl-3-vinylimidazolium bro-mide), poly(N-vinyl-2-pyrrolidone) (PVP, Mn =55,000)	K_2PtCl_4, room temperature, 1 h	[271]
Ag	1–3	H_2, 60 °C, 2 MPa	Chitosan Mv = 140 kDa, deacetylation degree 98 %	Silver 1,5-cyclooctadiene -, 1,1,5,5,5-hexafluoroacetylacetonate, CO_2	[172]
Cu	1–40	H_2, 125 °C, 2 MPa	Chitosan Mv = 140 kDa, deacetylation degree 98 %	Cu 1,1,1,5,5,5- hexafluoroacetyla-cetonate hydrate, CO_2	[172]

Ag		Hydrazine hydrate ($2 \cdot 10^{-2}$ M)	PMMA	Silver oxide ammonium solution ($10^{-2} – 10^{-4}$ M), 1,2-dichloroethane	[162]
Ag	5–35	UV (λ = 320 nm)	PHEMA	Silver oxide ammonium solution ($10^{-2} – 10^{-4}$ M), 11,2-dichloroethane, 2-hydroxyethylmethacrylate	[162]
Ag	~20	Ethanol	Poly(vinylpyrrolidone)	$AgNO_3$, 76–79 °C	[146]
Au	2.5 ± 0.2,	Non-aqueous N_2H_4 (0.02 vol.%)	$(PS_{25}$-b-$P4VP_{17})_4$	0.5 eq. $HAuCl_4 \cdot 3H_2O$	[194]
	2.4 ± 0.3		$(PS_{25}$- b -$P4VP_{37})_4$	24 h	
	2.0 ± 0.2		$(ПС_{25}$- block -$P4VP_{63})_4$	0.2 mg/mL micelle solution	
Polymer as reducing agent					
Au	~ 5	Acrylamide, sodium sulphite	Copolymer of acrylamide, N,N-methylene-bisacrylamide, sodium sulphite	$HAuCl_4$, 0.25 мМ в воде pH 7.0, 70 °C	[291]
Ag	10–40	Poly(methylvinylether-co-maleic anhydride)	Poly(methylvinylether-co-maleic anhydride)	$AgNO_3$ ($10^{-4}–10^{-3}$ M), water, pH 8–9, NaOH	[279]
Ag	12	Poly(2,5-dimethoxyaniline)	Poly(styrenesulphonic acid)	$AgNO_3$, [Ag]: [monomer] = 2:1 (mol), pH 1.4	[320]
Au	18–40	Carboxymethylcellulose, methylhydroxyethylcellulose	Carboxymethylcellulose, methylhydroxyethylcellulose	$HAuCl_3$ $3H_2O$, pH – 5.7–9.0	[277]
Pt	2	Poly(N-vinyl-2-pyrrolidone), HCHO	Poly(N-vinyl-2-pyrrolidone), (PVP, Mw = 10 000)	$PtCl_4$, pH 12, NaOH	[308]
Ag	8.7–15.4	Poly(acrylo-uretane-*co*-styrene)	Poly(acrylo-uretane-*co*-styrene)	$AgNO_3$ (0.5 wt.-%), 65 °C, 3–4 h	[284]

deposition is performed by deposition of a metal film on one surface of the membrane, which acts as a cathode for galvanic coating. The respective metal ions are then electrochemically reduced and precipitate in pores of a template membrane [348–350]. Sizes and morphology of nanoparticles can be varied by changing parameters of electric precipitation, such as voltage, temperature, precipitating time, and origin of surfactants and polymers [348, 349, 351]. Nanoparticles are extracted from a template by physical-chemical means [352].

Electrochemical synthesis in the etched templates is efficient method of obtaining of multi-segmented nanowires of various metals [351, 353]. Sometimes the processes of chemical reduction in solution and electrochemical deposition are combined. Thus, for example, highly oriented triangle golden nanoparticles on conducting glass surfaces are used [354].

2.3.2 Direct Electric Reduction of Metal Ions in Electrolyte Solution

The essence of the method is in direct electric reduction of metal ions in aqua electrolytes [355]. The main difficulty with which one encounters in this case is competition between two opposite processes. On the one hand, it is formation of metal nanoparticles, on the other hand, it is electric precipitation of metal on the cathode surface [356], and the second process usually dominates.

Under conditions of the controlled potential, electric plating of Pt nanoparticles from H_2PtCl_6 water solutions takes place. It is necessary to use a stabilizer (tetraalkylammonium salts) to prevent deposition of particles on the cathode surface [8].

2.3.3 The Method of Anode Oxidizing Electrolysis

Reduction is conducted in electrolytic cell, with solution of anode as a metal source. Tetraalkylammonium salts are most often used as electrolyte and stabilizing surfactant. The general scheme of the process includes oxidization of a block metal of anode with formation of metal cations, which migrate to the cathode, where they are reduced to atomic metal. The clusters formed captured by a surfactant form stable colloids [357, 358] (Scheme 2.16).

Scheme 2.16 Electrochemical synthesis of transition metal colloids

$$\text{Anode: } M_{bulk} \rightarrow M^{n+} + ne^-$$

$$\text{Cathode: } M^{n+} + ne^- + \text{stabilizer} \rightarrow M_{colloid}/\text{stabilizer}$$

$$M_{bulk} + \text{stabilizer} \rightarrow M_{colloid}/\text{stabilizer}$$

Growth of the particles is predominantly by the coalescence mechanism of the formed colloid particle with ensembles of the reacting atoms [138].

The main parameters, which allow control over particle size in the electrochemical process, are polarity of medium, current density, space between electrodes, temperature.

2.3.3.1 The Effect of Current Density

It is known that the critical size r_{crit} of a cluster during formation of metal powder on the cathode is found by the equation [359]

$$r_{\mathrm{crit}} = \frac{2M\gamma}{nF\eta\rho} \tag{2.27}$$

where M is molecular weight, γ is surface tension, F is Faraday constant, η is overtension, ρ is density of the cluster, n is valence.

As is seen, r_{crit} in inverse dependence on overtension, and it is directly related to current density: the higher current density, the smaller particles. This tendency was found during production of Pd nanoparticles under action of current 2.16 and 5.41 mA/cm^2, while particle sizes were 2.56 and 1.39 nm, respectively [138]. Moreover, at longer time of electrolysis and lower current density nanoparticles had bimodal distribution, while at high current density the size distribution of particles was narrow. As the current density increases, the reaction yield also increases [360].

2.3.3.2 The Effect of Space Between Electrodes

The effect of space between electrodes is one of the important factors of control over dispersion of particles. Though the reasons are not quite clear, if the space between electrodes was, for example, 5 mm, small particles of constant size formed independently on time of electrolysis, while at the distance 0.75 mm particle sizes were greater and they increased during the reaction (Fig. 2.36) [138].

Probably, the crucial role in this case play transport velocities, and electrophoretic mobility in a system, i.e. parameters which depend directly of a field strength.

Thus, chemical reactions of reduction are the most widespread and available methods of production of nanocomposite materials. A wide range of reducing agents (from vapor compounds such as H_2, CO, etc., hydrides and salts, such as borohydride and sodium citrate, to oxidized solvents such as alcohols and polymer molecules), vast variety of synthetic and natural polymers and surfactants used as matrix and stabilizing media, multivariate control over conditions of the reactions (temperature, molar ratios of reagents, pH of a medium, etc.) provide development of materials with given structures and properties. Kinetic and mechanical studies [361] show that formation of monodisperse nanoparticles conforms with classical mechanisms of nucleation and growth of nanoparticles including fast nucleation

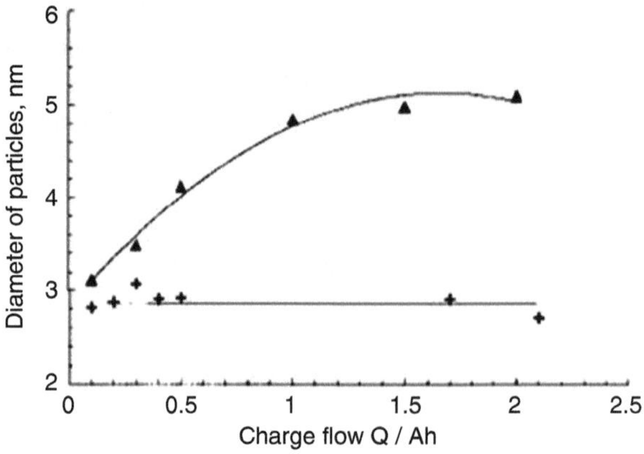

Fig. 2.36 Dependency of particle size of electrochemically prepared $(n\text{-}C_8H_{17})_4N^+Br^-$-stabilized Pd colloids on electrode distance (D_E) at different charge flow Q: $D_E = 0.75$ mm (×), $D_E = 5.0$ mm (▲) ($T = 20$ °C, $I_E = 5.41$ mA cm^{-2}, THF) [138]

with formation of a great amount of nuclei and their growth due to molecular incorporation [362–364]. The problem in the case of aggregation of fine nanocrystallites is described by alternative models, according to which the initial nucleation and growth origin from nanocrystalline aggregates of critical sizes. Growth of nanoparticles is achieved by coalescence of these aggregates, and the formed particles are characterized by polycrystalline structure, bimodal size distribution at the early stages and sigmoid growth kinetics [365–367]. Synthesis of anisotropic particles in solutions often demands deviation from thermodynamically controlled paths of reaction towards kinetically controlled, for example, by decrease in the rate of precursor decomposition or use of a weaker reducing agent, etc. A typical approach to synthesis of anisotropic particles is "seed" method, incorporation of a small amount of earlier prepared seed grains of nanoparticles into solution [368]. It should be noted that though chemical methods allow production of nanoparticles with controlled composition, sizes and homogeneous size distribution, use of organic solvents and hazardous chemical reagents are dangerous for environment. Moreover, target products are often contaminated by traces of reducing agent or intermediate products of reactions. Some of these problems are successfully solved with use of physical-chemical methods of production of nanocomposites.

References

1. M. Niederberger, G. Garnweitner, Chem. Eur. J. **12**, 7282 (2006)
2. S. Iijima, Nature **354**, 56 (1991)
3. C.B. Murray, D.J. Norris, M.G. Bawendi, J. Am. Chem. Soc. **115**, 8706 (1993)
4. X.G. Peng, L. Manna, W.D. Yang, J. Wickham, E. Scher, A. Kadavanich, A.P. Alivisatos, Nature **404**, 59 (2000)
5. C. Burda, X. Chen, R. Narayanan, M.A. El-Sayed, Chem. Rev. **105**, 1025 (2005)
6. Z.R. Dai, Z.W. Pan, Z.L. Wang, Adv. Funct. Mater. **13**, 9 (2003)
7. Y.N. Xia, P.D. Yang, Y.G. Sun, Y.Y. Wu, B. Mayers, B. Gates, Y.D. Yin, F. Kim, Y.Q. Yan, Adv. Mater. **15**, 353 (2003)

8. B.L. Cushing, V.L. Kolesnichenko, C.J. OSConnor, Chem. Rev. **104**, 3893 (2004)
9. T. Dwars, E. Paetzold, G. Oehme, Angew. Chem. Int. Ed. **44**, 7174 (2005)
10. U. Jeong, Y. Wang, M. Ibisate, Y. Xia, Adv. Funct. Mater. **15**, 1907 (2005)
11. J. Park, J. Joo, S.G. Kwon, Y. Jang, T. Hyeon, Angew. Chem. Int. Ed. **46**, 4630 (2007)
12. N. Pinna, M. Niederberger, N. Pinna, M. Niederberger, Angew. Chem. Int. Ed. **47**, 5292 (2008)
13. C. Feldmann, H. Goesmann, Nanoparticulate functional materials. Angew. Chem. Int. Ed. **49**, 1362 (2010)
14. Y. Yin, R.M. Rioux, C.K. Erdonmez, S. Hughes, G.A. Somerjai, A.P. Alivisatos, Science **304**, 711 (2004)
15. P.R. Selvakannan, M. Sastry, Chem. Commun. **1684** (2005)
16. Y.W. Jun, J.S. Choi, J. Cheon, Angew. Chem. Int. Ed. **45**, 3414 (2006)
17. S.E. Skrabalak, J. Chen, L. Au, X. Liu, X. Li, Y. Xia, Adv. Mater. **19**, 3177 (2007)
18. D. Kim, J. Park, K. An, N.K. Yang, J.G. Park, T. Hyeon, J. Am. Chem. Soc. **129**, 5812 (2007)
19. X. Wang, H. Fu, A. Peng, T. Zhai, Y. Ma, F. Yuan, J. Yao, Adv. Mater. **21**, 1636 (2009)
20. A.D. Pomogailo, V.N. Kestelman, *Metallopolymer Nanocomposites* (Springer, Berlin, 2005)
21. C.H. Kuo, M.H. Huang, Langmuir **21**, 2012 (2005)
22. S. Chen, Z.L. Wang, J. Ballato, S.H. Foulger, D.L. Carroll, J. Am. Chem. Soc. **125**, 16186 (2003)
23. J. Turkevich, P.C. Stevenson, P.C.J. Hiller, Discuss. Faraday Soc. **11**, 55 (1951)
24. B.J. Hornstein, R.G. Finke, Chem. Mater. **16**, 139 (2004)
25. A. Cacciuto, S. Auer, D. Frenkel, Nature **428**, 404 (2004)
26. U. Gasser, E.R. Weeks, A. Schofield, P.N. Pusey, D.A. Weitz, Science **292**, 258 (2001)
27. V.K. LaMer, R.H. Dinegar, J. Am. Chem. Soc. **72**, 4847 (1950)
28. Y. Yin, A.P. Alivisatos, Nature (London) **437**, 664 (2005)
29. X. Peng, J. Wickham, A.P. Alivisatos, J. Am. Chem. Soc. **120**, 5343 (1998)
30. Z.A. Peng, X. Peng, J. Am. Chem. Soc. **124**, 3343 (2002)
31. L.C. Ciacchi, M. Mertig, W. Pompe, S. Meriani, A. De Vita, Platin. Met. Rev. **47**, 98 (2003)
32. L.C. Ciacchi, W. Pompe, A. De Vita, J. Am. Chem. Soc. **123**, 7371 (2001)
33. L.C. Ciacchi, W. Pompe, A. De Vita, J. Phys. Chem. B **107**, 1755 (2003)
34. L.C. Ciacchi, M. Mertig, R. Seidel, W. Pompe, A. De Vita, Nanotechnology **14**, 840 (2003)
35. E.E. Finney, R.G. Finke, J. Colloid Interface Sci. **317**, 351 (2008)
36. S. Kinge, M. Crego-Calama, D.N. Chem, Phys. Chem. **9**, 20 (2008)
37. I. Haas, S. Shanmugam, A. Gedanken, J. Phys. Chem. B **110**, 16947 (2006)
38. I. Pastoriza-Santos, A. Sánchez-Iglesias, B. Rodríguez-González, L.M. Liz-Marzán, Small **5**(4), 440 (2009)
39. B.G. Ershov, Ros. Khim. Kh. (Zh. Ros. Khim. Obshch. D.I. Mendeleev) **45**, 21 (2001)
40. S. Kéki, L. Nagy, G. Deák, M. Zsuga, L. Somogyi, A. Lévai, J. Am. Soc. Mass Spectrom. **15**, 879 (2004)
41. J.A. McLean, K.A. Stumpo, D.H. Russell, J. Am. Chem. Soc. **127**, 5304 (2005)
42. X. Li, A.E. Kuznetsov, H.F. Zhang, A.I. Boldyrev, L.S. Wang, Science **291**, 859 (2001)
43. L.A. Peyser, A.E. Vinson, A.P. Bartko, R.M. Dickson, Science **291**, 103 (2001)
44. J. Zheng, R.M. Dickson, J. Am. Chem. Soc. **124**, 13982 (2002)
45. J.T. Petty, J. Zheng, N.V. Hud, R.M. Dickson, J. Am. Chem. Soc. **126**, 5207 (2004)
46. A. Henglein, Chem. Phys. Lett. **154**, 473 (1989)
47. T. Linnert, P. Mulvaney, A. Henglein, H. Weller, J. Am. Chem. Soc. **112**, 4657 (1990)
48. B.G. Ershov, E. Janata, A. Henglein, J. Phys. Chem. **97**, 339 (1993)
49. B.G. Ershov, N.L. Sukhov, D.A. Troitskii, Radiat. Phys. Chem. **39**, 127 (1992)
50. E. Janata, A. Henglein, B.G. Ershov, J. Phys. Chem. **98**(42), 10888 (1994)
51. V.V. Sviridov, T.N. Vorob'eva, T.V. Gaevskaya, L.I. Stepanova, in *Khimicheskoi osazhdenie metallov iz vodnykh rastvorov*, ed. by V.V. Sviridov (Izd. Universitetskoe, Minsk, 1987), p. 270
52. Z. Wu, J. Chen, R. Jin, Adv. Funct. Mater. **21**, 177 (2011)

53. D.V. Talapin, A.L. Rogach, M. Haase, H. Weller, J. Phys. Chem. B **105**, 21278 (2001)
54. D.V. Talapin, A.L. Rogach, E.V. Shevchenko, A. Kornowski, M. Haase, H. Weller, J. Am. Chem. Soc. **124**, 5782 (2002)
55. Q. Zhang, J. Xie, J. Liang, J.Y. Lee, Adv. Funct. Mater. **19**, 1387 (2009)
56. D.K. Lee, S.I. Park, J.K. Lee, N.M. Hwang, Acta Mater. **55**, 5281 (2007)
57. A.C.S. Samia, J.A. Schlueter, J.S. Jiang, S.D. Bader, C.J. Qin, X.M. Lin, Chem. Mater. **18**, 5203 (2006)
58. N.R. Jana, L. Gearheart, C.J. Murphy, Chem. Mater. **13**, 2313 (2001)
59. J. Park, E. Lee, N.M. Hwang, M. Kang, S.C. Kim, Y. Hwang, J.G. Park, H.J. Noh, J.Y. Kim, J.H. Park, T. Hyeon, Angew. Chem. **117**, 2932 (2005)
60. C.J. Murphy, N.R. Jana, Adv. Mater. **14**, 80 (2002)
61. H. Yu, P.C. Gibbons, K.F. Kelton, W.E. Buhro, J. Am. Chem. Soc. **123**, 9198 (2001)
62. J.P. Wilcoxon, P.P. Provencio, J. Am. Chem. Soc. **126**, 6402 (2004)
63. D.V. Talapin, A.L. Rogach, A. Kornowski, M. Haase, H. Weller, Nano Lett. **1**, 207 (2001)
64. J. Hambrock, R. Becker, A. Birkner, J. Weiss, R.A. Fischer, Chem. Commun. 68 (2002)
65. N.R. Jana, X. Peng, J. Am. Chem. Soc. **125**, 14280 (2003)
66. C. de Mello Doneg, P. Liljeroth, D. Vanmaekelbergh, Small **1**, 1152 (2005)
67. J. van Embden, P. Mulvaney, Langmuir **21**, 10226 (2005)
68. T.F. Tadros, *Applied Surfactants* (Wiley-VCH, Weinheim, 2005)
69. C.S.S.R. Kumar, *Biofunctionalization of Nanomaterials* (Wiley-VCH, Weinheim, 2005)
70. Y. Wang, X. Xu, Z. Tian, Y. Zong, H. Cheng, C. Lin, Chem. Eur. J. **12**, 2542 (2006)
71. Y. Song, J. Hormes, C.S.S.R. Kumar, Small **4**, 698 (2008)
72. Z. Wu, J. Suhan, R. Jin, J. Mater. Chem. **19**, 622 (2009)
73. Z. Wu, E. Lanni, W. Chen, M.E. Bier, D. Ly, R. Jin, J. Am. Chem. Soc. **131**, 16672 (2009)
74. Z. Wu, D. Jiang, E. Lanni, M.E. Bier, R. Jin, J. Phys. Chem. Lett. **1**, 1423 (2010)
75. T.K. Sau, A.L. Rogach, Adv. Mater. **22**, 1781 (2010)
76. Y. Xia, Y. Xiong, B. Lim, S.E. Skrabalak, Angew. Chem. Int. Ed. **48**, 60 (2009)
77. P.-F. Ho, K.-M. Chi, Nanotechnology **15**, 1059 (2004)
78. Y. Xiong, A.R. Siekkinen, J. Wang, Y. Yin, M.J. Kim, Y. Xia, J. Mater. Chem. **17**, 2600 (2007)
79. I. Washio, Y. Xiong, Y. Yin, Y. Xia, Adv. Mater. **18**, 1745 (2007)
80. Y. Xiong, I. Washio, J. Chen, H. Cai, Z.-Y. Li, Y. Xia, Langmuir **22**, 8563 (2006)
81. B. Lim, P.H.C. Camargo, Y. Xia, Langmuir **24**, 10437 (2008)
82. Y. Xiong, J.M. McLellan, J. Chen, Y. Yin, Z.-Y. Li, Y. Xia, J. Am. Chem. Soc. **127**, 17118 (2005)
83. Y. Sun, Y. Xia, Adv. Mater. **15**, 695 (2003)
84. Y. Sun, B. Mayers, Y. Xia, Nano Lett. **3**, 675 (2003)
85. H. Partridge, C.W. Bauschlicher Jr., S.R. Langhoff, Chem. Phys. Lett. **175**, 531 (1990)
86. T. Leisner, S. Vajda, S. Wolf, L. Woste, J. Chem. Phys. **111**, 1017 (1999)
87. Y. Xiong, I. Washio, J. Chen, M. Sadilek, Y. Xia, Angew. Chem. Int. Ed. **46**, 4917 (2007)
88. M.-P. Pileni, Nat. Mater. **2**, 145 (2003)
89. S. Brown, M. Sarikaya, E. Johnson, J. Mol. Biol. **299**, 725 (2000)
90. S.S. Shankar, A. Rai, A. Ahmad, M. Sastry, Chem. Mater. **17**, 566 (2005)
91. Y. Sun, B. Mayers, T. Herricks, Y. Xia, Nano Lett. **3**, 955 (2003)
92. Z.S. Pillai, P.V. Kamat, J. Phys. Chem. B **108**, 945 (2004)
93. M.H. Rashid, R.R. Bhattacharjee, A. Kotal, T.K. Mandal, Langmuir **22**, 7141 (2006)
94. M.H. Rashid, T.K. Mandal, J. Phys. Chem. C **111**, 16750 (2007)
95. H. Rashid, T.K. Mandal, Adv. Funct. Mater. **18**, 2261 (2008)
96. M. Liu, P. Guyot-Sionnest, J. Phys. Chem. B **109**, 22192 (2005)
97. T.K. Sau, C.J. Murphy, J. Am. Chem. Soc. **126**, 8648 (2004)
98. M. Faraday, Philos. Trans. R. Soc. Lond. **151**, 183 (1857)
99. B.V. Enüstün, J. Turkevich, J. Am. Chem. Soc. **85**, 3317 (1963)
100. J. Turkevich, G. Kim, Science **169**, 873 (1970)

101. Y. Wang, J. Ren, K. Deng, L. Gui, Y. Tang, Chem. Mater. **12**, 1622 (2000)
102. B.K. Park, S. Jeong, D. Kim, J. Moon, S. Lim, J.S. Kim, J. Colloid Interface Sci. **311**, 417 (2007)
103. B. Wiley, Y. Sun, B. Mayers, Y. Xia, Chem. Eur. J. **11**, 454 (2005)
104. F. Kim, S. Connor, H. Song, T. Kuykendall, P. Yang, Angew. Chem. Int. Ed. **43**, 3673 (2004)
105. S.I. Cha, C.B. Mo, K.T. Kim, Y.J. Jeong, S.H. Hong, J. Mater. Res. **21**, 2371 (2006)
106. E.I. Suvorova, V.V. Klechkovskaya, V.V. Kopeikin, P.A. Buffat, J. Cryst. Growth **275**, e2351 (2005)
107. H. Bonnemann, R. Richards, Eur. J. Inorg. Chem. 2455 (2001)
108. S. Kinge, H. Bonnemann, Appl. Organomet. Chem. **19**, 750 (2005)
109. H. Bonnemann, K.S. Nagabhushana, *Encyclopedia of Nanoscience and Nanotechnology* (Marcel Dekker, New York, 2004), pp. 739–760
110. Y. Yang, Y. Yan, W. Wang, J.R. Li, Nanotechnology **19**, 175603 (2008)
111. R.B. Vasil'ev, D.N. Darin, A.M. Gas'kov, Uspekhi Khimii **80**, 1190 (2011)
112. S. Komarneni, D. Li, B. Newalkar, H. Katsuki, A.S. Bhalla, Langmuir **18**, 5959 (2002)
113. J. Yang, T.C. Deivaraj, H.-P. Too, J.Y. Lee, Langmuir **20**, 4241 (2004)
114. R.O. Hutchins, K. Learn, B. Nazer, D. Pytlewski, A. Pelter, Org. Prep. Proced. Int. **16**, 335 (1984)
115. S.B. Kalidindi, U. Sanyal, B.R. Jagirdar, ChemSusChem **4**, 317 (2011)
116. D.V. Goia, J. Mater. Chem. **14**, 451 (2004)
117. M.T. Reetz, M. Maase, Adv. Mater. **11**, 773 (1999)
118. C.M. Welch, R.G. Comp, Anal. Bioanal. Chem. **384**, 601 (2006)
119. A. Troupis, T. Triantis, A. Hiskia, E. Papaconstantinou, Eur. J. Inorg. Chem. 5579 (2008)
120. Y. Ma, J. Zeng, W. Li, M. McKiernan, Z. Xie, Y. Xia, Adv. Mater. **22**, 1930 (2010)
121. T.K. Sau, C.J. Murphy, Philos. Mag. **87**, 2143 (2007); X. Jiang, Q. Zeng, A. Yu, Nanotechnology **17**, 4929 (2006)
122. N.R. Jana, Small **1**(8–9), 875 (2005)
123. P. Qiu, C. Mao, J. Nanopart. Res. **11**, 885 (2009)
124. I. Lisiecki, M.P. Pileni, J. Am. Chem. Soc. **115**, 3887 (1993)
125. I. Lisiecki, F. Billoudet, M.P. Pileni, J. Phys. Chem. **100**, 4160 (1996)
126. J. Tanori, M.P. Pileni, Adv. Mater. **7**, 862 (1995)
127. N. Pinna, K. Weiss, H. Sack-Kongehl, W. Vogel, J. Urban, M.P. Pileni, Langmuir **17**, 7982 (2001)
128. M. Maillard, S. Giorgio, M.P. Pileni, J. Phys. Chem. B **107**, 2466 (2003)
129. V. Germain, J. Li, D. Ingert, Z.L. Wang, M.P. Pilen, J. Phys. Chem. B **107**(34), 8717 (2003)
130. C. Salzemann, I. Lisiecki, J. Urban, M.P. Pileni, Langmuir **20**, 11772 (2004)
131. C. Salzemann, I. Lisiecki, J. Urban, M.P. Pileni, Adv. Funct. Mater. **15**, 1277 (2005)
132. X. Ji, X. Song, J. Li, Y. Bai, W. Yang, X. Peng, J. Am. Chem. Soc. **129**, 13939 (2007)
133. X. Liu, R. Huang, J. Zhu, Chem. Mater. **20**, 192 (2008)
134. D.O. Yener, J. Sindel, C.A. Randall, J.H. Adair, Langmuir **18**, 8692 (2002)
135. G. Wei, H. Zhou, Z. Liu, Y. Song, L. Wang, L. Sun, Z. Li, J. Phys. Chem. B **109**, 8738 (2005)
136. J. Zhang, H. Liu, Z. Wang, N. Ming, Adv. Funct. Mater. **17**, 3295 (2007)
137. S.A. Tikhomirov, M.A. Alymov, I.V. Tregubova, V.S. Shustov, Ross. Nanotekhn. **6**(3-4), 105 (2011)
138. M.T. Reetz, M. Winter, R. Breinbauer, T. Thurn-Albrecht, W. Vogel, Chem. Eur. J. **7**(5), 1084 (2001)
139. A. Miyazaki, Y. Nakano, Langmuir **16**, 7109 (2000)
140. X.C. Jiang, W.M. Chen, C.Y. Chen, S.X. Xiong, A.B. Yu, Nanoscale Res. Lett. **6**, 32 (2011)
141. Y. Wang, A.V. Biradar, G. Wang, K.K. Sharma, C.T. Duncan, S. Rangan, T. Asefa, Chem. Eur. J. **16**, 10735 (2010)
142. C. Salzemann, A. Brioude, M.P. Pileni, J. Phys. Chem. B **110**, 7208 (2006)
143. D.B. Pedersen, S. Wang, S.H. Liang, J. Phys. Chem. C **112**, 8819 (2008)
144. V.V. Klimov, *Nanoplazmonika* (Fizmatlit, Moscow, 2009)

145. S.P. Chandran, M. Chaudhary, R. Pasricha, A. Ahmad, M. Sastry, Biotechnol. Prog. **22**, 577 (2006)
146. A.K. Popov, J. Brummer, R.S. Tanke, G. Taft, M. Loth, R. Langlois, A. Wruck, R. Schmitz, Laser Phys. Lett. **3**(11), 546 (2006)
147. K. Ishizu, H. Kakinuma, K. Ochi, S. Uchida, M. Hayashi, Polym. Adv. Technol. **16**, 834 (2005)
148. H. Hirai, N. Yakura, Polym. Adv. Technol. **12**, 174 (2001)
149. L. Durán Pachón, G. Rothenberg, Appl. Organomet. Chem. **22**, 288 (2008)
150. L.S. Ott, B.J. Hornstein, R.G. Finke, Langmuir **22**, 9357 (2006)
151. A. Sutton, G. Franc, A. Kakkar, J. Polym. Sci. Part A Polym. Chem. **47**, 4482 (2009)
152. R. Hourani, M.A. Whitehead, A.K. Kakkar, Macromolecules **41**, 508 (2008)
153. R. Hourani, M.A. Whitehead, A.K. Kakkar, J. Mater. Chem. **15**, 2106 (2005)
154. H. Tsunoyama, H. Sakurai, Y. Negishi, T. Tsukuda, J. Am. Chem. Soc. **127**, 9374 (2005)
155. J.A. Widegren, R.G. Finke, J. Mol. Catal. A Chem. **191**, 187 (2003)
156. X. Mu, D.G. Evans, Y. Kou, Catal. Lett. **97**, 151 (2004)
157. I. Pastoriza-Santos, L.M. Liz-Marzán, Adv. Funct. Mater. **19**, 679 (2009)
158. P. Jiang, S.-Y. Li, S.-S. Xie, Y. Gao, L. Song, Chem. Eur. J. **10**, 4817 (2004)
159. J.L. Elechiguerra, L. Larios-Lopez, C. Liu, D. Garcia-Gutierrez, A. Camacho-Bragado, M.J. Yacaman, Chem. Mater. **17**, 6042 (2005)
160. N. Toshima, Macromol. Symp. **235**, 1 (2006)
161. J. Widoniak, S. Eiden-Assmann, G. Maret, Coll Surf A Physicochem. Eng. Asp. **270–271**, 340 (2005)
162. P.A. Muzalev, I.D. Kosobudskii, N.M. Ushakov, L.G. Panova, Perspect. Mater. **3**, 84 (2011)
163. Y. Park, P. Taranekar, J.Y. Park, A. Baba, T. Fulghum, R. Ponnapati, R.C. Advincula, Adv. Funct. Mater. **18**, 2071 (2008)
164. W. Wang, G. Wang, X. Wang, Y. Zhan, Y. Liu, C. Zheng, Adv. Mater. **14**, 67 (2002)
165. P.S. Kumar, I. Pastoriza-Santos, B. Rodríguez-González, F.J. García de Abajo, L.M. Liz-Marzán, Nanotechnology **19**, 015606(1–5) (2008)
166. D. Cabane, R. Duplessix, J. Physique **48**, 651 (1987)
167. D. Li, S. Komarneni, J. Am. Ceram. Soc. **89**(5), 1510 (2006)
168. E.N. Sobol, V.N. Bagratishvili, S.M. Khoudl, Dokl. AN **356**(6), 777 (1997)
169. J. Watkins, T. McCarthy, Chem. Mater. **7**, 1991 (1995)
170. A. Kameo, T. Yoshimura, K. Esumi, Colloids Surf. A **215**, 181 (2003)
171. S. Moisan, V. Martinez, P. Weisbecker, F. Cansell, S. Mecking, C. Aymonier, J. Am. Chem. Soc. **129**, 10602 (2007)
172. E. Said-Galiev, A.I. Gamzadze, T.E. Grigor'ev, A.R. Khokhlov, N.P. Bakuleva, I.G. Lyutova, E.V. Shtykova, K.A. Dembo, V.V. Volkov, Ross. Nanotekhn. **6**(5–6), 94 (2011)
173. M. Chatterjee, Y. Ikushima, Y. Hakuta, H. Kawanami, Adv. Synth. Catal. **348**, 1580 (2006)
174. A.D. Pomogailo, A.S. Rozenberg, I.E. Uflyand, *Metal Nanoparticles in Polymers* (Khimiya, Moscow, 2000)
175. T. Hasell, J. Yang, W. Wang, P.D. Brown, S.M. Howdle, Mater. Lett. **61**, 4906 (2007)
176. V. Sankaran, J. Yue, R.E. Cohen, R.R. Schrock, R.J. Silbey, Chem. Mater. **5**, 1133 (1993)
177. K. Holmberg, *Surfactants and Polymers in Aqueous Solution*, 2nd edn. (Wiley, Chichester, 2003)
178. R. Savic, L. Luo, A. Eisenberg, D. Maysinger, Science **300**, 615 (2003)
179. S. Főrster, M. Konrad, J. Mater. Chem. **13**, 2671 (2003)
180. S. Sun, C.B. Murray, D. Weller, L. Folks, A. Moser, Science **287**, 1989 (2000)
181. L. Luo, A. Eisenberg, Langmuir **17**, 6804 (2001)
182. L. Luo, A. Eisenberg, J. Am. Chem. Soc. **123**, 1012 (2001)
183. S. Mossmer, J.P. Spatz, M. Moller, T. Aberle, J. Schmidt, W. Burchard, Macromolecules **33**, 4791 (2000)
184. J.X. Zhao, C. Allen, A. Eisenberg, Macromolecules **30**, 7143 (1997)

185. R. Nagarajan, K. Ganesh, J. Colloid Interface Sci. **184**, 489 (1996)
186. N. Singh, A. Karim, F.S. Bates, M. Tirrell, K. Furusawa, Macromolecules **27**, 2586 (1994)
187. X. Qiu, Z.-G. Wang, J. Colloid Interface Sci. **167**, 294 (1994)
188. Y. Kang, T.A. Taton, Macromolecules **38**, 6115 (2005)
189. P. Spatz, S. Mőssmer, C. Hartmann, M. Mőller, Langmuir **16**, 407 (2000)
190. L.M. Bronstein, D.M. Chernyshov, I.O. Volkov, M.G. Ezernitskaya, P.M. Valetsky, V.G. Matveeva, E.M. Sulman, J. Catal. **196**, 302 (2000)
191. Y.J. Ho, M.K. Park, J. Locklin, R. Advincula, J.H. Yang, J. Mays, Langmuir **18**, 2457 (2002)
192. S.N. Sidorov, L.M. Bronstein, Y.A. Kabachii, P.M. Valetsky, P.L. Soo, D. Maysinger, A. Eisenberg, Langmuir **20**, 3543 (2004)
193. M. Filali, M.A.R. Meier, U.S. Schubert, J.F. Gohy, Langmuir **21**, 7995 (2005)
194. J. Li, L. Shi, Y. An, Y. Li, X. Chen, H. Dong, Polymer **47**, 8480 (2006)
195. M. Gauthier, J.M. Li, J. Dockendorff, Macromolecules **36**, 2642 (2003)
196. R. Zhang, J. Liu, B. Han, B. Wang, D. Sun, J. He, Polymer **46**, 3936 (2005)
197. F. Calderara, G. Riess, Macromol. Chem. Phys. **197**, 2115 (1996)
198. M. Moffitt, L. McMahon, V. Pessel, A. Eisenberg, Chem. Mater. **7**, 1185 (1995)
199. C.W. Wang, M.G. Moffitt, Langmuir **20**, 11784 (2004)
200. N. Duxin, F. Liu, H. Vali, A. Eisenberg, J. Am. Chem. Soc. **127**, 10063 (2005)
201. K. Gatsouli, S. Pispas, E.I. Kamitsos, J. Phys. Chem. C **111**, 15201 (2007)
202. J.-Y. Kim, D.-H. Shin, K.-J. Ihn, J. Appl. Polym. Sci. **97**, 2357 (2005)
203. A. Biffis, N. Orlandi, B. Corain, Adv. Mater. **15**, 1551 (2003)
204. C. Aymonier, D. Bortzmeyer, R. Thomann, R. Mulhaupt, Chem. Mater. **15**, 4874 (2003)
205. S. Xu, J. Zhang, C. Paquet, Y. Lin, E. Kumacheva, Adv. Funct. Mater. **13**, 468 (2003)
206. Y.Y. Li, F. Cunin, J.R. Link, T. Gao, R.E. Betts, S.H. Reiver, V. Chin, S.N. Bhatia, M.J. Sailor, Science **299**, 2045 (2003)
207. M. Han, X. Gao, J.Z. Su, S. Nie, Nat. Biotechnol. **19**, 631 (2001)
208. A.N. Shipway, I. Willner, Chem. Commun. 2035 (2001)
209. Y. Lu, P. Spyra, Y. Mei, M. Ballauff, A. Pich, Macromol. Chem. Phys. **208**, 254 (2007)
210. Y.M. Mohan, K.J. Lee, T. Premkumar, K.E. Geckeler, Polymer **48**, 158 (2007)
211. C. Wang, N.T. Flynn, R. Langer, Mater. Res. Soc. Symp. Proc. **820**, R2. 2. 1. (2004)
212. V. Thomas, M.M. Yallapu, B. Sreedhar, S.K. Bajpai, J. Colloid Interface Sci. **315**, 389 (2007)
213. A.N. Shipway, I. Willner, Chem. Commun. **20**, 35 (2001)
214. V. Pardo-Yissar, R. Gabai, A.N. Shipway, T. Bourenko, I. Willner, Adv. Mater. **13**, 1320 (2001)
215. P. Saravanan, M. Padmanabha Raju, S. Alam, Mater. Chem. Phys. **103**, 278 (2007)
216. Y. Murali Mohan, T. Premkumar, K.J. Lee, K.E. Geckeler, Macromol. Rapid Commun. **27**, 1346 (2006)
217. E.M. Ahmed, F.S. Aggor, J. Appl. Polym. Sci. **117**, 2168 (2010)
218. C. Wang, N.T. Flynn, R. Langer, Adv. Mater. **16**, 1074 (2004)
219. D. Suzuki, H. Kawaguchi, Langmuir **21**, 12016 (2005)
220. Y. Mei, Y. Lu, F. Polzer, M. Ballauff, M. Drechsler, Chem. Mater. **19**, 1062 (2007)
221. S.H. Anastasiadis, M. Vamvakaki, Int. J. Nanotechnol. **6**, 46 (2009)
222. C. Boyer, M.H. Stenzel, T.P. Davis, J. Polym. Sci. Part A Polym. Chem. **49**, 551 (2011)
223. Y. Dong, Y. Ma, T. Zhai, F. Shen, Y. Zeng, H. Fu, J. Yao, Macromol. Rapid Commun. **28**, 2339 (2007)
224. J.G. Zhang, N. Coombs, E. Kumacheva, J. Am. Chem. Soc. **124**, 14512 (2002)
225. J. Zhang, S. Xu, E. Kumacheva, J. Am. Chem. Soc. **126**, 7908 (2004)
226. D. Palioura, S.P. Armes, S.H. Anastasiadis, M. Vamvakaki, Langmuir **23**, 5761 (2007)
227. J.-T. Zhang, G. Wei, T.F. Keller, H. Gallagher, C. Stőtzel, F.A. Műller, M. Gottschaldt, U.S. Schubert, K.D. Jandt, Macromol. Mater. Eng. **295**, 1049 (2010)
228. C.-W. Chen, K. Arai, K. Yamamoto, T. Serizawa, M. Akashi, Macromol. Chem. Phys. **201**, 2811 (2000)

229. H. Hirai, N. Toshima, Polymer-attached catalysts, in *Tailored Metal Catalyst*, ed. by Y. Iwasawa (D. Deidel Publishing Company, Dordrecht, 1986), p. 87
230. M.L. Wang, C.H. Wang, W. Wang, Mater. Chem. Phys. **104**, 162 (2007)
231. K.A. Malini, M.R. Anantharaman, S. Sindhu, C.N. Chinnasamy, N. Ponpandian, A. Narayanasamy, B. Balachandran, V.N. Shivasankarapillai, J. Mater. Sci. **36**, 821 (2001)
232. E. Veena Gopalan, K.A. Malini, G. Santhoshkumar, T.N. Narayanan, P.A. Joy, I.A. Al-Omari, D. Sakthi Kumar, Y. Yoshida, M.R. Anantharaman, Nanoscale Res. Lett. **5**, 889 (2010)
233. B. Domènech, M. Muñoz, D.N. Muraviev, J. Macanás, Catal. Today **193**(1), 158 (2012)
234. R.C. Advincula, W.J. Brittain, K.C. Caster, J. Ruhe, in *PolymerBrushes: Synthesis, Characterization, Applications*, ed. by R. Advincula, W.J. Brittain, K.C. Caster, J. Ruhe (Wiley, Weinheim, 2004)
235. T.M. Birshtein, Soros. Obrazov. Kh. **5**, 42 (1999)
236. S.G. Boyes, W.J. Brittain, X. Weng, S.Z.D. Cheng, Macromolecules **35**, 4960 (2002)
237. A.M. Granville, S.G. Boyes, B. Akgun, M.D. Foster, W.J. Brittain, Macromolecules **37**, 2790 (2004)
238. B. Zhao, W.J. Brittain, W. Zhou, S.Z.D. Cheng, J. Am. Chem. Soc. **122**, 2407 (2000)
239. R. Iwata, H. Li, H. Zhang, Y. Xu, K. Zhang, P. Ai, X. Jin, J. Wang, Mater. Chem. Phys. **90**, 90 (2005)
240. K. Ohno, T. Morinaga, K. Koh, Y. Tsujii, T. Fukuda, Macromolecules **38**, 2137 (2005)
241. R. Djalali, S.Y. Li, M. Schmidt, Macromolecules **35**, 4282 (2002)
242. M. Yu, Y. Lu, M. Schrinner, F. Polzer, M. Ballauff, Macromol. Symp. **254**, 42 (2007)
243. M. Zhang, M. Drechsler, A.H.E. Müller, Chem. Mater. **16**, 537 (2004)
244. M. Schrinner, S. Proch, Y. Mei, R. Kempe, N. Miyajima, M. Ballauff, Adv. Mater. **20**, 1928 (2008)
245. D.G. Shchukin, G.B. Sukhorukov, Adv. Mater. **16**, 671 (2004)
246. T.C. Wang, M.F. Rubner, R.E. Cohen, Langmuir **18**, 3370 (2002)
247. M. Fang, P.S. Grant, M.J. McShane, G.B. Sukhorukov, V.O. Golub, Y.M. Lvov, Langmuir **18**, 6338 (2002)
248. T.C. Wang, B. Chen, M.F. Rubner, R.E. Cohen, Langmuir **17**, 6610 (2001)
249. S. Moya, O. Azzaroni, T. Farhan, V.L. Osborne, W.T.S. Huck, Angew. Chem. Int. Ed. **44**, 4578 (2005)
250. N. Ayres, S.G. Boyes, W.J. Brittain, Langmuir **23**, 182 (2007)
251. J. Chen, S.K. Spear, J.G. Huddleston, R.D. Rogers, Green Chem. **7**, 64 (2005)
252. A.S. Karakoti, S. Das, S. Thevuthasan, S. Seal, Angew. Chem. Int. Ed. **50**, 1980 (2011)
253. M.D. Howard, M. Jay, T.D. Dziublal, X.L. Lu, J. Biomed. Nanotechnol. **4**, 133 (2008)
254. M.J. Joralemon, S. McRae, T. Emrick, Chem. Commun. **46**, 1377 (2010)
255. P.K. Sudeep, Z. Page, T. Emrick, Chem. Commun. 6126 (2008)
256. N. Uekawa, M. Endo, K. Kakegawa, Y. Sasaki, Phys. Chem. Chem. Phys. **2**, 5485 (2000)
257. S. Kuchibhatla, A.S. Karakoti, S. Seal, Nanotechnology **18**, 075303 (2007)
258. M. Popa, T. Pradell, D. Crespo, J.M. Calderon-Moreno, Colloids Surf. A **303**, 184 (2007)
259. M. Zhang, Z.H. Wang, G.C. Xi, D.K. Ma, R. Zhang, Y.T. Qian, J. Cryst. Growth **268**, 215 (2004)
260. U.I. Tromsdorf, N.C. Bigall, M.G. Kaul, O.T. Bruns, M.S. Nikolic, B. Mollwitz, R.A. Sperling, R. Reimer, H. Hohenberg, W.J. Parak, S. Forster, U. Beisiegel, G. Adam, H. Weller, Nano Lett. **7**, 2422 (2007)
261. L. Gu, Z. Shen, C. Feng, Y.G. Li, G.L. Lu, X.Y. Huang, G.W. Wang, J.L. Huang, J. Mater. Chem. **18**, 4332 (2008)
262. M.L. Ji, W.L. Yang, Q.G. Ren, D.R. Lu, Nanotechnology **20**, 075101 (2009)
263. K. Naka, A. Nakahashi, M. Bravo, Y. Chujo, Appl. Organomet. Chem. **24**, 573 (2010)
264. J.-J. Park, X. Bulliard, J.M. Lee, J. Hur, K. Im, J.-M. Kim, P. Prabhakaran, N. Cho, K.-S. Lee, S.-Y. Min, T.-W. Lee, S. Yong, D.-Y. Yang, Adv. Funct. Mater. **20**, 2296 (2010)

265. A. Dokoutchaev, J.T. James, S.C. Koene, S. Pathak, G.K.S. Prakash, M.E. Thompson, Chem. Mater. **11**, 2389 (1999)
266. C.W. Chen, T. Serizawa, M. Akashi, Chem. Mater. **14**, 2232 (2002)
267. F. Wen, W. Zhang, G. Wei, Y. Wang, J. Zhang, M. Zhang, L. Shi, Chem. Mater. **20**, 2144 (2008)
268. Y. Mei, G. Sharma, Y. Lu, M. Ballauf, M. Drechsler, T. Irrgang, R. Kempe, Langmuir **21**, 12229 (2005)
269. Y. Lu, Y. Mei, M. Drechsler, M. Ballauf, Angew. Chem. Int. Ed. **45**, 813 (2006)
270. S. Li, J. Wang, Y. Kou, S. Zhang, Chem. Eur. J. **16**, 1812 (2010)
271. J. Yang, L. Qiu, B. Liu, Y. Peng, F. Yan, S. Shang, J. Polym. Sci. Part A Polym. Chem. **49**, 4531 (2011)
272. Y.S. Gu, X.M. Hou, H.Y. Hua, B. Yu, L.X. Wang, F. Zhou, Mater. Chem. Phys. **116**, 284 (2009)
273. S.F. Zheng, J.S. Hu, L.S. Zhong, L.J. Wan, W.G. Song, J. Phys. Chem. C **111**, 11174 (2007)
274. C.J. Murphy, J. Mater. Chem. **18**, 2173 (2008)
275. A.I. Loskutov, O.Ya. Uryupina, V.V. Vysotskii, M.P. Kiselev, Nanotekhnika 39 (2010)
276. O.Ya. Uryupina, V.V. Vysotskii, A.I. Loskutov, V.V. Matveev, A.V. Cherkasova, V.I. Roldugin, Kh. Prikl. Khim. **86**. P. 1268 (2013)
277. A.I. Loskutov, O.Y. Uryupina, V.V. Vysotskii, A.V. Gusel'nikov, Nanotekhnika **1**, 62 (2011)
278. R.R. Bhattacharjee, M.H. Rashid, T.K. Mandal, J. Nanosci. Nanotechnol. **8**, 3610 (2008)
279. D. Maity, M.K. Bain, B. Bhowmick, J. Sarkar, S. Saha, K. Acharya, M. Chakraborty, D. Chattopadhyay, J. Appl. Polym. Sci. **122**, 2189 (2011)
280. Z. Shervani, Y. Ikushima, M. Sato, H. Kawanami, Y. Hakuta, T. Yokoyama, T. Nagase, H. Kuneida, K. Aramaki, Colloid Polym. Sci. **286**, 403 (2008)
281. A. Kohut, A. Voronov, V. Samaryk, W. Peukert, Macromol. Rapid Commun. **28**, 1410 (2007)
282. A. Voronov, A. Kohut, W. Peukert, Langmuir **23**, 360 (2007)
283. T. Sakai, P. Alexandridis, Chem. Mater. **18**, 2577 (2006); T. Sakai, P. Alexandridis, Nanotechnology **16**, 8019 (2005)
284. J.-Y. Kim, D.-H. Shin, K.-J. Ihn, Macromol. Chem. Phys. **206**, 794 (2005)
285. C.H. Bamford, *Encyclopedia of Polymer Science and Engineering* (Wiley-Interscience, New York, 1988), p. 763
286. G.G. Odian, *Principles of Polymerization*, 3rd edn. (Wiley, New York, 1991), p. 220
287. Y.M. Shulga, O.S. Roshchupkina, G.I. Dzhardimalieva, A.D. Pomogailo, Izv. AN, Ser. Khim. **9**, 1565 (1993)
288. G.I. Dzhardimalieva, A.D. Pomogailo, Uspekhi Khimii **77**, 270 (2008)
289. A.A. Berlin, V.N. Kislenko, Prog. Polym. Sci. **17**, 765 (1992)
290. V.I. Kurlyankina, V.N. Shadrin, E.N. Kazbekov, V.A. Molotkov, M.K. Bukina, Zh. Obsch. Khim. **44**, 1593 (1974)
291. Y. Song, Z. Li, L. Wang, Y. Yao, C. Chen, K. Cui, Microsc. Res. Tech. **71**, 409 (2008)
292. I. Tarnavchyk, A. Voronov, A. Kohut, N. Nosova, S. Varvarenko, V. Samaryk, S. Voronov, Macromol. Rapid Commun. **30**, 1564 (2009)
293. J.D.S. Newman, G.J. Blanchard, Langmuir **22**, 5882 (2006)
294. C. Subramaniam, R.T. Tom, T. Pradeep, J. Nanopart. Res. **7**, 209 (2005)
295. M. Aslam, L. Fu, M. Su, K. Vijayamohanan, V.P. Dravid, J. Mater. Chem. **14**, 1795 (2004)
296. P.R. Selvakannan, A. Swami, D. Srisathiyanarayanan, P.S. Shirdue, R. Pasricha, A.B. Mandale, M. Sastry, Langmuir **20**, 7825 (2004)
297. J. Keilitz, M.R. Radowski, J.-D. Marty, R. Haag, F. Gauffre, C. Mingotaud, Chem. Mater. **20**, 2423 (2008)
298. M. Chen, Y.-G. Feng, X. Wang, T.-C. Li, J.-Y. Zhang, D.-J. Qian, Langmuir **23**, 5296 (2007)
299. K. Esumi, Top. Curr. Chem. **227**, 31 (2003)
300. I. Gitsov, C. Lin, Curr. Org. Chem. **9**, 1025 (2005)
301. S. Keki, J. Tőrők, G. Deák, L. Daróczi, M. Zsuga, J. Colloid Interface Sci. **229**, 550 (2000)
302. Y. Niu, L. Sun, R.M. Crooks, Macromolecules **36**, 5725 (2003)

303. M.J. Jasmine, M. Kavitha, E. Prasad, J. Lumin. **129**, 506 (2009)
304. X. Luo, T. Imae, J. Mater. Chem. **17**, 567 (2007)
305. F. Grőhn, Macromol. Chem. Phys. **209**, 2295 (2008)
306. I. Willerich, F. Grőhn, Chem. Eur. J. **14**, 9112 (2008)
307. M. Kavitha, M.R. Parida, E. Prasad, C. Vijayan, P.C. Deshmukh, Macromol. Chem. Phys. **210**, 1310 (2009)
308. J.-M. Lin, T.-L. Lin, U.-S. Jeng, Y.-J. Zhong, C.-T. Yeh, T.-Y. Chena, J. Appl. Cryst. **40**, s540 (2007)
309. D. Debnath, C. Kim, S.H. Kim, K.E. Geckeler, Macromol. Rapid Commun. **31**, 549 (2010)
310. M.L.C.E. Hoppe, I. Pardiñas-Blanco, M.A. López-Quintela, Langmuir **22**, 7027 (2006)
311. H.H. Huang, X.P. Ni, G.L. Loy, C.H. Chew, K.L. Tan, F.C. Loh, J.F. Deng, G.Q. Xu, Langmuir **12**, 909 (1996)
312. Y. Gao, P. Jiang, L. Song, J.X. Wang, L.F. Liu, D.F. Liu, Y.J. Xiang, Z.X. Zhang, X.W. Zhao, X.Y. Dou, S.D. Luo, W.Y. Zhou, S.S. Xie, J. Cryst. Growth **298**, 376 (2006)
313. N. Bicak, S. Sungur, N. Tan, F. Bensebaa, Y. Deslandes, J. Polym, Sci. Part A Polym. Chem. **40**, 748 (2002)
314. V.A. Tertykh, K.V. Katok, V.V. Yanishpolskii, Russ. J. Phys. Chem. **82**, 1438 (2008)
315. N. Ivashchenko, V. Tertykh, V. Yanishpolskii, S. Khainakov, A. Dikhtiarenko, Mater. Werkst. **42**, 64 (2011)
316. K.G. Neoh, T.T. Yong, N.T. Looi, E.T. Kang, Chem. Mater. **9**, 2906 (1997)
317. E.T. Kang, Y.P. Ting, K.L. Tan, J. Appl. Polym. Sci. **53**, 1539 (1994)
318. L.M. Huang, T.C. Wen, A. Gopalan, Thin Solid Films **473**, 300 (2005)
319. L.M. Huang, T.C. Wen, A. Gopalan, Mater. Lett. **57**, 1765 (2003)
320. L.-M. Huang, C.-C. Tsai, T.-C. Wen, A. Gopalan, J. Polym. Sci. Part A Polym. Chem. **44**, 3843 (2006)
321. A.K. Sra, R.E. Schaak, J. Am. Chem. Soc. **126**, 6667 (2004)
322. B.M. Leonard, N.S.P. Bhuvanesh, R.E. Schaak, J. Am. Chem. Soc. **127**, 7326 (2005)
323. Y. Vasquez, A.K. Sra, R.E. Schaak, J. Am. Chem. Soc. **127**, 12504 (2005)
324. Y. Vasquez, Z. Luo, R.E. Schaak, J. Am. Chem. Soc. **130**, 11866 (2008)
325. N.H. Chou, R.E. Schaak, J. Am. Chem. Soc. **129**, 7339 (2007)
326. J.F. Bondi, R. Misra, X. Ke, I.T. Sine, P. Schiffer, R.E. Schaak, Chem. Mater. **22**, 3988 (2010)
327. B. Lim, J. Wang, P.H.C. Camargo, C.M. Cobley, M.J. Kim, Y. Xia, Angew. Chem. Int. Ed. **48**, 6304 (2009)
328. K.E. Elkins, T.S. Vedantam, J.P. Liu, H. Zeng, S. Sun, Y. Ding, Z.L. Wang, Nano Lett. **3**, 1647 (2003)
329. S. Sun, S. Anders, T. Thomson, J.E.E. Baglin, M.F. Toney, H.F. Hamann, C.B. Murray, B.D. Terris, J. Phys. Chem. B **107**, 5419 (2003)
330. X. Huang, H. Zhang, C. Guo, Z. Zhou, N. Zheng, Angew. Chem. Int. Ed. **48**, 4808 (2009)
331. J.B. Wu, J.L. Zhang, Z.M. Peng, S.C. Yang, F.T. Wagner, H. Yang, J. Am. Chem. Soc. **132**, 4984 (2010)
332. D. Wang, Y. Li, Adv. Mater. **23**, 1044 (2011)
333. Y. Sun, B.T. Mayers, Y. Xia, Nano Lett. **2**, 481 (2002)
334. W.G. Menezes, V. Zielasek, G.I. Dzhardimalieva, S.I. Pomogailo, K. Thiel, D. Wohrle, A. Hartwig, M. Baumer, Nanoscale **4**, 1658 (2012)
335. C. Wang, H. Yin, R. Chan, S. Peng, S. Dai, S. Sun, Chem. Mater. **21**, 433 (2009)
336. H. Song, F. Kim, S. Connor, G.A. Somorjai, P. Yang, J. Phys. Chem. B **109**, 188 (2005)
337. S.M. Humphrey, M.E. Grass, S.E. Habas, K. Niesz, G.A. Somorjai, T.D. Tilley, Nano Lett. **7**, 785 (2007)
338. N. Zettsu, J.M. McLellan, B. Wiley, Y. Yin, Z.-Y. Li, Y. Xia, Angew. Chem. Int. Ed. **45**, 1288 (2006)
339. Y. Xiong, J. Chen, B. Wiley, Y. Xia, S. Aloni, Y. Yin, J. Am. Chem. Soc. **127**, 7332 (2005)
340. D. Seo, J.C. Park, H. Song, J. Am. Chem. Soc. **128**, 14863 (2006)

341. Y. Zhang, M.E. Grass, S.E. Habas, F. Tao, T. Zhang, P. Yang, G.A. Somorjai, J. Phys. Chem. C **111**, 12243 (2007)
342. T.M. Whitney, J.S. Jiang, P.C. Searson, C.L. Chien, Science **261**, 1316 (1993)
343. S. Dominguez-Dominguez, J. Arias-Pardilla, A. Berenguer-Murcia, E. Morallon, D. Cazorla-Amoros, J. Appl. Electrochem. **38**, 259 (2008)
344. S. Kim, Y. Jung, Y.S. Lee, S.J. Park, Diffus. Defect Data Pt. B **124–126**, 1821 (2007)
345. C.R. Martin, Chem. Mater. **8**, 1739 (1996)
346. J.-S. Yu, J.Y. Kim, S. Lee, J.K.N. Mbindyo, B.R. Martin, T.E. Mallouk, Chem. Commun. 2445 (2000)
347. C.K. Preston, M. Moskovits, J. Phys. Chem. **97**, 8495 (1993)
348. M. Wirtz, C.R. Martin, Adv. Mater. **15**, 455 (2003)
349. E.K. Payne, K.L. Shuford, S. Park, G.C. Schatz, C.A. Mirkin, J. Phys. Chem. B **110**, 2150 (2006)
350. J. Wang, J. Mater. Chem. **18**, 4017 (2008)
351. J.-G. Wang, M.-L. Tian, T.E. Mallouk, M.H.W. Chan, Nano Lett. **4**, 1313 (2004)
352. S.J. Lee, J.M. Baik, M. Moskovits, Nano Lett. **8**, 3244 (2008)
353. J.J. Mock, S.J. Oldenburg, D.R. Smith, D.A. Schultz, S. Schultz, Nano Lett. **2**, 465 (2002)
354. P.R. Sajanlal, T. Pradeep, Adv. Mater. **20**, 980 (2008)
355. L. Delplancke, J. Dille, J. Reisse, G.J. Long, A. Mohan, F. Grandjean, Chem. Mater. **12**, 946 (2000)
356. L. Rodrigues-Sanchez, M.C. Blanco, M.A. Lopez-Quintela, J. Phys. Chem. B **104**, 9683 (2000)
357. M.T. Reetz, W. Helbig, J. Am. Chem. Soc. **116**, 7401 (1994)
358. J. Becker, R. Schäfer, R. Festag, W. Ruland, J.H. Wendorff, J. Pebler, S.A. Quaiser, W. Helbig, M.T. Reetz, J. Chem. Phys. **103**, 2520 (1995)
359. Southampton Electrochemistry Group, *Instrumental Methods in Electrochemistry* (Ellis Horwood, Chichester, 1990)
360. A.B. Isaev, N.A. Zakargalieva, Z.M. Aliev, Ross. Nanotekhnol. **6**, 88 (2011)
361. T.S. Sabir, D. Yan, J.R. Milligan, A.W. Aruni, K.E. Nick, R.H. Ramon, J.A. Hughes, Q. Chen, R.S. Kurti, C.C. Perry, J. Phys. Chem. C **116**, 4431 (2012)
362. A.E. Saunders, M.B. Sigman, B.A. Korgel, J. Phys. Chem. B **108**, 193 (2003)
363. D.T. Robb, V. Privman, Langmuir **24**, 26 (2007)
364. H. Zheng, R.K. Smith, Y.-W. Jun, C. Kisielowski, U. Dahmen, A.P. Alivisatos, Science **324**, 1309 (2009)
365. S.P. Shields, V.N. Richards, W.E. Buhro, Chem. Mater. **22**, 3212 (2010)
366. T.O. Drews, M.A. Katsoulakis, M. Tsapatsis, J. Phys. Chem. B **109**, 23879 (2005)
367. V.N. Richards, N.P. Rath, W.E. Buhro, Chem. Mater. **22**, 3556 (2010)
368. J. Millstone, G. Métraux, C. Mirkin, Adv. Funct. Mater. **16**, 1209 (2006)

Chapter 3
Physical-Chemical Methods of Nanocomposite Synthesis

On the whole, methods of production of nanostructural materials can be divided in two groups: physical and chemical. However, this separation is conditional, because, for example, all physical methods contain considerable chemical component, and it is often difficult to draw a distinct boundary between different methods. Special character of physical methods, prevailingly vapor phased, is in formation of crystal nanoparticles with vast open surface, which advances formation of strong aggregates, difficultly parted into primary particles. Moreover, it is often impossible to obtain complex phases because of their segregation at high temperatures typical of gas phase processes. A special place has methods, in which nanoparticles form as a result of different "physical" impacts, for example, under ultrasonic or microwave radiation. This effect stimulates different processes in a reaction mixture, first of all, chemical reactions, which brings to formation of nanostructural material with a definite composition, structure and properties. Therefore, these methods are often related to intermediate physical-chemical group.

Most physical-chemical methods of production of nanoparticles are based on homogeneous nucleation in vapor phase or heterogeneous nucleation in contact with surface, followed by condensation and coagulation. A necessary condition for condensation from the vapor phase is supersaturation, which can be reached by physical or chemical methods. Depending on a character of the heating processes (resistive, laser, plasma, electric arc, induction, ionic) and cooling, different methods of production of nanomaterials are distinguished, such as flame pyrolysis (see Chap. 7), synthesis in a flow reactors, laser induced evaporation and pyrolysis, thermal and microwave plasma methods, laser ablation. This group also includes solvo-thermal synthesis, pyrolysis of aerosols, and most methods of growth of nanoparticles and films from vapor phase, for example, chemical deposition from vapor phase (CVD), etc. Many of these methods are rather well described in literature, there are voluminous reviews and monographs [1–5]. Below only most typical methods are considered, which, in our opinion, are of keen interest for production of nanocomposite materials.

A.D. Pomogailo, G.I. Dzhardimalieva, *Nanostructured Materials Preparation via Condensation Ways*, DOI 10.1007/978-90-481-2567-8_3

3.1 Synthesis of Nanocomposites by Microwave Irradiation

Prospects of using of microwave energy in development of nanocomposite materials are determined by a possibility of bulk heating of materials. As is shown in Chap. 2, methods of chemical reduction often include complex and continuous reactions of molecular precursors in presence of solvents, ligands and/or surfactants at elevated temperatures. Microwave heating has some advantages as compared to conventional methods of heating of condensed matters, in particular, high rate and low inertia of heating, uniformity of heating of material over the volume, and, as a consequence, decrease in thermal stress and high homogeneity of microstructure of the obtained composites, etc. Owing to this, microwave radiation is widely used in synthesis of nanocrystalline materials. General information about microwave heating of dielectric materials, criteria of estimation of efficiency of its usage in different processes of organic synthesis, polymer chemistry, materials science, nano- and biotechnology, etc. are described in reviews [6–9].

3.1.1 Basic Principals

Microwave radiation is non-ionizing electromagnetic radiation at frequencies from 300 MHz to 30 GHz. Most experimental studies are performed at the frequency of electromagnetic radiation 2.45 GHz (Fig. 3.1).

If a substance is under impact in the range of microwave frequencies, its dipoles are oriented along applied electric field. Oscillation of the field causes re-orientation of bipolar molecules, which try to follow alternate electric field, which is accompanied by energy loss due to molecular friction and dielectric loss.

Scattering of microwave irradiation power per unit volume in a material can be expressed by equation:

$$P = c|E|^2 f\varepsilon'' = c|E|^2 f\varepsilon' \tan\delta \tag{3.1}$$

where c is the constant, E is the intensity of electric field, f is the radiation frequency, ε'- imaginary part of equivalent dielectric permeability of a substance, and ε'' – real part, $\tan\delta = \varepsilon''/\varepsilon'$ – dielectric loss tangent.

There is optimal frequency range of dielectric loss coefficient, which determines absorption ability of a matter. On the whole, media with high tgδ are characterized by high absorption ability and, respectively, efficiency of heating (Table 3.1).

In the synthesis of nanocrystalline materials a wide range of solvents is used: with strong (for example, ionic liquids, ethylene glycol), intermediate (water, methyl pyrrolidone) absorption abilities, and also media almost transparent for microwave radiation (non-polar alkanes or alkenes), i.e. the substances should have either high coefficient of dielectric loss tangent (they should have mobile dipoles with rather high dipole momentum), or high ion, electron, or hole conductivity.

Water, alcohols, DMF, ethylene glycol have high dielectric loss tangent and are ideal for microwave heating (Table 3.2).

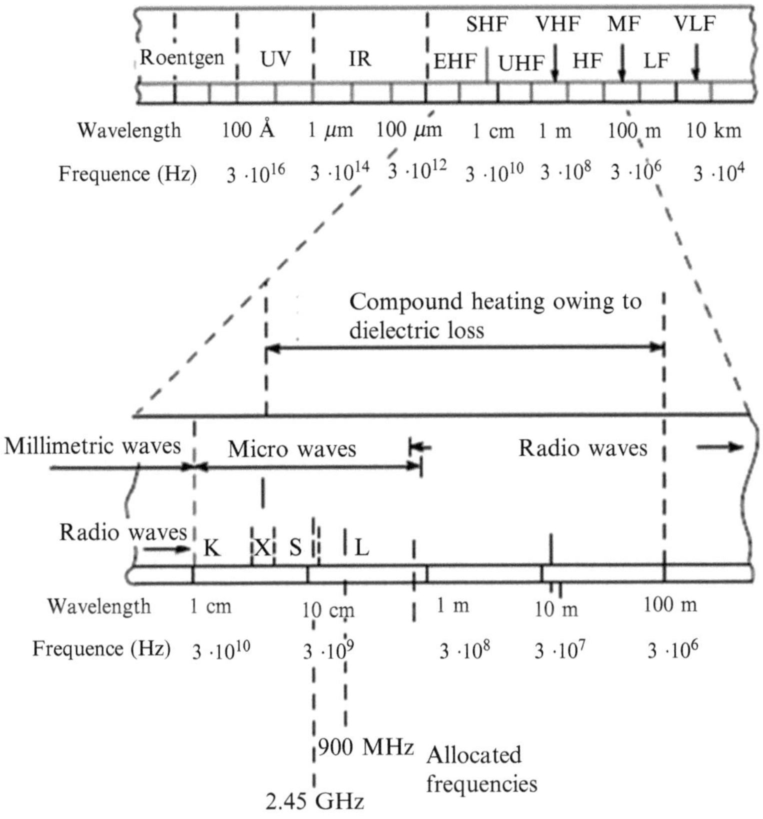

Fig. 3.1 Microwave electromagnetic irradiation scale

Table 3.1 Dielectric loss tangent (tanδ) for different solvents [6]

Solvent	tanδ	Solvent	tanδ
Ethylene glycol	1.350	DMF	0.161
Ethanol	0.941	1,2-dichloroethane	0.127
DMSO	0.825	Aqua	0.123
2-Propanol	0.799	Phenyl chloride	0.101
Formic acid	0.722	Acetonitrile	0.062
Methanol	0.659	Acetone	0.054
1,2- dichlorobenzene	0.280	Tetrahydrofuran	0.047
N-methyl-2- pyrrolidone	0.275	Dichloromethane	0.042
1-butyl-3- methylimidazolium hexafluorophosphate	0.185	Toluene	0.040
Acetic acid	0.174	Hexane	0.02

The processes in microwave field may be intensified by non-thermal effects of microwave radiation. Main approaches to grounding of the mechanism of non-thermal effect consider acceleration of diffusion processes [10, 11], some authors relate these phenomena to surface polarization [12].

Table 3.2 Physical parameters of typical solvents used for microwave heating [7].

Solvent	T_{boil}, °C	ε′	ε″	tanδ
Aqua	100	78.3	12.3	0.157
Methanol	65	32.7	20.9	0.639
Ethanol	78	24.3	6.08	0.200
N,N-dimethylformamide	153	36.7	–	–
Ethylene glycol	198	41.0	41.0	1.00
N-methyl-2-pyrrolidone	202	32.0	8.855	0.277

3.1.2 Microwave Heating in Synthesis of Colloid Nanocrystals

It should be noted that most experiments with microwave radiation are confined to simple "one-pot" schemes with one-step heating, when all reacting substances presence in the initial reaction mixture and are heated to a given temperature. In this case it is impossible to use, for example, such well-known procedures as "hot-injection" for separation of nucleation and growth stages, or to conduct consequent addition of selective reagents to control particle size distribution and shape. Under microwave radiation so called "heating-up" technique is used, when reagents mixed at environmental temperature are rapidly heated to high temperature. This scheme is especially efficient in colloid systems in which super saturation point is a limiting stage [13]. Monomer precursors can form in shape of intermediates during slow reaction or during decomposition of quite stable monomer-surfactant complexes. Thus, complex formation of poly(vinylpyrrolidone) with Ag^+ ions before the beginning of microwave heating advances reduction predominantly on polymer chains, and a polymer molecule additionally inhibits cluster aggregation at early stages of the reaction, which on the whole favors formation of narrow dispersed distribution of nanoparticles [14]. Specific impact of microwave irradiation can manifest itself in increase in reduction ability of a reducing agent, as was observed in polyethylene glycol [15] or glutathione [16] and $AgNO_3$ system. It is interesting to note that under comparable conditions, but at ordinary convection heating for 24 h no reaction products were found. Analogous effect of microwave irradiation was found in carboxymethylcellulose-Na_2PtCl_6, $HAuCl_4$ $3H_2O$ or $PdCl_2$, systems, in which noble metals are reduced only under conditions of microwave heating at 100 °C [17], no reducing agents or stabilizers were required.

Formation of different polygonal nanoplates (triangular, square, pentagonal and hexagonal) was observed at microwave heating (198 °C) during reduction of $HAuCl_4$ in ethylene glycol and in presence of PVP [18]. At the same time heating in oil bath to the same temperature causes formation of prevailingly spherical particles of 100—190 nm in diameter and a small fraction of rod-shaped nanoparticles. This difference in character of formation of nanostructures is, probably, caused by explosive nucleation owing to fast homogeneous dielectric heating, which cannot be reached under conditions of ordinary heating. Thus, microwave heating allows conducting of hydrothermal synthesis of ferrite [19] or perovskite [20–22] nanoparticles at lower temperatures and shorter times, which manifests itself in substantial decrease in sizes

Table 3.3 The effect of reaction conditions on sizes of ferrite nanoparticles [19]

Sample	Reaction conditions	Grain size by XRD data, nm	Nanoparticle size by TEM data, nm	Morphology
$NiFe_2O_4$, $CoFe_2O_4$, $MnFe_2O_4$	Ordinary hydrothermal heating, 250 °C, 1 h; the ratio oleic acid/iron nitrate 3:1, 6:1	7 ± 1	9 ± 1	Spherical
	Ordinary hydrothermal heating, 250 °C, 2 h; the ratio oleic acid/iron nitrate 3:1, 6:1	7 ± 1	9 ± 1	Spherical
	Ordinary hydrothermal heating, 250 °C, 2 h; the ratio oleic acid/iron nitrate 3:1, 6:1	7 ± 1	9 ± 1	Spherical
	Microwave heating, 160°C, 1 h	3.5 ± 1	5 ± 2	Spherical
	Microwave heating, 160 °C, 2 h	6.5 ± 1	9 ± 1	Spherical
γ-Fe_2O_3	Ordinary hydrothermal heating, 120 °C, 1 h; the ratio oleic acid/iron nitrate 3:1	9 ± 1	10 ± 1	Spherical and cubic
	Ordinary hydrothermal heating, 120 °C, 1 h; the ratio oleic acid/iron nitrate 3:1	4 ± 1	5 ± 1	Spherical

of formed nanoparticles and in increase in crystallization rate by one-two orders, than under conditions of ordinary hydrothermal synthesis (Table 3.3). In the case of perovskite dielectric ceramic materials this happens due to high absorption of electromagnetic radiation by these materials having high dielectric constants [23].

It is shown that advantages of microwave – polyolic synthesis of Ni nanoparticles as compared to traditional polyolic technique bring to increase in efficiency of the process and thus facilitate its scaling [24]. Microwave polyolic synthesis is very efficient for producing two- and three-component nanometer chalcogenides (selenides and tellurides) and can serve as a general way of their fabrication [25–28]. As it was noted in Sect. 2.2, polyols (such as ethylene glycol and other glycols) on the one hand, play the role of reducing agent and solvent in the processes of synthesis of metal nanoparticles, and, on the other hand, are very sensitive to microwave radiation due to high dipole moment and high dielectric loss tangent. Moreover, reducing properties of ethylene glycol are most efficient at high temperatures and, as is known, high-boiling solvents are less susceptible to carbonizing due to appearance of arc discharge. Besides, it should be noted that this phenomenon may be used as simpler and quite efficient method for production of metal/carbon composites with core-shell structure in organic solvents. Under conditions of microwave arc discharge, as a rule, highly organized carbon structures form, for example, graphene layers and Fe/C core-shell in nanocomposite obtained from toluene solution $Fe(CO)_5$-$C_{60/70}$ [29].

Increase in energy of microwave radiation effectively decreases reaction time and sizes of formed structures, which, probably, is caused by the fact that at higher energies a greater number of nuclei form, and this, provides a precursor concentration is the same, brings to a decrease in diameters of the particles [30], a decrease in role of side effects, an increase in reaction yield, and improvement of reproducibility of the reactions [31]. And the observed increase in reaction rate is purely thermal/kinetic effect, i.e. it is a consequence of high temperature of the reaction, super-heated solvent above boiling temperature. Especially effective these processes go in media, highly absorbing microwave radiation, in particular, in ionic liquids. It is interesting that just impact of microwave heating causes decomposition of 1-butyl-3-methylimidazolium tetrafluoroborate used as ionic solvent during synthesis of Cu and Ni with formation of carbon, which in turn serves as a reducing agent and protective shell for the formed metal particles [32]. In this case, depending on irradiation time, the products of reduction for copper nitrate were Cu_2O or Cu for 5-min or 10-min microwave treatment in solution of ionic liquid, respectively. It should be noted that advantages of microwave heating, which provide uniform heating and equal rates of achievement of necessary temperature are very attractive for implementation of various reduction processes in hetero-metal systems, for example, for producing of core-shell structures or nanoparticles of metal alloys [33].

It is assumed that non-isothermal effects caused by absorption of long-chained molecules of polymer surfactant can advance formation of coarse anisotropic structures [7] (Scheme 3.1):

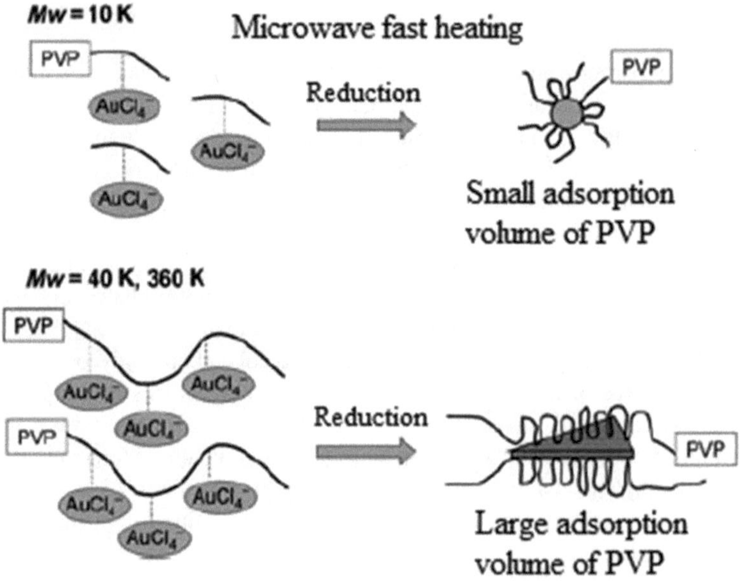

Scheme 3.1 Formation scheme of Au nanostructures using PVP with different chain lengths

Moreover, under conditions of microwave radiation transformation processes of nanoparticle shape are also possible. For example, coherent surface heating of silver particles stabilized by PVP brings to transformation from spherical Ag crystals into coarse prismatic particles, probably by Ostwald ripening mechanism [34].

3.2 Photo Chemical Reduction

Producing of metal nanoparticles under impact of high energies on chemical system is associated with generation of high-active strong reducing agents like electrons, radicals, excited particles. Typical wavelengths and energy ranges of the used electromagnetic radiations are shown in Table 3.4.

Methods of production of nanoparticles can be divided by the type of impact of input energy on chemical system as: photochemical (photolysis, < 60 eV), radiolytic (radiolysis, $60–10^4$ eV), irradiation by fast electron flow ($>10^4$ eV). Ionizing radiation is generated by a source of X-rays and gamma rays, ultraviolet and visible irradiations are usually generated by a mercury vapor lamp, or by xenon lamp. Among the special features of chemical processes induced by high-energy irradiation there are the following: high volume and surface energy density; non-equilibrium state, which manifests itself in non-Maxwell, non-Boltzmann functions of particle size distribution and population of quantum energy levels; approaching and overlap of typical τ times of physical, physical-chemical and chemical processes; the leading role in chemical transitions of highly reactive particles; multichannel character and instability of the processes in a reacting system (Table 3.5).

A number and size of formed nanoparticles can be controlled due to regulation of intensity of flow of electromagnetic radiation quanta, which are able to efficiently and rapidly stimulate reduction reactions.

3.2.1 *The Main Approaches to the Synthesis of Matrix-Stabilized Nanoparticles by Photochemical Irradiation*

Photo and radiation-chemical reduction has some advantages as compared to chemical reduction. Stimulated by radiation reduction reactions can be carried out in different media, including solid ones (for example, polymer matrices, films) and at low temperatures with good reproducibility. In parallel to this it is

Table 3.4 Scale of electromagnetic waves

Radiation	λ, nm	E, eV
Fast electrons	–	$>10^4$
Gamma and X-ray:		
Hard	0.2–100	$6.2 \cdot 10^3–12.4$
Soft radiation	>100	<12.4
Ultraviolet:	10–400	124–3.1
Deep (vacuum)	10–200	124–62
Near radiation	200–400	62–3.1
Visible light	400–700	3.1–1.67

Table 3.5 Some physical-chemical characteristics of irradiation impact

Characteristic	Methods Photolysis	Radiolysis	Fast electrons
Active reactive particles	Electron-vibrating-excited molecules, ions, and electrons	Electrons, ions, electron-excited particles	Electrons, ions, electron excited particles
Initial processes	$h\nu + M \rightarrow M^* \rightarrow$ Products	$e^-, h\nu + M \rightarrow M^+, e^-, (M^+)^* \rightarrow$ Products	$e^- + M \rightarrow M^+, M^*, (M^+)^* \rightarrow$ Products
Hierarchy of typical times of a process	$\tau_v < \tau_{ch} \leq \tau_e$	$\tau_{ch} \approx \tau_{dif}$	$\tau_{ch} \approx \tau_{tr}$
Generation rate of reactive particles ($v \cdot 10^{-15}$, cm^3/s)	10^{-2}–10^{-3}	~10^{-3}	~$2 \cdot 10^4$

Legend: *ch* chemical reaction, *v* vibrating relaxation, *e* electron relaxation, *tr* transport, *dif* diffusion

possible to obtain quite chemically pure target products without impurities, in particular accompanying chemical reduction methods. Duration of irradiation can be a simple experimental criterion for control over sizes of the formed particles [35], though in some cases increase in exposure time causes an increase in yield of nanoparticles, and has insignificant effect on their size [36].

At the same time, methods combining chemical and photochemical reduction, when a reducing agent also presents in a system [37, 38], are quite widely spread. A good example is photochemical reduction of $HAuCl_4$ in water media containing ethylene glycol of different concentrations and in presence of poly-4-vinylpyrrolidone. Mechanism of this reaction is studied in detail in [39]. As it was mentioned in Sect. 2.3, ethylene glycol is widely used as a reducing agent and viscous solvent slowing down diffusion in polyol processes of nanoparticles fabrication. However, these reactions as a rule are performed at high temperatures (160–280 °C), in order to reach reduction of metal ions. Continuous UV radiation of the ion metal salt under conditions of the considered reaction caused formation of nanoparticles at room temperature. Increase in rate of Au^{3+} extinction (absorption band at 323 nm in UV/visible range of the spectrum) during irradiation and in formation rate of nanoparticles (up to 0.4 mol fraction of glycol), as well as regeneration of Au^{3+} and its repeated disappearance some time later, and increase in intensity of a plasmon resonance band in a dark phase signs that ethylene glycol participates in the reduction reaction. The proposed mechanism took into account the role of ethylene glycol not only in reduction of excited Au^{3+}, but, probably, in reactions of Au^{2+} with Au^+:

$$Au^{3+}Cl_4^- \xrightarrow{h\nu} \left[Au^{3+}Cl_4^-\right]^* \tag{3.2}$$

$$[Au^{3+}Cl_4^-]^* \xrightarrow{glycol} Au^{2+}Cl_3^- + Cl^- \tag{3.3}$$

$$2Au^{2+}Cl_3^- \rightarrow Au^{1+}Cl_2^- + Au^{3+}Cl_4^- \tag{3.4}$$

$$4\,Au^{1+}Cl_2^- \rightarrow 2Au^{2+}Cl_3^- + 2Au^0 + 2Cl^- \tag{3.5}$$

$$3Au^{1+}Cl_2^- \rightarrow Au^{3+}Cl_4^- + 2Au^0 + 2Cl^- \tag{3.6}$$

$$Au^{m+} \xrightarrow{glycol} Au^0(m+\ :\ 3+,\ 2+,\ 1+) \tag{3.7}$$

$$nAu^0 \rightarrow Au^0 + (Au^0)_n \tag{3.8}$$

As a mole fraction of ethylene glycol increases, the solution becomes more viscous, and velocities of diffusion-controlled reactions, for example, disproportioning, slow down. It should be noted that special features of reduction of $AuCl_4^-$ and reactions of its disproportioning are intensely studied by many researchers [40–42]. As in the considered example, it is assumed that Au^{3+} transits to excited state, then reduces to Au^{2+}, which is unstable and disproportions with formation of Au^+ and Au^{3+}, then Au^+ reduces to Au^0 by other absorbed photon. The assumptions are made that nanoparticles can catalyze reactions of disproportioning [43].

Presence of UV or light radiation makes it possible to more flexible regulate characteristics of nanoparticles. For example, ammonium silver complex can be reduced by glucose at activation under light radiation at wavelength 365 nm at room temperature with stable dispersion of Ag nanoparticles with diameter 4–7 nm [44]. It is interesting to note that reduction of silver under the same conditions, but at heating and without UV impact, causes formation of nanoparticles with wide size distribution (4–100 nm). For the synthesis of anti-bacterial nanocomposite fibers, cellulose acetate (polyacrylonitrile, polyvinyl alcohol)/silver nanoparticles in DMF, which is simultaneously a solvent and a reduction agent, it is very efficient to apply UV radiation to increase yield of formed nanoparticles [45].

In the system of branched polyethyleneimine: 4-(2-hydroxoethyl)-1-piperazineethanesulfonic acid (HEPES): $AgNO_3$ photo-radiation causes oxidizing decomposition of polyethylene amine with following formation of positively charged Ag nanoparticles according to the proposed scheme (Scheme 3.2) [36]:

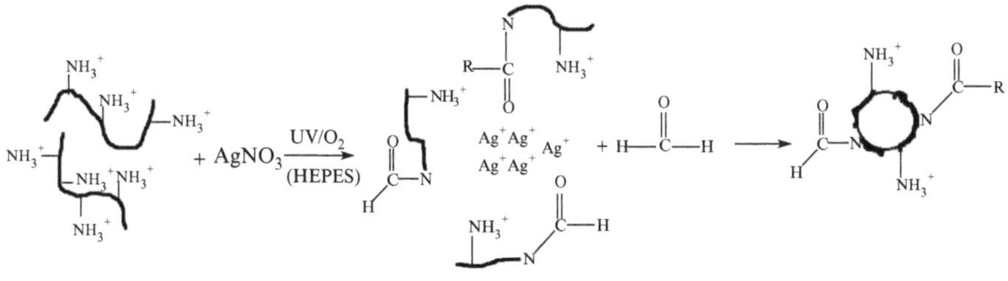

Scheme 3.2 The formation of branched PEI/Ag nanocomposite

It is assumed that formaldehyde formed as result of chemical transformation of PEI under UV treatment reduces silver ions, and HEPES has a catalytic effect on the process on the whole.

Important is the method of producing of nanocomposites, when polymerization and reduction processes induced by ultraviolet radiation are superimposed. Radicals formed under UV impact and being electron donors are able simultaneously polymerize monomers and reduce metal salts. Production of Ag nanoparticles and stabilizing polymer matrix is conducted in situ under irradiation ($\lambda = 320$ nm) of the mixture of monomer of 2-hydroxyethylmethacrylate and ammonium solution of silver oxide [46]. Often in these systems the main reactions are preceded by complex formation of metal ions with monomer molecules, as, for example, during formation of silver/polyacrylonitrile nanocomposite [47–49]. These processes can schematically be presented as:

$$Ag^+ + 2AN \rightleftharpoons [Ag(AN)_2]^+ \quad (3.9)$$

$$IN \xrightarrow{h\nu} IN^* \quad (3.10)$$

$$[Ag(AN)_2]^+ + NO_3^- \xrightarrow{h\nu} Ag^0 + 1/2O_2 + NO_2 + -(-AN-)_n- \quad (3.11)$$

$$IN^* + AN \rightarrow IN - AN^* \quad (3.12)$$

$$IN - AN^* + nAN \rightarrow IN - AN^* - (AN)_n \quad (3.13)$$

$$IN^* + Ag^+ \rightarrow IN^+ + Ag^0 \quad (3.14)$$

$$IN^+ + AN \rightarrow IN - AN^+ \quad (3.15)$$

$$IN - AN^+ + nAN \rightarrow IN - (AN)_n - AN^+ \quad (3.16)$$

where AN is an acrylonitrile molecule, IN is photoinitiator molecule.

The formed molecule of polyacrylonitrile plays a stabilizing role preventing aggregation of nanoparticles. The factors which allow control over nanoparticle sizes can be concentration of initial salt and content of a photoinitiator (Fig. 3.2).

Quite efficient control over nanoparticle size in such systems can be accomplished by usage of metal-containing monomers for precursors [50]. It can be demonstrated on the example of polystyrene/polyacrylic acid-Ag structured as polyelectrolyte brushes (see Sect. 2.2.3) [51]. Silver nanoparticles and polyacrylic chains are formed in situ during photopolymerization of silver acrylate on the surface of polystyrene particles coated by a thin layer of photoinitiator (2-[*p*-(2-hydroxy-2-methylpropiophenone)]-ethyleneglycol methacrylate) (HMEM) (Fig. 3.3). Radicals formed during UV irradiation initiate grafting polymerization of acrylate monomer and simultaneously reduce silver ions to metal Ag. Under these conditions local concentration of Ag^+ ions stays low, which causes expected ultrafine nanoparticle sizes (3 ± 1.2 nm).

Polyoxometalates of Keggin type (see Sect. 2.2.1) can serve as efficient matrix media for formation of metal nanoparticles due to their multielectron redox-chemistry and ability to accept and donor a certain number of electrons at some stages without decomposition [52, 53]. For example, under UV irradiation ($\lambda > 280$ nm) of the

Fig. 3.2 Dependence of average grain size of silver nanoparticles on concentration of $AgNO_3$ (*1*) and photoinitiator (*2*) [47]

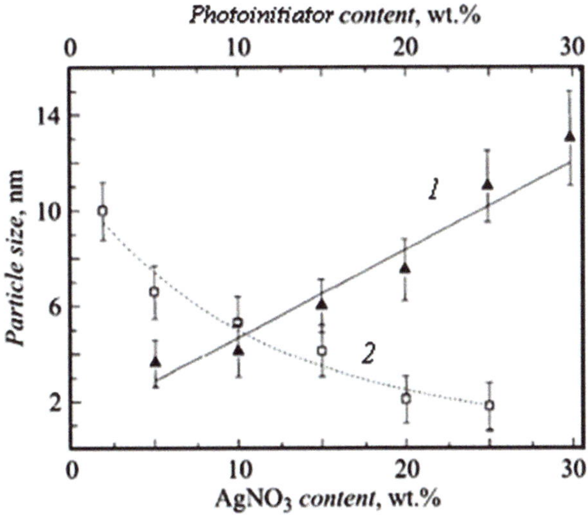

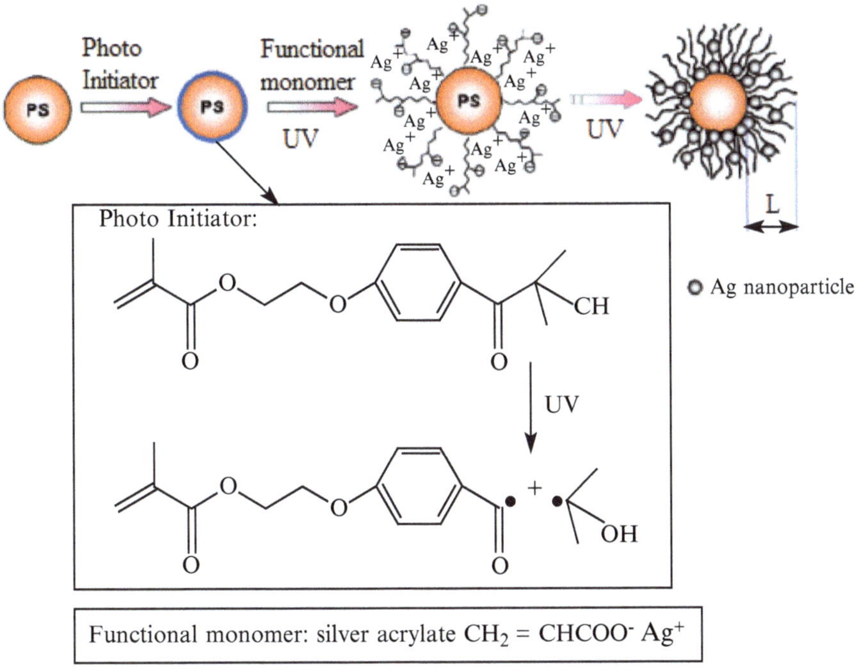

Fig. 3.3 Scheme of fabrication of PS-PAA-Ag composite particles *in situ* [51]

reaction mixture of $AgNO_3$ water solutions and $H_4(SiW_{12}O_{40})$ in presence of 2-propanol the following reactions take place [54]:

$$2(SiW_{12}O_{40})^{4-} + (CH_3)_2CHOH \rightarrow 2(SiW_{12}O_{40})^{5-} + (CH_3)_2C = O + 2H^+ \quad (3.17)$$

$$(SiW_{12}O_{40})^{5-} + Ag^+ \rightarrow (SiW_{12}O_{40})^{4-} + Ag \quad (3.18)$$

By the similar scheme photo-reduced interacts with $(SiW_{12}O_{40})^{5-}$ Pd^{2+}, $AuCl^{4-}$, and $PtCl_6^{2-}$ ions forming metal nanoparticles electrostatic and stoichiometric stabilized by Keggin ions [52]. In principle, the repeated irradiation can also reduce bound Keggin ions, and in this case polyoxometalates can be efficiently used for synthesis of bimetal nanoparticles as UV-switched reducing agents. This possibility is shown at production of Au-Ag nanoparticles with core-shell structure [53] in presence of phosphotungstic acid $H_3(PW_{12}O_{40})$ in propanol-2. Under irradiation by UV light (>280 nm, 450-W intermediate pressure lamp) single-electron reduction takes place during 4 h with formation of $(PW_{12}O_{40})^{3-}$ ions, which take immediate part in reduction of $HAuCl_4$ with formation of stable Au nanoparticles bound with Keggin ions. It is important that UV activation and reduction reaction of $AuCl_4^-$ do not touch the basic structure of the latter. Additional UV irradiation of Au-PTA solution brings to reduction of Keggin ions on the surface of nanoparticles and, in turn, addition of Ag_2SO_4 water solution completes by formation of Ag nanoparticles for 15 min. In the UV/visible spectrum of this solution there is weakening and blue shift of the plasmon resonance band of Au, while simultaneously appears a distinct absorption band at 415 nm, which confirms formation of Ag shell around Au core coated by PTA ions. The reducing agent bound this way and switched under irradiation provides reduction of the second metal just on the surface of the metal core, which makes this approach very promising in production of bimetallic structures, especially in catalysis.

It should be noted that thus obtained nanoparticles have long-term stability and, on the whole, the process is easily scaled.

3.2.2 Photochemical Reduction in Solutions

Photochemical reduction in the solutions is conducted in presence of electron-donor reagent and is most often used for synthesis of noble metal particles. Solutions of metal salts or their complexes in water, alcohol, and organic solvents are used as precursors and a medium for obtaining nanoparticles. In these media subjected to light exposure active particles form in form of solvated electrons:

$$H_2O \rightarrow aq\, e + H + OH, \quad (3.19)$$

or free radicals, to which, for example, belong ketyl, α-hydroxybenzene, benzoyl radicals, formed in reactions of photo-decomposition Norrish Type I [55, 56] of benzoin [57, 58], benzophenone [59–61] and other photoactivators. A typical representative widely used is 1-[4-(2-hydroxyethoxy)phenyl]-2-hydroxy-2-methyl-1-propan-1-on (Irgacure-2959,I-2959)[1]:

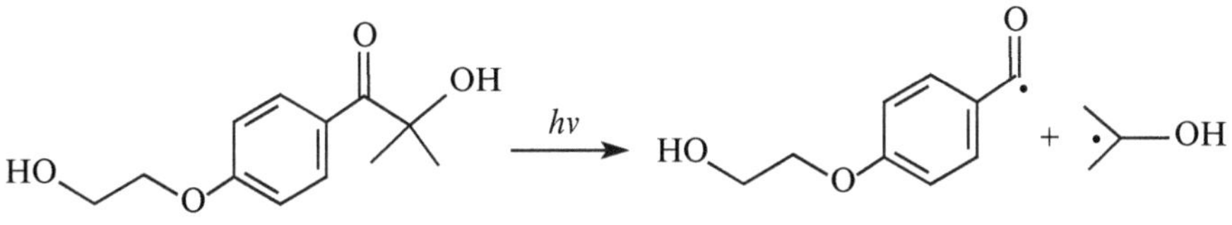

(3.20)

Active particles, in turn, via interaction with a metal ion, for example, Ag^+, reduce it to metal:

$$Ag^+ + aq\,e \rightarrow Ag^\circ \tag{3.21}$$

or

(3.22)

Three electron reduction of Au(III) ions can include several consequent stages of reduction and disproportioning [63, 64].

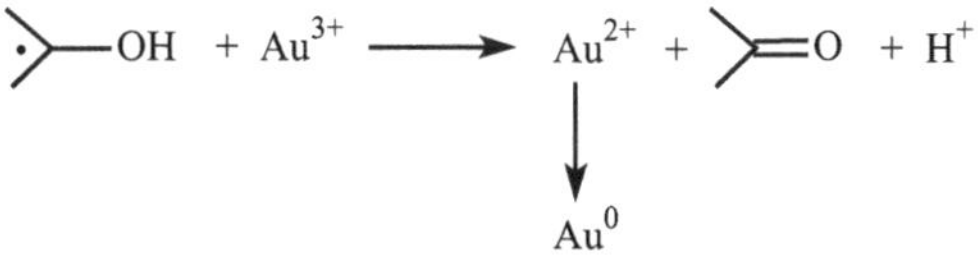

(3.23)

Efficiency of the photo process is such that under exposure of UV light >320 nm gold nanoparticles form during several minutes [64]. Redox potentials of ketyl radicals make all stages of reduction of Au^{3+} ions thermodynamically preferable [42].

[1] Irgacure-2959 has sufficient solubility in solvents with different polarities, such as water, toluene, THF, etc. Quantum yield of its photodecomposition is 0.29 and the processes of extinguishing of triplet state by ion metals almost do not go since times are short, which, on the whole, provides high efficiency of reduction reactions [62].

Photoinitiator 1-hydroxycyclohexylphenyl ketone forms two free radicals during photolysis:

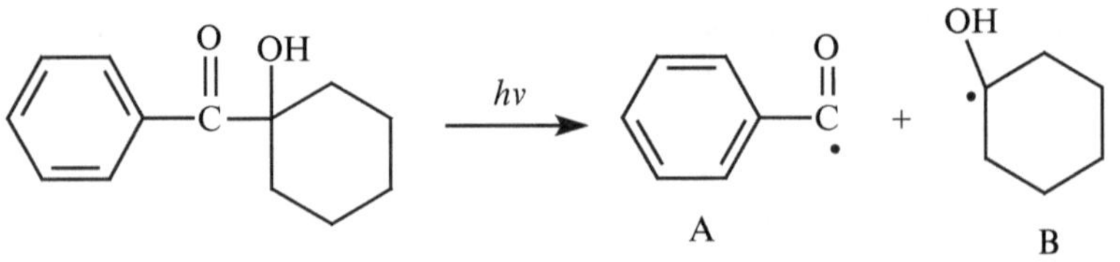

(3.24)

Chromatography and mass-spectral analysis were used to identify products of photolysis during the reduction reaction of coordination compound $CuCl_2$ with diethanolamine in the ethanol medium [65] and main ways of photochemical transformations in the system are found. Free radical A can split a hydrogen atom from ethanol molecule or from free radical B with formation of benzaldehyde (Scheme 3.3). In the other way A and B can recombine the initial photoinitiator, or two free A radicals interacting with each other form a couple:

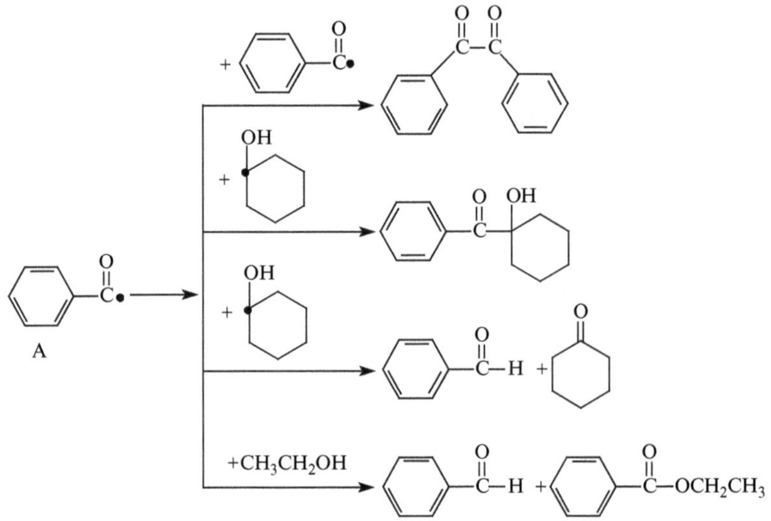

Scheme 3.3 The routes of a primary radical transformation at photoreduction of $CuCl_2$ in the diethanolamine-ethanol system

In no one of these transformations free radical A loses electrons, which points to the fact that it does not take part in reduction of copper ions. But, taking into account that in other products of the photochemical reaction cyclohexanol and cyclohexanone are found, corresponding to B radical by their chemical structure, probably, the main channels of the reduction reaction of copper ions proceed with participation of B radical, as is shown in the following scheme (Scheme 3.4):

Scheme 3.4 The routes of the $CuCl_2$ transformations in the diethanolamine-ethanol system

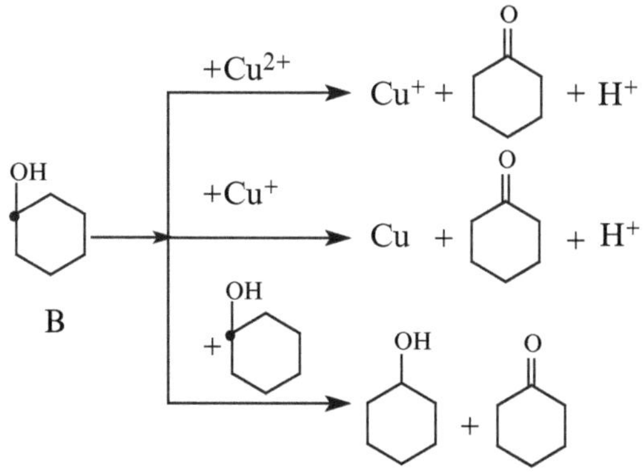

Color molecules can play a double role as photosensitizers: they can initiate protecting function for nanoparticles and modify their optical properties. Experimental data [35] confirm that, for example, thionine molecule

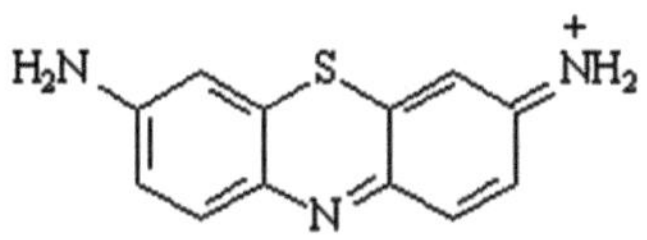

Thionine (TH^+)

under action of light transforms into singlet and triplet states:

$$TH^+ + h\nu \rightarrow {}^1TH^+ \rightarrow {}^3TH^+ \quad (3.25)$$

The formed triplet in absence of donors or acceptors of electrons can take part in disproportioning reaction with a non-excited thionine molecule with formation of oxidized and reduced products with electron transfer in non-polar medium at the rate 7.3 ($\pm$0.3)$\cdot 10^7$ M^{-1} s^{-1}:

$$^3TH^+ + TH^+ \rightarrow TH^\bullet + TH^{\bullet 2+} \quad (3.26)$$

$$TH^\bullet + Ag^+ \rightarrow TH^+ + Ag^0 \quad (3.27)$$

Rate of the direct reaction of extinguishing of triplet state by silver ions is small (1.30 ($\pm$0.06)$\cdot 10^6$ M^{-1} s^{-1}), which makes this reaction of electron transfer with formation of metal nanoparticles barely probable, as was discussed above. It should be noted that realization of photosensitized reduction of metal ions in water medium is possible only at participation of electron donors, for example, triethanolamine.

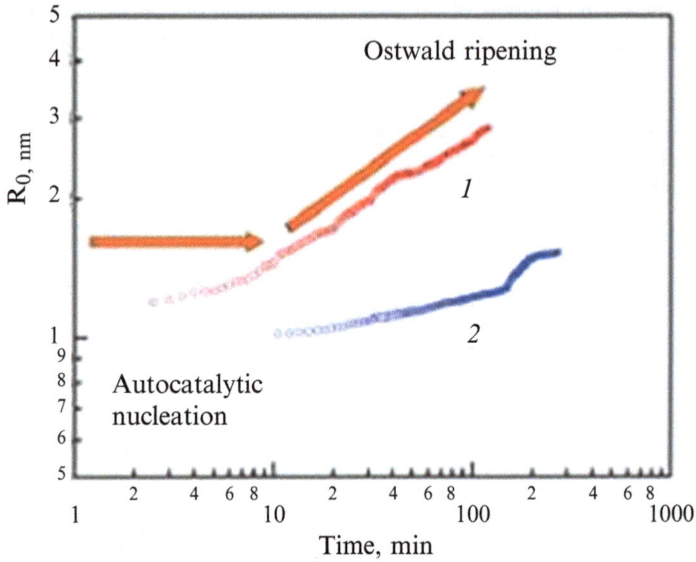

Fig. 3.4 Average radius (R_0) as a function of the reduction time for the metallic Pd (*1*) nd Rh (*2*) nanoparticles produced during the photoirradiation [67]

SAXS measurements in situ with time resolution made it possible not only to find mechanism of formation of Ag [66], Rh and Pd [67, 68] nanoparticles at photoreduction in the system of ionic precursors and surfactant of poly(N-vinyl-2-pyrrolidone), which includes consequent conjugate stages of autocatalytic reduction, nucleation, nucleation-growth, Ostwald ripening and dynamic coalescence, but also compare rates of individual reactions. Typically, depending on a kind of metal and its concentration, prevailing processes can be autocatalytic reduction-nucleation with the following growing stage in combination with coalescence, as in the case of Rh nanoparticles, or rapid reduction-nucleation with high velocities at the early stage and, then, via Ostwald ripening, to diffusion-limited growth of nanoparticles, as in the case of Pd (Fig. 3.4).

It should be noted that metal nanoparticles, especially noble, are very photoactive and under action of UV irradiation strongly polarized. This brings to positive charge of metal aggregates consisting of small clusters and, as a consequence, to their instability and decomposition by components [69, 70]. Since free metal ions present in a solution in substantial concentrations, at the initial state this causes formation of a great amount of small clusters, which rapidly coagulate in great aggregates. Nuclei with less than critical size are dissolved due to thermodynamic instability and provide growth of coarser particles. In turn, coarser aggregates are photo-chemically unstable and decompose under UV irradiation. On the whole, mechanism of formation of nanoparticles under UV irradiation can be considered as quasi-equilibrium [71], accompanied in parallel by growth and fragmentation processes, which in turn causes thermodynamically stable particles with relatively narrow size distribution.

Efficiency of photo-chemical synthesis of metal nanoparticles can, in the first place, decrease due to intense extinguishing of excited state of photoinitiator or photosensitizer by metal ions which are precursors of nanoparticles [64]. Thus, Ag (I) ions can suppress excited states of carbonyl compounds with the rate constants close to the values typical for diffusion-limited reactions [72]. In principle, these problems can be avoided if systems are used in which photo-induced processes required for metal ion reduction can be separated in time or space. For example, strategy of efficient photochemical synthesis, not only for Au nanoparticles, but with regard to other metals, can be put down to two main approaches proposed in [55]. This is, first of all, achieved by usage of monomolecular precursors (for example, benzoin derivatives such as already mentioned I-2559) with short lifetime of triplet state (10 ns or less) and effective absorption in UV region (315–400 nm). Another approach is in spatial segregation of radical-generated reaction from the stage of electron transport, which causes reduction of metal ion. To do this, one can use numerous self-assembling systems, for example, micelles such as sodium dodecyl sulfate, etc. The main demand to them is transparency in the region of UV absorption and absence of structural groups capable of barren extinguishing of triplet state.

Yield of ketyl radicals increases significantly at addition to the system of effective hydrogen donor. In [55] for this purpose 1,4-cyclohexadiene was used, its hydrophobic character is especially favorable for micellar systems. The process with participation of cyclohexadiene is illustrated by the following scheme (Scheme 3.5).

Scheme 3.5
Transformations of ketyl radicals in the presence of 1,4-cyclohexadiene

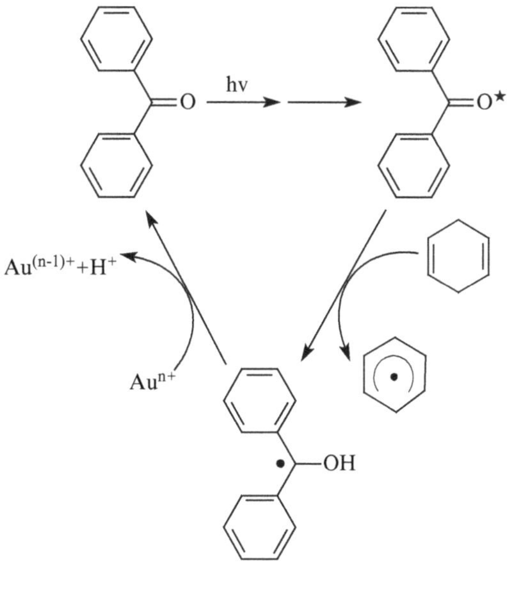

Cyclohexadiene radical is also a good reducing agent and can contribute to reduction of metal ions.

3.2.3 Photo Reduction on Solid Substrates

Chemical interaction of a solid substance with electromagnetic irradiation goes not only on a surface, but in surface layers of about several microns. Formation of nanoparticles in radiation stimulated solid phase processes is predominantly caused by reactions of photochemical decomposition. As regards other mechanisms, the considered method can have some limitations, because not all used matrices can be photochemically active, i.e. able to generate free electrons in the system under action of UV light, as it is typical, for example, for TiO_2. The TiO_2 gel films have intense absorption in the range of 250 nm, and under exposure at 254-nm irradiation, couples form: hole-free electron. This approach is successfully used for fabrication of Pt nanoparticles encapsulated in microporous carbon hollow shell [73]. For precursor the Pt(IV) compound and phenol served absorbed on the surface of titanium oxide powder. Photo generated electrons of metal oxide matrix causes reduction of Pt(IV) and oxidization of phenol with formation of phenol polymer on the surface of TiO_2 particles with Pt (~3 nm) particles encapsulated in organic matrix. The following carbonizing at 700 °C and removal of TiO_2 by dissolution causes formation of hollow carbon spheres with 3–5-nm thickness with incorporated metal particles. It is interesting that formed nanocomposites show effective catalytic properties in hydration of cyclohexene with higher TOFs (turnover frequencies), than for a catalyst obtained by impregnation.

Combination of photo irradiation with ultrasonic treatment causes penetration of metal salt into pores of a carrier and deposition of metal nanoparticles not only on surface, but also in inner pores of TiO_2 matrix [74].

On the whole, the process of photochemical deposition of nanoparticles can be implemented in two ways. In the first way, matrix on a solid substrate is immersed in a solution of a metal precursor, and this system is irradiated. In the second way first the solution of metal salt is subjected to photolysis, and then a substrate, for instance, film, is immersed in it. These procedures of reduction were used for plating of TiO_2 surface or indium-tin oxidized film (ITO) by Pt, Au or Ag nanoparticles [70]. A number of particles of deposited metal is regulated either by concentration of a metal salt and/or by holding time of a film in irradiated solution. It should be noted that nanocomposite materials of this type can be used as photo catalytic films in devices with self-cleaning surfaces or for water and air purification from contaminants [75, 76], and also as new electrode materials in electrochemistry [70].

Polysilanes are known as photosensible materials [77], UV irradiation causes break of Si-Si bond due to formation of silyl radicals, which can play a role of reducing agent [78]. Metal nanoparticles are formed on a surface of polystyrene particles containing incorporated molecules of poly(methylphenylsilane) (PMPS) by the following scheme (Scheme 3.6) [79]:

Scheme 3.6 Reduction of metal ions with the participation of polymethylphenylsilanes on the surface of polystyrene

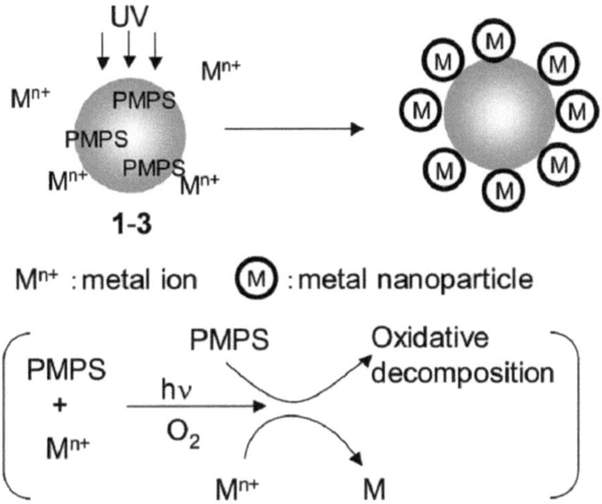

Under action of UV radiation photo oxidized decomposition of PMPS proceeds, which is accompanied by reduction of metal ion.

For carries used for immobilizing of metal ions most often are used polymers [80–82] or inorganic matrices such as TiO_2 [83–85], SiO_2 [86, 87]. Film nanocomposites can be fabricated in two ways. One of them provides films of 20–200 μm thickness by pouring and centrifuging from a polymer solution containing a metal precursor compound [81], after UV irradiation heating to various temperatures can be applied, which do not exceed glass transition temperature of a polymer. Silver nanoparticles in a matrix of linked oligourethane methacrylates are synthesized by impregnation of polymer plates with 2,2-dimethyl-6,6,7,7,8,8,8-heptafluoro-3,5-octanedione Ag metal organic complex in the medium of supercritical CO_2 [82] and the following radiation by filtrated radiation at $\lambda = 365$ nm from a mercury lamp. The size distribution function for nanoparticles and their concentration are controlled by different factors, such as structure of a matrix of impregnated polymer, type of solvent, temperature and other conditions of impregnation. It is interesting that this approach makes a possibility for designing in polymer matrices various optical microstructures consisting of nanoparticles, which would have varying refraction and absorption coefficients. In the other way metal containing precursor can be introduced in the initial polymer system, however, in this way the process is often complicated by interaction reactions of a metal compound with monomer or initiator molecules, inhibiting of polymerization reactions, etc. Only specially found conditions by type and concentration of the initial state of a metal, type of initiator, and temperature of reaction are suitable for production of nanocomposites, for example, as in MMA system and tetrachloroaurate (III) N-cetylpyridinium (Scheme 3.7) [88]:

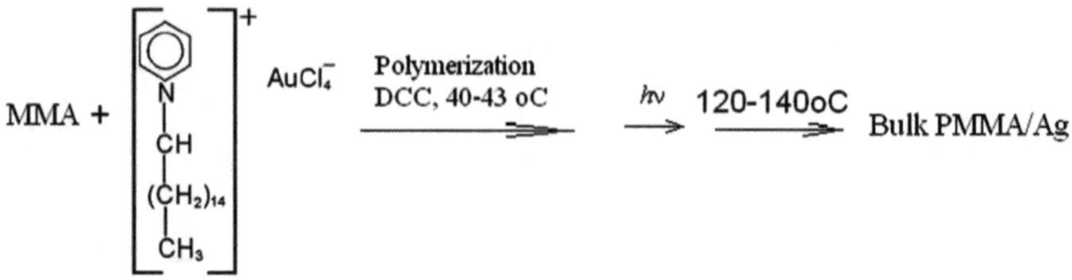

Scheme 3.7 Synthesis of a nanocomposite by polymerization in situ followed with photoreduction. *DCC* dicyclohexyl peroxydicarbonate

Character of interaction between metal ions and functional groups on a surface of polymer particles has a substantial effect on formation and immobilizing of metal nanoparticle. Opposite results were obtained in the polystyrene/PMPS and $HAuCl_4$ $4H_2O$, $AgNO_3$, Na_2PdCl_4 salts system depending on a type of surface carrier (cation or anion) and type of functional group of polymer particle (Table 3.6) [79].

As is seen from Table 3.6, interaction between cation polymer particle and metal ion (Ag^+, Pd^{2+}) impedes formation of metal nanoparticle on a surface of the polymer particle, while interaction between anion polymer particle and anion complex ion $AuCl^{4-}$, $PdCl_4^{2-}$ has no such effect and brings to formation of nanoparticle/polymer

Table 3.6 Preparation of Metal Nanoparticle/Polymer Hybrid Particles [79].

Polystyrene particles (surface charge and coordinative group[a])	Metal salt	Metal particles on the polymer surface: Morphology	Size, nm
PS (anionic)	$HAuCl_4{\cdot}H_2O$	Sphere	10–20
PS (cationic)	$HAuCl_4{\cdot}H_2O$	Sphere	10–20
PS-co-NIPAM (90:10) (anionic, NIPA)	$HAuCl_4{\cdot}H_2O$	Sphere	10–20
PS-co-NIPAM (90:10) (cationic, NIPA)	$HAuCl_4{\cdot}H_2O$	Sphere	10–20
PS (anionic)	$AgNO_3$	Sphere	10–20
PS (cationic)	$AgNO_3$	[b]	
PS (anionic)	Na_2PdCl_4	Sphere	<10
PS (cationic)	Na_2PdCl_4	Sphere	<10
PS-co-NIPAM (90:10) (anionic, NIPA)	Na_2PdCl_4	Shapeless	>10
PS-co-NIPAM (90:10) (cationic, NIPA)	Na_2PdCl_4	Sphere	<10
Poly(St-co AAEM) (99:1) (anionic, AcAc)	Na_2PdCl_4	Shapeless	>10
Poly(St-co AAEM) (99:1) (cationic, AcAc)	Na_2PdCl_4	Sphere	<10
PS-co-NIPAM (90:10) (anionic, NIPA)	$PdCl_2$	Shapeless	>10
PS-co-NIPAM (90:10) (cationic, NIPA)	$PdCl_2$	[b]	

[a]*NIPA* is N-isopropylacrylamide group; *AcAc* is acetoacetoxy group
[b]Metal particles were not formed

hybrid composite. The mechanism of photoreduction of silver on SiO_2 is supposed [89] which includes the donation of electron from silica to silver ions fixed on the oxide surface under UV irradiation. It is shown [90] that there are many differences in formation of Ag nanoparticles depending on whether there is reduction of free Ag^+ ions or ions absorbed on a surface of colloid silica gel. In presence of oxide carrier formation and aggregation of nanoparticles was predominantly on its surface due to absorption of Ag^+ ions on negatively charged surface of silica gel. Probably, here SiO_2 surface plays a double role: attracts and bonds positive silver ions, and also under irradiation supplies by electrons, which reduce metal ions.

By changing size of the surface of colloid silica gel ($S_{spec} = 140–345\ cm^2/g$), density of silver nanoparticles on silica gel particle can be controlled.

On the whole, photochemical deposition of metal nanoparticles minimizes usage of chemical reagents, solvents, and, consequently, is ecologically more attractive than chemical ways considered above. However, problems of control over nanoparticle size and their distribution on a solid substrate in many cases are not solved. It is significant, for example, that hybrid nanocomposites based on noble metals (Ag, Au, Pd, Pt) obtained in thin TiO_2 films by the following Scheme 3.8 [91] have

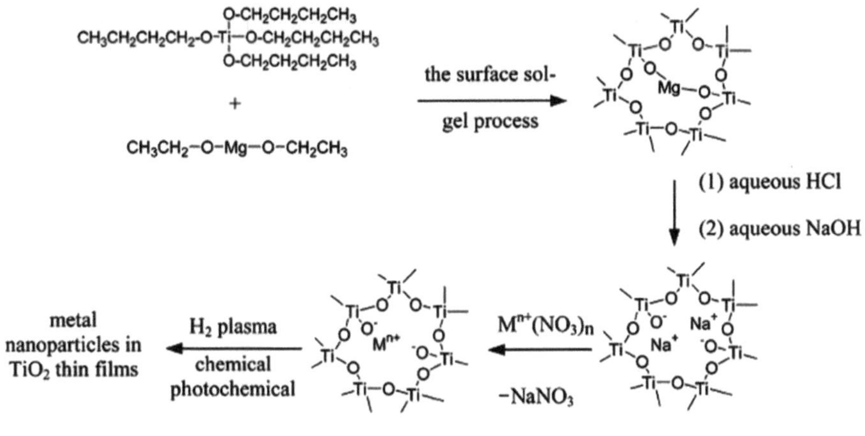

Scheme 3.8 Hybrid nanocomposites in TiO_2 thin films

different sizes of nanoparticles depending on reduction method. Average diameter of Ag nanoparticles and standard deviation (dispersion) are 6.0 ± 1.8 nm at usage of $NaBH_4$, 8.6 ± 3.0 nm at reduction by H_2 plasma and 7.6 ± 3.0 nm under UV irradiation.

3.2.4 Photo Induced Processes in Nanoparticles

Apart from photo reducing reactions, which have as a result formation of nanoparticles under light of UV irradiation the formed nanoparticles can be subjected to structural or morphological changes due to their high photochemical activity. Two main effects can be shown. On the one hand, individual particles or their aggregates can be subjected to photo fragmentation [70, 71, 92]. This mechanism can be caused by photo induced generation of electrons in a nanoparticle, which are injected in environment, consequently nanoparticles accept positive charge. Electron ejection brings, in turn, to destabilization of a nanoparticle or a nanoaggregate with the following fragmentation and formation of metal clusters. On the other hand, UV and light excitation can cause photo induced increase in temperature of a nanoparticle, which promotes partial melting of aggregates [93], accompanied by coalescence of neighbor particles, and, as a result, by formation of coarser individual nanoparticles. In accordance with these mechanisms, disappearance of Pt nanoparticles with 20 nm sizes and appearance of particles in H_2PtCl_6/TiO_2 system is, probably, caused by photo induced fragmentation of initially formed polycrystalline aggregates under UV exposure for 15 h [70]. Similar processes are described for gold and silver nanoparticles [92, 94]. Photochemical activity of metal nanoparticles promotes photo induced generation of charges and the following formation of charged metal particles. In turn, electrostatic repulsion forces raised between these particles can stop particle growth after quite long exposure to UV radiation. In these systems it is possible to obtain stable suspensions of nanoparticles for their long-time storage as it has been shown for Au particles [70]. No changes were observed in UV-visible spectrum range after 2 weeks of storage of $HAuCl_4$ solution subjected to photolysis during 3 h (Fig. 3.5).

As is seen, the band at 225 nm corresponding to charge carrier metal-ligand for the initial chloraurate complex significantly decreases in the first minutes of UV

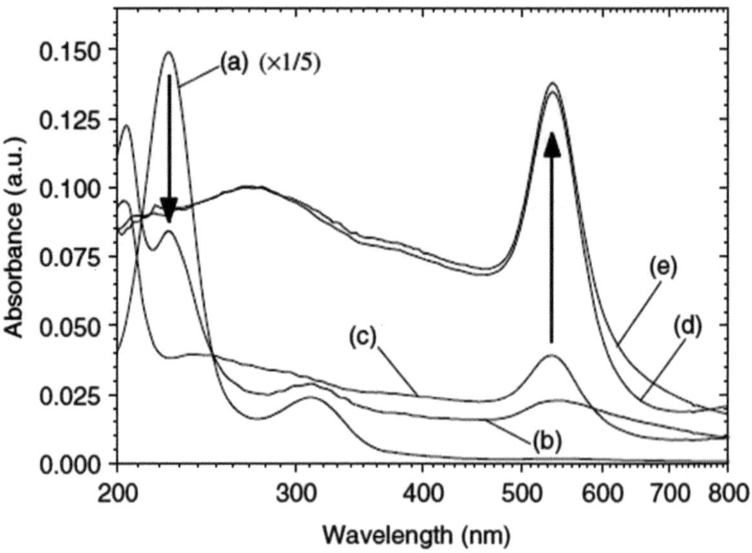

Fig. 3.5 UV-visible spectra of a 250-mM $HAuCl_4$ solution exposed to UV light for 0 (*a*), 2.5 (*b*), 45 (*c*) and 180 min, before (*d*) and after ageing in ambient conditions for 2 weeks (*e*) [70]

exposure, which points to photolytic reduction of salt. Simultaneous appearance of a band at 205 nm is, probably, caused by previously considered consequent reduction from Au^{3+} to Au^{1+} and further on to Au^0.

Thus, action of light produces photo effect, which can substantially change properties of immediate surrounding of metal nanoparticles. Respectively, this causes instability of colloid, for example, eliminates stabilizing impact of polymer shell, induces changes in size and shape of nanoparticles, and initiates their clustering, i.e. formation of aggregates consisting of hundreds to thousands superfine metal nanoparticles. Generation of particles with different shape and sizes and a change in distance s between particles in aggregates is expressed in appearance of additional absorption peaks and inhomogeneous broadening of resonance absorption peak. It has been found [95] that as after irradiation by mercury lamp plasmon resonance band of Au at 532 nm broadens and disappears, and a band appears at 840 nm, which is related to dipole-dipole interaction of particles in aggregates. It is found that aggregation rate depends on a solvent type and light wavelength. The effect of light wavelength is caused by increase in Van-der-Waals forces and by a change in Coulomb interaction of surface charges. Exchange by electric charges generated due to photoemission is caused by a dependence of Fermi energy on particle size, passes through dispersion medium and brings to equalization of potential of particles of different sizes. As a result of exchange, long-range electric forces appear, which promote approach of the particles to distances corresponding to Van-der-Waals forces, which cause aggregation.

Distances between particles are in inverse relation to usually observed spectral shift [96]. While polychrome light can increase aggregation rate [97, 98], laser impact is accompanied by photo modification of the aggregates [99]. As a rule, aggregation rate and properties of the formed aggregates depend on method of synthesis of nanoparticles. Coarse aggregates usually have loose dendrite-like structure with fractal properties.

Irradiation of colloid silver obtained by citrate method in the $AgNO_3/NaBH_4/$ sodium citrate/bis(p-sulfonatophenyl) phenylphosphine dihydrate dipotassium system by fluorescent lamp (7 mW/cm^2) during 20 h is accompanied by significant modification in shape of nanoparticles and formation of nanoprisms [100]. At the same time, usage of laser on argon ions ($\lambda = 514.5$ nm) in the colloid system based on $AgNO_3$/ethanol/poly(vinylpyrrolidone) (PVP) is efficient for structuring of large aggregates with the structure of statistical fractals with ~5,000 nm length.

3.3 Radiation-Chemical Reduction

In radiation-chemical reduction usually γ-radiation is used (for example, on ^{60}Co sources). Radiolysis is conducted in stationary conditions (constant radiation flow) or in alternate conditions (intermittent flow with pauses or pulse, when once for a short time high doze is applied) [101]. Radiation-chemical yield for each system depends on a great number of physical and chemical factors (time and absorbed energy, concentration of reacting substances, temperature, mass transport, pH of a medium, etc.), which change during irradiation. In a common case radiation yield being a number of active particles (molecules, ions, radicals, etc.) per 100 eV of absorbed energy is a function of absorbed radiation doze. Initial yield of intermediate products of radiolysis (electrons, ions, excited particles) is a value of about 10. Observed values of radiation yields are in wide range from 10^{-6} to 10^{8} particles per 100 eV [102]. Thus, for active particles which generate during radiolysis of water, the highest yield is typical of hydrated electrons $e^{-}{}_{aq}$ (2.6), OH^{-} (2.7), H^{+} (2.6), for H, H_2 and H_2O_2 it is 0.6, 0.45 and 0.7, respectively [103]. The formed $e^{-}{}_{aq}$ and hydrogen atom have high reduction potentials (−2.9 and −2.3 V, respectively), which allows, for example, in to reduce Cu^{2+} ions in triple polyelectrolyte PAA-PEI- Cu^{2+} complex to metal copper Cu^0, while in this system Cu_2O-based nanocomposite is obtained in presence of strong chemical reducing agent [104].

Oppositely, hydroxyl radical has high oxidizing ability (1.9 V). In order to avoid side reactions with participation of OH^{-} radicals, such as oxidation of metal ions, linking reactions, and destruction of macromolecules, organic substances are added to the system, for this purpose alcohols and salts of organic acids are used. In the case, for example, of isopropyl alcohol or formate-ions, as a result of reactions

$$(CH_3)_2CHOH + OH\,(H) \rightarrow (CH_3)_2COH + H_2O\,(H_2) \tag{3.28}$$

$$HCOO^{-} + OH\,(H) \rightarrow CO_2{}^{-} + H_2O\,(H_2), \tag{3.29}$$

initiated by impact of ionizing radiation on water solution, only reducing particles form in it, in particular $(CH_3)_2COH$ or $CO_2{}^{-}$, whose reduction potentials are −1.4 and −1.9 V [103].

Radiolysis in condensed matter is conventionally divided in three stages [102]. The first, physical stage is associated with formation of primary intermediate products and is rather fast (less than $3{\cdot}10^{-16}$ s). Second, physical-chemical stage is associated with spatial distribution of primary intermediate products and homogenization of a system as a result of transport process (for example, diffusion). In liquids with medium viscosity it is completed for 10^{-8}–10^{-7} s. During radiolysis, in contrast to photolysis, distribution of obtained intermediate particles is more homogeneous and provides synthesis of narrowly distributed particle sizes. The third, chemical stage includes processes of chemical transformation in volume of homogenized system. Duration of this stage is determined by average time of chemical transformation and can be infinitely long. In gas phase the second stage usually does not manifest itself due to high diffusion rate of primary products of radiolysis. On the contrary, in solids (especially at low temperatures), where diffusion rate is low, radiolysis can delay at the second stage.

3.3.1 *Kinetics and Mechanism of Radiation-Induced Synthesis of Nanocomposites*

Usage of pulse γ-radiolysis in combination with spectrophotometry made it possible to study experimentally on example of reduction of silver [103, 105], gold [103, 106], and copper ions [107] mechanism and kinetics of initial stages of formation of metal colloids in water, water-alcohol, and non-water solutions, whose basic principles have, probably, general character. One of advantages of radiolytic method is homogeneous and instant generation of a great number of atoms during irradiation, which provides favorable conditions for formation of monodisperse particles.

At early stages of transformation after fast one-electron reduction of metal ions follows disproportioning with formation of simple atomic and ionic clusters (Ag_2^+, Ag_2, Ag_4^{2+}, Ag_4^+, etc.), or clusters and intermediates (in the case of formation of copper colloids in presence of formate ions, $CuCO_2$, Cu^+, Cu^{2+}). Interaction of these associates causes formation of subcolloid particles containing one-two tens atoms (for example, 8–16 Ag atoms with maximum absorption at 325 nm [108, 109], 10 Pd atoms with maximum absorption at 320 nm) [103]. These processes proceed for times no less than 1 s after γ-pulse. In some cases subcolloid forms (for example, clusters Ag [110], Pd [111]) can be stable during several days.

Mechanism of transformation of reduced metal ions to nanoparticles on the example of formation of subcolloid particles [112] can be presented by the following scheme (Scheme 3.9):

$$Cu^{2+} \xrightarrow{CO_2^-,\ e_{aq}^-} CuCO_2 \xrightarrow{Cu^{2+}} Cu_2^+ \xrightarrow{x\ Cu_2^+} Cu_j^{m+} \longrightarrow Cu_n^0$$

Scheme 3.9 Formation of subcolloid Cu particles

or in general form (Scheme 3.10):

Scheme 3.10 The main steps of the formation of colloidal particles

Active species generation:

$$H_2O \xrightarrow{h\nu} H^\bullet,\ OH^\bullet,\ e_s^-,\ H_3O^+,\ H_2,\ H_2O_2$$

$$OH^\bullet(H^\bullet) + RH \longrightarrow R^\bullet + H_2O\ (H_2)$$

Metal reduction:

$$M^{n+} + nR^\bullet \longrightarrow M^0 + nR^\bullet + nH^+$$

$$M^{n+} + ne_s^- \longrightarrow M^0$$

Growth of colloids:

$$nM^0 \longrightarrow M^0_{colloid}$$

Formation of metal colloid during γ-radiolysis is accompanied by narrow plasmon band in absorption spectrum, which is typical of spherical particles with similar size distribution and weak interaction with a solvent. Changes in electron state from clusters with atomic-molecular characteristics to nanoparticles are well tracked in optical spectra. Individual narrow optical bands typical of clusters of different complexity, as cluster grows, transform into wide optical bands typical of quasi-metal particles, and then into plasmon absorption band, corresponding to metal state. The $Au_{18}Ag_{20}$ и Au_{55} particles show obvious surface plasmon resonance, whereas for clusters consisting of 8–13 atoms atomic-molecular electron state is typical [103]. Properties of intermediate sized particles are approximately quasi-metallic and, probably, they serve nuclei of metal phase and have tendency to growth. Presence of indifferent electrolyte $NaClO_4$, and OH^-, SO_4^{2-}, $HCOO^-$ ions, which are prone to specific sorption on a metal, accelerates nucleation process of silver or other metal.

Unusual reduction kinetics is shown by cobalt and nickel ions under γ-irradiation of de-aerated $2{\cdot}10^{-5}$–$5{\cdot}10^{-2}$ *M* solutions of $Co(ClO_4)_2$ and $Ni(ClO_4)_2$, containing HCOONa. Formation of metal is preceded by induction period (Fig. 3.6) and reduction is developed in autocatalytic regime, formation of metal soles and increase in their concentration accelerates the process.

These behavior of reduction of metal ions is explained by the ratio of reduction potentials of their intermediate oxidized forms and redox potentials of formed organic radicals $(CH_3)_2COH$ in the system with isopropyl alcohol and CO_2^- in presence of formate ions. Potentials of E^0 $Ni^{2+}/Ni^0 = -2.2$ V and $Co^{2+}/Co^0 = -2.33$ V pairs are more negative than for massive metals, and $E^0(Co^{2+}/Co^+)$ and E^0 (Co^+/Co^0) are -1.8 V and -2.9 V, respectively. Since that, hydrated electron (-2.9 V) can

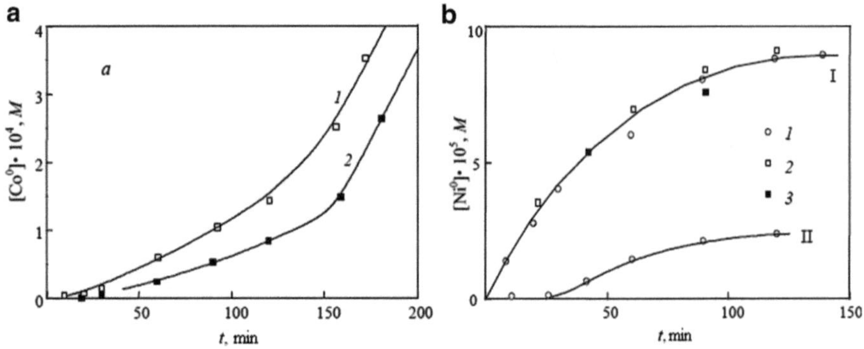

Fig. 3.6 Concentration of colloidal cobalt (**a**) and nickel (**b**) in the reaction solution as a function of the γ-irradiation time. (a) Solutions of $Co(ClO_4)_2$ with concentrations of $5{\cdot}10^{-3}$ *M* (*1*) and $1{\cdot}10^{-3}$ *M* (*2*); the admixtures are HCOONa (1×10^{-2} *M*) and polyacrylate (1×10^{-3} *M*). *Radiation dose* rate is 1 kGry/h, for 1 min of irradiation $4{,}4{\cdot}10^{-6}$ *M* e^-_{aq} and 5.6×10^{-6} *M* CO_2^- are formed. (b) Solutions of $Ni(ClO_4)_2$ with concentration of 1×10^{-4} *M*; the admixtures are polyethylene imine (*1*), polyacrylate (*2*), polyvinyl alcohol (*3*), HCOONa (I) or isopropyl alcohol (II). *Radiation dose* rate is 0. kGry/h, for 1 min of irradiation 2.2×10^{-6} *M* e^-_{aq} and 2.8×10^{-6} *M* CO_2^- are formed [103]

efficiently reduce Co^{2+} ion. Method of pulse radiolysis has confirmed the reaction realization ($k = 1.1 \cdot 10^{10}$ l·mol^{-1}·s^{-1}) (ref. by [103]):

$$Co^{2+} + e^{-}{}_{aq} \rightarrow Co^{+} \tag{3.30}$$

3.3.2 *Fabrication of Polymer Nanocomposites* In Situ

Initial stages of radiation-chemical reduction of metal ions should conduct under anaerobic conditions. To obtain stable nanoparticles and prevent coalescence of atoms, radiation reduction is performed in presence of stabilizing agents, among which most often are used various polymers such as polyethylene amine, polyvinyl sulfate [103], polyvinyl alcohol [113], poly(*N*-vinylcarbazole) [114], polyacrylonitrile [115], dendrimers, etc. Functional groups of polymers with high affinity to metal provide fixation of a molecule on a particle surface, and polymer chains in parallel to such factors as electrostatic repulsion and steric obstacles prevent particles from sticking to each other.

The considered method of producing of nanocomposites in situ includes solution, and often, a complex formation of metal salt with a polymer in water, water-alcohol, and non-polar media and the following reduction of metal ion under γ-irradiation. There are interesting approaches when the processes of formation of a nanoparticle and polymer matrix are joined. Gamma-induced polymerization of *N*- vinylcarbazole, particularly, its co-polymerization, since acrylic acid was added to the system, and formation of silver nanoparticles, cadmium sulfide and hetero-metal nanoparticles of Pd_{50}–Ag_{50} and Pt_{50}–Ru_{50} alloys (considered below) occurred simultaneously under action of γ-irradiation of 30 kGy at the doze power 1.0×10^4 Gy/h [114]. The same method was used to fabricate Ag/poly aniline nanocomposites with core-shell structure. Water solution of aniline, free radical initiator and silver salt were irradiated by γ-rays [116]. Reduction of silver salt in water aniline brings to formation of Ag nanoparticles, which in turn, catalyze oxidation of aniline to polyaniline.

Variation of in situ is fabrication of hydrogel nanocomposites, which are obtained by preliminary linking of polymer gel by gamma-radiation, then swelling of the linked polymer with metal salt and alcohol (isopropyl) water solution, and following by metal ion reduction. Silver nanocomposites stabilized in matrices polyvinyl alcohol/cellulose acetate/gelatin [117] and polyvinylpyrrolidone/alginate [118] were synthesized with usage of γ-radiation of no less than 25–40 kGy doze. Functional groups of the polymer system such as –OH, C = O and –NH work as templates for binding metal ion and provide homogeneous distribution of formed nanoparticles in hydrogel matrix. Additional advantages of the radiolysis here are in a possible combination in one process of polymerization, linking, and sterilization of hydrogel (the latter is important, for example, for using of such composites as dressing for wounds[2]) due to the fact that primary OH

[2] Doze 10 kGy provides sufficient sterilization of these materials [119].

Table 3.7 Mechanical properties, surface area and mean pore diameter of diisocyanate cross-linked silica aerogel composites

Metal loading	Density, g/cm^3	Load at fracture, kg	Elasticity modulus, MPa	Specific surface, m^2/g	Average pore diameter, nm
Au	0.56	19.15	77.35	164	19.6
Ag	0.58	16.07	62.80	164	19.6
Without metal	0.57	14.7	67	150	19.4

Diisocyanate concentration in the bathing solution before curing was 50 % w/w. Metal concentration was 3×10^{-4} mol $\cdot$ L^{-1} in both cases. Metal-loaded samples were irradiated with a total dose of 7 kGy. The plain sample was non-irradiated. Typical inaccuracy of the experimental data is ±10 % [123]

radicals interacting with hydrogel molecule generate polymer radicals, which participate in linking of chains. A degree of linking is well controlled by doze of radiation irradiation.

As in other methods of reduction, multi parametric control may be performed over dispersion and shape of particles. Average diameter of Ag nanoparticles and their distribution increased as a precursor concentration increased. Impregnation of hydrogel system: polyvinyl alcohol/cellulose acetate/gelatin with $AgNO_3$ with concentrations 5.10 and 20 mmol brought to formation of nanoparticles with average diameters 38.6, 56.8, and 60 nm, respectively [117].

In order to obtain mechanically strong hydrogel nanocomposites, approaches are developed for encapsulation of metal nanoparticles in aerogels with high strength characteristics [120–122]. In particular, silica gel hydrogels were produced by reactions of based-catalyzed catalysis and linking with di-isocyanate [123]. The linked hydrogel is multiply rinsed (by 12 h each time) by metal salt solution and bombarded by gamma-radiation. It should be noted that incorporation of metal nanoparticles and used irradiation doses do not have undesirable effect on mechanical properties and chemical composition of composite aerogel (Table 3.7).

3.3.3 *γ – Radiolysis in Synthesis of Semiconductor Nanomaterials*

Radiation-chemical method is one of the most efficient methods of producing of metal chalcogenide nanoparticles. Radiation-induced formation of monodisperse CdS, ZnS, PbS nanoparticles is usually performed in presence of thiol compounds [124–128], and on the whole the process goes on by one-type mechanism [129]. Irradiation of solution containing a polar solvent causes formation of active particles. Among them a hydrated electron having most strong reducing properties interacts with thiols by the following scheme:

$$HOCH_2CH_2SH + e_{aq}^{-} \rightarrow {}^{\bullet}CH_2CH_2OH + HS- \quad (3.31)$$

$$HOCH_2CH_2SH + OH^{\bullet} \rightarrow {}^{\bullet}SCH_2CH_2OH + H_2O \quad (3.32)$$

$$Pb^{2+} + OH^{\bullet} \rightarrow Pb^{3+} + OH^{-} \quad (3.33)$$

with the rate constants $k_1 = 1.2 \times 10^{10}$ M^{-1} s^{-1}, 6.8×10^{9} M^{-1} s^{-1} и $< 2 \times 10^{8}$ M^{-1} s^{-1}, respectively.

Monomer sulfide forms when Pb^{2+} reacts with HS^{-}:

$$Pb^{2+} + HS^{-} \rightarrow PbS + H^{+} \quad (3.34)$$

with the following coalescence and formation of a nanoparticle with radiation yield 2.7×10^{-7} mol·J^{-1} [129]:

$$nPbS \rightarrow (PbS)_n \quad (3.35)$$

When $Na_2S_2O_3$ is used as a sulfiding agent at producing of chitosan-ZnS and chitosan-ZnS:Mn^{2+} [130] the reduction reaction of $S_2O_3^{2-}$ can proceed with participation of e_{aq}^{-} or atomic hydrogen:

$$S_2O_3^{2-} + 2e_{aq}^{-} \rightarrow SO_2^{-3} + S^{2-}, \quad (3.36)$$

$$S_2O_3^{2-} + 2H^{\bullet} \rightarrow SO_3^{2-} + 2H^{+} + S^{2-}, \quad (3.37)$$

$$S^{2-} + Zn^{2+} + Mn^{2+} \rightarrow ZnS : Mn^{2+}, \quad (3.38)$$

$$n(ZnS : Mn^{2+}) \rightarrow (ZnS : Mn^{2+})_n. \quad (3.39)$$

The absorbed dose and irradiation power are important factors of control over sizes of formed nanoparticles and, finally, over their optical properties. ZnS nanoparticles were synthesized by γ-irradiation of water solution containing Zn^{2+} and thiol (RSH) [129]. Under low irradiation doses monodisperse ultra fine ZnS particles with diameter 1.5 nm form and absorption band appears at the wavelength 238 nm. Coarser nanoparticles form under higher doses with absorption at $\lambda > 260$ nm.

By change in irradiation dose, not only sizes of nanoparticles change, but, respectively, photoluminescent properties of the nanocomposites. Thus, the photoluminescent peak in ZnS-chitosan emission spectrum is barely seen under irradiation by 5 kGy dose, but at the dose 10 kGy quantum yield is 16.6 %, and emission peak appears at 442 and 584 nm. As irradiation dose increases, sizes of nanoparticles increase and photoluminescent emission peak reaches 22.5 %, and as the dose increases further on to 30 Gy, the yield begins to decrease, which is caused by breakage of a crystal structure of chitosan.

It should be noted that obtained by radiolysis nanocomposites of the considered type (quantum dots in biocompatible polymer matrix) are, as a rule, well dissolved in water. In combination with efficient luminescent properties these materials are interesting for different biomedical applications [131, 132]. The CdS polyamidoamine dendrimer with terminal amino- and carboxyl groups obtained from $CdCl_2$ solution and a polymer in presence of sodium thiosulfate under action of ^{60}Co-γ-source is tested for detection of ascorbic acid in pharmaceutical drugs [128]. Ability to detect ascorbic acid in micromolar concentrations is caused by efficient extinguishing of luminescence of a nanocomposite biosensor by the acid. However, even more than 100-fold concentration excess of ordinary interfering agents such as tartaric, uric, citric, acetic acids had no significant effect on fluorescence of the nanocomposite.

3.3.4 Synthesis of Bimetallic Nanocomposites

The method of radiolytic reduction is especially attractive for production of heterometallic nanocomposites. Nanoparticles consisting of two or more different metals are of a special interest for development of materials with new properties, because at the nanoscale intermetallic compounds and alloys can be obtained, which do not form in the case of compact metals. Radiation reduction of salt solutions was used to obtain nanoparticles including two [133–135] and three [136] metals. General strategy for producing these nanocomposites is analogous to the above mentioned and includes incorporation of metal ions in a solution or suspension of a matrix polymer and the following reduction by γ-irradiation. Chemical bonding of metal ions with matrix promotes yield and homogeneous distribution of nanoparticles in volume on a substrate surface, as it is shown in the system multiplayer carbon nanotubes-sulfided polyaniline alloy – Fe-Pd nanoparticles (Scheme 3.11) [134]:

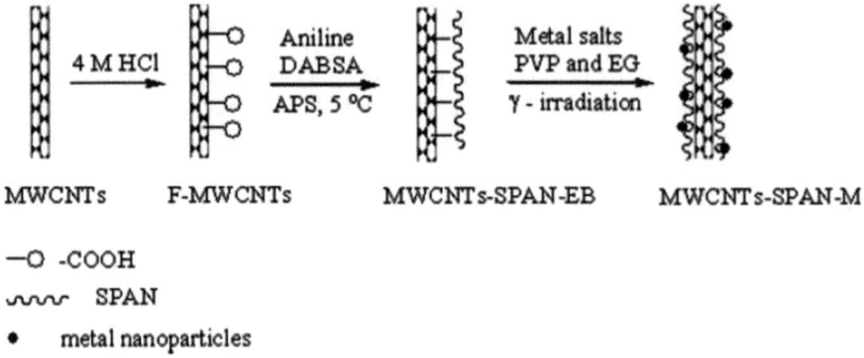

2,5-Diaminobenzene sulfonic acid (DABSA), carboxylic acid functionalized MWCNTs is F-MWCNTs, ammonium peroxydisulfate [$(NH_4)_2S_2O_8$)] or APS.

Scheme 3.11 Preparation of multilayered carbon nanotube-sulfonated polyaniline-Fe-Pd alloy nanocomposites

It should be noted that conductivity of this composite was 1.5 S cm^{-1}, which was far higher than conductivity of SPAN ($2.46 \cdot 10^{-4}$ S cm^{-1}).

Advantage of radiolytic method of production of heterometallic nanocomposites is that using of powerful irradiation dose makes it possible to reduce a metal ion very rapidly, and thus to prevent electron transport between metals. Therefore, it is possible to obtain desirable structures of core-shell or homogenous alloy nanoparticles. This strategy is successful in production of nanocomposite alloys, which are difficult to obtain from non-mixed metals in bulk state, as in case, for example, of Ag and Ni. Radiolytic synthesis of Ag-Ni alloy of different stoichiometric compositions and of $Pd_{0.5}$-$Ni_{0.5}$ alloy was performed successfully at room temperature [137]. Oppositely to thermodynamically stable core-shell structure of Ag-Ni nanoparticles, kinetically preferable phase with average particle diameter 5.8 nm and narrow distribution forms in $Ag_{0.5}$-$Ni_{0.5}$ under high irradiation dose (power 300 rad/s) [138]. Properties of the formed bimetallic nanoparticles and nanoalloys are predominantly conditioned by their structures, which are far more complex than those of monometallic particles. These structures include core-shell and subcluster-segregated alloys ordered or heterogeneously mixed nanoalloys, etc. Their particular feature is wide variety of shapes. Apart from spherical nanoparticles, under radiolysis conditions bimetallic nanostructures with high aspect ratio can form. Whiskers nanocrystalline structures of several microns in length and from 3 to 20 nm in diameter were produced by radiation reduction of Ag^+ и $PtCl_6^{2-}$ ions in water solution and in presence of the stabilizer, polyvinyl alcohol with a different degree of hydrolysis [139]. The crucial role in formation of nanowires has molar ratio of two metals and formation of metal-polymer complexes. It is assumed that presence of hydrophilic alcohol and hydrophobic acetate PVA groups can have an effect on anisotropy of formed structures during radiolysis reduction.

3.3.5 *In Situ Radiolysis*

Original possibility of using *in situ* radiation source (oppositely to exterior source), on the one hand, and production of radioactive nanoparticles, on the other hand, was demonstrated on the example of $H^{198}AuCl_4$ served as a precursor for synthesis of Au nanoparticles in polyethylene glycol [140]. Products of PEG and H_2O radiolysis formed under impact of γ-energy of ^{198}Au (411 keV) stimulate reduction of a metal ion with the following stabilization of PEG by the similar mechanism considered above for the cases of external irradiation. It should be noted that the obtained radioactive nanoparticles have high energy of γ-radiation ($\tau_{1/2} = 2.69$ day) and can be used *in vivo* and *in vitro* for different biomedical purposes (see Chap. 7), even more that the described method allows production of nanoparticles with rather high concentrations. It is interesting that *in situ* radioactivity of ^{198}Au isotope was also efficient in synthesis of Au-Pd bimetal nanoparticles with core-shell structure [141], despite essentially low radiation doses (0.01 mCi), as compared, for example, with very high irradiation power density up to $9 \cdot 10^2$ Gy/h for ^{60}Co γ-beams also used for production of bimetal nanoparticles [135].

3.4 Sonochemical Reduction

Sonochemistry is based on the acoustic cavitation phenomenon, which includes nucleation, growth and collapse of bubbles in liquid medium. Almost adiabatic collapse of microbubbles in a solution brings to rise in local temperatures up to (>5,000 K), pressure (>20 MPa), and cooling rate ($>10^7$ K s^{-1}) in micro-environment from 0.3 to 150 μm depending on ultrasonic frequency (5 MHz, 20 kHz) [142]. On the gas/liquid interface free radicals form, which induce sonochemical reactions. Different types of particles and reducing agents form during sonolysis of water and organic compounds (2-propanol,[3] ethylene glycol, surfactants, polymer molecules, etc.). Extreme conditions formed under impact of ultrasonic wave are widely used to produce different metal-containing nanoparticles and nanocomposites [143–146]. Sonochemical reduction of metals includes three main stages: formation of active particles, reduction of metal, and growth of these particles. These processes can take place in different media: in gas medium of cavitation bubbles under impact of high temperature and pressure water undergoes pyrolysis, as a result, H and OH radicals form on the interface between the bubbles and the solution, and finally in the solution (Scheme 3.12) [147].

Scheme 3.12 The main stages of sonochemical reduction

Generation of active species:

In cavitation bubbles:

$$H_2O \xrightarrow{\text{Ultrasound}} H^{\bullet} + OH^{\bullet}$$

At bubbles/solution interface:

$$RH \xrightarrow{\text{Ultrasound}} R_1^{\bullet} + H^{\bullet}$$

$$OH^{\bullet}(H^{\bullet}) + RH \longrightarrow R_2^{\bullet} + H_2O\ (H_2)$$

Metal reduction:

In solution:

$$M^{n+} + nR_1^{\bullet}(R_2^{\bullet}) \longrightarrow M^0 + nR_1^{\bullet}(R_2^{\bullet}) + nH^+$$

Growth of colloids:

In solution:

$$nM^0 \longrightarrow M^0{}_{\text{colloid}}$$

Due to low vapor tension of metal salts, reduction proceeds mainly on the bubbles/ solution interface. The important factors, which influence efficiency of cavitation and physical-chemical properties of the products, are power and frequency of ultrasonic action, temperature of the initial solution, character of a solvent, etc. It has been

[3] This is most often used organic addition is involved in sonochemical reactions as a polar solvent having lower boiling temperature and surface tension than water, but with higher vapor tension to change parameters of gas/liquid interface and to increase energy yield.

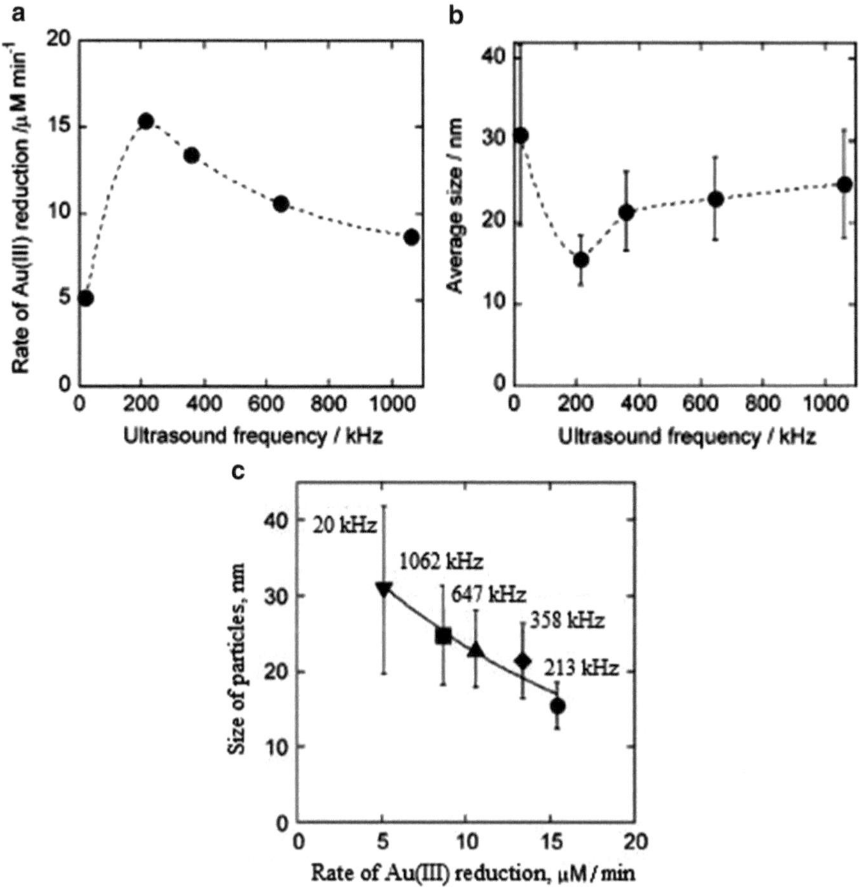

Fig. 3.7 Dependence of the reduction rate of Au(III) on ultrasonic frequency (**a**) and average size of the particles as a function of ultrasonic frequency (**b**) and of the reduction rate of Au(III) (**c**). Conditions: Au(III): 0.2 mM, 1-propanol: 20 mM, Ar, power of ultrasound is 0.1 W ml^{-1} [148]

shown that during sonochemical reduction of Au(III) in water solution containing 1-propanol, the initial rate of the reaction depends significantly on frequency of ultrasound [148]. As is seen from Fig. 3.7, the reduction rate decreases as ultrasonic frequency increases. It should be noted that size of the formed Au particles is predominantly determined by the reaction rate and almost does not depend on frequency of ultrasonic field. This confirms that sonochemical nucleation and growth processes of metal clusters and nanoparticles are slightly influenced by mechanical forces formed during cavitation [145].

However, in the case of Zn, Cr, Ni, and Mo metal particles with the sizes from 5 to 10 μm there were agglomeration/melting processes and necks formed between the particles due to local melting induced by high velocities of collisions between particles under impact of intense shock wave at 20 kHz ultrasonic frequency [149]. Moreover, 20-min impact of highly intense (40 W cm^{-2}) ultrasonic wave of the same frequency on the previously obtained citrate-preserved Au nanoparticles with the average diameter 25 ± 7 nm was enough to cause melting of nanoparticles in the zone of contacts and formation of dumb-bell like structures [150].

Comparison of the boundary layer thickness δ, tangential velocity near the bubble surface υ_t and the resulting shear stress Ψ found from the expressions (3.40, 3.41 and 3.42):

$$\delta = \sqrt{\frac{2\eta}{\rho\omega}} \tag{3.40}$$

$$\upsilon_t \approx \frac{\omega\xi^2 R_0^4}{r^5} \Rightarrow \frac{\omega\xi^2}{R_0} \tag{3.41}$$

$$\psi = \eta\frac{\partial\upsilon_t}{\partial r} \approx \frac{\eta\omega\xi^2}{R_0\delta} \tag{3.42}$$

where η and ρ are viscosity and density of water at room temperature, respectively, R_0 is equilibrium radius of the bubble, ξ is displacement amplitude and ω is frequency of ultrasonic wave has shown that tangential velocity υ_t and shear stress Ψ for the system under action of 200 kHz ultrasonic wave are by 3–4 orders of magnitude lower than in the case of the ultrasonic wave at 20 kHz frequency. This confirms that physical effects caused by ultrasound should be far more prominent in the latter case than at frequency 200 kHz and above. We shall also note that these conclusions agree well with the abovementioned examples of Au nanoparticles with the average size from 15 to 25 nm non-melted under ultrasonic impact in the frequency range from 213 to 1,062 kHz [148].

The method of sonochemical synthesis is also successfully used to obtain bimetallic systems. Morphological core-shell structures can be produced by consequent reduction as, for example, in the case of Pt-Rh nanostructure [151], or by simultaneous treatment in ultrasonic field of solution of mixture of metal ions as was shown for Au and Ag [152, 153]. The decisive role in core-shell formation of the structure plays reduction rate of a metal ion under conditions of sonochemical reaction and the character of stabilizing of a polymer matrix and formation of polymer-metal ion complex.

The approaches are intensely studied, which combine advantages of sonochemical reactions with other methods. Among them sono-electrochemical methods can be distinguished, their advantages are acceleration of mass transport, cleaning and degassing of the surface of electrodes, increase in reaction rates, etc. Pulse sono-electrochemical method has been successfully used for production of nanostructural materials, it is based on alternating impact of sonic and electric pulses on a reacting system [154, 155]. Initially metal or semiconductor particles were electro-deposited, then shock ultrasonic wave was applied to remove the particles from the electrode surface. This way fine Pd particles of different morphologies and sizes were obtained [156], as well as Ag [157] and Cu [158] in

Table 3.8 Effect of various deposition parameters on Cu particle size [158].

Variable parameter	Conditions	Particle size, nm
Current density, mA/cm^2	55	29 ± 2
	70	24 ± 4
	100	10 ± 2
Reaction duration, min	10	23 ± 2
	30	29 ± 2
	60	38 ± 4
PVPr concentration, %	0.2	55 ± 10
	1	33 ± 2
	2	29 ± 2
Temperature, °C	15	17 ± 3
	25	29 ± 2
	50	62 ± 5
Ultrasound power, W	35	29 ± 2
	50	29 ± 2
	76	29 ± 2

presence of poly (N-vinylpyrrolidone) (PVP). Sizes of the nanoparticles and their morphologies are controlled by change in reaction conditions (Table 3.8).

It draws attention that in many sonochemical reactions often poly (N-vinylpyrrolidone) is used as a stabilizing agent. Moreover, it is assumed that presence of shear stresses of a polymer in water solution can also induce sonochemical reaction in flowing medium [159]. The stabilizing mechanism of PVP action is based on specific structure of its molecule including polar functional groups, which can donate unpaired electron forming coordination bond with a metal ion, and which after reduction can also form bonds with clusters and metal nanoparticles preventing further aggregation of both as is shown in the sonochemical synthesis of Cu nanoparticles (Scheme 3.13) [158]:

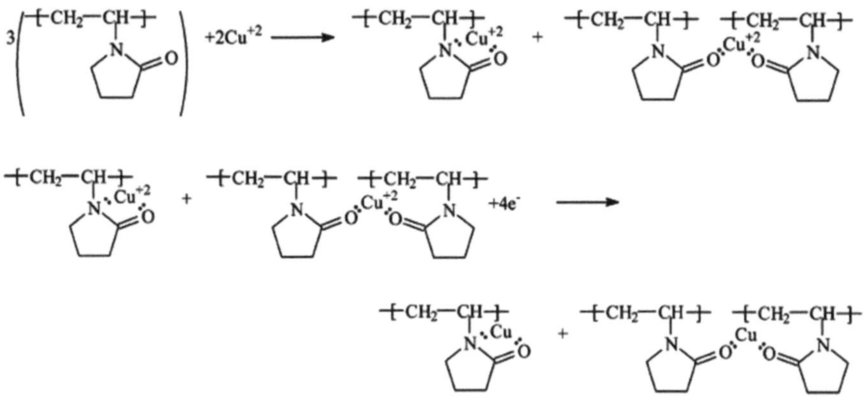

Scheme 3.13 Forming of complexes of Cu^{2+} ions and Cu nanoparticles with poly (vinylpyrrolidone)

Table 3.9 Characteristics of the Pd/PVP nanocomposite obtained by a sonochemical reaction [163].

Concentration of Pd(II) × 10^3, mol.	Average size of particles, nm	Calculated number of particles	Global particle surface, nm^2	$PVP_n/N_{particles}$	$PVP_n/S_{particles}$, molecules/nm^2
0.66	3.5	$0.26 \cdot 10^{18}$	$9.87 \cdot 10^{18}$	11.74	0.31
1.33	4.5	$0.24 \cdot 10^{18}$	$15.46 \cdot 10^{18}$	12.38	0.19
2.0	5.5	$0.19 \cdot 10^{18}$	$17.74 \cdot 10^{18}$	16.12	0.17
2.66	2.5	$2.84 \cdot 10^{18}$	$55.67 \cdot 10^{18}$	1.06	0.05

A large body of FT-IR spectral research [160–162] has shown that in the range of valence vibrational modes of the carbonyl group of poly (4-vinylpyrrolidone) 1,665–1,640 cm^{-1} there is displacement to low frequency region, which points to the interaction between C = O and surface of nanoparticles. The absorption band of the carbonyl group in the Pd(II)- poly (4-vinylpyrrolidone) system depends on concentration of a metal ion, i.e. on a number of particles in suspension [163]. A change in the initial concentration of Pd(II) from $0.66 \cdot 10^{-3}$ to $2 \cdot 10^{-3}$ mol brings to increase in size of Pd nanoparticles from 3–4 nm to 5–6 nm. The respective increase in the molar ratio Pd(II)/PVP is reflected in a decrease in number of nanoparticles, and for the maximum ratio Pd(II)/PVP there is maximum number of particles, therefore PVP cannot cover every particle (Table 3.9).

The active medium for sonochemical reactions are ionic liquids such as 1-n-alkyl-3-methylimidazolium- bis(trifluoromethanesulfonyl)imide, etc. [164, 165]. Non-traditional class of solvents containing organic cations such as imidazolium, pyridinium, pyrrolidinium, quaternary ammonium, sulfonium, and inorganic anion attract attention by their chemical stability at high temperatures, low volatility, low toxicity and low flammability, i.e. they mostly meet demands of green chemistry. By varying length of alkyl chain and duration of ultrasonic action it is possible to control efficiently morphology and sizes of nanoparticles, for example, to obtain ZnO in form of nanorods or nanosheets [166]. It should be noted that sonochemical synthesis, as a rule, brings to products with lower crystallinity than traditional heating methods. Therefore sometimes sonochemical reactions are performed at high temperatures. Thus, increase in temperature from room temperature to 40–100 °C for production of Mn_3O_4 colloid nanoparticles leads to considerable improvement of crystallinity of the product, though in the absence of ultrasonic impact completely amorphous colloid nanoparticles are formed (Fig. 3.8) [167].

Therefore, sonochemical approach has wide application due to low cost and high efficiency of the procedure of ultrasonic cavitation. On the other hand, impact waves and turbulent flows generated by multiple collapses of microbubbles, make nanoparticles move with velocities of hundreds meters per second and undergo non-elastic impact with change in shape and crystallinity, i.e. ultrasonic melting of inorganic materials provides production of metal particles at the nanoscale with different morphologies and properties.

Fig. 3.8 X-ray diffraction patterns of the samples of Mn_3O_4 colloid nanoparticles synthesized at: (*a*) 100 °C, (*b*) room temperature, (*c*) at room temperature without ultrasonic impact [167]

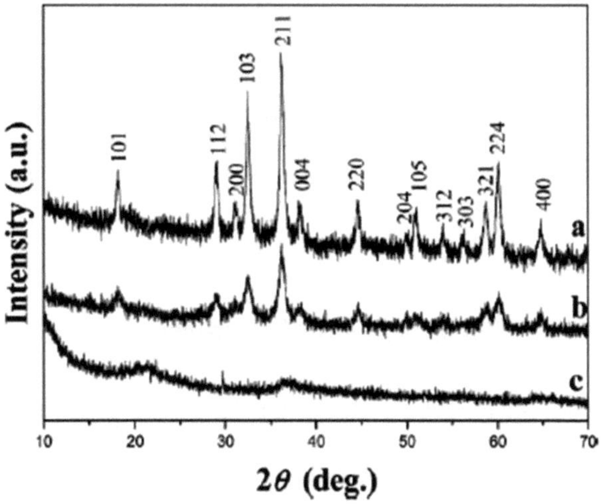

Sonochemical methods include unique reaction ways for development of different kinds of composite materials [148, 168–171]. Sonochemical intercalation of Au nanoparticles in interlayer space of Na-montmorillonite can be a general approach to synthesis of thermally stable metal-clay nanocomposites with high dispersion ability of metal nanoparticles, including catalytically active [172]. Important advantage of ultrasonic treatment is its multifunctional character, which displays itself in a possibility to perform different procedures under its action: dispersion, emulsification, organic synthesis, and polymerization. Combination of dispersion and polymerization processes *in situ* were used to obtain polyaniline/Fe_3O_4 [173] and poly (n-butyl methacrylate)/γ-Al_2O_3 nanocomposites [174]. In the latter case encapsulated polymerization induced by ultrasonic irradiation (at frequency $2{\cdot}10^4$–10^9 Hz) proceeds with high conversion in presence of sodium dodecyl sulfate in the system, which points to the fact that radicals which induce emulsion polymerization of butyl methacrylate form during dissociation of a surfactant.

On the whole, sonochemical method of nanocomposite production can be considered as one of the most attractive and in this view it serves as an alternative to ordinary solvothermal or hydrothermal synthesis, sol-gel processes, co-deposition, electrochemical and pyrolytic methods. Most of these synthetic methods need high temperatures, involve metal-organic precursors sensitive to oxidizing media and temperature, and often use environment-polluting agents, etc. [175].

3.5 Physical and Chemical Deposition from the Vapor Phase

Different variants of techniques of deposition from vapor phase are widely used for production of metal polymer nanocomposites [4, 176, 177]. Their essence is in simultaneous or consequent deposition of metal and organic components and formation of metal nanoparticles in growing composite films. Active metal atoms during collision with polymer surface can diffuse in volume of a polymer matrix, be captured by surface defects, and collide with each other, which results in aggregation and formation of cluster particles. The volume fraction of metal nanoparticles in a polymer film can be controlled by relationship between deposition rates of metal and polymer components [178]. We shall consider some ways of production of nanocomposites.

3.5.1 Plasmochemical Synthesis

In this method low temperature (4,000–10,000 K) nitrogen, ammonium, hydrogen, hydrocarbon, or argon plasma is used, which is generated by arc, glow, high-frequency or microwave discharge. The main conditions for obtaining nanoparticles by this method are the reaction process being far from equilibrium, high rate of the solid-state phase nucleation and low growth rate. Plasma chemical synthesis provides high rates of formation and condensation of a compound and quite high efficiency. Due to these advantages plasma technology is widely applied in production of nanostructural materials [179, 180].

The method of plasma condensation is one of universal methods for production of metal polymer nanocomposites based on spraying of a metal target and simultaneous polymerization of organic precursors in plasma [5, 176, 181]. This approach supplies production of homogeneously distributed nanoparticles in a polymer matrix and provides variation of degree of filling in wide range. Most often fluorocarbon [182], hydrocarbon [182–185], organosilicic [186, 187] precursors are used for production of polymer matrix. The Ti- [184] and Pt- [185] hydrocarbon nanocomposite coatings were obtained by one-stage plasma polymerization method.

Content, size and distribution of metal clusters depend on parameters of plasma chemical process, and can be regulated by power of plasma supply and flow velocity of a monomer (Fig. 3.9).

Among plasma-mediated methods there are interesting approaches based on use of monodisperse metal clusters generated in separate reactors by mechanism of gas aggregation in connection with filtration by mass [188–191]. Formation of clusters carries on in an aggregation chamber during thermal evaporation of a metal into flow of a carrier gas (Ar or He) under high pressure (~100 Pa). Metal atoms are condensed on gas molecules, and the generated clusters grow during homogeneous nucleation-condensation and evaporation of atoms. The formed clusters are transferred with the gas-carrier flow through the outlet nozzle to substrate in the reactor, where a polymer condenses gradually during plasma polymerization (Fig. 3.10).

It is interesting that the obtained by this strategy Pd cluster structures on PMMA display far higher adhesion than the metal films synthesized by thermal evaporation [192]. This method provides development of 3D nanocomposites with homogeneous size distribution of nanoparticles in volume of a polymer matrix, and

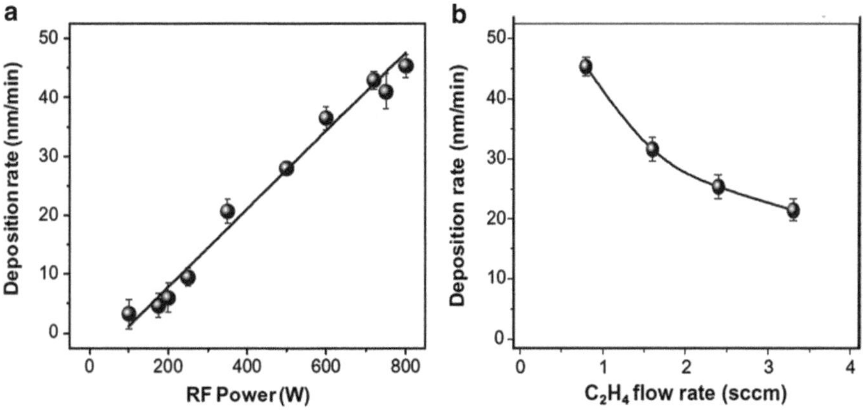

Fig. 3.9 Deposition rate of Pt/CH_x nanocomposite film vs. input power (**a**) and ethylene flow rate (**b**) [185]

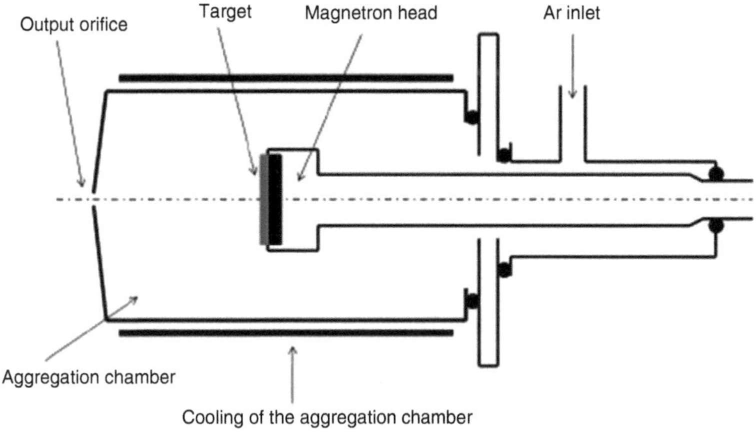

Fig. 3.10 Schematic drawing of reactor for synthesis of monodispersed metallic cluster by thermal evaporation and gas aggregation [188]

development of layered nanocomposite structures. In the latter case necessary concentrations are achieved by variations of deposition time, while for bulk deposition it is important that the deposition rate of a polymer was higher than the deposition rate of the clusters, like in Ag/$SiO_xC_yH_z$ [191].

3.5.2 Evaporation and Spraying Methods

The condensation (deposition) method from the vapor phase used for production of metal polymer nanocomposites is usually reduced to evaporation of a monomer and

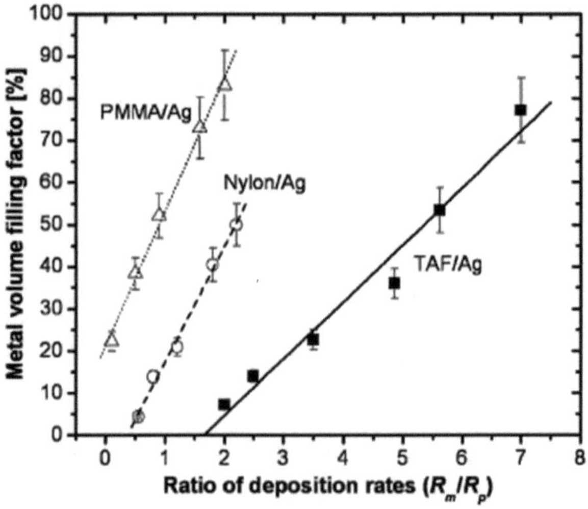

Fig. 3.11 The volume fraction of Ag in nanocomposites versus ratio of evaporation rates Rm/Rp of Ag to TAF (1), Nylon (2), and PMMA (3), respectively [178]

poly-condensation on a substrate, thermal disintegration of some polymers, for example, Teflon, and partial repolymerization of fragments on a substrate [4, 177, 193]. Nanocomposite materials with different content of metal are obtained by simultaneous deposition of noble metals and polymers (nylon 6, polytetrafluoroethylene, PMMA) [178]. Increase in particle size with increase in metal concentration is typical of vapor deposition, in which temperature can be used as additional controlled parameter. Volume fraction of a metal in nanocomposite depends almost linearly on the ratio of metal/polymer deposition rates (Fig. 3.11).

The considered technique can also be used to obtain nanocomposite films containing bimetal nanoparticles. The co-evaporation and co-deposition methods were used to synthesize Au_xAg_{1-x} [194] and Cu_xAg_{1-x} [195] in Teflon polymer matrix. Components of the formed nanocomposite were evaporated simultaneously from separate targets in a chamber under residual pressure 2.4×10^{-5} Pa. Despite of a complicated character of the processes in these systems (absorption and reemission of metals, surface diffusion, etc.), coefficients of condensation can reach 60 %, nanocomposite films are characterized by homogeneous size distribution of particles in the matrix (Fig. 3.12) in a wide range of their compositions and volume fractions. The nanodisperse phase has prevailingly a core-shell structure, which forms, probably, due to a considerable difference between coefficients of condensation, for example, between copper and silver atoms.

Methods of simultaneous and consequent spraying and condensation include radio frequency (RF) magnetron spattering of polymers from a polymer target and cathode spattering of a metal target with direct current (DC) energy supply. Under RF impact a polymer is fragmented and linked, which can bring to improvement of mechanical properties of the obtained nanocomposite materials. The most often used polymers in this technique are polytetrafluoroethylene and nylon [196, 197]. It should be noted that this approach is especially efficient for production of alloys of

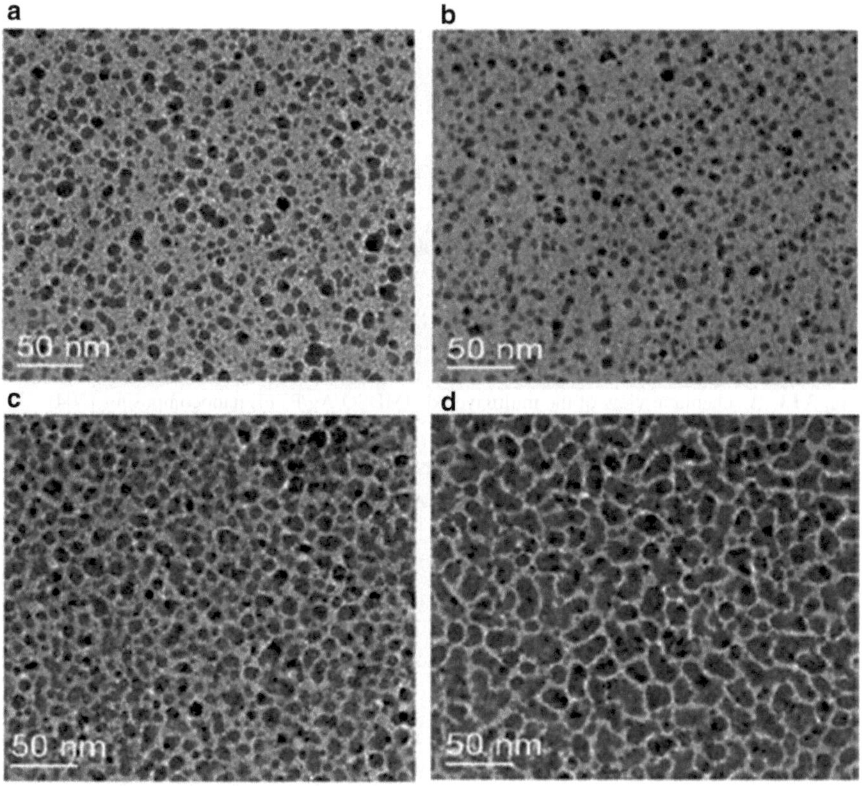

Fig. 3.12 TEM micrographs of bimetallic Cu_xAg_{1-x}/teflon nanocomposites with different composition (x) and volume fraction (f) of nanodispersed phase: (**a**) x = 0.2, f = 17 %; (**b**) x = 0.31, f = 17 %; (**c**) x = 0.34, f = 42 %; (**d**) x = 0.35, f = 54 % [195]

bimetallic nanoparticles on the surface of polymers, as has been shown in the case of Au-Ag/PTFE nanocomposites [194, 198], while consequent deposition processes, as a rule, bring to formation of core-shell structures, and in this case additional treatment is needed for production of alloyed particles, for example, annealing [199] or laser-induced melting [200]. A possibility of use of targets made of pure metals [201] and their alloys [202] is shown. The TiO_2/FeCo nanocomposite was produced one stage by simultaneous DC spraying of FeCo target and TiO_2-RF magnetron spattering [203].

An interesting approach was realized in multilayered nanocomposite material based on Ag-composite [204]. Hydrophilic polymer layer, composition and thickness of which was controlled by O_2 flow during polymerization, was formed by plasma polymerization of hexamethyldisiloxane (HMDSO) on the surface of Ag nanoparticles of 2D ensemble in a thin layer of RF-spattered PTFE (Fig. 3.13).

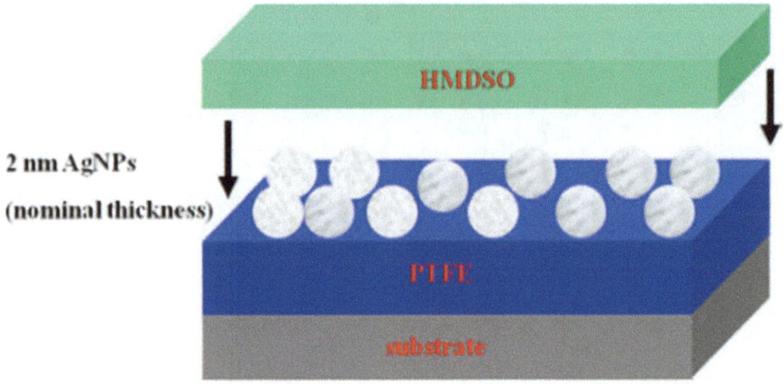

Fig. 3.13 A schematic view of the multilayered HMDSO/Ag/PTFE nanocomposites [204]

Microstructure and thickness of plasma-polymerized polymer film have a significant effect on properties of multilayer composite material, in particular, on release of silver ions.

3.5.3 Laser Ablation

Laser ablation of solids in liquid attracts more and more interest in production of nanoparticles [4, 205–207]. The essence of this method is in focusing laser beam through a transparent liquid to a metal target surface. Interaction of a powerful laser irradiation with absorbing media causes ablation of the latter and spattering of their material in air. At high absorption coefficients thickness of the removed layer is small, from units to tens nanometers. Nanoparticles form due to collision of molecules of a substance with each other during adiabatic expansion and recombination of plasma flame in rarefied gas. Mean free path of molecules is in the case of laser ablation of solids in liquids by several orders of magnitude lower than that in vacuum, because evaporated material of the target collides with vapors in the liquid near the target. High pressure of liquid vapors at the temperature of the target has effect on size distribution of the particles. Moreover, the function of size distribution of the particles can vary during irradiation due to absorption at the wavelength of laser irradiation. One of advantages of this method is the fact that synthesis of metal colloids can be realized in a solution in absence of chemical reagents, i.e. high purity grade products can be obtained.

Laser ablation in presence of surfactants and polymers brings to formation of nanocomposite materials. Microstructure of the obtained nanocomposites is substantially affected by parameters of laser interaction. Granular magnetoresistive Co/PTFE nanocomposites are obtained by laser ablation (Nd-YAG laser at the temperature 355 nm) [208]. Colloids of silver nanoparticles obtained by laser irradiation in water solution of polyvinylpyrrolidone were more stable than colloidal water solutions [209]. Method of pulse laser ablation make it possible to carry out functioning of nanoparticles by direct incorporation of functional agents in ablation medium before laser irradiation, thus favoring simultaneous generation of a nanoparticle and covalent bonding of a ligand, especially this approach is important for production of bionanocomposite materials (see Chap. 7) [210, 211]. In situ bonding of laser generated Au nanoparticles by functional biomolecules is shown for dextran-coated nanoparticles used as biosensors for lectins [212]. Another similar example relates to conjugation of Au nanoparticles with thermo-sensitive co- poly(N-isopropylacrylamide) polymer with terminal thiol groups in the processes of in- and ex-situ laser ablation [213]. Undoubtedly, in systems containing thermo-sensitive biomolecules it is necessary to take into account degrading effect of laser on them [214].

Theoretical studies in the field of applications of low temperature plasma in nanotechnologies are intensely carried on. The theoretical model is developed to estimate typical sizes of nanostructures formed on the surface of solids during melting under laser irradiation [215]. Solution of Stefan problem of melting and crystallization of the surface layer in combination with the theory of formation of crystal nuclei displays a dependence of cooling rate and typical sizes of the seeds on pulse duration and energy. At duration 10^{-8} s and energy density 4 J/cm^2 the calculation gives typical sizes of the structures about interatomic distances (5×10^{-10} m), while at duration 10^{-6} s and energy 6 J/cm^2 the typical size is ~100 nm.

Resuming, it can be noticed that classical solution methods for production of nanocomposite materials, such as sol gel synthesis (see Chap. 4), hydrothermal methods, etc., become more and more differentiated, and preparative instrumentation is supplemented by innovation approaches, such as microwave, laser, sonic-chemical techniques, etc. For example, metal nanoparticles incorporated in dielectric nanofibers are obtained in a sandwich-like microreactor by combined application of H_2 plasma of high-frequency microwave power and high temperature of substrate [216]. In order to develop inorganic n-p-hetero-junctions based on GaN/InGaN 1D heterostructures, the method of plasma-mediated molecular-beam epitaxy was successfully used [217]. In similar semiconductor structures TiO_2 shell was formed by plasma spattering on the surface of ZnO nanorods, preliminary obtained by hydrothermal method [218].

A tendency of development will, probably progress towards combination of new synthetic strategies and hybrid approaches with individually adapted instrumentation for design of technologically important nanostructural materials.

References

1. C. Boissiere, D. Grosso, A. Chaumonnot, L. Nicole, C. Sanchez, Adv. Mater. **23**, 599 (2011)
2. G.R. Patzke, Y. Zhou, R. Kontic, F. Conrad, Angew. Chem. Int. Ed. **50**, 826 (2011)
3. I. Marozau, A. Shkabko, G. Dinescu, M. Dobeli, T. Lippert, D. Logvinovich, M. Mallepell, C.W. Schneider, A. Weidenkaff, A. Wokaun, Appl. Surf. Sci. **255**, 5252 (2009)

4. F. Faupel, V. Zaporojtchenko, T. Strunskus, M. Elbahri, Adv. Eng. Mater. **12**, 1177 (2010)
5. A. Heilmann, *Polymer Films with Embedded Metal Nanoparticles* (Springer, Berlin, 2003)
6. M. Baghbanzadeh, L. Carbone, P.D. Cozzoli, C.O. Kappe, Angew. Chem. Int. Ed. **50**, 11312 (2011)
7. M. Tsuji, M. Hashimoto, Y. Nishizawa, M. Kubokawa, T. Tsuji, Chem. Eur. J. **11**, 440 (2005)
8. A.S. Vanetsov, Yu.D. Tret'yakov, Uspekhi Khim. **76**, 435 (2007)
9. Y. Bykov, K.I. Rybakov, V.E. Semenov, Russ. Nanotechnol. **6**, 60 (2011)
10. Y.I. Bokhan, I.A. Shkreb, Pis'ma v ZHETF **20**, 24 (1994)
11. Y.V. Bykov, K.I. Rybakov, V.E. Semenov, J. Phys. D **34**, R55 (2001)
12. H.M. Kingston, S.J. Haswell (eds.), *Microwave-Enhanced Chemistry* (American Chemical Society, Washington, DC, 1997)
13. J. Park, J. Joo, S.G. Kwon, Y. Jang, T. Hyeon, Angew. Chem. **119**, 471 (2007)
14. I. Pastoriza-Santos, L.M. Liz-Marzan, Langmuir **18**, 2888 (2002)
15. M.N. Nadagouda, R.S. Varma, Cryst. Growth Des. **8**, 291 (2008)
16. B. Baruwati, V. Polshettiwar, R.S. Varma, Green Chem. **11**, 926 (2009)
17. M.N. Nadagouda, R.S. Varma, Biomacromolecules **8**, 2762 (2007)
18. M. Tsuji, M. Hashimoto, Y. Nishizawa, T. Tsuji, Chem. Lett. **32**, 1114 (2003)
19. B. Baruwati, M.N. Nadagouda, R.S. Varma, J. Phys. Chem. C **112**, 18399 (2008)
20. X. Zhu, J. Wang, Z. Zhang, J. Zhu, S. Zhou, Z. Liu, N. Ming, J. Am. Ceram. Soc. **91**, 2683 (2008)
21. L.T. Guo, H.J. Luo, J.Q. Gao, L.Z. Guo, J.F. Yang, Mater. Lett. **60**, 3011 (2006)
22. B.L. Newalkar, S. Komarneni, H. Katsuki, Mater. Res. Bull. **36**, 2347 (2001)
23. W. Sutton, Ceram. Bull. **68**, 376 (1989)
24. D. Li, S. Komarneni, J. Am. Ceram. Soc. **89**, 1510 (2006)
25. H. Grisaru, O. Palchik, A. Gedanken, V. Palchik, M.A. Slifkin, A.M. Weiss, Y. Rozenfeld Hacohen, Inorg. Chem. **40**, 4814 (2001)
26. H. Grisaru, O. Palchik, A. Gedanken, V. Palchik, M.A. Slifkin, A.M. Weiss, Inorg. Chem. **42**, 7148 (2003)
27. J. Zhu, O. Palchik, S. Chen, A. Gedanken, J. Phys. Chem. B **104**, 7344 (2000)
28. H. Grisaru, V.G. Pol, A. Gedanken, I. Nowik, Eur. J. Inorg. Chem. **2004**, 1859 (2004)
29. Y.-L. Hsin, C.-F. Lin, Y.-C. Liang, K.C. Hwang, J.-C. Horng, J.-A. Ho, C.-C. Lin, J.R. Hwu, Adv. Funct. Mater. **18**, 2048 (2008)
30. B. Hu, S.-B. Wang, K. Wang, M. Zhang, S.-H. Yu, J. Phys. Chem. C **112**, 11169 (2008)
31. C.O. Kappe, Angew. Chem. Int. Ed. **43**(46), 6250 (2004)
32. D.S. Jacob, I. Genish, L. Klein, A. Gedanken, J. Phys. Chem. B Lett. **110**, 17711 (2006)
33. T. Yamauchi, Y. Tsukahara, K. Yamada, T. Sakata, Y. Wada, Chem. Mater. **23**, 75 (2011)
34. T. Yamamoto, H. Yin, Y. Wada, T. Kitamura, T. Sakata, H. Mori, S. Yanagida, Bull. Chem. Soc. Jpn. **77**, 757 (2004)
35. P.K. Sudeep, P.V. Kamat, Chem. Mater. **17**, 5404 (2005)
36. S. Tan, M. Erol, A. Attygalle, H. Du, S. Sukhishvili, Langmuir **23**, 9836 (2007)
37. Y. Niidome, K. Nishioka, H. Kawasaki, S. Yamada, Chem. Commun. **2376** (2003)
38. O.R. Miranda, N.R. Dollahon, T.S. Ahmadi, Cryst. Growth Des. **6**, 2747 (2006)
39. S. Eustis, H.-Y. Hsu, M.A. El-Sayed, J. Phys. Chem. B **109**, 4811 (2005)
40. K. Kurihara, J. Kizling, P. Stenius, J.H. Fendler, J. Am. Chem. Soc. **105**, 2574 (1983)
41. E. Leontidis, K. Kleitou, T. Kyprianidou-Leodidou, V. Bekiari, P. Lianos, Langmuir **18**, 3659 (2002)
42. E. Gachard, H. Remita, J. Khatouri, B. Keita, L. Nadjo, J. Belloni, New J. Chem. **22**, 1257 (1998)
43. R. Kissner, G. Welti, G. Geier, J. Chem. Soc. Dalton Trans. **10**, 1773 (1997)
44. A.-T. Le, P.T. Huy, T.Q. Huy, P.D. Cam, A.A. Kudrinskii, A.Y. Olenin, G.V. Lisichkin, Y.A. Krutyakov, Russ. Nanotechnol. **5**, 125 (2010)
45. N.L. Lala, R. Ramaseshan, L. Bojun, S. Sundarrajan, R.S. Barhate, L. Ying-jun, S. Ramakrishna, Biotechnol. Bioeng. **97**, 1357 (2007)

46. P.A. Muzalev, I.D. Kosobudskii, N.M. Ushakov, L.G. Panova, Perspekt. Mater. **84** (2011)
47. M.A. Kudryashev, A.I. Mashin, A.S. Tyurin, A.E. Fedosov, G. Chidichimo, G. De Filpo, Zh. Tekhn. Fiz. **81**, 95 (2011)
48. M.A. Kudryashev, A.I. Mashin, A.S. Tyurin, G. Chidichimo, G. De Filpo, Poverkhnost. Rentgenovskii, sinkhotronnye I neitronnye issledovaniya, **84** (2010)
49. Z. Zang, L. Zang, S. Wang, W. Chen, Y. Lei, Polymer **42**, 8315 (2001)
50. A.D. Pomogailo, G.I. Dzhardimalieva, V.N. Kestelman, *Macromolecular Metal Carboxylates and Their Nanocomposites* (Springer, Heidelberg, 2010)
51. Y. Lu, Y. Mei, M. Schrinner, M. Ballauff, M.W. Moller, J. Breu, J. Phys. Chem. C **111**, 7676 (2007)
52. A. Troupis, A. Hiskia, E. Papaconstantinou, Angew. Chem. Int. Ed. **41**, 1911 (2002)
53. S. Mandal, P.R. Selvakannan, R. Pasricha, M. Sastry, J. Am. Chem. Soc. **125**, 8440 (2003)
54. L. Yang, Y. Shen, A. Xie, J. Liang, S. Li, Q. Zhang, Eur. J. Inorg. Chem. **4658** (2006)
55. M.L. Marin, K.L. McGilvray, J.C. Scaiano, J. Am. Chem. Soc. **130**, 16572 (2008)
56. K.G. Stamplecoskie, J.C. Scaiano, Photochem. Photobiol. **88**, 762 (2012)
57. K. Esumi, T. Matsumoto, Y. Seto, T. Yoshimura, J. Colloid Interface Sci. **284**, 199 (2005)
58. C.S. Colley, D.C. Grills, N.A. Besley, S. Jockusch, P. Matousek, A.W. Parker, M. Towrie, N.J. Turro, P.M.W. Gill, M.W. George, J. Am. Chem. Soc. **124**, 14952 (2002)
59. S. Kapoor, T. Mukherjee, Chem. Phys. Lett. **370**, 83 (2003)
60. N. Kometani, H. Doi, K. Asami, Y. Yonezawa, Phys. Chem. Chem. Phys. **4**, 5142 (2002)
61. S. Eustis, G. Krylova, A. Eremenko, N. Smirnova, A.W. Schill, M. El-Sayed, Photochem. Photobiol. Sci. **4**, 154 (2005)
62. S. Jockusch, M.S. Landis, B. Freiermuth, N.J. Turro, Macromolecules **34**, 1619 (2001)
63. C.M. Gonzalez, Y. Liu, J.C. Scaiano, J. Phys. Chem. C **113**, 11861 (2009)
64. K.L. McGilvray, M.R. Decan, D. Wang, J.C. Scaiano, J. Am. Chem. Soc. **128**, 15980 (2006)
65. X. Zhu, B. Wang, F. Shi, J. Nie, Langmuir **28**, 14461 (2012)
66. M. Harada, E. Katagiri, Langmuir **26**, 17896 (2010)
67. M. Harada, N. Tamura, M. Takenaka, J. Phys. Chem. C **115**, 14081 (2011)
68. M. Harada, Y. Inada, Langmuir **25**, 6049 (2009)
69. P.V. Kamat, in *Semiconductor Nanoclusters – Physical, Chemical and Catalytic Aspects*, ed. by P.V. Kamat, D. Miesel (Elsevier Science, Amsterdam, 1997), pp. 237–259
70. D. Riassetto, F. Roussel, L. Rapenne, H. Roussel, S. Coindeau, O. Chaix, F. Micoud, M. Chatenet, M. Langle, J. Exp. Nanosci. **5**, 221 (2010)
71. S.K. Ghosh, S. Kundu, M. Mandal, S. Nath, T. Pal, J. Nanopart. Res. **5**, 577 (2003)
72. J.C. Scaiano, C. Aliaga, S. Maguire, D. Wang, J. Phys. Chem. B **110**, 12856 (2006)
73. Y.H. Ng, S. Ikeda, T. Harada, T. Sakata, H. Mori, A. Takaoka, M. Matsumura, Langmuir **24**, 6307 (2008)
74. X. Wang, J.C. Yu, H.Y. Yip, L. Wu, P.K. Wong, S.Y. Lai, Chem. Eur. J. **11**, 2997 (2005)
75. O. Carp, C.L. Huisman, A. Reller, Prog. Solid State Chem. **32**, 33 (2004)
76. D. Hufschmidt, D. Bahnemann, J.J. Testa, C.A. Emilio, M.I. Litter, J. Photochem. Photobiol. A **148**, 223 (2002)
77. R.D. Miller, Chem. Rev. **89**, 1359 (1989)
78. M. Sakamoto, T. Tachikawa, M. Fujitsuka, T. Majima, Chem. Mater. **20**, 2060 (2008)
79. T. Tamai, M. Watanabe, Y. Hatanaka, H. Tsujiwaki, N. Nishioka, K. Matsukawa, Langmuir **24**, 14203 (2008)
80. N.O. Yakimovich, N.V. Sapogova, L.A. Smirnova, A.P. Aleksandrov, T.A. Gracheva, A.V. Kirsanov, N.M. Bityurin, Khim. Fiz **27**, 61 (2008)
81. L.A. Smirnova, A.P. Aleksandrov, N.O. Yakimovich, N.V. Sapogova, A.V. Kirsanov, L.V. Soustov, N.M. Bityurin, Dokl. Akad. Nauk **400**, 779 (2005)
82. A.O. Rybaltovskii, A.A. Aksenov, V.I. Gerasimova, V.V. Zosimova, B.K. Popov, A.B. Solov'eva, P.S. Timashev, V.N. Bagratishvili, Sverkhkriticheskii fluidy: teoriya I praktika **3**, 50 (2008)

83. M. Kohsuke, Y. Miura, S. Shironita, M. Tomonari, N. Mimura, H. Yamashita, Stud. Surf. Sci. Catal. B **170**, 1319 (2007)
84. P. He, M. Zhang, D. Yang, J. Yang, Surf. Rev. Lett. **13**, 51 (2006)
85. K.P. Yu, W.Y. Yu, M.C. Kuo, Y.C. Liou, S.H. Chien, Appl. Catal. B **84**, 112 (2008)
86. S. Kundu, M. Mandal, S.K. Ghosh, T. Pal, J. Photochem. Photobiol. A **162**, 625 (2004)
87. N. Luo, L. Mao, L. Jiang, J. Zhan, Z. Wu, D. Wu, Mater. Lett. **63**, 154 (2009)
88. N.A. Agareva, A.P. Aleksandrov, L.A. Smirnova, N.M. Bityurin, Perspectivnye Materk **5** (2009)
89. M. Muniz-Miranda, J. Raman Spectrosc. **35**, 839 (2004)
90. J.C. Vinci, P. Bilski, R. Kotek, C. Chignell, Photochem. Photobiol. **86**, 806 (2010)
91. J. He, I. Ichinose, T. Kunitake, A. Nakao, Langmuir **18**, 10005 (2002)
92. P.V. Kamat, J. Phys. Chem. B **106**, 7729 (2002)
93. M.A. El-Sayed, Acc. Chem. Res. **34**, 257 (2001)
94. P.V. Kamat, M. Flumiani, G.V. Hartland, J. Phys. Chem. B **102**, 3123 (1998)
95. K. Kimura, J. Phys. Chem. **98**, 11997 (1994)
96. R.C. Jin, Y. Cao, E. Hao, G.S. Metraux, G.C. Schatz, C.A. Mirkin, Nature **425**, 487 (2003)
97. S.V. Karpov, V.V. Slabko, G.A. Chiganova, Colloid J. **64**, 425 (2002)
98. S.V. Karpov, A.L. Bas'ko, A.K. Popov, V.V. Slabko, Opt. Spectrosc. **95**, 241 (2003)
99. W.D. Bragg, V.A. Markel, W.T. Kim, K. Banerjee, M.R. Young, J.G. Zhu, R.L. Armstrong, V.M. Shalaev, Z.C. Ying, Y.E. Danilova, V.P. Safonov, J. Opt. Soc. Am. B **18**, 698 (2001)
100. A.K. Popov, J. Brummer, R.S. Tanke, G. Taft, M. Loth, R. Langlois, A. Wruck, R. Schmitz, Laser Phys. Lett. **3**, 546 (2006)
101. B.G. Ershov, Izv. AN, Ser. Khim. 1 (1999)
102. L.T. Bugaenko, M.G. Kuz'min, *Khimiya vysokikh energii* (Khimiya, Moscow), 1 (1988)
103. B.G. Ershov, Ros.khim. zh. (Ros. Khim. Zh. Ob-va im. D.I. Mendeleeva) **45**, 20 (2001)
104. A.B. Zezin, V.B. Rogacheva, S.P. Valueva, N.I. Nikonorova, M.F. Zansokhova, A.A. Zezin, Russ. Nanotechnol. **191** (2006)
105. A. Henglein, M. Gutierrez, E. Janata, B.G. Ershov, J. Phys. Chem. **96**, 4598 (1992)
106. A. Henglein, D. Meisel, Langmuir **14**, 7392 (1998)
107. B.G. Ershov, E. Janata, A. Henglein, Radiat. Phys. Chem. **39**, 123 (1992)
108. B.G. Ershov, E. Janata, A. Henglein, J. Phys. Chem. **97**, 339 (1993)
109. B.G. Ershov, N.L. Sukhov, D.L. Troitskii, Radiat. Phys. Chem. **39**, 127 (1992)
110. D.A. Troitskii, N.L. Sukhov, B.G. Ershov, A.V. Gordeev, Khim. Vysok. Energii **29**, 218 (1994)
111. M. Michaelis, A. Henglein, J. Phys. Chem. **96**, 4719 (1992)
112. B.G. Ershov, E. Janata, M. Michaelis, A. Henglein, J. Phys. Chem. **95**, 8996 (1991)
113. A. Radosavljević, D. Božanić, N. Bibić, M. Mitrić, Z. Kačarević-Popović, J. Nedeljković, J. Appl. Polym. Sci. **125**, 1244 (2012)
114. Y.-O. Kang, S.-H. Choi, A. Gopalan, K.-P. Lee, H.-D. Kang, Y.S. Song, J. Appl. Polym. Sci. **100**, 1809 (2006)
115. S.-H. Choi, M.S. Choi, K.P. Lee, H.D. Kang, J. Appl. Polym. Sci. **91**, 2335 (2004)
116. M.R. Karim, K.T. Lim, C.J. Lee, M.T. Bhuiyan, H.J. Kim, L.S. Park, M.S. Lee, J. Polym. Sci. Part A Polym. Chem. **45**(24), 5741 (2007)
117. H.L. Abd El-Mohdy, J. Polym. Res. **20**, 177 (2013)
118. R. Singh, D. Singh, J. Mater. Sci. Mater. Med. **23**, 2649 (2012)
119. Z. Ajji, I. Othman, J.M. Rosiak, Nucl. Instrum. Methods Phys. Res. B **229**, 375 (2005)
120. N. Leventis, C. Sotiriou-Leventis, G. Zhang, A.-M.M. Rawashdeh, Nano Lett. **2**, 957 (2002)
121. G. Piccaluga, A. Corrias, G. Ennas, A. Musinu, *Sol-Gel Preparation and Characterization of Metal-Silica and Metal Oxide-Silica Nanocomposites*. Materials Science Foundation (Trans Tech Publications, Switzerland, 2000)
122. J.F. Hund, M.F. Bertino, G. Zhang, C. Sotiriou-Leventis, N. Leventis, A. Tokuhiro, J. Farmer, J. Phys. Chem. B **107**, 465 (2003)

123. M.F. Bertino, J.F. Hund, G. Zhang, C. Sotiriou-Leventis, A.T. Tokuhiro, N. Leventis, J.Sol-Gel Sci. Technol. **30**, 43 (2004)
124. M. Mostafavi, Y.P. Lin, P. Pernot, J. Belloni, Radiat. Phys. Chem. **59**, 49 (2000)
125. A.H. Souici, N. Keghouche, J.A. Delaire, H. Remita, M. Mostafavi, Chem. Phys. Lett. **422**, 25 (2006)
126. A.L. Pan, J.G. Ma, X.Z. Yan, B.S. Zou, J. Phys. Condens. Matter **16**, 3229 (2004)
127. H.L. Su, J. Han, Q. Dong, D. Zhang, Q.X. Guo, Nanotechnology **19**, 025601 (2007)
128. S. Ghosh, S.C. Bhattacharya, A. Saha, Anal. Bioanal. Chem. **397**, 1573 (2010)
129. A.H. Souici, N. Keghouche, J.A. Delaire, H. Remita, A. Etcheberry, M. Mostafavi, J. Phys. Chem. C **113**, 8050 (2009)
130. S.-q. Chang, B. Kang, Y.-d. Da, H.-x. Zhang, D. Chen, Nanoscale Res. Lett. **6**, 591 (2011)
131. X. Michalet, F.F. Pinaud, L.A. Bentolila, J.M. Tsay, S. Doose, J.J. Li, G. Sundaresan, A.M. Wu, S.S. Gambhir, S. Weiss, Science **307**, 538 (2005)
132. S. Ghosh, A. Priyam, A. Chatterjee, A. Saha, J. Nanosci. Nanotechnol. **8**, 5952 (2008)
133. N.A. Melosh, A. Boukai, F. Diana, B. Gerardot, A. Badolato, P.M. Pelroff, J.R. Heath, Science **300**, 112 (2003)
134. K.R. Reddy, K.-P. Lee, A.I. Gopalan, M.S. Kim, A.M. Showkat, Y.C. Nho, J. Polym, Sci. Part A Polym. Chem. **44**, 3355 (2006)
135. A. Henglein, J. Phys. Chem. B **104**, 2201 (2000)
136. Q. Wan, C.L. Lin, X.B. Yu, T.H. Wang, Appl. Phys. Lett. **83**, 124 (2004)
137. Z. Zhang, T.M. Nenoff, K. Leung, S.R. Ferreira, J.Y. Huang, D.T. Berry, P.P. Provencio, R. Stumpf, J. Phys. Chem. C **114**, 14309 (2010)
138. Z. Zhang, T.M. Nenoff, J.Y. Huang, D.T. Berry, P.P. Provencio, J. Phys. Chem. C **113**, 1155 (2009)
139. C.M. Doudna, M.F. Bertino, F.D. Blum, A.T. Tokuhiro, D. Lahiri-Dey, S. Chattopadhyay, J. Terry, J. Phys. Chem. B **107**, 2966 (2003)
140. K. Roy, S. Lahiri, Green Chem. **8**, 1063 (2006)
141. K. Roy, S. Lahiri, Anal. Chem. **80**, 7504 (2008)
142. K.S. Suslick, D.A. Hammerton, R.E. Cline, J. Am. Chem. Soc. **108**, 5641 (1986)
143. K.S. Suslick, S.B. Choe, A.A. Cichowlas, M.W. Grinstaff, Nature **353**, 414 (1991)
144. N.A. Dhas, C.P. Raj, A. Gedanken, Chem. Mater. **10**, 1446 (1998)
145. K. Okitsu, A. Yue, S. Tanabe, H. Matsumoto, Y. Yobiko, Y. Yoo, Bull. Chem. Soc. Jpn. **75**, 2289 (2002)
146. R.A. Caruso, M. Ashokkumar, F. Grieser, Langmuir **18**, 7831 (2002)
147. A. Roucoux, J. Schulz, H. Patin, Chem. Rev. **102**, 3757 (2002)
148. K. Okitsu, M. Ashokkumar, F. Grieser, J. Phys. Chem. B **109**, 20673 (2005)
149. T. Prozorov, R. Prozorov, K.S. Suslick, J. Am. Chem. Soc. **126**, 13890 (2004)
150. D. Radziuk, D. Grigoriev, W. Zhang, D. Su, H. Mo¨hwald, D. Shchukin, J. Phys. Chem. C **114**, 1835 (2010)
151. K. Vinodgopal, Y. He, M. Ashokkumar, F. Grieser, J. Phys. Chem. B **110**, 3849 (2006)
152. S. Anandan, F. Grieser, M. Ashokkumar, J. Phys. Chem. C **112**, 15102 (2008)
153. Y. Mizukoshi, E. Takagi, H. Okuno, Y. Maeda, Y. Nagata, Ultrason. Sonochem. **8**, 1 (2001)
154. J. Reisse, T. Caulier, C. Deckerkheer, O. Fabre, J. Vandercammen, J.L. Delplancke, R. Winand, Ultrason. Sonochem. **3**, S147 (1996)
155. A. Durant, J.L. Delplancke, R. Winand, J. Reisse, Tetrahedron Lett. **36**, 4257 (1995)
156. X.F. Qiu, J.Z. Xu, J.M. Zhu, J.J. Zhu, S. Xu, H.Y. Chen, J. Mater. Res. **18**, 1399 (2003)
157. L.P. Jiang, A.N. Wang, Y. Zhao, J.R. Zhang, J.J. Zhu, Inorg. Chem. Commun. **7**, 506 (2004)
158. I. Haas, S. Shanmugam, A. Gedanken, J. Phys. Chem. B **110**, 16947 (2006)
159. J. Rae, M. Ashokkumar, O. Eulaerts, C.V. Sonntag, J. Reisse, F. Grieser, Ultrason. Sonochem. **12**, 325 (2005)
160. G. Cárdenas-Trivinõ, R.A. Segura, J. Reyes-Gasga, Colloid Polym. Sci. **282**, 1206 (2004)
161. I. Szaraz, W. Forsling, Langmuir **17**, 3987 (2001)
162. C.M. Koo, H.T. Ham, M.H. Choi, S.O. Kim, I.J. Chung, Polymer **44**, 681 (2003)

163. A. Nemamcha, J.-L. Rehspringer, D. Khatmi, J. Phys. Chem. B **110**, 383 (2006)
164. T. Alammar, A.-V. Mudring, J. Mater. Sci. **44**, 3218 (2009)
165. T. Alammar, A. Birkner, O. Shekhah, A.-V. Mudring, Mater. Chem. Phys. **120**, 109 (2010)
166. T. Alammar, A.-V. Mudring, ChemSusChem **4**, 1796 (2011)
167. S. Lei, K. Tang, Z. Fang, H. Zheng, Cryst. Growth Des. **6**, 1757 (2006)
168. J. Zhang, J. Du, B. Han, Z. Liu, T. Jiang, Z. Zhang, Angew. Chem. Int. Ed. **45**, 1116 (2006)
169. R. Abu-Much, A. Gedanken, Chem. Eur. J. **14**, 10115 (2008)
170. L. Xu, L.-P. Jiang, J.-J. Zhu, Nanotechnology **20**, 045605 (2009)
171. B.M. Teo, S.W. Prescott, M. Ashokkumar, F. Grieser, Ultrason. Sonochem. **15**, 89 (2008)
172. V. Belova, H. Mo¨hwald, D.G. Shchukin, Langmuir **24**, 9747 (2008)
173. G. Qiu, Q. Wang, M. Nie, J. Appl. Polym. Sci. **102**, 2107 (2006)
174. Y. Liao, Q. Wang, H. Xia, Polym. Int. **50**, 207 (2001)
175. S. Manickam, in *Theoretical and Experimental Sonochemistry Involving Inorganic Systems*, ed. by Pankaj, M. Ashokkumar (Springer, Heidelberg, 2011), pp. 191–211
176. H. Biederman, *Plasma Polymer Films* (Imperial College Press, London, 2004)
177. A. Biswas, Z. Marton, J. Kanzow, J. Kruse, V. Zaporojtchenko, F. Faupel, T. Strunskus, Nano Lett. **3**, 1 (2003)
178. H. Takele, H. Greve, C. Pochstein, V. Zaporojtchenko, F. Faupel, Nanotechnology **17**, 3499 (2006)
179. J. Zheng, R. Yang, L. Xie, J. Qu, Y. Liu, X. Li, Adv. Mater. **22**, 1451 (2010)
180. A. Borras, A. Barranco, F. Yubero, A.R. Gonzalez-Elipe, Nanotechnology **17**, 3518 (2006)
181. A.D. Pomogailo, V.N. Kestelman, *Metallopolymer Nanocomposites* (Springer, Berlin/Heidelberg, 2005)
182. A. Heilmann, J. Werner, M. Kelly, B. Holloway, E. Kay, Appl. Surf. Sci. **115**, 365 (1997)
183. C. Laurent, E. Kay, J. Appl. Phys. **65**, 1717 (1989)
184. A. Choukourov, P. Solar, O. Polonskyi, J. Hanus, M. Drabik, O. Kylian, E. Pavlova, D. Slavinska, H. Biederman, Plasma Process. Polym. **7**, 25 (2010)
185. E. Dilonardo, A. Milella, F. Palumbo, G. Capitani, R. d'Agostino, F. Fracassi, Plasma Process. Polym. **7**, 51 (2010)
186. F. Fracasssi, R. d'Agostino, F. Palumbo, F. Bellucci, T. Monetta, Thin Solid Films **264**, 40 (1995)
187. H. Biederman, L. Martinu, in *Plasma Deposition, Treatment, and Etching of Polymers*, ed. by R. d'Agostino (Academic, San Diego, 1990), p. 269
188. H. Biederman, O. Polonskyi, M. Drabik, O. Kylian, J. Kousal, J. Hanus, A. Choukourov, P. Solar, A. Serov, D. Slavinska, 30th ICPIG, August 28th–September 2nd 2011, Belfast, Northern Ireland, UK
189. P.-G. Reinhard, E. Suraud, *Introduction to Cluster Dynamics* (Wiley-VCH, Mörlenbach, 2004)
190. S.H. Baker, S.C. Thornton, K.W. Edmonds, M.J. Maher, C. Norris, C. Binns, Rev. Sci. Instrum. **71**, 3178 (2000)
191. T. Peter, S. Rehders, U. Schu¨rmann, T. Strunskus, V. Zaporojtchenko, F. Faupel, J. Nanopart. Res. **15**, 1710 (2013)
192. L. Ravagnan, G. Divitini, S. Rebasti, M. Marelli, P. Piseri, P. Milani, J. Phys. D. Appl. Phys. **42**, 082002 (2009)
193. A. Kubono, N. Okui, Prog. Polym. Sci. **19**, 389 (1994)
194. H.T. Beyene, V.S.K. Chakravadhanula, C. Hanisch, M. Elbahri, T. Strunskus, V. Zaporojtchenko, L. Kienle, F. Faupel, J. Mater, Science **45**, 5865 (2010)
195. H.T. Beyene, V.S.K. Chakravadhanula, C. Hanisch, T. Strunskus, V. Zaporojtchenko, M. Elbahri, F. Faupel, Plasmonics **7**, 107 (2012)
196. H. Biederman, Vacuum **59**, 594 (2000)
197. U. Schu¨rmann, H. Takele, V. Zaporojtchenko, F. Faupel, Thin Solid Films **515**, 801 (2006)
198. N. Alissawi, V. Zaporojtchenko, T. Strunskus, I. Kocabas, V.S.K. Chakravadhanula, L. Kienle, D. Garbe-Schönberg, F. Faupel, Gold Bull. **46**, 3 (2013)

199. K. Baba, T. Okuno, M. Miyagi, Appl. Phys. Lett. **62**, 437 (1993)
200. M. Schierhorn, L.M. Liz-Marzan, Nano Lett. **2**, 13 (2002)
201. G. Gonzalo, D. Babonneau, C.N. Afonso, J.-P. Barnes, J. Appl. Phys. **96**, 5163 (2004)
202. E. Cottancin, J. Lerme, M. Gaudry, M. Pellarin, J.-L. Vialle, M. Broyer, Phys. Rev. B **62**, 5179 (2000)
203. A. Kulkarni, V.S.K. Chakravadhanula, V. Duppel, D. Meyners, V. Zaporojtchenko, T. Strunskus, L. Kienle, E. Quandt, F. Faupel, J. Mater. Sci. **46**, 4638 (2011)
204. N. Alissawi, T. Peter, T. Strunskus, C. Ebbert, G. Grundmeier, F. Faupel, J. Nanopart. Res. **15**, 2080 (2013)
205. A.V. Simakin, V.V. Voronov, G.A. Shafeev, Trudy institute obshchei fiziki im. A.M. Prokhorova **60**, 83 (2004)
206. T.X. Phuoc, Y. Soong, M.K. Chyu, Optics Lasers Eng. **45**, 1099 (2007)
207. I.V. Kavetskaya, T.V. Voloshina, V.A. Karnavskii, V.I. Krasovskii, Kondensirovannye sredy I mezhfaznye granitsy **11**, 53 (2009)
208. H.Y. Kwong, Y.W. Wong, K.H. Wong, J. Appl. Phys. **102**, 114303 (2007)
209. T. Tsuji, D.H. Thang, Y. Okazaki, M. Nakanishi, Y. Tsuboi, M. Tsuji, Appl. Surf. Sci. **254**, 5224 (2008)
210. S. Barcikowski, F. Devesa, K. Moldenhauer, J. Nanopart. Res. **11**, 1883 (2009)
211. S. Petersen, S. Barcikowski, Adv. Funct. Mater. **19**, 1 (2009)
212. S. Besner, A.V. Kabashin, F.M. Winnik, M. Meunier, J. Phys. Chem. C **113**, 9526 (2009)
213. S. Salmaso, P. Caliceti, V. Amendola, M. Meneghetti, J.P. Magnusson, G. Pasparakis, C. Alexander, J. Mater. Chem. **19**, 1608 (2009)
214. Y. Takeda, T. Kondow, F. Mafune, J. Phys. Chem. B **110**, 2393 (2006)
215. S.I. Mikulitskii, V.Y. Khomich, V.A. Shmakov, V.A. Yamshchikov, Russ. Nanotechnol. **6**, 65 (2011)
216. M.S. Hu, H.L. Chen, C.H. Shen, L.S. Hong, B.R. Huang, K.H. Chen, L.C. Chen, Nat. Mater. **5**, 102 (2006)
217. W. Guo, M. Zhang, A. Banerjee, P. Bhattacharya, Nano Lett. **10**, 3355 (2010)
218. M.L. Wang, C.G. Huang, Y.G. Cao, Q.J. Yu, W. Guo, Q.L. Liu, J.K. Liang, M.C. Hong, Nanotechnology **20**, 285311 (2009)

19. K. Baba, T. Ogino, M. Miyagi: Appl. Phys. Lett. 62, 437 (1993)
20. [illegible]

Chapter 4
Physics and Chemistry of Sol-Gel Nanocomposites Formation

This method called *sol-gel* or *dip-* and *spin-on-glass process, spin-spray-coating, sol – gel glasses*, which was widely used in the second half of the twentieth century [1][1] is one of the most universal condensation ways of production of nanoparticles stabilized by inorganic oxide or polymeric matrices. A special interest is attracted to materials obtained by combination of sol-gel chemistry and aerosol or spray processes, and combination of sol-gel synthesis with intercalation. This also includes combinations of sol-gel processes with different types of thermolysis, and approaches met in nature in biomineralization processes, etc., which are new aspects of integrative chemistry approach. Traditional sol-gel concept is based on hydrolysis and condensation of metal alkoxides and many metalloids including different ways of their modification [2, 3]. Main reactions proceed at relatively low temperatures with usage of prepared in advance or synthesized in parallel polymers, and are convenient techniques for preparation of organic-inorganic nanocomposites compared with commercial silicate-intercalation technique. Sizes of formed nanoparticles in a composite can be reduced to 10 nm by choice of relative conditions. Polymer-inorganic materials have high mechanical strength and thermal stability in combination with optimal heat transport properties. They are widely used in practice due to unique physical and chemical properties: incorporation of nanometric inorganic component into polymeric matrix improves mechanical properties of material [4–7], permeability of polymers [8]. Light-sensitive materials are used in optic and electronic industry, printed-circuit boards, photoconductive cells, as key elements in solid coatings, packing materials, as cover materials they have good transparency [9–11]. For example, sol-gel method was used to produce various polymer/TiO_2 composites with high refractive index [12]. Such type of composites are used for chromatographic carriers, membrane materials, these are the main class of plastics for different applications, including airspace. Controlled properties of the surface of these colloid

[1] We refer here to the recent review devoted to celebration of 40th anniversary of innovation researches in polymer-inorganic ceramics produced by sol-gel technique [1].

A.D. Pomogailo, G.I. Dzhardimalieva, *Nanostructured Materials Preparation via Condensation Ways*, DOI 10.1007/978-90-481-2567-8_4

materials make reason for their wide use in pharmaceutical, cosmetic, food, biologic, and medical applications. Nanocomposite materials combining organic and inorganic phases are subjects for novel nanotechnologies, due to their combination of the best properties of metal oxides and polymers or biopolymers. They have already found their place in fabrication of various contact lenses, optic wave guides, thermally resistant coatings, memory devices and printers, chemical filters, solid electrolytes, biosensors, semiconductors, catalysts of many reactions, dip-pen lithography, ink jet printing, electrospinning, etc.

Technology of introduction of inorganic fillings in situ, sol-gel process makes it possible to obtain homogeneous products of constant chemical composition, high purity, with good reproducibility. Homogeneity of the initial solution provides good control over sizes and microstructures of inorganic filler, and conservation of these features in a polymer [13]. In turn, strict control over nanostructure of organic/inorganic nanocomposites is an important factor of formation of highly functioning self-organized materials, though thus produced nanocomposites, are subjected to change in structure and morphology during thermal transformations. It is important that sol-gel method is zero-discharge and ecological.

In order to obtain hybrid composite materials either with covalent bond between polymer and inorganic component or without it, many polymers were tested modifying morphology of inorganic networks. Especially convenient way is combination of inorganic "hosts" (precursors are silica titanium oxides, aluminum, vanadium, molybdenum oxides, etc.) for intercalation of polymer matrix ("guests" [14]) and polymeric nets at molecular level. For materials and their formation by gel formation, transformation from sol to gel is used in the systems including a polymer or its precursor, known as polymer sol-gel synthesis [15–19].

This approach allows, on the one hand, studying in more detail mechanism of formation of nanoparticles, control over their sizes, and, on the other hand, construction of the class of the novel materials having synergetic properties of the initial components: inorganic phase can capture into "a trap," oxide network, not only nanoparticles, but monomers or polymer links. The leading role in these non-equilibrium self-organized systems have interphase interactions between inorganic and polymer components. Nanometric building blocks (among them can be silica nanotubes or sheets, ceramics of layered silicates, including heterometallic ceramics, etc.) give improved properties to composites as compared to conventional microcomposites due to maximal interphase adhesion.

In this chapter we briefly focus on these problems from uniform position of "bridge building" between "inorganic" sol-gel synthesis of precursor nanoparticles and organic polymer phase taking part some way or other in formation of a nanocomposite.

4.1 General Characteristics of Sol-Gel Reactions

Sol-gel synthesis includes the following main stages: preparation of solution → gel formation → drying → heat treatment (xerogel).

Using precursor compounds, which do not introduce impurities in content of the final products, makes it possible to exclude stages of rinsing, which decreases ecological burden of the process. Precursors in this case are metal alkoxides (in the case of silicon derivatives there are silicic acids, chalcogenides, and alkoxide compounds), reaction proceeds in the medium of organic solvents, mostly alcohols, and water is a reagent whose addition causes hydrolysis $M(OR)_n$ ($M = Si$, Ti, Zr, V, VO, Ta, Zn, Al, Sn, Ce, Mo, W, also lanthanides, etc. $R = C_nH_{2n+1}$, alkyl aromatic and unsaturated ligands), also chelated alkoxides or metal salts, chlorides, sulfates, nitrates, etc. are used. Chemistry of sol-gel process is based on reactions of polymerization (poly-condensation) of inorganic compounds. They go by multi-stage mechanisms, first of which is hydroxylation of metal alkoxides and metal salts going via hydrolysis of alkoxy groups and deprotonation of molecules of coordinated water. Just after hydroxyl-groups have formed, branched oligomer or polymers form with nuclei having metal framework for a core.

Intermediates are also highly active products, which are involved in reactions of poly condensation due to residual hydroxyl and alkoxy groups, taking part in olation or oxolation, respectively bringing to gel formation. Acid hydrolysis and the following condensation can be represented by the formal schemes:

$$Si(OR)_4 + 4H_2O \rightarrow Si(OH)_4 + 4ROH; \quad (4.1)$$

$$mSi(OH)_4 \rightarrow (SiO_2)_m + 2mH_2O, \quad (4.2)$$

of (4.3) alkoxolation – alcohol release, (4.4) oxolation – water release:

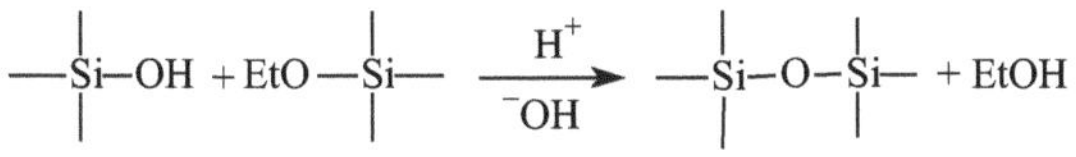

(4.3)

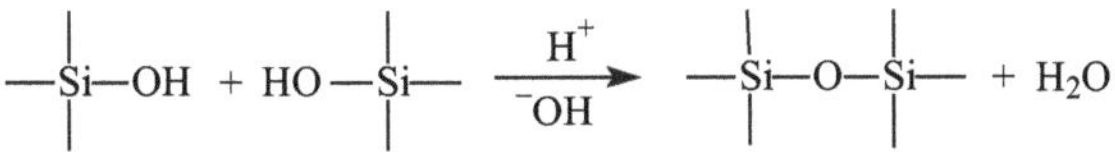

(4.4)

Sols of silicic acid stabilized by alkali condense at 30 °C to the sizes 5–10 nm during several months [20]; increase in temperature accelerates their growth.

For deposition of silica layers on different surfaces often well studied consequence of reactions is used:

$$SiCl_4 + 4H_2O \rightarrow Si(OH)_4 + 4HCl \quad (4.5)$$

$$Si(OH)_4 \rightarrow Si_nO_{4n-x}H_{4n-2x} + xH_2O \quad (4.6)$$

The last reaction is basic, it determines the following transformations of silicic acid. Less often its derivatives are used as precursors. Thus, dissociation of catecholate complex completes almost instantly in the range of neutral pH:

$$[K_2Si(C_6H_4O_2)_3] + 2H^+ + 4H_2O \rightarrow Si(OH)_4 + 3C_6H_4(OH)_2 + 2K^+ \quad (4.7)$$

$(RO)_4Si$ Precursor $\xrightarrow{\text{Hydrolysis}}$ $(SO_2)_n$ Nanosol $\xrightarrow{\text{Coating}}$ $(SO_2)_n$ Liogel film $\xrightarrow{\text{Drying}}$ Porous xerogel film $\xrightarrow{T>500°C}$ Glass oxide film

Fig. 4.1 Sol-gel synthesis of SiO_2 coatings [22]

Advantage of inorganic precursor (sodium silicate) is that it is more available (it also concerns its cost), and silicon acid forms as a result of its hydrolysis in aqua medium, in contrast to, for example, hydrolysis of triethoxysilane (TEOS) [21].

Thus, sol-gel process includes polymerization of gel-precursor (chemically controlled condensation), most often of alkali metal silicates, tetramethoxysilane (TMOS) or TEOS, which form silica-gel structure (*host*) around a dopant (*guest*) as though confined in a special cell-trap. In gels, as a rule, condensation reactions, monomer and oligomer molecules redeposit, phase transformations between solid and liquid phases go on.

Gel formation is sintering of the neighbor silica particles. Increase in concentration of precursor and pH reaction accelerates condensation.

In order to get powder or film ceramics, drying and consolidation (often by sintering) are conducted via sequence of stages (Fig. 4.1) [22].

Study of condensation of silicon acid has a long history and is related not only to producing of silica-based nanocomposite materials, but also to problems of bioinspiration and biomimetics (see Chap. 7). The first stage of condensation of silicon acid is formation of small oligomer particles with Si–O–Si bonds, then they are bound via the same bonds into branched chains and then concentrate to gel and a rigid frame. In the case of acid pH values particle growth is determined by aggregation of tetrahedrons, condensation goes due to one Si–OH bond to the degree of condensation 300, after that particle growth goes by Ostwald ripening due to dissolution of small particles and coarsening of coarser to 2–4 nm [23]. Gigantic 3D molecule forms from polymerizing branched oligomers, a special cluster [24], which coexists with sol. Control over gel formation of alkali silicon compounds and the main methods of control over these compounds are considered in [25].

At the final stage it is important to remove a solvent (as a rule, it is ethanol) at 70 °C, aerosol drying at temperature in the heating zone at 350–500 °C (predominantly at 400 °C). Duration of holding is several seconds, which prevents decomposition of organic component of the product. Its calcination is carried out at 350–700 °C during several hours in air. Ceramic components are sintered by powder during several hours in air. Sintering of the ceramic components by powder technology (mixing with the following compression, casting without melting) causes formation of pure metal oxides with high melting temperature.

The same method is used to obtain non-silicate meso-structured ceramics, for example, anatase and rutile modifications of TiO_2 [26, 27].[2] The particles form as a result of reaction of hydrolysis and co-condensation in the following way.

[2] It is more difficult to obtain polymetal-organic siloxanes of linear structure $(SiR_2\text{–O–}M)_n$, где M = Cu, Ni, Co, Mn, where R is carbohydrate radical by hydrolysis of alkyl trichlorsilane, alkali splitting of the formed polysiloxane with the following exchanging reaction of organo silanolate with metal chalcogenide (see for example [27]).

Hydrolysis:

$$\mathrm{Ti\,(OR)_4 + H_2O \rightarrow Ti\,(OR)_3OH + ROH} \tag{4.8}$$

Condensation of two Ti-OH groups goes through oxolation, while condensation of one Ti-OH with Ti-OR goes through alkoxalation with formation of Ti-O-Ti oxo-bridges by the following reactions (4.9) and (4.10):

Oxolation (splitting of water molecule):

$$\mathrm{Ti\,(OR)_3OH + Ti\,(OR)_3OH \leftrightarrow (OR)_3Ti - O - Ti\,(OR)_3 + H_2O} \tag{4.9}$$

Alkoxalation (splitting of alcohol molecule of alcohol):

$$\mathrm{Ti\,(OR)_3OH + Ti\,(OR)_4 \leftrightarrow (OR)_3Ti - O - Ti\,(OR)_3 + ROH} \tag{4.10}$$

However, these stages are difficult to identify (see, for example, [28]). Reaction capability of $M(OR)_4$ alkoxides increases in the line: $Si(OR)_4 << Sn(OR)_4$, and $Ti(OR)_4 < Zr(OR)_4 < Ce(OR)_4$. It is interesting that in the same line increase ionic radius of the central atom (respectively 0,04, 0,06, 0,064, 0,087 and 0,102 nm) and its coordination number KN (4, 6, 6, 7, 8), and its unsaturation, a difference between KN and valence (0, 2, 2, 3 and 4). As KN of the central atom (Si, Ti, Zr) increases, deposits form during shorter time for the same ligands. However, the primary important is the ratio $\gamma = H_2O/M(OR)_n$. In particular, in the case of $VO(OPr)_3$ uniform transparent gel with alkoxide polymer network in *n*-propanol is reached at $\gamma = 3$, while when $\gamma > 100$, the formed gel has a structure which cannot form compounds of a filler. Kinetics of growth of TiO_2 nanoparticles at high ratio between water and titanium alkoxide is specially studied [29]. Metal oxo alcoholates, among which are poly-nuclear $Ti_xO_y(OR)_{4x-2y,}$ can form as intermediated forms during hydrolysis of oxo alcoholates, many of these are characterized quite well. As a result, formation of TiO_2 solid phase monodisperse powder follows some consecutive stages: hydrolysis → condensation → nucleation → growth of particles. In traditional sol-gel process oxo- oligomers or polymers (sometimes with relatively high molecular mass), and also cross-linked polymers form. Most probably, particles grow by diffusion mechanism; the diffusion coefficient is about 10^{-9} $cm^2 \cdot s^{-1}$. Nevertheless, Fourier transformation infrared (FTIR) spectroscopy and small-angle X-ray scattering have shown that the main stage of hydrolysis of $Zr(OBu)_4$ goes at early stages (during the first second), and even within this time range condensation already takes place [30]. Silicate and non-silicate precursors, as a rule, demonstrate absolutely different kinetics of sol-gel condensation. Rate constants of these reactions are very high. One of the most convenient ways of studies of such processes is study of kinetics of formation of Ti–O–C bonds by IR spectroscopy (FTIR) during reaction [31]. Area of the absorption band of the respective Ti–O–C bond at the initial moment

Table 4.1 Experimental reaction rate constants and the energy of activation for hydrolysis of Ti $(OC_4H_9)_4$ of different concentrations in the Squalane medium [32]

	k, min^{-1} (180 °C)	k, min^{-1} (220 °C)	k, min^{-1} (250 °C)	E, kJ mol^{-1}
$Ti(OC_4H_9)_4$	0.29	0.50	0.83	47
$Ti(OR)_4$ 10 wt.%	0.10	0.61	0.91	65
$Ti(OR)_4$ 20 wt.%	0.07	0.37	0.75	66
$Ti(OR)_4$ 30 wt.%	0.06	0.28	0.87	79

(t = 0) is designated as A_0, and for time t (min) it is A_t. Then the degree of transformation of Ti–O–C bonds p is calculated by equation [32]:

$$p = \left(1 - \frac{A_t}{A_0}\right) \times 100 \tag{4.11}$$

Linear dependence lnp on f(t) is observed according to the first-order reaction. Kinetic constants at different temperatures and concentration of precursor in the medium of high-boiling Squalane solvent and activation energies are listed in Table 4.1. These data are compatible with absorption bands obtained for the melt.

Heat effects of hydrolysis reaction of $Ti(OBu)_4$ in ROH (R = Et, Pr^i, Bu) at different ratios in the range of $\gamma = 0.2 \div 70$ increase to $\gamma \approx 1$ (average value is 19.3 kJ mol^{-1}) and thereafter almost do not change; in other words, the thermal effect corresponds to the reaction of substitution of one alkoxy group. In this case optimization of conditions of the process is very important, as is using of catalysts (including nucleophilic type NH_4F, trifluoroacetic and even polymer acid, for example, polystyrene sulfonic acid), and origin of metal and its alkoxy group (because the hydrolysis rate of $Ti(OBu)_4$ is almost by 150 times slower than the hydrolysis rate of $Ti(OEt)_4$). A significant effect has presence of other ligands in the case of alkoxides of mixed type (especially chelate: β-diketonate, α- or β-hydroxy acid, polyolic, etc.) in $Zr(OSiMe_3)_2(acac)_2$ compounds applied in processes of chemical metal-organic deposition from gas phase (MOCVD).

An important role has also a degree of association of alkoxides (for example, for $[Ti(OEt)_4]_n$ $n = 2$; 3). Rate of hydrolysis of formed in the process oxo- and alkoxo clusters, such as $Ti_{18}O_{22}(OBu)_{26}(acac)_2$ is far lower than the hydrolysis rate of the initial $Ti(OR)_4$. Presently a wide diversity of methods for production of metal oxides (for example, commercial TiO_2) is known, nevertheless, sol-gel process looks the most promising. As is known, sols are thermally unstable, constantly changing systems with high free surface energy, and they can exist only in presence of stabilizing agents. One of the efficient ways of their stabilizing is electrostatic, which is controlled by pH of a system. Stabilizing of highly concentrated soles is a difficult problem, because disturbance of the system causes uncontrolled agglomeration of particles. There are several chemical strategies, which can be used for stabilizing of inorganic precursor in a solution and production of nanostructured materials by aerosol process, in particular, usage for precursors chlorine derivatives or complex forming agents (acetic acid or chlorine-fluorine hydrogen acid), which are strongly connected to the surface of the SiO_2, ZrO_2, TiO_2or Al_2O_3 particles [33].

Reaction ability of titanium alkoxides with respect to water is very high, so that it even causes uncontrolled formation of a deposit, which limits their usage in sol-gel synthesis. These problems are substantially overcome by using of organic ligands such as acetylacetone (AcAc), diols, carbonic acids acting as bidentate ligands [34, 35]. The structuring agents for many of them are polyethylene oxide (PEO) and its copolymers, which are used for formation of mesopores in a product (within 4–6 nm). Crystallization of metal oxide matrices is conducted by thermal activation, and interesting properties can be achieved, such as catalytic activation of TiO_2 (anatase crystals), providing aluminum γ-oxide with acidity, etc.

For mesostructural films of these composites critical nucleation and collapse temperatures are found (for example, for anatase it is 350 °C and 700 °C, respectively) [36].

Oxopolymers synthesized by sol-gel synthesis method have superfine porous net with pore size 1–10 nm like zeolites have, sometimes they are called nanoperiodical (mesostructured) materials, their specific surface changes from 130 to 1,260 $m^2 \cdot g^{-1}$ depending on synthesis conditions. A special interest (also for catalysis) have titanium containing silica mesoporous molecular sieves like hexagonal MCM-41, Ti-MCM-41 and cubic MCM-48 materials (graded by Mobil corporation). Significant size of their pores (2–3 nm and more) gives great opportunities for modification of inner surface for control of hydrophobic-hydrophilic and acid properties, design of catalytically active centers. Drying regime, during which volatile components are removed, determines texture of the product: during long drying in air coarse xerogels can form due to coagulation of gel particles in air. If drying conditions prevent action of capillary forces, highly disperse aerogels form, xerogels with a structure of wet gel, whose pores are filled with air instead of removed liquid phase (in recent years CO_2 in supercritical state is often used for it). Formation of structure and texture of the product completes at the stage of heat treatment. Its properties are also influenced by atmosphere of pyrolysis: inert, vacuum or oxidizing (air) [[37–43][3]].

Annealing of material is also accompanied by many physical-chemical processes: destruction of organic fragments, restructuring of inorganic polymer, its crystallization, and sintering.

---

[3] Sol-gel process with usage of highly porous silica-gel in polar solvent can also be applied for microcapsulation of different large molecules such as pigments, photochromic fluorescent substances, scintillators, porphyrins, phthalocyanines, ferments, proteins, etc. (see Chap. 7) enclosed into sol-gel matrix obtained by combined hydrolysis of $Si(OEt)_4 - Zr(OBu)_4$, [37]. Methods of their preparation and structure are similar to metal polymer nanocomposites. In particular, MMA, in which perillene organic dye can be solved, is introduced in xerogel, and after thermal or UV polymerization co-impregnated composition consisting of SiO_2-PMMA-dye is obtained [38, 39]. Another variant is topochemical nanometically-precise imprinting of a dye in radial positions in silica [40, 41]. These materials are often called organically doped or organically modified sol-gel materials [42, 43].

General scheme of hydrolysis and condensation of tantalum alkoxides is similar to that of silicon, titanium, and vanadium oxides and can be reduced to the following [44–46]:

$$\mathrm{Ta\,(OR)_5 + xH_2O \rightarrow Ta\,(OR)_{5-x}(OH)_x + \ xROH} \tag{4.12}$$

$$\begin{aligned} &\mathrm{(OR)_{5-x}(OH)_{x-1}Ta - OH + \ XO - Ta(OH)_x(OR)_{4-x} \rightarrow} \\ &\quad \mathrm{(OR)_{5-x}(OH)_{x-1}Ta - O - Ta\,(OR)_{4-x}(OH)_x + \ XOH} \\ &\quad \text{где } \mathrm{R = C_2H_5; X = H, R} \end{aligned} \tag{4.13}$$

As a result of hydrolysis of $Ta(OR)_5$ unstable intermediate hydroxyalkoxides $Ta(OH)_x(OR)_{5-x}$ form, whose polycondensation causes formation of a net by oxolation or alkoxolation. Oleic acid forms micellar structure, which limits further growth of Ta_2O_5 nanoparticles. It acts as surfactant and chelating element, transforming Ta_2O_5 from hexagonal to orthorhombic form at increase in temperature of baking from 700 to 750 °C. Usually recrystallization temperature of orthorhombic blocked Ta_2O_5 is 800–1,000 °C. Consequently, a decrease in sizes of grains of orthorhombic Ta_2O_5 to nanometric scale decreases temperature of phase transition below 750 °C. This highly disperse Ta_2O_5 can be used as active filler for polymeric matrices, and addition of HEMA (see below) as solidifying agent brings to improvement of mechanical and dielectric properties, thermal stability.

Interesting direction of development of sol-gel synthesis $VO(OPr^i)_3$ – is production of multilayered (from 2 to 30) vanadium oxide nanotubes [47] with the following hydrothermal reaction causing formation of nanoparticles of diameter from 15 to 150 nm.

Sol-gel process is a convenient way for production of fine ceramic materials often called ceramers. Organic groups introduced in $R'_nSi(OR)_{4-n}$ based materials can fulfill two functions: modifying ormosils (ORMOSIL – ORganically MOdified SILicates) and forming a net ORMOCER – ORganically MOdified CERamics. Hybrid TiO_2/ormosil films are efficient for application in photonics [48].

Classification of sol-gel technologies should consider not only structure and sources for the formed nets, but also type of bonding (Table 4.2) [49]. The first group has materials in which organic and inorganic components are connected by strong chemical bonds, in the second one there are only weak physical bonds (captured by the nets). It is important that many of these products can be modified not only at the stage of formation, but at the stage of following chemical transformations of a composite by known ways.

Oxide-based ceramics is called white ceramics, while ceramic based on carbides (SiC, B_4C, TiC, etc.) or nitrides (Si_3N_4, AlN, BN, TiN, TiAlN, ZrN, GaN, InN), or carbonitrides (SiCN, BCN, TiCN) is called black.

Table 4.2 Classification of the main methods of sol-gel technology

Structure	The bond type	Precursor
(1) Colloid	*The particles* interconnected by Van-der-Waals or hydrogen bonds	Oxide or hydroxide sols
(2) Metal oxide polymer	*Inorganic polymers* interconnected by Van-der-Waals or hydrogen bonds	Hydrolysis and condensation of metal alkoxides
(3) Carbon chain polymer	*Organic polymers* interconnected by Van-der-Waals or hydrogen bonds	Covalent polymers largely used in organic chemistry
(4) Metal complexes	*Associates* weakly interconnected by Van-der-Waals or hydrogen bonds	The concentrated solutions of metal complexes (citrate method)
(5) Polymer complexes		
In-situ polymerable complexes	Organic polymers interconnected by coordinate, Van der Waals or hydrogen bonding	Polymerization between α-hydroxycarboxylic acids and polyhydroxyalcohols in the presence of metal complexes; e.g. Pechini method
Polymer- complex solution		Coordinating polymer, e.g. polyvinyl alcohol, and metallic salt solutions

4.2 Combination of Reactions of Polymerization and Sol-Gel Synthesis of Nanocomposites *in-situ*

Hybrid nanocomposites can be structured in different ways, which differ by prehistory of introduction of organic and inorganic phase in material. Thus, an organic component, which can be monomer or oligomer, is introduced as a precursor; linear polymers are introduces from a solution, melt, emulsion; polymer nets are physically or chemically sewed. In turn, mineral ingredient also can be included in composition of hybrid material in different ways: as metal oxide monomer, in preparation of nanoparticles, in form of nanoporous structures (for example, aerogel).

Alternative method is sol-gel synthesis in presence of organic monomers with the consequent or competitive polymerization or solidification. One of advantages of this approach in sol-gel process is strict stabilization of unstable particles or their precursors due to bonding of the sol surface with special molecules, monomers [50–52]. If bifunctional molecules are used, which apart from hydrolyzed silane or M-OR groups also include double bond, for example, methacrylic acid (MAA) and zirconium alkoxides, then after controlled hydrolysis the respective precursors can be obtained, which include ZrO_2 particles with the sizes ~2 nm, which can co-polymerize due to double MAA bond, which also plays the role of surface modifier (Scheme 4.1).

Moreover, zirconium alkoxide has high reaction ability to hydrolysis, and in the direct process ZrO_2 aq is deposited, which cannot be used for preparation of homogeneous composite material. Binding with MAA dramatically decrease

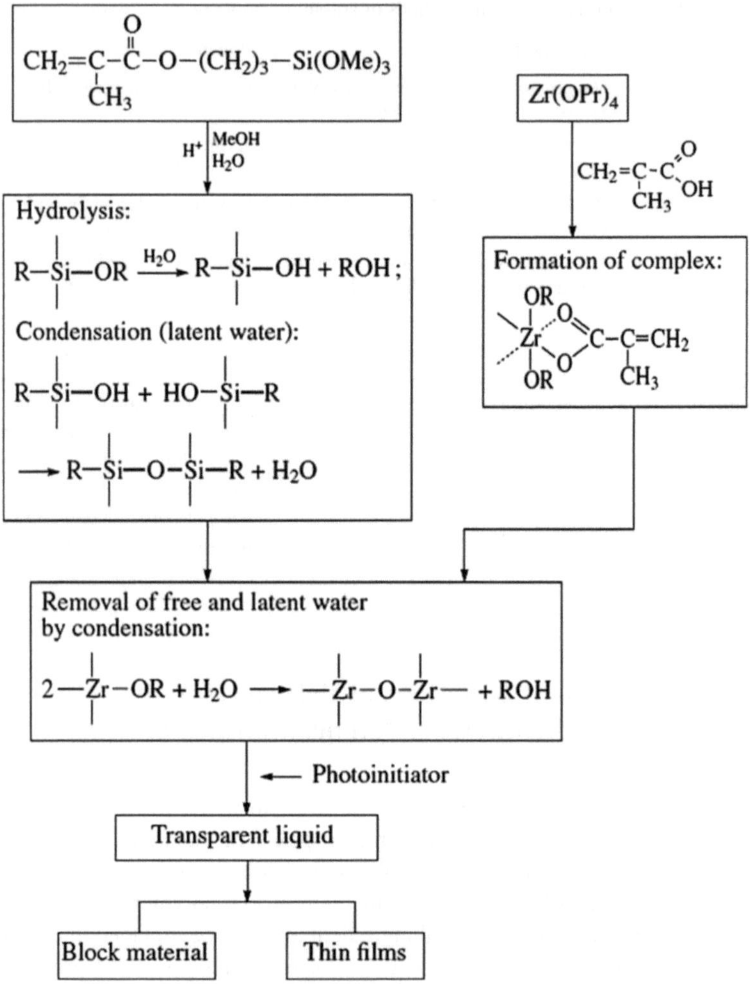

Scheme 4.1 Block-scheme of combination of sol-gel synthesis and polymerization of organic monomer

ability of $Zr(OR)_4$ to hydrolysis, therefore in presence of latent water in hydrolyzed and condensed silane well dispersed ZrO_2 nanoparticles form.

Practical realization of this combined reaction is difficult because sol-gel process is fast as compared to polymerization, which can last several hours; not always there is a common solvent, the growing polymer can become a template with different shapes, sizes, etc.

Nevertheless, the method based on polymerization transformations of "hybrid" monomers, compounds including precursor of inorganic component polymer forming group (single-source precursors) is a convenient way for obtaining hybrid nanocomposites. These compounds are called metal containing (element containing) monomers (a situation related to their synthesis and polymerization transformations is considered in the study [53]). Synthesis of titanium(trialkoxide) methacrylate monomers $CH_2{=}C(CH_3)COO\text{-}Ti(OR)_3$, in which R = Bu, Pr^i, *t*-Bu, *t*-amyl or 2-ethyl-hexyl can be displayed as an example.

Titanium alkoxides (as other metal alkoxides) are prone to association based on nucleophilic connection of OR group with negative charge to positively charged metal atom. A degree of this association is defined by reaction conditions (especially by a solvent origin and by temperature), and by alkyl groups. NMR ^1H, ^{13}C and IR spectroscopy were used to find existence of different structures in which methacrylate group forms a bridge bond:

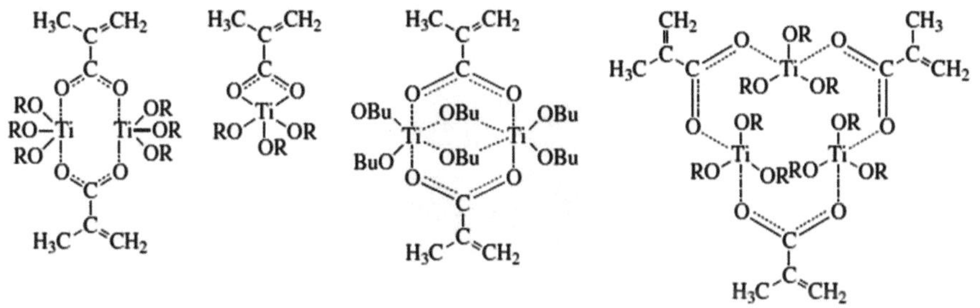

In principle, in order to obtain ormosil type materials vinyl derivatives can be used, for example, $CH_2{=}CHSi(OEt)_3$ (as is known, R–Si are non-hydrolyzed groups) (see, for example [54]), including those for functional mesoporous silica MCM-41. Quite often formation of inorganic polymer in these systems is preceded by cluster formation. Thus, for these purposes metal clusters of $[(RO)_nM]_xY$ type are used, where Y is polymer-forming organic group $x \geq 2$ [55]. Methacrylate substituted tetra-nuclear titanium, zircon, and tantalum oxide clusters of $Ti_4O_2(OPr^i)_6(OMc)_6$, $Zr_4O_2(OMc)_{12}$, $Ta_4O_4(OEt)_8(OMc)_4$ types are synthesized. Mechanism of formation of these structures is complicated and it is reduced to substitution of one or more alkoxide ligands by methacrylate groups, and released alcohol reacts with excess of acid, forming ether and latent water. The latter by hydrolyzing of non-reacted alkoxide radicals brings to formation of oxide and hydroxide groups in a cluster. Taking into account the fact that these processes are relatively slow, there is a possibility of strict control over growth of carboxylate substituted oxometallate clusters. Their nuclear ability and shape are controlled by the ratio of initial components and nature of OR groups in alkoxide. This was the way to produce cluster monomers of other metals, for example, $Hf_4O_2(OMc)_{12}$, $Nb_4O_4(OPr^i)_8(OMc)_4$, and also clusters of higher nuclear ability (Ti_6, Zr_6, Ti_9), including a series of different metal titanium/zirconium oxide clusters. Structure of these formations is studied in detail. In Fig. 4.2, for example, the Ti_4 cluster consists of 4 octahedral links more condensed than Ta_4, because two central octahedrons are bound to this face via μ_3-oxygen [56]. Functions of six methacrylate groups are reduced to balance of charges and coordination places of metal atoms.

Structure of Zr_4 is similar to titanium, except that their central atoms are seven-coordinated, whereas the exterior ones are eight-coordinated [57]. Therefore, a degree of substitution by bidentante carboxyl groups is higher than in Ti_4.

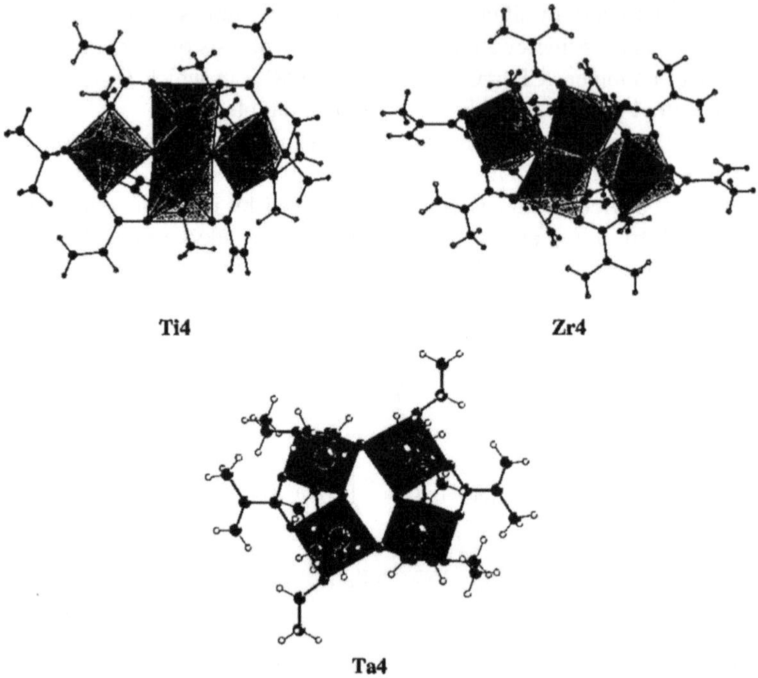

Fig. 4.2 Structure of polynuclear methacrylate complexes

Molecular structure of Ta_4 is a central symmetric cycle consisting of four octahedrons connected by angles. Methacrylate groups form square planar surrounding normal to cluster frame. Sol-gel synthesis from hydroxypropyl cellulose and Ti (O-*i*-Pr)$_4$, modified by methacrylic acid was used to obtain the respective nanocomposites. Raman spectroscopy has found complexes of octahedral coordination of titanium atom in them [TiO_6] surrounded by HPC, in spite of chelate bond of [TiO_6] with MMA [58]. During calcinations efficient control over sizes of TiO_2 inorganic phase is done. This is interesting approach to creation of hybrid materials based on assembling of strictly organized nanostructural blocks, which are preliminary formed calibrated objects preserving their integrity in a final product.

Numerous silicon-based oxo clusters of polynuclear structure and titanium (like $Ti_7O_4(OEt)_{20}$, $Ti_{16}O_{16}(OEt)_{32}$, $Ti_6O_4(OEt)_{14}(OOCPh)_2$, etc.) [57, 59–61], zirconium [62], may also be potential building blocks of inorganic phase of hybrid composites. Usually they are obtained in solution by controlled under-stoichiometric (i.e. at less than one mol-equivalent of OH^- groups per one mol equivalent of alkoxy groups) hydrolysis of metal alkoxides or their complexes M $(OR)_{n-x}(LZ)_x$ (where LZ is organic ligand). Stability of clusters increases with their nuclear ability and in presence of organic ligands. In the system of (tetra-*n*-propyl) or (tetra-isopropyl) orthotitanates and methacrylic acid, depending on molar ratios of the components, temperature of reaction, and time of holding of the reaction products, crystal or amorphous Ti_9O_8(O-*н*-Pr)$_4$ (MMA)$_{16}$ oxo titanium alkoxide form, which contains nine $[TiO_6]$ ring-shaped octahedral links with six μ_2- and two μ_3-oxide bridges, and yellow crystal compound, oligomer oxo titanium alkoxide carboxylate with the formula $Ti_4O_2(O\text{–}iPr)_6(MAA)_6$ [63]. Synthesis of metal and metal oxide nanoparticles in silica matrices with usage of metal coordinated organo-functional trialkoxy silanes is considered in detail in recent review [64].

Sources including hybrid monomer, precursor of $(RO)_3Si\text{–}X\text{–}M$ type, in which organic group X is connected to $Si(OR)_3$-group and metal-containing unit (metal-ion or metal alkoxide group), have advantages in control over dispersion of particles in producing MOx/SiO_2 or M/SiO_2 nanocomposites.

The controlled post-synthetic treatment of gel causes formation of metal or metal oxide nanoparticles as a result of degradation of organic groups; this method is convenient for powder formation, for thin films or coatings preparation.

In recent years not only unsaturated carbon acids but also unsaturated alcohols $Ti(OR)_3(OR')$, $VO(OR)_{3-n}(OR')_n$ are used for production of sol-gel products by polymerization method [65, 66], which are produced by reaction:

$$M(OR)_n + mR'OH \rightarrow M(OR)_{n-m}\left(OR'\right)_m + R'OH \tag{4.14}$$

where OR' is residual of the unsaturated alcohol ($R = Pr^i$, $R' = CH_2C{\equiv}CH$, $CH_2CH{=}CH_2$, $(CH_3)_2CC{\equiv}CCH{=}CH_2$ and $(CH_2)_2OC(O)C(CH_3){=}CH_2$) [67].

Under condition of removal of alcohol released in the reaction, for example, in form of azeotropy with a solvent, a range of polymerized alkoxides of Ti^{4+} and V^{5+} is received. Iso-evgenol, 2-metoxi-4-propenylphenol were used to synthesize potential monomers, the respective titanium alkoxides and more complicated products like $\{Zr(OPr^i)_2[(OC_6H_3)(OMe_2)(CH_2CH{=}CH_2)_4](\mu\text{-}OPr^i)\}_2$, their crystallographic structure is studied. Hybrid precursor is synthesized also by reaction of metal amides, for example, $Ti(NMe_2)_4$ with hydroquinone. Alcoholysis of $Ti(OPr^i)_4$ by hydroquinone brings to formation of covalent 3D Ti^{4+}- quinone networks, the same relates to $Zr(NMe_2)_4$ with 2,6- dimethylphenol or with 2-methoxy-4-propylenephenol and to condensation of yttrium-iso-propoxide with diols (Scheme 4.2).

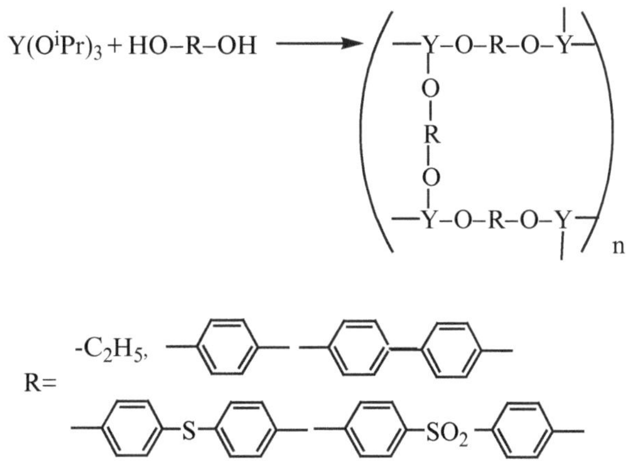

Scheme 4.2 Reaction of polycondensation alcoholysis

Polymer silanes are easily synthesized by condensation of polymers with end carboxyl groups with 3-triethoxysilyl propylamine and radical polymerization of 3-(trimethoxysilyl)propyl thiol as chain transfer agent (Scheme 4.3).

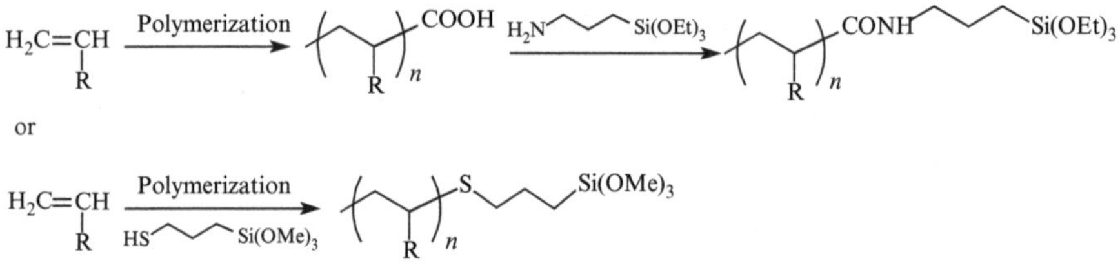

Scheme 4.3 A scheme for synthesis of polymer silanes

A special interest for functioning of alkoxy derivatives of different metals or related nanoparticles has 2-hydroxyethyl methacrylate (HEMA), in particular, due to its OH-groups, HEMA is a good co-solvent for TEOS and water, and viscosity of the composition is quite low to suit good blending.

In this sol-gel process two main reactions proceed: an ordinary one (hydrolysis and condensation of TEOS, during which –SiOSiO– net forms) and condensation of hydroxyl groups of silanol with 2- hydro methyl methacrylate in presence of acid [68]. Depending on reaction conditions (ratio between organic and inorganic reagents, pH of a medium, origin and concentration of a solvent), size of formed nanoparticles changes significantly, and the smallest achievable size is 20 nm. As content of inorganic component and solvent increases, pH decreases, thermal stability of the synthesized composites improves significantly. The highest T_g and T_d values are 110 and 313 °C, respectively, which is by 50° and 70° higher than for pure poly (HEMA), thermal expansion coefficient is far lower than for pure poly(HEMA).

Free radical co-polymerization of (methacrylate)titanium-tri-alkoxides with MMA is conducted at the temperatures 40–60 °C in toluene under action of AIBN [69], total concentration of monomers is 0.5–2 mol·L^{-1}. In the case of polynuclear clusters 3D structures form (Scheme 4.4):

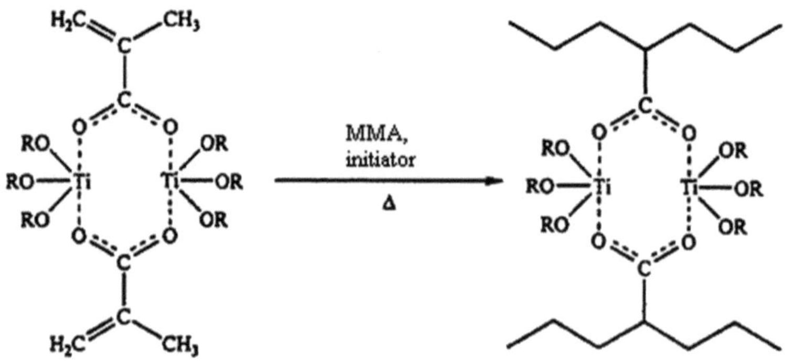

Scheme 4.4 Formation of three-dimensional structures during polymerization of titanium alkoxy derivatives

Fig. 4.3 A scheme of interfacial polarization on the surface of Ta_2O_5/acrylic chain

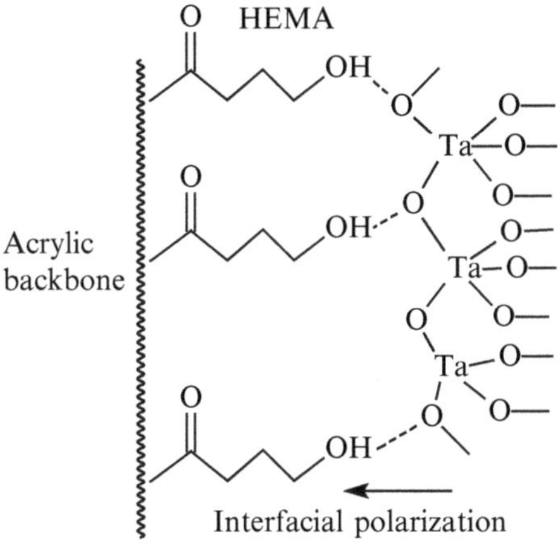

The formed active inorganic surface of nanometer sizes provides great opportunities for control of polymerization process. Thus, strictly uniform hybrid nanocomposites can be obtained by integration of nanoparticles synthesized by controlled/living radical polymerization. In particular, on the surface of SiO_2 nanoparticles (75 nm in diameter) macrocatalyst is immobilized and styrene is polymerized in presence of CuCl/4,4′-di(5-nonyl)-2,2′-dipyridyl by atom transfer radical polymerization (ATRP) [70, 71]. Properties of thus synthesized nanocomposite are far better than those obtained by traditional grafting polymerization of respective monomers. The analogous method (binding of CuX/pyridyl methane imine on SiO_2 particles) is used for ATRP methylmetacrylate [72].

Nanocomposite obtained by photocuring of acrylic monomers (MMA, HEMA, trimethylolpropane triacrylate) in presence of Ta_2O_5 nanoparticles was taken for example to show strong interphase interaction between organic-inorganic components (Fig. 4.3) [73].

Sol-gel condensation of bimethacrylate silane monomer obtained by reaction of attachment of aminopropyltrimethoxysilane to ethylene glycol acrylate methacrylate in triethylene glycol dimethacrylate (TGDMA) causes formation of methacrylate-functioned nanoparticles suspended in TGDMA as in a reactive solvent (Fig. 4.4) [74]. Oppositely to ordinary acrylic dispersions of nanoparticles, in which strong interparticle interactions determine their thixotropic properties, efficient steric stabilization of nanoparticles presents and viscosity of the formed composites does not depend on a shift rate. The best results were found at a filler concentration 1–4 wt% in TGDMA. These nanocomposite dispersions displaying improved mechanical properties and low shrinking deformation are of interest for dentistry and technologies with reactive composite particles.

It also should be noted that for producing nanocomposites by polymerization, hybrid macromers may be used (see, for example [75, 76]). Macromers based on polyhedral oligosesquioxane (POSS) containing almost spherical inorganic core (Si_8O_{12}) with diameter about 1.4 nm are most often used for these purposes (Fig. 4.5).

Methacryloyl-functioned POSS (POSS-MA) is a new class of monomers, which are actively studied for producing hybrid materials. They are easily co-polymerized

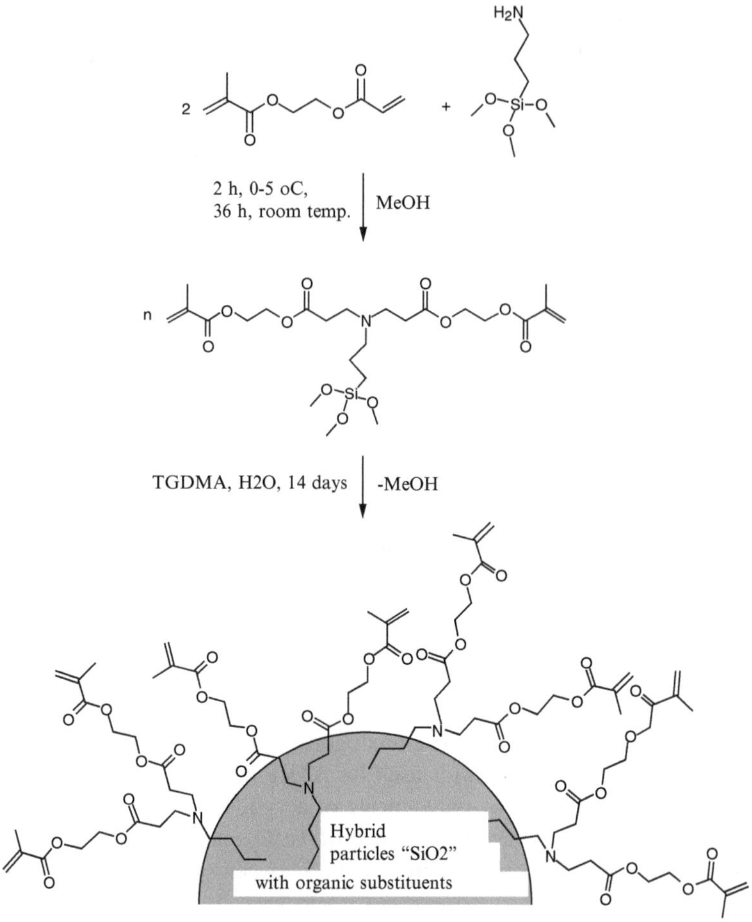

Fig. 4.4 A scheme of synthesis of hybrid organic-inorganic nanocomposites according to the reaction of condensation of alkoxysilanes

with other methacrylate monomers, such as 1.6-hexamethylenedimethacrylate, triethylene glycol dimethacrylate, HEMA, etc. (Fig. 4.6). These products are promising as potential alternative for usage in dental prosthetics, because the materials based on multi-methacrylates have low shrinking deformation[4] [77]. Really, including of just 5 wt% of POSS-MA almost restores shrinking of standard materials.

One of interesting examples of formation of hybrid nanocomposites is demonstrated by combination of radical polymerization reaction of N, N-dimethylacrylamide (DMAAm) and acid catalysis of hydrolyze-condensation of

[4] Significant decrease in volume (shrinkage) during polymerization of methylmethacrylate monomers, which causes stress in matrix of dental tissue (material) and low conversion of double bonds are two factors, which limit clinical application of dental materials based on methylmethacrylate.

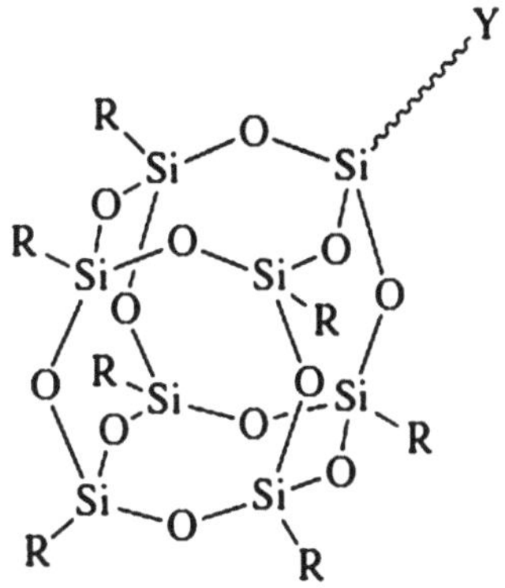

Fig. 4.5 Structure of POSS macromers

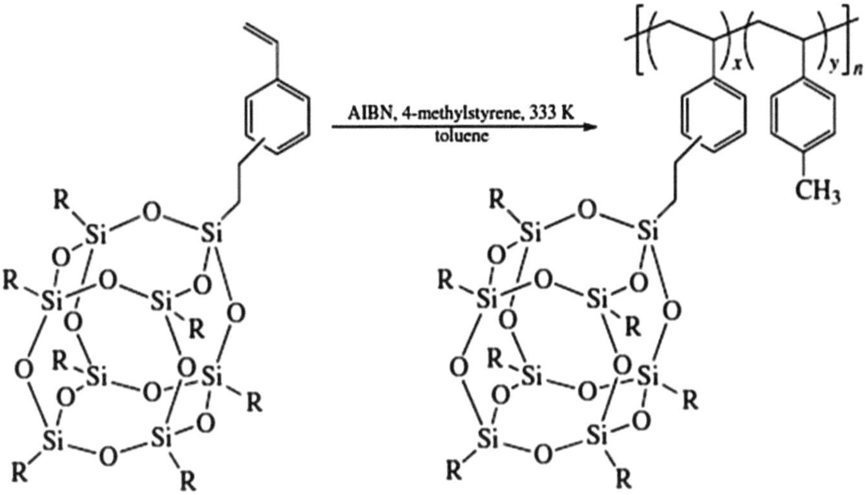

Fig. 4.6 Principal scheme of the preparation of new class of POSS-based hybrid materials

TMOS (Fig. 4.7) [78]. Polymer obtained in the mixture of DMAAm/TMOS = 1:2 and concentration of catalyst 1 % had average molecular mass 76,000. The composite had high degree of homogenization and was suspended in aggregates with domain sizes more than 30 nm. The value of specific surface in the nanocomposites after baking of hybrid polymers at 600 °C was dependent on production conditions, especially on amount of acid, and reached 365 $m^2 \cdot g^{-1}$. At that, two types of mutually penetrating polymer networks (PPNs) formed: one was formed by spontaneous polymerization of N,N′-dimethylacrylamide and N,N′-methylene-bis-acrylamide in methanol solution of TMOS. The second type of PPN is obtained by co-polymerization of styrene with divinylbenzene in presence of TMOS. It is assumed that siloxane network is so hard that it even does not change at this temperature. Nevertheless, pore size is expectable and comparable with domain sizes of a polymer in the initial hybrid, which confirms its dendrimer structure [77].

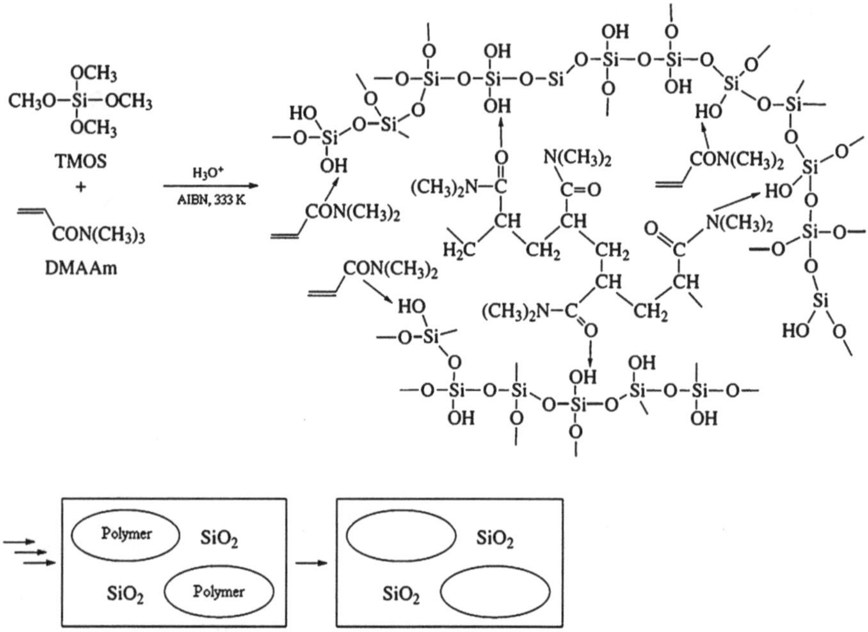

Fig. 4.7 Typical route of the preparation of hybrid nanocomposite

If the first group of monomers for producing nanocomposites by sol-gel synthesis includes compounds tending to opening of a multiple bond, the second one no less widely spread, includes polymerization transformations of compositions like $(RO)_3M$–X–A, in which X is a connecting bridge (organic spacer), and A-group is able to be subjected to opening of the cycle, polymerization (polycondensation) or coupling reaction [28]. Often net-like epoxy group is used for A, for example, in the case of 3-(glycidoxypropyl) it is binding for compatibility of TEOS with polyacids:

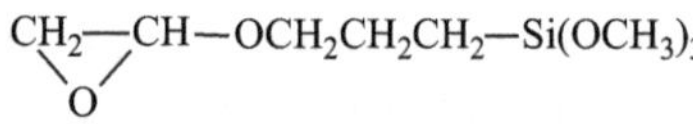

and more rarely with PEO[5]79] taking part in network formation. For this alkoxy derivatives of different metals, for example, aluminum, can be used (Fig. 4.8): epoxy silanes transform into cross-linked polymers, which is provided by catalytic effect of γ-particles of aluminum oxide.

For modification of sol-gel products and obtaining of hybrid materials isocyanate derivatives of alkoxides, for example, 3-(trimethoxysilyl)propylisocyanate and precursors with alkoxy groups substituted by others, for example, hexa (methoxymethyl)melamine are used.

[5] PEO is used for solvation of small cations such as lithium; in combination with oxopolymers it is used in sol-gel synthesis for producing of polyelectrolytes with ionic conductivity (ormalites). Classification of gel electrolytes and the recent data on application of metal-polymer nanocomposites in this field are described in review [79].

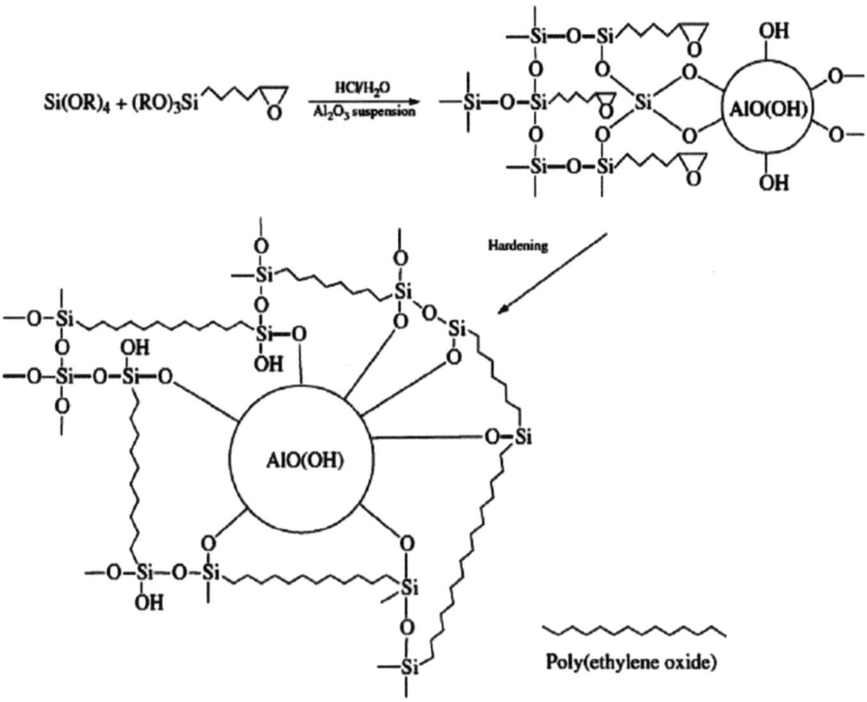

Fig. 4.8 Reactions and structural models of network systems based on Al^{3+}

One of extraordinary methods for production of hybrid materials of this type is thermal polymerization with opening of the cycle and formation of a new class of poly (ferrocenylsilanes), high-molecular silaferrocenophanes [80]. Material with interesting magnetic properties can be obtained by living anion polymerization with opening of the cycle under action of BuLi. Polysiloxane block (Pt^0 catalyst, toluene) is connected to this polymer with formation of block-co-polymer [81] (Scheme 4.5):

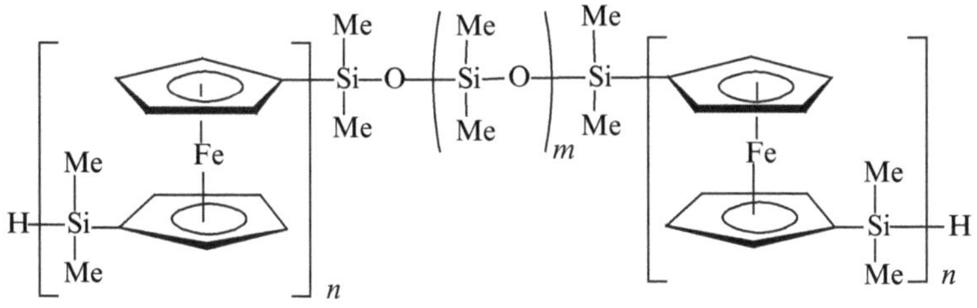

PFS-b-PDMS-b-PFS

Scheme 4.5 Structure of poly(ferrocenesilanes)

During solution in hexane this poly(ferrocenylsilanes)-block-poly (dimethylsilane), as many other block-polymers, is self-organized in cylinder,

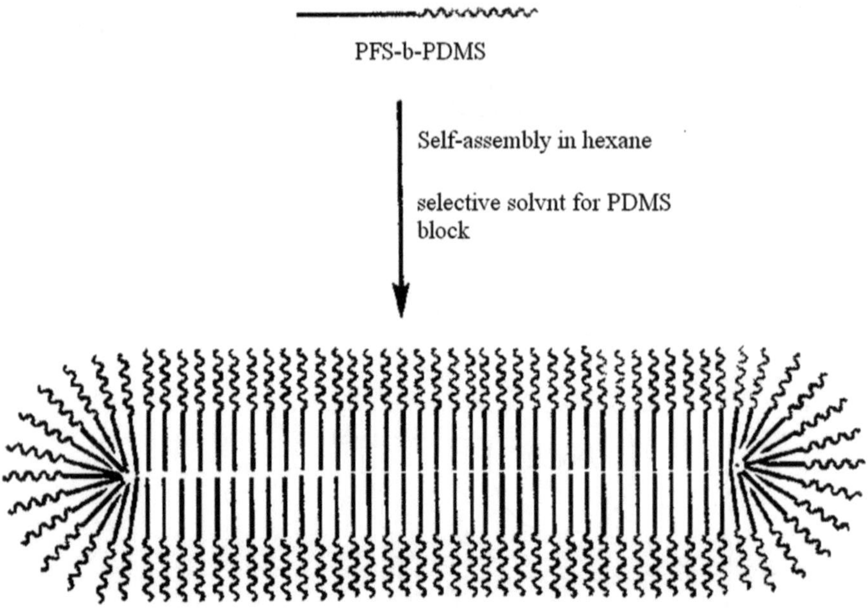

Fig. 4.9 Self-organization of PFS-b-PDMS block-copolymer in hexane with the formation of cylinder micelles

wormlike micelles, built as a core – poly(ferrocenylsilanes) blocks and shells – poly (dimethylsilane) crowns [82] (Fig. 4.9).

A large group of monomers for sol-gel synthesis are derivatives of trimethoxysilyl ferrocene. We do not consider here precursors of more complicated content, which could potentially be used for these purposes, for example, silylated chalcogenes such as $E(R)SiMe_3$ ($E = S$, Se, Te), etc. However, interesting are sol-gel products of multicomponent co-polymerization of CO_2 with allyl glycidyl ether (AGE) and oxide of cyclohexene (CHO) (poly(AGE-CHO-carbonate)) (Scheme 4.6) of the relevant molecular mass [83] catalyzed by the coordination system $Y(CF_3COO)_3$-$Zn(C_2H_5)_2$-pyrogallol in 1,3-dioxolane:

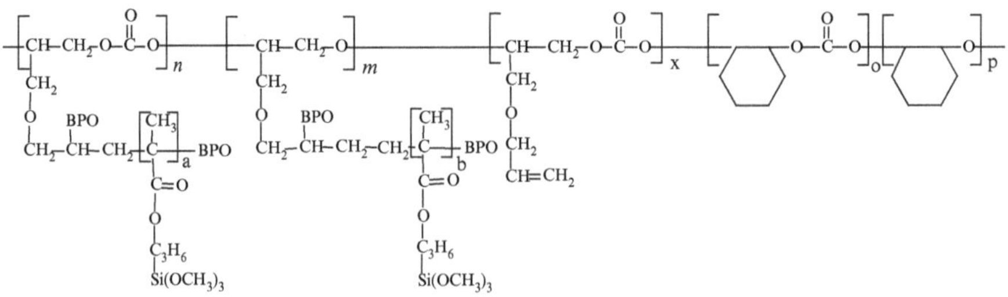

Scheme 4.6 Structure of the multicomponent copolymer of poly(allylglycidyl ether – cyclohexene oxide-carbonate) – precursor of SiO_2-nanocomposites obtained by sol-gel synthesis

Their radical co-polymerization (catalyst is benzoyl peroxide (BPO)) with 3-(trimethoxysilyl)propylmethacrylate results in formation of poly (AGE-CHO-carbonate)-SiO_2 nanocomposites, whose thermal and mechanical properties are better than those of the initial material.

Studies of co-polymerization of monomers one of which contains trialkoxysilyl groups have been quite detailed; however, hydrolytic instability of trialkoxysilyl groups implanted in polymers is a drawback of the obtained products. Anyway, their properties are useful. Let us consider a typical example. Methacryloxypropyltrimethoxysilane is hydrolyzed and condensed at adding of 0.5 N HCl, the product of interaction of equimolar amounts of $Zr(OR)_4$ and MAA are mixed with controlled amount of condensed silane in required proportions and calculated amount of H_2O is added. Optimal amount of photocatalyst and alcohol is taken as a solvent for viscosity control. Thin film of photosensitive material is applied for coating by immersing. The structure of this nanocomposite is shown in Fig. 4.10.

Even more convenient technique for synthesis of these nanocomposites is dispersion polymerization of the relative polymers, which is static co-polymerization of silicon containing monomers (Fig. 4.11). As a result beads of 100–500 nm form, which work as scaling templates at heating to 1,000 °C for removal of organic component; the final particles have size about 100 nm. Thus, the general scheme of production of nanoceramics via the stage of polymerization of the respective monomer can be reduced to the following (Fig. 4.12).

Here we have briefly analyzed the main approaches to formation of hybrid nanocomposites by sol gel method in cases when organic and inorganic polymers form interpenetration networks (IPN) *in situ* according to classification shown in Table 4.2. Alternative variant is usage of prepared and characterized polymers for these purposes.

4.3 Formation of Inorganic Precursor in Presence of Organic Polymers: Polymer Sol-Gel Synthesis

Thus obtained chemical materials are widely spread, most promising, and fast-developed organic-inorganic nanocomposites, and the basic studies in this field have been performed in the recent years, (for example [84–87]). At the stage of synthesis of nanocomposites by sol-gel method nanoparticles of metal oxides can be obtained and implanted into previously prepared organic polymers (see, for example, [88–91]). As the composites obtained in other ways, these nanocomposites are classified by the nature of interaction between organic and inorganic phases: nanocomposites with interphase covalent bond [92–94] and nanocomposites with weak (for example, hydrogen) bonds between the phases [95–97]. There are quite many examples in which these types of nanocomposites are described and characterized, we shall confine just to some of them [98–110]. Class 5 by classification [49] from Table 4.2 belongs to this type of materials

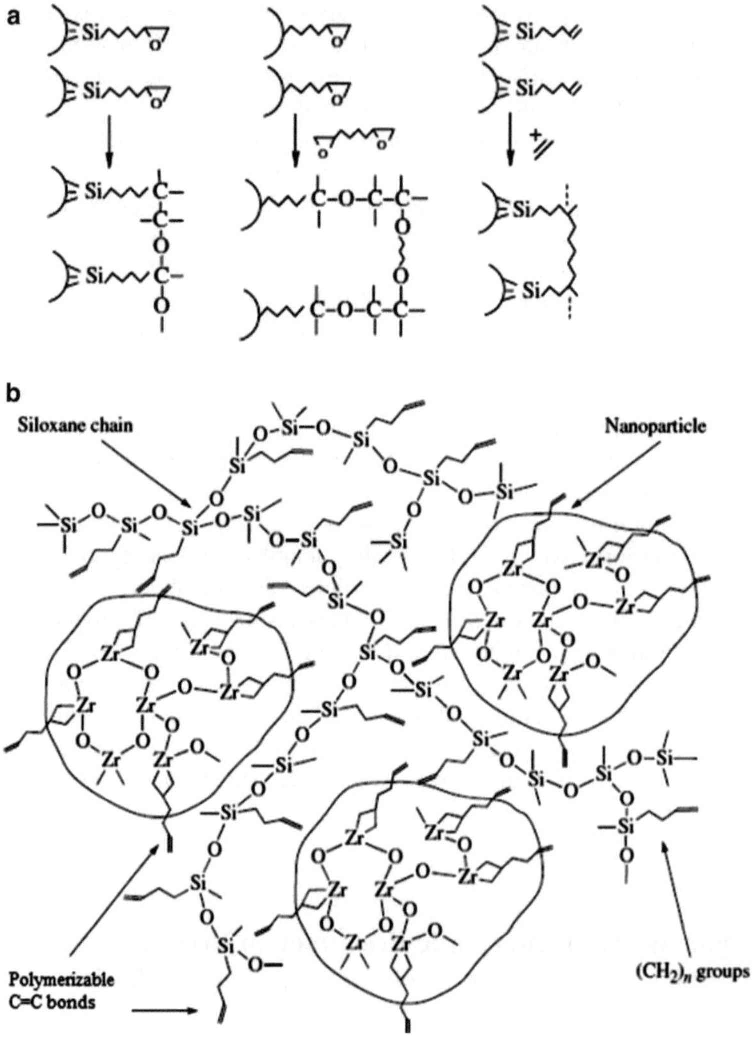

Fig. 4.10 General routes of organic-inorganic cross-linking by functionalized silanes (**a**) and model of the ZrO_2 – methacrylic acid – methacryloxysilane system (**b**)

They include hard metal oxide polymer and organic polymer capable of binding metal ions, thus forming metal-polymer complex. Presently many variants of sol-gel process with organic ingredients employing surface modification by polymers, including absorption, different coatings of grafting with colloid particles, etc. are used. In sol-gel synthesis most often used are the polymer/TEOS systems based on various macromolecules: poly(methyl methacrylate) (PMMA), poly(n-butyl acrylate), polyvinyl alcohol (PVA), PEO, polyvinyl acetate (PVA), PVP, poly(N, N-dimethyl acrylamide), silicon containing polymers, poly(amide-ethylene oxide), poly(ethylene oxide-co-epichlorohydrin), Nafion, polyoxazoline, polyphosphazene, etc. Assortment of hybrid composites based on TiO_2 is no less impressive.

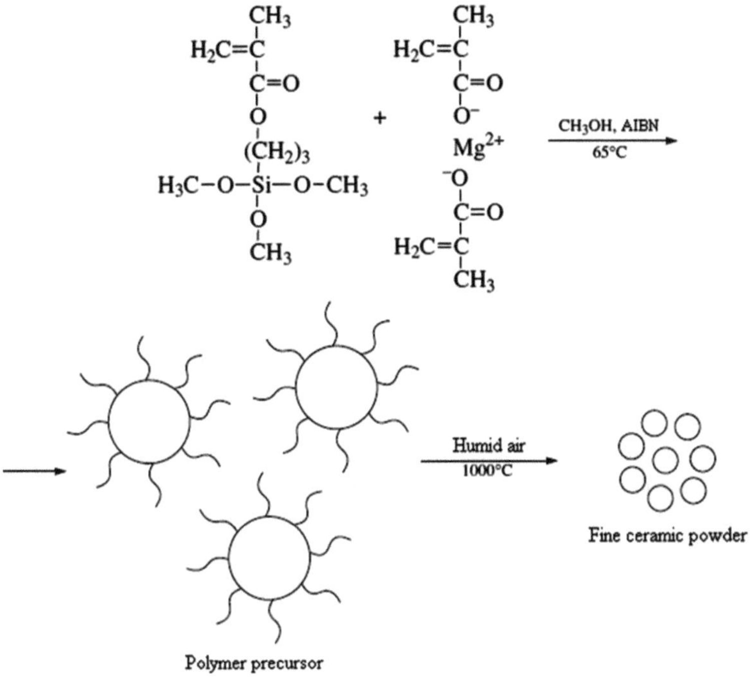

Fig. 4.11 Nanocomposites obtained by copolymerization of prepolymers

Fig. 4.12 Scheme for the preparation of nanoceramics

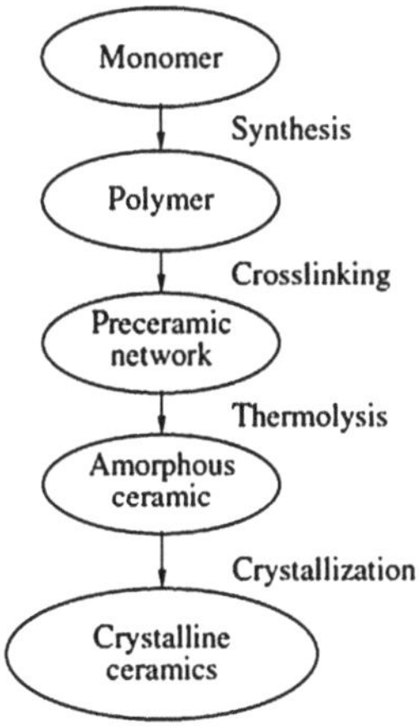

Microstructure of these nanocomposites is controlled by formation of weak Van-der-Waals bonds, hydrogen bonds or by hydrophilic-hydrophobic interactions. Alternative to this is implantation of a polymer (or its precursor) in oxygel during mixing of metal alkoxides or their impregnation (intercalation) into pores of the network of oxide xerogel. Many variants of coating of nanoparticles by polymer

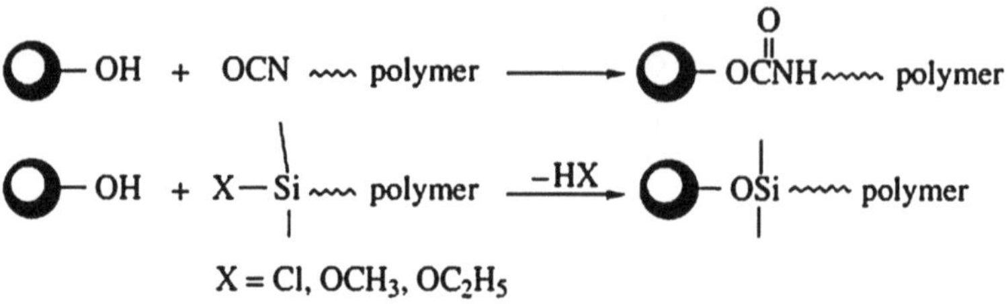

Fig. 4.13 The scheme of interactions between functional groups of ingredients

latex during sol-gel synthesis are known. Even more the reverse process, coating of latex particles by formed nanoparticles is known, for example, due to hydrolysis of $Ti(OBu)_4$ [111]. The required coating thickness of the coating is reached by concentration relations, and after calcinations (250–900 °C) hollow TiO_2 nanoparticles form, and as temperature increases, their morphology changes in the line: amorphous → rutile → anatase. Unique by their homogeneity mesoporous materials (having 2D and 3D hexagonal pores with diameter 2.7 and 3.1 nm) with specific surface 750 and 1,170 $m^2 \cdot g^{-1}$ are formed in the cases when polymer forming and inorganic fragments are in one molecule, for example, in 1, 2-di (trimethoxysilyl)ethane $(CH_3O)_3Si\text{-}CH_2\text{-}CH_2\text{-}Si(OCH_3)_3$.

Taking into account that the systems polymer matrix-inorganic component are, as a rule, thermodynamically incompatible, in conventional cases of production of hybrid nanocomposites surface of a filler particles are dressed by special agents before mixing. Hybrid materials produced by sol-gel synthesis are formed by the mechanism of interaction of surface hydroxyl (or specially implanted) groups of inorganic particles with chemically active polymer groups, including end groups [112]. The general scheme of these interactions (polymer as intermediate link) can be presented as in Fig. 4.13.

In these systems non-covalent interactions between side functional groups take place [113, 114]. Moreover, amphiphile polyorganosiloxanes (hydrophilic inner part and hydrophobic shell) can be themselves stabilizing agents (specific nanoreactors) for nanoparticles, for example, noble metals [115]. Polysilane shell cross-linked by PMMA is a polymer micelle stabilizing Au nanoparticles (average diameter 25 nm) and Pd (20 nm), the latter has effective catalytic activity in Heck reaction [116].

Hierarchic structure of pores in these structures is directly related to bimodal packaging formed in the mixture of two colloids with different sizes of particles, when coarser particles are structuring agents for finer particles via using of latex polystyrene particles as template for colloid silica gel [117–119]. As SiO_2 colloid with particle sizes from 5 to 25 nm is mixed with PS latex (particle sizes from 42 to 178 nm), critical ratio of diameters of SiO_2/PS latex is found, at which spontaneous packaging of fine colloid particles around closely packed PS latex particles is observed. Maximum diameter d of SiO_2 particles which can fill space between latex particles with diameter D is $d = (\sqrt{3}/3 - 1)D$. The following technological scheme of treatment consists of several heating zones: 1st stage is drying of liquid

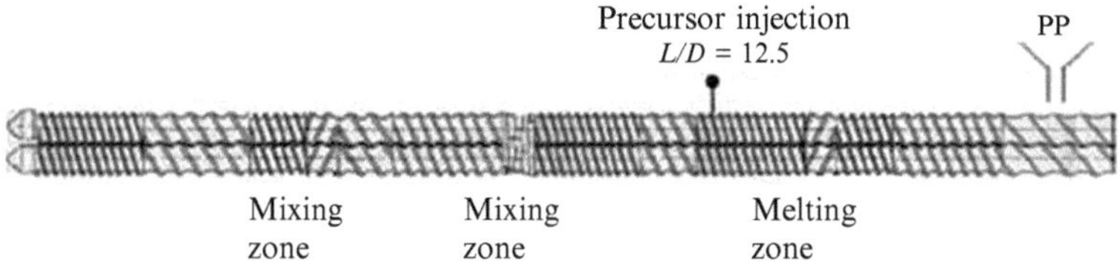

Fig. 4.14 Profile of the screw and the point of Ti(O*n*-C_4 H_9)$_4$ injection

drops of suspended colloid at 100–200 °C, 2nd stage is calcinations of PS latex at the temperature 450–500 °C, and 3rd stage is baking at 1,500 °C (ideal case is 900 °C) for formation of SiO_2 structure.

Very strong interactions in composites of this type are observed in the case of polyvinyl butyral (PVB), which has hardness and flexibility and high impact strength at low temperatures. Moreover, PVB has good adhesive properties to many materials, such as glass, metal, plastic, wood. Really, PVB/TiO_2 composites obtained in sol-gel process have good compatibility, strong interactions between components and high mechanic properties.

We shall consider the typical example of sol-gel synthesis, grafting of silica gel nanoparticles, using vinyltriethoxysilane as a precursor, to the surface of polypropylene (PP) whose pores work as a specific reactor [120]. Chemical bond between PP and SiO_2 phases is established by different physical chemical methods.

Inorganic nanoparticles improve mechanical properties of the composite, for example, cause two-time increase in impact strength. This points to the fact that the chemical bonding between particles of a filler and PP matrix brings to brittle fracture, and it confirms hypothesis that breakage of the bond is needed for improvement of impact strength of PP by inorganic fillers.

Interesting method is production of PP/TiO_2 nanocomposites in a special twin-screw extruder (Fig. 4.14) [32], which provides *in situ* synthesis in PP melt (the length/radius ratio is L/D = 34.5). Right-handed screw was a feeding zone, while left-handed was the PP melt zone before spraying of Ti(O*n*-C_4 H_9)$_4$ and the following two mixing zones, one of which was a stirring block with five-disc spreader and the left-handed was a die exit. Software controlled continuous injection of liquid titanium precursor was performed at a constant rate through the feeding mechanism and mixed with the PP melt.

Hybrid nanocomposites based on polyacrylamide (PAAm, molar mass 10,000) and TEOS are prepared using transparent water solution of PAAm by sol-gel synthesis [121]. Morphology of the hybrid material is particle-matrix structure with in- and intermolecular hydrogen bonds between the phases (Fig. 4.15). Depending on conditions of the synthesis (pH of medium), a number of charged Si-O particles on a polymer surface is different. Their presence, as well as presence of uncharged Si-OH particles, causes interaction with polymers due to formation of ionic and hydrogen bonds.

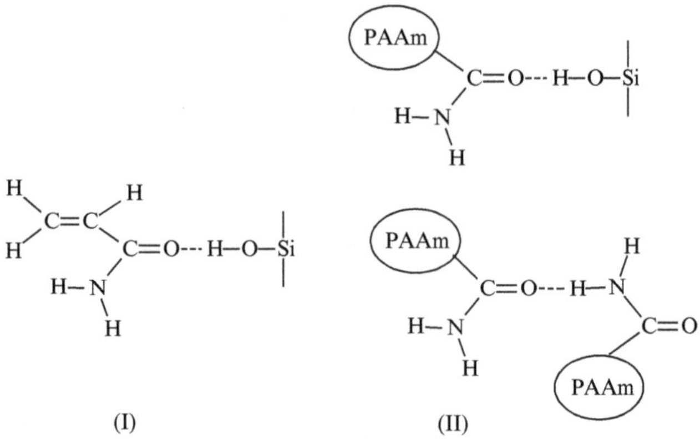

Fig. 4.15 A proposed model of the formation of hydrogen bonds between TEOS and AAm monomer (I) and PAAm polymer (II)

At low concentration of TEOS highly disperse SiO_2 particles form, as its concentration increases, roughness of the surface and aggregation of silica gel particles becomes stronger.

Spin coating is used to prepare flawless nanocomposite films with smooth surface. Clusters of SiO_2 formed after baking are more compacted, linked into silica gel microstructured net.

Relatively often for preparation of organic-inorganic composites polymer amines are used, and increase in degree of polymerization of polyamines causes an increase in the condensation rate of silicic acid [122–125]. The driving force of interaction between them is not formation of hydrogen bonds, but electrostatic interaction between protonated amino- and silanol groups. A possibility is not excluded of formation of microemulsion advantaging formation of non-aqua micromedia around these particles, which would bring to contribution of donor-acceptor interaction between them. Composites deposited through condensation of silicic acid in presence of polyamines include 30-nm nanoparticles.

It is interesting that polyamino-amide dendrimers inhibit the condensation process due to association and stabilization of oligo-silicates, which impedes further growth of particles [126]. However, because of impossibility of complete inhibiting, colloid particles deposit.

Polymers of 2-(dimethylamino)ethyl methacrylate with different degree of quaternization catalyze condensation of silicic acid at neutral pH [127], in the case of co-polymer wit acrylamide spherical particles about 50 nm with positive ζ-potential form [128].

Polyvinylimidazole (PVI) has a specific effect on condensation: it is able to stabilize formed SiO_2 particles in the case of high molecular fractions of PVI, and at the same time they bring to association of particles followed by their deposition at neutral pH and excess of silicic acid [129, 130]. Phosphorylated chitosan inhibits

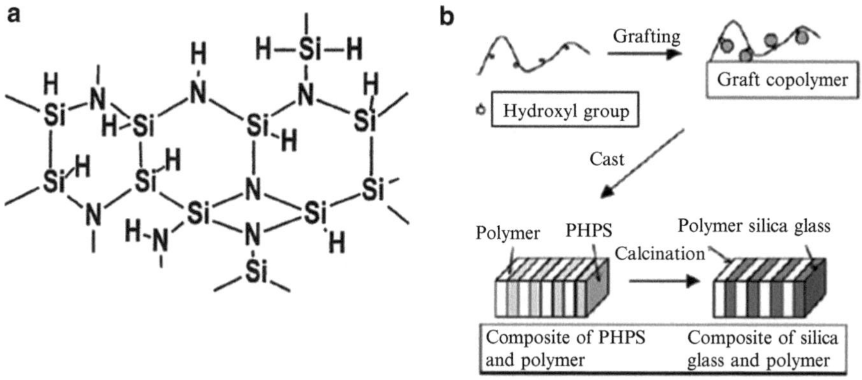

Fig. 4.16 Structure of perhydropolysilazane (PHPS) (**a**) and a grafted polymer on its base (**b**)

condensation of silicic acid and also TMOS [131, 132]. Transparent PS/SiO_2 nanocomposites are synthesized by split coating of mixture of solution of hydroxyl containing derivatives of PS (such as poly(styrene-co-4-vinylphenol), poly(styrene-co-hydroxymethylstyrene), etc.) and perhydropolysilazane (PHPS) (Fig. 4.16) with molar mass 700 with the following calcinations of the mixed film at 100 °C [133].

Silazane groups are very reaction active with respect to hydroxyl groups, which favors grafting of polysilazane to hydroxyl containing polymer and formation of transparent preceramic component in organic solvent. Organo/silicate or organo/silicate/cobalt nanocomposites were prepared by mixing of PHPS and/or cobalt acetate with PMMA, poly-2-vinylpiridine, poly-4-vinylpyridine [134], PS [133–135]. Thermal stability of the composites did not change during hybridization of PHPS due to formation of nanostructure. Leakage current decreased during hybridization, however, refraction index, dielectric constant remained unchanged.

Thickness of the silicon containing coating is tens nanometers, surface hardness of these composites increases dramatically (from 0.22 to 0.55 GPa) with increase in silicon fraction to 33.7 wt.%, morphology of the nanocomposite is microphase separated.

Therefore, condensation of silicon containing precursors in presence of polymers proceeds with formation of stable organic-silicon particles including oligosilicate core and polymer shell.

We shall consider formation of similar nanocomposites in the systems of polyimides (PI)-polyamides (PA), important engineering plastics having high thermo-oxidizing stability and perfect mechanic, adhesion, and electric properties, good resistance to organic solvents. A great number of PI is widely studied for application as suitable polymer matrices, adhesive materials, highly efficient coatings for microelectronics [136–142]. They are obtained by the scheme (Fig. 4.17) [143].

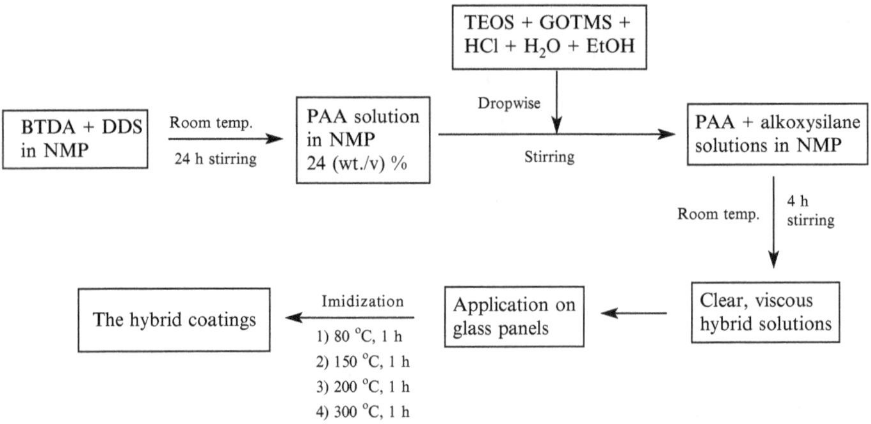

Fig. 4.17 A scheme of synthesis of the hybrid material on the base of polyamic acid (PAmA) – 3,3′,4,4′-benzophenone dianhydride tetracarboxylic acid (BTDA) and 3,3′ diaminodisulphonephenyl (DDS) – SiO_2 in N-methyl-2-pyrrolidone (NMP)

Thus, properties of PA-66 semi-crystal thermoplastic can be improved by silica gel nanoparticles (10–100 nm) [144–147]. However, preparation of nanoparticles with the size less than 10 nm counters some considerable difficulties. In this regard interest is drawn to polyoxometalates (POM) $[X^{n+} M_{12}O_{40}]^{(8-n)-}$, $[X^{n+}M_{12}O_{42}]^{(12-n)-}$ or $[X^{n+}M_6O_{24}]^{(12-n)-}$ (X = P(V), As(V), Si(IV), Ge(IV), Ti(IV), Ce(IV)), in which metal (M = W, Mo, V) is bound by oxygen atoms [148–150]. POM is interesting class of materials for electrochromic, photochromic, catalytic, energy-efficient, and biomedical applications, but their drawback is low workability [151–158]. In [142] it is reported about synthesis of nanocomposites based on PA-66 and trioxide tungsten (WO_3) clusters formed by sol-gel synthesis of 1–10-nm nanoparticles [159].[6] Combination of WO_3 (10 wt%) with PA-66 brings to intriguing results, for example, dynamic elastic modulus (accumulation modulus) at 258 °C increases 6 times [160].

Involving of water-solved polyimides in sol-gel synthesis makes it possible to significantly improve compatibility of organic-inorganic components of PI/SiO_2 nanocomposite [161–164]. Often small amounts of doping agents, such as 3-glycidyloxypropyl trimethoxysilane, are introduced in these systems to provide covalent bonding of nanoparticles with polymer chain [165, 166] (Scheme 4.7):

[6] PA-66 was solved in 99 % formic acid at room temperature, tungsten trioxide was solved in sodium hydroxide at pH > 10, which was slowly added to PA-66 at mixing at room temperature. Acid character of PA-66 provided formation of polyoxymetalate WO_2^{4-}. Since in acid medium the formed POM took negative charge, cluster particles were bound to protonized chains PA-66. PA-66/POM nanocomposite was deposited with water addition. Variation in content of the initial tungsten trioxide (0, 1, 3, 5, and 10 wt %) can be used to obtain nanocomposites with different properties.

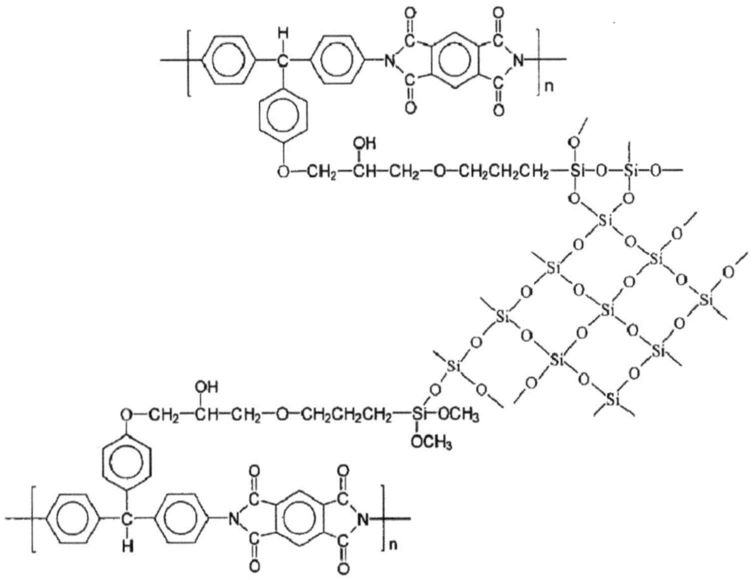

Scheme 4.7 Covalent binding of nanoparticles with a polymer chain by 3-glycidiloxytrimethoxy silane

This provides good transparency of polyimide nanocomposite films, and their improved thermal and mechanical properties (ultimate strength and tension elongation reach 91 MPa and 9.86 % at SiO_2 concentration 15 %).

Sol-gel method was used [167] to obtain nanocomposites based on bis-maleimide novolak resin/silsesquioxane (BMI-PN/$SiO_{3/2}$) by the scheme (Fig. 4.18).

Thermal stability (temperature of the beginning of their decomposition is above 550 °C) is their specific feature, while mechanical properties are improved by implanting of 10 or 5 wt% of SiO_2 nanoparticles. Variation in compatibility between the components is due to different types of diamines. At high content of inorganic phase mechanical and thermal properties deteriorate, which is caused by aggregation of silica gel particles in the hybrid. These polyamides are promising candidates for usage in microelectronics [168], other types of sol-gel products of similar composition are promising for production of membranes with high gas transporting properties [169], ceramics [170], nonflammable nanocomposites [171], materials with low dielectric permeability [172, 173] and with stable optical properties for non-linear optics, [174], etc. Also widely used are polyamide-silica gel hybrid coatings (0–20 wt% SiO_2), which are obtained, for example, from PI, TEOS, glycidyloxypropyl trimethoxysilane and polyamic acid (synthesized from 3, 3, 4 benzophenone tetracarboxylic dianhydride (BTDA) and 3,3′-Diaminodiphenyl sulfone)(DDS) [143]), combination of sol-gel technique with

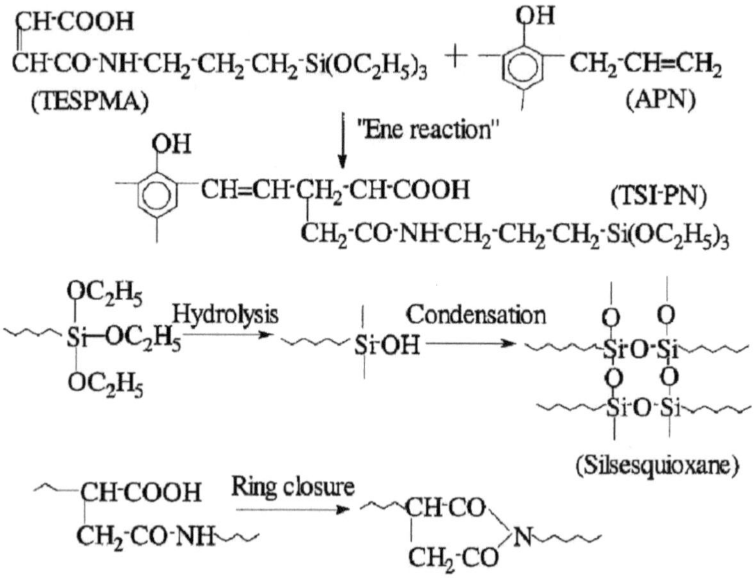

Fig. 4.18 The main reactions of the synthesis of novolak resin/silsesquioxane nanocomposite (TESPMA is N-triethoxysilylpropylmaleamic acid, APN is allylnovolak resin, TSI-PN is prepolymer with triethoxysilylpropylic groups)

thermal imidization. Hardness of these coatings increases with increase in SiO_2 content, tensile tests shows significant improvements, inorganic particles are distributed homogeneously in PI matrix.

The method of production of nanocomposites based on novolak resins modified by 4,4′-bismaleimide diphenylmethane in composition with alko-derivatives of titanium is interesting [175]. TiO_2 particles dispersed in novolak resin matrix have size less than 150 nm, however, implanting of titanium phase, while increases elasticity modulus at temperatures below 200 °C, does not improve T*g* of material and does not decrease its thermal stability because of incomplete removal of AcAc (Fig. 4.19) which is coordinated with titanium tetrabutoxide.

Polyimide materials, containing SiO_2 and TiO_2 nanoparticles forming 3D inorganic nets, have high mechanical hardness. Interesting variant for synthesis of such nanocomposites is using precursors for producing polyimide- polysilsesquioxane (PI-POSS) composites [176]. In particular, during condensation of 1,1-bis (4-aminophenyl)-1-phenyl-2,2,2- trifluoroethane and derivatives of pyromellitic anhydride (PA) with aminophenyltrimethoxysilane (with simultaneous imidization at 250–350 °C) films of organic-inorganic hybrids containing 32–70 % of homogeneously distributed SiO_2 with particle size 0.5–7 nm are formed (Scheme 4.8).

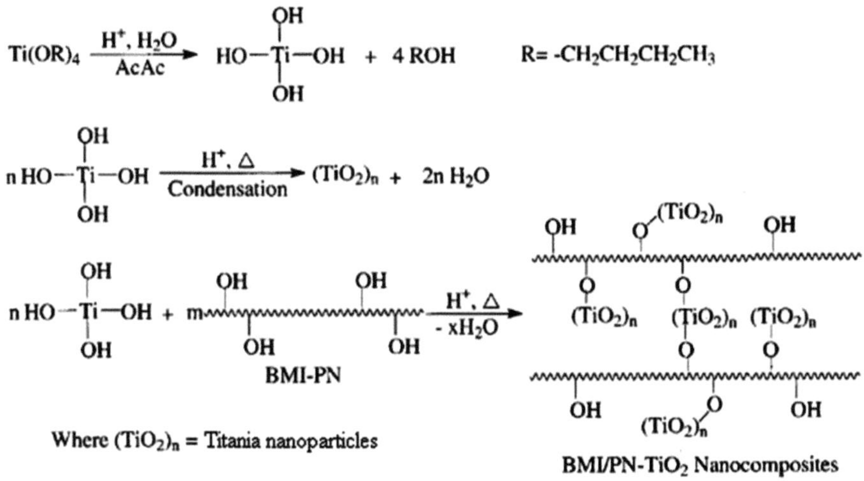

Fig. 4.19 A scheme of synthesis of nanocomposite on the base of novolak resin modified with bismaleimidediphenylmethane and TiO_2 nanoparticles

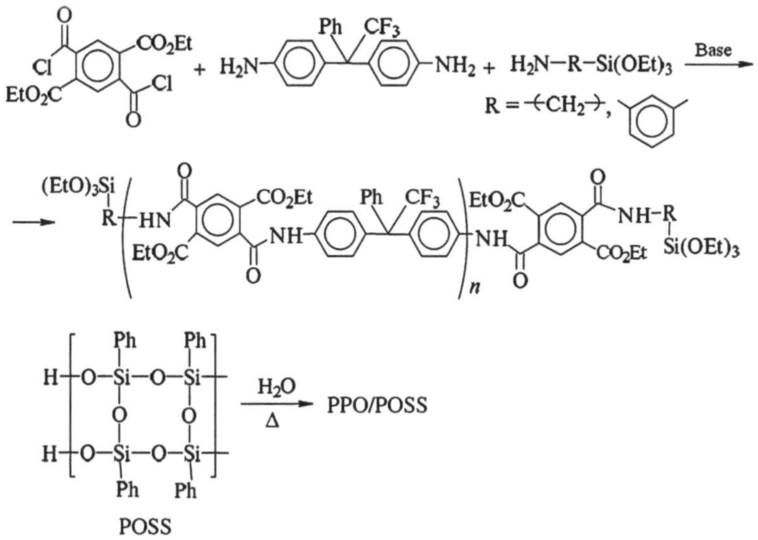

Scheme 4.8 Synthesis of polyimide-polysilsesquioxane (PPO/POSS) composites

Polyimide –SiO_2 nanocomposites are synthesized and characterized also in non-aqua sol-gel processes [177] (Fig. 4.20).

Nanoparticles are dispersed homogeneously in matrix and this is a way for development of reinforced composites. Morphology of a nanocomposite is controlled by sol-gel stages and by imidization.

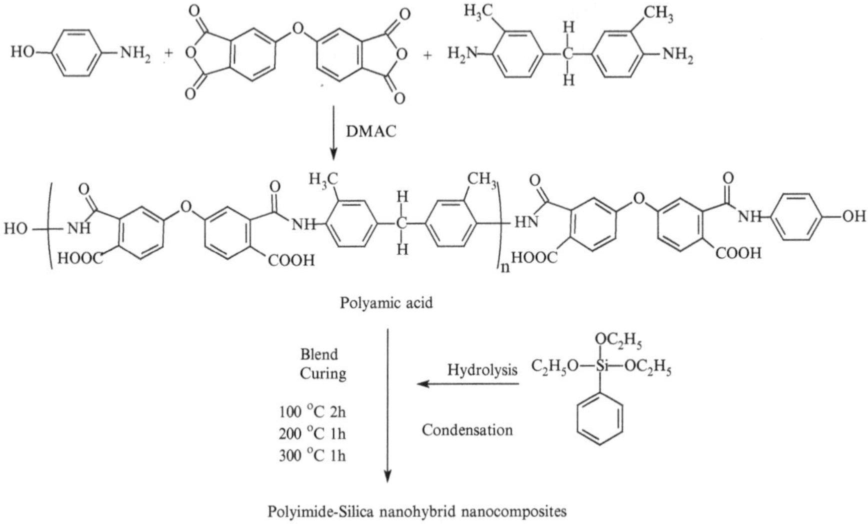

Fig. 4.20 Sol-gel synthesis of hybrid PI- SiO_2 nanocomposite in N,N-dimethylacetoamide

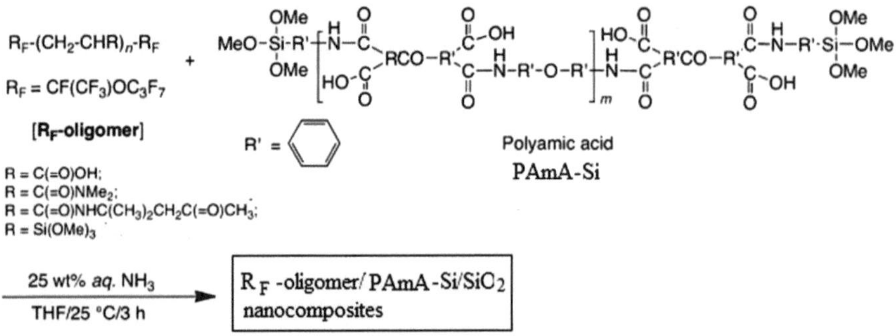

Fig. 4.21 Formation of fluorine oligomer nanocomposites

Fluorinated oligomers and polyamic acids with trimethoxysilyl groups under alkali conditions form PAmA/SiO_2 nanocomposites by sol-gel synthesis reaction [178]. These matrices well stabilize fine nanoparticles not only in aqua solutions, but also in conventional organic solvents. Imidization of these oligomers smoothly goes at 270 °C with formation of fluorinated oligomer/PAmA/SiO_2 nanocomposites by the scheme (Fig. 4.21).

Nanocomposites of this type are also obtained on the basis of polyimide analogues, polyoxazolines, among which are functioned triethoxysilane groups.

We shall briefly consider the main ways of production of hybrid nanocomposites based on two cross-linked polymers, whose organic and inorganic components form strong covalent or ionic chemical bonds. Most widespread are two approaches to synthesis of these nets. The first one is formation of secondary nets in preliminary formed primary nets, which should be functionalized in a special manner; more rarely functionalized inorganic macromers and oxopolymers are used. The second method is simultaneous formation of two different nets (including mutually penetrating) of molecular precursors having both organic and inorganic functionality and reacting by different mechanisms (polyaddition, polycondensation, hydrolysis-condensation, etc.).

At once we shall focus on complication of the problem and its conjugation with many difficulties; studies in this direction just start and, as far as we know, there is no clarity here, and a number of reliably implemented variants is scarce. Primary attempts were based on usage of polymer solutions, however, phases separated uncontrolled during gel formation, and heterogeneous material formed. We shall note that the similar method is used for preparation of thin interphase layer between oxidized Al, Si surfaces, porous Si via interaction with alkoxy zirconium complexes, and also between these surfaces and poly(ethylene-co-acrylic acid) through alkoxy zirconium [112]. Sol-gel materials, probably close to the required type, are obtained by solution of HO-PES-OH (PES is polyether sulfone) and TEOS or TMOS in DMF [179]: acid catalyzed hydrolysis-condensation of tetra alkoxysilanes causes formation of products by the scheme:

$$\equiv \mathrm{Si} - \mathrm{OR} + \mathrm{HO} - \mathrm{PES} - \mathrm{OH} \rightarrow \equiv \mathrm{Si} - \mathrm{O} - \mathrm{PES} - \mathrm{OH} + \mathrm{ROH} \quad (4.15)$$

Deep interaction between inorganic and organic chains is confirmed by different methods: cross-linking of polyether sulfone links by alkoxysilane bridges and formation of nanoparticles by condensation of alkoxysilane with these linking groups is proved (Fig. 4.22).

In the cases when polymer has low surface energy and is not porous, for example, polytetrafluoroethylene films (PTFE) activation of their surface is

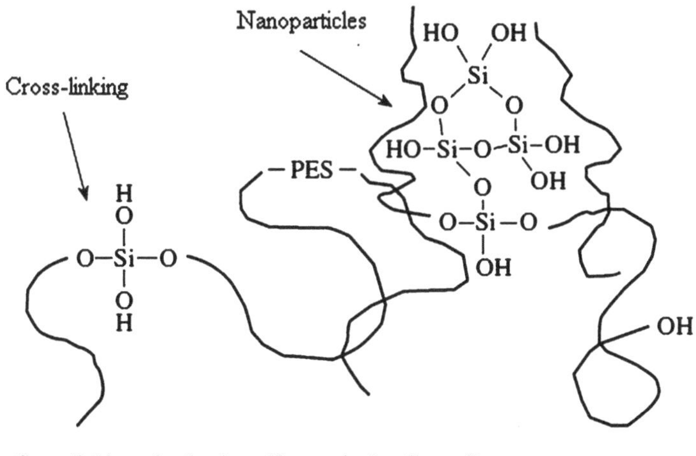

Fig. 4.22 Cross-linking of polyethersulfone units by alkoxysilane groups

provided by deposition of silicon dioxide formed *in situ* from $SiCl_4$ and water without chemical destruction or modification of a polymer block [180]. This technique increases significantly adhesion of silica to fluoride polymer due to $SiCl_4$ and water diffusion to surface layers of the polymer. Chemical interaction between organic and inorganic polymer components with formation of covalent bond was found during sol-gel process in the systems containing styrene polymer with maleic anhydride, TEOS in presence of a bonding agent, (3-aminopropyl)triethoxysilane (see, for example, [28]):particles of less than 20 nm formed (Scheme 4.9). Organic co-polymer worked as compatibilizer of organic-inorganic composite. $Ti(OBu)_4$ or allyl acetyl acetone (3-allyl-2,4-pentadione) can be used as a coupling agent. Organic-inorganic chemical hybrid filler for dental use [181] is produced by sol-gel reaction of poly[methyl methacrylate-co-3-(trimethoxysilyl)propyl methacrylate].

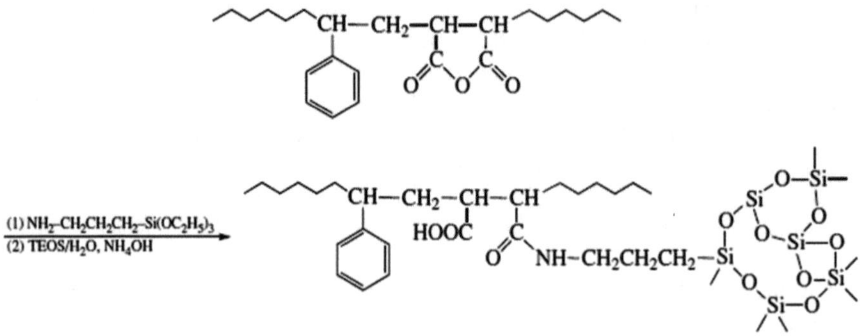

Scheme 4.9 Sol-gel synthesis of poly[methylmethacrylate-co-3-(triethoxysilyl)propylmethacrylate]

It is assumed that in this hybrid nanocomposite polymethacrylate chains are covalently bound and homogeneously distributed in silicon nets, on molecular level, without macroscopic separation of organic-inorganic phases. Oppositely to already mentioned standard dental materials based on 2,2-bis(*n*-2-hydroxy-3-methacryloxypropoxyphenyl)propane/triethylene glycol dimethacrylate filled with fine SiO_2, there is stronger interaction in the sol-gel produced nanocomposites. The surface of thus fabricated hybrid filler includes initially polymer component, whose structure is similar to the matrix. Most probably, one of the most important conditions for homogeneity of the formed hybrid, polystyrene (PS) – the sol-gel reaction product of phenyltrimethoxysilane is π-π-interaction between benzene rings of PS and modified silica gel. Probably, even more complicated is structural arrangement of composites based on poly(isoprene-block-ethylene oxide), in which phase-separated block-copolymers were mixed with alkoxide (3-glycidil oxypropyl)trimethoxysilane and aluminum butoxide in sol-gel process. Fixing of PEO macromolecules on the surface of metal oxides due to H-bonds of oxygen atoms with OH-groups of the surface conducts even through layers of coordination-coupled with these groups water [26]. As a rule, strong hydrogen bonds usually form between organic and inorganic phases. Sometimes for producing of hybrid gels such as SiO_2/PEO both components are modified, mineral precursor and polymer. Thus, for synthesis of efficient luminescent materials 3-isocyanotriethoxysilane is used with O,O′-di(2-aminopropyl)- polyethylene oxide, which while swelling in ethanol solution of terbium nitrate and 2,2-dipyridyl couples both lanthanide ion and ligand. The same way is used to obtain hybrid materials including not only luminescent but also redox-active or catalytic active centers [182], for example, oxygen-selective organic-inorganic hybrid membranes contain salcomine as oxygen carrier, etc.

In composites polyamidoimide (PAI)/TiO_2 prepared *in situ* in sol-gel process (area sizes of TiO_2 change from 5 to 50 nm at increasing of its composition from 3.7 to 17.9 wt %), formation of hydrogen bond between amido-groups of PAI and hydroxyl groups of inorganic oxide is found [17]. Composite materials obtained in sol-gel process, based on inorganic structures and polymers and having high thermal and mechanical characteristics, as compared to the initial ingredients, are also used for production of optic waveguides. They are formed by deposition of polymer – TiO_2 coating on a glass plate with the following heat treatment (30 min at 300 °C in N_2 atmosphere). Advantage of production of nanocomposites in polymer solution is a possibility to form transparent films due to avoiding formation of yellow colored titanium complexes. Thermal stability of these materials (containing 4 % TiO_2) is almost the same as that of the initial polymer: mass loss is 1 % in N_2 atmosphere at 461 and 463 °C, respectively. This points to the fact that impregnation of TiO_2 almost does not change thermal characteristics of host-polymer. Thus, in some cases just mixing of components, polyimide solution with sol-gel-precursor, which causes phase separation, is needed for formation of polymer-inorganic composite. Phase separation in these IPN is, as a rule confined to nanometric areas of different phases.

Optimal variant for synthesis of SiO_2 – polymer nanocomposites with minimal shrinkage is metathesis polymerization with cycle opening and free radical connection of cyclic alcohols [183]. Synchronic formation of mutually penetrating cross-links was due to competitive polymerization and hydrolysis with condensation of silicon alkoxide under action of nucleophilic catalyst NaF (Fig. 4.23).

In the PVP-TiO_2 composites the latter fast crystallizes at the stage of calcinations due to strong interaction with polymer. The same effect is caused by addition of organic gelling agents, organogelators, at the stage of sol-gel reaction. In the case of MAA modified silanes, as was already mentioned, these monomers are co-polymerized, silicon containing polymer is used as matrix material for nanoparticles.

Titanium and silicon alkoxides are coupling reagents for many natural polymers, such as polysaccharides, cellulose materials, derivatives of vegetable oils, etc. [184]. They contain highly active hydroxyl groups, which can form oxo polymers. Co-condensation of macromers containing trialkoxysilyl groups with known polymers such as functioned PS, polyoxazoline, PI, PEO, polyether ketones, PMMA, etc., including end functioning groups, is a convenient method for production of telechelic cross-links [185]. The same relates to poly tetramethylene oxide, whose end links are

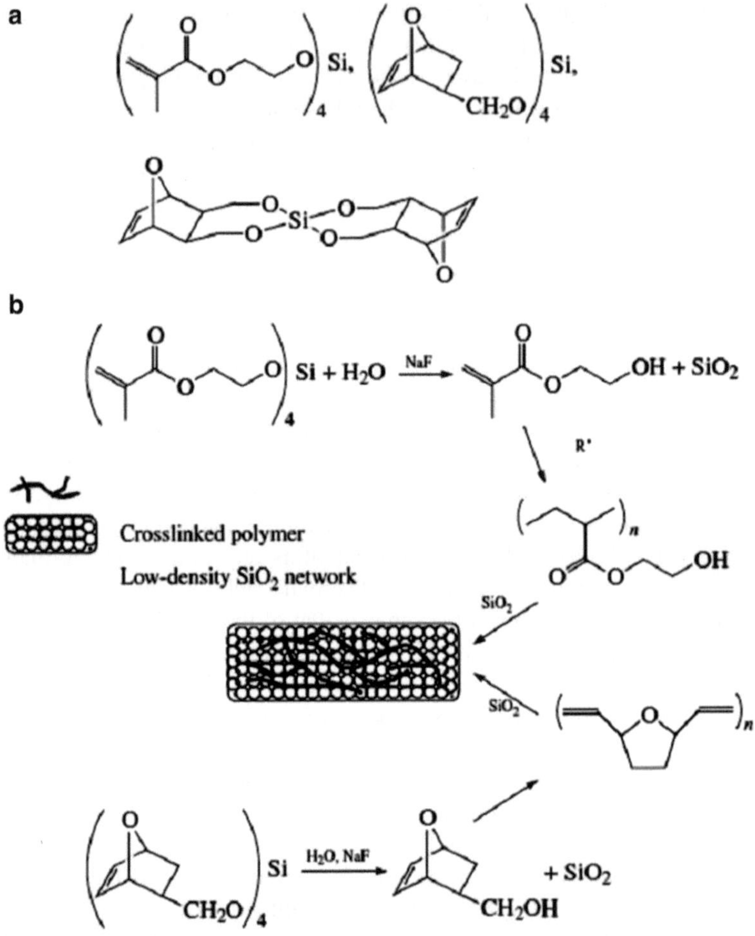

Fig. 4.23 Modified precursors of Si alkoxides (**a**) and scheme of the formation of organic-inorganic interpenetrating networks with the participation of precursors (**b**)

modified by triethoxysilyl groups $(EtO)_3Si–(CH_2)_3–O–[–(CH_2)_4–O–]_n–(CH_2)_3–Si(OEt)_3$.

For the abovementioned method of co-polymerization of titanium alkoxy derivatives with acrylic acid there is alternative way which includes polymer-analogue transformations etherification reaction of PAA and $Ti(OR)_4$ (Scheme 4.10):

Scheme 4.10 Etherification reaction of polyacrylic acid with $Ti(OR)_4$

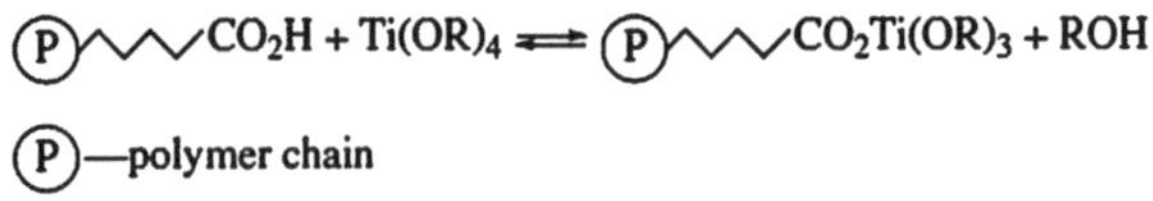

In the cases when molar ratio Ti(IV)/COOH < 1, i.e. at deficiency of $Ti(OR)_4$, 3D structures of clusters reinforcing polymers form (Scheme 4.11):

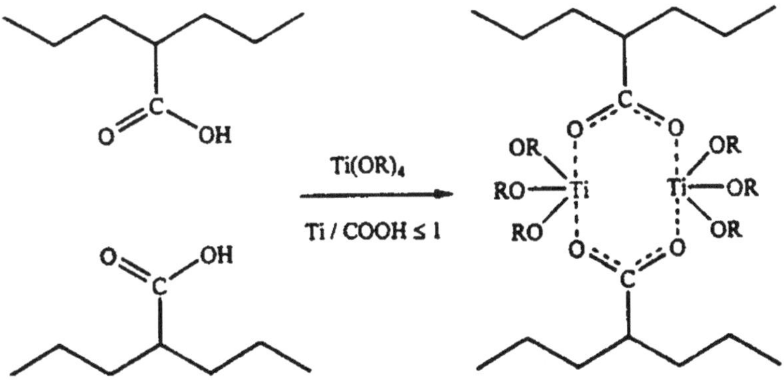

Scheme 4.11 Formation of cross-linking structures with the participation of $Ti(OR)_4$

Their structure is similar to the composites produced at the stage of co-polymerization of the respective monomers (compare with the Scheme 4.4). TiO_2 – based nanocomposites are synthesized by hydrolysis of $Ti(OPr^i)_4$ solved in inverse micelles formed by iso-octane and sodium salt of bis(2-ethylhexyl) sulfosuccinic acid (AOT) and containing calculated amount of water. At that, Ti $(OPr^i)_4$ diffused slowly in micelles, hydrolyzed and condensed in them as in microreactors, forming nanoparticles. The latter were extracted and dispersed in fluorinated polyimide solution.

We shall also note one interesting possibility of application of sol-gel polymer synthesis. The case in point is formation of inorganic coatings on the surface of polymer materials. This approach can be demonstrated on example of formation of multicomponent microspheres Ni/PS/TiO_2 [186]. Previously synthesized Ni nanoparticles were coated by PS shell during radical polymerization of styrene, then inorganic coating was fabricated by hydrolysis of $Ti(OBu)_4$. This multilayered coating material had high sensitivity (response) to electric and magnetic fields [187–190]. Nanoparticles of noble metals coated by TiO2 are used in photocatalysis [191], the same relates to TiO_2/Ag nanoparticles [192, 193]. These hetero elemental composites will be considered in the next chapter.

It needless to say that the presented brief analysis is not exhaustive, regarding diversity of hybrid nanocomposites obtained by polymeric sol-gel technique. Most probably, this listing just gives an idea about possibilities of this method.

4.4 Hybrid Nanocomposites Based on Heteroelemental Oxide/Oxide Ceramics

Sol-gel synthesis is used to produce films or powders consisting of binary and ternary oxides (non-silicate mesostructured ceramics). The following heteroelements are usually impregnated in silicate matrices: aluminum [194, 195], zirconium [196, 197], calcium [33, 88, 198], iron [199], palladium [200], phosphate-ions [201], etc. However, most popular is already mentioned ceramic mixture of Si-Ti, Ti-Zr oxides [202–213]. Sol-gel synthesis was performed at low temperatures and makes it possible to realize strict control over ratio of elements and their homogenous distribution in formed nanocomposites. Thus, SiO_2/TiO_2-nanocomposites (with sizes $\approx 10 \div 50$ nm) were obtained on the basis of petroleum sulfonate ionomers, and TiO_2/ZrO_2 nanocomposites were formed from $Ti(OR)_4$, Zr $(OR)_4$ in matrices of co-polymer styrene with 4-vinyl phenol, etc. This way is also convenient for synthesis of Ti, Ta, and Nb bimetal nanocomposites, which are interesting for catalytic application (see, for example [214, 215]) and even for ternary compounds like Ti-Zr-Sn, Ti-Zr-Al, and also complex perovskite oxides $Ba_{1-x}Sr_xTiO_3$, Ga-Si-O-materials including acid centers, serving good precursors for new catalysts, Brönsted acids and many others. The most important value have zircon $ZrSiO_4$ (or $SiO_2 \cdot ZrO_2$), obtained from $ZrOCl_2 \cdot 8H_20$ or $Zr(OPr^i)_4$ and Si $(OEt)_4$ with the following heat treatment, and V- or Ni-zircons. Various methods are used for synthesis of $SrTiO_3$ [216] (Fig. 4.24a) and zircon with 12 mol % CeO_2 additives (Fig. 4.24b).

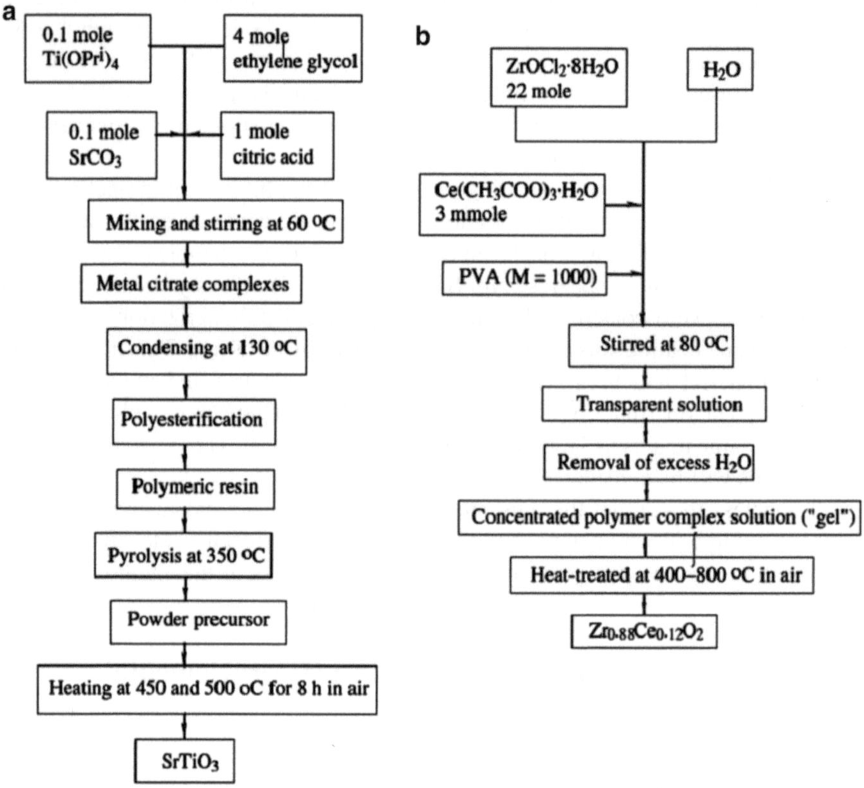

Fig. 4.24 Block-scheme of synthesis of $SrTiO_3$ (**a**) and $Zr_{0.88}Ce_{0.12}O_2$ (**b**)

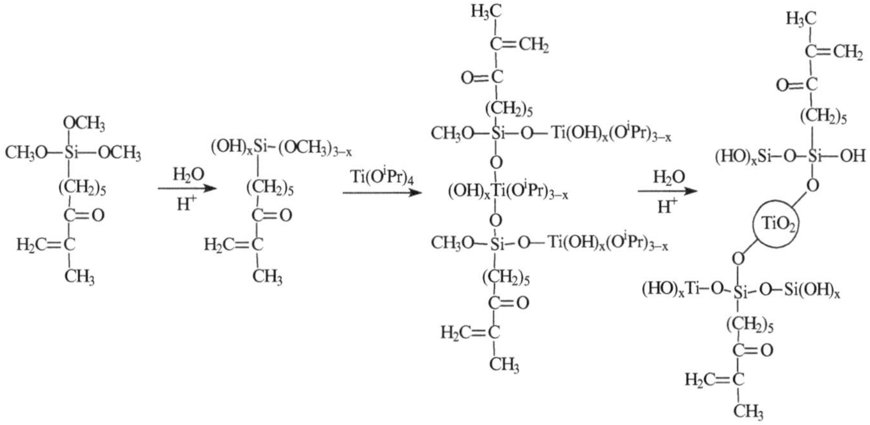

Fig. 4.25 Formation scheme of hybrid Si-Ti- nanocomposite

Polytitanium siloxanes are also obtained in one stage by combined controlled hydrolysis of $Si(OEt)_4$ and $Ti(OPr^i)_2(acac)_2$. Ladder polymer is structured, which contains chain forming links Si–O–Si and Si–O–Ti, the ratio between them depends on synthesis conditions and can reach 10. It determines time of gel formation and a possibility of fiber fabrication by spinning, while ceramic fibers are obtained by annealing of material at 500–900 °C. Hybrid film material is prepared [217] in the process including phase separation of TiO_2 by hydrolysis and condensation of tetraisopropoxide and 3-methacryloxypropyl trimethoxysilane (Fig. 4.25) with excess of aqua solution of the acid at elevated (80 °C) temperature, different molar ratios Ti/Si, and different aging times. Nanoparticles containing nanocrystalline TiO_2 have no cracks, their dielectric constant is 5.0–7.1 and depends on content.

For production of mixed metal oxides, for example, $Zr_xTi_{(1-x)}O_2$, $Zr_xCe_{(1\ x)}O_2$, 10%CuO/TiO_2, doped by Nb_2O_5 and others also aerosol process is widely used [88]: spray-drying is a convenient way for synthesis of powders with high loading of heteroelements, which causes wide variation of properties, such as improvement of hydrothermal stability (Al or Zn incorporation), biological activity (introduction of Ca and/or phosphates), catalytic properties (addition of Pt, Au, Zr, and Al), thermal stability (implantation of Zr), electric conductivity (Nb_2O_5-doped Ru), etc. Localization of heteroelements, incorporated in the initial solution as molecular precursor, during spray-drying is very sensitive to their reaction ability, interaction with structuring agent and silicon oligomers [218–221].

When concentration of heteroelement in solution increases, captured volume of a structuring agent and silicon oligomer play a crucial role in final localization of metal centers. At low concentrations heteroelement is usually homogenously distributed in the final particles. At high concentrations mesostructures of pure silica are observed, which have no effect on incorporation of small amount of metal. However, as concentration of heteroelement increases, there is a tendency to reorganization of a structure. Limiting amount of one heteroelement is predetermined by its ability to interaction with silicon oligomer or with a structuring agent. This interaction prevents inorganic-inorganic or organic-inorganic phase separation until there is radial concentration gradient caused by drying of drops. If attraction is weak or capability of capture of silicon oxide oligomers and/or surfactant is suppressed (for example, by introduction of high concentration of heteroelements), heterocenters can migrate faster than silica gel oligomers along radial gradient of concentration, and concentration gradient of heteroelement or its phase separation can be in the center of a particle. Concentration gradient sometimes can serve for formation of particles with core-shell structure, for example, for delivery of drags or for nucleation of functional nanoparticle in the center of such particle as magnetite.

Variety of the method is gas-flame spray-pyrolysis, for example, for production of nanocomposites on the basis $MgO–ZrO_2$,$Y_2O_3–MgO$, $CoO_x–Al_2O_3$ [222–225].

The main obstacle for effective usage of oxide/oxide nanoceramics is precise control over synthesis conditions and high cost of metal organic precursor, which causes its rise in price. Less expensive are etherized derivatives, which are used in many areas of sol-gel techniques. For example, multicomponent nanocomposites $NiAl_2O_4/SiO_2$ or Co^{2+} doped by $NiAl_2O_4/SiO_2$, nanocomposite materials containing 5 % NiO – 6 % Al_2O_3 – 89 % SiO_2 or 0.2 % CoO – 4.8 % NiO – 6 % Al_2O_3 – 89% SiO_2 are produced by the following block-schemes (Scheme 4.12 and Fig. 4.26) [226]:

$$Si(OC_2H_5)_4 + Ni(NO_3)_2{\cdot}6H_2O + Al(NO_3)_3{\cdot}9H_2O \xrightarrow[60\ ^{\circ}C]{nH_2O,\ nC_2H_5OH}$$

$$\longrightarrow NiOAl_2O_3SiO_2 + nC_2H_5OH + nH_2O\ \text{(Gel)}$$

$$NiO{-}Al_2O_3{-}SiO_2\ \text{(Gel)} \xrightarrow{600\ ^{\circ}C} NiO{-}Al_2O_3{-}SiO_2\ \text{(Glass)}$$

$$NiO{-}Al_2O_3{-}SiO_2\ \text{(Glass)} \xrightarrow{800\ ^{\circ}C} NiAl_2O_4/SiO_2$$

Crystalline/glass : nanocomposite

Scheme 4.12 Schemes of synthesis of Ni-doped $Al_2O_3–SiO_2$ ceramic nanocomposites

Nanoparticles grow in SiO_2 amorphous matrices during heating of dried gel in the temperature range from 60 to 800 °C at the rate 1 deg/min. According to TEM data, $NiAl_2O_4$ and its cobalt-doped nanocrystals are homogeneously dispersed in SiO_2 amorphous matrix, its average size is ~5–8 nm.

Materials with structure of oxide spinels attract great attention in materials science, because has interesting properties, such as magnetism, semiconducting, catalytic properties, heat-resistance in wide temperature range, they belong to dyed ceramics. In metal aluminates with spinel structure and general formula MAl_2O_4 (M = Ni, Co, Mg, Zn), two metal cations M^{2+} and Al $^{3+}$, have tetrahedral and octahedral configurations, respectively. In the spinel structure distribution of different cations along available tetrahedral and octahedral sets is determined by

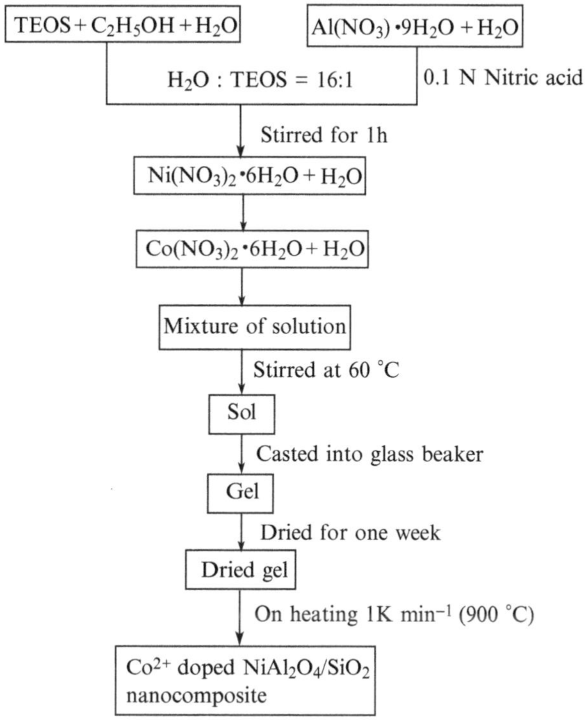

Fig. 4.26 Block-scheme of sol-gel synthesis of Co^{2+} – doped $NiAl_2O_4/SiO_2$ nanocomposite

respective energy preference, which depends on Madelung potential, ion size, polarization of anion, temperature of calcinations, etc. [227–233].

Based on polymer-derivatives structural ceramics are produced for high temperature applications [234–242]. For example, silicon carbonitride (SiCN) shows oxidizing resistance at high temperatures, creep at compression appears only at high temperatures [243], silicon oxycarbide (SiOC) synthesized by transformation of polymer-ceramics based on polysiloxane is high-temperature resistant material. The same relates to hafnium doped silicon oxycarbide ceramics (Fig. 4.27) [244, 245].

A special place have oxide ceramic materials with incorporated rare metal ions (Er^{3+}, Sc^{3+}, Y^{3+}, La^{3+} etc.), which are efficient for use in solid lasers, photo and luminescent applications, optic amplifiers, active planar waveguides, etc. High demands are made to telecommunication segments, to new materials with special optic properties, and to improvement of materials properties [246]. Preparation by sol-gel method, structure, morphology and luminescent properties of Eu^{3+} doped SiO_2/Ta_2O_5 nanocomposite is reported (Si/Ta molar ratio is 90:10, 80:20, 70:30, 60:40, 50:50). Sol was held at 60 °C before xerogel formation by baking for 2 h at 900–1,100 °C. The samples with low Ta content (10 mol %) baked at 1,100 °C had a structure with separated phases, and nanocrystals had size 3.8 nm. As tantalum concentration increased, phases sharply separated and crystallized depending on content and baking temperature. As OH^- group was displaced, SiO_2 as well as Ta_2O_5 cross-linked nets formed, as a result, glassed ceramic formed containing

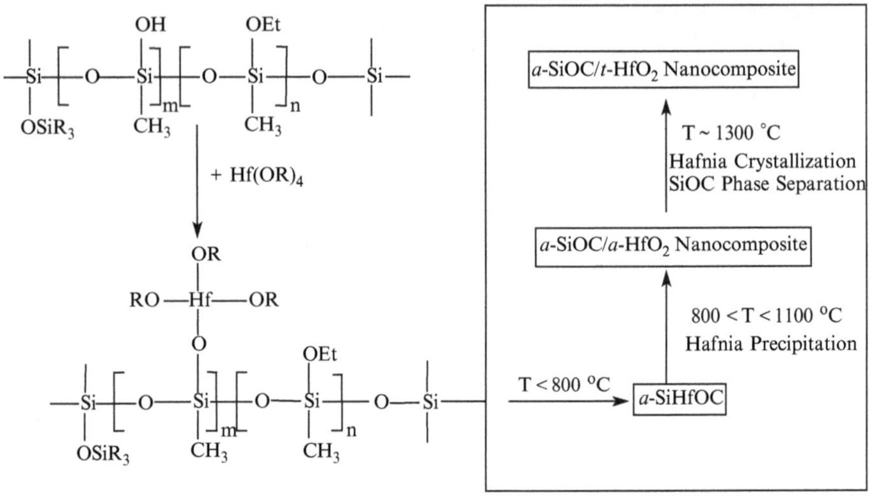

Fig. 4.27 Evolution of ceramization of polymethylsilsesquioxane modified by hafnium alkoxide ($Hf(OBu)_4$, 1 vol. %.) [245]

Ta_2O_5 nanocrystals dispersed in SiO_2 amorphous matrix. Nanocomposites baked at 1,100 °C for 2 h show wide band emission at 1,535 nm, which increases by 64 and 92 nm as tantalum concentration increases (molar ratio Si/Te is 60:40 and 50:50). Lifetime of active state points out that Er^{3+} ions are predominantly localized in areas enriched by Ta_2O_5 nanoparticles.

Many heteroderivatives based on silica materials obtained by sol-gel method are described, among them are Sc_2SiO_5, Y_2SiO_5, HfO_2–SiO_2, etc. (see, for example, [247–256]).

During co-synthesis, one-pot procedure, different types of organic and hybrid components, for example, fluorescent colors, rhodamine B, spiropyran or spirooxazine [257], phenanthroline Eu^{3+} Tb^{3+} complexes [258] and other agents, which provide transparent photochromic films with 97–100 % transmission, can be impregnated into mesostructured microparticles.

Stable homogeneous MgO–Y_2O_3 fine-grained nanocomposites and phase domains were obtained by new "foam" method. This process is realized with usage of non-expensive precursors under soft conditions in low-dense organic foams. Their transformation with respective oxides is reached by low temperature calcinations followed by high temperature post-calcinations. Formation of amorphous intermediate during calcinations is critical for formation of fine structure of oxide/oxide nanocomposites [259]. Implementation of this procedure can be reduced to the follows. The MgO–Y_2O_3 nanocomposites were prepared from respective nitrates by etherification reaction of ethylene glycol and citric acid. A foam forms during emanation of nitrogen oxide, and during calcinations extremely fine nanocomposite powder forms from it. Surprisingly, this microstructure remains unchanged during post-calcinations and consolidation at heat treatment under pressure and is a result of competitive crystallization and phase separation.

Homogenous separation of domains of submicron phase is found in $MgO–ZrO_2$ [222] and in $MgO–Y_2O_3$ [260]. A composite containing about 90 mol % of MgO is prepared from respective amounts of 0.5 M $Y(NO_3)_3$ (7 mmol) and 0.5 M Mg $(NO_3)_2$ (33 mmol) water solutions, citric acid (40 mmol) and ethylene glycol (13 mmol), and was placed after mixing in oven heated to 200 °C. After grinding colorless composite material containing MgO and Y_2O_3 phases formed.

Stabilized yttrium-zirconium ceramics (YSZ) is produced from $ZrOCl_2$ $8H_2O$ and YCl_3 $6H_2O$, YSZ contained 3 mol % Y_2O_3 [261]. Coating of carbon nanofibers by yttrium zirconium ceramics forms superconducting ceramics (120 vol % YSZ per carbon fiber). This material has perfect mechanical, thermal, and electric properties, it is used in many applications.

The most serious problem in preparation of composites of this type is difficulties in homogenous phase distribution with different properties of nanocarbon fibers and zirconyl caused by weak interaction between them. Therefore they are prepared by plasma sintering of powder mixtures or by coating nanofibers surface by wet chemical method [262–267].

Therefore, properties of products fabricated by sol-gel method depend also on additionally implanted modifying agents, and highly disperse fillers, which can be metal nanoparticles or their precursors. Among numerous examples of this type we shall highlight just several ones [28]: $SiO_2{\cdot}x$M (M = Ni, Cu, Co, Cu/Ni, Co/Ni, Co/Cu) xerogels, metal acetates/TEOS M = Pt, Gd/Pt, Au, Pd, Cr_2O_3, Co, Zn, Cu, which are obtained by system hydrolysis, stable nanofilms $Pb(Zr,Ti)O_3$, etc. [268–272].

Based on TEOS, $Ti(OPr^i)_4$, SAS and NaOH, using chromium salts or vanadyl, the already mentioned template materials M-MCM-48 (S_{sp} for M = Ti ~ 1,000 m^2. g, d_{por} ~ 2,5 nm), and also M = Cr have been obtained [17]. Their special property is an increase in condensation degree of silica frame (by ~25 %) at inclusion of heterometal ion. Similar procedure is also used for growth of ultra thin barium-ferrite powders, and also LiZn ferrite thin films, $CoFe_2O_4$ cobalt-ferrite and many others.

Anyway, the polymerization method used for synthesis of heterometallic ceramics is most widely applied for perovskites with ABO_3. Found in nature $CaTiO_3$ perofskite has a pseudo cubic crystal lattice, in which big cations are in corners of the cell, and small ones are in its center, oxygen ions are in the face centers. Traditional way of their production, for example, $PbTiO_3$ ($A^{2+}B^{4+}O_3$-ceramics) is solid phase mixing in mills of PbO и TiO_2 followed by baking at temperatures above 600 °C (*ex situ*-production). However, this process was accompanied by formation of undesired and toxic PbO. Sol-gel synthesis (Scheme 4.13a), in which complex formation of components with citric acid and polyetherification by ethylene glycol followed by polymerization (at 160 °C) and pyrolysis are used, is free of these drawbacks. In this case thin films with properties of blocked materials can be obtained.

Barium stannate with perovskite structure $BaSnO_3$ is obtained by polymerization, pyrolysis of preliminary produced $BaSn(C_6H_5O_7)_2$ citrate complex (Scheme 4.13c), whose grain size is 80–100 nm; this method is used to produce

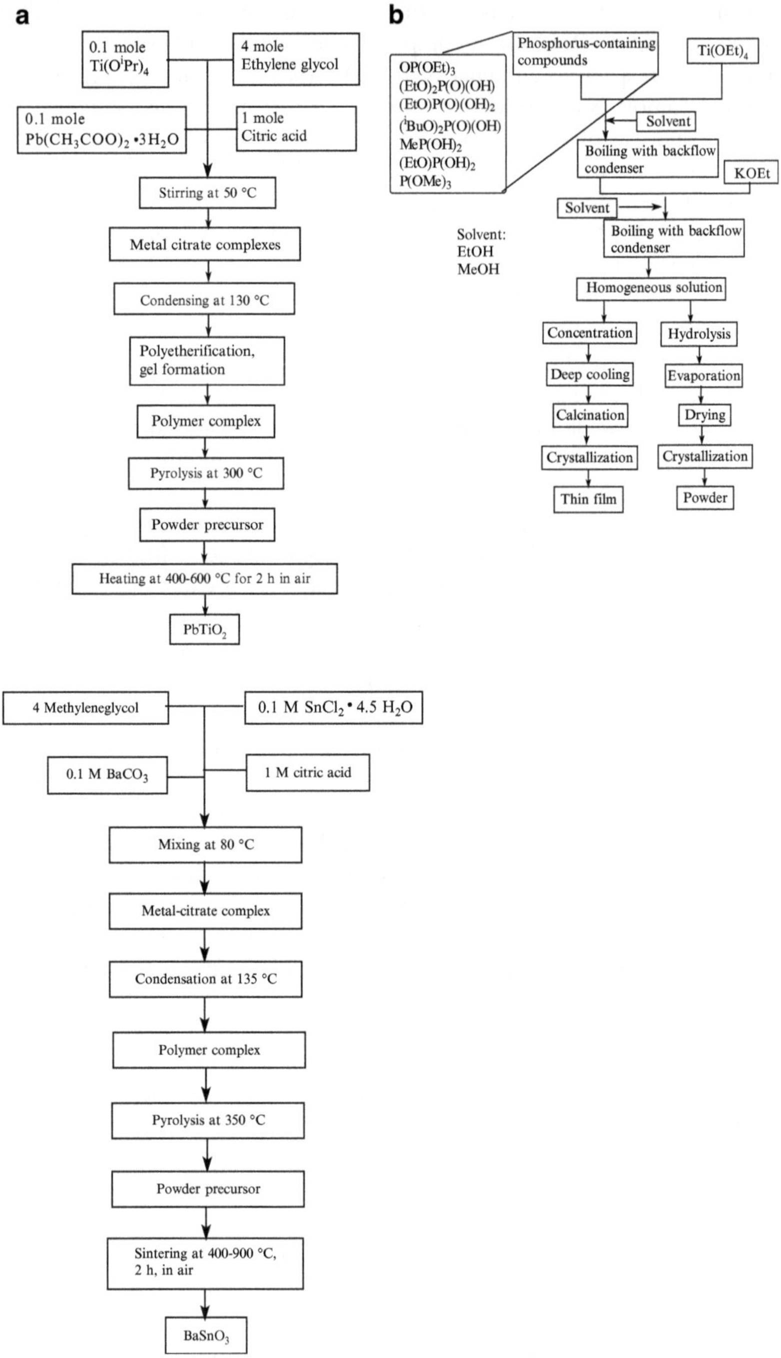

Scheme 4.13 Block scheme of synthesis of $A^{2+}B^{4+}O_2$ (**a**, **b**) and $A^{2+}B^{4+}O_3$ (**c**) ceramics

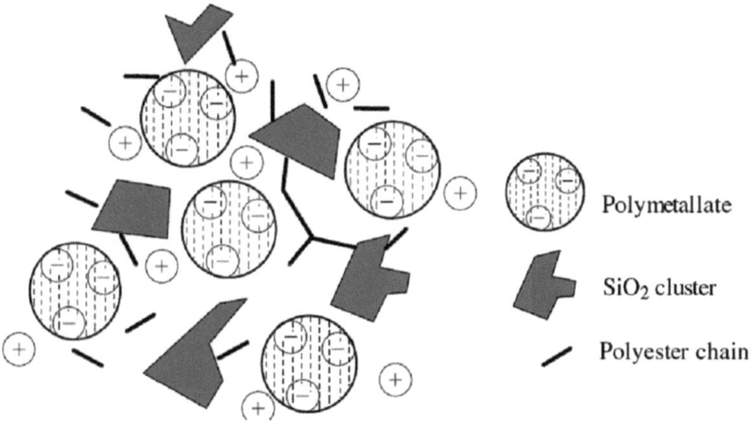

Fig. 4.28 Schematic representation of the structure of organo–inorganic nanocomposites based on polymetallates

other perofskite ceramics (see, for example, [28]): monodisperse $BaTiO_3$, (Ba,Sr) TiO_3, $Pb(Zr,Ti)O_3$, $NdAlO_3$ ($A^{3+}B^{3+}O_3$-ceramics), ZrO_2/X ($X = CeO_2$, Y_2O_3, Y_6WO_{12}), $SrBi_2Ta_2O_9$ nanoparticles, and even high temperature superconducting ceramics $YBa_2Cu_3O_{7-\delta}$ [273]. As a result of polymer synthesis with the following decomposition of organic phase at 300–400 ºC and calcinations at 400–900 ºC oxides of molecular degree of homogeneity and high purity are obtained. Here we shall just notice a possibility of fabrication of one phase thin films $KTiOPO_4$ (having high thermal stability and impressive optical properties used for non-linear optics) by using $Ti(OEt)_4$, KOEt and various phosphorous sources for precursors (Scheme 4.13b), among which the optimal one is $(n\text{-}BuO)_2P(O)(OH)$ (see [15]).

To this group of nanocomposites also belong heteropolymetallates such as Keggin acids ($H_3PW_{12}O_{40}$, $H_4SiW_{12}O_{40}$ – spheres with diameter ~1 nm) included in organic-inorganic structures with good electrochemical properties for potential usage in holography. Nanocomposites are obtained in two ways: by mixing with TEOS (at the ratio $W/Si_{alkoxide} = 0.2 \div 0.6$) and tetraethylene glycol or by their impregnation in organosilanes. Idealized structure of these materials is shown in Fig. 4.28.

Prospective method is polymerized transformations of specific clusters, oxometalates, whose W-O-Si links are included in side chains of a polymer. They are synthesized by interaction of substituted trichlorosilane $RSiCl_3$ (R is vinyl, allyl, methacryl, styril, etc.) with $K_4SiW_{11}O_{40}$ polyanion. The formed cluster-containing monomers are built from metal-oxo core with polymer forming ligands, set on periphery (Fig. 4.29a). Metal-oxo cores are used for formation of a cluster net and formation of nanohybrides, which are attached ensembles of exhaustively characterized structural blocks. Yield of polymer under action of radical initiators and chain lengths depended on reaction ability of unsaturated groups and decreased in sequence: $SiW_{11}O_{40}Si(vinyl) << SiW_{11}O_{40}Si(allyl)_2 < SiW_{11}O_{40}Si(methacryl)_2 < SiW_{11}O_{40}Si(styril)_2$. Principal scheme of these polyoxometalate polymers are shown in Fig. 4.29b.

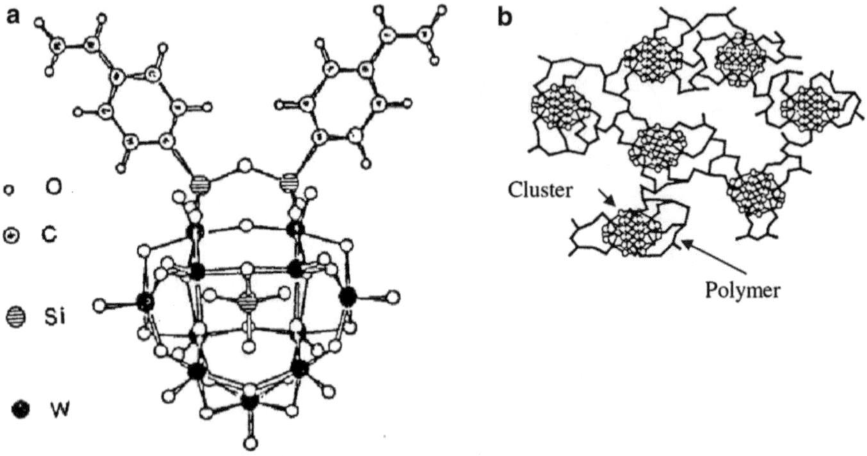

Fig. 4.29 The structure of $[SiW_{11}O_{40}Si(C_6H_5CH{=}CH_2)_2]^{4-}$ polyanion (**a**) and schematic representation of the structure of hybrid polymer-inorganic material based on such anion (**b**)

We shall just note that by the same mechanism, bonding in a complex with bi-functioning ligand containing alkoxo- or hydroxy silyl for reactive group, Os_3, PtSn, Fe_2P, Co_2P, Ni_2P or Ge clusters are impregnated in the composite, and Pd^{2+}, Cu^{2+} and other ions are implanted in poly(ethyltrimethylsilane).

We shall especially highlight usage of sol-gel technique for performance of template syntheses, which are considered as processed of formation of purposeful products by assembling of simple components, carried on under conditions of strict stoichiometric orientation of reagents. These conditions, first of all, small reaction volume, due to which closer attraction of reacting molecules as compared with processes in solution or in solid phase advantage so called "soft" template process. The result of this is that assembling has place under normal conditions (sometimes just at room temperature). For example, already analyzed mesoporous molecular sieves (like MCM-41) are highly organized substances, their formation includes self-organization of SAS molecules in solution, and sol-gel processes of frame formation around micellar structures, which play function of templates. Synthesis with usage of inverse micelles is introduction of surface-active substances (in particular, didodecyl-dimethylammonium bromide, AOT, etc.) in toluene, as a result, inverse micelles form, in which metal salts, for example, $AuCl_3$ and gel precursor TEOS are impregnated. The reducing agent is $LiBH_4$/THF. Hydrolysis of TEOS and condensation cause formation of impregnated in wet gel Au nanoparticles with average sizes dependent on the ratio H_2O/Si and SAS concentration (at $H_2O{:}Si = 1{:}1 \div 4{:}1$, average particle size is 5–7 nm). To obtain closely packed and organized 2D gold layers (about 10-nm thick), its colloid particles immobilized on the surface of highly disperse SiO_2 by bi-functional $(CH_3O)_3Si(CH_2)_3NHCH_2{-}CH_2NH_2$ amino silane stabilize SAS by dodecanethiol (see, for example, [28]). The following treatment of RSH particles makes them movable by breaking bonds with SiO_2, repetition of the immobilization-stabilization acts (Fig. 4.30a) caused growth of highly ordered 2D particles with interlinked grains (Fig. 4.30b).

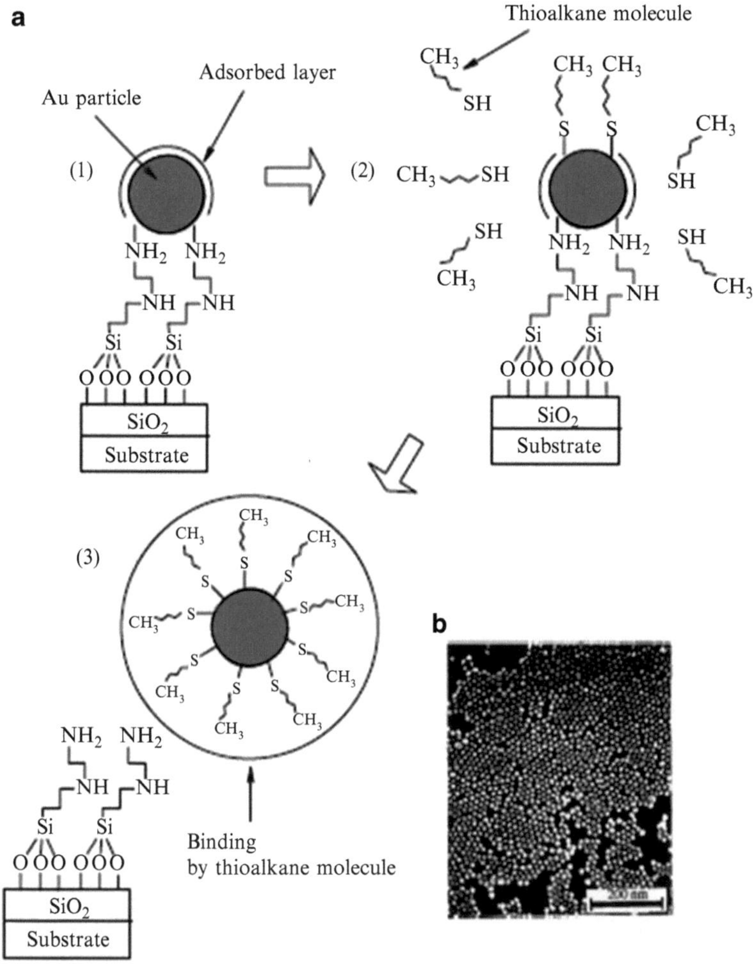

Fig. 4.30 Formation mechanisms of hybrid-phase materials based on inverse micelles (**a**) and electronic micrograph of highly organized 2D Au particle latex obtained by scheme (**b**): *1* immobilization of colloidal Au particles, *2* exchange reactions of alkanthiol molecules and breakage of Au–NH_2 bonds, *3* fast restoration of the layer at complete binding of Au particles

Silver and gold nanoparticles are synthesized in sol-gel process in thin SiO_2, TiO_2 and ZrO_2 films [274] by the method based on coordination chemistry of respective ligands for stabilizing Ag^+и Au^{3+} ions in sol-gel systems. As is seen from Fig. 4.31, size of gold nanoparticles depends on precursor (Ti or Zr) and baking temperature for the ceramics.

Convenient method of incorporation of metal nanoparticles in these inorganic polymers Pt/SiO_2 is combination of nanocluster formation reactions by reduction of respective salts in inverse micelles with sol-gel synthesis (Fig. 4.32).

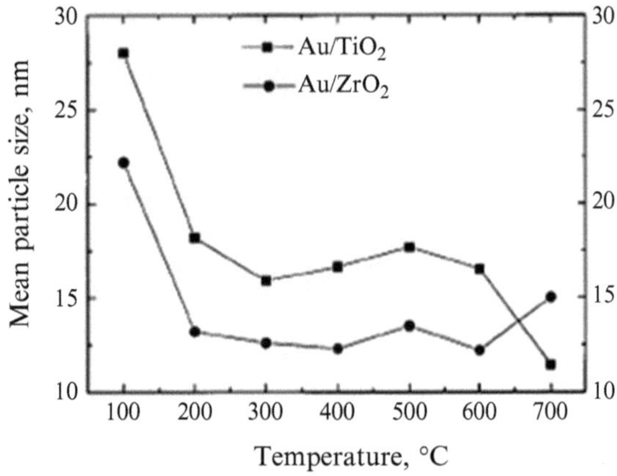

Fig. 4.31 Dependence of the mean size of particles on the temperature of sintering for the Au/TiO_2 and Au/ZrO_2 systems [274]

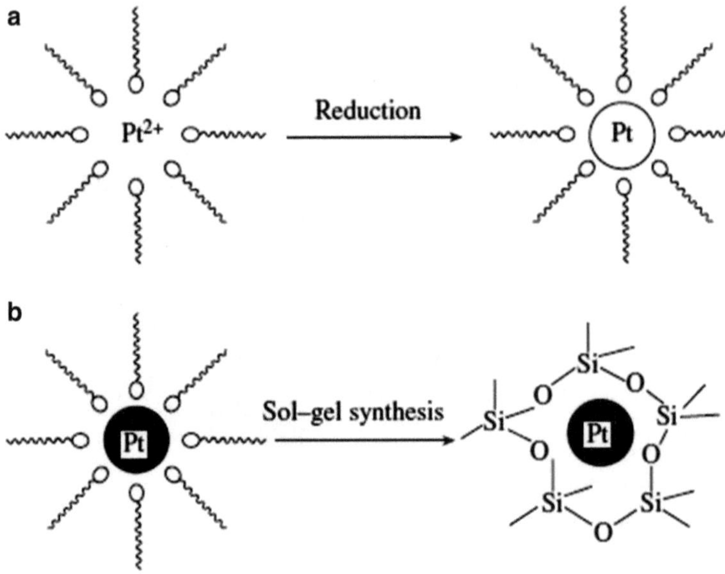

Fig. 4.32 Reduction of a metal salt combined with sol-gel processing. (**a**) formation of nanoclusters by the inverse micelle technique, (**b**) formation of the gel around nanoclusters

This method can be used for other noble metals (Pd, Rh, Ru, Ir). At sintering metal nanoparticles are incorporated in xerogel structures without agglomeration. Assortment of these transformations is doubtless far wider. It includes almost all chain forming elements of Periodic Table of Elements.

For example, formation of Pt/Fe binary systems occurs during thermal reduction of acetylacetonate complexes incorporated in TiO_2 matrix at the stage of xerogel formation [275]. It is important that combination of sol-gel synthesis with thermal reduction in this system favors formation of isolated $L1_0$ FePt nanoparticles at lower temperatures. Moreover, presence in the system of complex forming ligand gives stability to metal ion and thus improves control over its reaction ability under conditions of sol-gel synthesis during formation of metal nanoparticles. General approach in this method is demonstrated in [274] during formation of Au and Ag nanoparticles in SiO_2, TiO_2, ZrO_2 in sol-gel process, when respective variation of coordinating ligands, matrix origin, temperature allows conduction of controlled synthesis of thin filmed nanocomposites (Table 4.3).

Table 4.3 The influence of nature of coordinating ligands on the nanocomposite formation in sol-gel process

Nanocomposite	Without adding of ligand	Diene	CH_3CN	DMSO	Pyridine
Ag-SiO_2	Slow precipitation of brown compound		Stable sol without precipitation	Slow formation of brown compound	Stable sol during 3 days without precipitation
Ag-TiO_2	Formation of black precipitate in a day	Stable sol with slow formation of black precipitate	Formation of black precipitate in a day	Formation of black precipitate in a day	Stable sol during 1 week, at low ration pyridine/Ag sol becomes red, then a black precipi-tate forms
Ag-ZrO_2	Very quickly formation of black precipitate (during several min)	Stable sol during 3 days	Quickly for-mation of black precipitate	Turbidity at the addition of DMSO, sol becomes red, then a black precipitate forms	Turbidity at the addition of pyridine, sol becomes red, then a black precipitate forms
Au-TiO_2	Formation of black precipitate in a day	Formation of black precipitate in a day		Slow formation of small amount of black precipitate (more than for 1 week)	
Au-ZrO_2	Formation of black precip-itate in sev-eral hours	Formation of black precipitate in several hours		Slow formation of small amount of black precipitate (more than for 1 week)	

4.5 Morphology and Fractal Models of Hybrid Nanocomposites

Many special features of morphology of nanocomposites formed by sol-gel method were repeatedly considered. Here we shall consider interesting examples. Interest is drawn to photo-curable polymer nanocomposites due to their different applications and interesting morphology [276]. Thus, photo-curable polyacrylate nanocomposite with silica gel nanoparticles is prepared by sol-gel synthesis. In contrast to the ordinary sol-gel process (using water for liquid), in this work hydrolysis of the precursor was conducted through air moisture in order to confine size of silica gel nanoparticles to nanometer scale and protect the composite from residual water molecules [277]. Morphology of crystals in polyurethane cast films (PU) with different gold particles content 0.1, 0.3 and 1.0 % is studied (Fig. 4.33) [278]. Crystals of pure PU are very fine and not seen. Cross wheel of these spherical crystals was observed in PU-Au, size of spherical crystals decreased as gold content increased. In the samples with higher gold content there are many spherolites coagulating due to nucleation and crystallization. Gold nanoparticles become the center of hard segment in PU and star-like coupling of PU appears.

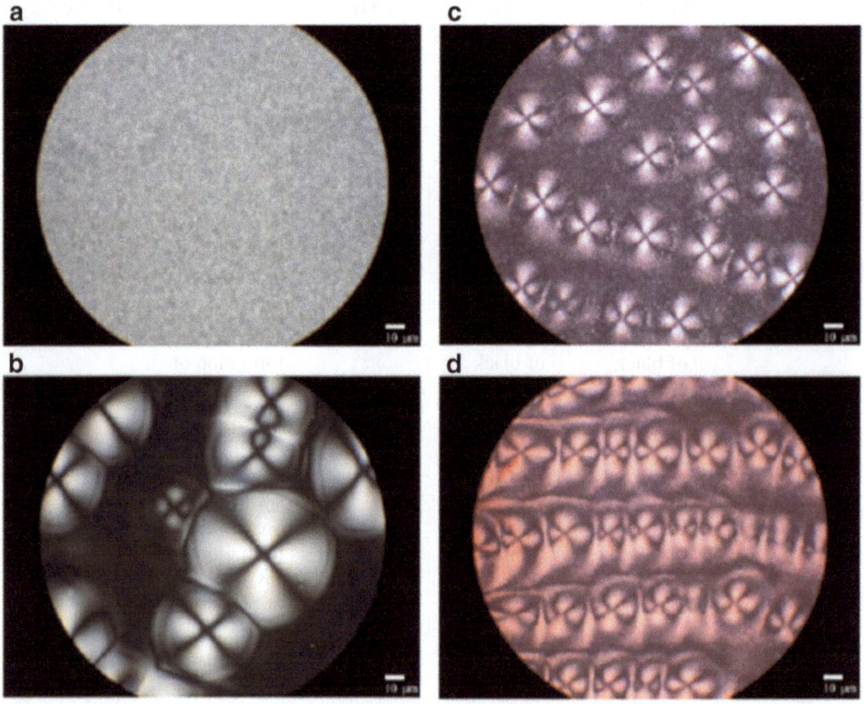

Fig. 4.33 Crystal morphology of polyurethane (**a**) and polyurethane containing 0.1 (**b**), 0.3 (**c**) and 1.0 wt.% (**d**) Au by the optical polarization microscopy data [278]

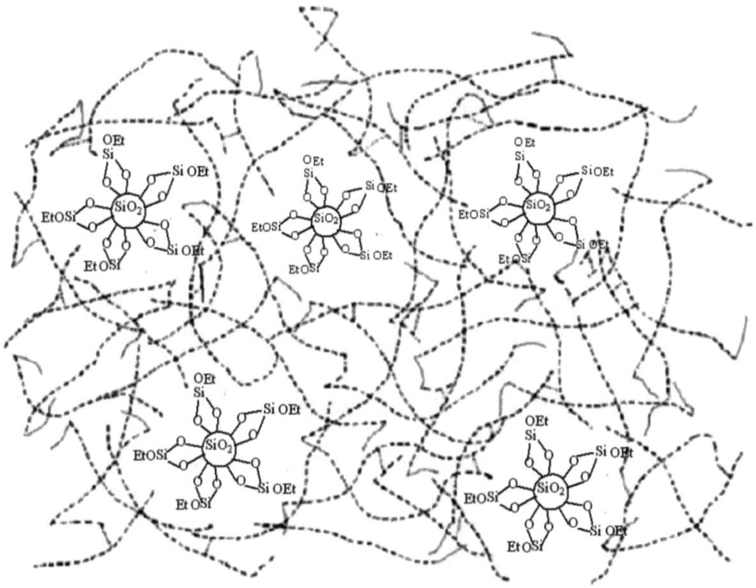

Fig. 4.34 Schematic illustration of nanocomposite microstructure

The simplified scheme of hybrid composite, in particular, containing cluster fragments, can be represented as followed (Fig. 4.34) [279].

Sol-gel synthesis is a perfect method for illustration of theory of fractals in the percolation theory as applied to structure of hybrid materials. Detailed analysis and theoretical ground of the theory were recently considered in the monograph [280]. We shall reduce the limit just to basic information. In materials of sol-gel synthesis at different degrees of porosity bulk and surface fractals are observed [281]. Bulk fractals form in two extreme cases: formed by porous medium or by frame of a porous material, in intermediate cases product can show properties of a surface crystal. As is known, fractal aggregates are characterized by fractal dimensionality D_m ($1 \leq D_m \leq 3$), which can be found from the relation:

$$N \approx R^D, \tag{4.16}$$

where N is mass of an aggregate, or a number of primary particles in it; R is typical size equal, for example, to inertia radius of an object.

Self-similarity or scale invariance of a fractal aggregate means that at different scales of its length its structure does not change overall, and any of its part is similar to the whole aggregate [24, 25, 28].

Fractal surface is characterized by fractal dimensionality D_s, which correlates surface of an object **S** and its radius r:

$$\mathrm{S} = r^{\mathrm{D_s}} \tag{4.17}$$

where $2 \leq D_s \leq 3$. The higher D_s, the rougher is surface of a fractal object: $D_s = 2$ corresponds to smooth surface, while $D_s = 3$ corresponds to maximum rough surface.

Products of hydrolysis of $Ti(OBu)_4$ are multilevel systems, whose structural element is a complex oligomer molecule of titanium oxobutyrate (see, for example [28]). Under drying transformation of polydisperse particles to homogenous particles with fractal dimensionality $D_s = 3$ takes place. This respects to the case of surface fractal with maximum developed surface. Role of the coupling agent can play compounds including labile, vinyl, methacrylic, epoxy and other groups.

At the same time fractal dimensionality of polymer and colloid formations of aerogel TiO_2 with high S_{spec} (prepared by sol-gel technique with CO_2 drying in supercritical state) is 2.6–2.8. Fractal surface of particles is just a little less regular than that of polymer aerogels. This similarity in dimensionalities means that morphology at nanometer level is predetermined at early stages of synthesis, contribution from the process is insignificant. However, morphology at mesoscale depends on conditions of synthesis. Thus, extent of poly-ε-caprolactone incorporated in the cross-link of silicon oxide depends on molar relations HCl:TEOS and H_2O:TEOS. Structure of interphase material is a bulk fractal with respect to changing morphology of both phases, it is more open when acid concentration is below and water concentration is higher than average (D_m changes from 1,4 to 2,0). This provides a possibility of control over parameterization of the surface. For example, in [282, 283] morphology of two types of nanocomposites is compared: **A** (30 wt % SiO_2) obtained by block polymerization of HEMA, preliminary functioned with silicon dioxide particles (~13 nm) and **B** based on synthesized *in situ* SiO_2 (as a result of hydrolysis of TEOS hydrolysis) during radical HEMA polymerization.

In the first case (**A**) nanocomposite has ordinary particle-matrix morphology in which particles of mineral phase have a tendency to aggregation in continuous poly-HEMA phase (Fig. 4.35a).

Another material (**B**) has finer structure. Its mineral part obtained *in situ* has almost perfect morphology consisting of very open mass-fractal silicate formations in bicontinuous structures at molecular level. This fine morphology is as though frozen in vitrified organic phase. Thus, composite **B** can be called nanocomposite

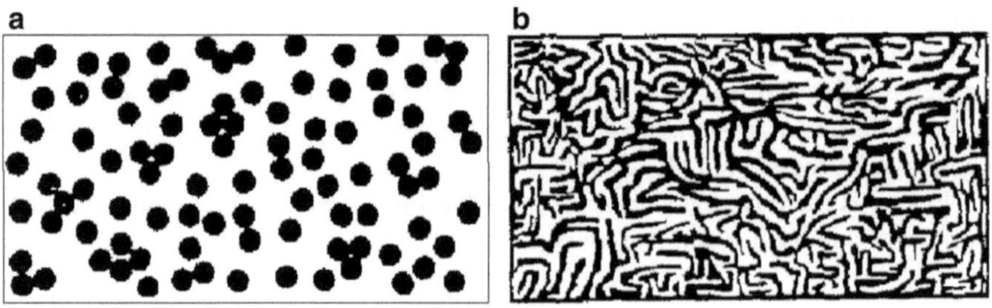

Fig. 4.35 Morphology of hybrid nanocomposites formed by different procedures (see text) [283]

with molecular level of morphology. At that in **A** material two types of size of nanoparticles are identified by SAXS data: compact primary which form smooth and distinct interphase with polymer matrix (their average diameter is ~14 nm) and secondary – aggregation of the primary nanoparticles. These aggregates show mass-fractal geometry, whose fractal dimensionality is lower than 2 ($D_s = 1.6$). Their equivalent spherical diameter is close to 30 nm, which confirms that they include just few primary microparticles. Typical fractal models fulfilled at condensation of TEOS in solution are not found during polymerization in the matrix (polysiloxane cross-links). High molecular masses of the matrix and formed polymer bring to a decrease in entropy of the mixture and, consequently, to phase separation. Not separated by phases systems are obtained just under definite conditions, however, at concentration of SiO_2 above 5 %, phase separation takes place, which increases mechanical strength of the composites. It is interesting that morphology and structure of hybrid PEO-TEOS composite depends also on nature of a catalyst used in sol-gel synthesis: in materials obtained in acid medium (oppositely to NH_4F), links of a polymer chain were bound with SiO_2 particles forming ideal composite. In the case of *in situ* deposition of TiO_2 (up to 5 % filling) extremely coarse domains were observed, strength properties of these composites are inferior to the systems with SiO_2. Mechanism of phase separation in hybrid materials prepared by sol-gel method is keenly studied (see, for example [24]). Two mechanisms of phase separations are considered: nuclei formation (domain formation) and growth (NG), and also spinodal decomposition (SD). Mechanism NG is preferable for extinguishing (quenching) of metastable phase bimodal and spinodal curves. Nucleation is initiated by local density and fluctuation concentration with formation of nuclei with distinct interface. This requires activation energy depending on interphase energy for formation of these centers. Evolution process followed by diffusion of macromolecules in nucleus domains is spontaneous. Composition of nuclei stays constant for entire phase separation. The observed morphology during NG phase separation at early and late stages has dropped/matrix type. On the other hand, SD mechanism of phase separation prevails in extinguishing of unstable regions in spinodals. Extended concentration of fluctuation is spontaneously initiated by prevailing wavelength, which is constant at the very early stages and depends on intensity of extinguishing. Fluctuation of wavelength increases at intermediate stages and phase coalescence is observed at late stages. If it is possible to avoid coalescence, 3D spatial continuous morphology is observed. When a solvent is removed from hybrid material poly(ethylenoxide -co-epichlorohydrin)-TEOS, spontaneous growth of inorganic net begins, mechanism of phase separation is spinodal decomposition; the same morphology was observed in hybrid films of poly(ethylenoxide-amide-6) including sol-gel prepared SiO_2 and TiO_2 as inorganic fillers, however, titanium films are harder [284]. Although correlations between changes in properties of sol-gel products and their fractal dimensionality are still not found, studies in this direction are promising.

As it follows from the above brief analysis, using of sol-gel technique for combination of polymer with inorganic material has several directions and allows purposeful using of a complex of methods for development of novel materials with unique properties. First of all, usage of alkoxides modified by two different functional groups is promising: one of the groups reacts with monomer link of a macromolecule, another one reacts with sol-gel precursor, as a result strong chemical bond forms between a polymer and sol-gel cross-links. Secondly, interesting is to use special bonding agents, which also form chemical bond between these components. For example, modifying surfaces of highly disperse SiO_2 with aminobutyric acid, then dispersing of this product in ε-capramide and catalyzed co-polymerization. Uniqueness of the considered method of obtaining of hybrid nanocomposites is that at the stage of formation of the products of sol-gel synthesis different modifying dopants of organic or inorganic type can be incorporated, for example, 1,10-phenantrolinhydrochloride and butyl amine for absorption in UV region (<350 nm) and transparency in visible light. In sol-gel process also can be involved alumoxanes, carboxylates with general formula $[Al(O)_x(OH)_y(OOCR)_z]_n$, at this stage ceramics can be doped by additional incorporation of acetylacetonate complexes $M(acac)_n$ ($M = Ca^{2+}$, Mn^{2+}, Y^{3+}, etc.) and metal chelates of other origin including transuranium, various heterometallic compounds.

There may be advantageous sol-gel process, which causes formation of power-intensive materials [285–289]. For example, polymer sols resorcinol – formaldehyde and resorcinol – furfural were mixed with water solutions of inorganic salts NH_4ClO_4,$Mg(ClO_4)_2$ and NH_4NO_3. After gel formation and drying under normal pressure hybrid nanocomposite showed properties of power-intensive material, which after beforehand heating in flame boisterously burned with loud noise and sparks. In these composites including organic gel-oxidant, important role have a solvent and a possibility of dry gel production containing 50 % oxidant (with respect to polymer matrix) with balanced ratio oxidant/fuel without breaking of homogeneity of the system.

Sol-gel process is very flexible and multisided by type of products produced, for example, for optical and communication technologies (Fig. 4.36) [290]. Motivation for this field is a decrease in optic scattering on interface, when phase domain is smaller than wavelength of electromagnetic photon.

We shall notice actively developed method of heterophase sol-gel synthesis, and the deposition-precipitation method with usage of products of sol-gel synthesis for production of highly active heterogeneous catalysts. These are independent research fields, which also should be analyzed. This is the way to obtain porous monolithic products with meso- and macro- architecture of pores, and components may be incorporated at the stage of SiO_2 formation, which abruptly increases its absorption ability with respect to water (for example, up to 30 % of waterless $CaCl_2$) [291].

In order to increase compatibility of organic and inorganic ingredients, different co-polymers will be used as compatibilizers. Thus, already analyzed styrene block-copolymer and 4-vinyl phenol (50:50) was very efficient for these purposes. Probably, expectation will be realized [69] of polymers based on acrylate titanium alkoxides, oppositely to tin-organic ones will be self-polished non-toxic coatings for ships, which emanate biocides, absolutely innocuous for sea medium.

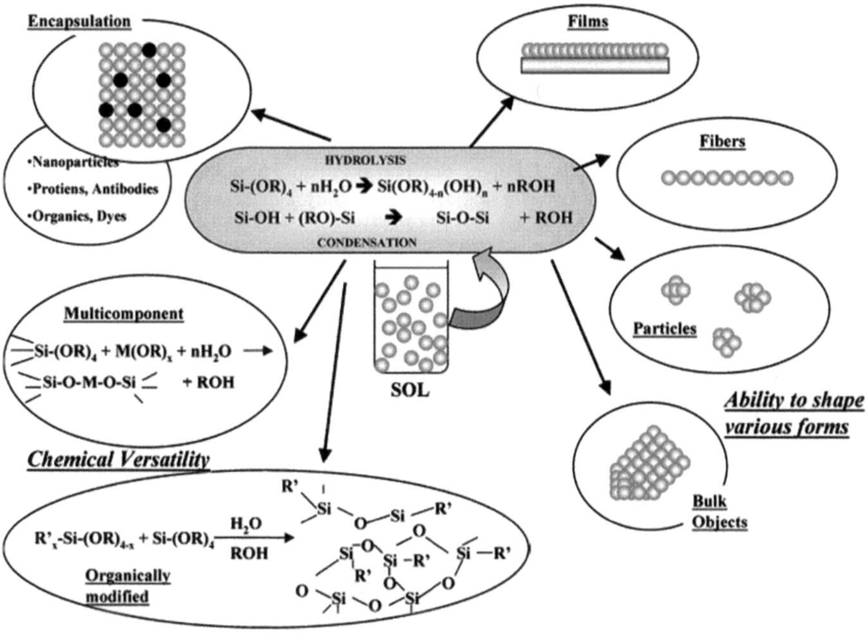

Fig. 4.36 The *circles* represent the basic moieties of sol-gel products used in optical communication technology [290]

There are all grounds to believe that development of new nanocomposite materials based on carbon nanofibers in composition with ceramics (including heteroelemental), having good mechanic, thermal and electric properties and interesting for a range of industrial applications will progress. Among them stands out ceramic material based on zirconium dioxide, which has unique mechanical properties. The best illustration of preparation of such a product, is a material, which is difficult to prepare, with homogeneous phase distribution of carbon fibers and ZrO_2 with different properties due to weak interaction between them. These composites are prepared by the powders mixture and sintering in spark plasma [261, 263, 265–267].

The considered organic-inorganic hybrid materials are biocompatible and biodegradable, in particular, this relates to the system based on TEOS-poly(ε-caprolactam). These nanobiocomposites will be analyzed in Chap. 7.

References

1. P. Colombo, G. Mera, R. Riedel, D. Soraru, J. Am. Ceram. Soc. **93**, 1805 (2010)
2. P. Gómez-Romero, C. Sanchez, *Functional Hybrid Materials* (Wiley-VCH, Weinheim, 2004)
3. G. Kickelbick, *Hybrid Materials* (Wiley-VCH, Weinheim, 2006)
4. T. Liu, K.P. Lim, W.C. Tjin, K.P. Pramoda, Z.K. Chen, Polymer **44**, 3529 (2003)

5. J.C. Huang, C.B. He, Y. Xiao, K.Y. Mya, J. Dai, Y.P. Siow, Polymer **44**, 4491 (2003)
6. A.F. Zimmerman, G. Palumbo, K.T. Aust, U. Erb, Mater. Sci. Eng. A **328**, 137 (2002)
7. I. Isik, U. Yilmazer, G. Bayram, Polymer **44**, 6371 (2003)
8. A.P.V. Pereira, W.L. Vasconcelos, R.L. Orerfice, J. Non-Cryst. Solids **273**, 180 (2000)
9. J. Chen, R. Chareonsak, V. Puengpipat, S. Marturunkakul, J. Appl. Polym. Sci. **74**, 1341 (1999)
10. S.D. Burnside, E.P. Giannelis, J. Polym. Sci. Part B Polym. Phys. **38**, 1595 (2000)
11. J.P. Jolivet, *Metal Oxide Chemistry and Synthesis* (Wiley, Chichester, 2000)
12. S. Wu, G. Zhou, M. Gu, Opt. Mater. **29**, 12 (2007)
13. E.O. Oh, K. Chakrabarti, H.Y. Jung, C.M. Whang, Mater. Sci. Eng. B **90**, 60 (2002)
14. A.D. Pomogailo, Russ. Chem. Rev. **69**, 53 (2000)
15. A.D. Pomogailo, Colloid J. **67**, 658 (2005)
16. A.D. Pomogailo, Vysokomolek. Soed. C **48**(7), 1318 (2006)
17. A.D. Pomogailo, A.S. Rozenberg, I.E. Uflyand. Metal nanoparticles in polymers (Khimiya, Moscow, 2000)
18. A.D. Pomogailo, Inorg. Chem. **41**(1), S47 (2005)
19. E.V. Tarasyuk, O.A. Shilova, A.M. Bochkin, A.D. Pomogailo, Glass Phys. Chem. **32**(4), 439 (2006)
20. A.M. Herring, J. Macromol. Sci. Part C Polym. Rev. **46**, 245 (2006)
21. M. Meyer, A. Fischer, H. Hoffmann, J. Phys. Chem. B **106**, 1528 (2002)
22. R. Riedel, A. Gurlo, E. Ionescu, Chem. Unserer Zeit **44**, 208 (2010)
23. G.A. Icopini, S.L. Brantley, P.J. Heaney, Geochim. Cosmochim. Acta **69**(2), 293 (2005)
24. O.A. Shilova, V.V. Shilov, Nanosyst. Nanomater. Nanotechnol. **1**, 9 (2003)
25. A.I. Maximov, V.A. Moshnikov, Y.M. Tairov, O.A. Shilova, *The Basis of Sol-gel Technology of Nanocomposites* (Elmor, St. Petersburg, 2008)
26. B.V. Eremenko, T.N. Bezuglaya, A.N. Savitskaya, M.L. Malysheva, I.S. Kozlov, L.G. Bogodist, Colloid J. **63**, 194 (2001)
27. V.V. Smirnov, M.M. Levitskii, I.G. Tarkhanova, B.G. Zavin, F.I. Bilyachenko, Kinet. Catal. **44**, 625 (2003)
28. A.D. Pomogailo, V.N. Kestelman, *Metallopolymer Nanocomposites* (Springer, Berlin/Heidelberg/New York, 2005)
29. G. Oskam, A. Nellore, R. Pen, P. Searson, J. Phys. Chem. B **107**, 1734 (2003)
30. M.Z.-C. Hu, J.T. Zielke, C.H. Byers, J. Mater. Sci. **35**, 1957 (2000)
31. C. Boissiere, D. Grosso, H. Amenitsch, A. Gibaud, A. Coupe, N. Baccile, C. Sanchez, Chem. Commun. **2798** (2003)
32. W. Bahloul, O. Oddes, V. Bounor-Legarer, F. Merlis, P. Cassagnau, B. Vergnes, AICHE J. **57**, 2174 (2011)
33. D. Arcos, A. Lopez-Noriega, E. Ruiz-Hernandez, O. Terasaki, M. Vallet-Regi, Chem. Mater. **21**, 1000 (2009)
34. M. Adachi, Y. Murata, J. Takao, J. Jiu, M. Sakamoto, F. Wang, J. Am. Chem. Soc. **126**, 14943 (2004)
35. F.X. Perrin, V. Ngujen, J.L. Vernert, J. Sol-Gel Sci. Technol. **28**, 205 (2003)
36. E.L. Crepaldi, G.J.A.A. Soler-Illia, D. Grosso, F. Cagnol, F. Ribot, C. Sanchez, J. Am. Chem. Soc. **125**, 9770 (2003)
37. L.M. Ellerby, C.R. Nishida, F. Nishida, S.A. Yamanaka, B. Dunn, V.J. Selverstone, J.I. Zink, Science **255**, 113 (1992)
38. R. Reisfeld, D. Brusilovsky, M. Eyal, E. Miron, Z. Burstein, J. Ivri, Chem. Phys. Lett. **160**, 43 (1989)
39. E.J.A. Pope, A. Asami, J.D. Mackenzie, J. Mater. Res. **4**, 1018 (1989)
40. Y. Hu, J.D. Mackenzie, Mater. Res. Soc. Proc. **271**, 681 (1992)
41. A. Blaaderen, A. Vrij, J. Colloid Interface Sci. **156**, 1 (1993)
42. E.M. Moreno, D. Levy, Chem. Mater. **12**, 128 (2000)
43. M.T. Reetz, R. Wenkel, D. Avnir, Synthesis **781** (2000)

44. P.D. Cozzoli, A. Kornowski, H. Weller, J. Am. Chem. Soc. **125**, 14539 (2003)
45. C. Bartic, H. Jansen, A. Campitelli, S. Borghs, Org. Electron. **3**, 65 (2002)
46. X. Wang, J. Zhuang, Q. Peng, Y. Li, Nature **437**, 121 (2005)
47. F. Krumeich, H.-J. Muhr, M. Niederberger, F. Bieri, B. Schnyder, R. Nesper, J. Am. Chem. Soc. **121**, 8324 (1999)
48. W. Que, X. Hu, Q.Y. Zhang, Chem. Phys. Lett. **369**, 354 (2003)
49. M. Kakihana, M. Yoshimura, Bull. Chem. Soc. Jpn. **72**, 1427 (1999)
50. K.T. Nam, D.-W. Kim, P.J. Yoo, C.-Y. Chiang, N. Meethong, P.T. Hammond, Y.-M. Chiang, A.M. Belcher, Science **312**, 885 (2006)
51. N.C. Bigall, M. Reitzig, W. Naumann, P. Simon, K.-H. van Pée, A. Eychmüller, Angew. Chem. Int. Ed. **47**, 7876 (2008)
52. S.-M. Lee, E. Pippel, U. Gösele, C. Dresbach, Y. Qin, C.V. Chandran, T. Bräuniger, M.K.G. Hause, Science **324**, 488 (2009)
53. A.D. Pomogailo, V.S. Savost'yanjov, *Synthesis and Polymerization of Metal-Containing Monomers* (CRC Press, Boca Raton, 1994)
54. C. Rottman, G. Grader, D. Avnir, Chem. Mater. **13**, 3631 (2001)
55. S. Gross, V. Di Noto, G. Kickelbick, U. Schudert, Mater. Res. Soc. Symp. Proc. **726**, Q4.1, 1 (2002)
56. L. Long, E. Ruckenstain, J. Appl. Polym. Sci. **67**, 1891 (1998)
57. B. Moraru, N. Hüssing, G. Kickellbick, U. Schubert, P. Fratzl, H. Peterlik, Chem. Mater. **14**, 2732 (2002)
58. T. Halamus, P. Wojciechowski, Polym. Adv. Technol. **18**, 411 (2007)
59. H.S. Nalwa, *Handbook of Organic–Inorganic Hybrid Materials and Nanocomposites*, vol. 1 (American Scientific Publishers, Los Angeles, 2004)
60. G. Fornasieri, L. Rozes, S. Le Calve, B. Alonso, D. Massiot, N. Rager, M. Evain, K. Boubekeur, C. Sanchez, J. Am. Chem. Soc. **127**, 4869 (2005)
61. I. Mijatovic, G. Kickelbick, U. Schubert, Eur. J. Inorg. Chem. **1933** (2001)
62. S. Gross, V.D. Noto, U. Schubert, J. Non-Cryst. Solids **322**, 154 (2003)
63. G. Kickelbick, U. Schubert, Eur. J. Inorg. Chem. **2**, 159 (1998)
64. U. Schubert, Polym. Int. **58**, 317 (2009)
65. A.D. Pomogailo, N.D. Golubeva, Izv. Ross. Akad. Nauk, Ser. Khim. **12**, 2139 (1994)
66. E.I. Klabunovskii, E.I. Karpeiskaya, G.I. Dzhardimalieva, N.D. Golubeva, A.D. Pomogailo, Russ. Chem. Bull. **48**, 1717 (1999)
67. A.D. Pomogailo, G.I. Dzhardimalieva, V.N. Kestelman, *Macromolecular Metal Carboxylates and Their Nanocomposites* (Springer, Berlin/Heidelberg, 2010)
68. D.-J. Lin, C.-C. Chen, Y.-C. Su, S.-H. Huang, L.-P. Cheng, J. Appl. Polym. Sci. **94**, 1927 (2004)
69. M. Camail, M. Hubert, A. Margaillan, J.L. Vernet, Polymer **25**, 6533 (1998)
70. F. Aldinger, S. Prinz, N. Janakiraman, R. Kumar, M. Christ, M. Weinmann, A. Zimmermann, Int. J. Self-Proparg High-Temp. Synth. **10**, 249 (2001)
71. K. Matyjaszewski, T.P. Davis, *Handbook of Radical Polymerization* (Wiley Int, Hoboken, 2002)
72. T. von Werner, T.E. Patten, J. Am. Chem. Soc. **123**, 7497 (2001)
73. H.-C. Huang, T.-E. Hsieh, J. Appl. Polym. Sci. **117**, 1252 (2010)
74. E. Müh, H. Frey, J.E. Klee, R. Mülhaupt, Adv. Funct. Mater. **11**, 425 (2001)
75. M. Yoshihara, H. Oie, A. Okada, H. Matsui, S. Ohshiro, Macromolecules **35**, 2435 (2002)
76. K. Dahmouche, L.D. Carlos, C.V. Santilli, V.Z. Bermudes, A.F. Craievich, J. Phys. Chem. B **102**, 4377 (2002)
77. F. Gao, Y. Tong, S.R. Schricker, B.M. Culbertson, Polym. Adv. Technol. **12**, 355 (2001)
78. J.D. Lichenhan, in *Polymeric Materials Encyclopedia*, ed. by J.C. Salamone (CRC Press, Boca Raton, 1996)
79. M.M.E. Jacob, E. Hackett, E.P. Giannelis, J. Mater. Chem. **13**, 1 (2003)

80. S.C. Hong, J.-F. Lutz, Y. Inoue, C. Strissel, O. Nuyken, K. Matyjaszewski, Macromolecules **36**, 1075 (2003)
81. T. Werne, T.E. Patten, J. Am. Chem. Soc. **121**, 7400 (1999)
82. J. Massey, K.N. Power, I. Manners, M.A. Winnik, J. Am. Chem. Soc. **120**, 9533 (1998)
83. C.-S. Tan, T.-W. Kuo, J. Appl. Polym. Sci. **98**, 750 (2005)
84. J. Xie, Y. Zheng, J.Y. Ying, J. Am. Chem. Soc. **131**, 888 (2009)
85. V. Berry, S. Rangaswamy, R.F. Saraf, Nano Lett. **4**, 939 (2004)
86. V. Berry, A. Gole, S. Kundu, C.J. Murphy, R.F. Saraf, J. Am. Chem. Soc. **127**, 17600 (2005)
87. M. Nogi, H. Yano, Adv. Mater. **20**, 1849 (2008)
88. C. Boissiere, D. Grosso, A. Chaumonnot, L. Nicole, C. Sanchez, Adv. Mater. **23**, 599 (2011)
89. S.-K. Lee, D.S. Yun, A.M. Belcher, Biomacromolecules **7**, 14 (2006)
90. V. Berry, R.F. Saraf, Angew. Chem. Int. Ed. **44**, 6668 (2005)
91. B. Samanta, H. Yan, N.O. Fischer, J. Shi, D.J. Jerryc, V.M. Rotello, J. Mater. Chem. **18**, 1204 (2008)
92. P.M. Ajayan, L.S. Schadler, P.V. Braun (eds.), *Nanocomposite Science and Technology* (Wiley-VCH, Weinheim, 2004)
93. M. Grzelczak, M.A. Correa-Duarte, V. Salgueirinõ-Maceira, R.D.M. Giersig, L.M. Liz-Marzán, Adv. Mater. **18**, 415 (2006)
94. M.A. Correa-Duarte, J. Pérez-Juste, A. Sánchez-Iglesias, M. Giersig, L.M. Liz-Marzán, Angew. Chem. Int. Ed. **44**, 4375 (2005)
95. S. Zhang, W. Ni, X. Kou, M.H. Yeung, L. Sun, J. Wang, C. Yan, Adv. Funct. Mater. **17**, 3258 (2007)
96. M. Grzelczak, M.A. Correa-Duarte, V. Salgueirinõ-Maceira, B. Rodriguez-González, J. Rivas, L.M. Liz-Marzán, Angew. Chem. Int. Ed. **46**, 7026 (2007)
97. H.-S. Qian, M. Antonietti, S.-H. Yu, Adv. Funct. Mater. **17**, 637 (2007)
98. T. Li, B. Ye, Z. Niu, P. Thompson, S. Seifert, B. Lee, Q. Wang, Chem. Mater. **2046** (2009)
99. M. Gensheimer, M. Becker, A. Brandis-Heep, J.H. Wendorff, R.K. Thauer, A. Greiner, Adv. Mater. **19**, 2480 (2007)
100. A. Singh, S. Hede, M. Sastry, Small **3**, 466 (2007)
101. M.N. Tahir, M. Eberhardt, H.A. Therese, P.T.U. Kolb, W.E.G. Müller, H.-C. Schröder, W. Tremel, Angew. Chem. Int. Ed. **45**, 4803 (2006)
102. M.-S. Hu, H.-L. Chen, C.-H. Shen, L.-S. Hong, B.-R. Huang, K.-H. Chen, L.-C. Chen, Nat. Mater. **5**, 102 (2006)
103. S.K. Dixit, N.L. Goicochea, M.-C. Daniel, A. Murali, L. Bronstein, M. De, B. Stein, V.M. Rotello, C.C. Kao, B. Dragnea, Nano Lett. **6**, 1993 (2006)
104. M. Liu, Q. Zhao, C. Wang, Z. Mo, S. Cao, Polymer **44**, 2537 (2003)
105. R. Agrawal, N.S. Saxena, K.B. Sharma, S. Thomas, M.S. Sreekala, Mater. Sci. Eng. A **277**, 77 (2000)
106. K.M. Bromley, A.J. Patil, A.W. Perriman, G. Stubbs, S. Mann, J. Mater. Chem. **18**, 4796 (2008)
107. H. Dong, D. Wang, G. Sun, J.P. Hinestroza, Chem. Mater. **20**, 6627 (2008)
108. A. Sugunan, P. Melin, J. Schnürer, J.G. Hilborn, J. Dutta, Adv. Mater. **19**, 77 (2007)
109. G. Antoniadis, K.M. Paraskevopoulos, D. Bikiaris, K. Chrissafis, Thermochim. Acta **493**, 68 (2009)
110. A.I. Wallbank, J.F. Corrigan, Can. J. Chem. **80**, 1592 (2002)
111. H. Shiho, N. Kawahashi, Colloid Polym. Sci. **278**, 270 (2000)
112. T. Sugimoto, *Fine Particles. Synthesis, Characterization and Mechanism of Crown* (Marcel Dekker, New York, 2000)
113. J.B. Carroll, A.J. Waddon, H. Nakade, V.M. Rotello, Macromolecules **36**, 6289 (2003)
114. K. Yoshinaga, Bull. Chem. Soc. Jpn. **75**, 2349 (2002)
115. N. Jungmann, M. Schmidt, M. Mascos, Macromolecules **36**, 3974 (2003)
116. T. Sanji, Y. Nakatsuka, M. Tanaka, H. Sakurai, Chem. Lett. **32**, 980 (2003)
117. F. Iskandar, A. Mikrajuddin, K. Okuyama, Nano Lett. **1**, 231 (2001)

118. F. Iskandar, A. Mikrajuddin, K. Okuyama, Nano Lett. **2**, 389 (2002)
119. A. Thill, O. Spalla, J. Colloid Interface Sci. **291**, 477 (2005)
120. S. Jain, J.G.P. Goossens, M. van Duin, Macromol. Symp. **233**, 225 (2006)
121. J. Jang, H. Park, J. Appl. Polym. Sci. **83**, 1817 (2002)
122. D.J. Belton, S.V. Patwardhan, C.C. Perry, J. Mater. Chem. **15**, 4629 (2005)
123. S.V. Patwardhan, N. Mukherje, S.J. Clarson, Silicon Chem. **1**, 47 (2002)
124. D.J. Belton, S.V. Patwardhan, V.V. Annenkov, E.N. Danilovtseva, C.C. Perry, PNAS **105**, 5963 (2008)
125. V.V. Annenkov, S.V. Patwardhan, D. Belton, E.N. Danilovtseva, C.C. Perry, Chem. Commun. **1521** (2006)
126. K.M. Hawkins, S.S. Wang, D.M. Ford, D.F. Shantz, J. Am. Chem. Soc. **126**, 9112 (2004)
127. Y. Jia, G.M. Gray, J.N. Hay, Y. Li, G.-F. Unali, F.L. Baines, S.P. Armes, J. Mater. Chem. **15**, 2202 (2005)
128. X. Li, T. Yang, Q. Gao, J. Yuan, S. Cheng, J. Colloid Interface Sci. **338**, 99 (2009)
129. V.V. Annenkov, E.N. Danilovtseva, E.A. Filina, Y.V. Likhoshway, J. Polym. Sci. Part A Polym. Chem. **44**, 820 (2006)
130. V.V. Annenkov, E.N. Danilovtseva, Y.V. Likhoshway, S.V. Patwardhan, C.C. Perry, J. Mater. Chem. **18**, 553 (2008)
131. K.D. Demadis, A. Ketsetzi, K. Pachis, V.M. Ramos, Biomacromolecules **9**, 3288 (2008)
132. B. Leng, Z. Shao, P.H.H. Bomans, L.J. Brylka, N.A.J.M. Sommerdijk, G. de With, W. Ming, Chem. Commun. **46**, 1703 (2010)
133. R. Saito, S. Kobayashi, T. Hosoya, J. Appl. Polym. Sci. **97**, 1835 (2005)
134. R. Saito, T. Tobe, J. Appl. Polym. Sci. **93**, 749 (2004)
135. R. Saito, T. Tobe, Polym. Adv. Technol. **16**, 232 (2005)
136. V. Ratta, A. Ayambem, J.E. McGrath, G.L. Wilkes, Polymer **42**, 6173 (2001)
137. M.A. Wahab, I.I. Kim, W.J. Cho, C.S. Ha, Mol. Cryst. Liq. Cryst. **417**, 127 (2004)
138. J.H. Lee, J.S. Im, K.W. Song, J.O. Lee, K. Yoshinaga, J. Macromol, Sci. Part A Pure Appl. Chem. **41**, 1345 (2004)
139. C. Chang, G.P. Wang, H.C. Tsai, Y.S. Hong, Int. J. Polym. Anal. Charact. **8**, 157 (2003)
140. Y.H. Zhang, Y. Li, S.Y. Fu, J.H. Xin, W.A. Daoud, L.F. Li, Polymer **46**, 8373 (2005)
141. C. Yen, W.C. Chen, D.J. Liaw, H.Y. Lu, Polymer **44**, 7079 (2003)
142. N. Cheval, F. Xu, N. Gindy, R. Brooks, Y. Zhu, A. Fahmi, Macromol. Chem. Phys. **212**, 180 (2011)
143. S. Karatas, N. Kayaman-Apohan, H. Demirer, A. Gungor, Polym. Adv. Technol. **18**, 490 (2007)
144. L. Shen, I.Y. Phang, T. Liu, K. Zeng, Polymer **45**, 8221 (2004)
145. X. Xu, B. Li, H. Lu, Z. Zhang, H. Wang, Appl. Surf. Sci. **254**, 1456 (2007)
146. X. Liu, Q. Wu, L.A. Berglund, Polymer **43**, 4967 (2002)
147. D.L. Long, E. Burkholder, L. Cronin, ChemInform **38**, 15 (2007)
148. M.T. Pope, *Polyoxometalate Chemistry: From Topology Via Self-Assembly To Applications* (Springer, Heidelberg, 2001)
149. D.L. Long, L. Cronin, Chem. Eur. J. **12**, 3698 (2006)
150. P. Gomez-Romero, Adv. Mater. **13**, 163 (2001)
151. T. He, J. Yao, Prog. Mater. Sci. **51**, 810 (2006)
152. S. Liu, D.G. Kurth, H. Möhwald, D. Volkmer, Adv. Mater. **14**, 225 (2002)
153. W. Qi, H. Li, L. Wu, J. Phys. Chem. B **112**, 8257 (2008)
154. M. Jiang, E. Wang, G. Wei, L. Xu, Z. Li, J. Colloid Interface Sci. **275**, 596 (2004)
155. C.L. Hill, J. Mol. Catal. A Chem. **262**, 2 (2007)
156. I. Kozhevnikov, *Catalysts for Fine Chemical Synthesis: Catalysis by Polyoxometalates*, vol. 2 (Wiley-VCH, Weinheim, 2002)
157. N. Muradov, A.T. Raissi, J. Sol. Energy **128**, 326 (2006)
158. P. Gómez-Romero, K. Cuentas-Gallegos, M. Lira-Cantú, N. Casañ-Pastor, J. Mater. Sci. **40**, 1423 (2005)

159. Y. Han, Y. Xiao, Z. Zhang, B. Liu, P. Zheng, S. He, W. Wang, Macromolecules **42**, 6543 (2009)
160. N. Cheval, F. Xu, N. Gindy, R. Brooks, Y.Q. Zhu, A. Fahmi, Key Eng. Mater. **450**, 169 (2010)
161. A. Arkhireeva, J.N. Hay, J. Mater. Chem. **13**, 3122 (2003)
162. A. Arkhireeva, A.M. Groth, J.N. Hay, Polym. Int. **56**, 350 (2007)
163. J. Chiefari, B. Dao, A.M. Groth, J.H. Hodgkin, High Perform. Polym. **15**, 269 (2003)
164. J.H. Hodgkin, B.G. Laycock, T.C. Morton, D.G. Hawthorn, *Aqueous polyimide process* (US Patent 6333391, 2001)
165. B.-K. Chen, T.-M. Chiu, S.-Y. Tsay, J. Appl. Polym. Sci. **94**, 382 (2004)
166. C.-L. Chiang, C.-C.M. Ma, D.-L. Wu, H.-C. Kuan, J. Polym. Sci. Part A Polym. Chem. **41**, 905 (2003)
167. G. Lu, Y. Huang, Y. Yan, T. Zhao, Y. Yu, J. Polym, Sci. Part A Polym. Chem. **41**, 2599 (2003)
168. B.-K. Chen, Y.-T. Fang, J.-R. Cheng, S.-Y. Tsay, J. Appl. Polym. Sci. **105**, 1093 (2007)
169. C.J. Cornelius, E. Marand, J. Membr. Sci. **202**, 97 (2002)
170. Z. Ahmad, J.E. Mark, Chem. Mater. **13**, 3320 (2001)
171. J. Liu, Y. Gao, F. Wang, M. Wu, J. Appl. Polym. Sci. **75**, 384 (2000)
172. V.Y. Kramarenko, T.A. Shantalil, I.L. Karpova, K.S. Dragan, E.G. Privalko, V.P. Privalko, D. Fragiadakis, P. Pissis, Polym. Adv. Technol. **15**, 144 (2004)
173. K. Kyriakos, K.N. Raftopoulos, P. Pissis, A. Kyritsis, F. Naether, L. Haeussler, D. Fischer, A. Vyalikh, U. Scheler, U. Reuter, J. Appl. Polym. Sci. **128**, 3771 (2013)
174. H.-L. Lin, T.-Y. Chao, Y.-F. Shih, S.A. Dai, W.-C. Su, R.-J. Jeng, Polym. Adv. Technol. **19**, 984 (2008)
175. G. Lu, Y. Huang, Y. Yan, T. Zhao, Y. Yu, J. Appl. Polym. Sci. **102**, 52 (2006)
176. K. Kuraoka, Y. Chujo, T. Yamazawa, Chem. Commun. **2477** (2000)
177. G.-H. Hsiue, J.-K. Chen, Y.-L. Liu, J. Appl. Polym. Sci. **76**, 1609 (2000)
178. Y. Gotoa, N. Otsukab, H. Sawada, Polym. Adv. Technol. **23**, 290 (2012)
179. K. Yoshinaga, Bull. Chem. Soc. Jpn. **5**, 2349 (2002)
180. K. Nakane, J. Ohashi, F. Suzuki, J. Appl. Polym. Sci. **71**, 185 (1999)
181. N. Uangvanich, K.A. Mauritz, J. Appl. Polym. Sci. **67**, 1799 (1998)
182. S.M. de Paul, J.W. Zwanziger, R. Ulrich, U. Wiesner, H.W. Spiess, J. Am. Chem. Soc. **121**, 5727 (1999)
183. Z.-K. Zhu, Y. Yang, J. Yin, Z.-N. Qi, J. Appl. Polym. Sci. **73**, 2977 (1999)
184. M.-P. Zheng, Y.-P. Jin, G.-L. Jin, M.-Y. Gu, J. Mater. Sci. Lett. **19**, 433 (2000)
185. S. Kobayashi, K. Hanabusa, M. Suzuki, M. Kimura, H. Shirai, Bull. Chem. Soc. Jpn. **73**, 1913 (2000)
186. H. Guo, X. Zhao, G. Ning, G. Liu, Langmuir **19**, 4884 (2003)
187. B. Wu, C. Guo, N. Zheng, Z. Xie, G.D. Stucky, J. Am. Chem. Soc. **130**, 17563 (2008)
188. X. Chen, S.S. Mao, Chem. Rev. **107**, 2891 (2007)
189. D.V. Bavykin, J.M. Friedrich, F.C. Walsh, Adv. Mater. **18**, 2807 (2006)
190. C. Su, B.-Y. Hong, C.-M. Tseng, Catal. Today **96**, 119 (2004)
191. L. Gang, B.G. Anderson, J. Grondelle, R.A. Santen, Appl. Catal. D **40**, 101 (2002)
192. L. Zhang, J.C. Yu, H.Y. Yip, Q. Li, K.W. Kwong, A.-W. Xu, P.K. Wong, Langmuir **19**, 10372 (2003)
193. H. Li, G. Zhao, Z. Chen, B. Song, G. Hanz, J. Am. Ceram. Soc. **93**, 445 (2010)
194. S. Pega, C. Boissiere, D. Grosso, T. Azais, A. Chaumonnot, C. Sanchez, Angew. Chem. Int. Ed. **48**, 2784 (2009)
195. M.T. Bore, M.P. Mokhonoana, T.L. Ward, N.J. Coville, A.K. Datye, Microporous Mesoporous Mater. **95**, 118 (2006)
196. M. Colilla, M. Manzano, I. Izquierdo-Barba, M. Vallet-Regi, C. Boissiere, C. Sanchez, Chem. Mater. **22**, 1821 (2010)
197. M.T. Bore, R.F. Marzke, T.L. Ward, A.K. Datye, J. Mater. Chem. **15**, 5022 (2005)

198. F. Goettmann, C. Boissiere, D. Grosso, F. Mercier, P. Le Floch, C. Sanchez, Chem. Eur. J. **11**, 7416 (2005)
199. T.H. Zheng, J.B. Pang, G. Tan, J.B. He, G.L. McPherson, Y.F. Lu, V.T. John, J.J. Zhan, Langmuir **23**, 5143 (2007)
200. J.E. Hampsey, S. Arsenault, Q.Y. Hu, Y.F. Lu, Chem. Mater. **17**, 2475 (2005)
201. C. Coelho, T. Azais, C. Bonhomme, C. Bonhomme-Coury, C. Boissiere, G. Laurent, D. Massiot, Compte Rendus de Chimie **11**, 387 (2008)
202. C. Sanchez, H. Arribart, M.M. Giraud Guille, Nat. Mater. **4**, 277 (2005)
203. R. Backov, Soft Matter **2**, 452 (2006)
204. C. Sanchez, L. Rozes, F. Ribot, C. Laberty-Robert, D. Grosso, C. Sassoye, C. Boissiere, L. Nicole, Comptes Rendus Chimie (CRC) **13**, 3 (2010)
205. T. Yogo, T. Yamamoto, W. Sakamoto, S.H. Hirano, J. Mater. Res. **19**, 3290 (2004)
206. L. Nicole, L. Rozes, C. Sanchez, Adv. Mater. **22**, 3208 (2010)
207. S.I. Hirano, T. Yogo, W. Sakamoto, S. Yamada, T. Nakamura, T. Yamamoto, H. Ukai, K. Bannp, T. Nakafuku, Y. Ando, J. Sol-Gel Sci. Technol. **26**, 35 (2003)
208. G. Kwak, S.Y. Kim, M. Fujiki, T. Masuda, Y. Kawakami, T. Aoki, Chem. Mater. **16**, 1864 (2004)
209. R.A. Caruso, M. Antonietti, M. Giersig, H.-P. Hentze, J. Jia, Chem. Mater. **13**, 1114 (2001)
210. E. Prouzet, S. Ravaine, C. Sanchez, R. Backov, New J. Chem. **32**, 1284 (2008)
211. K. Okuyama, I.W. Lenggoro, Chem. Eng. Sci. **58**, 537 (2003)
212. A.M. Ruiz, G. Dezanneau, J. Arbiol, A. Cornet, J.R. Morante, Chem. Mater. **16**, 862 (2004)
213. P.H. Mutin, V. Lafond, A.F. Popa, M. Granier, L. Markey, A. Dereux, Chem. Mater. **16**, 5670 (2004)
214. R.A. Zoppi, S. Das Neves, S.P. Nunes, Polymer **41**, 5461 (2000)
215. J.-B. Tang, R.-G. Qiu, Y. Wei Chin, J. Chem. **19**, 198 (2001)
216. K. Kosuge, P.S. Sing, Chem. Lett. **9** (1999)
217. H. Kim, N.C. Pramanik, B.Y. Ahn, S.I. Seok, Phys. Status Solid A Appl Mater. Sci. **203**, 1962 (2006)
218. S. Miraux, E. Duguet, C. Sanchez, J. Mater. Chem. **17**, 1563 (2007)
219. J.W. Liu, X.M. Jiang, C. Ashle, C.J. Brinker, J. Am. Chem. Soc. **131**, 7567 (2009)
220. E. Ruiz-Hernandez, A. Lopez-Noriega, D. Arcos, I. Izquierdo-Barba, O. Terasaki, M. Vallet-Regi, Chem. Mater. **19**, 3455 (2007)
221. M.H. Sorensen, R.W. Corkery, J.S. Pedersen, J. Rosenholm, P.C. Alberius, Microporous Mesoporous Mater. **113**, 1 (2008)
222. C.K. Muoto, E.H. Jordan, M. Gell, M. Aindow, J. Am. Ceram. Soc. **93**, 3102 (2010)
223. C.K. Muoto, E.H. Jordan, M. Gell, M. Aindow, J. Am. Ceram. Soc. **94**, 372 (2011)
224. T. Jiang, A.K. Mukherjee, J. Am. Ceram. Soc. **93**, 769 (2010)
225. J. Azurdia, J. Marchal, R.M. Laine, J. Am. Ceram. Soc. **89**, 2749 (2006)
226. I. Prakash, P. Muralidharan, N. Nallamuthu, N. Satyanarayana, M. Venkateswarlu, D. Carnahan, J. Am. Ceram. Soc. **89**, 2220 (2006)
227. T. Shiono, K. Shiono, K. Miyamoto, G. Pezzotti, J. Am. Ceram. Soc. **83**, 235 (2000)
228. X.L. Duan, D. Yuan, X. Cheng, C. Luan, Z. Sun, X. Wei, S. Guo, D. Xu, M. Lv, Inorg. Chem. Commun. **7**, 62 (2004)
229. Y.S. Han, J.B. Li, X.S. Ning, B. Chi, J. Am. Ceram. Soc. **88**, 3455 (2005)
230. T. Suzuki, K. Horibuchi, Y. Ohishi, J. Non-Cryst. Solids **351**, 2304 (2005)
231. J. Merikhi, H.O. Jungk, C. Feldmann, J. Mater. Chem. **10**, 1311 (2000)
232. P. Jeevanandam, Y. Koltypin, A. Gedanken, Mater. Sci. Eng. B **90**, 125 (2002)
233. N. Rajic, M. Ceh, R. Gabrovsek, V. Kaucic, J. Am. Ceram. Soc. **85**, 1719 (2002)
234. M. Belmonte, Adv. Eng. Mater. **8**, 693 (2006)
235. K.N. Lee, Surf. Coat. Technol. **133–134**, 1 (2000)
236. N.P. Padture, M. Gell, E.H. Jordan, Science **296**, 280 (2002)
237. B. Gleeson, J. Propuls. Power **22**, 375 (2006)
238. S. Direr, R. Ceccato, F. Babonneau, J. Sol-Gel Sci. Technol. **34**, 53 (2005)

239. X.Q. Cao, R. Vassen, D. Stoever, J. Eur. Ceram. Soc. **24**, 1 (2004)
240. S. Tian, C.Z. Chen, D.Y. Wang, Q. Ji, Surf. Rev. Lett. **12**(3), 369 (2005)
241. Y.R. Vassen, G. Kerkhoff, D. Stoever, Mater. Sci. Eng. A **303**, 100 (2001)
242. L.B. Chen, Surf. Rev. Lett. **13**, 535 (2006)
243. R. Riedel, G. Mera, R. Hauser, A. Klonczynski, J. Ceram. Soc. Jpn. **114**, 425 (2006)
244. E. Ionescu, B. Papendorf, H.-J. Kleebe, R. Riedel, J. Am. Ceram. Soc. **93**, 1783 (2010)
245. E. Ionescu, B. Papendorf, H.-J. Kleebe, F. Poli, K. Muller, R. Riedel, J. Am. Ceram. Soc. **93**, 1774 (2010)
246. J.L. Ferrari, K.O. Lima, L.J.Q. Maia, S.J.L. Ribeiro, R.R. Goncalves, J. Am. Ceram. Soc. **94**, 1230 (2011)
247. R.M. Almeida, A.C. Marques, S. Pelli, G.C. Righini, A. Chiasera, M. Mattarelli, M. Montagna, C. Tosello, R.R. Goncalves, H. Portales, S. Chaussedent, M. Ferrari, L. Zampedri, Philos. Mag. **84**, 1659 (2004)
248. R.R. Gonçalves, G. Carturan, L. Zampedri, M. Ferrari, A. Chiasera, M. Montagna, G.C. Righini, S. Pelli, S.J.L. Ribeiro, Y. Messaddeq, J. Non-Cryst. Solids **322**, 306 (2003)
249. R. Balda, J. Fernandez, J.M. Fernandez-Navarro, Opt. Express **17**, 8781 (2009)
250. L.J.Q. Maia, V.R. Mastelaro, S. Pairis, A.C. Hernandes, A. Ibanez, J. Solid State Chem. **180**, 611 (2007)
251. R.R. Gonçalves, G. Carturan, L. Zampedri, M. Ferrari, M. Montagna, A. Chiasera, G.C. Righini, S. Pelli, S.J.L. Ribeiro, Y. Messaddeq, Appl. Phys. Lett. **81**, 28 (2002)
252. L. Zampedri, C.G. Righini, H. Portales, S. Pelli, L. Zampedri, G.C. Nunzi, M. Montagna, M. Mattarelli, R.R. Gonçalves, M. Ferrari, A. Chiasera, M. Bouazaoui, C. Armellini, J. Non-Cryst. Solids **345–346**, 580 (2004)
253. R.R. Gonçalves, G. Carturan, M. Montagna, M. Ferrari, L. Zampedri, S. Pelli, G.C. Righini, S.J.L. Ribeiro, Y. Messaddeq, Opt. Mater. **25**, 131 (2004)
254. J.L. Ferrari, K.O. Lima, L.J.Q. Maia, S.J.L. Ribeiro, A.S.L. Gomes, R.R. Gonçalves, J. Nanosci. Nanotechnol. **10**, 1 (2010)
255. R.R. Gonçalves, J.J. Guimaraes, J.L. Ferrari, L.J.Q. Maia, S.J.L. Ribeiro, J. Non-Cryst. Solids **354**, 4846 (2008)
256. A. Polman, F.C.J.M. van Veggel, J. Opt. Soc. Am. B **21**, 871 (2004)
257. N. Andersson, P. Alberius, J. Ortegren, M. Lindgren, L. Bergstrom, J. Mater. Chem. **15**, 3507 (2005)
258. L. Li, C.K. Tsung, T. Ming, Z.H. Sun, W.H. Ni, Q.H. Shi, G.D. Stucky, J.F. Wang, Adv. Funct. Mater. **18**, 2956 (2008)
259. C.-H. Chen, J.K.M. Garofano, C.K. Muoto, A.L. Mercado, S.L. Suib, M. Aindow, M. Gell, E.H. Jordan, J. Am. Ceram. Soc. **94**, 367 (2011)
260. C.K. Muoto, E.H. Jordan, M. Gell, M. Aindow, J. Am. Ceram. Soc. **93**, 31029 (2010)
261. A. Borrell, V.G. Rocha, R. Torrecillas, A. Fernarndez, J. Am. Ceram. Soc. **94**, 2048 (2011)
262. S. Maensiri, P. Laokul, J. Klinkaewnarong, V. Amornkitbamrung, Mater. Sci. Eng. A **447**, 44 (2007)
263. J. Sun, L. Gao, M. Iwasa, T. Nakayama, K. Niihara, Ceram. Int. **31**, 1131 (2005)
264. K. Hirota, H. Hara, M. Kato, Mater. Sci. Eng. A **458**, 216 (2007)
265. S. Kobayashi, W. Kawai, Compos. Part A **38**, 114 (2007)
266. S.Y. Lee, H. Kim, P.C. McIntyre, K.C. Saraswat, J.S. Byun, Appl. Phys. Lett. **82**, 2874 (2003)
267. J. Dusza, G. Blugan, J. Morgiel, J. Kuebler, F. Inam, T. Peijs, M.J. Reece, V. Puchy, J. Eur. Ceram. Soc. **29**(15), 3177 (2009)
268. K. Waada, K. Yamada, T. Kondo, T. Mitsudo, Chem. Lett. **12** (2001)
269. Z. Liu, Y. Sakamoto, T. Ohsuma, K. Higara, O. Terasaki, C.H. Ko, H.J. Shin, R. Rioo, Angew. Chem. Int. Ed. **39**, 3107 (2000)
270. Y.-H. Han, J.M. Kim, G.D. Stucki, Chem. Mater. **12**, 2068 (2000)
271. H.J. Shin, R. Rioo, Z. Liu, O. Terasaki, J. Chem. Soc. **123**, 1246 (2001)
272. J.M. Watson, U.S. Oskan, J. Catal. **210**, 295 (2002)

273. A.D. Pomogailo, V.S. Savostuanov, G.I. Dzhardimalieva, A.V. Dubovitskii, A.N. Ponomarev, Russ. Chem. Bull. **44**, 1056 (1995)
274. M. Epifani, C. Giannini, L. Tapfer, L. Vasanelli, J. Am. Ceram. Soc. **83**, 2385 (2000)
275. J. Sort, S. Surinach, M.D. Baro, D. Muraviev, G.I. Dzhardimalieva, N.D. Golubeva, S.I. Pomogailo, A.D. Pomogailo, W.A.A. Macedo, D. Weller, V. Skumryev, J. Nogues, Adv. Mater. **466** (2006)
276. K.K. Baikerikar, A.B. Scranton, Polymer **42**, 431 (2001)
277. Y.-C. Chou, Y.-Y. Wang, T.-E. Hsieh, J. Appl. Polym. Sci. **105**, 2073 (2007)
278. C.-C. Chang, C.-H. Chang, Polym. Int. **59**, 910 (2010)
279. U. Schubert, J. Sol-Gel Sci. Technol. **26**, 47 (2003)
280. V.I. Roldugin, Uspekhi Khim. **72**, 1027 (2003)
281. J. Park, Fractals **8**, 301 (2001)
282. B.K. Coltrain, C.J.T. Landry, J.M. O'Reilly, A.M. Chamberlain, G.A. Rakes, J.S.S. Sedita, L.W. Kelts, M.R. Landry, V.K. Long, Chem. Mater. **5**, 1445 (1993)
283. P. Hajji, L. David, J.F. Gerard, J.P. Pascault, G. Vigier, J. Polym, Sci. Part B Polym. Phys. **37**, 3172 (1999)
284. D. Tian, S.W. Blacher, R. Jerome, Polymer **40**, 951 (1999)
285. T.M. Tillotson, L.W. Hrubesh, R.L. Simpson, R.S. Lee, R.W. Swansiger, L.R. Simpson, J. Non-Cryst. Solids **225**, 358 (1998)
286. T.M. Tillotson, A.E. Gash, R.L. Simpson, L.W. Hrubesh, J.H. Satcher Jr., J.F. Poco, J. Non-Cryst. Solids **285**, 338 (2001)
287. A. Pivkina, P. Ulyanova, Y. Frolov, S. Zavyalov, J. Schoonman, Propellants Explos. Pyrotech. **29**, 39 (2004)
288. U. Teipel (ed.), *Energetic Materials – Particle Processing and Characterization* (Wiley-VCH, Weinheim, 2005)
289. S. Cudziło, W. Kicinrski, Propellants Explos. Pyrotech. **34**, 155 (2009)
290. S. Bhandarkar, J. Am. Ceram. Soc. **87**, 1180 (2004)
291. A. Martino, A.G. Sault, J.S. Kawola, E. Boespflug, M.L.F. Phillips, J. Catal. **187**, 30 (1999)

Chapter 5
Physical Chemistry of Intercalated System

The driving force of development of intercalation chemistry is significant improvement of properties of fabricated nanocomposites and design of materials with new properties. Forming hybrid structures (sometimes called "ship-in-the-bottle") define important functional characteristics: improved mechanical strength (increase in modulus), enhanced stiffness, heat resistance (decrease in thermal expansion coefficient), heat stability, and other thermal physical properties, water resistance, interesting barrier properties for gas separation, high flame and fire resistance, electric and electrochemical behavior, size stability, chemical stability, different from the simple additive properties [1–20].

Hybrid-phase nanocomposites [21] are not only of academic, but of commercial interest, because they posses improved mechanical and thermal properties as compared with the same content of conventional fillers (such as carbon soot or deposited silica gel). The methods of production of hybrid nanocomposites from polymer solutions or polymerization in situ in combination with delamination and exfoliation were already studied at the end of the last century (for example, [22, 23]). However, there were technological drawbacks: these methods had low correlation with polymer production and demanded great amounts of organic solvents. At the beginning of 1990s more convenient technique appeared on the basis of PEO (polyethylene-oxide) and MMT (montmorillonite), which advantageously demonstrated the intercalation process in the polymer melt, as well as potential applications of the obtained products in solid phase electrolytes of recharged lithium batteries.

However, highly promising commercial strategy in this problem was formulated after the group of researchers of Toyota company [24–26] had found extraordinary strengthening of mechanical properties of polymer-layered nylon-based nanocomposites, which was caused by extremely large surface contacts of the ingredients and high aspect ratio reached in the intercalation/exfoliation process, and by homogenous dispersion of silica plates in the polymer matrix [27, 28]. On the one hand, these functional materials have a relation to nanocomposites via

A.D. Pomogailo, G.I. Dzhardimalieva, *Nanostructured Materials Preparation via Condensation Ways*, DOI 10.1007/978-90-481-2567-8_5

nanosized galleries of layered silicates (1—5 nm), distances between networks and layers formed by polymer and inorganic components, on the other hand, via the value of their exfoliated fragments [29].

Most widely known hybrid organic-inorganic nanocomposites are polymer-clay ones, which combine organic polymers with smectites. MMT is the most convenient layered silicate for application into intercalation systems, because their natural reserves are unlimited, substitution of their inlayer inorganic cations by organic ones, such as ammonium groups with long hydrocarbon chains, is easy. Such a modification provides better compatibility of a silicate with a polymer matrix [30, 31]. Polymer is inserted into interlayer space of a smectite via penetration from a solution, melt with following delamination (layering) or exfoliation (peeling) with in situ formation of mineral nanoparticles, which are uniformly distributed in the polymer. Intercalation of inorganic component into organic materials having limited interlayer distance with preserving of their layered structure is, on the one hand, a perfect way of structuring of organic-inorganic nano-ensembles [9], and, on the other hand, it provides additional possibilities for studying physical-chemical properties of these systems. These are supramolecular formations with original molecular architecture (see, for example, [9]).

Rigid crystalline matrices of these materials (*hosts*) with regulated system of percolation pores and nanosized layers can be filled with atomic or molecular *guest* structures: molecules of organic matters, not only monomer links, but also solvents, alkyl amines, crown ethers, cryptands, etc., and also clusters, nanoparticles, inorganic coordination polymers of $(CdS)_x$ type, large molecules, like C_{60} fullerenes. For example, intercalation of buckminsterfullerene functionalized with ethylene diamine into mica-like silicate of hectorite type was found [32].

Most substantial are other problems, whose solutions are closely connected to intercalation physical chemistry of periodic mesoporous structures: sorption and phase transitions; ion exchange and complex formation; formation of metal, oxide, sulfide, and semiconductor clusters and nanowires; covalent inoculation of ligand and functional groups to a *host*; finding of a composition and structure of hybrid materials, obtained via in situ assembling, condensation, polymerization, or insertion of primarily prepared macromolecules in the interlayer space, finding of topochemistry of individual stages of reactions, etc. Numerous studies concerning intercalation of organic metal organic and inorganic compounds, and properties of the formed products are published (previous studies on the subject were considered in our reviews [33, 34]. The great surge of the studies relates to the last decade and concerns mostly “bridges constructing” between inorganic materials having intercalation properties and intermediates forming, as well as to characteristics of hybrid-phase nanocomposites.

5.1 Composition, Structure, and Intercalation Properties of Layered Materials

From four thousands of known minerals hundreds are of zeolite type (amphoteric silicates) [35, 36]. Layered silicates or smectites, smectite clays are groups of clay rock minerals, 2D layered aluminosilicate, phyllosilicate, such as MMT are crystal formations of natural origin and produced by various synthetic methods (including hydrothermal, sol-gel synthesis, etc.). This is the most important type of solids with interesting functional and structural properties, whose special feature is a weak bond between individual layers. Natural Na-MMT clay of intercalation purity modified by long-chain ammonium cations, for example, trimethyl octadecyl ammonium chloride has exchange capacity 1.2 mg-eq/g, contains 56.3 mass.% of inorganic component, basal distance is $d_{001} = 2.07$ nm, aspect ratio, length to thickness ratio, is AR = l/c from 80 to 1,120, average AR is 320. For comparison, we shall consider oxide content of the native montmorillonite (%): SiO_2 62.9, Al_2O_3 19.6, TiO_2 0.009, FeO 0.32, Fe_2O_3 3.35, MnO 0.006, MgO 3.05, CaO 1.68, Na_2O 0.153, K_2O 0.53, exchange capacity is 0.76 mg-eq/g. For non-modified MMT thickness of one structural layer is about 1 nm (0.96 nm), surface of the plates reaches 800 m^2/g, aggregates consist of tens of such plates, their size in nm is: saponite 50–60, MMT 100–150, hectorite 200–300, for the latter AR is also high [30].

Most widely known are also other natural materials used due to their prominent intercalation abilities: hectorite, laponite, nontonite, beidelite, volkonskoite, sepiolite, stevensite, sauconite, swinfordite, kenyaite, bentonite, in particular, bentonite clay containing predominantly MMT, $(Na,Ca)_{0.33}(Al,Mg)_2Si_4O_{10}(OH)_2 \cdot (H_2O)_n$, $d_{001} = 1.206$ nm, product of volcanic ash erosion, layered mica silicates, including vermiculite, illite and close to them hydro-mica tubular attapulgite, a kind of palygorskite, the mineral with tubular structure formed owing to corrugation of structural layers, etc. The structure and composition of the layered silicates, their physical-chemical properties, the origin of active surface have been studied in detail for a long time, their models and properties of intercalation systems were considered in many reviews (see, for example, just one of them [37]). These are specific materials, whose interlayer space includes sandwich OH- groups of octahedron aluminum hydroxide and oxygen atoms of tetrahedron silica sheets: the crystals consist of altering layers of cations and negatively charged silicates in the 2:1 ratio. Their main structuring blocks are $Si(O, OH)_4$ tetrahedrons and $M(O, OH)_6$ octahedrons, in which $M = Mg^{2+}$, Al^{3+}, $Fe^{2+,3+}$. In kaolin (M = Fe and Mg) tetrahedrons and octahedrons are condensed in a monolayer with the ratio 1:1. In many minerals there is isomorphism, which is due to substitution of Si^{4+} ions in tetrahedral or Al^{3+} or Mg^{2+} in octahedral layers. In these cases a lattice can acquire charge depending on a valence state – from electrically neutral, in the mineral groups of pyrophyllite, talc to $x + y = 0.25$–0.6 (layered silicates with such a charge are particularly called smectites), 0.6–0.9 in vermiculites, synthetic clays of hectorite

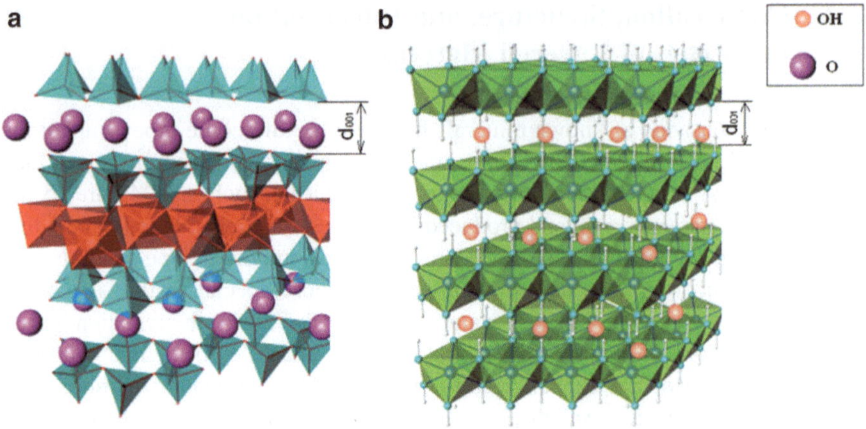

Fig. 5.1 The chemical general structures are respectively for: 2:1 phyllosilicates (**a**) and LDH (**b**), with the schematic representation of the tetrahedra (T) and octahedra (O) occupancies, substitution elements (Y), and interlayer ions (Z) in natural and synthetic layered nano-fillers. (a) Talc (T = Si, O = Mg (occupancy 3/3), Z none; hectorite: T = Si, O = Mg (occupancy 3/3); Y = Li, Z = Na^+/Mg^{2+}; saponite: T = Si, O = Mg (occupancy 3/3), Z = Na^+/Mg^{2+}, Y = Al; montmorinollite: T = Si, O = Al (occupancy 2/3), Z = Na^+/Mg^{2+}, Y = Mg; beidellite: T = Si, O = Al (occupancy 2/3), Z = Na^+/Mg^{2+}, Y = Al. б) brucite: O = Mg (occupancy 3/3), T = Z = Y none; gibbcite: O = Al (occupancy 2/3), T = Z = Y none; hydrotalcite: O = Mg/Zn etc., Z = Cl^-, CO_3^-, NO_3^-, T = Al

type, as $MgO(SiO_2)_s(Al_2O_3)_a(AB)_b(H_2O)_x$ (AB is a ionic pair of NaF type) and even 1 for mica. Oxide composition of Na-hectorite is $Na_{0.33}\,Mg_{2.67}Li_{0.33}Si_3O_{10}F_2$ and that of potassium tetrasilicate mica is $KMg_{2,5}Si_4O_{10}F_2$.

Idealized 2:1 structure of the layered silicate is presented by two tetrahedral sheets, which are condensed with octahedral sheets (the ratio is between tetrahedral and octahedral layers). The native silica gallery is usually filled with hydrated cations of alkali metal. The parameter, which characterizes intercalation activity of clay minerals can change depending on natural form of smectite clays, which is a crystal structure of aluminum silicate, a size of exchanging anions and aqua portion in the interlamellar space (Fig. 5.1).

In synthetic layered silicates (LHD) only octahedral and framing elements (Al, Mg, Zn, etc.) exist. General formula of dioctahedral minerals (pyrophyllite, vermiculite, and others) formed by Al-octahedrons, can be presented as follows $(Si_{8-x}M_y)^4(Al_{4-}{'}_y)^6(OH)_4O_{20}M''_{(x+y)/n}(H_2O)_w$ (content of oxides in vermiculite is, mass.%): SiO_2 38.8, Al_2O_3 13.8, TiO_2 1.97, FeO 1.11, Fe_2O_3 6.25, MgO 18.3, foreign oxides MnO 0.03, Cr_2O_3 0.01, ZnO 0.01, NiO 0.12, CoO 0.01, cation capacity of this silicate is 1.36 mg-eq/g), and for Mg-octahedron based trioctahedral minerals (talc, saponite, biotite, hectorite, etc.) the formula is $(Si_{8-x}M_x)^4(Mg_{6-y}M'_y)^6(OH)_4O_{20}M''_{(x+y)/n}(H_2O)_w$, where x is a portion of isomorphic substitution.

Some characteristics of the surface and porosity of the most widely known alumosilicates[1] [38] are presented in Tables 5.1 and 5.2.

In the presence of water molecules in the MMT gallery it is difficult to obtain clear evidence of intercalation process. In the system dry MMT/PEO, MMT-B34 the intercalation process, for example, with PEO, is found by various physical chemical methods, for example, by X-ray powder diffractometry (XRD),[2] wide-angle X-ray scattering (WAXS), differential scanning calorimetry (DSC), etc.

We shall demonstrate this using as an example magadiite modified with long chained phosphorus-containing [39] surfactant (tributylhexadecylphosphonium bromide), which by its intercalation properties is close to organomodified MMT (OMMT)(Fig. 5.2).

In majority of layered systems basal (interlayer) space is no more than ~5 nm, a layer thickness in perovskite, for example, varies from 0.5 to 2.2 nm, dehydrated Na-MMT the space between the neighbor layers is approximately 1.0 (0.96 nm). It can increase from 1.25 nm to 1.50–1.55 or even to 1.80–1.90 nm, when mono-, bi-, and trimolecular water layers are inserted into the lamellar space. Layered

---

[1] As a rule, for scientific researches Swy-1 County Creek, Wyoming USA standard clays, or highly iron-concentrated smectite, Gafsa, Tunisia, supplied by Source Clay Repository, Clay Minerals Society or other suppliers (see, for example, [38]), Na-MMT - University of Missouri-Columbia (USA), Cloisite organic silicates, Benahavis (Málaga, Spain) vermiculite are used. MMT- Closite 30B is supplied by Southern Clay Products Inc (Texas, USA), Tokyo Kasei Industries Ltd, Na^+-MMT is supplied by Zhe-Jiang Fenghong Clay Company(China), modifier, N- diamine octadecyl ammonium chloride supplied by Zhejiang Chemi-520 Agent Company (China). It should be noted that also widely used are clays from Angrensk, Glukhovets national deposits, vermiculite from Kovdorsk deposit, etc. In the vermiculite from Kovdorsk deposit between mica packages $(Si_3{}^{4+}Al)$ $O^{2-}{}_{10}\,(Mg^{2+})_6(OH^-)^{4-}$ hydrated layer is inserted, Mg ions are surrounded by 12 water molecules forming double-layer shell of Mg^{2+}. If an interlayer magnesium cation is replaced by univalent cations, the initial mineral is split into separate layers, bond strength between them depends on water state, its molecular mobility. First Mg^{2+} ions are substituted by Na^+ (via boiling in saturated NaCl solution), at the second stage sodium ions are substituted by $RNH_3{}^+$ (during boiling in 20–40 % aqua solution of respective alkyl ammonium chloride), cation exchange capacity is 1.7 mg-eq/g, after intense one-dimensional swelling, the mineral is dried in air. Water content in the dry product is 15–20 mass.%.

[2] Powder samples of organic silica were scanned to reflect in the angular range $2\theta = 2$–$10°$. The interlayer distance of the organic silicate is calculated using Wulf-Bragg equation by position of the d_{001} peak in the diffraction pattern $\lambda = 2d \sin\theta$, $\lambda = 0.1542$ nm is wavelength for Ni-filtered Cu K_α-radiation, d is interplanar distance, θ diffraction angle. After subtraction of background scattering, intensity of scattering distribution is implanted by azimuth angle and presented as dispersion function of q-vector in space:

$$q = \frac{4\pi}{\lambda}\sin\frac{2\theta}{2} \quad \text{and} \quad \begin{aligned} d_{001} &= \frac{2\pi}{q} \\ d_{001} &= \frac{1}{2}\frac{\lambda}{\sin(2\theta/2)} \end{aligned}$$

Table 5.1 Characteristics of some aluminum silicates [395]

Silicate	Annealing temperature, °C	S_{sp} BET, m^2/g	V_{gen} porous, cm^3/g
Montmorillonite	Initial	50.0	0.081
	450	368	0.392
	500	382	0.402
	550	360	0.382
Vermiculite	Initial	23	0.043
	450	484	0.427
	500	471	0.396
Silicate			
Minerals	450	517	0.340
	500	600	0.398
	550	584	0.388

Table 5.2 Characteristics of some modified MMT [396]

Organosilicate[a]	Modifier ammonium cation[b]	2θ, deg	D_{001}, nm	Concentration of modifier, mg-eq/g
Cloisite® 10A	$(CH_3)_2(HT)(CH_2C_6H_5)$ N^+	4.38	2.01	1.25
Cloisite® 6A	$(CH_3)_2(HT)_2$ N^+	2.38, 4.34	3.7	1.40
Cloisite® C20A	$(CH_3)_2(HT)_2$ N^+	2.42		
Cloisite® 30B	$(CH_3)_2(T)$ $(CH_2CH_2OH)_2$ N^+	4.72	1.87	0.90

[a]Cation exchange capacity of the non-modified silicate is 0.926 mg-eq/g
[b]T(tallow) are amines of fatty acids of hydrated tall oil (~65 % C_{18}, ~30 % C_{16}, 5 % C_{14})

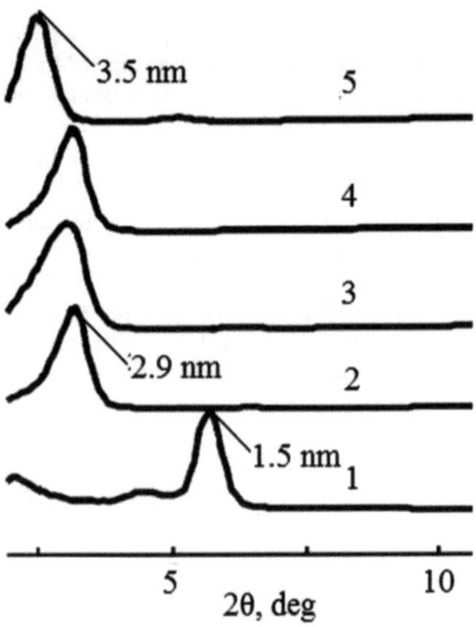

Fig. 5.2 WAXS of the magadiite powder (peak at 2θ = 5.7° corresponds to a basal spacing of 1.5 nm) (*1*); organomodified magadiite (peak at 2θ = 3.0° corresponds to 2.9 nm) (*2*); organomagadiite after melt annealing at 160 °C for 30 min (*3*); organomagadiite in PS microcomposites (*4*); and organomagadiite in PCL nanocomposites (peak at 2.58 corresponds to 3.5 nm) (*5*) [39]

perovskites are easily exfoliated into separate single nanosheets. From this point of view exfoliation can be regarded as the ultimate case of intercalation.

The longitudinal size of disc particles of clay is governed by the methods of its preparation: clays obtained by grinding have a typical plate shape with longitudinal size 0.1–1.0 μm. Taking into account the fact that their colloid sizes are characterized by extraordinary high length-to-thickness ratio, for example, AR = 25–27 for synthetic laponites, the exfoliated clays have extremely developed surface. For the AR to be AR > 25 is a mandatory demand for good intercalating properties of materials, which are of interest for the technology of fabrication of polymer nanocomposites (thickness of exfoliated plates is 0.7–2.5 nm) with far better characteristics of materials. In contrast to mica, cations in natural forms of smectite clays easily exchange as a result of isomorphic substitution of Al^{3+} by Mg^{2+} in $Si_8(Al_{3.34}\ Mg_{0.67})(OH)_4O_{20}M^+{}_{0.67}$ MMT, including replacement by the transition metal ions.

Exchange cation capacity depending on a crystal structure of aluminum silicate reaches 0.64–1.50 mg-eq/g (for MMT it is 1.10, for hectorite it is 1.20, for saponite it is 0.87 mg-eq/g, etc) [40]. A negative charge appears in the octahedral layer, it is compensated by metal cations in the interlayer space, which may be Na^+, Ca^{2+}, Mg^{2+} or Al^{3+}. Apart from non-stoichiometric octahedral substitution the natural MMT can also have tetrahedral isomorphic substitutions (in particular, Al^{3+} instead of Si^{4+}), and also excess negative charge on the side surfaces of the crystals and OH^- groups of the basic character. When M^{2+} cations are substituted by M^{3+} (Fe, Al, Ga) or by Zr^{4+}, Sn^{4+} in the layered materials, for example, LDH, exchange anion capacity can reach significant value, 2–4 mg-eq/g (Fig. 5.3) [41].

One important property of MMT is their tendency to swelling in polar reagents of ionic (water) and non-ionic (monoatomic and polyatomic alcohols, amines, nitro

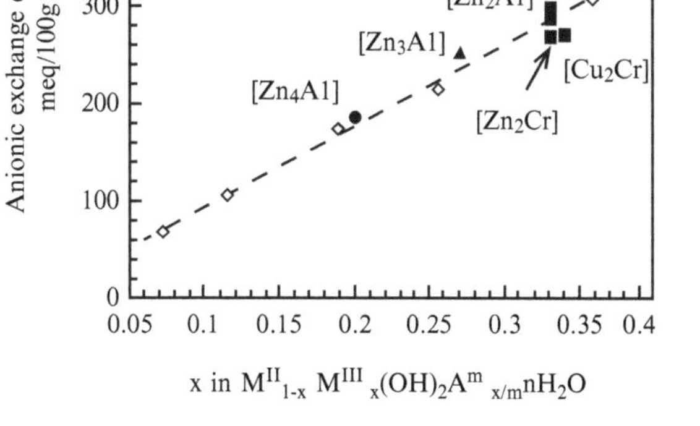

Fig. 5.3 Anionic exchange capacity *versus* layer charge for some LDH compositions [43]

methane, acetonitrile, ketones, aromatic hydrocarbons, etc.). They insert between aluminum-silica-oxygen layers and easily draw them apart. These molecules, though they form ion-dipole, Van der Waals, hydrogen bonds, may be substituted by others which can penetrate into interlayer distances. In the natural MMT interlayer distance is filled with metal cations, which impede dispersion of silicates in a polymer matrix. Substitution of metal ions by alkyl ammonium cations is a widely used method for enhancement of silicate dispersion in a polymeric matrix and for improvement of compatibility with the matrix.

Interlayer compensating cations of the layered silicates can be also replaced by various organic cations, the most interesting are ammonium cations with hydrocarbon chain C_1–C_{18}. Most often cation surfactants are used, such as octadecylamine (modification of MMT surface 15–30 mass.%), hexadecylamine, decylamine, and dodecylamine, hexadecyltrimethylammonium chloride, and also ammonium cations of amines of higher fatty acids of hydrated tallow oil, such as methylenbis (2-hydroxyethyl)tallow (2M2HT), alkylammonium $[CH_3(T)N(C_2H_4OH)_2^+/MMT]$, consisting of ~65 % C_{18}, 30 % C_{16} and 5 % C_{14}, intergallery cation with exchange capacity 0.9–1.25 mg-eq/g, tetralkyl ditallow ammonium modifier B34 in bentonite increases *d*-basalt space by 25 % [42] (sometimes the latter named by trade name Nanometer 1.30 T). We should also note di(2-oxyethyl)-12alkano-3-methyl-aminochloride,

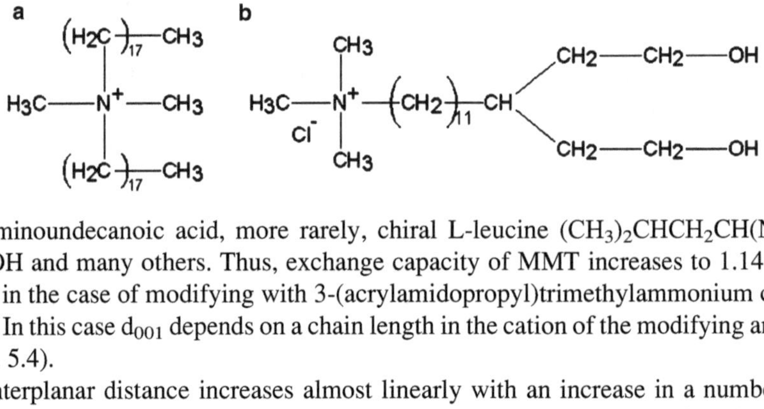

aminoundecanoic acid, more rarely, chiral L-leucine $(CH_3)_2CHCH_2CH(NH_2)$ COOH and many others. Thus, exchange capacity of MMT increases to 1.14 mg-eq/g in the case of modifying with 3-(acrylamidopropyl)trimethylammonium chloride. In this case d_{001} depends on a chain length in the cation of the modifying amine (Fig. 5.4).

Interplanar distance increases almost linearly with an increase in a number of carbon atoms in the primary amine $C_nH_{2n=1}NH_2 \cdot HCl$, intercalated in a layered silicate (Fig. 5.4), in this case the correlation equation is approximated as follows: d_{001} (nm) = 0.998 + 0.052n, the correlation score is $r = 0.980$ [43]. In other words, each CH_2- group increases d by 0.1 nm.

Degree of intercalation of a modifier is calculated taking into account interplanar distance D_F in a composite, with account for widening of the gallery (from 3.2 to 4 nm), and the magadiite sheet thickness (L = 1 nm) [39].

$$D_F = \frac{[(d_{001}(\text{organomagadiite}) - L) - (d_{001}(\text{magadiite}) - L)]}{[(d_{001}(\text{magadiite}) - L)]} \tag{5.1}$$

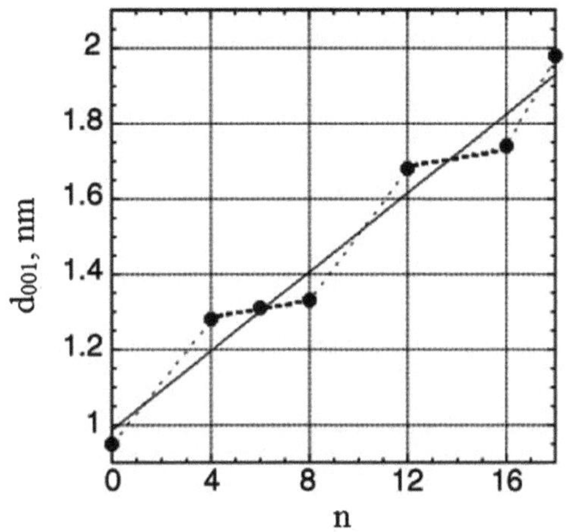

Fig. 5.4 The dependence of the interlayer spacing on the number of carbon atoms (n) in the amine intercalated into a layered silicate [43]

Table 5.3 Characteristics of organic silicate

Designation[a]	Interlayer distance, d_{001}, nm
1.40C	3.47
1.25C	2.36
0.95C	1.89

[a]The values are cation exchange capacity in the organic silicate, mg-eq/g

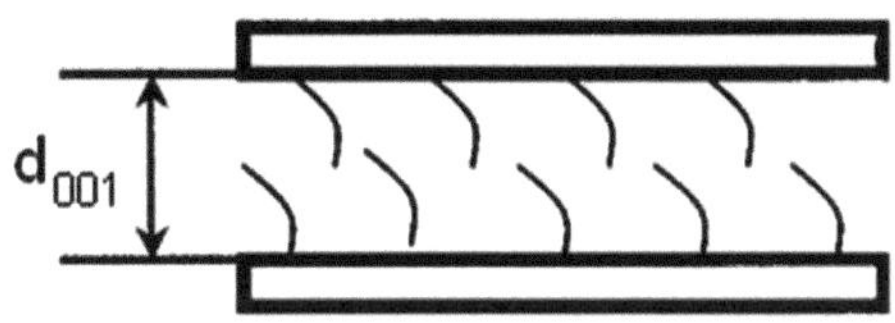

Fig. 5.5 Orientation of the modifying agent in silicate layers

The interplanar distance also depends on a modifying degree of a silicate (its cation exchange capacity), and on orientation of a modifying ammonium cation (Table 5.3).

The modifying agent can be parallel to the silicate layers, normal to them or placed in other way (Fig. 5.5).

Exchange mainly occurs in the layers containing solvated potassium cations. Taking into account the fact that hydrogen bonds form between layers, the character of small polar molecules inserted directly between them is limited and includes such ones as dimethylformamide, dimethyl sulfoxide, glycerol. Adsorption, its constant, and ΔH increase as a length of hydrocarbon radical increases. Interplanar distance also depends on orientation of cations in a layer, on how the layers are close-packed. These parameters, in turn, are defined by a surface charge: in low-charged silicates alkyl chains are between layers, in high-charged they are vertical to the surface [44]; for this type of substitution organic silicates gain hydrophobic properties due to blocking of absorption centers, and ability to swell

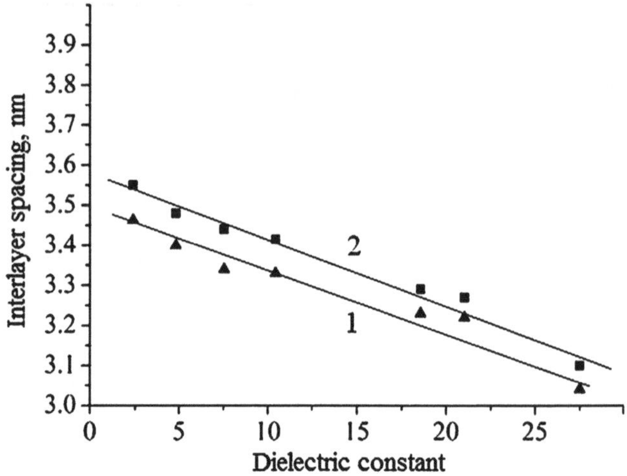

Fig. 5.6 Interlayer spacing of the MMT modified by 30 % (*1*) and 35 % (*2*) stearylbenzyldimethylammonium as a function of the dielectric constant of the solvent [44]

in organic solvents reaches maximum in the case of alkyl ammonium modifiers with a number of carbon atoms in a chain 12–18 (Fig. 5.6).

Molecular mobility of a modifying surfactant (taking for example tributylhexadecyl phosphonium bromide or dodecyl trimethyl ammonium bromide) in layers of synthetic magadiite [39]. In this case tributylhexadecyl phosphonium bromide was bound to imineoxyl radical to obtain spin-labeled surfactants such as

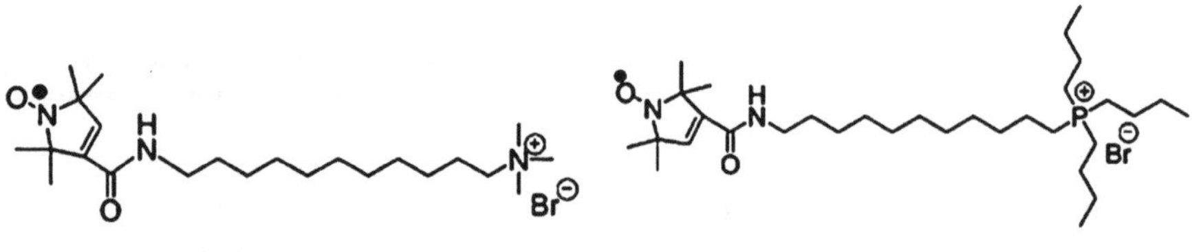

which made it possible to use EPR, NMR ^{31}P, and other techniques. Existence of two types of modifiers was found, bound to magadiite and individual, just one type of ligand was found as the temperature increased. Amine modifier enhances intercalation more efficiently than P-containing, due to smaller volume of its functional group. Behavior of these spin labels (rotation activation energy is 6–60 kJ/mol) resembles polymer-immobilized ones. The layered silicates modified with quarter amines with two long aliphatic chains have mixed layered structure consisting of packages with different interplanar distances. In their interplanar spaces ordered adsorption layers of a modifier are formed, which substantially draw apart plates of the layered silicate [45].

As a rule, mechanical properties of intercalated nanocomposites, such as ultimate tension strength, relative elongation in tension, elastic modulus and fracture energy are improved. Ultimate strength for tensile increases as a length of amine chain increases depending on character of a clay: Na-MMT > bentonite > K-MMT, which corresponds with an increase in interplanar distance. Thus, an amine defines organophilic properties of a silicate, this, in turn, improves interaction between polymer and filler and properties of a composite [46]. Improvement of compatibility is proved by different methods, including solving and mechanical mixing. A tendency to improvement of properties is followed in the range: organophilic clay > Na hectorite > K bentonite > deposited SiO_2, which reflects their intercalation and organophilic abilities. It should be noted that not only ammonium but also alkylphosphonic salts [47], including tributylhexadecyl phosphonium bromide are suggested as organic exchanging cations, intercalating agents. Phosphor containing surfactants have higher thermal stability, than their nitrogen containing analogues, especially, in compositions on the basis of layered materials [48, 49]. Modifiers can be ranged in order of their intercalation, and finally, exfoliation activity: aqua-soluble sulfonates, sulfates, carboxylites. Modification of MMT by intercalation with various polyhydrocomplexes and other additives makes it possible to significantly increase its adsorption-structural characteristics and thermal stability.

Synthetic phyllosilicate clays, silicic acids, acid potassium phyllosilicate, magadiite, layered double hydroxides (LDHs) (Fig. 5.7), bilayered ($HLaNb_2O_7$) and trilayered ($HLa_2Nb_3O_{10}$) perovskites, zirconium phosphates (ZrPs), dichalcogenides are obtained by different, relatively simple ways [50]. Their main advantage is in chemical purity (absence of amorphous sand contaminators, arsenic, iron, heavy metals), they are transparent, which provides glossy color of a product, they have a wide rang of the aspect ratio from 20 to 6,000 [51]. The most widely applied is hydrothermal technique with usage of templates [52–57].

Coprecipitation of the respective cursors is used to obtain $Zn_{0.67}Al_{0.33}(OH)_2Cl_{0.33} \cdot 0.75H_2O$ or $Cu_{0.68}Cr_{0.32}(OH)_2Cl_{0.33} \cdot 0.92\ H_2O$, and magadiite is obtained via hydrothermal synthesis from two components: silica oxide, potassium hydroxide, and water in the ratio 3:1:200, its calculated formula is $Na_2Si_{14}O_{29} \cdot 11H_2O$, the exchange capacity is 1.82 mg-eq/g two potassium ions are substituted by one molecule, the total absorption of the modifier – long chained amine is 0.85 mmol/g. Layered nanohybrid materials are synthesized via interaction

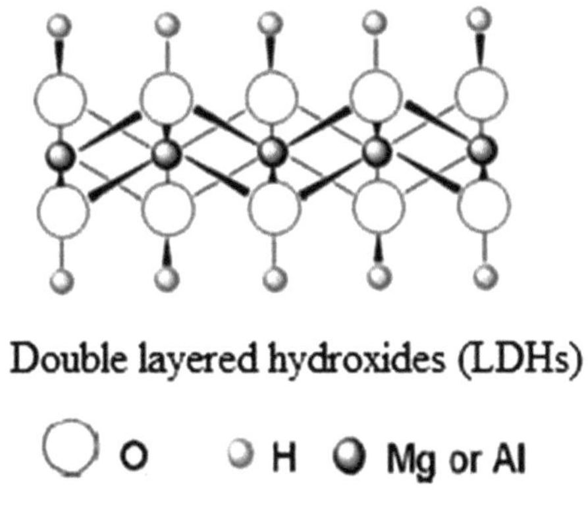

Fig. 5.7 The schematic illustration of LDH layers (The interlayer space is 0.50–0.60 nm) [50]

of $Zn(OH)_2$, ZnO or Zn/Al with carboxylic acids or with oxychlorides; the *guest/host* ratio in these composites is from 1.5:0.5 to 1.8:0.2. Depending on a character of the reacting components, morphology of the particles changes from fibers to plates, interlayer distance increases from 1.61 to 2.01 nm. Hydrotalcite materials[3] ($[AlMg_2(OH)_2(OH)_4]^+[0,5CO_3, OH, Cl]^-$) LDH [9] consist of layers of positively charged $[Mg_6Al_2(OH)_{16}]^{2+}$ cations neutralized by CO_3^{2-} ions in the interlayer area [4, 43, 58–61].

For modifying of LDH the following four approaches can be taken: anion exchange, in LDH, direct synthesis via precipitation, rehydration and calcinations of LDH, and thermal reactions. Styrene sulfonate RSO_3^-, alkyl benzene sulfates, dodecyl benzene sulfate are relatively stable modifiers of LDH, break of the tail-head bond in a surfactant happens at the temperatures 100–200 °C. However, these modifiers has not been widely applied for nanocomposites.

It should be especially noted that some metal oxides prepared by "wet" chemical synthesis (hydrothermal, solvothermal [62, 63]) have anisotropic structure, which should cause variation in their properties along different crystallographic directions. It relates, for example, to α-MoO_3, which is one-dimensional rod [64] to layered double hydroxides (LDHs) $M^{2+}{}_{(1-x)}M^{3+}{}_x(OH)_2]^{y+}(A^{n-\,y/n})m\ H_2O$, for example, $Mg_6Al_{3\text{-}4}(OH)_{18\text{-}8}(CO_3)_{1\text{-}7}H_2O$ or $Zn_6Al_2(OH)_{16}CO_3nH_2O$. Materials with layered-columnar structure made on the basis of montmorillonite clays are widely used in various areas, including sorption technique and catalytic processes. This is especially related to widely used in catalysis Ti(IV) columnar montmorillonite materials (see for example [65–70]). They include centers of acid and Lewis types (three-coordinate Al^{3+}cations substituting Si^{4+} in tetrahedral layers), as well as Brőnsted centers (containing Si-OH-Al- groups). Residual water has a significant effect on their acid capacity; they inflate when filled with organic solvents, monomers, or polymers. Inflation of interior crystalline cavities, which is sometimes called characteristic basis of clay swelling, depends on a nature of cations in layers and also on density of negative charge on them. Moreover, inorganic acids have surface hydroxyl groups, which develop strongly polar surface active in bonding of various metal ions. However, main properties of these materials are governed by characteristics of porous open systems and especially by pore sizes. In the compounds of insertion *guest*-molecules are in crystallographic hollow sites of the *host*-matter. The layered "sandwich" structures and one-dimensional channel structures mostly belong to these compounds. Metal ions included in these structures can easily reduce under action of various chemical reducers, or thermally, less often by photochemical process, and form nanoparticles [71].

---

[3] Synthetic hydrotalcite $Ex_{0.66}[Li_{0.66}\ Mg_{5.34}Si_8O_{20}(OH, F)_4]$ (Ex = Li, R) is obtained, from the precursor of the composite 0.32 R, 1.0 LiF, 5.3 $Mg(OH)_2$, 8 $SiO_2 \cdot nH_2O$, in which R is organic salt of univalent metal and Ex is substituted cation. The process begins from solution of 0.72 mmol of organic salt in water and addition of 4.8 mmol of LiF during stirring. Separately 24 mmol of $MgCl_2 \cdot 6H_2O$ is solved in water and mixed with 32 ml 2 N NH_4OH to formation of fresh Mg $(OH)_2$.

The layered materials are distinguished as two types by their architecture and properties. First of them is characterized by stiff pores with a constant volume, parallel insulation of the lattice channels and communicating channels of the network. Localization, concentration and spatial distribution of *guests* are defined by the topology, chemical origin, and reaction ability of the inner surface of the *host*, whose matrix can additionally be subjected to functionalization. Moreover, *guests* types are limited by minimal conjugated channels, which causes selected intercalation behavior of *guests* in matrices (like molecular "sieves"). The second type is presented by low-dimensional lattice of a *host*, i.e. layered or chained structure. This provides "flexible" pores, whose dimensionality can adapt to dimensionality of a *guest*. Matrix lattice of a *host* can have no effect on intercalation-deintercalation (in the case of lattices of insulators, zeolites, layered aluminum silicates, metal phosphates, etc.). For them intercalation behavior is predominantly characterized by acid-base and exchange properties.

Host lattices with electron conductivity (semiconductors based on layered metals, their chalcogenides, etc.), undergoing redox reactions during intercalation with electron or ion transport accompanied by significant change in physical properties of a host matrix take a special place. Thus, a great intercalation capacity of xerogel on the base of V^{5+} oxide is known: this is $H_2V_{12}O_{31} \cdot nH_2O$ polyvanadium acid, in which negative charge of vanadium-oxygen layers is distributed along V_2O_5 fibres [72]. The fibers are plane ribbons up to 100 mm in length and 10 nm in width (Fig. 5.8) and include water molecules, chemically bound with vanadium atoms via different types of bonds. Intercalation can develop by dipole absorption, ion exchange or oxidation-reduction processes. Vanadyl phosphates are relatives to intercalation gels. Insertion of quite large $VOPO_4 \cdot H_2O \cdot EtOH$ molecules in $VOPO_4 \cdot 2H_2O$ layered structure, interlayer distance can increase from 0.75 to 1.03 nm.

Other phosphates such as: α-$Zr(HPO_4)_2 \cdot H_2O$, $HUO_2PO_4 \cdot 4H_2O$, layered aluminum phosphates, such as, for example, $AlPO_4$, berlinite, $Al_4(PO_4)_3(OH)_3 . 9H_2O$,

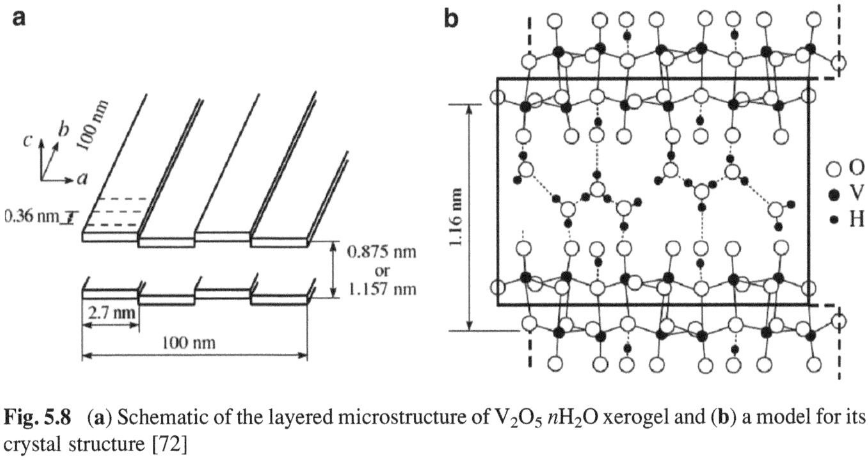

Fig. 5.8 (**a**) Schematic of the layered microstructure of V_2O_5 nH_2O xerogel and (**b**) a model for its crystal structure [72]

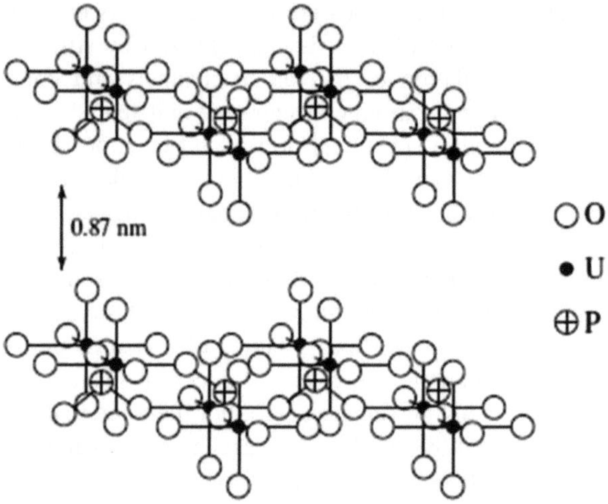

Fig. 5.9 Structure of $HUO_2PO_4\ 4H_2O$ [34]

vantasselite, produced by hydrothermal synthesis from H_3PO_4 and $Al(OH)_3$ with structural-governing agent (template), M^{4+} phosphates or phosphonates ($M^{4+} = Ti$, Zr, Sn), for example, α-form $Zr(HPO_4)_2 \cdot 2H_2O$, γ-form $ZrPO_4O_2P(OH)_2 \cdot 2H_2O$, λ-form $ZrPO_4XY$ (where X, Y is anion or neutral ligand), phosphorus-molybdenum acid $H_3PMo_{12}O_{40}$, etc. For example, uranyl hydrophosphate is layered bright yellow material with interplanar distance 0.87 nm. The layers include dumbbell shaped UO_2^+-ions with additional coordination of uranium with four equatorial oxygen atoms of PO_4^{3-} tetrahedrons forming 2D sheets (referred by [34]) (Fig. 5.9).

An important place in intercalation processes have cerium(IV)hydrophosphates, belonging to acid salts of tetravalent metals, in which inorganic network is built from $M^{4+}O_6$ conjugated octahedrons (M = Zr, Ti, Ce, Sn, etc.) and XO_4 tetrahedrons (X = P or As), structured as 2D and 3D structures [73]. Some layered acids, such as kanenite, makatite, octasilylsilicate, magadiite, kenyaite have interesting intercalating properties.

Among the materials, which are obtained by sol-gel synthesis [74] we shall note once again TiO_2 (anatase) pillared clays. They consist of saponite, natural purified montmorillonites (MMT-TiO_2), mica (mica-TiO_2). Nanoparticles TiO_2 intercalated into silicate clays widen their interlayer distance, while crystal structure and sizes of anatase remain almost identical for various types of clays [75]. Intercalation of TiO_2 in MMT or in hectorite can go in two ways [76]: preabsorption of precursors, titanium alkoxides, on mineral plates with the following hydrolysis and calcinations or heterocoagulation of titanium dioxide and the mineral [77, 78]. It is especially typical of SiO_2 intercalation, which depends on the character of alkoxysilane and initial clay, modified by longchained amines; most efficient is tetramethoxysilane (TMOS) with varied (from 1:0.5 to 1:5 mass.%) ratio clay/silane. A different ability of silanes to penetrate interlayer space is caused by facilitated diffusion, steric hindrances, which are weaker for TMOS than for $Si(OC_2H_5)_4$, etc.

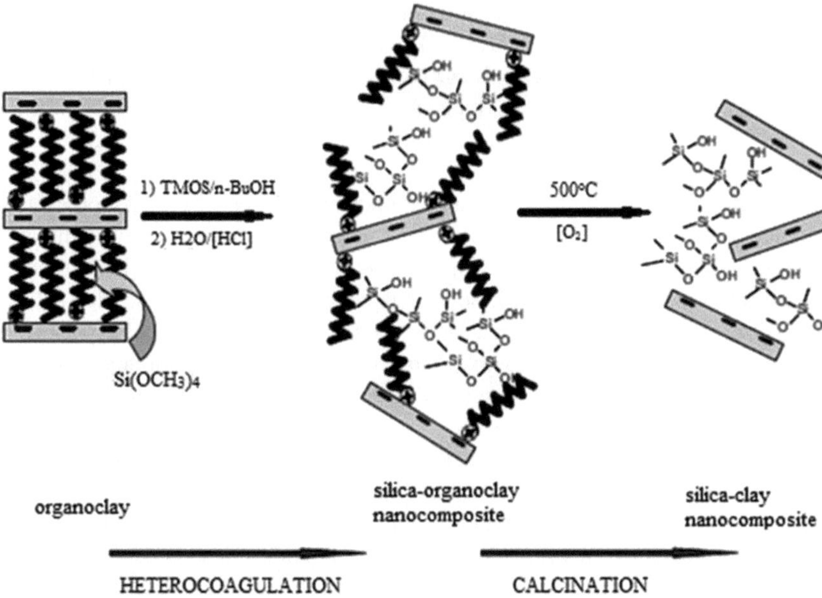

Fig. 5.10 Combined sol-gel-intercalation synthesis using a modified silicate [395]

General scheme of this combined sol-gel-intercalation process can be briefly described as follows [39, 79] (Fig. 5.10).

This process may be influenced by different density of electric charges in silicate, which is defined by a number of alkyl ammonium cations per unit area of its interlayer distance. After annealing at 500 °C SiO_2^- inorganic composite forms. For alkoxides of various metals (Ti, Al, Zr, Sn) this way can be used to obtain layered titanites, zirconites, etc.

An interesting field is production of vanadium oxide nanotubes, nanotubular forms of intercalating compounds. The case in point is sol-gel synthesis with participation of $VO(OPr^i)_3$ in presence of primary or α,ω-diamines followed by hydrothermal reaction in autoclave at 180 °C for 2–7 days. After washing and drying vanadium oxide of mixed valence had a nanorod shape with the composition $V_2O_5 \cdot 0.3\ H_2O$, diameter from 15 to 150 nm (inner diameter from 5 to 50 nm), its length reached 15 μm [80]. The tubes consisted of 2–30 layers of crystal oxides of vanadium and amine molecules, intercalated between them. A distance between layers is 1.7–3.8 nm proportional to the amine length, which acted as structure governing template. In contrast to mesoporous oxides of metals with lamellar structure these nanocomposites are characterized by tubular morphology and well-structured walls. "Curls" of vanadium-oxygen layers are very flexible, they can be stretched in line and take part in many exchange reactions. Structural flexibility provides possibilities of further formation of nanotubular structure and modifying of mechanical, electrical, and chemical properties of this material. Vanadium oxides, as will be shown below, are perfect *hosts* for obtaining of composites on the basis of conjugated polymers. For intercalation far rarely are used layered minerals, oxides and hydroxides, such as brucite $Mg(OH)_2$; gibbsite $Al(OH)_3$; also $LiMn_2O_4$; $LiCoO_2$; $LiNiO_2$; $H_2Si_2O_3$; V_6O_{13}; $HTiNbO_3$; potassium titanate whiskers $K_2Ti_6O_{13}$; $W_{0.2}$ $V_{2.8}O_7$; Cr_3O_8; $MoO_3(OH)_2$; $CaPO_4CH_3 \cdot H_2O$; $MnHAs \cdot H_2O$; $Ag_6Mo_{10}O_{33}$; monoclinic structures (type m-WO_3), etc. [62–64].

Last advances in this field are described in recent reviews, concerning exfoliation of layered oxides [81, 82], etc.

As the methods of obtaining of nanocomposites will be analyzed, some other inorganic materials will be also discussed, among them the abovementioned phosphorus-molybdenum acid $H_3PMo_{12}O_{40}$, and layered compounds like graphite, graphene, chlorides $FeCl_3$, FeOCl, some chalcogenides TiS_2, MoS_2, MoS_3, $Cr_{0.5}$ $V_{0.5}S_2$, $(PbS)_{1.8}(TiS_2)_2$. Thus, layered MoS_2 structure consists of quite thin packs less than 5 nm in size (several layers) with a distance between layers (periodicity) 3.4 nm. Presence of negative charges in monolayer dispersions of MoS_2 separated by solvent molecules (which is ionic system $(MoS_2)^{x-}$ – alkali metal-hydroxide-anion [83] give a possibility of their modifying by various cations, including formation of layered compounds with the cations of quarter alkyl ammonium, etc. [84].

At the same time layered halogenides and thiophosphates of transition metals, oppositely to the abovementioned layered metal oxides (and considered below metal chalcogenides) did not attract much attention for intercalation. Maybe, a reason for this is low stability of majority of them that undergo various chemical transitions, hydrolysis, solution, decomposition, etc. one of exceptions is α-$RuCl_3$, which has a lamellar structure (Fig. 5.11).

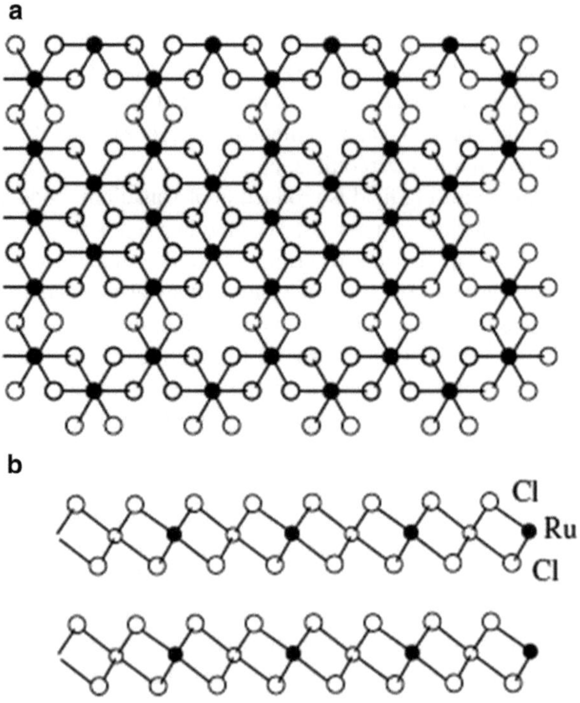

Fig. 5.11 Layered structure of α-$RuCl_3$: (**a**) octahedral coordination of Ru atoms; (**b**) projection along the *c* axis [34]

5.2 General Characteristics of Intercalated Systems

As it has already been noted, layered materials interesting for usage in technology of polymer composites, should form flakes 0.7–2.5 nm in thickness and having a definite size of interspace distance [85–87].

Nanocomposites are usually products, which consist of at least two different materials with a clear interphase boundary between them, at least one of them should have nano- (1–100 nm) sizes in no less than at least one direction. The same relates to the distances between networks and layers formed by polymer and inorganic gradients [71]. The last type includes materials in which monomeric or polymeric molecules are inserted as guests in a host structure, which possesses intercalation properties. Under effect of the forming polymer particles of this "host" disperse into individual layers which results in formation of exfoliated nanocomposite (Fig. 5.12). The "peeling" technique is quite well developed. Peeling lamellas of MMT can act on polymeric chains as nanosized "hosts" and effect structure and orientation of intercalated macromolecules.

The driving force of intercalation process is a decrease in free energy in the system: a change in enthalpy is due to intermolecular interactions, and change is entropy is due to configuration interactions. Enthalpy is a dominant factor, which governs intercalation process, at least because the liquid-plate interaction characterizes behavior of liquid molecules with respect to surface of a solid and is crucial in this case.

Analysis of thermodynamic parameters [88, 89] shows that nanostructures are formed if free energy of a change in interlayer volume (ΔF_v) is negative, it can be written as follows

$$\Delta F_v = \Delta E_v - T\Delta S_v \tag{5.2}$$

where ΔE_υ, ΔS_υ are enthalpy and entropy of a change in volume during intercalation, T is temperature.

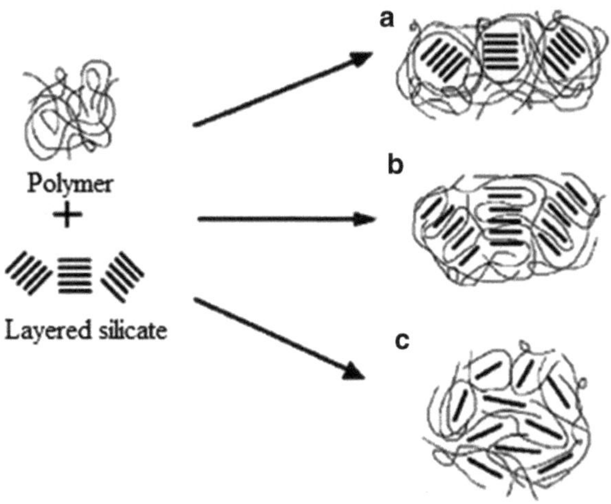

Fig. 5.12 Morphology structure of layered silicates in (**a**) conventional composites, (**b**) intercalated nanocomposites and (**c**) exfoliated nanocomposites [85]

Taking into account the fact that a small increase in gallery space has insignificant effect on a total change in entropy, the driving force of the process is a change in total enthalpy, which is expressed as follows:

$$\Delta E_v = \varphi_1 \varphi_2 \frac{1}{Q}\left(\frac{2}{h_0}(\varepsilon_{\mathrm{sp}} - \varepsilon_{\mathrm{sa}}) + \frac{2}{r}\varepsilon_{\mathrm{ap}}\right) \quad (5.3)$$

where φ_1 and φ_2 are volumes of the intercalated polymer and bonded chains of a surfactant, Q is a number of links, h_o is initial size of organosilicate gallery, r is radius of intercalation surface of the chains of the bonded surfactant, ε_{sp}, ε_{sa} and ε_{ap} is intercalation energy of interaction between a layered silicate and polymer, layered silicate and intercalating (modifying) agent and intercalating agent and polymer, respectively.

Exfoliation can appear only for positive value of Flory-Huggins parameter between polymers and thin discs modeling silicate layers, though a mixture of polymers and clay can delaminate, while intercalation is expected if the interaction parameter is negative. Thus a polymer diffuses through energetically preferable gallery, which enhances its contact with two limiting silicate layers. The diffusing polymer molecule sticks two neighbor surfaces, making them kinetically captured. Therefore, even though interaction energy is required for diffusion of the polymer into the gallery of silicate layers, its increase would cause predominant intercalation, despite predominant exfoliation [90–92]. Adhesion behaviors of polar polymers and hydrophilic silicate layers is sometimes called sticking effect [93].

This effective way of obtaining of various types of nanocomposites, for example, based on PS, PE, PP [94], polyamide (PA) [95, 96], polyethylene terephthalate (PETP) [97–99] and others [100, 101]. As was already mentioned, exfoliation is accompanied by formation of monolayers of nanometer thickness with highly anisotropic shape, uniformly distributed in a polymeric matrix [9, 12, 85, 88, 102–109].

Thus, in this process three main hybrid-phase composites can form (taking for example dispersion of OMMT particles in polymer matrices): microcomposite with separated phases (traditional material), intercalated (including folliculated material) and exfoliated nanocomposites. Traditional microcomposites with separated phases include a clay tactoid with layers aggregated in non-intercalated state. A silicate is finely dispersed as a separate phase, these composites are often called delaminated (delaminated suspension). Depending on nature of the reacting components and conditions, a mixed composite can form, which contains the described structures in various proportions. Intercalation state is less stressed, its morphology is characterized by preserved order in the silicate layers, including one or more polymer chains intercalated in gallery space with a certain structure, which are tend to increased basal space of a layered silicate. Exfoliation of the silicate in a nanocomposite is a result of homogenous dispersion of nanoplates of delaminated silicate in a continuous phase. In other words, exfoliated or delaminated nanocomposites are materials in which packed silicate layers are absolutely fractured and single layers are uniformly distributed in polymeric matrix. In this case a collapsed structure consisting of elongated silicate plates can be absolutely random or can have some order. In other words exfoliated nanocomposites can be formed as a result of ordered (by topography of silicate nanoplates) and disordered exfoliation [110]. Ratio between depends on such factors as capacity of the layers to swelling under conditions of the process, maximum separation, depending on volume fraction of exfoliated layers. Moreover, intermediate morphologies appear owing to rotation of partly exfoliated associates dispersed in layers in combination with small thickness of their bunches [111]. Widening, a distance between the separated layers is, as a rule, 10 nm and more and depends on a nature of inorganic component, its aspect ratio, concentration of a polymer in a composite [1, 110, 112, 113].

Exfoliated nanocomposite contains low concentrations of a silicate (which plays a role of nanorods), which does not cause a significant increase in density and deterioration in recycling ability of a material, which is a solid structure, its properties are defined by a nanoconfined polymer. One of the aspects of delamination is a possibility of introduction of polymers by the mechanism *guest-host* with intermediate reaction time and following precipitation of the intercalated systems after removal of a solvent. Taking into account directions of modified agents between layers of a silicate (Fig. 5.13) localized macromolecules between these layers can be presented as follows:

Moreover, morphology of the nanocomposites depends on polarity of the silicate, way of its modification, polarity of intercalated polymer, interaction between components, OH-groups in the silicate. The effect of these factors is schematically shown in Fig. 5.14 [114]. In a non-modified silicate non-polar polymer is localized, mainly, on side surfaces of the silicate, in a modified intercalated structures form, while polar polymers have mostly exfoliated morphology.

A significant difference in characteristics of these materials is shown, for example, for their barrier properties.

One more moment should be highlighted. The matter is about obtaining of concentrates of the composites (the method of dosed introduction of components, masterbatch). As a rule, they are obtained at the stage of polymerization or from a melt of polymer, taking high (higher than 21 mass.%) concentrations of a silicate with the following extrusion (including introduction of additional ingredients at this stage) for production of a necessary homogenous nanoconcentrate, which is then

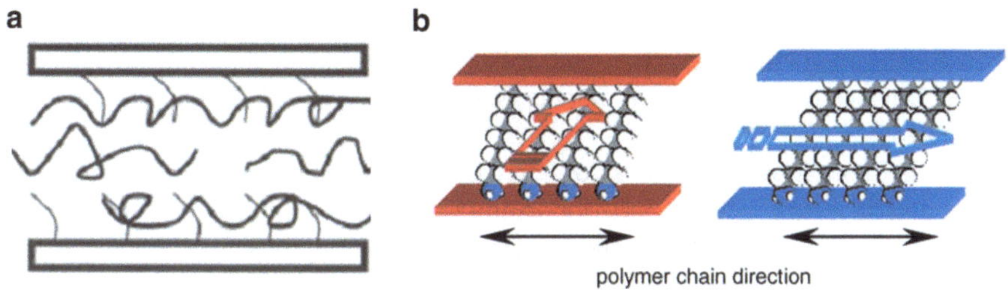

Fig. 5.13 Model of oriented structure of long-chain alkylamine (**a**) and polymer chain (**b**) in an interlayer space of MMT [42]

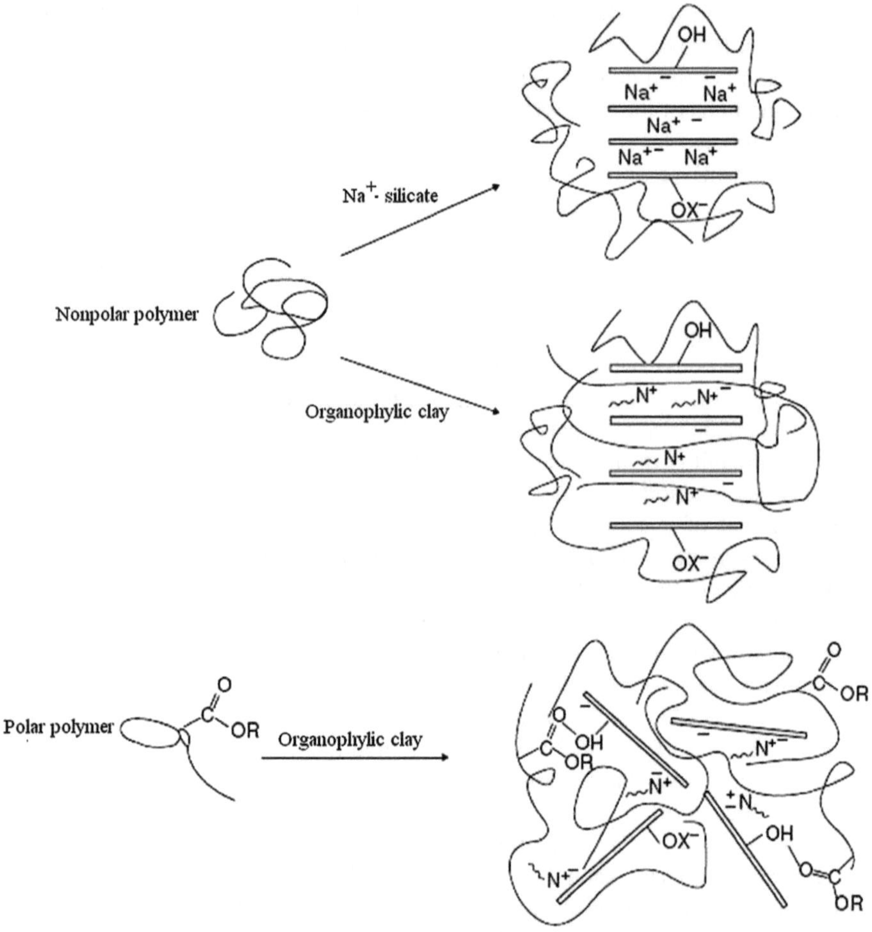

Fig. 5.14 Effect of organophilic and polar interactions on the morphologies of composites [114]

used for production of nanocomposites of a requires content. Other types of silicates are less active in intercalation of polymers due to their low intercalation ability. This relates, for example, to a family of crystal alkaline-earth polysilicates, whose oxygen ions are neutralized by sodium cations, such as kanemite, makatite, ortosilicate, magadiite, kenyaite, and hollandite, though for these purposes sometimes MnO_2 (KMn_8O_{16} type) nanocrystals are used with hollandite or mica structure.

The field of physical chemistry of intercalated systems and its lines are relatively often and in detail generalized, for example, composites based on intercalated polyolefins [115] and functionalized thermoplastic materials were considered in the informative review: [116] main data on PE intercalation, including linear low-density polyethylene (LLDPE), syndiotactic polypropylene, polybutadiene, copolymers poly(ethylene-co-propylene), poly(ethylene-co-vinyl acetate), their preparation, thermal, mechanical, rheological, barrier and plasma-retarding properties, intercalated polyvinyl chlorides, polyamides, organic-inorganic materials for energy-efficient technologies in [15, 33, 34, 106, 117], etc.

Typical methods of production of intercalated composites, which are here analyzed, can be confined to following:

- in-situ polymerization (solution mixing)
- mixing in suspension
- latex compounding
- direct intercalation in a polymer melt (reactor mixing).

5.3 Special Features of Polymerization of Monomers in Interlayer Space

Probably, this is one of the most interesting types of intercrystalline chemical reactions, inserting of monomer molecules and initializing particles in pores of a *host* with the following interior transformation into polymer, oligomer, or hybrid sandwich products (postintercalation transitions). Often this method is called "ship-in-the-bottle" polymerizing approach or in situ polymerization of pre-intercalated monomers. A layered silicate in this process plays a role of nanoreactor. This is rather exhaustively investigated field, many studies of this way performed nanocomposite formation may be referred to [11, 118].

A special attention is drawn to monomers and polymer links synthesized via topotactic in-situ process [119, 120] characterized by low activation energy of diffusion controlled by structural organization of insertion compounds. The obtained nanocomposites are, as a rule, metastable, and cannot be synthesized in other ways, for example, by thermal synthesis or mixture of components (because of exfoliation of a polymer from inorganic component).

From thermodynamic point of view *in situ* intercalation polymerization has an advantage in formation of finer dispersion. Formation of hybrid systems takes place if free energy of a chain ΔG caused by separation of silicate layers and insertion of a polymer chain from a block is negative. Entropy to enthalpy ratio defines intercalation of a monomer molecule and its yield: the greater decrease in internal energy, the higher tendency to formation of intercalated/exfoliated hybrids. Since a great amount of heat is emitted during in situ intercalation polymerization, it is efficiently compensated by loss in entropy of the hybrid and promotes penetration of polymer in silicate layers.

Intercalation was modeled by molecular dynamics (MD) [121]. The calculations confirm non-equilibrium process of intercalation in layers with low aspect ratio at low concentration of the component stronger bound to the layers. As the latter increases, initial stages of exfoliated structures are generated [122]. Also intercalation of a solvent and polymer in the silicate gallery were modeled by Monte-Carlo technique [123].

A matrix polymer in interlayer space of aluminum silicate is synthesized almost by any known mechanism; polymerization in the interlayer space can be initialized by conventional methods. Since polymerization goes in presence of an incentive, various additives, a necessity appears to study intercalation processes of a monomer, incentive substance and additives modifying a silicate surface.

5.3.1 Emulsion Polymerization in Interlayer Space

The simplest method of intercalation of polymers in inorganic structures, one staged methods of emulsion or suspension polymerization of traditional monomers (most often styrene, methyl methacrylate, acrylonitrile, vinyl acetate, etc.) in presence of various organophilic minerals.

Though detailed mechanism of the process is not completely studied, say nothing about topochemistry of elemental stages, physical picture of intercalation can be considered as follows. Characteristic basis of a clay swelling in aqua systems containing 2–10-nm monomer micelles allows their deep penetration in inflated OMMT layers. At the same time, large monomer drops (10^2–10^4 nm) form during polymerization in a solution, therefore they usually easily absorb or are connected to exterior surface of silicate particles. A typical emulsion polymerization of latex particles takes place in ordered 2D colloid ensembles. This principle is used to produce PMMA/MMT, PVC/MMT, PS/MMT and others, their copolymers [124–129].

PS-organosilicate composites are obtained by suspension polymerization of intercalated styrene, an effect of a length of alkyl chain of ammonium modifier and its concentration on properties of the composite is found [130]. Non-extracted nanocomposites, styrene-acrylonitrile SAN/MMT is formed in two ways by intercalation technique: one staged emulsion co-polymerization in water or from a copolymer solution in cyclohexane and OMMT. Interlayer distance of the emulsion substance increases dramatically, by 0.76 nm, whereas in the solute substance it widens only by 0.39 nm. Emulsion polymerization of aniline in situ in presence of TiO_2 causes incapsulation of the nanocomposite, the process is interesting because of use of functionalized protonic acid, dodecylbenzene sulfonic acid, which is simultaneously surfactant for water emulsion polymerization and a doping agent for a forming polymer [131]. Emulsion polymerization has ecological advantages before the solution techniques, because it does not require organic solutions.

5.3.2 Interlayer Blocked Radical Polymerization and Copolymerization

Often during polymerization in situ intercalation of monomers in layered double hydroxides is used (see, for example, [132–134]). Thus, in LDH of the $Zn_{0.67}Al_{0.33}(OH)_2Cl_{0.33}$ 0.75H_2O or $Cu_{0.68}Cr_{0.32}(OH)_2Cl_{0.33} \cdot 0.92\ H_2O$ composition chlorine is substituted by acid monomer during interaction with vinylbenzene-sulfonate (VBS) sodium salt or aminobenzenesulfonate (ABS) before in situ polymerization. Interstitial monomers increase interlayer distance from 0.776 to 1.80 nm (VBS) or 1.54 nm (ABS) [132].

Lamellar aluminum silicates are often used as *hosts* for polymerization of acryl monomers [1, 113]. The type of forming nanocomposites polymer/clay is determined by a character of a monomer, conditions of its formation.

During radical polymerization of vinyl monomers a significant functional groups of long-chained amines-modifiers of layered silicates taking part in complex formation with polar monomers can play an important role. One of most obvious examples is synthesis of the functional nanocomposites via copolymerization of pre-intercalated itaconic acid with cations of N,N-(dimethyl)dodecyl ammonium (DMDA) modifying MMT surface, and n-butyl methacrylate (used as interior plasticizing comonomer) in the MEC solution [135–138]. In fact, they deal with a new type of radical polymerization, interlamellar complex-radical copolymerization (Fig. 5.15). Comparative analysis of various parameters points to formation of interlayer hydrogen bond with oxygen atom of flexible n-butyl methacrylate ether link (BMA) and its crucial role in interlamellar copolymerization and intercalation/exfoliation in presence of radical initiators (AIBN):

In future this approach was broadened, details of its mechanism were studied [139].

A special place in this problem should be assigned to orientation control of a *guest* molecule in intercalation systems of a *host* and, as a result, to formation of stereo-regular polymers: during intercalation the *host* identifies the *guest* type and reversibly accepts it in 2D layered structure [12] (Fig. 5.16).

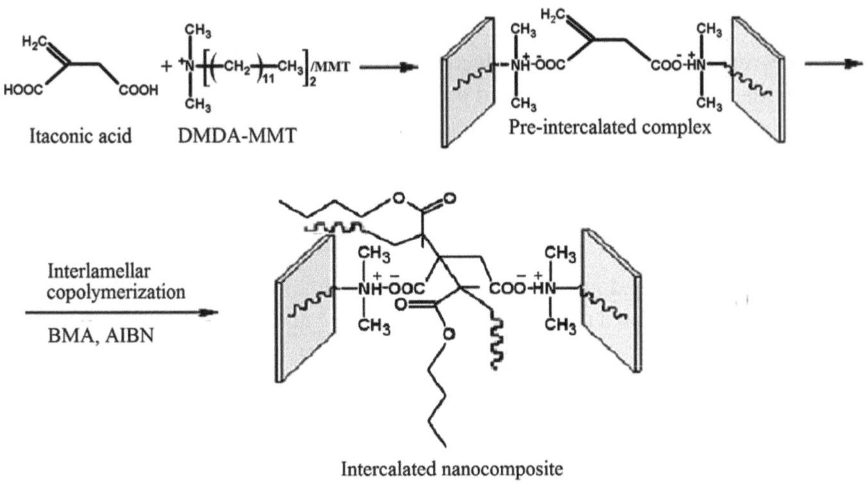

Fig. 5.15 Interlamellar complex-radical copolymerization of pre-intercalated itaconic complex with n-butyl methacrylate [135]

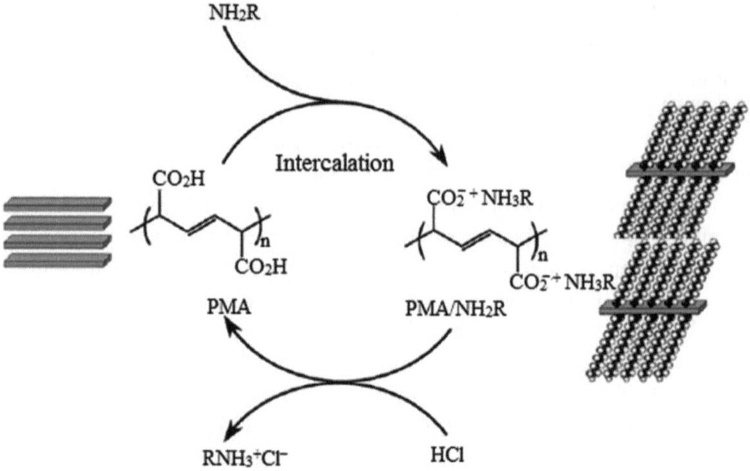

Fig. 5.16 Intercalation polymerization of muconic acid and orientational control of guest molecule

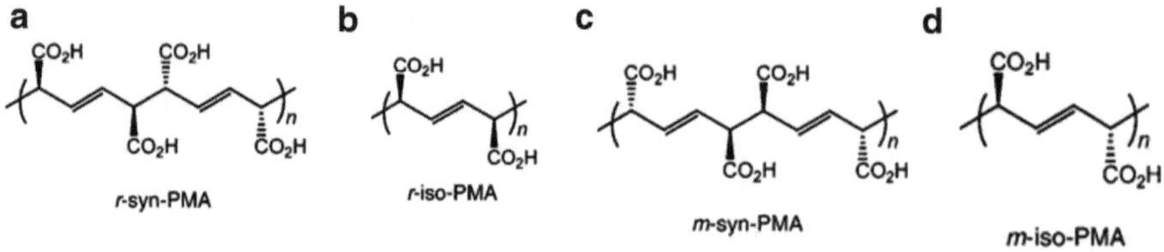

Fig. 5.17 Stereostructures of polymuconic acids obtained via intercalation polymerization of muconic acid: r-syn-tactic (**a**), r-iso-tactic (**b**), m-syn-tactic (**c**), and m-iso-tactic (**d**)

Stereo-regular polymers are produced via topochemical polymerization (Z, Z)- or (E, E)-derivatives of muconic acid *m*-iso-tactic, *r*-*syn*-tactic, *m*-*syn*-tactic, *r*-*iso*-tactic (Fig. 5.17) having translation and alternative stacked structure, respectively, in a solid state under UV-radiation of acid salt crystals and amines with different length of hydrocarbon chain in silicate layers [140–148]. Hydrolysis (HCl или H_2SO_4) converts them into stereo-regular polymuconic acids (PMA) in silicate layers.

Both *m*-*iso*-PMA and *r*-*iso*-PMA crystals show inclined line of the dependence of interlayer distance on a number of carbon atoms in alkylamine, *d* increases by 1Å with each carbon atom (Fig. 5.18). Most probably, *m*- *r*-*syn*-PMA (A, C, straight line 2) have different structures, as is also true for *iso*-PMA (B, D – straight line 1), which differ by a layer structure and by different packages of alkyl ammonium groups. Alkyl chain of a *guest* has a regular structure, corresponding to a *trans*-planar conformation with a partial *gauche*-conformation with different positions of carboxylic groups in the plane of polymer and by 2D structure of intercalated ammonium complex.

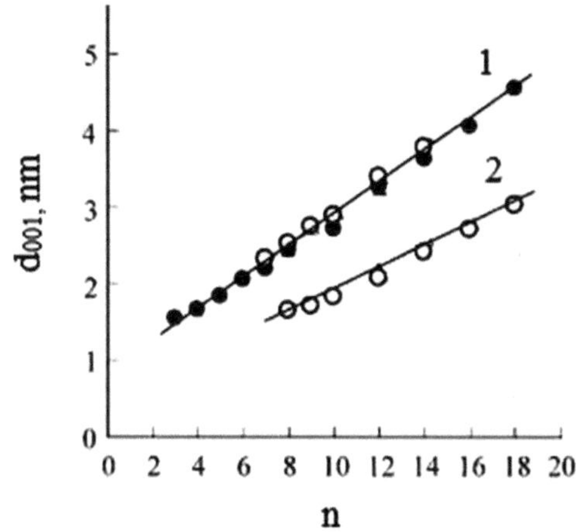

Fig. 5.18 Change in basal space value (d) as a function of the carbon number of n-alkylamines (n). Closed and open symbols indicate isotactic and syndiotactic polymers, respectively: *m-r-iso*-PMA (*1*), *m-r-syn*-PMA (*2*) [142]

The *host* provides control over variation of the *guest* orientation (tactics) and direction, normal or parallel to a polymer chain in isotactic or syndiotactic polymer sheets, respectively. Orientation control of the *host*-molecule (alkylamine in a polymeric crystal) opens a way to design of finely regulated layered materials with modern functions.

5.3.3 Interlamellar Catalytic Polymerization

An important stage in increase in intensity of intercalation processes and exfoliation of a layered silicate is intercalation of Ziegler-Natta catalyst with the following polymerization and co-polymerization of olefin monomers. Reviews of anion-coordination polymerization of olefins in presence of mineral fillers are referred to [33, 149, 150]. Then metallocenes catalysis was studied: after protection of the inner surface of a silicate with aluminum organic compound (most often methylalu-minoxane) and introduction of zirconocene, the formed system causes catalytic polymerization of ethylene its copolymerization with higher olefins (for production of low-density polyethylene, LDPE), intercalated stereo-regular or oligomer poly-propylene. More convenient approach is usage of single-component catalysts of olefins polymerization. In particular, this concerns the Pd^{2+} chelate complex [[(dppp)Pd(OTs)(H_2O)]OT[(dppp)Pd(OTs)(H_2O)]OT]

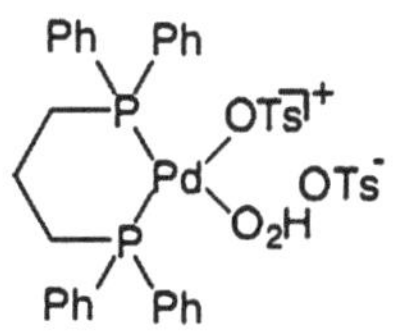

(where dppp = 1.3 Bis(diphenylphosphino)propane, OTs – *n*-toluene sulfonate) which is intercalated in synthetic hectorite modified by tetra(decyl)ammonium cations and performs gas phase polymerization of ethylene at 295 K with formation of high molecular polyethylene (PE extracted with toluene, $M_n = 159{,}000$, $M_w = 262{,}000$). After 2-h absorption of ethylene a dramatic increase in the silicate/catalyst size was observed followed by its collapse: in the composite produced after 24 h of polymerization there are absolutely no diffraction peaks of the silicate. Thus, at the initial stage of in situ formation of hybrid nanocomposites intercalated polymer forms, while at later stages exfoliation of a silicate takes place.

As it was mentioned in Sect. 5.2, almost all natural silicates contain as impurity constitutionally bound compounds of transition metals, components of metal complex catalysts (copper, titanium, vanadium chromium, zirconium, molybdenum, etc.) In this connection those are of a special interest that include impurity ions of metals initiating polymerization of an intercalated monomer. Thus, hectorite, whose sodium ions are substituted by Cu^{2+}, initializes styrene polymerization, in pores as well as on the surface. The polymer has a "brush" structure, which points to orientating effect of inorganic surface, which decreases as the chain growth distances from the surface. Compulsory neighboring in limited basal space has an effect on higher ordering of the polymer, its optical and mechanical properties. Intercalated PS exists in two forms, first of which is similar to blocked polystyrene on the surface, and another one is stiffer, maybe because of high ordering of the polymer. Redox-properties also have Fe^{2+} and Ru^{3+} cations exchanging in clay minerals: many monomers in their coordination sphere have a trend to intercalation radical polymerization. In the already mentioned review [35] crystal chemistry of the frame-forming silicate materials, in which impurity atoms of some transit metals, such as Ti, Nb, Ta, W, Fe, Mn (naturally immobilized metal ions) and others with coordination numbers 5, 6, as well as tetrahedral coordination atoms, are analyzed. Probably, they may serve as starting elements for hybrid materials, however, presently these data are limited, in the systems where naturally immobilized clays activated by aluminum organic compounds are used, activity of these elements in polymerization of olefins is low.

It should be noted that there is a possibility of obtaining molecular polymer-polymer nanocomposites [151] using spatial limitations which arise during catalytic formation of various chains in mesoporous systems [152]. Experimentally this approach is realized by treatment of silicates by methylalumoxane followed by bonding of two catalysts Cp_2TiCl_2 (Cp- cyclopentadienyl ligand) – the catalyst of ethylene polymerization and $CpmTi(OMe)_3$ (Cpm-pentamethylcyclopentadienyl ligand) – the catalyst of syndiotactic styrene polymerization. As a result, a composite forms with improved properties due to uniform (at nanometric level) distribution of polymers in a reactor mixture.

Oppositely to polar polymer matrices, in the case of polyolefin nanocomposites it is impossible to reach significant improvements of a complex of properties at low filling factor owing to, on the one hand, limited compatibility of non-polar polymers with layered silicates, and, on the other hand, difficulties in reaching exfoliation of a layered ingredient in single nanolayers in polyolefin matrix.

We shall especially highlight catalytic polymerization of MMA and formation of a stereo complex syndiotactic/isotactic PMMA in the layered silicate [153]. To do this, methyl-bis(2-hydroxyethyl)alkyl of fatty acids [$Me(T)N(C_2H_4OH)_2^+$/MMT] modified by ammonium salt is activated by toluene adduct *tris*-(pentafluoride phenyl)alane $Al(C_6F_5)_3(C_7H_8)_{0.5}$ (A activator) or $Al(CH_3)_3$ with the following treatment $B(C_6F_5)_3$ (B activator) and then combine with zirconium containing catalyst for production of iso- and syndiotactic PMMA [154]. Combination of thus activated (A and B) OMMT with decyl metallocene complexes of various symmetry, including achiral C_{2v}-symmetric Cp_2ZrMe_2, chiral C_2-symmetric *ras*-(EBI)$ZrMe_2$ (EBI = $C_2H_4(Ind)_2$, Ind- indenyl), pro-chiral Cs-symmetric $CGCTiMe_2$ (CGC = $Me_2Si(Me_4C_5(t$-BuN), with in situ MMA polymerization brings to formation with high yield of PMMA/OMMT nanocomposite with intercalated morphology (Fig. 5.19). In this case the forming polymer chain is subjected to stoichiometric control, realized as *atactic*, *iso-* or *syndiotactic* polymer from *iso-syndio-stereoblock* governed by metallocene symmetry (like polymerization of abovementioned muconic acid).

A range of stereo regulated systems of the considered type is significantly widened, see for example [155–157].

5.3.4 Intercalation Assembling

Molecules of a monomer intersticed by displacement form hydrogen and other bonds with a host, because kaolin-methanol intercalated compounds are comfortable intermediates for substitution by *guests* such as *n*-nitroaniline, ε-caprolactam, alkylamines, vinylpyrrolidone and others by relatively large molecules. This is a convenient way for formation of clays and nanocomposites based on condensation type of polymers, for example, ω- amino acids with a short carbon chain, β-alanine [H_2N-$(CH_2)_2COOH$], 6-aminocaproic acid [H_2N-$(CH_2)_5COOH$]. Taking into account that direct formation of compounds of 6-aminocaproic acid insertion is hindered by its relatively large dimensions and low intercalation ability it is more often intercalated in vermiculite or indirect way is used, *guest*-substitution reaction, when kaolin-methanol intercalate is used for intermediate [27]. Polymerization is performed via heating during 1 h at 250 °C in nitrogen flow. As is seen from Fig. 5.20, basal space of the matter of acid intercalation is 1.23 nm (intensive peak at 10.6° (0.84 nm)) corresponds to crystal structure of 6-aminocaproic acid absorbed on surface), it is greater than for kaolin (0.72 nm) or for the intercalated compound kaolin/methanol (1.11 nm). It is interesting that the basal space of thermally treated substance is 1.16 nm, which is shorter than the initial one (1.23 nm).

Probably, the reason for this decrease is reaction of polycondensation accompanied by spontaneous reorganizing of hydrogen bonds between OH-groups of *guest* molecules.

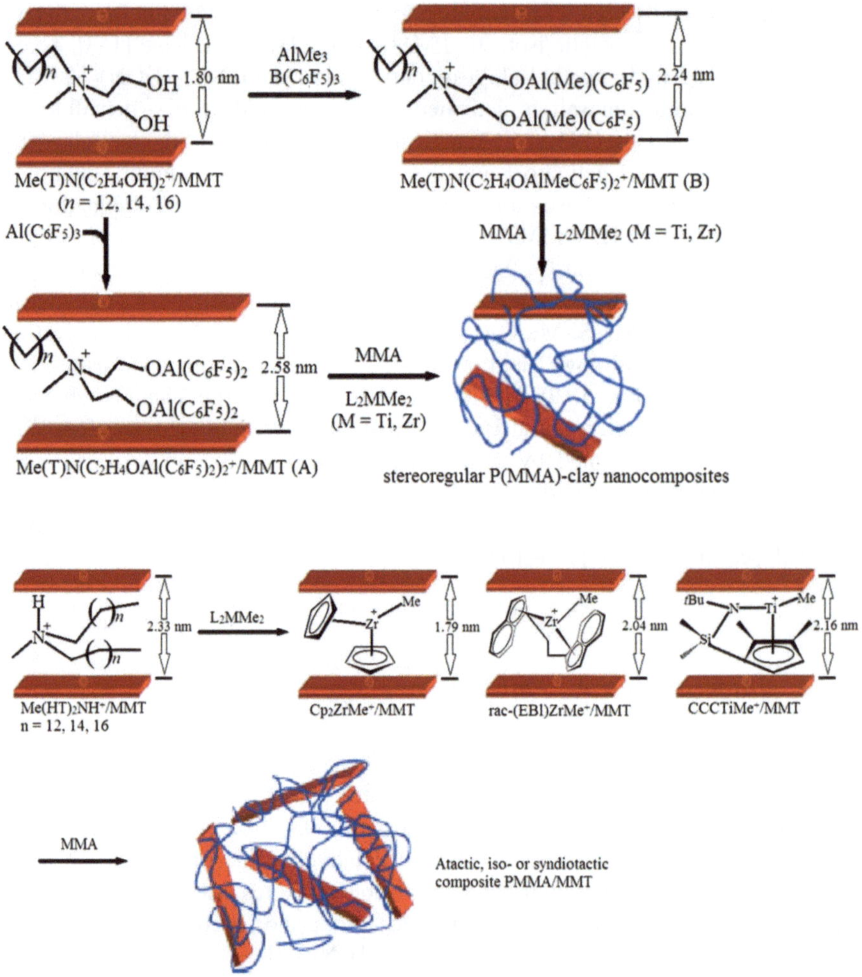

Fig. 5.19 Intercalated zirconocene cations for stereochemical control of PMMA chain growth [153]

Formation of aramid composite [158] can be presented schematically (Fig. 5.21), from which it is seen that interstitial co-monomer is simultaneously MMT modifier:

Intercalation technique [159] is applied to produce PA by polymerization in N,N (dimethylacetate amide) using 4,4(oxydianiline) (ODA) and isophthaloyl chloride (IPC) by the scheme (Fig. 5.22).

However, more widely used for formation of this type of composites are melt-intercalation techniques (see Sect. 5.4.4).

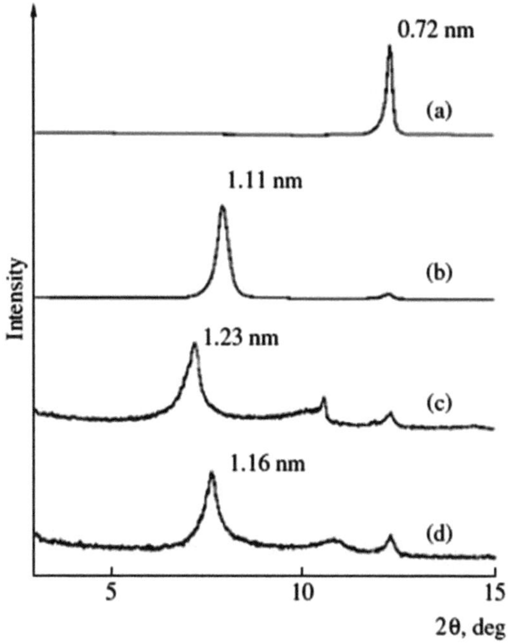

Fig. 5.20 X-ray diffraction patterns of (**a**) kaolinite, (**b**) a kaolinite–methanol intercalation compound prepared wetchemically, and (**c**, **d**) a kaolinite-6-aminohexanoic acid intercalation compound before and after heat treatment at 520 K, respectively [27]

5.3.5 Polymerization with a Ring Opening

This is a limited group of nanocomposites, only few examples of their synthesis are known. Hybrid polybenzoxazine/MMT nanocomposites are obtained via OMMT polymerization with a precursor ring opening (270–280 °C). The surface of OMMT has catalytic properties in this reaction [160]. In nanocomposites containing up to 10 mass.% OMMT there is distortion and breakage of its layered structure. In other words, nanocomposites with high content of organo-modified MMT have a mixed morphology, exfoliated in addition to intercalated one. The mechanism of the composite formation via ring-opening polymerization has much in common with intercalating polymerization of vinyl type of monomers [161, 162].

5.3.6 Redox-Intercalation Polymerization

Apart from vinyl monomers in mineral matrices also acetylene monomers are polymerized, for example, 2-ethylpyridine (EPy), implanted in MMT galleries [163, 164].

Probably, it is possible to realize photochemical polymerization of diacetylene-3,5-octadiine in the layers of metal phosphates (M = Mg, Mn or Zn). It can be assumed that under respective conditions monomers fill almost all volume of pores or interlayer distance. The next oxidizing polymerization is governed by molecular

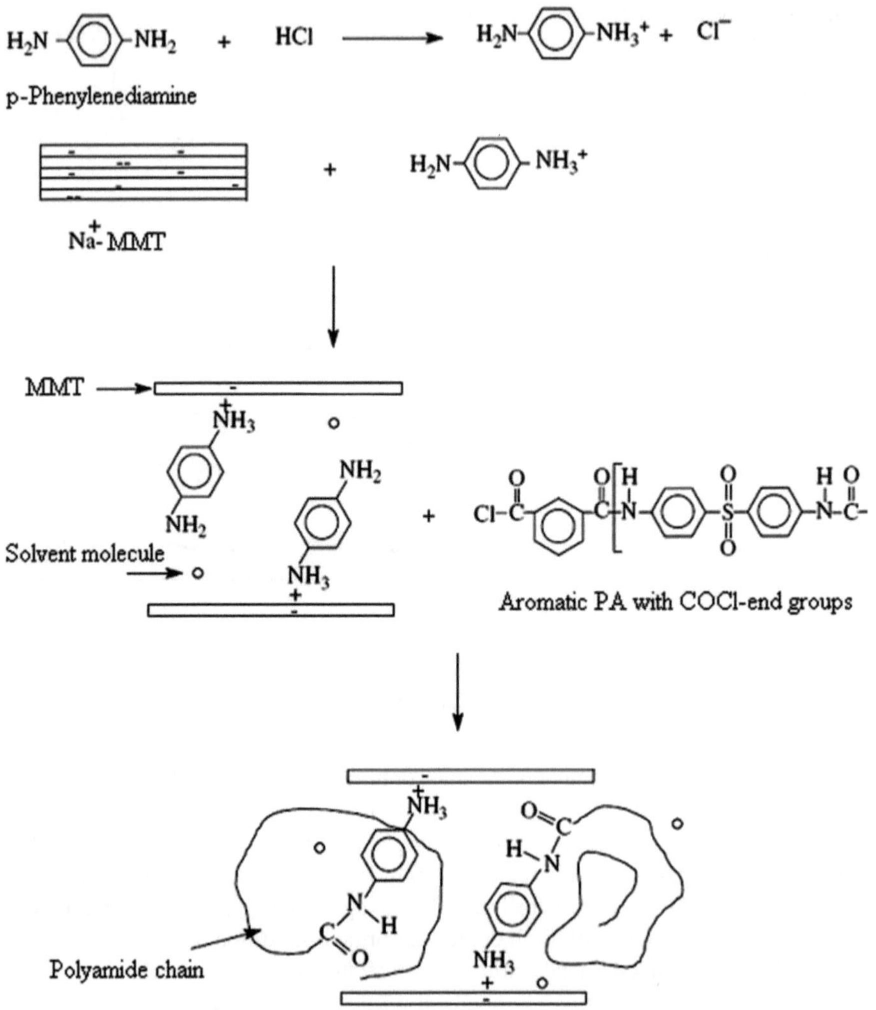

Fig. 5.21 Schematic representation for formation of aromatic PA/MMT nanocomposite [158]

oxygen acting as electron acceptor in presence of redox-active *host*, catalyst of electron transport, for example, Fe^{3+}, Cu^{2+}. Widely known are hybrid nanocomposites based on polyconjugated conductive polymers, such as polyaniline (PAn), poly(2-ethylaniline), poly-n-phenylene, polythiophene, polypyrrole (PPy) and polyacrylonitrile (followed by pyrolysis), and on various mineral matrices. Thus, interchannel reactions of polymerization of preadsorbed acrylonitrile in a confined volume result in formation of a fiber polymer. As a result of this pyrolysis, carbonized conductive material forms in channels of a *host*. Postintercalation polymerization of aniline (associated, as a rule, with amino groups proton treatment) is performed in air at 130 °C in phosphate layers: $Zr(HPO_4)_2$ zirconium

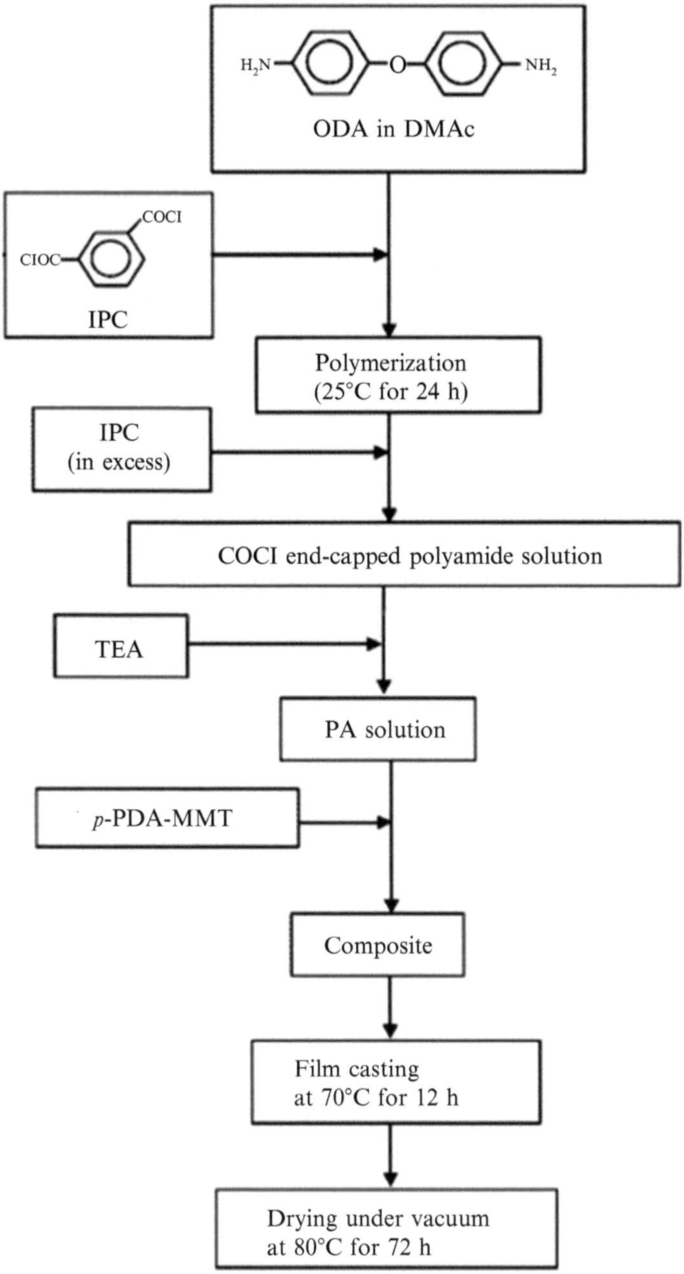

Fig. 5.22 Flow chart diagram for the synthesis of aramid/p-PDA-MMT nanocomposites [159]

phosphate, $VOPO_4$ vanadium phosphate, uranium HUO_2PO_4 hydrophosphate, Mg-Al layered binary hydroxides, $HMWO_6 \cdot H_2O$ (M = Nb, Ta), in layered acid zirconium-copper phosphates, etc. (see, for example [73, 165–167]).

Redox-intercalation aniline (An) polymerization proceeds with loss of two electrons and two protons for each monomer link:

$$nC_6H_5NH_2 \rightarrow (-C_6H_4NH-)_n + 2nH^+ + 2ne \tag{5.4}$$

It is important that oxidation polymerization of aniline, pyrrole, dithiophene monomers, intercalated in layered aluminum silicates leads to formation of highly oriented *host-guest* layers [117].[4]

Oxidation polymerization of pyrrole, dithiophene, tetrahydrofuran, acrylonitrile in FeOCl lattice has been known fairly long. Intercalated aniline forms hydrogen bonds with chlorine atoms of a lattice, polymerization proceeds along its diagonal (Fig. 5.23).

This lattice proves to be convenient for oxidation polymerization of An introduced from aprotonic solution. Gross content of the obtained substance is expressed by the formula $(An)_{0,28}FeOCl$, single crystal was grown from this. Zigzag polymer chains with $M_w = 6{,}100$, commensurable with FeOCl lattice are directed along the *host* crystal and hydrogen bonds of NH-groups with chlorine atoms of the lattice layers. In this case the ratio Fe^{2+}/Fe^{3+} is ~1:9, and polymer intercalate behaves as *p*-type semiconductor (specific conductivity of a single crystal is $1{,}5 \cdot 10^{-2}$ S/cm); during its long oxidation in air a mixture forms consisting of PAn and β-FeOOH. Other layered chlorides $FeCl_3$, CdI_2, $CdCl_2$ are seldom used in intercalation practice. At the same α-$RuCl_3$/PAn nanocomposites were obtained in situ via oxidation

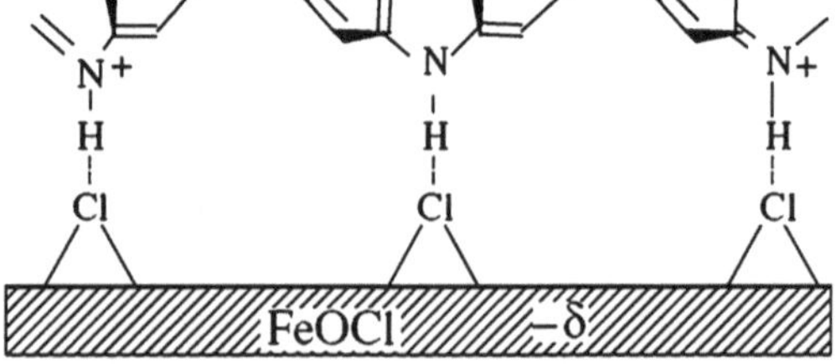

Fig. 5.23 Schematic representation of polyaniline intercalation into FeOCl lattice

[4] Very interesting is a title of one of the recent reviews [117], which considers energy efficient hybrid materials *Hybrid organic–inorganic materials: from child's play to energy applications*.

intercalation polymerization [168]. Moreover, α-$RuCl_3$ behaves as a perfect intercalation *host* polymer.

We should also notice that the materials obtained in polymerizing systems from N-vinylcarbazole and iron chloride impregnant with MMT (referred to [33]).

Hybrid nanocomposites based on PAn have special properties – variability of structures doping mechanisms, proper stability and recycling ability are widely used as materials for electronics. The most widely used variants are: PAn intercalations in V_2O_5, TiO_2, MoO_3, SnO_2, SiO_2, $BaSO_4$, HUO_2PO_4, Fe_2O_3, etc. [73, 131]. Thus, intercalation of aniline in $HMWO_6$ Bronsted layer acids proceeds at 130 °C in air and is accompanied by polymerization according to the scheme:

$$(C_6H_5NH_3)_xH_{1-x}MMoO_6 \times H_2O \xrightarrow{O_2} PAn \times H_{1-x}MMoO_6 + (1+x)H_2O \quad (5.5)$$

Lattice parameters of the monomer intercalate (x = 0.31) increase by ~0.69 nm as compared with non-aqueous $HMMoO_6$, while the corresponding polymeric only by 0.5 nm (Fig. 5.24). Probably, the polymer chain in PAn/$HMWO_6$ is so oriented that C_2 axis is parallel to inorganic lattice. Similar situation takes place in the V_2O_5-based nanocomposites.

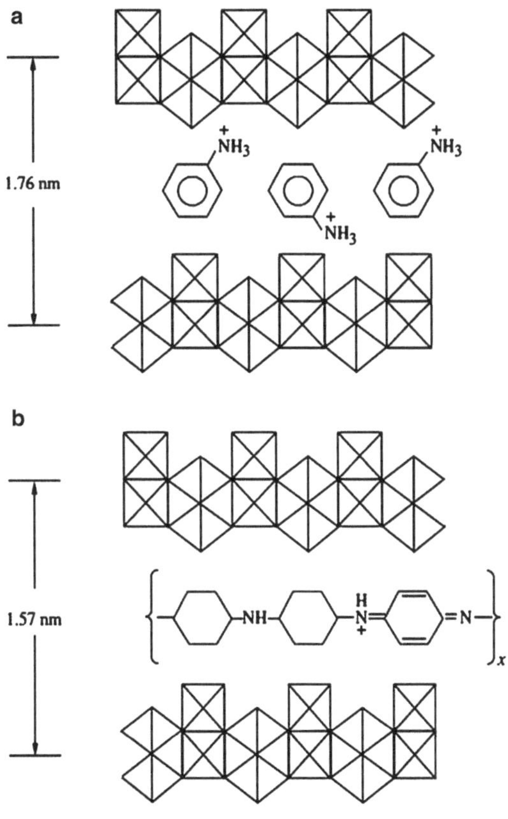

Fig. 5.24 Configurations of intercalated (**a**) aniline and (**b**) PAn in $HMMoO_6$. (M = Mg, Mn, Zn)

The hybrid material produced by polymerization of pyrrole in this fiber matrix is a solid conductive nanocomposite. Inorganic phase based on V_2O_5 xerogel is more expanded and is used for synthesis of hybrid materials with conductive polymers. Various intercalated types of conductive polymers, such as PAn, poly (2,2-dithiophen) or PPy are rather well investigated [169, 170]. Their redox-intercalation polymerization in V_2O_5 xerogel is an efficient method for obtaining of layered materials with a conductive polymer layer formed by polyconjugated anisotropic laminar microstructures [171, 172]. Their formation is based on oxidation-reduction reaction in which a monomer polymerizes after oxidation and V^{5+} is partially reduced to tetravalence. Intercalations (for example, in the case of aniline formation $(C_6H_4NH)_{0.44} \cdot V_2O_5 \cdot 0{,}5H_2O$) deep-blue complex with metal gloss is formed) are accompanied by polymerization. Growth of polymeric chains are realized first inside the xerogel, which is critical for formation of intercalation polymer, and are related to molecular oxygen transport, meanwhile, V_2O_5 xerogel also acts as a catalyst. Conductive polymer with different PAn/V_2O_5 ratios is formed with participance of mixed valence (V^{4+}/V^{5+}) lamellas of the *host* ordered in one direction. The material consists of altering vanadium oxide and polymer layers, its conductivity is by 4 orders of magnitude higher than the conductivity of initial $V_2O_5 \cdot nH_2O$ xerogel and is at room temperature ~$0.5\,\Omega^{-1} \cdot cm^{-1}$. Typically, depending on molar ratio of the components, two phases form with the composition $(C_6H_4NH)_{0.6} \cdot V_2O_5 \cdot nH_2O$ и $(C_6H_4NH)_{1.2} \cdot V_2O_5 \cdot nH_2O$. For the first one the interlayer distance is 1.4 nm, which corresponds to intercalation of one monolayer in the xerogel, for the second one (d = 2.0 nm), which corresponds to two monolayers. These structures allow doping of intercalated PAn accompanied by a small increase of electric conductivity of the produced materials, redox-active both in acid media and in aprotonic organic electrolytes. Physical chemical studies prove formation of a salted form of PAn. Aging of the material in air induces partial oxidation of inorganic frame and oxidizing binding of PAn in interlamellar space accompanied by further polymerization of aniline oligomers in xerogel layers. The formed polymer is as if frozen in them, which is caused by binding interaction organic and inorganic components due to formation of hydrogen bonds –N–H··· O–V–.

It is interesting that microstructure and properties of another conducting material, PPy/V_2O_5 aerogel composite depend on the synthesis method: it can be obtained as a result of combined polymerization of pyrrole and sol gel transformation of vanadyl alkoxide precursor ($VO(OPr)_3$ + pyrrole + water + acetone mixture) or mixturing of V_2O_5 gel with PPy). The best conductivity has a tablet nanocomposite synthesized by "post-gelatin" polymerization including consecutive polymerization of inorganic and organic phases. There is almost nothing known about co-intercalation of different types of monomers for synthesis of hybrid materials including two polymer components (by polymer-polymer type). Apart from the abovementioned catalytic way of formation of PE/PS composite, the method should be found based on mixing of methanol solution of 2,5-dimercapto-1, 3, 4-thiadiazole and aniline with $V_2O_5 \cdot nH_2O$ gel in molar ratio 1:2:2 [173]. Competitive polymerization and insertion of aniline takes place

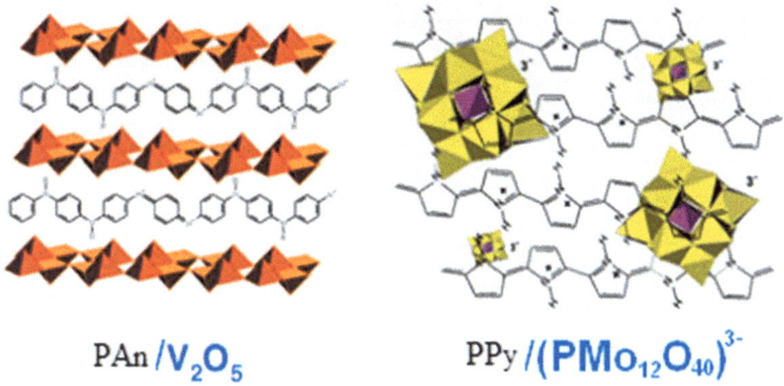

Fig. 5.25 Structure of conducting PAn/V_2O_5 and PPy/$(PMo_{12}O_{40})^{3-}$ composites [117]

also in the case of MoO_3 with formation of $(PAn)_{0,24} \cdot MoO_3$ nanostructures, including usage of low temperature intercalation technique. Polyaniline chains widen layers and change a potential surface, decreasing polarizing ability of the lattice. Further oxidation of polyaniline chains intercalated in MoO_3 is due to $(NH_4)_2S_2O_8$. Chemical synthesis of hybrid materials can realize via simple addition of pure aniline to the solid fluorite-molybdenum acid $H_3PMo_{12}O_{40}$. Its gallery includes a water layer bound with hydrogen bonds. Cation H^+-galleries can be substituted not only by other compensating cations, but also by aniline molecules with its following polymerization.

Heteropolyanions compose redox-frame of PAn, PPy [174], poly(3, -4-ethylenedioxythiophene) [175, 176] (Fig. 5.25).

The structure of conducting composites, ranged from the nucleus-shell complex to nanostructures and 100-nm nanoparticles, intercalation hybrids or smaller than 1 nm molecular structures are distinguished by a type of layered structures. Various inorganic phases with developed surface such as RuO_2, MnO_2, etc. are used as components of components of conductive systems [177].

Integration of polyoxometalates in conductive polymers is unusual class of electrically active hybrids, the result of a new idea of application of these in electrochemical supercapacitors. Figure 5.26 shows electrochemical characteristics of $(PMo_{12}O_{40})^{3-}$/PAn hybrid-based solid phased symmetric supercapacitor gauges as active electrodes [178, 179], in this case electric capacity evolves with increase in cycling acts up to 300 cycles (up to 120 F/g).

Polyoxometalates, which are ideal model electrodes for molecular batteries, are also active intercalating elements. In the forming hybrid organo-inorganic nanomaterials the conductive phase is polyaniline matrix and fluorine-molybdate anion $(PMo_{12}O_{40})^{3-}$ is active inorganic redox-component. Advantages of hybrid electrodes can be described as follows. First of all, a possibility of synthesis of hybrid materials exclusively with doped centers is shown. Next, they show stable behavior in terms of cyclic anion/cation integration. Thirdly, they can be used as blocked solid materials with molecular centers of fluorine-molybdate-anion.

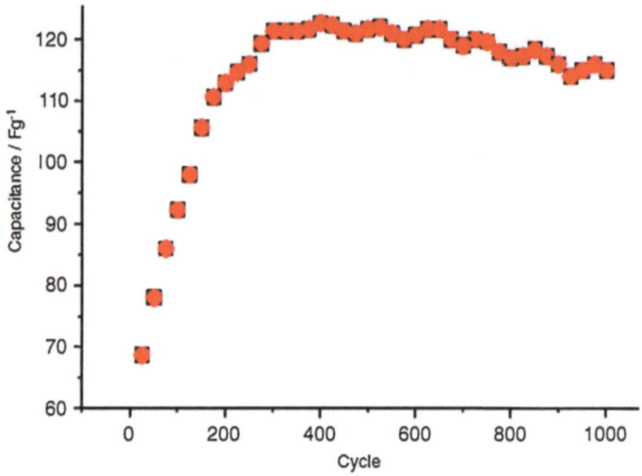

Fig. 5.26 Capacitance as a function of cycle number for a symmetrical supercapacitor cell with the hybrid $(PMo_{12}O_{40})^{3-}$/PAn as electrode material [117]

Very interesting intercalated agent for production of conductive materials is molybdenum disulfide (Fig. 5.27). X-ray diffraction patterns of MoS_2 and PPy intercalated in them show intense peaks at $2\theta = 7.63$ deg (Fig. 5.27b, 2) and 8.96 deg (Fig. 5.27b, 3) are associated with reflection (002, c axis) from the plane of layered molybdenum disulfide, interlayer distance in MoS_2 with intercalated polypyrrole 1.158 for PPy/MoS_2-1 and 0.986 nm for PPy/MoS_2-2, respectively [180].

Taking into account that the interlayer distance in the initial MoS_2 is estimated as 0.615 nm (Fig. 5.27b, 1) and it increases by 0.543 and 0.371 nm in two composites, which confirms intercalation of PPy between the layers of MoS_2. On the other hand, appearance of two different interplanar distances shows that the polypyrrole chains between MoS_2 layers are in two different conformations (orientations) (Fig. 5.27a). Most probably, smaller interplanar distance corresponds to a *guest* polymer molecule lying on a plane between layers of MoS_2, greater interplanar distance is responsible for the polymer molecules perpendicular to MoS_2. Intercalation of PPy in MoS_2 causes significant increase in electric conductivity of both conformations (PPy/MoS_2-2 far more than PPy/MoS_2-1). Probably, this molecule of the previous polymer lies in the plane of π-electrons parallel to layers of MoS_2, which facilitates electron transport between organic and inorganic components.

Despite great facilities of the intercalation method carried out via in situ polymerization, it has drawbacks. The main drawbacks of the intercalation polymerization method are the problems of specifying molecular – mass characteristics of produced polymers, which impedes investigation of kinetic parameters of intercalation polymerization and finding of its special features as compared to traditional processes. Applications of these methods are confined to thin films production, coats, etc.

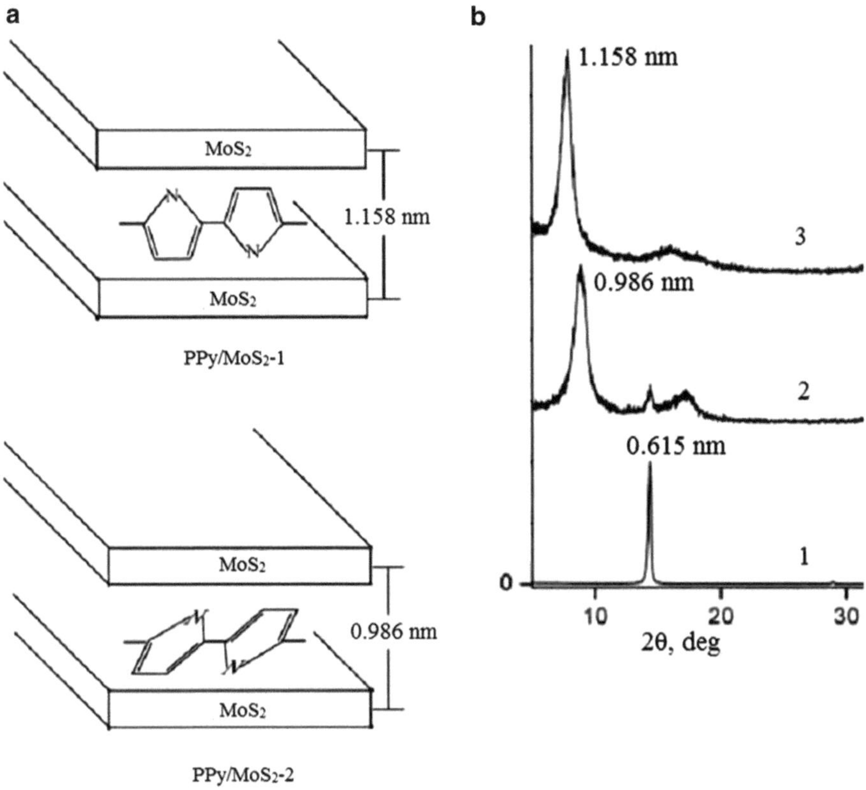

Fig. 5.27 (**a**) Schematic representation of PPy intercalation models into MoS_2 : PPy/MoS_2-1 и PPy/MoS_2-2 with perpendicular and flate orientation of polymer molecule, respectively (**b**) X-ray diffraction patterns of MoS_2 (*1*) and PPy/MoS_2 -1 (*2*), MoS_2-2 (*3*) intercalated nanocomposites [180]

5.4 Hybrid Nanocomposites Produced by Direct Intercalation of Macromolecules in *Host* Layers

Though intercalation from polymer solution of melt has being studied for rather long time, commercial applications have presently only composites on Nylon 6 and OMMT basis [39]. It seems promising to use polymer/silicate hybrids obtained this way in automotive industry, packing industry, and aerospace industry, as well as materials with high thermal, gas-separating, and mechanical properties.

Matrices are made of PS, polycaprolactam (PCL), PEO ($T_m = 328$ K) polyvinylidene fluoride ($T_m = 444$ K), polymethylmethacrylate (PMMA) ($T_m = 527$ K), poly-4-vinylpyridine ($T_m = 473$ K), etc. Degree of intercalation of a polymer does not correlate with its dielectric constant, most probably, presence of carbonyl group and low T_g of the polymer encourages its intercalation. Aminomodifier increases intensity of intercalation (as compared with a fluoride containing one) due to smaller volume of its functional group and formation of stronger hydrogen bonds of surface hydroxyl groups of silicate with polycaprolactam. In this case mobility of spin-labeled modifier is increased in presence of polymer, which is probably caused by the fact that T_g of the polymer in the surfactant layer is lower than the order-disorder temperature in pure organo-magadiite. This phenomenon is similar to polymer plastification.

Direct methods (especially, melt technique) are of versatile interest. Firstly, they are relatively simple, as compared to polymerization in interlayer space, and are therefore, more practical for structuring of inorganic/organic polylayered composites [118]. Next, for this can be used well characterized polymer. And the last, the process is interesting due to its unusual intercalation physical chemistry. It displays in acquiring of electron conductivity by a system (in polyconjugated systems) for application, for example, in reversible electrodes, in improved physical-mechanical properties of many nanocomposites. Studying of these systems also can give important information on special features of absorption of a polymer on nanometer materials. Thus, absorption isotherms of polyvinylpyrrolidone in CeO_2 suspension (synthesized under hydrothermal conditions, particles size about 9 nm) for various pH is well studied (see, for example [33]. Most widely used are two types of direct intercalation of polymers: from solution or from melt.

In discussion main properties of nanocomposites obtained by comparative methods are analyzed, for example [181]. Nevertheless, still remain many problems as regards the effect of various parameters on properties of the forming nanocomposites.

5.4.1 Intercalation of Polymers from Their Solutions

Nanosilicate composite produced in this process includes three components: silicate layers, solvent and polymer. The controlling parameters are: composition, concentration and distribution of components, molecular mass of a polymer, temperature; distribution of the components plays important role in achievement of enhanced characteristics.

The most important factors are: the origin and polarity of a solvent, a silicate modifier, a character of interaction between the components, precipitation, etc. For example, precipitation is successfully used for formation of the systems based on ABS [182], polyvinylidene fluoride (PVdF) [183], PE [184], however, there is no intercalation from the copolymer solution of ethylene with acrylic acid into MMT, though it easily proceeds from the melt [185]. The effect of a solvent depends also on the origin of a modifier: when a modifier with one long alkyl chain is used, interlayer distance decreases linearly with increase in dielectric constant, while if a modifier with two long chains is applied, linear dependence of d on polarity of the solvent is observed under condition that the solvent has low (below 10) value of dielectric constant. Other listed parameters play less important role [44].

The experimental procedure is as follows. The layered silicates modified by organic cations are swelled in some solvent with the following intercalation of a polymer solved in the same solvent. Various approaches are possible to obtain these hybrids: mixing of a solution (for example, salt solutions of polyanilinehydrochloride in acid-methanol solution, sulfonic salt of poly-*n*-phenylelene soluble in many solvents, etc.), suspension, confinement of polymer in gel or xerogel, etc.

Layered silicates are dispersed to small aggregates (packs) consisting of several layers with high aspect ratio and intercalated between them and the polymer parallel to it consisting of one-two layers of macromolecules. The driving force of intercalation from the solution is an increase in entropy during desorption of a solvent made of active swelling polymer, which overpowers a decrease in entropy during sorption of swelled macromolecules. Though usage of great amounts of a solvent makes this process less attractive, the solution intercalation is applied in the fields where thin films are required, for example, in coatings [186], membranes [187], etc.

Most widely used are the following types of nanocomposites produced this way: PS thermoplastics [188], styrene-acrylonitrile copolymer, PMMA [189], thermoplastic elastomers [190–193], fluorine elastomers [194], PEO-MMT, nylon-layered silicates, composites based on hectorite and polyaniline, polythiophene or polypyrrole, TiO_2 and polyaniline shaped as nanotubes [195], etc. More rarely this method is used to form reactive plastics – poly(glycidyl methacrylate-co-methylmethacrylate) [196], epoxycomposites [197]. These hybrids can be used as nanofilms or nanocoatings formed in continuous (flow) regime during surface modification of various thermoplastic films, as reaction-capable compatibilizers, fillers for thermoplastic polymer compositions, especially based on acryl polymers, components of nanomaterials, made via extrusion, masterbatches, etc. [86, 195, 198–201].

Though intercalation of polyolefins is impeded by their solution, there are solutions adapted for this purpose, in particular, solutions of low density linear PE and PE-grafted-MA (PE-*gr*-MA) in xylene in combination with binary Zn_3Al-hydroxides for producing of exfoliated nanocomposites [202].

Generally, in this type of nanocomposites also hybrid structures of two types can form: intercalated, in which a polymer or a part of its links are implanted between silicate layers, or delaminated with dispersed individual silicate layers in organic matrix, and the composite structure is microscopically isotropic. Layered silicates are mostly used like in the case of polymerization, most often Na-MMT, including those modified by long chained amines.

To implant PEO from water solution, a convenient inorganic matrix is V_2O_5, in this case its interplanar distance increases to 1.32 nm, the same is observed in the case of PVP, PPO, etc. PEO is also implanted in lamellar networks of $V_2O_5 \cdot nH_2O$ and $CdPS_3$. Thus, PEO water solution (molar mass 10^5) with $V_2O_5 \cdot nH_2O$ gel (interplanar distance 1.155 nm) after water removal forms a nanocomposite of xerogel with the general formula $(PEO)_xV_2O_5 \cdot nH_2O$ (referred by [33]. Interplanar distances in the matrix increases from 1.32 to 1.68 nm at $x = 0.5–1.0$, while at $1 < x < 3$ it increases to 1.83 nm. It is interesting that the composites with $x < 1$ contain a monolayer, while at $x \geq 1$ bilayer of closely packed PEO molecules completely filling the interlayer distance of xerogel with fiber mono- and bilayers.

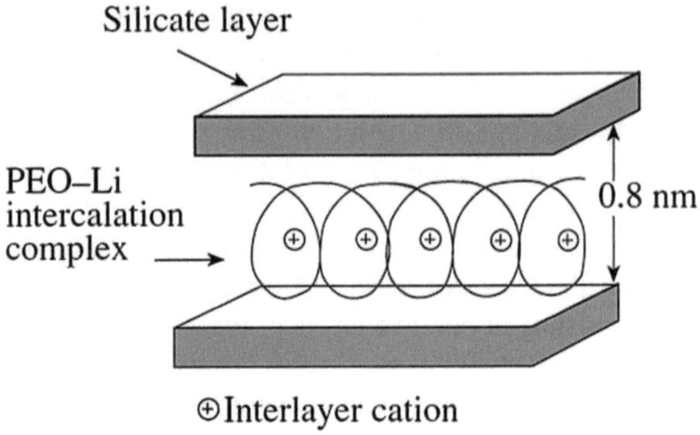

Fig. 5.28 Structure of PEO intercalated into a monoionic silicate layer [34]

The one-phased nanocomposite is formed at $x < 0.8$, it easily forms flexible thin film and can participate in oxidation-reduction intercalation. As a result of these processes solid electrolytes and materials with electrochromic properties form:

$$y\mathrm{LiI} + (\mathrm{PEO})_x\mathrm{V_2O_5} \times n\mathrm{H_2O} \rightarrow \mathrm{Li}_y(\mathrm{PEO})_x\mathrm{V_2O_5} \times n\mathrm{H_2O} + (y/2)\mathrm{I_2} \qquad (5.6)$$

Alkali metal ions and PEO form interstitial compounds that also can implant into silicate layers, for example, in MMT (Fig. 5.28). Intercalated salt complexes PEO/Li^+ – MMT are apart at 0.8 nm distance, in this case PEO chain has a little stressed helicoid conformation. To compare, we shall note that broadening in PEO-MMT nanocomposite is 0.81 nm, which is wider than in nylon-6-clay nanocomposite (0.6 nm).

The PEO-LiX (X – halogen, alkoxy-, aryloxy- groups, etc.) systems are widely used as flexible (soft) electrolytes, new ionic conductive materials. For these systems it is interesting to find the effect of value and number of charges on mesomorphic behavior and ion conducting properties [203]. Though lithium salts are used in majority of practical applications, other cations are also well studied. The most intensely studied during the last decade are PEO/Zn salt systems. The compounds containing different numbers of PEO links are considered and characterized: $(–CH_2–CH_2–O–)_nZnBr_2$ ($n = 6$, 8, 15). Local neighboring of macrocomplexes has a significant effect on properties of polymer electrolytes, which is demonstrated taking $(–CH_2–CH_2–O–)_n\ [(ZnBr_2)_{1-x}(LiBr)_x]$ ($n = 20$–80, $x = 0$–0,5) systems for example [204].

As in other already analyzed cases, a cation can be intercalated by the ion exchanging and by reduction, while neutral polar molecules can be intercalated only via substitution by molecules of a solvent. For water-soluble polymers PEO, PVP, polyethylene imine (PEI) the method of incapsulated precipitation from solution is used. As it was already mentioned, α-$RuCl_3$, having lamellar structure, has high intercalation properties. In the inert atmosphere as a result of interaction between α-$RuCl_3$, and $LiBH_4$ Li_xRuCl_3 ($x \approx 0,2$) is obtained, which, in turn, forms intercalation compounds with $Li_x(PEO)_yRuCl_3$ type of polymer (Fig. 5.29).

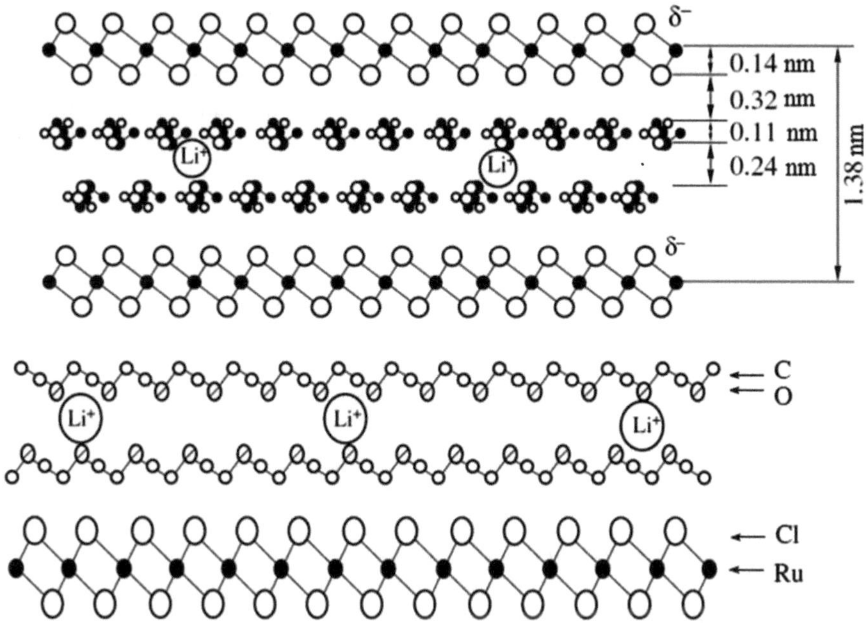

Fig. 5.29 Structural model for $Li_x(PEO)_yRuCl_3$ and schematic 1D electron-density map (projection along the *c* axis)

These products have ionic conductivity comparable with the best polymer electrolytes [168].

Comparative with the melt method, in situ intercalation polymerization causes homogenous dispersions and is accompanied by substantial improvements of properties [111]. However, taking into account the fact that intercalation of some polymers has special features (for example, requires their solubility) it has been studied for few polymers, in contrast to more conventional melt technique.

5.4.2 Intercalation Composites Produced from Polymer Melts

This method is a direct mixing of a polymer melt with organic silicate under action of shear loads at temperatures higher than the temperature of the matrix melt. Exfoliation of layered silicates or minerals caused by mixing with a polymer, when both components, polymer and inorganic material in the interphase layer, are exposed to thermomechanical stress under the treatment, combined with a filler dispersion into uniformly distributed nanoparticles forming a nanostructured system, significantly improve properties of the composites.

This preparation technique also called reactor mixing, is very convenient, it is used for various types of polymers (non-polar PE, PP, PS, natural resin; polar polyamides, polylactic acids, etc.), for its realization compaction, injection formation, extrusion can be used, which reduces production time and cost of materials, the process is ecological [205–216]. The experiments on the polymer/silicate hybrid materials show that intercalation of polymer chains causes 25 % increase in space between silicate layers. This increase corresponding to radius of inertia of the polymer melt assumes plane conformation of the chains in the galleries. Non-disturbed radius of inertia in the melt is calculated from the formula $R = u(N/6)^{1/2}$, where u is length of a segment of a polystyrene chain based on polymerization degree N, a u is valued 5 Å from the table data [217].

Theoretical studies of melt intercalation receive much attention. Thus, molecular dynamic techniques (elastic spheres method) is applied for promotion of intercalate and exfoliate formation in nanocomposite material. The effect of many factors is taken into account, among them the most important are temperature, structure of polymer and co-polymer chains, including functioning of end links, content and degree of interaction of polymer-silicate layers [218].

Commercial importance of this widely used versatile intercalation technique as ecologically preferable and technically easily realized has already been discussed. Lastly, it should be noted that intercalation is often enforced by stimulating power actions, for example, ultrasonic treatment (Fig. 5.30).

Intercalation of polystyrene from melt to organic clays is controlled by mass transport in primary particles and is not specifically controlled by diffusion of polymer chains in the space of a silicate gallery. Activation energy of hybrid

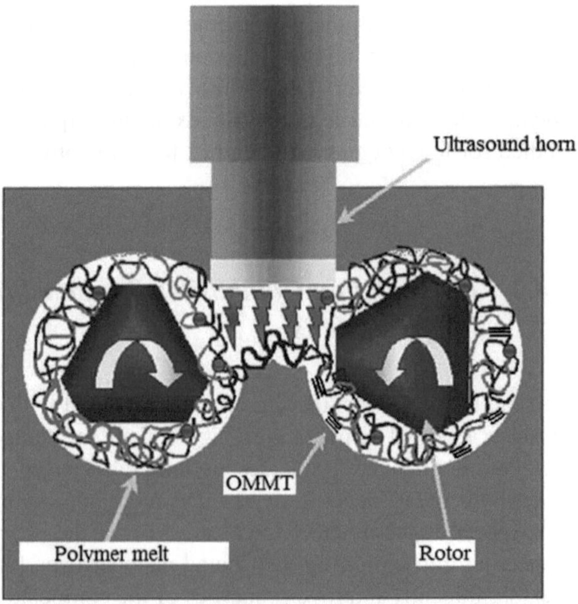

Fig. 5.30 Schematic diagram of mixing in melt activated by high intensive ultrasound irradiation (20 kHz)

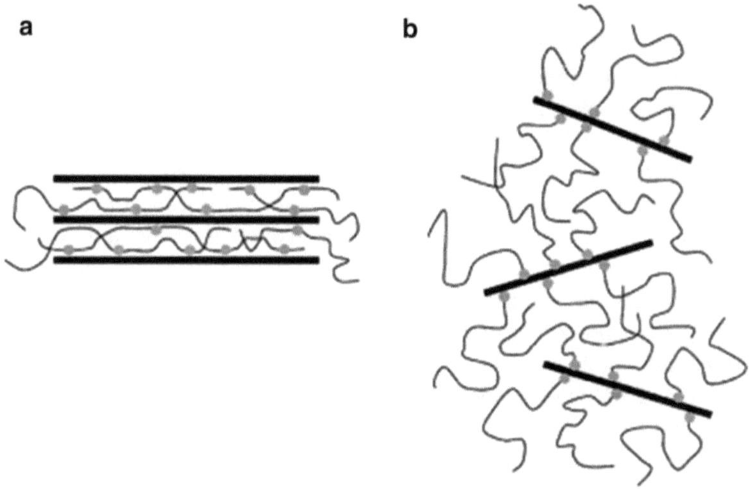

Fig. 5.31 Molecular structure of chain-end-functionalized polyolefin (**a**) and side-chain-functionalized polyolefin between the layers and related morphologies of nanocomposites (**b**) [219]

formation is analogue to activation energy of self-diffusion of polystyrene in blocked melt. Dynamic processes of hybrid formation on the basis of non-polar polymer/organic clay nanocomposites were studied using rheological techniques [97]. In contrast to this, PP functioned in the side chain or block-co-polymers have multiple contacts with each surface of the silicate, which provides not only equalizing of the polymer chains with respect to the silicate surface, but formation of sequent bridges between its plates to promote intercalation structures (Fig. 5.31) [219].

5.4.3 Nanocomposites Based on Non-polar Thermoplastics Produced by Reactor Mixing

Since this area of investigations develops very intensely, we will just briefly analyze here the most widely applied and typical, in our opinion, variants and laws. The polymer melts (PELD, PEHD, CEP, LPELD, PP, PS, etc.) are used to prepare polymer/clay composites, including screw, extruder, polymer fibers [184, 220–227].

In order to refine nanoclays, not only silicate chemical modification, but also those of polymers and implanted compatibilizer are used in the melt of non-polar polymer. In the case of PE with statistically distributed or end dimethyl-ammonium-chloride groups or PE-block-PMAA it is shown [228, 229] that one end ammonium group is, probably, not enough to change a structure of even organophilic clay, contrary to the results obtained for PP. The best results are achieved for PE containing more than one ammonium group in a side chain, and due to increase in PE-block-PMMA concentration [122, 230].

Also various polymer-modifiers and increasing compatibility of ingredients additives are implanted into the melt, the most widely used is maleic anhydride (MA) applied as compatibilizer, however, it is more often introduced in PP melt in the composition PP-gr-MA, these composites are called PP-MA/OMMT or PP/PP-pr-MA-MMT. The silicate gallery (2–3-nm width) one or more stretched PP-MA chains are implanted. They simultaneously change rheology of the composition, interparticle structure, and chemistry of hybrid materials. Polyolefin-MA-MMT composites have improved mechanical properties, thermal stability, fire resistance, barrier properties, corrosion resistance, they favor non-isothermal crystallization, etc. [45, 231–244].

Properties of polyolefins induced by the intercalation technique are analyzed recently in the review [116]. Silicate layers of about 150 nm in length and 4 nm in thickness are fined in PP-MA-matrix, hierarchical structure of the intercalated PP-MA correlated with particle sizes of the clay (30–50 nm), which is comparable with the inertia rotational radius of macromolecules. The crystallized lamellas have thickness 7–15 nm; they can form a spherulite texture [245] with 10-μm diameter particles (Fig. 5.32). The clay particles are used as nuclei-agents for the PP-MA matrix (like in the case of PP crystallization in presence of carbon nanotubes) [246].

Crystallization of polymers is one of efficient methods of control over extension of an intercalated polymer chain in the silicate gallery. It is assumed that crystallization is impeded due to an interlayer space limited to 2-3-nm, however, one can expect that heat should accelerate diffusion and define orientation of a polymer in the silicate gallery [247]. The one-staged method of nanocomposite preparation is proposed using isotactic PP or its oligomer and organically modified (with dodecylamine) MMT with 0.15–1.22 mol% MA as compatibilizer (inoculated in double screw extruder initiated by dicumile peroxide): *iso*-PP was used a matrix polymer, oligo (*iso*-PP-*pr*-MA) as a reaction-capable compatibilizer and dodecylamine was used as MMT surface modifier. Under conditions of extrusion controlled destruction of *iso*-PP takes place and inoculation of MA links to *iso*-PP chains in the melt. In this case nanostructured morphology of PP/oligo(PP-gr-MA)/OMMT forms via creation of hydrogen bonds and the following reactions of

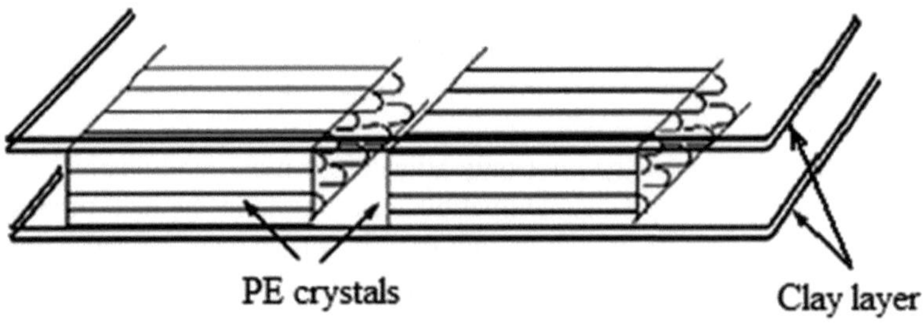

Fig. 5.32 Schematic of PE crystals intercalated between MMT layers [245]

Fig. 5.33 Schematic representation of in situ reactions in the formation of nanostructural architectures in the PP/PP-gr-MA-OMMT system [198]

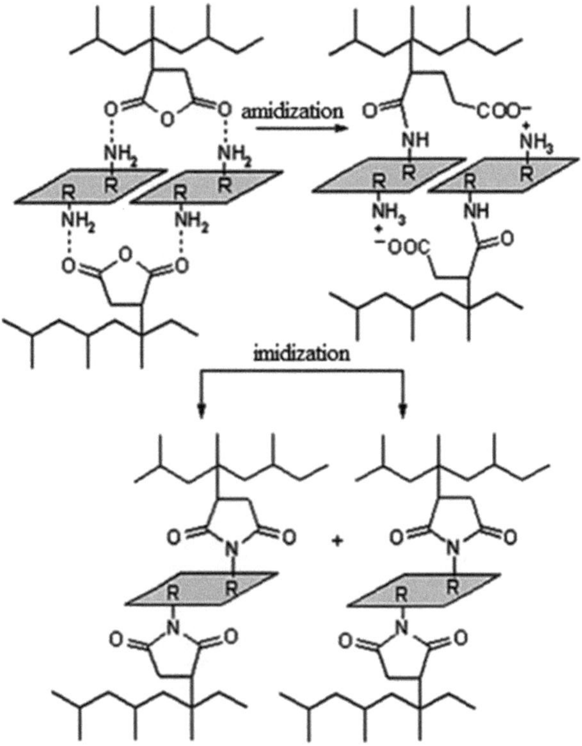

amidation/imidation with participation of anhydride cycles of exfoliated *iso*-PP and modifying OMMT alkylamine (Fig. 5.33).

It has been found that if intercalation is improved, the complete exfoliation of a composite may be not achieved [248–251]. Usage of inoculated MA as a compatibilizer and a modifying agent brought to good results as regards intercalation and exfoliation, however, high concentration of MA causes phase separation, which means a disadvantage in achievements of good mechanical properties of the composites [252]. Therefore numerous attempts are made to decrease MA concentration due to usage of other molecules favoring intercalation (secondary intercalated molecules) [253–255]. Relatively often PVA additives or its polymers are implanted [252, 256–260].

Fineness of a silicate depends on a matrix and an intercalation agent. MMT modified with octadecyl ammonium in the ethylene-vinyl acetate (EVA) matrices collapses owing to significant increase in the interlayer distance [260]. Good results are also achieved via intercalation of different components into basalt space of silicates (Fig. 5.34). Thus, mixing in a melt was used to obtain composites including PP, PP-pr-MA and a long chain amide, AM (docosanoic acid C_{21}, unsaturated erucic acid C_{22}) and OMMT in the ratios 95.5: 2: 0.5: 2 [261]. In fact, this amid increases interplanar distance, however, without PP-gr-MA it is difficult to reach homogenous dispersion of a silicate during exfoliation. Double system of PP-gr-

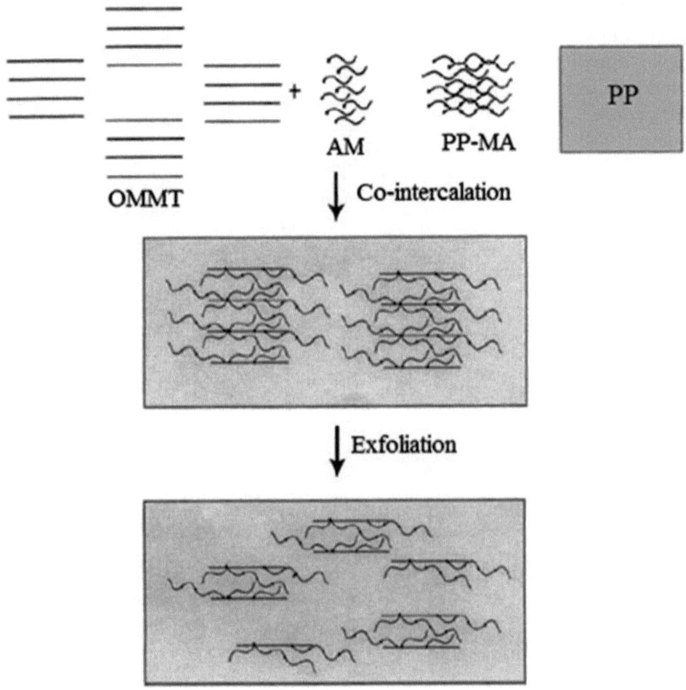

Fig. 5.34 Schematic illustration of preparation of polypropylene-clay nanocomposite by the co-intercalation of 13-*cis*-docosenamide (AM) and maleic anhydride grafted polypropylene (PP–MA) [261]

MA and AM improves properties of the produced composites, in particular, their thermal stability. Meantime, MMT modified by ammonium cations including links of carboxylic acids is not suitable for formation of these composites [262].

Comparative analysis of properties of the nanocomposites PE/PE-gr-MA-OMMT obtained by the solution (dimethyl benzene, 130 °C) and the melt techniques has shown [191] that the intercalation effect is expressed in improvement of nanocomposite structure and its physical mechanical properties caused by an increase in PE-gr-MA concentration, increase in the chain length of the modifying ammonium cation. A degree of crystal order of a composite and thickness of formed crystallites decrease as concentration of PE-gr-MA increases [263], crystallites formed in the solution technique are far smaller than those obtained via direct intercalation from melt (Table 5.4). Therefore, it is concluded that OMMT and PE-gr-MA are heterogeneous nuclei of PE crystallization from melt, maximum ultimate tensile strength (23.3 MPa) is achieved at the PE-gr-MA concentration 6 wt%, strength increases as a fraction of PE-gr-MA increases to 9 wt% and reaches 122.2 J/m.

Third polar component is introduced to achieve better refining of MMT in the nanocomposites [250, 264–267].

Table 5.4 Comparative analysis of basal space during formation of MMT-organosilicate exfoliated nanocomposite by various techniques [263]

	In situ polymerization		Intercalation from melt		Method of dosed integration of components (masterbatch)	
OMMT, wt %	2θ°	d_{001}, nm	2θ°	d_{001}, nm	2θ°	d_{001}, nm
0.73	2.43	3.63	2.65	3.30	2.72	3.25
1.60	2.40	3.68	2.65	3.30	2.74	3.24
2.40	2.69	3.28	2.64	3.32	2.65	3.30
3.36	2.65	3.30	2.60	3.35	2.69	3.28

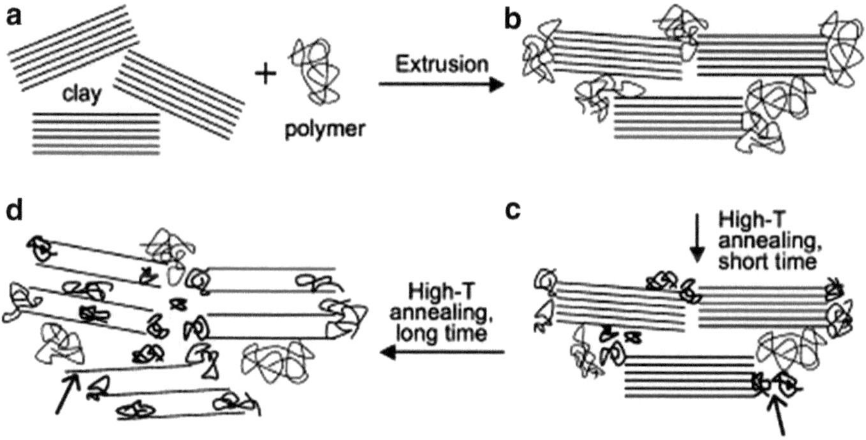

Fig. 5.35 A scheme depicting the evolution of nanocomposite morphology upon extrusion and subsequent annealing: **a**, **b**, **c**, **d** [268]

AIBN was introduced in organic-modified MMT, after which PS was intercalated from melt. Some increase in interplanar distance can be related with exothermal process of decomposition of the initiator and forming of free volume due to gas emission, which accelerates exfoliation [268]. Extrusion of a polymer with aggregated silicate (Fig. 5.35a) is accompanied by intermediate formation of agglomerate of polymer chains with potentially suitable intercalation places (Fig. 5.35b). Low temperature short-time annealing during t_1 interval or annealing in inert atmosphere change the structure insignificantly (Fig. 5.35c), whereas a long high temperature annealing during t_2 or annealing in presence of O_2 is associated with chains breakage and formation of bridge-structures with the following diffusion of short chains into layers (allow labeled). As the time increases, short PS chains generated by chains breaks insert into interlayer distance and cause fracture of the plates, the latter draw apart (Fig. 5.35d, pointed to by an arrow). The total molecular mass of this PS is up to 30,000 (initial mass is $M_w = 330{,}000$). The calculation are made and these processes are theoretically grounded (see for example, [269]).

These results confirm the well known fact that diffusion of long-chain PS to the silicate gallery is impeded more than that of short-chained. Double-screw extrusion of high-molecular PS and organically modified silicate with profile optimization of the screw compensates [270, 271] diffusion difficulties, which would otherwise need static annealing at high temperature [272, 273].

5.4.4 Composites Formed in Melt of Polar Matrices

A possibility of intercalation of polar polymers between the silicate layers is governed by total enthalpy of intercalation including adverse effect of interaction between the polymer and aliphatic chains of the modifying agent and favorable effect of polar interaction between a polymer and the silicate layers [274]. In other words, a decrease in entropy caused by integration of macromolecules into the interlayer space is compensated by increase in conformation freedom of alkyl radicals of the modifier at drawing apart the silicate layers or a polymer melt with a mineral. Drawing layers apart, especially exfoliation depends on their position, preferable interaction between the polymer and the surface of a silicate and further decrease in energy of the system.

An ethylene co-polymer with vinyl acetate (EVA) with various numbers of polar groups controlled by its composition is intercalated in the melt in OMMT. The intercalation/exfoliation behavior and properties are influenced by VA links content, concentration of MA inoculated to HDPE. Fineness of MMT and its exfoliation are improved as HDPE-MA content increases, inoculated MA amount decreases and vinyl acetate concentration increases, which, probably, is due to the synergetic effect of the polar groups. In this case PEVA-MA additive increases the elastic modulus (Fig. 5.36). One of remarkable and extensively studied

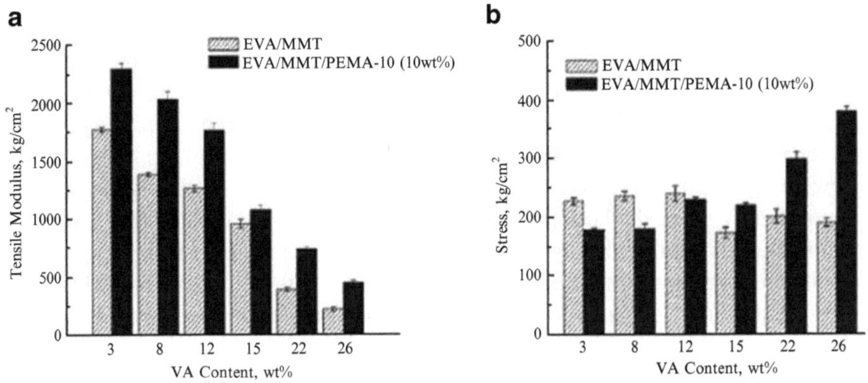

Fig. 5.36 Dependence of tensile modulus (**a**) and tensile stress at break (**b**) on the vinylacetate content in EVA/MMT/PEMA-10 (10 wt% maleic anhydride) composites [265]

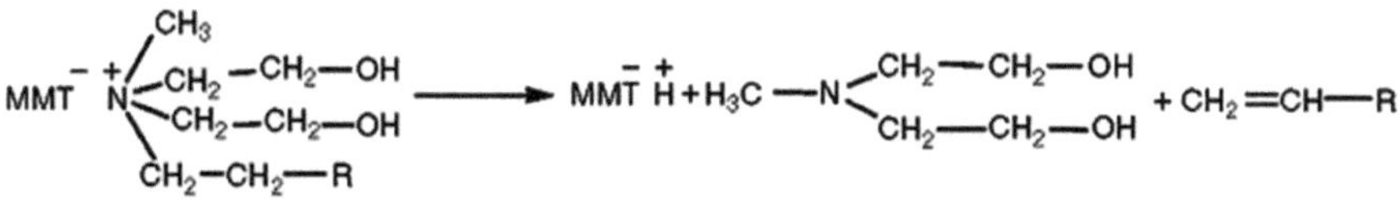

Fig. 5.37 Degradation of the OMMT ammonium salt with the formation of an olefin, an amine and a protonated silicate [277]

properties of the nanocomposites of this type is manifestation of flame-inhibiting ability[5] (see, for example, [243, 275, 276]).

Deacetylation of EVA polymer in the nanocomposite occurs at lower temperature, than in the initial polymer, which may be caused by catalytic effect of strongly acid centers on thermal decomposition of the modified silicate, for example, by Hoffman elimination or S_N 2 nucleophilic substitution at temperature below 155 @C, causing acid activation of OMMT [277, 278], which, in turn, has an effect on inflammability of the nanocomposite (Fig. 5.37).

The main part of decrease in permeability id owing to extension of real diffusive paths of gases, which flow in presence of a filler, layered silicates. Their sheet-like morphology causes maximization of this path length. The factor of convolution (τ) is defined as a ratio of a real distance (d'), which should be passed by a penetrant to the shortest distance (d), corresponding to a distance without a layered silicate (L, W, and Φ are length, width, and volume fraction of these plates, respectively, $W \sim 1$ nm, $L = 50$–1,000 nm and $\Phi = 0.05$):

$$\tau = \frac{d'}{d} = 1 + \left(\frac{L}{2W}\right)\Phi \tag{5.7}$$

As a rule, $W \sim 1$ nm, $L = 50$–1,000 nm and $\Phi = 0.05$.

The effect of convolution on permeability is expressed [279] as

$$\frac{P_s}{P_p} = (1 - \Phi)\tau, \tag{5.8}$$

in which P_s and P_p are permeability of the penetrant in the nanocomposite and in pure matrix, respectively. This idea is displayed in Fig. 5.38.

Actually, low flammability is caused by many factors, one of which is formation of barrier carbonated layer on the surface of a fractured composite during combustion [280].

[5] Often additives are introduced to decrease inflammability. In particular, 0.5 wt% cyclo [dineopentyl(diallyl)] pyrophosphate dineopentyl (diallyl) zirconate or Tamol 20011 (trade mark KZTPP1, Rohm and Haas Co, Philadelphia, PA, USA), hydrophobic highly efficient pigment, acid copolymer based on acrylic acid (un-neutralized acrylic-based acid copolymer) is used to prevent agglomeration of silicates.

Fig. 5.38 Schematic representation of the gas path in a nanocomposite (path of the penetrant = d') and in a pure polymer (path of the penetrant = d)

Fig. 5.39 Variation of the burning rate of the ABS with lower (*1*) and higher (*2*) acrylonitrile content and the corresponding ABS/Clay nanocomposites with the different Cloisite Clays, all nanocomposites at 4 wt% loading [281]

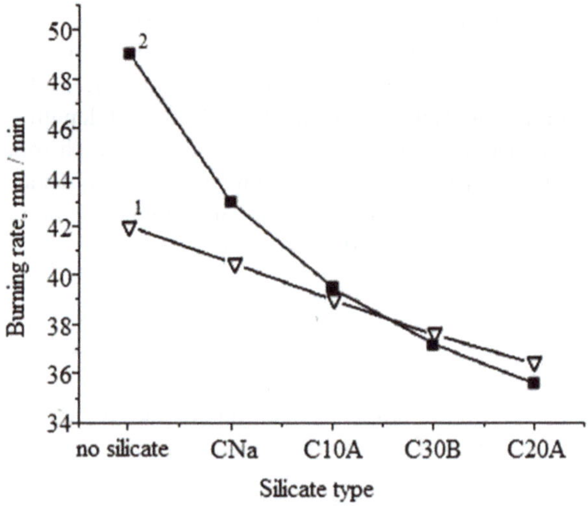

Under industrial conditions modified concentrated clays are also used (masterbatches), with their help it is easier to control dispersing of organosilicates in melt of components. Moreover, in combination with nanoclays as enforcing agents for a polymer composite, containing nanofillers such as carbon nanotubes, multifunctional materials with improved mechanic, thermal, and electric properties can be obtained.

The best characteristic of thermal stability and inflammability have Cloisites20A [281] (a decrease in a burning rate of plastic is 27 and 13 % with high and low concentration of AN, respectively) and 30B (24 and 11 %, respectively) (Fig. 5.39).

We shall highlight some more moments.

Intercalation of syndiotactic and isotactic PMMA from melt into organomodified bentonite [42] increase d-basalt space by 25 % and leads to improvement of physical mechanical properties: [161, 282, 283] polymer chains

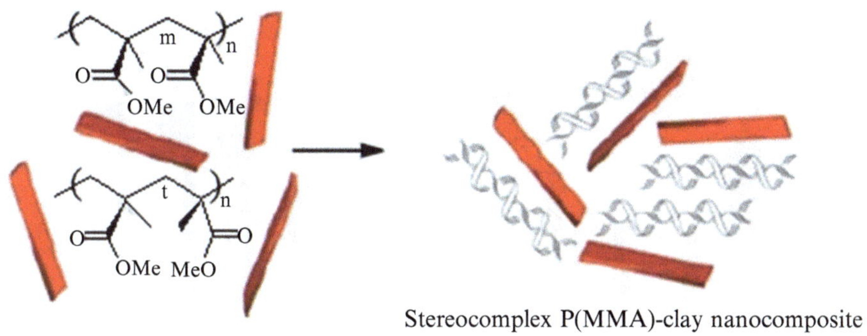

Fig. 5.40 Schematic illustration for the formation of isotactic/syndiotactic stereocomplex PMMA-clay nanocomposites. Modifying agent is methyl bis(2-hydroxyethyl)tallow alkyl ammonium [153]

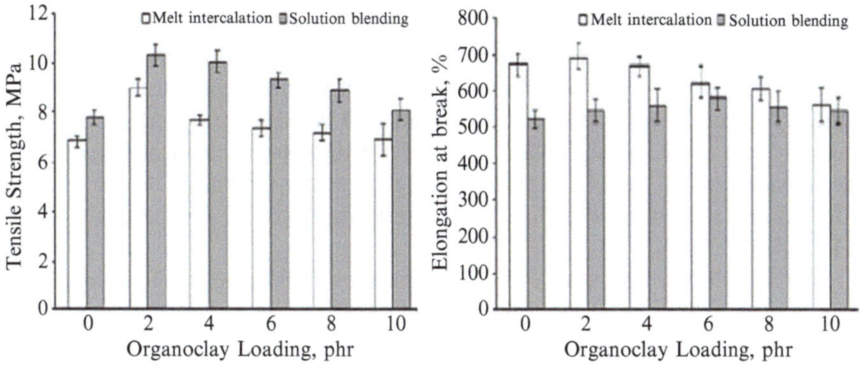

Fig. 5.41 Tensile strength and elongation at break for neat poly(ethylene-co-vinylacetate)/natural rubber and its nanocomposites obtained by solution blending (*dark column*) and melt intercalation (*white column*) [284]

with modified cation are associated (Fig. 5.40), thermal stability of the composite as compared to the initial polymer is increased, T_g of polymers increases [157, 284].

The solution (from toluene) and the melt methods of production and properties of intercalated composites based on poly(ethylene-co-vinyl acetate) mixtures (15 % VA)/natural rubber (0–10 mass parts per 100 parts of the mixture)/OMMT are compared.

The composites obtained via solution mixing show higher ultimate tensile stress (Fig. 5.41) and thermal stability due to better fineness of OMMT in the polymer matrix, than the respective value of the composites obtained from melt. At the same time the way of preparation of a nanocomposite has no effect on its flammability. Though relative elongation during tension of the composite produced by reactor mixing is higher than that of the one produced by solution method, it decreases as OMMT content increases [285].

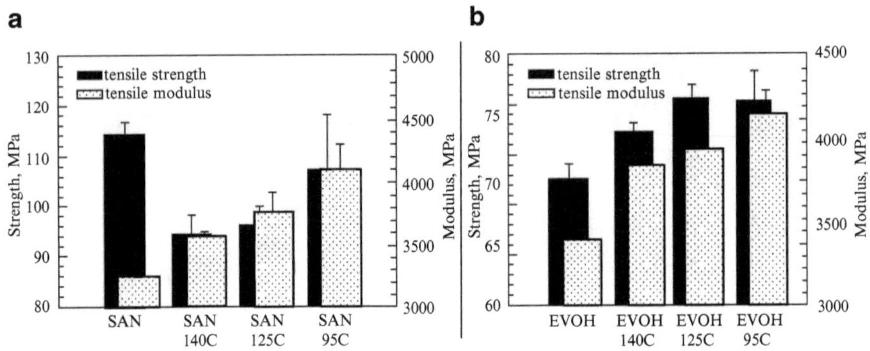

Fig. 5.42 The tensile properties of the intercalated poly(St-*co*-AN/OMMT and poly(E-*co*-VA)/OMMT nanocomposites prepared by dynamic melt intercalation [285]

Interphase interaction in the silicate-polymer systems appears in different ways for polar, copolymer of ethylene with vinyl alcohol (poly(E-*co*-VA), EVOH) forming strong hydrogen bonds with the surface of silicate and less polar bonds with lower interaction energy in the copolymer of styrene with acrylonitrile (poly (St-*co*-AN, SAN). Though poly(E-*co*-VA) more efficiently penetrates into a silicate gallery due to its hydrophilic nature, exfoliation is easier in the poly(St-co-AN) based system (with weaker polar interaction). However, mechanical properties of the poly(St-*co*-VA) based systems have higher values, probably, due to stronger interaction between nanoparticles and interphase layer of the polymer matrix (Fig. 5.42).

Intercalated poly(St-co-AN)/OMMT reveals lower ultimate tensile strength than the initial poly(St-*co*-AN); poly(E-*co*-VA)/OMMT shows a significant increase in ultimate strength and modulus of elasticity, probably, due to increase in interphase adhesion.

It should be specially noted that the functioned nanocomposites, for example, hydrogels [286] based on N-isopropylacrylamide, acrylic acid, neutralized by sodium hydroxide and MMT, modified by (3-acrylamidopropyl) trimethylammonium chloride have exfoliation properties both in dry and swelled states, in the latter case the products are potential agents of purposeful delivery of drugs.

5.4.5 Melt Intercalation of PVC

The method of intercalation from melt is widely used for production of PVC-based nanocomposites (see, for example, [287–294]).

In practice, apart from a polymer and a modified silicate also necessary amounts of the ingredients are introduced into the melt, including stabilizers (calcium stearate, tin-organic compounds, for example, di-n-octyl bis(isoctylthioglycolate) tin), plasticizers (dioctyl phthalate), etc. Two method of production of hybrid nanocomposites are used: intercalation from PVC solution and in situ intercalation polymerization of vinyl chloride. In case of polymerization in interlayer distance neither type of a silicate, nor nature of the used ammonium cation has an effect of morphology of the formed nanocomposite, in contrast to that prepared by melt method. When the values polarity of the silicate and the polymer are almost equal, silicate layers are well dispersed in the matrix. The nanocomposites based on PVC and layered silicate (5 wt% of bentonite) are prepared using extrusion followed by roll milling [295]. KZTPP1 (cyclo[dineopentyl(diallyl)] pyrophosphate dineopentyl (diallyl) zirconate) or Tamol 2001 (un-neutralized acrylic-based acid copolymer) (0.5 wt%) introduced into the composites as inflammatory agents prevent agglomeration of silicates. The PVC-bentonite-KZTPP1 is anisotropic composite, which assumes compelled flow of nanoparticles with orientation in 001 direction, in other words, macromolecules are intercalated in silicate particles. In the PVC-bentonite-Tamol 2001 composite just insignificant isotropy was observed, which pointed to exfoliation of the silicate. The glass transition temperature of PVC and nanocomposites based on it increased in the range of PVC (1) <KZTPP (2) <Tamol 2001(3) (Fig. 5.43). The changes in mechanical modulus and activation energy obeyed the same relations.

These results confirm the fact that even small additives of nanosilicate have a significant effect on microstructure of the extruded composite and on relaxation of polymer chains.

5.4.6 *Nanocomposites PEO-Silicate Obtained by Melt Technique*

One of the most well studied polymers of direct intercalation is integration of PEO in layers of plane silicates, such as mica, by interaction between a melted polymer with Na^+- or NH_4^+-substitutional *host* lattice [78, 296–299].

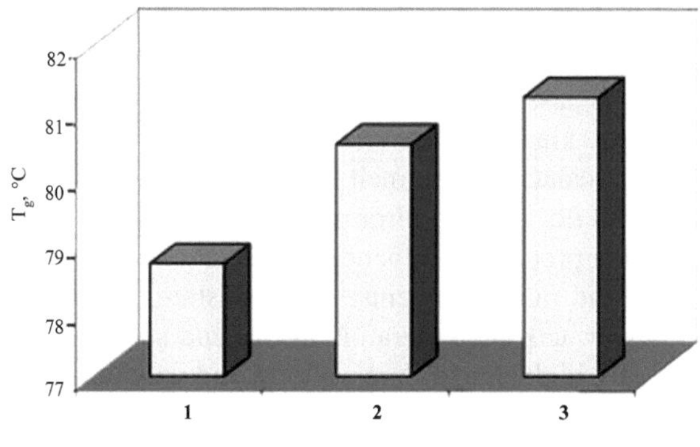

Fig. 5.43 Dependence of glass transition temperatures of PVC composites on their compositions (see the text) [295]

Intercalation of PEO melt into the layered silicates is one-staged process in the case when interlayer distance in them is smaller than 0.8 nm (according to spiral structure of a molecule, diameter of PEO molecule is 0.8 nm in normal to the axis direction) and low concentration of PEO is used. In the PEO/MMT-B34 system interlayer space increases by 0.68 nm, and, most probably, the intercalated molecule has slightly distorted elongated spiral shape.

This doped hybrid nanocomposite has ionic electric conductivity: specific conductivity of PEO/Li^+-MMT (40 wt% PEO) $1.6 \cdot 10^{-6}$ S/m at 30 °C, activation energy 2.8 kcal/mol. Stability of polymer electrolyte depends on ceramic additives, grain sizes, polymer/lithium ratio, conductivity between the silicate and the polymer is synergetic. The general formula of PEO intercalate in the layered silicate magadiite obtained from solution in aprotonic solvent and in the polymer melt at 155 °C corresponds to the composition $H_2Si_{14}O_{29}(–OCH_2CH_2–)_3$ [300]. This agrees with a structure in which one unit cell of magadiite correlates with three oxyethylene links. In other words, interplanar distance is not completely saturated with oxyethylene links.

In the general case intercalation in MMT-B34 is influenced by saturation with the polymer (PEO/MMT-B34 21:79, 10:90), the temperature and annealing time, molecular mass of PEO. Though there are data on polymer extraction from layered nanocomposites, there are no data on degree of extraction and it is unknown whether only a polymer is extracted (there is no confidence about presence of the polymer in the nanocomposite, i.e. it is not clear whether the process is accompanied with destruction). In this connection we shall highlight information [301] on a rapid and quantitative method of PEO extraction from $K_x(C_2H_4O)_4$ $M_{1-x/2}PS_2$ (M = Mn, Cd) composite. The polymer is quantitatively extracted using aqua salt of tetra-ethyl-ammonium under conventional conditions. This process is analyzed in detail, its kinetic effects are found, etc. [302]. Intensity of the peak responsible for PEO intercalation increases with its concentration (Fig. 5.44).

As it was already mentioned during PS intercalation, kinetics of a hybrid formation straightly depends on molecular mass of PEO: lower molecular polymer faster penetrates interlayer space, which is advanced by lower viscosity of the melt and higher diffusion coefficient. This confirms that melt-intercalation is mass transport and depends on mobility and diffusion rate of polymer chains. Theoretical models and developments predict that an increase in length of a polymer chain causes a decrease in compatibility between a layered silicate and a polymer and in a tendency of the system to intercalation. Studies of a structure of intercalated systems based on polymers with various molecular masses prove that intercalation has only kinetic nature and does not affect structure of a finite material.

Intercalation from melt in the PEO-clay system (MMT, hectorite, laponite) is induced not only by ultrasonic, but also by microwave radiation [303]. The intercalation process is influenced by radiation time (optimal 10 min), power (525 W), mass and ratio of regents, their moisture, and salt additions. It is also noted that polymer acts agglomerating agent, and severe control of the process is required, because high temperatures in localized polymer-silicate zone can cause destruction of PEO.

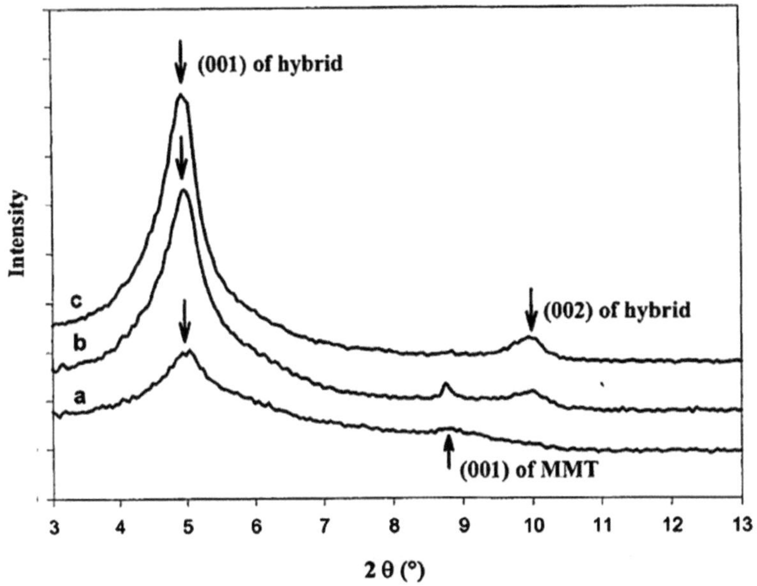

Fig. 5.44 XRD patterns of composites with PEO/MMT = 5:95, 10:90 and 15:85 [302]

We shall consider one more example of intercalated nanocomposites based on PEO or PPO as conducting materials. The obtained by sol-gel technique monoclinic structures (of m-WO_3 type) can be chemically composed with polypropylene glycol, end groups of which are modified by triethoxysilane [304] bringing to film formation with sensor properties. Often these systems are used in combination with ceramic fillers: $LiAlO_2$, Al_2O_3, TiO_2 (10 % TiO_2 with average grain size 13 nm or Al_2O_3 with the size 5.8 nm are dispersed in acetonitrile with $LiClO_4$ and PEO is implanted up to molar ratio $LiClO_4$: PEO links 1:8; conductivity of these polymer-inorganic composites is $10^{-4}\ S \cdot cm^{-1}$ at 50 and $10^{-5}\ S \cdot cm^{-1}$ at 30 °C).

5.4.7 Intercalated Nanocomposites Formed in Polyamide Melts

Polyamides and their aramid varieties are nanocomposites of intercalation type which are of great interest due to their special properties, such as various coatings, they are applied as structuring, inflammable and electronic materials [305–307], which caused not only research but also industrial interest [308, 309]. Properties of a traditional system containing poly-ε-caprolactam (PA-6)—MMT are quite completely studied [159, 310–315] and some of them represented in Table 5.5 [316].

Table 5.5 Some properties of PA/MMT compositions

MMT concentration, %	Maximum stress, MPa	Maximum strain, %	Initial modulus, MPa	Impact toughness, MPa	T_g, °C	Equilibrium water absorption, %
0.0	35.6	0.131	578.8	3.76	78.0	9.4[a]
2.0	47.5	0.095	943.4	3.23	83.7	8.6
4.0	50.4	0.085	950.9	2.96	89.1	8.1
6.0	50.5	0.066	1.240.7	2.31	90.3	7.6
8.0	48.8	0.064	1.125.7	2.04	92.1	7.1
10.0	48.3	0.062	1.117.3	2.03	94.9	6.7
12.0	48.1	0.061	1.113.1	1.94	97.4	6.1
14.0	47.7	0.059	1.082.7	1.71	99.2	5.7
16.0	47.5	0.056	995.4	1.63	105.0	2.1
20.0	47.4	0.051	977.6	1.34	100.0	0.0

[a]For 168 h

For such composites the mechanism of formation has been studied in detail as well as the dependencies of their properties and structure on the nature of polymer and silicate components [317–332].

Many properties of polyamides deteriorate because of water absorption (the initial polymer absorbs 9.4 % of water for 168 h): interaction between organic and inorganic phase decreases ability of polar groups to bind water, therefore, water absorption of these composites decreases as silicate concentration increases. Moreover, impenetrable clay layers make a trajectory convoluted for water penetration, as was in the case of gas penetration (see Fig. 5.38).

I.e. increased barrier characteristics, chemical resistance, low permeability of solvents, fire protection – all is a consequence of difficulties and complicated diffusion ways of gases through a layered nanocomposite.

5.4.8 Intercalated Polymer Mixtures

Mixtures of polymers with complementary properties are widely used [151]. Some of them have high compatibility, especially those formed in presence of compatibilizers. Composites based on polymer mixtures of PPO/PA-6 type formed in presence of intercalating agents are of interest [333]. Dispersed silicate plates play important role in control over morphology of the formed PPO/PA compositions, which is confirmed by selective localization of clay in PA phase. An example of LDPE/PA-66 composition reveals role of the silicate as nuclei agent, which causes a decrease in crystallite sizes [334], as a result compatibility between components is improved and morphology of the composite changes. It should be noted for comparison the tin composite of polybutylene terephthalate (PBT)/PE/Silicate silicate localizes in the PBT matrix and this has an effect on new morphology of the composite [218].

PBT is a typical engineering plastic with good mechanical properties and used, especially, for fiber formation or cast mould articles formation. Architecture of intercalated PBT/PE nanocomposites is similar to liquid crystal phase structure [218]. Studies of intercalation kinetics of PBT/PE (rheological technique) into organosilicate proves [218] that intercalation does not carry on during preparation of the sample, instead of it multilayered product forms with two PBT molecules and one PE/MMT layer, which alternately impose on each other and are compressed in a layered sheet. The entire characteristic of relative volume fraction of intercalated tactoid can be obtained from parameters of viscosity and accumulation modulus at lower frequencies, which could be used for definition of apparent diffusion coefficient for mass transport in primary particles at different temperatures. The data on calculated activation energy show that formation of PBT/clay composite depends on molecular mass, which can be conditioned by an increase in interaction of polar groups in PBT chains and on the surface of a silicate.

The coefficients of effective (s^{-1}) and observed ($cm^2 \cdot s^{-1}$) diffusion at temperatures 230, 240 and 250 °C are: [218] 0.0007 and $0.7 \cdot 10^{-13}$; 0.0013 and $1.3 \cdot 10^{-12}$; 0.0017 and $1.7 \cdot 10^{-12}$.

5.4.9 *Intercalated Network Composites*

Multicomponent nanocomposites in layered gallery are actually composed of ingredients. In the melt triple copolymer of ethylene-propylene-diene is vulcanized in layered silicates, and CEPD/silicate nanohybrid forms using organophilic clays (MMT, modified by hexadecyl trimethyl amine, octadecyl trimethyl amine or distearyl dimethyl amine) and accelerators of vulcanization [335] (Fig. 5.45). There is several times increase in strength properties of these hybrid materials [336].

After extrusion CEPD/MMT composite is subjected to roll milling in presence of vulcanizing components, and are then vulcanized [181].

Intercalated nanocomposites based on nitrile rubbers is relatively new area in chemistry of polymer materials. It includes a range of various approaches, such as coagulation of rubber latex with water dispersion of silicate, mixing of rubber solution with dispersed clay [205, 337–339] intercalation from melt [340–343].

Properties of the nanocomposites based on nitrile rubber (content of acrylonitrile is 45 %) and two types of modifiers Cloisite 15A (cation exchange-capacity is 1.25 mg-eq/g) and Cloisite 30B (cation-exchange capacity 0.9 mg-eq/g) dispersed by various procedures (melt mixture at 130 °C or in chloroform solution or in THF). The nanocomposite is solidified by the system dicumile peroxide in presence of phenylene bismaleimide as co-agent of solidification (Table 5.6).

The solution intercalation process is more effective than the melt process. In all cases elastic properties, such as ultimate tensile strength, modulus of elasticity improve, which points to good reinforcing properties of silicates. The best quality is reached for "solution" material (especially for OS30B), which is explained by high

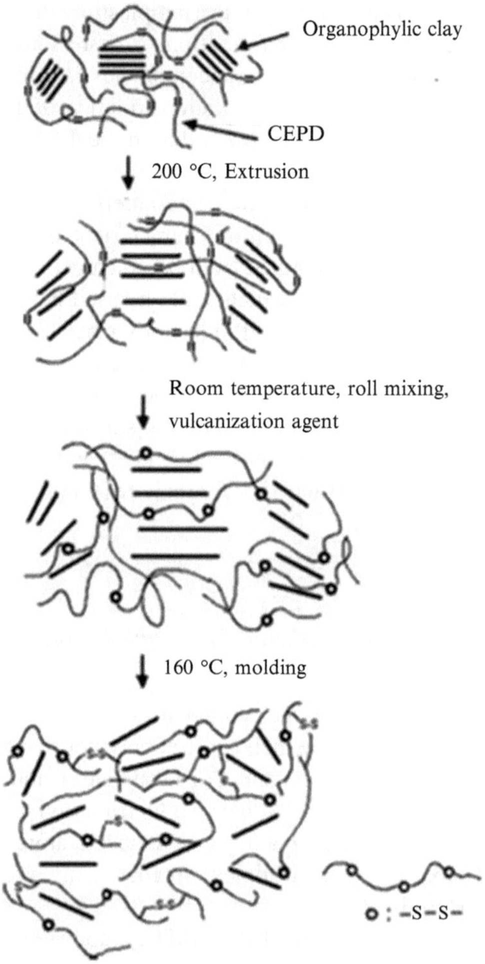

Fig. 5.45 A scheme of the vulcanization of CEPD and intercalation into clay gallery [336]

dispersion ability of the silicate as one of the most important characteristic for improvement of mechanical properties. These results confirm improvement of physical mechanical properties, including deformation, in the systems with high capacity of exfoliation/intercalation, which are reached during preliminary dispersion of the silicate in polymer solution.

At the same time, silicone rubbers (mixture of hydrated and containing end ethylene groups of siloxanes) in OMMT galleries modified with di(2-oxyethyl)-12-alkan-3-methyl-aminchloride are vulcanized at room temperature [344] with formation of intercalated or exfoliated structures (Fig. 5.46). This functioned silicon rubber has interesting physical chemical properties [345–350].

Nanocomposites based on unsaturated polyethers (glyptal resin), including the resin with styrene (up to 35 %) and modified in different ways silicates, solidified with cobalt octoate and peroxide are obtained by various methods: statistical,

Table 5.6 Some characteristics of intercalation systems and physical mechanical properties of intercalated CEPD

Organosilicate MMT-modifier	Intercalation procedure	Percentage of the modifier, %	d_{001}, nm	Total surface	Fraction of intercalated MMT, %	Ultimate tensile strength, MPa	Relative elongation in tension, %	Modulus at 10 % elongation, MPa
OS15A	From melt	2.5	4.2	1,676	43	4.1	730	2.8
OS15A	From melt	5.0	4.0	4,364	56	4.4	720	3.1
OS30B	From melt	2.5	>1.9	2,257	47	3.6	640	2.7
OS30B	From melt	5.0	>1.9	3,602	55	4.5	670	3.2
OS15A	From solution	2.5	4.0	1,450	15	4.3	750	3.0
OS15A	From solution	5.0	4.0	3,997	54	7.9	630	4.9
OS30B	From solution	2.5	>1.9	1.508	82	6.3	660	3.5
OS30B	From solution	5.0	>1.9	1,498	81	9.6	270	6.8
MMT-OS15A	–	100	3.0	2,526	–	–	–	–
MMT-OS30B	–	100	1.9	1,435	–	–	–	–
Initial CEPD	–	0	–	–	–	2.5	450	2.7

By the data [181]

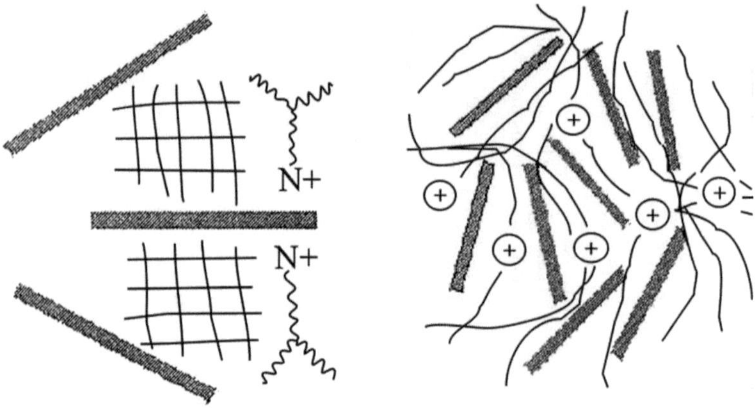

Fig. 5.46 Vulcanized silicone rubber in the gallery of the OMMT [344]

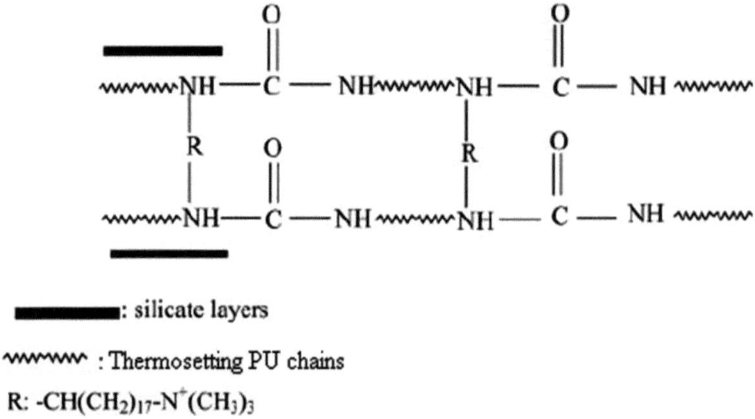

Fig. 5.47 A possible mechanism of the formation of layered PU/silicate composite [355]

mixing, and mixing with ultrasonic irradiation. Rheological properties of the composites correlate with temperature and time of mixing, intercalation space, delamination, and fineness of nanoparticles of exfoliated organic silicates [240, 306, 351–355].

Polyurethane (PU) composite is a polar polymer compatible with OMMT (modified with N-diamine octadecyl trimethyl ammonium chloride) is obtained in presence of 1, 2, 3, and 4.5 wt% OMMT and its treatment with PEG (MM 2000) at 80 °C and toluene diisocyanite, added by 3, 3′-dichloro-4,4′- 4 4′-diphenylmethane diamine (150 °C, 3 h). During mixing –NCO groups react not only with –OH, but also with $-NH_2$ groups of silicate modifying agent (Fig. 5.47). This effect is exothermal, it decreases Coulomb electrostatic attraction between the silicate and

Table 5.7 Elastic properties and temperature of thermal decomposition of SPU and SPU/OMMT composites [356]

Sample, silicate concentration, %	Ultimate tension strength, MPa	Relative elongation at fracture, %	T_{dec} (intense)	Carbon residue, %
SPU	6.85	500	299	3.5
SPU/OMMT-1	10.33	530	312	5.0
SPU/OMMT-2	15.68	570		
SPU/OMMT-3	20.19	590		
SPU/OMMT-4	24.79	630	320	6.7
SPU/OMMT-5	10.95	540	307	8.9

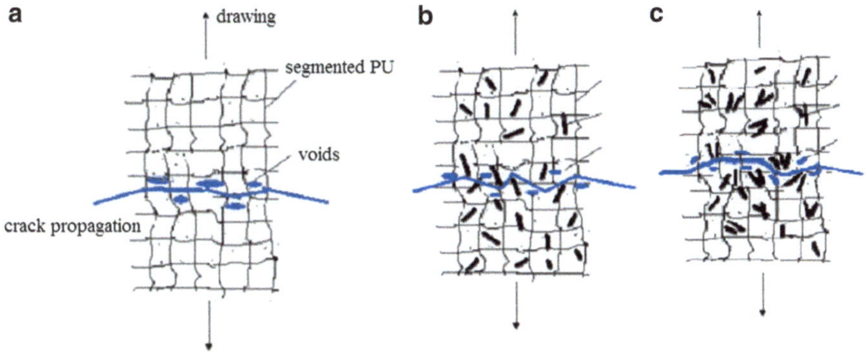

Fig. 5.48 Scheme of failure development during tensile drawing (**a**) thermosetting polyurethane TSPU; (**b**) TSPU/OMMT-4; (**c**) TSPU/OMMT-5 [355]

polymer chains, changing total enthalpy. This intercalate can be expressed as follows:

New nanocomposites have been synthesized, in particular, segmented SPU PU/clay based on PCL, methylene-diphenyl diisocyanite, butanediol and PCL-silicate-prepolymer [356]. This composite showed high elastic and thermostable properties as compared with pure PU matrix up to silicate/PCL concentration 4.2 wt %, however, when it was exceeded, properties of the composite deteriorated abruptly (Table 5.7), which confirmed transformation of PU/silicate from elastomer to thermoplastic material with increase in PCL/silicate portion.

Morphology of SPU/OMMT-4 nanocomposite displays presence of superfine mixture of intercalated and exfoliated silicate layers in the matrix. It is supposed to be the reason for the nanoenhancement effect in combination with a number and size of pores and increase in propagation paths of cracks (Fig. 5.48).

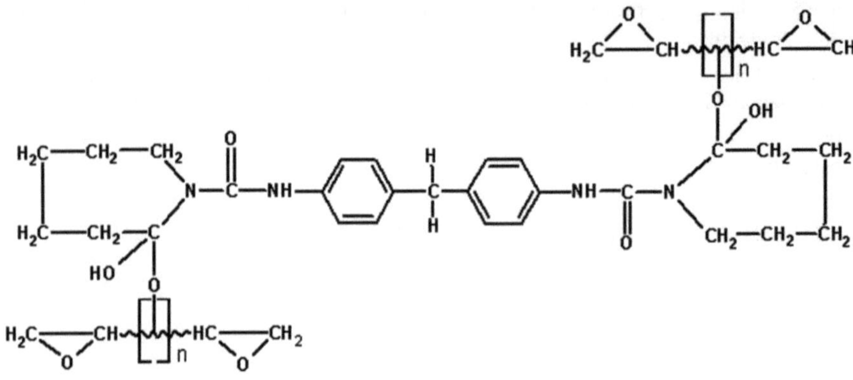

Fig. 5.49 A scheme of caprolactam-blocked methylenediphenyl diisocyanate-epoxy matrix [362]

The proposed mechanism of enhancement of mechanical properties on the example of SPU/OMMT-4 nanocomposites is associated with high dispersion ability of OMMT and an increase in crack widening in the nanocomposite. As OMMT portion increases, the character of crack development in material changes.

Far less comprehensively hybrid epoxy-silicate nanomaterials are studied [240, 357–359]. Interpenetrating networks based on enforced epoxy matrix with different content (5, 10, and 15 %) of caprolactam-blocked methylene diisocyanite (CDMI) are modified with 1, 3, 5 wt% of organophilic MMT (Fig. 5.49).

A decrease in T_g of epoxy systems filled with organic clay is registered as compared with initial epoxy resin, as well as a tendency to its thermal stability. Similarly, introduction of organic clay and CMDI into epoxy improves strength, elasticity and impact strength according to percentage of these additives [360–363].

Nanocomposites based on epoxy resins and layered silicates (ELS) are synthesized using pre-intercalated novolak resin layered silicate (NLS) nanocomposites. For this different approaches were used: melt, melt-US, solution (water, alcohol, toluene, o- xylene acetone, etc.) and solution-US methods of intercalation, various types of novolac epoxies, including benzene novolac, bis-phenol-A novolac, acryl phenol novolac, o-cresol-novolac, etc., and MMT modified with benzene octadecyl ammonium bis(2-hydroxyethyl)ethyl)methyl tallow ammonium (Cloisite 10A with exchanging capacity 1.25 mg-eq/g and Cloisite 30B – 0. 90 mg-eq/g).

NLS and ELS displayed stable structures at highly intense US irradiation, they intercalated and exfoliated. The material based on NLS and ELS showed far higher thermal and mechanical properties, for which chemical affinity between phenol epoxies and OMMT was very important.

To obtain exfoliated polymers of novolak epoxies hydroxyl modified MMT is more efficient than benzene-modified high efficiency in US dispersion and intercalation is observed in both melt and solution variants. Pre-intercalated NLS structures are very stable and do not fracture during solidification of epoxy resins, however, they have a trend to exfoliation during self-propagated intercalation and ELS solidification. Dramatic increase in mechanical properties, such as their mechanic rigidity, thermal properties, including T_g decomposition temperature, which is assigned to different morphologies of intercalation and exfoliation, and with special features of solidification reaction between epoxy and phenol resins conducted in presence of layered silicates. Most probably, to obtain optimal composites from thermally reactive resins, especially in the case of self-propagating intercalation in the solidified systems, a balance is needed between the intercalation and solidification rates. However, many problems of intercalation chemistry of thermosetting plastic are not solved yet.

5.5 Supramolecular Assembling in Nanolayered Materials

Substantial development have hybrid self-assembling nanocomposites formed as multilayers with the general formula $(P/M)_n$, where *M* и *P* are nanometer oppositely charged of inorganic component and polymer (referred by [71]). Versatile approaches are made to assembling of layered systems, for example, polyelectrolytes and clays, flaked zirconium phosphates, colloid metal particles. Positively charged poly (diallyldimethylammonium chloride) (PDMAC) and glass, quarts, silver, gold, and even teflon nanoplates were used to consider in detail mechanism of formation of these materials. Subsequent immersion of these plates in *P* solution or *M* suspension cause increase in a number *n* of the layers, and each action is accompanied by increase in thickness by 1.6 nm in the case of *P* and by 2.5 nm in the case of *M*.

Similarly self-organized layers form from TiO_2 cation nanoparticles (~3 nm) obtained by acid hydrolysis of $TiCl_4$. They are organized in layered structures on the surface of superthin (~1 nm) films of cation polymers of poly(sodium 4-styrenesulfonite) (PSS) or already mentioned PDMAC. Optically transparent organized molecularly ordered films of up to 120 layers thick (60 bi-layers, each thickness is evaluated 3.6 nm) are gathered on the surface of substrate (metal, silicon, polymer) ((Fig. 5.50). This strategy makes it possible to obtain various

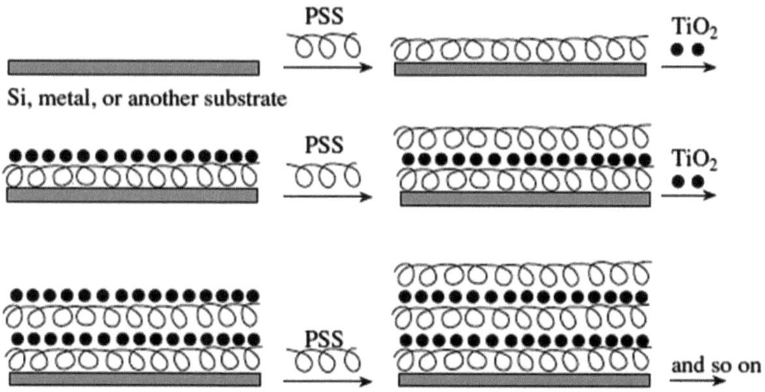

Fig. 5.50 Schematic representation of the fabrication of ultrathin TiO_2–polymer films [34]

combinations of semiconductor materials metal-insulator with nanometer sites *p-n, p-n-p, n-p-n* and others. Formation of multilayers includes the following stages. First, *P*-layer is absorbed at the surface of substrate via electrostatic and Van der Waals interactions. In this case structural hierarchy of *M* gives infinite possibilities of its usage as was used for a template for more versatility of molecules and clusters. They may be inserted between swelled layers and in *M* plates and on their surface, in planes of individual or coagulated @M. Secondly, @M-layer is strongly and irreversibly, which is confirmed by physical-chemical studies, adsorbed on the oppositely charged polymer electrolyte, displaying very dense plane orientation. Irregular *M*-layers cannot provide complete coating of intercalated *P*-layers, thus forming overlapped packs. Interphase roughness exceeds thickness of the *P*/*M* layer, they do not depend on nature of a substrate, control over this can be performed by applying external voltage during the self-assembling.

Polymers with inert surface (polyolefins, fluoropolymers, polyethers) "activate" PVC by spontaneous absorption and sewing with glutaric aldehyde [364]. Nanometer SiO_2 and TiO_2 layers are obtained by reaction of –OH groups (alcohol, silanol, or titanol) with $SiCl_4$ or $TiCl_4$ followed by hydrolysis (these cycles are repeated if necessary) by the scheme described below. In principle, this approach is similar to widely known and often used in practice "functional group lamination method" on inorganic surfaces.

We shall show some more typical samples. Particles TiO_2 form on the surface of photochemical electrode (from $TiCl_3$ solution at pH 2.5), size ~20 nm. Simultaneously, electrodeposition of PPy is registered, which is formed due to electrochemical polymerization of pyrrole (referred to by [71]). This is preceded by absorption in pores of an electrode coated by photosensitive ruthenium complex. These structurally controlled "templates" are analogue of self-assembling supramolecular aggregates. The supramolecular systems of molecular recognition, structured this way, including chromophores, semiconductors, cluster aggregates, and fulfilling specific optical and electronic functions, is a contemporary approach to modeling of ferments. If Al_2O_3-membranes are used as templates, matrix compounds, Au/PPy colloids with 1D morphology are synthesized. Fiber ZnO structures were obtained on the basis of zinc acetate and WO_3, electrode materials on V_2O_5 basis, MnO_2, Co_3O_4 oxide composites, and others. Fibrillate and tubular materials had a diameter corresponding with diameter of pores in Al_2O_3, however, their length was far more. In the case of TiO_2 metastable single crystals of anatase were observed. Tubular nanocomposites obtained via template sol gel synthesis in fibers or nanotubes, and in micro and nanoporous membranes are perfect photocatalysts [195].

At last, it should be noted that layered nanostructures are obtained using liquid-crystal polymers with ionic groups, and MMT or hydrotalcite are used as inorganic components. According to different estimations, average thickness of this pair *M*/*P* is 4.9 nm. It is assumed that such electrostatic assembly provides a close contact between components and strict molecular ordering, it can lead to formation of new types of liquid crystal structures with unusual properties.

A special place in the intercalation problem has layered interstitial compounds implanted in graphite (LGC) and graphite-like structures. Even though it seems that these problems are similar at first glance, graphite can be considered as a macromolecule of aromatic type with a number of aromatic rings about 1,000, period of identity (an interspace between neighbor planes) is 0.335 nm. Absence of chemical bonds between parallel carbon layers in graphite (energy of interlayer interaction is just 16.8 J/mol) makes possible insertion of monomolecular layers of various substances, including monomers, metal ions with formation of layered (laminated) graphite compounds. General methods of LGC production are confined to interaction between graphite and monomer vapors or solutions, metals in strongly ionizing solutions, volatile chlorides or cation metal complexes, accompanied by their implantation in interlayer distance of graphite lattice. Depending on a number of carbon layers, separating two neighbor layers of implanted matter, these compounds are classified as the products of the 1st, 2nd, etc. stages of implantation.

Nanocomposites PS/graphite are prepared as follows [365]. Graphite surface is modified by silane containing agent and subjected to ultrasonic treatment (for 0.5–1 h at 50 °C). Then PS/graphite nanocomposite is prepared by rolling, intercalation from THF solution, grinding of PS and graphite grains in a mill. The composite has high thermal conductivity at graphite concentration 34 vol.% (1.95 W/mK) (Fig. 5.51). In this case different methods of preparation of PS/graphite composites cause exfoliation, which increases in the range: intercalation via rolling < solution intercalation < combined grinding of PS and graphite grains. In the last case the structural chains of graphite are agglomerated, and the composite has high thermal conductivity and mechanical properties.

The origin of a bond between LGC and implanted metal depends on the latter. Thus, for Fe, Co, Ni, Mn, Cu the bond is provided by Van der Waals interactions, sometimes π-electron density is transported from graphite to the implanted metal layer. In this case carbon network of graphite is a kind of polymer ligand. For alkali metals this bond is formed due to transporting of electron from metal atoms to conducting zone of the neighboring graphite, i.e. due to electrostatic interaction of

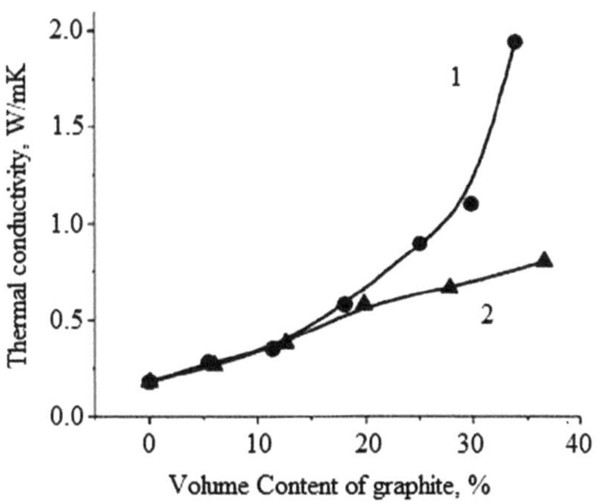

Fig. 5.51 Effect of the content of colloidal (*1*) and nano- (*2*) graphite on the thermal conductive properties of PS/graphite composites [365]

positive ions with free electrons of the conductive zone of graphite. The processes of reduction of interstitial metal ions can be accompanied by partial yield of them from layered packs and reduction on the exterior surface of graphite with formation of nanoparticles, for example, Ti included in its defective lattice. Under conditions of high pressure in combination with shear deformation many different atoms can be implanted in graphite. Due to interplanar distances being extremely constricted, as compared to silicates, there are very scarce data on intercalation of monomers and polymer links and exfoliated composites in graphite. Conducting properties of the PE-graphite nanoplates composite produced by in situ polymerization can be highlighted [366]. Nanographite plates are uniformly coated with the forming PE layer, the product is characterized by low (2.7 vol.%) percolation threshold and high dielectric permeability.

A special interest is drawn to graphemes, single layered carbon atoms [367] which attract increasing interest due to special conductive and transport properties and mass-less Fermi-Dirac interaction between graphene layers. Wide choice of intercalates with different physical properties allows achievement of high electric, thermal, and magnetic characteristics [368, 369]. Intercalation in few layer graphenes (FLG) is efficient method of their properties modification, for example, I_2, Br_2 intercalate in FLG [370, 371].

Intercalation distance increases dramatically in presence of intercalating agent, which abruptly changes properties of FLG. Thus, $FeCl_3$-FLG complex (Fig. 5.52) with homogeneous distribution of the intercalated agent is stable in environment [372]. Intercalation is easier than in blocked graphite (in chlorine atmosphere), $FeCl_3$ as a doping is completely intercalated in FLG. Fourier spectroscopy has shown that there is no electron interaction between adjusted graphene layers. These results confirm that FLG is a promising material not only modifying electron structure of graphene, but also changing electric, thermal, and magnetic properties during usage of different intercalating agents [368], for example, Ca-FLG is expected to display superconducting properties [373]. Most probably, this line of studies is expected to develop intensely.

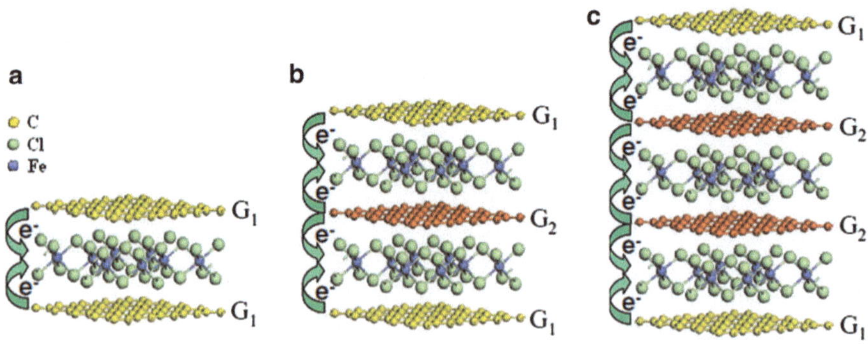

Fig. 5.52 Schematic crystal structures of FLG with two (**a**), three (**b**) and four (**c**) layers of grapheme. The model is constructed based on $FeCl_3 \cdot GIC$. The graphene layers flanked on one/both side(s) by $FeCl_3$ layer(s) [372]

5.6 Structure of Inclusion Nanocomposites Metal Chalcogenide-Polymer

Synthesis and characteristics of hybrid nanocomposites with periodic nanostructure of a superconductor arises a special interest due to their unusual optical, electrical, and other properties. Owing to this, they have great application prospects in microelectronics. Composites of this type are obtained, as a rule, by "aqueous" chemical methods, including sol gel method, in which CdS crystals are trapped in SiO_2 films with formation of Cd-S- SiO_2 systems, $Co_6S_8(PPh_3)_x$ intercalate in MoS_2, also CdS-Ag are described hybrids of self-assembling particles, semiconductors on Cd_3P_2 basis, etc.

A special feature of layered materials, dichalcogenides of metals, such as MoS_2 or TaS_2 with low density of a layer charge is ability to decompose into nanometer structural blocks under specific conditions and to form colloid solutions. If anion is easily polarized, and cation has strong polarizing properties, then MX_2 compound has a layered structure. This structure have molybdenum dichalcogenides, in which a layer of molybdenum is between two layers of chalcogen (X), as a result, triple-layered packages form. Bonds in them are far stronger than between two triple-layered packages, between them are Van der Waals cohesion forces. Monolayers of molybdenum disulfate display distorted octahedral configuration with (2×1) superlattice and coordination with unsaturated Mo-centers in prismatic sites of MoS_2. It is interesting that NbS_2 и TaS_2, and MoS_2 and WS_2, obtained in hydrogen flow during heating of MS_3, have multifaceted fullerene-like structure of nanotubes. For example, NbS_2 are individual nanotubes with hollow core of 4–15 nm in diameter [374]. Probably, the same structure have TiS_2, ZrS_2 и HfS_2 obtained in the same way. Exfoliation procedure in this case is well developed.

Interlayer distance of MoS_2 is small, $d_{001} = 0.615$ nm, the treatment in *n*-butyllithium in hexane leads to formation of $Li_{1.15}MoS_{2.216}$ complex, which is dispersed in aqua solution of PEO with exfoliation [375]. The purified dry product contains from 21 to 31 wt% PEG, $d_{001} = 1.45$ nm, which corresponds to double layer of polymer in the gallery. Thermal stability of the polymer increases as temperature increases to 255 °C, PEO intercalation increases conductivity of MoS_2 by 4–5 orders of magnitude [376]. However, this system is unstable: already at 90 °C MoS_2 catalyses decomposition of PEO, depending on its molecular mass and content of a composite causing a loss in conductivity of the system. A replacement of PEO by PAN causes formation of more stable $[Li_{0.6}MoS_2(PAN)_{1.2} \cdot 0.5H_2O]$ system [377] having mixed electronic conductivity [378, 379].

One of the best superconductors among layered chalcogenides is $NbSe_2$, its superconduction transition temperature is $T_c = 7.2\,K$. The general scheme of PVP, PEO, and PPO from aqua solutions in monolayers of suspended $NbSe_2$ can be expressed [380] as:

$$NbSe_2 + xLiBH_4 \rightarrow Li_xNbSe_2 + xBH_3 + x/2H_2 \tag{5.9}$$

Fig. 5.53 Structure of a PAn–MoS_2 nanocomposite

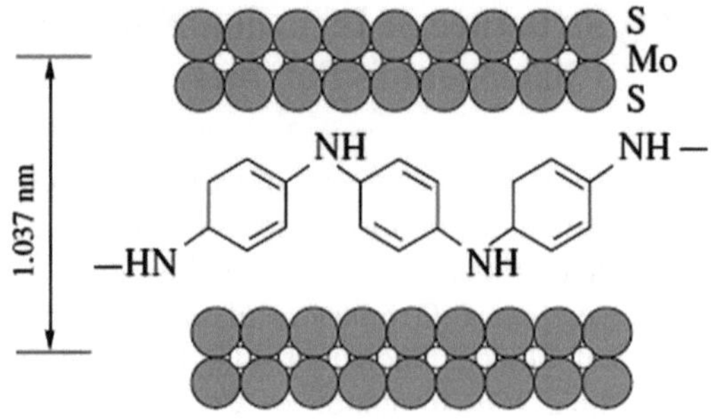

$$Li_xNbSe_2 + xH_2O \rightarrow NbSe_2(\text{monolayer}) + x/2H_2 + LiOH \quad (5.10)$$

$$NbSe_2(\text{monolayer}) + \text{polymer} \rightarrow NbSe_2/\text{polymer} \quad (5.11)$$

Principal scheme for these nanocomposites is shown in Fig. 5.53.

Polymers are implanted in WS_2 phase by the same mechanism. The promising application of these nanocomposites may be plastic superconducting electromagnetic materials. In Sect. 5.4 direct intercalation of polyaniline in interlayer space of MoS_2 ($d = 1{,}037$ nm) through colloid suspensions was already mentioned. It is important that PPy/MoS_2 is a product of in situ oxidation polymerization, which under kinetically limited conditions is *p*-type conductor; its electronic conductivity is by three orders of magnitude higher than that of the initial MoS_2. Similarly PEO is intercalated in delaminated TiS_2, Ta_{S2} и MoS_2 suspensions.

Interesting results are obtained during studies of intercalated polymer electrolytes alkali metal-PEO in layered $MnPS_3$ and $CdPS_3$ synthesized by two-staged a solvent ion-exchanging mechanism [72]:

$$CdPS_3 + 2xA^+{}_{aq} \rightarrow Cd_{1-x}PS_3A_{2x}(H_2O)_y + xCd^{2+}{}_{aq}(A = Li, K, Na, Cs) \quad (5.12)$$

$$Cd_{1-x}PS_3A_{2x}(H_2O)_y + PEO \rightarrow Cd_{1-x}PS_3A_{2x}(PEO) + yH_2O \quad (5.13)$$

Widening of the *host* lattice due to intercalation is $\Delta = 0.8$ nm, some of PEO links have trans-*gauche* and *gauche*-slanted configurations, spiral conformation in intercalated PEO disappears. The experiment proves planar zig-zag structure of intercalated PEO, its likewise conformation in complex with $HgCl_2$. Under equal conditions oligo oxyethylene ethers intercalate in CdS, CdSe and ZnS by direct template mechanism, whereas it is not true for the case of Ag_2S, CuS and PbS [381] and from aqua solution of linear PEI with layered $MoSe_2$, TiS_2, MPS_3 (M = Mn, Cd) the respective nanocomposites are synthesized, from the same polymer or from poly(styrene-4-sulfonate) in TiO_2/PbS layers hybrid materials with semiconductor properties are obtained.

A keen interest is in the inverse problem, implantation of metal chalcogenides in polymers, especially, 1D, self-organized semiconductor nanoparticles. The abovementioned hydrothermal polymerization and spontaneous sulfidizing [382] bring to formation of 1D nanocomposites with finely dispersed self-collecting CdS nanoparticles in polymer nanorods. These products are applied for plated semiconductor materials with superconducting properties and forms of superconductors such as wires, films with special properties, etc.

Nanocomposites of the considered type is a new class of hybrid organic-inorganic photoactive materials. The example of CdS/polyvinylcarbazole (PVC) is used to show photorefracting ability of nanocomposites in the visible spectrum. The composites including PbS or HgS nanocrystals (~50 nm) are photosensible at the wavelength 1.31 μm [383] nanocomposites PbS/polymer have high display properties. Different types of nanostructured systems based on PbS are studied including self-collecting layered lead sulfide ensembles in MMT and PbS nanocrystals dispersed in various polymer matrices. In ethanol (homogenous system) PbS nanoparticles are uniformly distributed in PVA matrix. In heterogeneous (aqua) mixture nanocomposites are structured as a nanocable with PbS core 30 nm in diameter and PVA shell with length from 80 nm to 10 μm [384].

Semiconductor properties have other layered materials, such as PbI_2, BiI_3, HgI_2, Bi_2S_3, Sb_2S_3. Thus, PbI_2 is an interesting material for X-ray digital visualization. Organic solution technique was used to synthesize CdS-wurtzite nanorods and nanowires.

Interesting approach is instabilizing of CdSe/CdS nanocrystals with the core/shell structure in dendrimers of third generation [385]. After removal of inorganic nanocrystal in the center of a dendrimer forms a nanometer cavity with very thin shell. These nanometer capsules are a new class of molecular containers, calibrated mesoporous structures.

Metal chalcogenides with a more complex structure $PbNb_2S_5$ or $SmNb_2S_4$ can also be exfoliated and intercalated. In polar solvents one-dimensional *host* phases with various chalcogenides like MMo_3Se_3 (M = Li, Na) form colloid systems with mono disperse negatively charged condensed cluster chains (Mo_3Se_3), they can also be of interest for intercalation.

A list of these examples and generalization require, of course, a special consideration.

As is seen from the performed analysis, the intercalation physical-chemical methods, chemistry of intercrystalline structures *guest-host* provide almost infinite possibilities for structuring of nanocomposites of hybrid type. Presently many processes accompanying formation of these materials are found, which are associated with formation of these materials, basically their structure and most important properties are studied. It becomes clear that intercalation of monomers and polymers in interlayer space of layered materials is the most important way for creation of hybrid phased nanocomposites, intensely and fruitfully developed line. At the same time, many problems, especially concerning implanting mechanism and *guest-host* interaction interphase processes, control for maximum intercalation, *guest-host* ratio followed by exfoliation – drawing layers apart, are not completely understood. In the intercalated nanocomposite inorganic layer preserves structural characteristics of matrix polysilicate (or other layered material), and organic polymer layer has a stressed structural organization, causing strong interaction with interlamellar surface of the host. Architecture of these formations can be very versatile, not only layered, but depending on the nature of organic component, which governs formation of hybrid phases of different structures. For example, for zinc oxide and sulfide nanocrystals three molecular blocks form with rod-like, dendritic, and coil-like architecture [386].

This property is widely used in the *host-guest* chemistry and for selective recognition. Probably, in future this line incited by increasing demands in many fields of materials science to hybrid nanocomposites will intensely develop. Probably, new types of these materials with required hierarchic structure of intercalated systems will appear. The composites of this type can be used for formation in them of metal nanoparticles, and electrically active polymers (polyaniline, polypyrrole, etc.) with a silicate – for binding of metal ions (gold, platinum, palladium). As much as polymers, hydrazine reduces those forming immobilized particles, catalysts of different reactions. At last, it should be noted that hybrid composites form at the stage of polymer reprocessing. From numerous examples we shall highlight just production of hybrid composites polyamide – 6 "potassium-titanium wickers" ($K_2Ti_6O_{13}$) in double-screw extruder with the following injection and forming [387]. For commercial applications of most polymers of intercalation type an important role plays their ecology, usage of available materials and reagents, low flammability. The last characteristic is determined by ability of a fracturing composite to form a barrier carbonized layer on the surface during burning [280].

Thus, intercalation polymerization has a great potential for structuring of hybrid polymer-inorganic nanocomposites by different ways [352] which are still far from implementation, and this governs intense development of studies in this direction. For example, the considered methods can be used to obtain composite materials of not only polymer, but also of metal types. Thus, one of the methods [352] is based on application of double-layered aluminum and lithium hydroxide, containing in the interlayer distance Ni, Co and Cu complexes with organic ligand, $[LiAl_2(OH)_6]_2[M(Edta)] \cdot nH_2O$, ethylenediaminetetraacetic acid. During annealing in vacuum at 400–450 °C materials form including resistant to oxidizing nanometer mono- and bimetal particles (3–4 to 40–50 nm in diameter): carbonized matrix reliably protects them from oxidizing not only from air oxygen, but also during holding in nitric acid.

Whilst intercalation of polymers followed by exfoliation is one many types of topochemical reactions, it is not the most widely used (Fig. 5.54). Topochemical intercalation procedures can include other lines of development with structural characteristics including ion exchange, reduction or oxidization (while implanting cations or anions, respectively), substitution reactions in layers, especially, H_2O. Taking into account ability of layered perovskites to exfoliate into individual nanosheets, exfoliation can be considered as the edged case of intercalation. Among other reactions we shall single out reduction intercalation, inoculation (modification type), formation of layered structures between neighboring layers, including columnar interlayer structures (pillars), etc. Thus, formation of structural columns between the neighboring layers, layered-columnar materials (pillaring) having far more developed surface as compared to the initial silicates, are widely currently used for preparing catalyst carriers of different types.

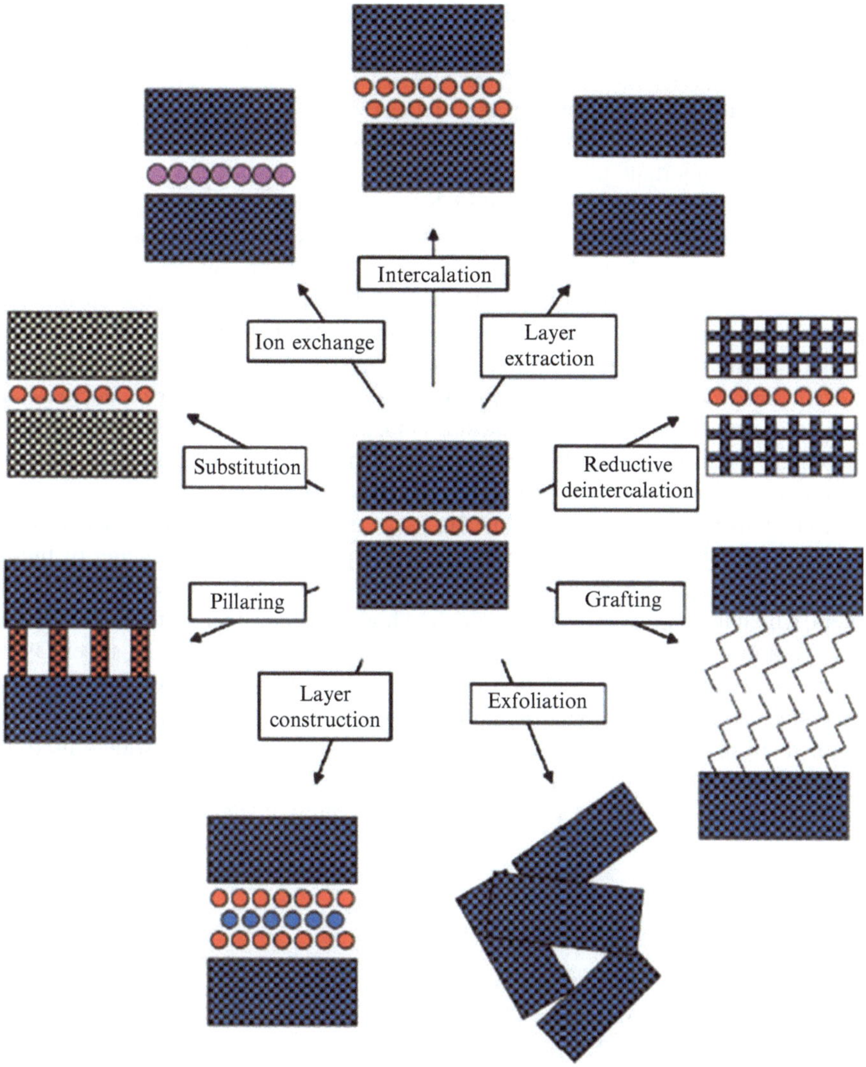

Fig. 5.54 Topochemical reactions in an interlayered space [388]

Topochemical syntheses are performed in a solid phase under relatively soft conditions (temperature below 500 °C), and the products are characterized by high stability. Many aspects of topochemical transformations in layers are very important (they are described in [388]), for example, ion or molecular stability, demands to redox-potential of intercalated compounds, thermal stability of reagents and products, etc. As was repeatedly mentioned, special demands are made to structuring of interlayer frame and its basal space. Topochemical methodology supposes steric control over the structure of intercalated component, which in the end brings to regulation of properties of the produced products. Ability to use this approach should take into account structural perfection of components and knowledge about mechanisms of the processes, and tools for their realization, physics and chemistry of intercalated systems.

A possibility of purposeful usage of these approaches together with routine design and synthesis of new materials with specific properties can give a wide approach to important materials [389], such as superconductors, ferroelectrics, thermoelectric materials, optical materials, catalysts, ionic conductors, microwaves absorbing materials, components of batteries and solar cells, various types of nanocomposites, including structuring ones. Synthetic approaches used in topochemical procedures, although limited as compared to molecular chemistry, are still efficient. In matrices of layered silicates, in particular in pillared DHL, protected nanoparticles of transition metals (Co, Ni, Cu and others) can be obtained by reduction reactions [390]. Encapsulation does not cause a significant decrease in magnetic properties of cobalt or nickel nanoparticles in these dielectric matrices stabilizing them from oxidation by air oxygen, after storage in nitric acid conservation of the particles was found.

Here a diversity of problems related to usage of porphyrin, natural polymers, etc. as intercalating agents are not considered. The promising for producing of these intercalating agents are also water-resistant microcrystal cellulose and other polymers, which in contrast to synthetic ones are highly porous fiber systems and can be used as a matrix for intercalation of metal nanoparticles, in particular, silver [391] There are no data on intercalation of new class of monomers containing metal, potential candidates for metal composites and catalysts based on them. Finally, methods of intercalation physical chemistry in the recent years are widely used in structuring of pharmaceutical [392], biological [393], including cellular, objects, for example, [394] which rest out of limits of the present study.

References

1. P. Gomez-Romero, Adv. Mater. **13**, 163 (2001)
2. R.S. Sinha, M. Okamoto, Prog. Polym. Sci. **28**, 1539 (2003)
3. S. Yariv, H. Cross (eds.), *Organoclay Complexes and Interactions* (Marcel Dekker, New York, 2002)
4. K.A. Carrado, Appl. Clay Sci. **17**, 1 (2000)
5. E. Ruiz-Hitzky, in *Organic-Inorganic Materials*, ed. by P. Gómez-Romero, C. Sanchez (Wiley-VCH, Weinheim, 2004)
6. E. Ruiz-Hitzky, P. Aranda, J.M. Serratosa, in *Handbook of Layered Materials*, ed. by S. Aucherbach, K.A. Carrado, P. Dutta (Marcel Dekker, New York, 2004)
7. D. Gournis, A. Lappas, M.A. Karakassides, D. Tobbens, A. Moukarika, Phys. Chem. Miner. **35**, 49 (2008)
8. R.K. Bharadwaj, R.A. Vaia, B.L. Farmer, in *PolymerNanocomposites*, ed. by R. Krishnamoorti, R. Vaia. ACS Symposium Series, vol. 804 (American Chemistry Society, Washington, DC, 2002)
9. T.J. Pinnavaia, G.W. Beall (eds.), *Polymer-Clay Nanocomposites* (Wiley, West Sussex, 2000)

10. E. Ruiz-Hitzky, A. Van Meerbeeck, Polymer-clay nanocomposites, in *Handbook of Clay Science*, ed. by F. Bergaya, B.K.G. Theng, G. Lagaly (Elsevier, Amsterdam, 2006)
11. M. Biswas, S.S. Ray, Adv. Mater. Sci. **155**, 167 (2001)
12. M.A. Alexandre, P. Dubois, Mater. Sci. Eng. **28**, 1 (2000)
13. S. Letaïef, E. Ruiz-Hitzky, Chem. Commun. **2996** (2003)
14. S. Yoda, Y. Nagashima, A. Endo, T. Miyata, S. Yanagishita, K. Otake, T. Tsuchiya, Adv. Mater. **17**, 367 (2005)
15. P. Gomez-Romero, C. Sanchez, *Functional Hybrid Materials* (Wiley-VCH, Weinheim, 2004)
16. K. Vermogen, K. Masenelli-Varlot, R. Seguela, J. Duchet-Rumeau, S. Boucard, P. Prele, Macromolecules **38**, 9661 (2005)
17. J. Liu, W.-J. Boo, A. Clearfield, H.-J. Sue, Mater. Manuf. Process **21**, 143 (2006)
18. T.G. Gopalkumar, J.A. Lee, M. Kontopoulou, J.S. Parent, Polymer **43**, 5483 (2002)
19. L. Jiankun, K. Yucai, Q. Zongneng, Y.J. Xiao-Su, Polym. Sci. B. Polym. Phys. **39**, 115 (2000)
20. S. Horsch, G. Serhatkulu, E. Gulari, R.M. Kannan, Polymer **47**, 7485 (2006)
21. A.D. Pomogailo, Russ. Chem. Rev. **69**, 53 (2000)
22. B.K.G. Theng, *Formation and Properties of Clay-Polymer Complexes* (Elsevier, New York, 1979)
23. H.V. Olephen, *An Introduction to Clay Colloidal Chemistry* (Wiley, New York, 1977)
24. A. Okada, Y. Kojima, M. Kawasumi, Y. Fukushima, T. Kurauchi, O. Kamigaito, J. Mater. Chem. **8**, 1179 (1993)
25. Y. Kojima, A. Usuki, M. Kawasumi, A. Okada, T. Kurauchi, O. Kamigaito, K. Kaji, J. Polym. Sci. Polym. Phys. **32**, 625 (1994)
26. Y. Kojima, A. Usuki, M. Kawasumi, A. Okada, T. Kurauchi, O. Kamigaito, K. Kaji, J. Polym. Sci. Polym. Phys. **33**, 1039 (1995)
27. A. Matsumura, Y. Komori, T. Itagaki, Y. Sugahara, K. Kuroda, Bull. Chem. Soc. Jpn. **74**, 1153 (2001)
28. L.A. Goettler, K.Y. Lee, H. Thakkar, Polym. Rev. **47**, 291 (2007)
29. F. Gao, Mater. Today **7**, 50 (2004)
30. S.S. Varghese, J.J. Karger-Kocsis, Appl. Polym. Sci. **91**, 813 (2004)
31. S. Pavlidou, C.D. Papaspyrides, Prog. Polym. Sci. **33**, 1119 (2008)
32. V. Mehrotra, E.P. Giannelis, R.F. Ziolo, P. Rogalskyj, Chem. Mater. **4**, 20 (1992)
33. A.D. Pomogailo, Polym. Sci. Ser. C **48**, 85 (2006)
34. A.D. Pomogailo, Inorg. Mater. **41**, S47 (2005)
35. N.V. Chukanov, I.V. Pekov, R.K. Rastsvetaeva, Usp. Khim. **73**, 227 (2004)
36. K.A. Carrado, S. Petit, F. Bergaya, G. Lagaly, Synthetic clay minerals and purification of natural clays, in *Handbook of Clay Science*, ed. by F. Bergaya, B.K.G. Theng, G. Lagaly (Elsevier, Amsterdam, 2006)
37. H.G. Karge, J. Weitkamp (eds.), *Molecular Sieves – Science and Technology* (Springer, Berlin/Heidelberg/New York, 2002)
38. S. Letaïef, B. Casal, N. Kebir-Ariguib, M. Trabelsi-Ayadi, E. Ruiz-Hitzky, Clay Miner. **37**, 517 (2002)
39. Q. Mao, S. Schleidt, H. Zimmermann, G. Jeschke, Macromol. Chem. Phys. **208**, 2145 (2007)
40. L.A. Utracki, *Clay-Containing Polymeric Nanocomposites*, vol. 1 & 2 (RAPRA, Shawbury, 2004)
41. F. Leroux, J.P. Besse, Chem. Mater. **13**, 3507 (2001)
42. Z. Shen, G.P. Simon, Y.-B. Cheng, J. Appl. Polym. Sci. **92**, 2101 (2004)
43. P. Reichert, H. Nitz, S. Klinke, R. Brandsch, R. Thomann, R. M lhaupt, Macromol. Mater. Eng. **275**, 8 (2000)
44. M. Huskic, M. Zigon, J. Appl. Polym. Sci. **113**, 1182 (2009)
45. V.A. Gerasin, PhD thesis, Institute of Oil Chemistry. Synthesis RAS, Moscow, 2005
46. R. Magaraphan, W. Thaijaroen, R. Lim-Ochakun, Rubber. Chem. Technol. **76**, 406 (2003)

47. J.M. Yeh, S.J. Liou, C.Y. Lin, C.Y. Cheng, Y.W. Chang, K.R. Lee, Chem. Mater. **14**, 154 (2002)
48. W. Xie, R. Xie, W. Pan, D. Hunter, B. Keone, L. Tan, R. Vaia, Chem. Mater. **14**, 4837 (2002)
49. E.M. Moujahid, J.P. Besse, F. Leroux, J. Mater. Chem. **12**, 3324 (2002)
50. J.H. Shi, X. Shu, M. Li, A.I.Ch.E. J **56**, 1352 (2010)
51. L.A. Utracki, M. Sepehr, E. Boccaleri, Polym. Adv. Technol. **18**, 1 (2007)
52. X. Duan, D.G. Evans (eds.), *Layered Double Hydroxides* (Springer, Berlin, 2006)
53. J.M. Miehe-Brendle, L. Delmotte, R. Le Dred, Micropor. Mesopor. Mater. **66**, 155 (2003)
54. G. Alberti, M. Casciola, U. Costantino, R. Vivani, Adv. Mater. **84**, 291 (2004)
55. M. Reinholdt, J. Miehe-Brendle, L. Delmotte, R. Le Dred, Clay Miner. **40**, 177 (2005)
56. J.M. Miehe-Brendle, C.R. Chimie **8**(2), 229 (2005)
57. J.M. Miehe-Brendle, L. Delmotte, R. Le Dred, Solid State Sci. **7**, 610 (2005)
58. A.I. Khan, D. O'Hare, J. Mater. Chem. **12**, 3191 (2002)
59. F. Leroux, C. Taviot-Gueho, J. Mater. Chem. **15**, 3628 (2005)
60. D.G. Evans, X. Duan, Chem. Commun. **485** (2006)
61. P. Ding, W. Chen, Q.U. Baojun, Prog. Nat. Sci. **16**, 573 (2006)
62. C. Pcholski, A. Kornowski, H. Weller, Angew. Chem. Int. Ed. **41**, 1188 (2002)
63. G.R. Patzke, F. Krumeich, R. Nesper, Angew. Chem. Int. Ed. **41**, 2446 (2002)
64. X.W. Lou, H.C. Zeng, J. Am. Chem. Soc. **125**, 2697 (2003)
65. K. Ebitani, T. Kawabata, K. Nagashima, T. Mizugaki, K. Kaneda, Green. Chem. **2**, 157 (2000)
66. K.V. Bineesh, D.R. Cho, S.Y. Kim, B.R. Jermy, D.W. Park, Catal. Commun. **9**, 2040 (2008)
67. G.K. Zhang, X.M. Ding, F.S. He, X.Y. Yu, J. Zhou, Y.J. Hu, J.W. Xie, Langmuir **24**, 1026 (2008)
68. P. Yuan, X. Yin, H. He, D. Yang, L. Wang, J. Zhu, Micropor. Mesopor. Mater. **93**, 240 (2006)
69. A.L. Villa, D.E. DeVos, F. Verpoort, B.F. Sels, P.A. Jacobs, J. Catal. **198**, 223 (2001)
70. K. Motokura, N. Fujita, K. Mori, T. Mizugaki, K. Ebitani, K. Kaneda, J. Am. Chem. Soc. **127**, 9674 (2005)
71. A.D. Pomogailo, A.S. Rozenberg, I.E. Uflyand, *Metal Nanoparticles in Polymers* (Khimiya, Moscow, 2000)
72. G.S. Zakharova, V.L. Volkov, Usp. Khim. **72**, 346 (2003)
73. C. Verissimo, O.L. Alves, J. Mater. Chem. **13**, 378 (2003)
74. A.D. Pomogailo, Colloid J. **67**, 658 (2005)
75. C. Ooka, H. Yoshida, K. Suzuki, T. Hattori, Chem. Lett. **32**, 896 (2003)
76. K. Mogyorosi, I. Dekany, J.H. Fendler, Langmuir **19**, 2939 (2003)
77. J. Njuguna, K. Pielichowski, S. Desai, Polym. Adv. Technol. **19**, 947 (2008)
78. M. Pluta, M.A. Paul, M. Alexandre, P. Dubois, J. Polym. Sci. B. Polym. Phys. **44**, 299 (2006)
79. H.Y. Zhu, Z. Ding, C.Q. Lu, G.Q. Lu, Appl. Clay Sci. **20**, 165 (2002)
80. F. Krumeich, H.-J. Muhr, M. Niederberger, F. Bieri, R. Nesper, Z. Anorg, Allg. Chem. **326**, 2208 (2000)
81. M.A. Bizeto, A.L. Shiguihara, V.R.L. Constantino, J. Mater. Chem. **19**, 2512 (2009)
82. M. Osada, T. Sasaki, J. Mater. Chem. **19**, 2503 (2009)
83. A.S. Golub, Y.V. Zubyavichus, Y.L. Slovokhotov, Y.N. Novikov, Uspekhi Khim. **72**, 138 (2003)
84. A.S. Golub, D.P. Rupasov, N.D. Lenenko, Y.N. Novikov, Z. Neorg, Zh. Neorg. Khim. **55**, 1239 (2010)
85. R. Sengupta, S. Chakraborty, S. Bandyopadhyay, S. Dasgupta, R. Mukhopadhyay, K. Auddy, A.S. Deuri, Polym. Eng. Sci. **47**, 1956 (2007)
86. Q.T. Nguyen, D.G. Baird, Adv. Polym. Technol. **25**, 270 (2006)
87. F. Santiago, A.E. Mucientes, M. Osorio, C. Rivera, Eur. Polym. J. **43**, 1 (2007)
88. E. Manias, L. Touny, K. Wu, B. Strawhecher, Chem. Mater. **13**, 3516 (2001)
89. A. Mousa, J. Karger-Kocsis, Macromol. Mater. Eng. **286**, 260 (2001)

90. S.-S. Lee, C.S. Lee, M.-H. Kim, S.Y. Kwak, M. Park, S.H. Lim, C.R. Choe, J.J. Kim, Polym. Sci. B: Polym. Phys. **39**, 2430 (2001)
91. S.-S. Lee, J. Kim, J. Polym. Sci. B: Polym. Phys. **42**, 246 (2004)
92. S.-S. Lee, J. Kim, J. Polym. Sci. B: Polym. Phys. **42**, 2367 (2004)
93. S.S. Ray, K. Okamoto, M. Okamoto, Macromolecules **36**, 2355 (2003)
94. E. Manias, H. Chen, R. Krishnamoorrti, J. Genzer, E.J. Kramer, E.P. Giannelis, Macromolecules **33**, 7955 (2000)
95. Z.G. Wu, C.X. Zhou, R.R. Qi, H.B. Zhang, J. Appl. Polym. Sci. **83**, 2403 (2002)
96. Z.G. Wu, C.X. Zhou, Polym. Test. **21**, 479 (2002)
97. G. Galgali, C. Ramesh, A. Lele, Macromolecules **34**, 852 (2001)
98. J. Li, C.X. Zhou, G. Wang, J. Appl. Polym. Sci. **89**, 3609 (2003)
99. J. Li, C.X. Zhou, G. Wang, D. Zhao, J. Appl. Polym. Sci. **89**, 318 (2003)
100. D.F. Wu, C.X. Zhou, F. Xie, D.L. Mao, B. Zhang, Polym. Polym. Compos. **13**, 61 (2005)
101. D.F. Wu, C.X. Zhou, F. Xie, D.L. Mao, B. Zhang, Eur. Polym. J. **41**, 2199 (2005)
102. I. Matsubara, K. Hosono, N. Murayama, W. Shin, N. Izu, Bull. Chem. Soc. Jpn. **77**, 1231 (2004)
103. M.R. Bockstaller, R.A. Mickiewicz, E.L. Thomas, Adv. Mater. **17**, 1331 (2005)
104. M.J. Kawasumi, Polym. Sci. A Polym. Chem. **42**, 819 (2004)
105. R.A. Vaia (ed.), *ACS Symposium Series*, vol. 804 (American Chemical Society, Washington, DC, 2002)
106. S.S. Ray, M. Okamoto, Prog. Polym. Sci. **28**, 1539 (2003)
107. D.D. Schmidt, D. Shah, E.P. Giannelis, Curr. Opin. Solid. State. Mater. Sci. **6**, 205 (2002)
108. C. Zilg, F. Dietsche, B. Hoffmann, C. Dietrich, R. Mülhaupt, Macromol. Symp. **169**, 65 (2001)
109. M. Zanetti, S. Lomakin, G. Camino, Macromol. Mater. Eng. **279**, 1 (2000)
110. L.S. Schadler, Polymer-based and polymer-filled nanocomposites, in *Nanocomposite Science and Technology*, ed. by P.M. Ajayan, L.S. Schadlerand, P.V. Braun) (Wiley-VCH, Weinheim, 2003)
111. V.V. Ginzburg, C. Singh, A.C. Balazs, Macromolecules **33**, 1089 (2000)
112. R.A. Vaia, Structural characterization of polymer-layered silicate nanocomposites, in *Polymer–Clay Nanocomposites*, ed. by T.J. Pinnavaia, G.W. Beall (Wiley, Chichester, 2000), p. 229
113. D.W. Kim, A. Blumstein, S.K. Tripathy, Chem. Mater. **13**, 1916 (2001)
114. E. Passaglia, R. Sulcis, F. Ciardelli, M. Malvaldi, P. Narducci, Polym. Int. **54**, 1549 (2005)
115. A.C. Balazs, J. Bicerano, V.V. Ginzburg, in *Polyolefin Composites*, ed. by D. Nwabunma, T. Kyu (Wiley, Hoboken, 2008). Chapter 15
116. F. Chiardelli, S. Coiai, E. Passaglia, A. Pucci, G. Rugg, Polym. Int. **57**, 805 (2008)
117. P. Gómez-Romero, O. Ayyad, J. Suárez-Guevara, D. Muñoz-Rojas, J. Solid State Electrochem. **14**, 1939 (2010)
118. A. Leszczyńska, J. Njuguna, K. Pielichowski, J.R. Banerjee, Thermochim. Acta. **453,** 75 (2007); **454**, 1 (2007)
119. M. Wark, Porphyrins and phtalocyanines encapsulated in inorganic host material, in *The Porphyrin Handbook*, ed. by K.M. Kadish, K.M. Smith, R. Guilard. Phtalocyanines: Properties and materials, vol. 17 (Elsevier Science, Amsterdam, 2003), pp. 247–283
120. T. Ogoshi, H. Itoh, K.-M. Kim, Y. Chujo, Macromolecules **35**, 334 (2002)
121. K. Anderson, A. Sinsawat, R. Vaia, B.L. Farmer, J. Polym. Sci. B Polym. Phys. **43**, 1014 (2005)
122. A. Sinsawat, K.L. Anderson, R.A. Vaia, B.L. Farmer, J. Polym. Sci. B Polym. Phys. **41**, 3272 (2003)
123. R.B. Pandey, B.L. Farmer, J. Polym, Sci. B Polym. Phys. **47**, 2487 (2009)
124. H. Li, Y. Yu, Y. Yang, Eur. Polym. J. **41**, 2016 (2005)
125. Y.S. Choi, H.T. Ham, I.J. Chung, Chem. Mater. **16**, 2522 (2004)
126. Y.S. Choi, M. Xu, I.J. Chung, Polymer **44**, 6989 (2003)

127. M.R. Moghbeli, N. Mehdizadeh, J. Appl. Polym. Sci. **123**, 2064 (2012)
128. L.B. de Paiva, A.R. Morales, F.R.V. Díaz, Appl. Clay Sci. **42**, 8 (2008)
129. Y.K. Kim, Y.S. Choi, K.H. Wang, I. Chung, J. Chem. Mater. **14**, 4990 (2002)
130. W. Xie, J.M. Hwu, G.J. Jiang, T.M. Buthelezi, W. Pan, Polym. Eng. Sci. **43**, 214 (2003)
131. W. Feng, E. Sun, A. Fujii, H. Wu, K. Niihara, K. Yoshino, Bull. Chem. Soc. Jpn. **73**, 2627 (2000)
132. E.M. Moujahid, F. Leroux, M. Dubois, J.P. Besse, C. R. Chimie **62**, 259 (2003)
133. P. Ding, B. Qu, J. Appl. Polym. Sci. **101**, 3758 (2006)
134. L. Vieille, E.M. Moujahid, C. Taviot-Guého, J. Cellier, J.P. Besse, J. Phys. Chem. Solids **65**, 385 (2004)
135. Z.M.O. Rzayev, A. Güner, E.A. Soÿlemez, S. Kavlak, Polym. Adv. Technol. **22**, 1349 (2011)
136. J. Rzayev, Macromolecules **42**, 2135 (2009)
137. K. Huang, J. Rzayev, J. Am. Chem. Soc. **131**, 6880 (2009)
138. L. Zhao, M. Byun, J. Rzayev, Macromolecules **42**, 9027 (2009)
139. Z.M.O. Rzayev, E.A. Sôylemez, B. Davarcioğlu, Polym. Adv. Technol **23**, 278 (2011)
140. A. Matsumoto, T. Odani, Macromol. Rapid Commun. **22**, 1195 (2001)
141. A. Matsumoto, Polym. J. **35**, 93 (2003)
142. S. Oshita, A. Matsumoto, Chem. Eur. J. **12**, 2139 (2006)
143. A. Matsumoto, K. Sada, K. Tashiro, M. Miyata, T. Tsubouchi, T. Tanaka, T. Odani, S. Nagahama, T. Tanaka, K. Inoue, S. Saragai, S. Nakamoto, Angew. Chem. **114**, 2612 (2002)
144. A. Matsumoto, T. Odani, M. Chikada, K. Sada, M. Miyata, J. Am. Chem. Soc. **121**, 11122 (1999)
145. A. Matsumoto, S. Nagahama, T. Odani, J. Am. Chem. Soc. **122**, 9109 (2000)
146. T. Tanaka, A. Matsumoto, J. Am. Chem. Soc. **124**, 9676 (2002)
147. S. Nagahama, T. Tanaka, A. Matsumoto, Angew. Chem. **116**, 3899 (2004); Angew. Chem. Int. Ed., **43**, 3811 (2004)
148. S. Oshita, T. Tanaka, A. Matsumoto, Chem. Lett. **34**, 1442 (2005)
149. P. Sozzani, S. Bracco, A. Comotti, R. Simonutti, I. Camurati, J. Am. Chem. Soc. **125**, 12881 (2003)
150. P. Sozzani, S. Bracco, A. Comotti, M. Mauri, R. Simonutti, P. Valsesia, Chem. Commun. **1921** (2006)
151. A.D. Pomogailo, Russ. Chem. Rev. **71**, 1 (2002)
152. S.-H. Chan, Y.-Y. Lin, C. Ting, Macromolecules **36**, 8910 (2003)
153. W. Mariott, N. Escude, E.Y.-X. Chen, J. Polym. Sci. A Polym. Chem. **45**, 2581 (2007)
154. W.R. Mariott, E.Y.-X. Chen, J. Am. Chem. Soc. **125**, 15726 (2003)
155. Y.H. Jin, H.J. Park, S.S. Im, S.Y. Kwak, S. Kwak, Macromol. Rapid. Commun. **23**, 135 (2002)
156. S.-Y.A. Shin, L.C. Simon, J.B.P. Soares, G. Scholz, Polymer **44**, 5317 (2003)
157. A. He, L. Wang, J. Li, J. Dong, C.A. Han, Polymer **47**, 1767 (2006)
158. S. Zulfiqar, I. Lieberwirth, Z. Ahmad, M.I. Sarwar, Polym. Eng. Sci. **48**, 1624 (2008)
159. S. Zulfiqar, Z. Ahmad, M. Sarwar, Polym. Adv. Technol. **19**, 1720 (2008)
160. T. Agag, T. Takeichi, Polymer **41**, 7083 (2000)
161. M. Okamoto, S. Morita, H. Taguchi, Y.H. Kim, T. Kataka, H. Tateyama, Polymer **41**, 3887 (2000)
162. C. Zeng, L.J. Lee, Macromolecules **34**, 4098 (2001)
163. D.W. Kim, A. Blumstein, H. Liu, M.J. Downey, J. Kumar, S.K. Tripathy, J. Macromol. Sci. Pure Appl. Chem. **38**, 1405 (2001)
164. S.K. Sahoo, D.W. Kim, J. Kumar, A. Blumstein, A.L. Cholli, Macromolecules **36**, 2777 (2003)
165. M. Casciola, A. Donnadio, M. Pica, V. Valentini, P. Piaggio, Macromol. Symp. **230**, 95 (2005)
166. G.D. Gatta, S. Masci, R. Vivani, J. Mater. Chem. **13**, 1215 (2003)

167. R. Singhal, M. Datta, J. Appl. Polym. Sci. **103**, 3299 (2007)
168. L. Wang, M. Rocci-Lane, P. Brazis, C.R. Kannewurf, Y. Kim, W. Lee, J.-H. Choy, M.G. Kanatzidis, J. Am. Chem. Soc. **122**, 6629 (2000)
169. I. Boyano, M. Bengoechea, I. DeMeatza, O. Miguel, I. Cantero, E. Ochoteco, H. Grande, M. Lira-Cantu, P. Gomez-Romero, J. Power. Sources **174**(2), 1206 (2007)
170. I. Boyano, M. Bengoechea, I. de Meatza, O. Miguel, I. Cantero, E. Ochoteco, J. Rodriguez, M. Lira-Cantu, P. Gomez-Romero, J. Power. Sources **166**, 471 (2007)
171. D.R. Rolison, B. Dunn, J. Mater. Chem. **11**, 963 (2001)
172. A. Gapeev, C.N. Yang, S.J. Klippenstein, R.C. Dunbar, J. Phys. Chem. **104**, 3246 (2000)
173. E. Shouji, D.A. Buttry, Langmuir **15**, 669 (1999)
174. P.J. Kulesza, M. Chojak, K. Miecznikowski, A. Lewera, M.A. Malik, A. Kuhn, Electrochem. Commun. **4**, 510 (2002)
175. L. Adamczyk, P.J. Kulesza, K. Miecnikowski, B. Palys, M. Chojak, D. Krawczyk, J. Electrochem. Soc. **152**, E98 (2005)
176. J. Vaillant, M. Lira-Cantu, K. Cuentas-Gallegos, N. Casañ-Pastor, P. Gómez-Romero, Prog. Solid. State. Chem. **34**, 147 (2006)
177. V. Subramanian, H.W. Zhu, R. Vajtai, P.M. Ajayan, B.Q. Wei, J. Phys. Chem. B **109**, 20207 (2005)
178. P. Gomez-Romero, M. Chojak, A.K. Cuentas-Gallegos, J.A. Asensio, P.J. Kulesza, N. Casañ-Pastor, M. Lira-Cantu, Electrochem. Commun. **5**, 149 (2003)
179. A.K. Cuentas-Gallegos, M. Lira-Cantu, N. Casañ-Pastor, P. Gomez-Romero, Adv. Funct. Mater. **15**, 1125 (2005)
180. T. Wang, W. Liu, J. Tian, X. Shao, D. Sun, Polym. Compos. **25**, 111 (2004)
181. B.G. Soares, M. Oliveira, S. Zaioncz, A.C. Gomes, A.A. Silva, K.S. Santos, R.S. Mauler, J. Appl. Polym. Sci. **119**, 505 (2011)
182. B. Pourabas, V. Raeesi, Polymer **46**, 5533 (2005)
183. D.R. Dillon, K.K. Tenneti, C.Y. Li, F.K. Ko, I. Sics, B.S. Hsiao, Polymer **47**, 1678 (2006)
184. L. Qiu, W. Chen, B. Qu, Polymer **47**, 9 (2006)
185. S. Filippi, E. Mameli, C. Marazzato, P. Magagnini, Eur. Polym. J. **43**, 1645 (2007)
186. D. Burgentzle, J. Duchet, J.F. Gerard, A. Jupin, B. Fillon, J. Colloid Interface Sci. **278**, 26 (2004)
187. M.K. Song, Y.M. Kim, Y.T. Kim, H.W. Rhee, A. Smirnova, N.M. Sammes, J.M. Fenton, J. Electrochem. Soc. **153 A**, 2239 (2006)
188. A.K. Ghosh, E.M. Woo, Polymer **45**, 4749 (2004)
189. T.L. Wang, W.S. Hwang, M.H. Yeh, J. Appl. Polym. Sci. **104**, 4135 (2007)
190. V. Sridhar, D.K. Tripathy, J. Appl. Polym. Sci. **101**, 3630 (2006)
191. G. Liang, J. Xu, S. Bao, W. Xu, J. Appl. Polym. Sci. **91**, 3974 (2004)
192. H. Acharya, S.K. Srivastava, A.K. Bhowmick, Polym. Eng. Sci. **48**, 837 (2006)
193. M. Frounchi, S. Dadbin, Z. Salehpour, M. Noferesti, J. Membr. Sci **282**, 142 (2006)
194. R. Valsecchi, M. Vigano, M. Levi, S. Turri, J. Appl. Polym. Sci. **102**, 4484 (2006)
195. L. Zhang, M. Wan, J. Phys. Chem. **107**, 6748 (2003)
196. Y. Someya, M. Shibata, Polymer **46**, 4891 (2005)
197. I.M. Daniel, H. Miyagawa, E.E. Gdoutos, J.J. Luo, Exp. Mech. **43**, 348 (2003)
198. Z.M.O. Rzayev, A. Yilmazbayhan, E. Alper, Adv. Polym. Technol. **26**, 41 (2007)
199. E. Söylemez, N.C. Aylak, Z.M.O. Rzayev, eXPRESS Polym. Lett. **2**, 639 (2008)
200. S.C. Tjong, Mater. Sci. Eng. **53**, 73 (2006)
201. H.K. Can, Z.M.O. Rzayev, A. G ner, J. Appl. Polym. Sci. **90**, 4009 (2003)
202. W. Chen, L. Feng, B. Qu, Chem. Mater. **16**, 368 (2004)
203. T. Ohtake, Y. Takamitsu, K. Ito-Akita, K. Kanie, M. Yoshizawa, T. Mukai, H. Ohno, T. Kato, Macromolecules **33**, 8109 (2000)
204. B.P. Grady, C.P. Rhodes, S. York, R.E. Frech, Macromolecules **34**, 8523 (2001)
205. D. Choi, M.A. Kader, B.H. Cho, Y. Huh, Y. Nuh, C. Nah, J. Appl. Polym. Sci. **98**, 1688 (2005)

206. A. Kelarakis, E.P. Giannelis, K. Yoon, Polymer **48**, 7567 (2007)
207. S.D. Wanjale, J.P. Jog, J. Appl. Polym. Sci. **90**, 3233 (2003)
208. S. Sadhu, A.K. Bhowmick, Rubber Chem. Technol. **78**, 321 (2005)
209. Y.-R. Liang, W.-L. Cao, X.-B. Zhang, Y.-J. Tan, S.-J. He, L.-Q. Zhang, J. Appl. Polym. Sci. **112**, 3087 (2009)
210. J. Karger-Kocsis, C.-M. Wu, Polym. Eng. Sci. **44**, 1083 (2004)
211. K.S. Santos, S.A. Liberman, M.A.S. Ovieda, R.S. Mauler, J. Polym. Sci. B Polym. Phys. **46**, 2519 (2008)
212. J.H. Joo, J.H. Shim, J.H. Choi, C.-H. Choi, D.-S. Kim, J.-S. Yoon, J. Appl. Polym. Sci. **109**, 3645 (2008)
213. L. Liu, D. Jia, Y. Luo, B. Guo, J. Appl. Polym. Sci. **100**, 1905 (2006)
214. N. Ristolainen, U. Vainio, S. Paavola, M. Torkkeli, J. Appl. Polym. Sci. **43**, 1892 (2005)
215. S.H. Kim, J.W. Chung, T.J. Kang, S.Y. Kwak, T. Suzuki, Polymer **48**, 4271 (2007)
216. D. Kang, D. Kim, S.-H. Yoon, D. Kim, C. Barry, J. Mead, Macromol. Mater. Eng. **292**, 329 (2007)
217. M. Kurata, Y. Tsunashima, in *Polymer Handbook*, ed. by J. Brandrup, E.H. Immergut, vol. Section VII (Wiley-Interscience, New York, 1989), p. 381
218. D. Wu, C. Zhou, H. Zheng, J. Appl. Polym. Sci. **99**, 1865 (2006)
219. Z.M. Wang, H. Nakajimia, E. Manias, T.C. Chung, Macromolecules **36**, 8919 (2003)
220. P. Ding, B. Qu, Polym. Eng. Sci. **46**, 1153 (2006)
221. S. Zhang, T.R. Hull, A.R. Horrocks, G. Smart, B.K. Kandola, J. Ebdon, P. Joseph, B. Hunt, Polym. Degrad. Stab. **92**, 727 (2007)
222. Y. Ji, B. Li, S. Ge, J.C. Sokolov, M.H. Rafailovich, Langmuir **22**, 1321 (2006)
223. F.R. Costa, M.A. Goad, U. Wagenknecht, G. Heinrich, Polymer **46**, 4447 (2005)
224. F.R. Costa, U. Wagenknecht, D. Jehnichen, M.A. Goad, G. Heinrich, Polymer **47**, 1649 (2006)
225. U. Costantino, A. Gallipoli, M. Rocchetti, G. Camino, F. Bellucci, A. Frache, Polym. Degrad. Stab. **90**, 586 (2005)
226. D. Wang, J. Zhu, C.A. Wilkie, Polym. Degrad. Stab. **80**, 171 (2003)
227. W. Lertwimolnun, B. Vergnes, Polym. Eng. Sci. **46**, 315 (2006)
228. M. Mainil, M. Alexandre, F. Monteverde, P. Dubois, J. Nanosci. Nanotechnol. **6**, 337 (2006)
229. M.A. Osman, J.E.P. Rupp, U.W. Suter, Polymer **46**, 8202 (2005)
230. K. Chrissopoulou, I. Altintzi, S.H. Anastasiadis, E.P. Giannelis, M. Pitsikalis, N. Hadjichristidis, N. Theophilon, Polymer **46**, 12440 (2005)
231. S. Parija, S.K. Nayak, S.K. Verma, S.S. Tripathy, Polym. Compos. **25**, 646 (2004)
232. K.H. Chen, S.M. Yang, J. Appl. Polym. Sci. **86**, 414 (2002)
233. Y. Güldoğan, S. Eğri, Z.M.O. Rzaev, E. Pişkin, J. Appl. Polym. Sci. **92**, 3675 (2004)
234. J.M. Yeh, S.J. Liou, C.Y. Lai, P.C. Wu, T.Y. Tsai, Chem. Mater. **13**, 1131 (2001)
235. H. Nasegawa, M. Okamoto, A. Usuki, J. Appl. Polym. Sci. **93**, 758 (2004)
236. A. Tidjani, O. Wald, M.-M. Pohl, M.P. Henschel, B. Schartel, Polym. Degrad. Stab. **82**, 133 (2003)
237. C.M. Koo, M.J. Kim, S.M.N. Choi, O. Kim, I.J. Chung, J. Appl. Polym. Sci. **88**, 1526 (2003)
238. W. Xu, G. Liand, W. Wang, S. Tang, P. He, W.-P. Pan, J. Appl. Polym. Sci. **88**, 3093 (2003)
239. Z.M.O. Rzaev, A. Güner, H.K. Can, A. Asıcı, Polymer **42**, 5599 (2001)
240. X. Kornmann, H. Lindberg, L.A. Berglund, Polymer **42**, 1303 (2001)
241. J. Zhang, C.A. Wilkie, Polym. Degrad. Stab. **80**, 163 (2003)
242. Y. Tang, Y. Hu, S.F. Wang, Z. Gui, Z. Chen, W.C. Fan, Polym. Degrad. Stab. **78**, 555 (2002)
243. J.W. Gilman, C.J. Jackson, A.B. Morgan, J.R. Harris, E. Manias, E.P. Giannelis, M. Wuthenow, D. Hilton, S.H. Philips, Chem. Mater. **12**, 1866 (2000)
244. M. Böhning, H. Goering, A. Fritz, K.-W. Brzezinka, G. Turky, A. Schönhals, B. Schartel, Macromolecules **38**, 2764 (2005)
245. J.T. Xu, Q. Wang, Z.Q. Fan, Eur. Polym. J. **41**, 3011 (2005)

246. L. Valentini, J. Biagiotti, M.A. López-Manchado, S. Santucci, J.M. Kenny, Polym. Eng. Sci. **44**, 303 (2004)
247. Z.M.O. Rzaev, A. Yilmazbayhan, E. Alper, Adv. Polym. Techn. **26**, 41 (2007)
248. F. Perrin-Sarazin, M.T. Ton-That, M.N. Bureau, J. Denault, Polymer **46**, 11624 (2005)
249. D. Garicia-López, O. Picqazo, J.C. Merino, J.M. Pastor, Eur. Polym. J. **39**, 945 (2003)
250. C.H. Jeon, S.H. Ryu, Y.W. Chang, Polym. Int. **52**, 153 (2003)
251. Y. Wang, F. Chen, Y. Li, K. Wu, Soc. Plast. Eng. Ann. Tech. Conf. **61**, 3670 (2003)
252. K.A. Nara, Soc. Plast. Eng. Ann. Tech. Conf. Proc. **61**, 3717 (2003)
253. X. Liu, Q. Wu, Polymer **42**, 10013 (2001)
254. Y. Zang, J. Lee, J.M. Rhee, K.Y. Rhee, Compos. Sci. Technol. **64**, 1383 (2004)
255. Y. Zang, J. Lee, H. Jang, C. Nah, Compos. Eng. **35**, 133 (2004)
256. H.R. Dennis, D.L. Hunter, S.K.D. Chang, J.L. White, J.W. Cho, D.R. Paul, Polymer **41**, 9513 (2001)
257. M. Zanetti, G. Camino, R. Mülhaupt, Polym. Degrad. Stab. **74**, 413 (2001)
258. M. Zanetti, G. Camino, R. Thomann, R. Mülhaupt, Polymer **42**, 4501 (2001)
259. A. Riva, M. Zanetti, M. Braglia, G. Camino, L. Falqui, Polym. Degrad. Stab. **77**, 299 (2002)
260. F. Bellucci, G. Camino, A. Frache, V. Ristori, L. Sorrentino, S. Iannace, X. Bian, M. Guardasole, S. Vaccaro, e-Polym. **014**, (2006)
261. U.N. Ratnayake, B. Haworth, D.J. Hourston, J. Appl. Polym. Sci. **112**, 320 (2009)
262. M. Alexandre, G. Beyer, C. Henrist, R. Cloots, A. Rulmont, R. Jerome, P. Dubois, Macromol. Rapid Commun. **22**, 643 (2001)
263. U. Yilmazer, G. Ozden, Polym. Compos. **27**, 249 (2006)
264. K.H. Wang, M.H. Choi, C.M. Koo, Y.S. Choi, I.J. Chung, Polymer **42**, 9819 (2001)
265. I.S. Suh, S.H. Ryu, J.H. Bae, Y.W. Chang, J. Appl. Polym. Sci. **94**, 1057 (2004)
266. J.T. Yoon, W.H. Jo, M.S. Lee, M.B. Ko, Polymer **42**, 329 (2001)
267. Q. Zhang, Q. Fu, L. Jiang, Y. Lei, Polym. Int. **49**, 1561 (2000)
268. D.J. Frankowski, S.A. Khan, R.J. Spontak, Adv. Mater. **19**, 1286 (2007)
269. M. Vacatello, Macromolecules **34**, 1946 (2001)
270. S. Tanoue, L.A. Utracki, A. Garcia-Rejon, J. Tatibouët, K.C. Cole, M.R. Kamal, Polym. Eng. Sci. **44**, 1046 (2004)
271. D.Z. Chen, H.Y. Yang, P.S. He, W.A. Zhang, Compos. Sci. Technol. **65**, 1593 (2005)
272. Y.Q. Li, H. Ishida, Macromolecules **38**, 6513 (2005)
273. B. Hoffmann, C. Dietrich, R. Thomann, C. Friedrich, R. Mülhaupt, Macromol. Rapid Commun. **21**, 57 (2000)
274. R.A. Vaia, E.P. Giannelis, Polymer **42**, 1281 (2001)
275. A.B. Morgan, R.H. Harris Jr., T. Kashiwagi, L.J. Chyall, J.W. Gilman, Fire Mater. **26**, 247 (2002)
276. O. Monticelli, Z. Musina, A. Frache, F. Bellucci, G. Camino, S. Russo, Polym. Degrad. Stab. **92**, 370 (2007)
277. M. Zanetti, Polym. Nanocomposites **253** (2006)
278. M. Lewin, E.M. Pearce, K. Levon, A. Mey-Marom, M. Zammarano, C.A. Wilkie, B.N. Jang, Polym. Adv. Technol. **17**, 226 (2006)
279. R.K. Bharadwaj, Macromolecules **34**, 9189 (2001)
280. S.M. Lomakin, G.E. Zaikov, Vysokomol. Soed. Ser. B **47**, 105 (2005)
281. A.P. Patio-Soto, S. Sanchez-Valdes, L.F. Ramos-Devalle, J. Polym, Sci Part Polym. Phys. **46**, 190 (2008)
282. G. Chen, X. Chen, Z. Lin, W. Ye, J. Mater. Sci. Lett. **18**, 1761 (2002)
283. J.M. Hwu, G.J. Jiang, Z.M. Gao, W. Xie, W.P. Pan, J. Appl. Polym. Sci. **83**, 1702 (2002)
284. Y. Munusamy, H. Ismail, M. Mariatti, C.T. Ratnam, J. Vinyl Addit. Technol. **15**, 244 (2009)
285. S.-S. Lee, M.H. Hur, H. Yang, S. Lim, J. Kim, J. Appl. Polym. Sci. **101**, 2749 (2006)
286. W.-F. Lee, L.-L. Jou, J. Appl. Polym. Sci. **94**, 74 (2004)
287. T. Pepmicek, J. Duchet, L. Kovarova, J. Malac, J.F. Gerard, J. Simonik, Polym. Degrad. Stab. **91**, 1855 (2006)

288. M.A. Souza, L.A. Pessan, N. Rodolfo Jr., Polimeros **16**, 257 (2006)
289. Y. Yoo, S.-S. Kim, J.C. Won, K.-Y. Choi, J.H. Lee, Polym. Bull. **52**, 373 (2004)
290. B. Dietrich, J. Vinyl Addit. Tech. **7**, 168 (2001)
291. C. Wan, X. Qiao, Y. Zhang, Y. Zhang, Polym. Test. **22**, 453 (2003)
292. X.L. Xie, R.K.Y. Li, Q.X. Liu, Y.W. Mai, Polymer **45**, 2793 (2004)
293. L. Kovarova, A. Kalendova, J.F. Gerard, J. Malac, J. Simonik, Z. Weiss, Macromol. Symp. **221**, 105 (2005)
294. H. Hu, M. Pan, X. Li, X. Shi, L. Zhang, Polym. Int. **53**, 225 (2004)
295. M.E. Romero-Guzman, A. Romo-Uribe, E. Ovalle-Garcıa, R. Olayo, C.A. Cruz-Ramos, Polym. Adv. Technol. **19**, 1168 (2008)
296. P. Jawahar, M. Balasubramanian, J. Nanosci. Nanotechnol. **6**, 3973 (2006)
297. J. Chandradass, M.R. Kumar, R. Velmurugan, Mater. Lett. **61**, 4385 (2007)
298. J. Kwiatkowski, A.K. Whittaker, J. Polym. Sci. B Polym. Phys. **39**, 1678 (2001)
299. I. Mironi-Harpaz, M. Narkis, A. Siegmann, Polym. Eng. Sci. **45**, 174 (2005)
300. M.-J. Binette, C. Detellier, J. Chem. **80**, 1708 (2002)
301. C.O. Oriakhi, M. Lerner, Chem. Mater. **8**, 2016 (1996)
302. Z. Shen, G.P. Simon, Y.-B. Cheng, Polym. Eng. Sci. **42**, 2369 (2002)
303. P. Aranda, Y. Mosqueda, E. Perez-Cappe, E. Ruiz-Hitzky, J. Polym. Sci. B Polym. Phys. **41**, 3249 (2003)
304. U. Opara-Krasovec, R. Jese, B. Orel, G. Drazic, Monat. Chem. **133**, 1115 (2002)
305. N. Hasegawa, H. Okamoto, M. Kato, A. Usuki, N. Sato, Polymer **44**, 2933 (2003)
306. T.D. Fornes, P.J. Yoon, H. Keskkula, D.R. Paul, Polymer **42**, 9929 (2001)
307. T.D. Fornes, P.J. Yoon, D.L. Hunter, H. Keskkula, D.R. Paul, Polymer **43**, 5915 (2002)
308. M.I. Sarwar, S. Zulfiqar, Z. Ahmad, Colloid Polym. Sci. **285**, 1733 (2007)
309. B. Lepoittevin, N. Pantoustier, M. Alexandre, C. Calberg, R. Jerome, P. Dubois, J. Mater. Chem. **12**, 3528 (2002)
310. D.L. VanderHart, A. Asano, J.W. Gilman, Chem. Mater. **13**, 3781 (2001)
311. G. Gorrasi, M. Tortora, V. Vittoria, E. Pollet, B. Lepoittevin, M. Alexandre, P. Dubois, Polymer **44**, 2271 (2001)
312. P.B. Messersmith, E.P. Giannelis, J. Polym. Sci. A Polym. Chem. **33**, 1047 (1995)
313. S. Bourbigot, M. LeBras, F. Dabrowski, J.W. Gilman, T. Kashiwagi, Fire Mater. **24**, 201 (2000)
314. M.K. Akkapeddi, Polym. Compos. **21**, 576 (2000)
315. A. Usuki, M. Kawasumi, Y. Kojima, A. Okada, T. Kurauchi, O. Kamigaito, J. Mater. Res. **8**, 1174 (1993)
316. S. Zulfiqar, M. Ishaq, M.I. Sarwar, Surf. Interface Anal. **40**, 1195 (2008)
317. T.D. Fornes, D.R. Paul, Polymer **44**, 3945 (2003)
318. S. Wu, F. Wang, C.M. Ma, W. Chang, C. Kuo, H. Kuan, W. Chen, Mater. Lett. **49**, 327 (2001)
319. M.I. Sarwar, S. Zulfiqar, Z. Ahmad, J. Sol-Gel Sci. Technol. **45**, 89 (2008)
320. S. Zulfiqar, Z. Ahmad, M.I. Sarwar, Colloid Polym. Sci. **285**, 1749 (2007)
321. K.H. Tyan, T. Wei, E. Hsieh, J. Polym. Sci. B Polym. Phys. **38**, 2873 (2000)
322. H. Fong, W. Liu, C.S. Wang, R.A. Vaia, Polymer **43**, 775 (2002)
323. D.P.N. Vlasveld, H.E.N. Bersee, S.J. Picken, Polymer **46**, 10269 (2005)
324. L. Li, L.M. Bellan, H.G. Craighead, M.W. Frey, Polymer **47**, 6208 (2006)
325. Y. Liang, X. Xia, Y. Luo, Z. Jia, Mater. Lett. **61**, 3269 (2007)
326. S. Zulfiqar, I. Lieberwirth, M.I. Sarwar, Chem. Phys. **344**, 202 (2008)
327. S. Zulfiqar, M.I. Sarwar, High Perform. Polym. **21**, 3 (2009)
328. D.P.N. Vlasveld, P.P. Parlevliet, H.E.N. Bersee, S.J. Picken, Appl. Sci. Manuf. **36**, 1 (2005)
329. T. Liu, K.P. Lim, W.C. Tjiu, K.P. Pramoda, Z.-K. Chen, Polymer **44**, 3529 (2003)
330. D.M. Lincoln, R.A. Vaia, Z.-G. Wang, B.S. Hsiao, Polymer **42**, 1621 (2001)
331. A.B. Morgan, J.W. Gilman, J. Appl. Polym. Sci. **87**, 1329 (2003)
332. V.E. Yudin, G.M. Divoux, J.U. Otaigbe, V.M. Svetlichnyi, Polymer **46**, 10866 (2005)
333. Y.J. Li, H. Shimizu, Polymer **45**, 7381 (2004)

334. M. Mehrabzadeh, M.R. Kamal, Polym. Eng. Sci. **44**, 1152 (2004)
335. H. Zheng, Y. Zhang, Z. Peng, Y. Zhang, J. Appl. Polym. Sci. **92**, 638 (2004)
336. A. Usuki, A. Tukigase, M. Kato, Polymer **43**, 2185 (2002)
337. S. Sadhu, A.K. Bhowmick, J. Polym. Sci. B Polym. Phys. **42**, 1573 (2004)
338. S. Sadhu, A.K. Bhowmick, J. Polym. Sci. B Polym. Phys. **43**, 1854 (2005)
339. M.A. Kader, K. Kim, Y.S. Lee, C.J. Nah, Mater. Sci. **41**, 734 (2006)
340. A. Das, R. Jurk, K.W. Stöckelhuber, G. Heinrich, Macromol. Mater. Eng. **293**, 479 (2008)
341. A. Das, R. Jurk, K.W. Stöckelhuber, G. Heinrich, Expr. Polym. Lett. **1**, 717 (2007)
342. R.R. Tiwari, K.C. Khilar, U. Natarajan, Appl. Clay Sci. **38**, 203 (2008)
343. S. Siengchin, J. Karger-Kocsis, R. Thomann, Exp. Polym. Lett. **2**, 746 (2008)
344. J. Wang, K. Yang, N. Xu, J. Appl. Polym. Sci. **123**, 1293 (2012)
345. L. Celine, M. Fabien, A. Michael, D. Philippe, J. Nanosci. Nanotechnol. **9**, 2731 (2009)
346. G. He, Y. Chen, M.M. Zhang, D.X. Wang, K. Cong, H.B. Fan, F. Zhou, J. Polym. Mater. Sci. Eng. **27**, 150 (2011)
347. W.S. Mark, T.S. Keith, L.O. Duan, J. Inorg. Organomet. Polym. **18**, 364 (2008)
348. A.O. Maged, M. Vikas, M. Massimo, Macromolecules **36**, 9851 (2003)
349. K.G. Gatos, R. Thomann, J. Karger-Kocsis, Polym. Int. **53**, 1191 (2004)
350. A.B. Morgan, Mater. Matters. **2**, 20 (2007)
351. A.-R. Lee, H.-M. Park, H. Lim, T. Kang, X. Li, W.-J. Cho, C.-S. Ha, Polymer **43**, 2495 (2002)
352. H. Ishida, S. Campbell, J. Blackwell, Chem. Mater. **2**, 1260 (2000)
353. Y.H. Hyun, S.T. Lim, H.J. Choi, M.S. Jhon, Macromolecules **34**, 8084 (2001)
354. B.J. Ash, L.S. Schadler, R.W. Siegel, Mater. Lett. **55**, 83 (2002)
355. J.-C. Wang, Y.-H. Chen, R.-J. Chen, J. Polym. Sci. B Polym. Phys. **45**, 519 (2007)
356. B. Han, A.M. Cheng, G.D. Ji, S.S. Wu, J. Shen, J. Appl. Polym. Sci. **91**, 2536 (2004)
357. J.Y. Lee, M.J. Shim, S.W. Kim, Polym. Eng. Sci. **39**, 1993 (1999)
358. C. Zilg, R. Thomann, J. Finter, R. Mulhaupt, Macromol. Mater. Eng. **280/281**, 41 (2000)
359. H. Zhang, Z. Zhang, K. Friedrich, C. Eger, Acta Mater. **54**, 1833 (2006)
360. C. Karikal Chozhan, M. Alagar, P. Gnanasundaram, Polym. Polym. Comp. **16**, 283 (2008)
361. C. Karikal Chozhan, M. Alagar, R. Josephine Sharmila, P. Gnanasundaram, J. Polym. Res. **14**, 319 (2007)
362. S. Premkumar, C. Karikal Chozhan, M. Alagar, Polym. Eng. Sci. **49**, 747 (2009)
363. S.M. Lee, T.R. Hwang, Y.S. Song, J.W. Lee, Polym. Eng. Sci. **44**, 1170 (2004)
364. M. Quarmyne, W. Chen, Langmuir **19**, 2533 (2003)
365. H. Tu, L. Ye, Polym. Adv. Technol. **20**, 21 (2009)
366. I.A. Chmutin, P.N. Brevnov, G.R. Sabirova, O.D. Nazirova, N.G. Ryvkina, L.N. Novokshonova, Nanotekhnika **3**, 33 (2011)
367. K.S. Novoselov, A.K. Geim, S.V. Morozov, D. Jiang, Y. Zhang, S.V. Dubonos, I.V. Grigorieva, A.A. Firsov, Science **306**, 666 (2004)
368. M.S. Dresselhaus, G. Dresselhaus, Adv. Phys. **51**, 1 (2002)
369. A. Reina, X.T. Jia, J. Ho, D. Nezich, H.B. Son, V. Bulovic, M.S. Dresselhaus, J. Kong, Nano Lett. **9**, 30 (2009)
370. Y.B. Zhang, Y.W. Tan, H.L. Stormer, P. Kim, Nature **438**, 201 (2005)
371. K.S. Novoselov, A.K. Geim, S.V. Morozov, D. Jiang, Y. Zhang, M.I. Katsnelson, I.V. Grigorieva, S.V. Dubonos, A.A. Firsov, Nature **438**, 197 (2005)
372. D. Zhan, L. Sun, Z.H. Ni, L. Liu, X.F. Fan, Y. Wang, T. Yu, Y.M. Lam, W. Huang, Z.X. Shen, Adv. Funct. Mater. **20**, 3504 (2010)
373. I.I. Mazin, A.V. Balatsky, Philos. Mag. Lett. **90**, 731 (2010)
374. M. Nath, C.N.R. Rao, J. Am. Chem. Soc. **123**, 4841 (2001)
375. M. Lerner, C. Oriakhi, Layered metal chalcogenides, in *Handbook of Layered Materials*, ed. by S. Auerbach, K.A. Carrado, P. Dutta (Marcel Dekker, New York, 2004)
376. E. Benavente, M.A. Santa Ana, G. Gonzalez, Phys. Status Solidi B-Basic Res. **241**, 2444 (2004)

377. N. Mirabal, P. Aguirre, M.A. Santa Ana, E. Benavente, G. Gonzalez, Electrochim. Acta **48**, 2123 (2003)
378. M.A. Santa Ana, E. Benavente, P. Gómez-Romero, G. González, J. Mater. Chem. **16**, 307 (2006)
379. G. Alberti, M. Casciola, M. Pica, G. Di Cesare, Ann. N.Y. Acad. Sci. **984**, 208 (2003)
380. H.-L. Tsai, J.L. Schindler, C.R. Kannewurf, M.G. Kanatzidis, Chem. Mater. **9**, 875 (1997)
381. T. Hanaoka, T. Tago, K. Wakabayashi, Bull. Chem. Soc. Jpn. **74**, 1349 (2001)
382. J.-H. Zeng, J. Yang, Y. Zhu, Y.-F. Liu, Y.-T. Qian, H.-G. Zheng, Chem. Commun. **15**, 1332 (2001)
383. J.G. Winiarz, L. Zhang, J. Park, P.N. Prasad, J. Phys. Chem. B **106**, 967 (2002)
384. Z. Qiao, Y. Xie, M. Chen, J. Xu, Y. Zhu, Y. Qian, Chem. Phys. Lett. **321**, 504 (2000)
385. W. Guo, J.J. Li, Y.A. Wang, X. Peng, J. Am. Chem. Soc. **125**, 3901 (2003)
386. L. Li, E. Beniash, E.R. Zubarev, W. Xiang, B.M. Rabatic, G. Zhang, S.I. Stupp, Na. Mater. **2**, 689 (2003)
387. S.C. Tjong, Y.Z. Meng, Polymer **40**, 1109 (1999)
388. K.G.S. Ranmohotti, E. Josepha, J. Choi, J. Zhang, J.B. Wiley, Adv. Mater. **23**, 442 (2011)
389. W.H. He, J. Huang, X.F. Sun, A.J. Frontier, J. Am. Chem. Soc. **130**, 300 (2008)
390. K.A. Tarasov, V.P. Isupov, B.B. Bokhonov, Y.A. Gaponov, B.P. Tolochko, M.R. Sharafutdinov, S.S. Shatskaya, J. Mater. Synth. Proc. **8**, 21 (2000)
391. N.E. Kotelnikova, G. Vegner, T. Paakari, R. Sepimaa, V.N. Demidov, A.S. Serebryakov, A.V. Shchukarev, A.V. Gribanov, Z. Obshch, Zh. Obshch. Khim. **73**, 447 (2003)
392. V. Uskokovic, D.P. Uskokovic, J. Biomed, Mater. Res. B Appl. Biomater. **96B**, 152 (2011)
393. J.A. Shellman, H.R. Reese, Biopolymers **39**, 161 (1996)
394. R.D. Snyder, D.E. Ewing, L.B. Hendry, Environ. Mol. Mutagen. **44**, 163 (2004)
395. S. Letaïef, M.A. Martín-Luengo, P. Aranda, E. Ruiz-Hitzky, Adv. Funct. Mater. **16**, 401 (2006)
396. H. Haiyan, P. Mingwang, L. Xiucuo, S. Xudong, Z. Liucheng, Polym. Int. **53**, 225 (2004)

Chapter 6
Thermolysis of Metallopolymers and Their Precursors as a Way for Synthesis of Nanocomposites

In this book we analyse all the types of metallopolymer used as precursors for obtaining nanocomposites via thermolysis. This method is an easy, reproducible and well-controlled route to prepare the nanocomposite materials. The research into the thermal decomposition of organic (metallo)polymers has long prehistory. Back during the Second World War, works on the design of polymers stable up to 870 K were started by the order of the USA air forces (cited from Ref. [1]). The attention was focused on the coordination polymers because the stability of organic compounds was known to increase upon their coordination to metal ions. Thus N-hydroxyethylethylenediamine rapidly decomposes under the action of hot nitric acid, while its complex with CoII withstands long-term (several hours) refluxing in concentrated nitric acid. Similarly, β-diketimine, which decomposes at moderate temperatures, starts to slowly decompose only at 620 K when complexed with copper. Although at that time the search for thermally stable coordination polymers has not met with success (in some cases, very short polymer chains were formed; in other cases, the organic groups connecting the metals were destroyed, etc.), they laid the foundation for the studies that recently resulted in the production of metallopolymer nanocomposites. Currently, various methods for the formation of the 'core -shell' type composites have been developed. These include, for example, sonochemical synthesis techniques [2], electrochemical methods [3], laser-induced reactions in the melt [4] and catalytic growth [5, 6], vapour-phase transport [7], IR pyrolysis [8, 9], polymer sol-gel synthesis [10], intercalation, 'wet' chemical methods [11, 12] etc. [13–15]. Controlled thermolysis was recognized as the simplest and most efficient, economical and versatile method. This process is usually carried out at temperatures from 500 to 1,300 K. It is environmentally clean and serves as a method for polymer waste disposal. An important application of the thermolysis is the preparation of carbon nanomaterials (for example, single- and multiwalled nanotubes, nanorods, nanospheres, etc.). Depending on the nature of the polymer and conditions of thermolysis, various carbon structures can be formed, including graphite- and diamond-like structures.

A.D. Pomogailo, G.I. Dzhardimalieva, *Nanostructured Materials Preparation via Condensation Ways*, DOI 10.1007/978-90-481-2567-8_6

The thermal stability of a polymer usually determines its upper operating temperature limit in the environment. It is related to kinetic parameters: the initial temperature and the rate of destruction [16].The knowledge of these parameters is necessary for the application and storage of polymers [17]. Thermolysis of polymers is a complex process involving a number of chemical reactions: destruction, cross-linking, transformations of functional groups and intramolecular rearrangements. For stabilizing a polymer and decreasing its reactivity, chemical or physical insertion of metal-containing particles is sometimes used. This affords new materials and composite systems. The simultaneous synthesis of finely dispersed particles (at their fairly high relative concentration) and the stabilizing polymer shell, together with process simplicity and controllability are the advantages of this approach. Although these studies have been carried out since the mid-twentieth century [18], let us note here the most interesting recent works [19–21] and references their]. The routes of thermolysis of polymers are versatile [22–24].

On the stage of thermolysis the different additives can be incorporated into polymers including metal-containing one which both influence the mechanism of thermolysis and lead to the quite number of interesting products, for example, polymer derivatives of ceramics. Thus, by thermolysis of polyvinyl alcohol (PVA) the polymer with conjugated bonds similar to polyacetylene is obtained. When the additives such as lead and bismuth formates, silver oxalate, potassium iodide or sodium hydroxide are introduced to the PVA a dopped conjugated polymer is formed [25]. The metal nanoparticles formed promote electron transfer between the polymer chains and thus polymer conductivity is enhanced. The additives of TiO_2, ZnO or γ-Al_2O_3 considerably influence the thermolysis at 340–390 K of acrylic and fluorine-containing polymers and copolymers [26].

For stabilizing a polymer and decreasing its reactivity, chemical or physical insertion of metal-containing particles is sometimes used. This affords new materials and composite systems. The simultaneous synthesis of finely dispersed particles (at their fairly high relative concentration) and the stabilizing polymer shell, together with process simplicity and controllability are the advantages of this approach.

Numerous methods for the preparation of metal-containing polymeric composites are known. The key method is thermolysis of metal compounds that are readily decomposed both in the pure state and in the polymer matrix (carbonyls, nitrosyls, thiolates, saturated carboxylic acid salts, organometallic and complex compounds) where destruction of the organic part affords the polymeric shell. Other methods are also widely used, in particular, introduction of metal-containing powders with various degrees of dispersion into the ready polymeric matrices and mixing them with oligomers; heteroadagulation or spraying of nanoparticles on polymer powders, films and fibres; evaporation of the solvent from a solution containing the polymer and the metal-containing precursor or hardening of their melts in liquid media; extraction substitution and soon. However, all of these methods have drawbacks. The main drawbacks include the statistical pattern of distribution and spatial inhomogeneity of the metal-containing particles in the matrix and the lack of control over the reaction at the dispersed phase – polymeric medium level. However, in all cases, the polymeric shell which separates the metal particles from one another and from the environment functions as a stabilizing agent, which may be represented by a natural (proteins, polysaccharides) or synthetic material obtained by polymerization or polycondensation. Of particular interest are the macromolecular metal complexes in which the metal is chemically bound to a polymer containing active functional groups (carboxylate, nitrile, amino, amido, imino and hydroxy groups and so on) but studies of the thermal transformations of such systems are only at the early stage of development.

Of particular interest are the macromolecular metal complexes in which the metal is chemically bound to a polymer containing active functional groups (carboxylate, nitrile, amino, amido, imino and hydroxy groups and so on) but studies of the thermal transformations of such systems are only at the early stage of development.

Relatively recently, a new approach has been developed [27] based on the use of metal-containing monomers as precursors. In this case, polymerization and synthesis of metal-containing nano-sized particles occur simultaneously during the thermal transformation. This field of research developed most intensely in recent years was called 'polymer-mediated synthesis' in our study [28] or it is called 'combined solid-state polymerization – thermolysis.' For the systems considered here, 'dry' methods for the composite formation are mainly used.

Note that quantitative studies of the thermolysis giving matrix-stabilized nanoparticles are few. These works are mainly qualitative and aim at describing the averaged characteristics and size distributions of the particles.

6.1 Kinetic Approaches to the Investigation of Thermolysis of (Metallo)polymer Systems

The kinetic approaches to the studies of thermolysis of (metallo)polymer systems provide quantitative knowledge about the course of transformations in each step for relatively slow processes at temperatures below 1,370 K. In recent years, thermogravimetric methods such as differential thermogravimetry (DTG) and differential thermal analysis (DTA), differential scanning calorimetry (DSC), thermodilatometry and thermomechanical analysis, dielectrical analysis, micro- and nanosized thermal analysis (including atomic force microscopy and scanning thermal microscopy) have been widely used, in particular, to study the thermolysis of (metallo)polymers. Recent achievements in the field of thermal analyses are summarized in the monograph [21].

Any thermal process is accompanied to a certain extent by the change in the inner heat content of the system, i.e., heat absorption or evolution. During the low-temperature thermolysis of most condensed polymer systems, the transformation rates are relatively low and the temperatures are low at rather intense heat transfer. Irrespective of the heat effect, these processes occur with time in the spatially isothermal mode [29] i.e., the temperature of the surface of a condensed sample (T_a) coincides with the ambient temperature (T_s). In this case, the isothermal methods of classical chemical kinetics are applicable and the process conditions are formulated in view of scaling requirements, in particular, the ratio of the characteristic times of heating (τ_h) and transformation (t_r).

The kinetics of thermal transformation of metallopolymer systems is generally described by the macrokinetic equation:

$$W(T,\eta) = k(T) \times \varphi(\eta), \tag{6.1}$$

where $W(T, \eta)$ is the reaction rate (s^{-1}), T is the temperature (K), $k(T)$ is the temperature-dependent rate constant determined by the reaction mechanism (s^{-1})[1], $\varphi(\eta)$ is a kinetic function. The last-mentioned value depends on the reaction mechanism and reflects the relationship between the concentrations of unreacted and reacted compound.

$$\varphi(\eta) = (1 - \eta)^n \tag{6.2}$$

where n is the order of the reaction, η is the degree of conversion ranging from 0 to 1.

Then the characteristic time of the reaction

$$\tau_r \approx W(T,\eta)^{-1}.$$

Since the transformations occur in the condensed phase, it is expedient to use as τ_h the characteristic time of sample warming-up traditional for macrokinetics: [30]

$$\tau_h \approx d^2/a,$$

where d is the sample diameter, a is the sample temperature conductivity. In this case, for obtaining reliable kinetic results, the following condition should be met: $\tau_h << \tau_p$.

As a rule, thermolysis of polymers is a multistage process. Its mechanism is studied using various types of kinetic models. In a pseudo-single-component model, the material is considered to be ordered and single-component, and the mass loss is described by studying the kinetics of the solid- or gas-phase reaction over the whole temperature range [31]. However, this model does not describe the possible changes in the kinetics and mechanism upon the change in the temperature of thermolysis.

The pseudo-multicomponent summary model analyzes the material as consisting of a number of pseudo-components each of them being a separate real component or mixture of several components and the polymer thermolysis is

[1] The rate constant for the reaction can be expressed by the Arrhenius dependence $k(T) = A\exp(-E_a/RT)$, where $A(s^{-1})$ is the pre-exponential factor (s^{-1}), $E_{a,\text{eff}}$ is the effective activation energy (kJ mol^{-1}), R is the universal gas constant.

modelled by superposition of decompositions of these components. However, this approach considerably complicates the kinetic studies of thermal destruction. For example, in the case of thermolysis of polyaryl sulfides, the weight loss takes place in two stages [32]. The pseudo-two-component first-order stage-separated model considers the material as being decomposed in two temperature regions, which are described by different kinetic models. The temperature corresponding to the minimum in the derivatogram is the separation point between the two stages of weight loss. This method is considerably complicated when the two peaks are not completely separated but overlap.

The non-isothermal method is widely used to describe the kinetics and the mechanism of reactions in the condensed phase. Since the rate of thermolysis depends on the temperature, the change in the sample weight can be presented as the dependence:

$$\frac{d\eta}{dt} = k(T)\varphi(\eta) = A\exp\left(-\frac{E_a}{RT}\right)\varphi(\eta) \tag{6.3}$$

When the sample temperature is controlled by a constant heating rate β ($\beta = dT/dt$), this equation has the form:

$$\frac{d\eta}{dt} = \frac{A}{\beta}\exp\left(-\frac{E_a}{RT}\right)(1-\eta)^n \tag{6.4}$$

In the integral form, expression (6.4) looks as follows:

$$g(\eta) = \int_0^{\eta} \frac{d\eta}{(1-\eta)^n} = \frac{A}{\beta}\int_0^{T} \exp\left(-\frac{E}{RT}\right)dT = \frac{AE}{\beta R}P(u) \tag{6.5}$$

and

$$\ln\left[\frac{g(\eta)}{T^2}\right] = \ln\frac{AR}{\beta E_a} - \frac{E_a}{RT}. \tag{6.6}$$

For $n = 1$

$$\ln\left[-\ln\frac{1-\eta}{T^2}\right] = \ln\frac{AR}{\beta E_a} - \frac{E_a}{RT} \tag{6.7}$$

For $n \neq 1$ in the most cases expression (6.7) is solved numerically.

Equations (6.5) and (6.6) are the fundamental expressions for calculating the kinetic parameters from thermogravimetric (TG) data. These are analyzed using the Kissinger, Flynn - Wall -Ozawa (FWO), Coat – Redfern, van Krevelen (vK), and Avramy-Erofeev methods in the differential and integral forms (see, for example [33–35]).

The dynamic parameter of thermal decomposition under linear heating is determined from the modified Kissinger equation at T_{max}, i.e., the temperature corresponding to the maximum conversion in the TG curves. The straight-line plot $\ln(\beta/T^2_{max}) - 1{,}000/T_{max}$ is used to calculate the activation energy and the pre-exponential factor. Equations (6.5) and (6.6) describe the thermolysis at 473 K of polymer networks and star-like polymers based on ethylene glycol dimethacrylate containing degradable acryl groups to give low-molecular-mass products [19]. Also the kinetics of nonisothermal decomposition of polypropylene filled with both mineral additives such as calcium carbonate, dioxide of silicone, zinc oxide, carbon black and fibers [36, 37] and organic ones as rice husk [38, 39] is well described by these equations. The degradation kinetics of the LDPE stabilized by cobalt stearate was investigated with the Flynn-Wall-Ozawa method [40, 41]. Several kinetic methods were used to investigate the thermal decomposition of chromium polyacrylate complexes [42, 43]. Under comparable conditions, the initial temperatures for the polymer and the metal complex were 628 and 591 K, respectively, and the final temperatures were 806 and 813 K; the temperature peaks in the derivatogram were observed at 713 and 690 K, the weight losses (%) were 53.40 and 65.56, and the activation energies of decomposition (at $0.05 < Z < 0.90$) were 187.8 and 122.5 kJ/mol. The orders of reaction differed from unity for all of the decomposition models. These parameters depend to some extent on the storage time of the aged (for 1–5 weeks) macrocomplex.

At relatively low temperatures (for instance, the decomposition of metal carbonyls such as $Cr(CO)_6$, $Fe(CO)_5$, $Co_2(CO)_8$ in toluene at 363 K [44]) the study of kinetics can be carried out with spectral methods, in particular, by FTIR analysis on the decreasing of the band intensity at 1,977 cm^{-1} (at 1,996 and 2,019 cm^{-1} for iron and cobalt carbonyls, respectively).

Nonisothermal kinetic modeling of thermolysis of natural fibers commonly used in industry and their composites [45] showed the good agreement between the calculated kinetic parameters and experimental data in the rather wide range of temperatures.

Formal kinetics method [46, 47] in combination with the Netzsch-Thermokinetics programme allows one to describe the behavior of the system analysed; the data obtained are most informative in the study of complex multistage processes as thermolysis (Table 6.1).

Table 6.1 The kinetic models and parameters for different stages of thermolysis

The kinetic model	$E_a/R \cdot 10^3$ K	log A	*N*
Prout Tompkins	14.17 ± 0.18	9.12 ± 0.14	1.37 ± 0.02
*n*th-order reaction	16.71 ± 0.16	9.85 ± 0.12	1.43 ± 0.02
Prout Tompkins	15.45 ± 0.12	8.18 ± 0.09	1.28 ± 0.04

E_a/R is the temperature coefficient of the reaction rate, A is the pre-exponential factor, *n* is the effective reaction order

Thermal treatment considerably changes the structure and properties of a polymeric material. This is due, first of all, to the physical and chemical processes that accompany the polymer destruction. The physical destruction processes (crystallization, recrystallization, etc.) are usually reversible and do not lead to cleavage or cross-linking of polymer chains. The chemical processes are irreversible and are accompanied by chemical bond cleavage, cross-linking of macromolecules, changes in the chemical structure or decrease/increase in the molecular mass [48, 49].

For the design of metallopolymer compositions, the thermal stability of polymer matrices is important. It can be estimated from the strength of the bond between atoms. According to the Boltzmann law, the bond strength is determined by the expression $S = E_b/RT^2$, where E_b is the bond energy (in kJ/mol). The dependence of S on $\gamma = (E_{X-Y} - E_{C-C})/E_{C-C}$, i.e., the change in the bond energy between the atoms X and Y (E_{X-Y}) relative to the average carbon–carbon bond energy (E_{C-C}) in organic polymer, shows that organic homo- and heterochain polymers are least thermally stable. Inorganic matrices (clays, hexagonal boron nitride, silicate glasses, zeolites, etc.) having no carbon backbone and graphite are distinguished by enhanced thermal stability and mechanical strength. The bond energies between most typical atoms of the organic and inorganic matrices are summarized in Table 6.2. Polymeric compounds have especially strong Si–O, B–N and B–O bonds (note that the probability of chemical bond cleavage depends also on temperature).

A different kinetic picture is observed at relatively high temperature. In the case of high-temperature thermolysis, the time τ_r related to the temperature T_a decreases and τ_h does not depend on T_a and remains constant. Then in the case of exothermic reaction at $\tau_h \approx \tau_r$, thermal explosion is possible, and at $\tau_h \gg \tau_r$, ignition occurs well before the studied compound has been warmed-up and has time to react in the area adjoining the

Table 6.2 Bond energy between the atoms in the polymer matrices [50, 51]

Bond	Bond energy, kJ/mol	Bond	Bond energy, kJ/mol
Carbon-containing matrices			
C–C	336	C–B	372
C–O	327	Al–N	363
C–N	277	Si–O	364
C–Si	241	B–O	500
C–Al	258	B–N	387
Inorganic polymer matrices			
C–C	715[a]	Ti–O	933
Al–O	1,027.6	B–O	1,047
Al–N	1,121	B–N	1,289
Si–O	586		

[a]For crystalline graphite

hot surface [29]. In the case of fast endothermic reactions,[2] no-self ignition takes place upon increase in T_a, and the temperature profile for the studied compound is not spatially isothermal. It depends on both the reaction coordinate and time, and the temperature assignment of the detected transformation becomes uncertain.

In particular, metal-polytetrafluoroethylene (M/PTFE) mixtures represent condensed high-energetic materials in which vigorous exothermic reactions take place according to the idealized scheme:

$$-(CF_2 - CF_2)_n - +4n\mathrm{M} \rightarrow\downarrow 2n\mathrm{C} + 4n\,\mathrm{MF}$$

The PTFE is chemically and thermally stable due to the high energy of the C–F bond (481 kJ/mol). The composites can be used as oxidation agents because in many cases the strength of metal-fluorine bond is even more than that for C–F bond. The rate of thermolysis depends on the size of the reacting particle and concentration of reduction agent in the mixture [52–55]. Such composites are widely used as igniters of solid propellants and pyrotechnics [55, 56]. Self-propagation heat waves, associated with the oxidation of metals at heating rates at 10^6 K/c can yield new materials resulting from these ultra-fast reactions. High temperatures of thermolysis (theoretically exceeding 3,000 K for Mg/PTFE [57]) and specific conditions for fast product growth favour the formation of nanostructures and polyynes $(–C{\equiv}C–)_n$.

Calorimetric studies (Fig. 6.1) showed that thermolysis proceeds in a self-sustaining high-temperature regime [58]. The heats of the reaction range from very low (less than 1,000 kJ/kg) for Zn/PTFE composites enriched with a metal to very high (more than 8,000 kJ/kg) for the reactants containing Al_3Mg_4 at a concentration close to the postulated stoichiometry. Nevertheless, these numbers account for only ca. 70 % of the theoretical exothermal effect calculated assuming the chemical equilibrium of the thermolysis.

Hence, the classical isothermal methods are inapplicable to the kinetics of fast high-temperature processes. These processes are to be studied using the approaches of nonisothermal chemical kinetics the foundations of which were laid in the fundamental works: N. N. Semenov, D. A. Frank-Kamenetsky, Ya. B. Zeldovich, A. G. Merzhanov and were considered in detail in the book [29].

6.2 General Characterization of the Principal Methods of Thermolysis of (Metallo)polymers and Their Precursors

Thermal transformation of (metallo)polymers and their precursors are studied using external (or internal) heating under isothermal (or non-isothermal) conditions in a closed (or open) system. Depending on the task, the degree of conversion of the

[2] An example is forced high-temperature destruction of the polymer binders of the thermal protection covers of space vehicles, which accompanies the combustion of solid rocket propellant components.

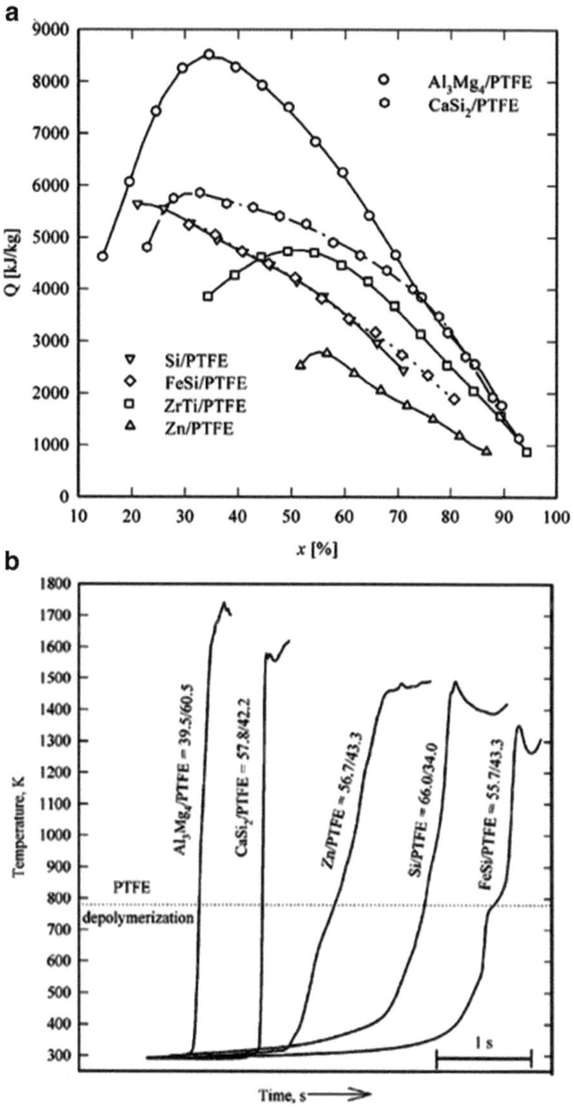

Fig. 6.1 Heats of reaction for M/PTFE mixtures as a function of metal (M) content (**a**) and change of temperature during ultra-fast burning reaction of the PTFE mixtures (**b**)

reactant is determined by a variety of methods. They are classified into gravimetric (thermogravimetric) and volumetric methods and determine the selection of the experimental gear.

Development of the instrumentation for measuring the kinetics of thermolysis has started since the first quarter of the twentieth century; the approaches to the design of instruments and protocols were analyzed in detail about 50 years ago in a monograph [59], which is still not out of date (see also the recently appeared review [21]). In recent years, integrated (synchronous) devices that combine several methods for monitoring the reactant conversion with automation and utilize the potential of modern computers within one instrument have started to appear. The specific features of the principal methods used to measure the kinetics of formation of nanoparticles are briefly analyzed below.

6.2.1 Thermogravimetric Methods

Thermogravimetric methods measure the relative change in the substance weight during the transformation as a function of temperature. Most often, the sample experiences programmed temperature change with a linear dependence on time. In some cases, thermogravimetry is combined with tensimetry and gas evolution is simultaneously measured. As a rule, these measurements are carried out under isothermal conditions in a dynamic vacuum and with recording in both continuous and discrete modes. For example, a study of the kinetics of consumption of volatile fractions formed in the pyrolysis of non-fractionated 230 kDa polystyrene was carried out using spring balance [59]. The resulting kinetic curves were S-shaped with a maximum transformation rate at $\eta = 0.35 - 0.55$. A modern thermobalance is suitable for continuous (automated) recording of the temporal variation of the sample weight (Δm), and with known law of time variation of temperature, these dependences can be displayed via a computer. The TG methods are mainly based on traditional non-isothermal variants of linear heating.

In recent years, thermogravimetric methods such as differential thermogravimetry (DTG) and differential thermal analysis (DTA) have been widely used, in particular, to study the thermolysis of (metallo)polymers. In the former case, the first-order derivative of thermogravimetric curve is found either with respect to time (dm/dt) or with respect to temperature (dm/dT). In the latter case, the temperature difference between the substance and the reference is measured as a function of temperature, the substance and the reference being programmed heated. The development of DTA toward increase in the accuracy of quantitative determination of process heats led to the advent of the new investigation technique called differential scanning calorimetry (DSC) [60].

The combined use of thermogravimetric analysis (TGA) and DTA is illustrated by the data on thermooxidative destruction of polytetrafluoroethylene (PTFE) [61]. By DTA measurements (Fig. 6.2), it was shown for the first time that the transformation of PTFE in air proceeds in three stages with a complex heat evolution pattern in the temperature range of 735–870 K; the last two stages with characteristic temperatures of 810–820 and 850–860 K are accompanied by the loss of 94–95 % of the sample weight. The exothermic character of each stage attests to partial oxidation of the products formed upon PTFE destruction with air oxygen. The average apparent activation energy of the main period (735–850 K) calculated from the data of a non-isothermal experiment using TG curves is 295 kJ/mol, being independent of the heating rate or the particle size of the polymer. Meanwhile, at 735–870 K, the results of thermooxidative destruction of PTFE in the isothermal

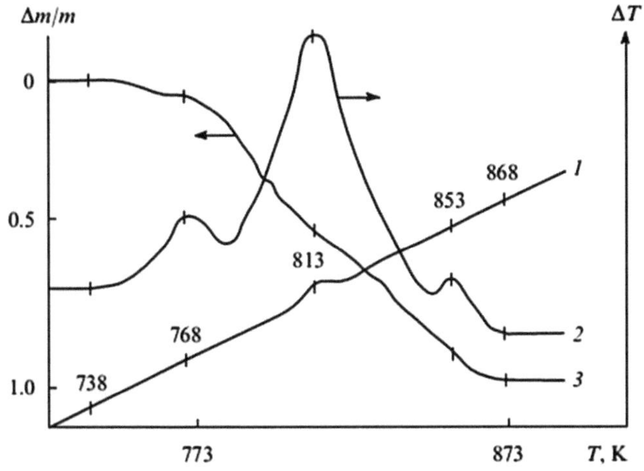

Fig. 6.2 Derivatogram of polytetrafluoroethylene [61]. (*1*) Temperature change, (*2*) DTA data, (*3*) TGA data

mode up to a degree of conversion of 0.9 are described satisfactorily by the 2/3-order reaction kinetics (the 'compressing' sphere equation [62]). In the temperature range of 735–870 K, the reaction rate increases 20-fold at $\eta = 0.35$ and sixfold at $\eta = 0.7$. The activation energies are 24 kJ/mol in the temperature range of 735–770 K and 102 kJ/mol in the temperature range of 770–870 K. These values are considerably (more than 2.5 times) lower than the values obtained in the dynamic mode, which raises doubts on the correctness of determination of the kinetic data from PTFE destruction experiments performed in the non-isothermal mode.

The vast majority of the procedures used to determine the kinetic parameters from the DTA- TGA experiments carried out in the linear heating mode are based on the solution of the mathematical problem with allowance for only one kinetic equation. The mathematical simplicity of this scheme accounts for its popularity. However, during the experiment, the system is self-heated, which may generally surpass the quasi-stationary temperature difference by a large factor or even by orders of magnitude [63]. In the case of programmed heating rate, the kinetic characteristics of the process include the temperature of the decomposition onset, the decomposition stages, and temperature ranges of stability of the intermediate compounds [60]. Ignoring the specific features of exo- and endothermic reactions and the effect of heat-exchange parameters (heat transfer coefficient, heating rate, reaction heat, sample weight, etc.) affects the legitimacy of using a particular method for kinetic data acquisition.

For reducing the distortions introduced by the noncontrolled pressure and by the permanently increasing temperature, a new TG experiment was developed and served as the basis for the quasi-isothermal quasi-isobaric thermogravimetry technique [60, 63]. This implies fine adjustment of heating in such a way that the rate of variation of either weight or pressure is maintained constant.

Table 6.3 Characteristics of thermoanalytical methods [64][a]

Method	Measured characteristics	Application
Common methods		
Thermogravimetry	Weight	Phase destruction, dehydration, oxidation
DTA, DSC	Temperature difference between the reference and the tested compound	Phase transitions (temperature and heat) and chemical reaction (heat capacity)
Thermomechanical analysis (TMA)	Deformation	Mechanical changes, deformations
Thermoptometry	Optical properties	Phase changes, surface reactions, colour changes
Dielectric thermal analysis	Dielectric constant	Phase changes, changes in polymers
Specific methods		
Synchronous thermal analysis	Combines two or more methods of sample investigation	
Thermal analysis with a specified process rate	The rate of the change of sample properties is maintained constant	

[a]Note: In some cases, these methods are used in a reducing atmosphere, for example, in a 92:8 (by volume) N_2:H_2 mixture

It is noteworthy that until recently all determinations of the kinetic parameters have been performed by 'discrete' type procedures using a small fraction of the 'continuous' information embedded in the set of DTA–TGA curves. For strict approximation to the real process, the kinetic equation should be considered together with the heat balance equation.

The best known thermoanalytical methods and their applications are indicated in Table 6.3.

6.2.2 *Linear Pyrolysis Method*

Linear pyrolysis (LP) is a steady-state one-dimensional propagation of the thermal reaction front of a condensed substance with heat supply from an external source [29]. Depending on the relationship between the effective (E_{eff}) and true (E_{tr}) activation energies, LP can occur in either of two regimes: kinetic (the macrokinetics of pyrolysis coincides with the true kinetics of decomposition, $E_{eff} \approx E_{tr}$) or internal diffusion regime (by heat, $E_{eff} \approx E_{tr}/2$). The development of this method was caused by the need for investigating the kinetics of fast high-temperature processes in condensed media where transition from one temperature region to another is accompanied by a change in the rate- limiting step of the reaction. In addition, the knowledge of the regularities of forced high-temperature decomposition of polymers

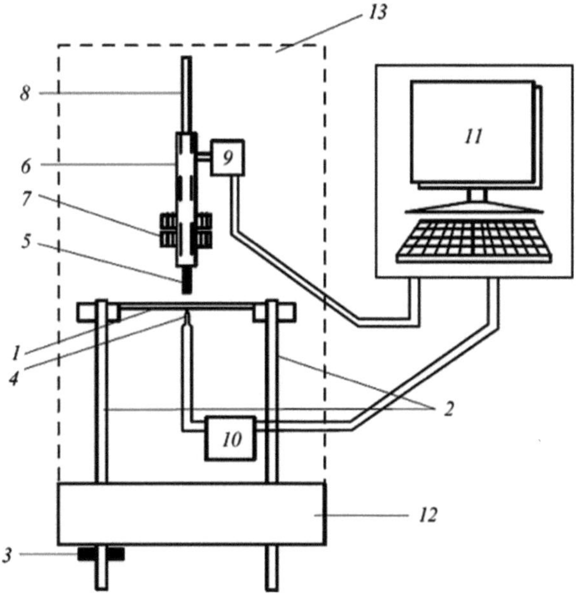

Fig. 6.3 Schematic diagram for linear pyrolysis. (*1*) Heater plate, (*2*) conducting planes, (*3*) current lead, (*4*) thermocouple, (*5*) sample, (*6*) sliding piston, (*7*) ring loads, (*8*) guide bar, (*9*) photodiode block for process rate recording, (*10*) control and heater plate temperature recording block, (*11*) computer block with a monitor, (*12*) massive vinyl plastic plate, (*13*) cover

is important for the development of modern space rocket facilities. They are necessary for solving problems related to combustion and decomposition of polymers, energetic compounds, ablation (evaporation) of the thermal protection cover of spacecrafts, etc. [29].

The principle of action of the devices used for LP experiments is presented in Fig. 6.3. A sample of the tested compound is pressed upon the surface of the heating plate at constant temperature. During the experiment, the temperature of the heater surface (T_0) and the velocity of sample migration (U) equal to the linear decomposition rate are recorded. When the rate of the linear pyrolysis related to the experiment parameters is below some limiting value (which is true for most of the real conditions), the temperature of the hot sample surface can be considered to be T_0 without considerable error. For a condensed substance with low thermophysical characteristics, LP is fairly fast.

Under conditions of low heat exchange with the environment, its rate is determined (for a zero-order reaction) by the Merzhanov relation [29]:

$$U = \sqrt{\frac{ak_0RT_S^2\exp(-E/RT_s)}{E[(T_s - T_0) \pm Q/2c]}} \tag{6.8}$$

where a and c are the temperature conductivity and the heat capacity of the compound, respectively, k_0 is the pre-exponential factor in the Arrhenius equation, T_s is the temperature of the environment and of the opposite surface of the sample, Q is the reaction heat.

Table 6.4 Kinetic parameters of high- and low-temperature destruction of PMMA from linear pyrolysis and isothermal method [59]

Polymer	T, K	E, kJ/mol	$\lg k_0$ [k_0, c^{-1}]
Linear pyrolysis			
PMMA	$723 \div 793$	179.7 ± 12.5	13.3 ± 1.0
	>803	$\geq 292.6 \div 364.4$	$\geq 22 \div 23$
PMMA + 2 % TEGM	$763 \div 863$	179.7 ± 12.5	12.5 ± 1.0
	>873	$\geq 292.6 \div 364.4$	$\geq 21 \div 22$
PMMA + 10 % TEGM	$763 \div 863$	179.7 ± 12.5	12.5 ± 1.0
	>873	$\geq 292.6 \div 364.4$	$\geq 21 \div 22$
Isothermal method			
Linear PMMA	$493 \div 523$	129.6	9.5
	$499 \div 529$	125.4	8.6
	$563 \div 613$	177.6	12.5
	$653 \div 703$	179.7	13.0
PMMA + 2 % TEGM	$583 \div 633$	181.8	12.1
PMMA + 10 % TEGM	$583 \div 633$	182.1	12.3

We emphasize that the data on low-temperature thermal decomposition kinetics of polymers are hardly applicable to the high-temperature pyrolysis. For example, decomposition of poly(methyl methacrylate) (PMMA) at temperatures ≤ 620 K follows first-order kinetics (for degree of conversion $\eta > 0.2$) with the activation energy $E_a \approx 167$ kJ/mol [59]. Then for LP rates from 10^{-3} to 10^{-2} m s^{-1}, Eq. (6.8) gives the temperature of the polymer surface ranging from 1,070 to 1,170 K. However, the calculated values from the results of thermocouple measurements far exceed the real values. These examples are quite numerous. Table 6.4 gives the results of measurements of the rate constants for high-temperature and low-temperature destruction of PMMA with added triethylene glycol methacrylate (TEGM). Their comparison demonstrates that extrapolation of the data obtained for PMMA transformation in the low-temperature region using usual kinetic methods to the high-temperature region (for example, to the polymer combustion region) should be done with caution.

6.2.3 *Volumetric Methods*

Volumetric methods are used to detect small amounts of a substance, i.e., to study the process at early stages of transformation. The time dependence of the volume of the evolved gas [$V(t)$] can be measured either discretely or continuously. The key drawback of pressure gauge units is the possibility of direct contact of gas evolution products, which are often chemically corrosive, with the corroding metallic parts and mercury. An obvious advantage of these setups is the use of glass membrane type pressure gauges devoid of metallic parts. They have short response times and

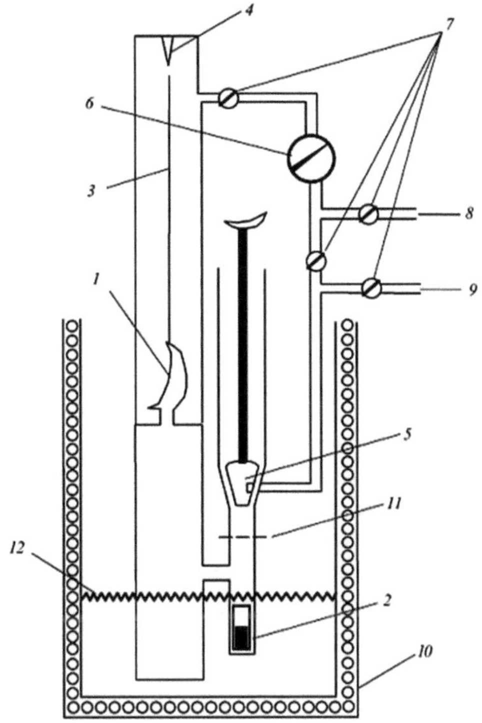

Fig. 6.4 Schematic diagram of the reaction vessel with a Bourdon pressure gauge for kinetic measurements. (*1*) Quartz crescent-shaped membrane, (*2*) tested compound, (*3*) rod, (*4*) zero indicator (zero reader), (*5*) vacuum valve, (*6*) recording mercury pressure gauge, (*7*) valves, (*8*) compensating gas inlet, (*9*) pumping out, (*10*) high-temperature thermostat, (*11*) vacuum sealing-off position, (*12*) position of the reaction vessel relative to the thermostat in the non-isothermal scheme

thus they are suitable for studying the rates of fast processes in a close reactor space by means of a Bourdon membrane-type pressure gauge (Fig. 6.4). A deformable highly sensitive membrane 1 separating the reaction space from the compensating space of the reactor serves as the pressure transducer. The gas pressure is determined either from the calibration curve which relates the membrane deformation value to the vapor pressure or by the compensation method (as a zero reader in the compensating circuit). Gas evolution during the transformation takes place in a self-generated atmosphere (SGA). Recently this method has been called RAPET (Reaction under Autogenic Pressure at Elevated Temperature) (see, for example, Ref. [65]). The reaction volume occurs under static isothermal conditions and is fully thermostated. The pressure created in the reaction chamber deforms membrane 1 and this induces bending of rod 3 welded to it. If a compensating gas is fed to the compensating chamber via valve 8 under pressure equal to the pressure of the gases evolved from the sample, rod 3 returns to the starting position, which is detected by pointer 4.

This reactor manufactured of quartz glass having a sealed-off reaction chamber is suitable for studying the kinetics of thermolysis of metallopolymers at fairly high temperatures (1,270–1,470 K) under fully isothermal conditions and for analyzing and interrupting gas evolution at any transformation stage. The number of moles of the evolved heated gas (n_{exp}) reduced to the number of moles (n_{Σ}) evolved at room temperature (T_r) is found from the equation

$$n_{exp}/n_{\Sigma} = (V_{exp}/V_{\Sigma})(T_r/T_{exp}).$$

Usually the reactor is designed in such a way that the ratio of the heated volume (V_{exp}) to the total reactor volume (V_{Σ}) is 1–5 %.

6.2.4 Thermolysis Induced by High-Energy Radiation

The thermal methods considered in the previous Section are based on the external heating of the tested sample. High-frequency (HF) electric fields can be used as an alternative source of heating [66–68]. On the one hand, polar groups and fragments of polymer molecule (dielectric material) placed in a variable electric field orient themselves following the change in the field polarity. On the other hand, thermal motion and non-polar groups and molecules prevent this orientation. The energy spent for overcoming the material disorientation is dissipated in the material and heats it. The intensity of heating increases with an increase in the oscillation frequency and electric field strength.

The key advantage of the HF heating is that heating occurs throughout the whole sample bulk without a contact. High-frequency treatment at 100–300 MHz (warming-up temperature may reach 510 K) of thermosetting oligomers containing finely powdered conductive fillers (iron, aluminium or copper powders, metal salts, complex organometallic compounds or soot) with particle size of less than 500 μm in an amount of up to 85 % of the weight produces homogeneous composite materials. They can be used for injection moulding, production of adhesives and electrically conducting articles for motor-vehicle construction, aircraft industry, microelectronics.

When amorphous titanium, zirconium and hafnium hydroxides are heated in a muffle microwave kiln or exposed to a 2,450 MHz electromagnetic field with a 700 W power, the warming-up temperature reaches 495–770 K [69]. The obtained metal oxide powders have well crystallized structure, highly developed surface (>500 m^2/g) and nanometre-range particle size.

A promising approach is the thermal treatment of metallopolymer materials under the action of lasers or accelerated electrons [70]. Laser ablation is also applicable. Carbon nanotubes were produced from polyacrylonitrile (PAN), PTFE, acrylonitrile and methyl methacrylate copolymers, etc. by a number of methods (arc discharge, laser or magnetron radiation, plasma treatment and so on) [71, 72]. In some cases, the polymer incorporates a metal- containing component, which affects considerably the nature of carbon structure: this gives carbon nanotubes with a metal nanoparticle shell. In particular, nanocrystalline Co particles with diameter of 4–7 nm and bimetallic Co–Pt particles in the carbon matrix are formed on exposure of (bi)metallic and carbon plates to an argon ion beam

[73]. Gold and ruthenium nanoparticles were obtained from metallized polymer precursors by electron beam treatment [74, 75].

Let's consider thermolysis of the PAN fiber in more detail. Commercial PAN fiber is a triple copolymer of acrylonitrile, methylacrylate and itaconic acid, the contents of the monomer units are equal to 93.0, 5.7 and 1.3 %, respectively. Thermolysis is carried out in an oxidizing atmosphere. Oxygen promotes the hydrogen elimination from the structure of the carbon chain polymer followed by the formation of intra- and intermolecular cross-linking and polyconjugated bonds. Cyclization of nitrile groups occurs at 150 °C yielded the conjugated C=N-bonds, but the system of conjugated C=N-bonds is formed at 200–220 °C. The gaseous products evaporate vigorously in the initial stage of carbonization (pre-carbonization). The loss of heteroatoms results of the formation of polyconjugates carbon structures accompanied by the hybridization of residual sp^3-carbon into sp^2-form. Further heating leads to the formation of ordered phase nuclei and ordered structures as well as a total polyconjugated system. These processes are approximated by Avramy-Erofeev equation of n-th order. The apparent activation energy is relatively low. Subsequent stages are accompanied with the formation of nanocrystalline structures of different dimensions and defects. At carbonization temperatures (1,300–1,400 °C) the layered nanocrystallites of graphene fragments are obtained [76, 77] (Fig. 6.5).

Considering its mechanism, IR thermolysis also belongs to this group of methods. However, in this case, the rates of chemical reactions are much higher, which is related to selective action of IR radiation on the vibrational energy of particular bonds of the macromolecule. For example, above mentioned thermolysis of PVA at 200 °C is accompanied by dehydration of macromolecules followed with the formation of polyconjugated bonds system. In this case anisotropic

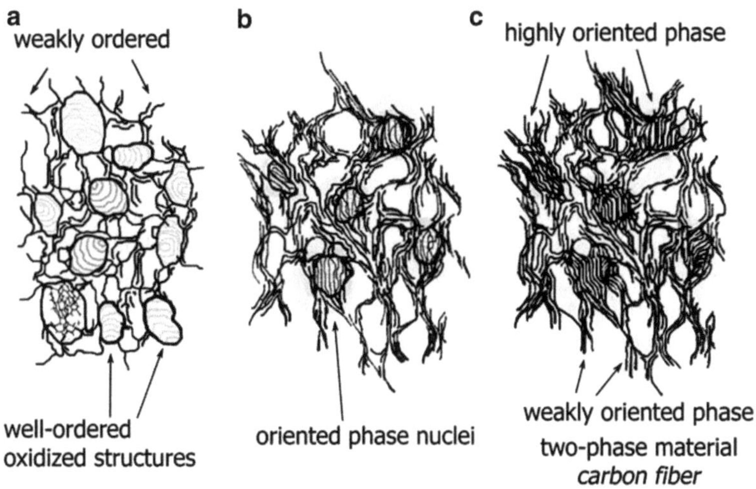

Fig. 6.5 A scheme of structural transformations of the PAN during carbonization. (**a**) an oxidized fiber; (**b**, **c**) – formation of ordered clusters in the volume of material

polarizability and decreasing of the vibration band in Raman spectrum associated with the vibration of atoms along a polymer chain and an increase of the band of deformation transverse vibrations are observed [78].

The IR thermolysis of PAN is best studied. At only 473 K, the polymer nitrile groups start to undergo cyclization caused by the mobility of the hydrogen atom at the tertiary carbon atom and its migration to the nitrile group to form an imine bond (>NH). The cyclization is facilitated by hydrogen bonding of the imine with the nitrile group [8]. At 708 K a vigorous IR-pyrolysis of PAN occur. This is accompanied by restructuring of the material yielded the formation of pyrolized conducting film, particularly, in the case of its doping [79]. It is of interest that in the case of IR pyrolysis of the PAN/Co metallopolymer system, the structure of the resulting carbon phase is perfected at the expense of the carbon present in the polymer. Hence, the cobalt nanoparticles catalyze the formation of crystalline graphite [80]. The hypothetical sequence of transformations is shown in Scheme 6.1.

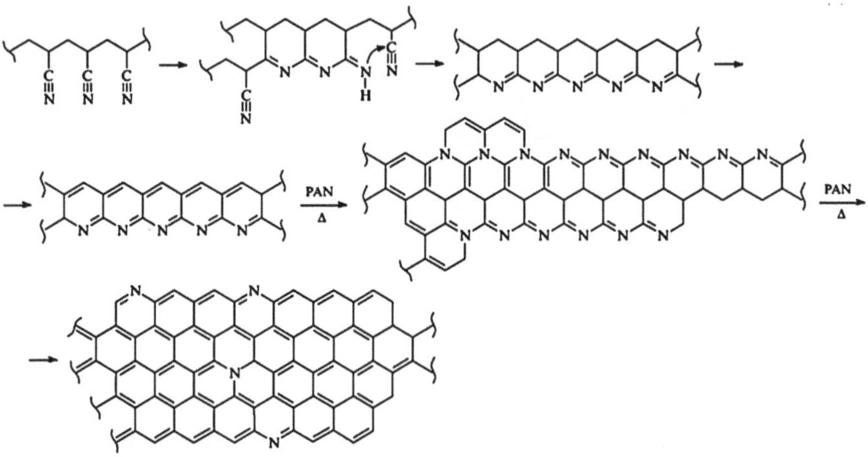

Scheme 6.1 A scheme of pyrolysis of PAN

Most likely, such structures are stacked into layers. Probably, carbonization of other polymers, for example, polybenzimidazole, polyphenylcarbyne, and so on, proceeds in a similar way. The preliminary annealing is carried out in air at 473 K and then under argon at 673–1,373 K. The precursor is a solution in DMF containing simultaneously three components: PAN, $PtCl_4$ and Ru (Re, Rh)Cl_3. Under these conditions, the salts form complexes with PAN and the products of its transformation. Already at 513 K the complexes are destroyed; however, the reduced metal atoms are surrounded by a polymeric structure, which restricts their mobility and prevents aggregation during the subsequent pyrolysis. This factor is responsible for the formation of rather small and uniform nanoparticles in the composite. The size of spherical nanoparticles is 2–18 nm, the presence of the $Pt_{13}Ru_{27}$ intermetallic and alloy formation throughout the whole homogeneity range of Pt solid solutions were also detected.

Other metal – carbon nanocomposites were obtained in a similar way using uranium nitrate, ammonium heptamolybdate, etc. IR-thermolysis of mixtures of PAN, $FeCl_3.6H_2O$ and $NiCl_2.6H_2O$ yields $FeNi_3/C$ nanocomposite with controlled electrophysical and magnetic properties [81, 82]. The particles were 10–80 nm in size and homogeneously distributed in a carbon matrix. Mechanism of the nanocomposite formation is multistage. Dehydration of the starting components takes place at 150 °C, the chemical bond of iron and nickel with nitrile groups of PAN appears up to 250 °C, the $NiFe_2O$ and $FeNi_3/C$ occur at 400 °C and 600 °C, respectively, and finally, γ-(Fe,Ni) forms at 800 °C.

6.2.5 Spray Pyrolysis

As regards the character and the equipment used, spray pyrolysis differs from the procedures considered above. This is an aerosol process that is used efficiently on a large scale to obtain metal and metal oxide nanoparticles and powdered ceramics and nanostructured materials [83, 84]. This technique has proved itself good in materials science and chemical and food industry.

Spray pyrolysis includes five key stages. The first stage is generation of the spray from a liquid precursor using an appropriate drop generator. Then it is followed by spray transport by an air or inert gas stream during which the solvent evaporates. The next stage is precipitation of the solute after the critical supersaturation limit has been reached in the drops. The key stage is thermolysis of the precipitated substance giving nanoporous particles followed by calcination to obtain dense structures. The extraction of particles from the gas stream completes the process. In some cases, sol-gel synthesis and spray pyrolysis techniques are combined. Other techniques such as ultrasound-, plasma-, and laser pyrolysis as well as flame spray-pyrolysis are also widely used [85–87]. The latter method has be proven to be more effective for obtaining zinc oxide in comparison with other known approaches such as CDV, sol-gel synthesis, etc. [88]. It is interesting note that ZnO[3] [89–95] is the third pigment on production volume (next of carbon black and titanium dioxide), its production exceeds 600,000 t/year. Introducing it into elastomers the materials with the developed specific surface can be obtained to apply for pharmaceutical, cosmetic, catalytic purposes, etc. [96]. Flame spray-pyrolysis synthesis of nanostructured ZnO (diameter of particles of 10–20 nm) can be represent by a principal scheme (Fig. 6.6) [97, 98].

The gas – particle conversion methods and the liquid–solid phase transformations followed by thermolysis and pulverization have important advantages. These

[3] In turn, nanocrystalline zinc oxide is one of widely investigated nanomaterial [89–95]. This is a large-band-gap semiconductor (with a band gap of 3.37 eV at 300 K and a large exciton binding energy of 60 meV) and a multifunctional material with the unique set of mechanical, electrical and luminescent properties. Quasi-onedimentional structures of ZnO are the significant building blocks of different nanodevices such as light-emmiting, sensor (including biosensor), piezoelectric, solarvaltaic arrays etc.

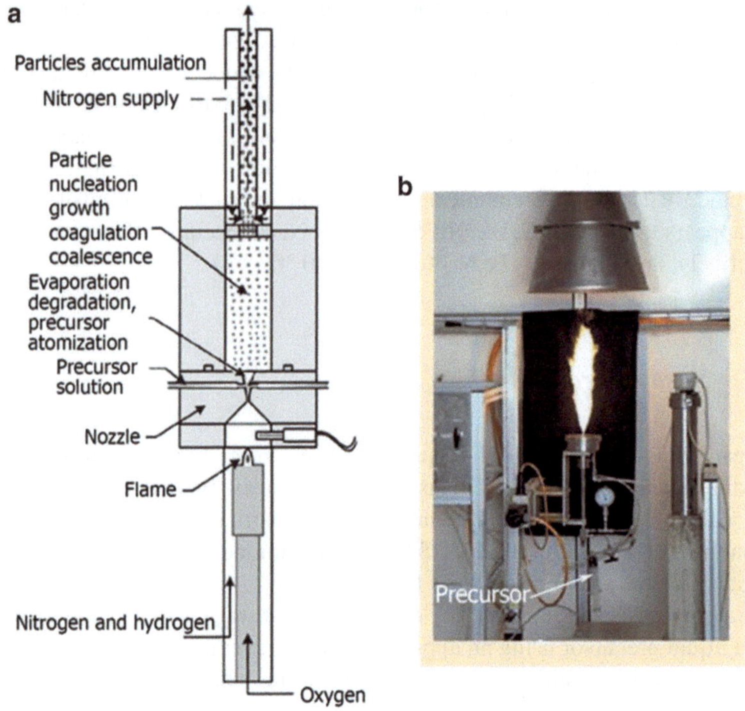

Fig. 6.6 A scheme of flame spray pyrolysis (a) and a view of the high-temperature stream reactor (**b**)

are the purity and the uniform chemical composition of the obtained material, narrow particle size and shape distributions, the possibility of synthesizing multicomponent systems, and the simplicity and scalability of the whole process. They are suitable for preparing particles of three different morphologies: [99] solid nanoporous, microporous and hollow ones (Fig. 6.7). The first (e.g., Al_2O_3) and third (e.g., MgO and ZnO) type particles are formed from the corresponding nitrates. Upon variation of the temperature of spray pyrolysis, ZrO_2 particles of all types were obtained.

Currently, some stages of spray pyrolysis[4] [100] are being under detailed investigation, for example, drop formation [101], evaporation and drying stages [102]. However, it is still unclear how the change of reaction conditions (component concentrations, flow rates, evaporation – precipitation or thermolysis) affects the mechanism and the possibility of control of nanoparticle formation. It was noted [103, 104] that the addition of a polymer, e.g., poly(ethylene glycol), to the initial solution of the precursor can be used to prepare nanoparticles of europium-doped yttrium oxide of controlled morphology.

[4] To consider the spray-pyrolysis in the more detail including the formation of bioceramics and processes with organic components the recent rather comprehensive review can be recommended [100].

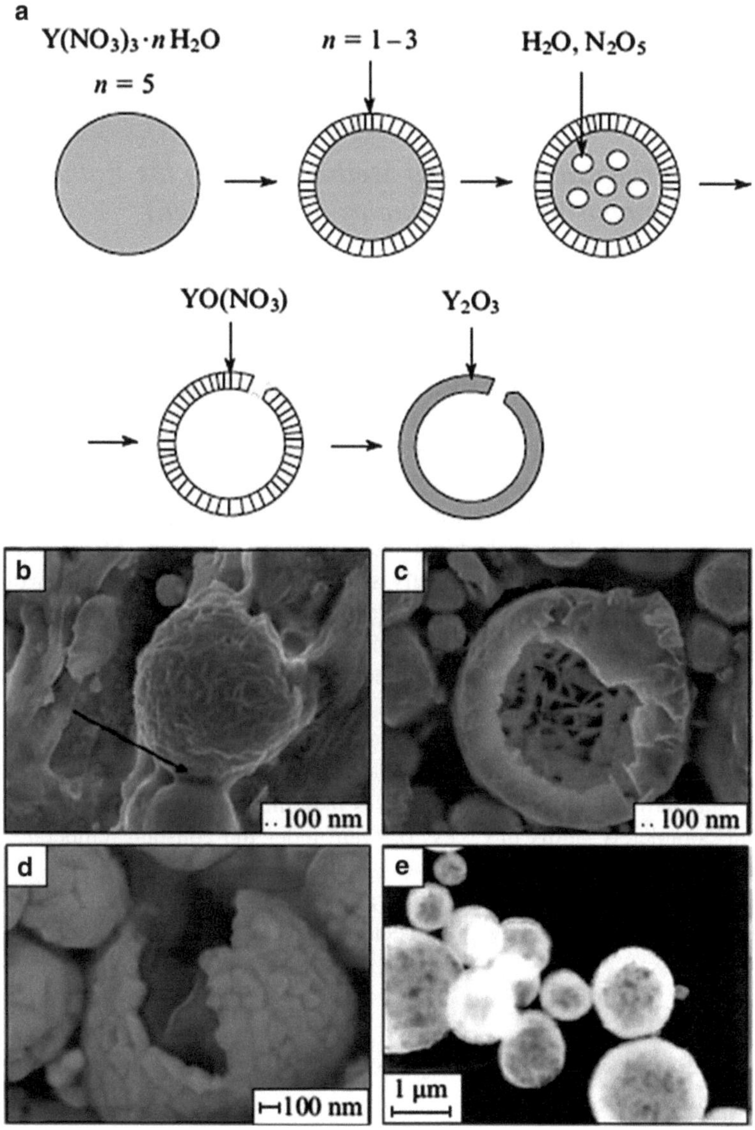

Fig. 6.7 Schematic view of the change of the morphology of $Y(NO_3)_3.nH_2O$ drops during spray pyrolysis (**a**); electron microscopic images of the products formed after drying and the first stage of thermolysis (390 K) (**b**), after thermolysis at 673 K (**c**), after thermolysis at 973 K and subsequent annealing at 1,673 (**d**) or 1,473 K (**e**) [99]

Nanoparticles and nancomposites on the base of metal oxides obtaining by spray-pyrolysis are quite well characterized (see, for instance [105–112]).

New gas phase methods are extensively developed to obtain new ceramic materials. Flame spray pyrolys combining with aqua and non-aqua sol-gel synthesis, hydro- and solvothermal methods, pyrolysis of polymer materials and high-pressure techniques has great synthetic potential. Probably, the same way is suitable for synthesis of copper nanoparticles stabilized with graphene [113].

6.2.6 *Calorimetric Bomb as a Method to Study the Enthalpy of Combustion (the Enthalpy of Formation) of Nanoparticles*

This is one of the important methods for studying of thermochemical data by the burning of material in calorimetric bomb. This method is widespread and has been brought to perfection. The bomb calorimeters are widely used in a thermoanalysis to measure the enthalpy of reaction for many compounds. Hear we have not analyzed the well-known method of Bartolo and its modifications as well as calorimetric bombs of Rossini and Prosen (the information can be found in the corresponding handbooks). Let's consider briefly a modified calorimetric bomb constructed in the thermochemical laboratory in Moscow State University (Fig. 6.8). The bomb is a hermetically sealed stainless steal vessel to withstand pressures of at least 40 atm. There are electrodes to carry an ignition current. The calorimetric bomb holds in a container (calorimeter) filled by 2–4 L of water, as a rule. Before a material with an unknown heat of combustion can be tested in a bomb calorimeter, the calorimeter must be calibrated with allowance of the energy contributed by the fuse. Benzoic acid is often used as a standard material with a known heat of combustion under controlled and reproducible operating conditions.

The standard enthalpy of combustion is the enthalpy change when 1 mol of a reactant completely burns in excess oxygen under the constant temperature and pressure yielded liquid water and gaseous dioxide carbon.

Thus, different methods for the preparation of nanocomposites and study of the kinetics of their formation have now been developed.

6.3 Thermolysis of Metalloprecursors

In this Section we consider the thermal decomposition of more particularly used precursors with a comparative analysis of their decomposition in polymer matrices. It should be noted that non-hydrolytic routes are the most popular to carry out the processes of thermolysis. This is a thermolysis of metal salts, metal oxysalts and their derivatives in high-boiling organic solvents. In addition, in many cases, it is difficult to compare the thermodynamic and kinetic data because the different conditions of experiments (inert and self-generated atmosphere or air, the different reaction rates, etc.).

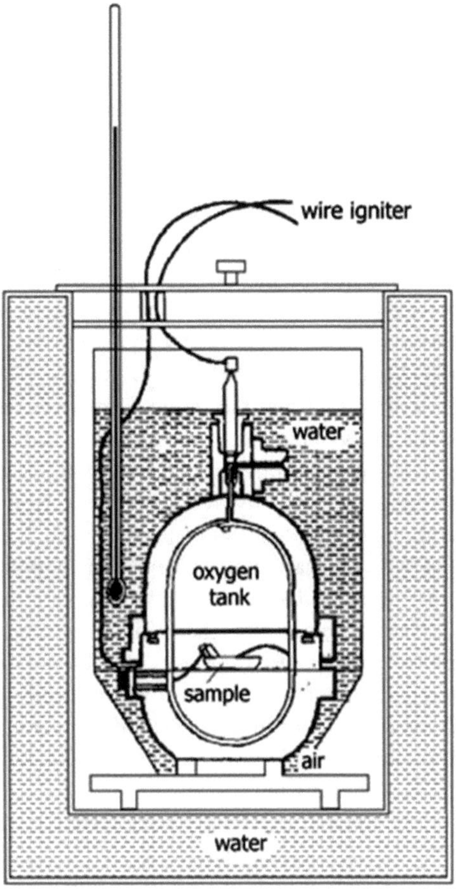

Fig. 6.8 A scheme of the calorimetric bomb

6.3.1 Metal Nitrates

Thermal decomposition of metal nitrates, as well as metal perchlorates, is used to prepare pure metal oxides, in particular, as parts of composites, for example, explosives [114]. Metal nanoparticles, for example, of aluminum in combination with ammonium nitrite and cyclotrimethylene trinitramine increase the rate and temperature of burning. Such energetic nanomaterials are of interest for a heterogeneous burning (see, for instance [115]). The mechanism of thermal decomposition of metal nitrates has been established; for actinides and lanthanides (except for cerium and samarium salts), it can be represented as follows:

$$\begin{aligned}&M\,(NO_3)_3\cdot 6H_2O\,(\text{ads.}) \rightarrow M(NO_3)_3(\text{ads.}) \rightarrow\\&\rightarrow MONO_3(\text{ads.}) \rightarrow M_2O_3 + \text{nitrogen oxides.}\end{aligned}$$

High energy thermolysis of nitrate complexes of bis-ethylenediamine $[M(EDA)_2](NO_3)_2$ (M=Cu, Co, Ni and Zn) is investigated in detail [116]. Their decomposition proceeds in two stages. First, one EDA molecule and monoethylene diamine are evolved; formation of high-dispersed metal oxides completes the reaction:

$$[M(EDA)_2](NO_3)_2 \xrightarrow{443-532\ K} [M(EDA)_2](NO_3) + EDA \xrightarrow{523-553\ K} MO + \text{gaseous products}$$

The kinetics of the process is described by second- and third-order equations: $1-(1-\eta)^{1/2} = kt$ and $1-(1-\eta)^{1/3} = kt$, respectively. The thermal stability and the activation energy of decomposition (88.8–108 kJ/mol) increases as follows: $[Cu(EDA)_2](NO_3)_2 < [Co(EDA)_2](NO_3)_2 < [Ni(EDA)_2](NO_3)_2 < [Zn(EDA)_2](NO_3)_2$. It is important that the initial complexes can be used as burning rate modifiers of solid propellants. By introducing 2 wt% of the complex into a composite of polybutadiene with terminal hydroxyl groups – ammonium perchlorate the rate of burning increases by 1.1–1.80 times. Still more activity is characteristic for the complexes of transition metal salts with 5-nitro 2,4-dihydro-3H-1,2,4-triazole-3-on [117]. In this case high-dispersed metal oxides are formed with the more developed surface that enhances the flame temperature.

The interesting mechanism is revealed under thermal transformations in the system of $Cu(NO_3)_2{\cdot}3H_2O$ – 1,2–ethandiol [118, 119]. At the comparative low temperatures and in an acid medium the spontaneous oxidation of 1,2–ethandiol occurs in a coordination sphere of metal caused by an equilibrium shift to the formation of the polynuclear coordinated product. The thermal conversion at 1273 K yields CuO.

Decomposition of anhydrous cerium nitrate to CeO_2 (in the temperature range of 500–633 K) corresponds to second-order kinetic equation $kt = [1/(1-\eta)]^{-1}$,, the reaction enthalpy is 111.1 kJ/mol and the apparent activation energy is 104 kJ/mol (for comparison, these values for thermolysis of anhydrous $Nd(NO_3)_3$, $Dy(NO_3)_3$ and $Yb(NO_3)_3$ are 33, 23 and 46 kJ/mol, respectively) [120]. Decomposition of the corresponding nitrates gave yttrium group rare earth metal sesquioxides (M_2O_3, M=Y, Dy, Ho, Er) with a controlled geometric structure as nanobelts [121–123] (Fig. 6.9).

The nanoparticle shape is considerably affected by the reaction conditions, specifically, the solvent composition, concentration of the precursor, temperature and time. Thus the mechanism of the hierarchical self-organization of cerium (IV) oxide nanoparticles [124] upon thermolysis of $(NH_4)_2Ce(NO_3)_6$ in oleic acid – oleylamine mixture includes two key stages. First, during hydrolysis of the initial salt at 413 – 493 K, CeO_2 cluster particles framed by surfactant and NO_3^- anion ligands are formed. The next stage (above 493 K) comprises spontaneous assembly of primary particles to flower-like structures (cubic, tetrapetalous, star-like and other structures) caused by the decrease in the concentration of the surface ligands as a result of fast decomposition of the precursor at high temperature.

Spray pyrolysis of cerium nitrate complexes gives rise to hollow spheres of CeO_2, [125] whereas Zn_2SiO_4 and ZnO form microporous spheres [126]. Synthesis

Fig. 6.9 Photomicrographs of lanthanide oxide nanobelts: Dy_2O_3 (**a**, **b**), Ho_2O_3 (**c**, **d**) and Er_2O_3 (**e**, **f**) from scanning (**a**, **c**, **e**) and transmission (**b**, **d**, **f**) electron microscopy (TEM) [121]

of non-doped and doped with Cu (0.1–5 %) ZnO was carried out using Zn $(NO_3)_2 \cdot 6H_2O$, $Cu(NO_3)_2 \cdot 3H_2O$ and glycine [127]. The products possessed a hexagonal phase and symmetry of wurtzite.

Quasi-1D belt-like nanostructures are convenient objects for studying the fundamental physical and chemical properties of closed systems. They are obtained by various methods including thermolysis [128, 129]. For example, the thermolysis of $M(NO_3)_3xH_2O$ (M = Dy, Ho, Er) in a dodecylamine–octadec-1-ene high-boiling solvent mixture is described [121]. Since the solubility of salts in this mixture is low, heat transfer between the liquid and solid phases takes place; dodecylamine plays the key role in controlling the structure formation in such 'solid-liquid' chemical systems with separated nucleation and particle growth stages. The ternary oxides such as $MAlO_3$ and $M_3Al_5O_{13}$ can be obtained in the same way. However, the authors did not consider the possibility of octadec-1-ene polymerization under these conditions (at temperatures of up to 600 K) or the formation of a protecting shell. It should be mentioned also thermolysis of silver nitrate [130].

6.3.2 *Thermolysis of Metal Carboxylates*

Metal salts of carboxylic acids have been widely used since ancient times. For example, calcium propionate is used as a preservative in food industry and in the soap and lubricant production. Its thermal behaviour has been studied in detail [131], as well as the behaviour of metal carboxylates with the number of carbon atoms (n) in the aliphatic chain ranging from 12 to 18. (These salts are present, for example, in human gallstones [132]). The products of their thermolysis (metal nanoparticles, metal oxides or carbides) are of interest owing to a wide application as catalysts of basic organic synthesis reactions, inorganic pigments and so on (see [133–135]).

Metal carboxylates usually exist as monohydrates, their dehydration at the temperature T_1 (>383 K) being accompanied by a phase transformation and considerable loss of crystallinity [136, 137]. At temperature T_2, which decreases by an exponential law following increase in the number of carbon atoms (Table 6.5), the salts start to decompose over a broad temperature range extending beyond 573 K starts (Fig. 6.10).

A typical scheme of the reaction in case of an anhydrous salt can be represented as follows:

$$Fe(NO_3)_3 + 3CH_3(CH_2)_7CH = CH(CH_2)_7COOH$$
$$\rightarrow Fe(C_{17}H_{33}COO)_3 + 3HNO_3 \uparrow$$

Synthesis of the monodispersed product proceeds through two separated stages: nucleation and growth of the crystal. It is supposed that under thermolysis of

Table 6.5 Temperatures of dehydration (T_1/K) and onset of the decomposition (T_2/K) of calcium carboxylates Ca $(n\text{-}C_nH_{2n+1}CO_2)_2 \cdot H_2O$

Carboxylate	T_1	T_2
C3	386.3	585
C4	392.8	547
C5	389.9	517
C6	394.8	500
C8	391.9	485
C10	391.6	458
C12	394.2	443
C13	391.7	439

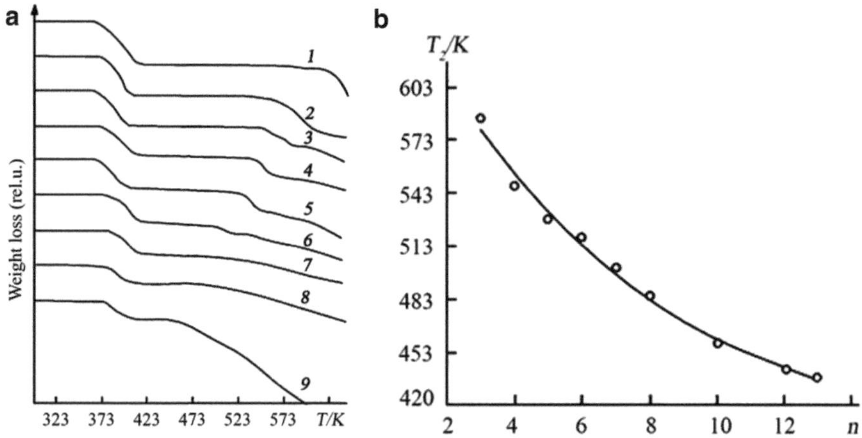

Fig. 6.10 Profiles of the thermograms of decomposition of Ca(II) carboxylates (**a**) and initial decomposition temperature (T_2) vs. the number of C atoms in the aliphatic chain (**b**) [136]. Heating rate 10 deg min^{-1}; n = 3 (*1*), 4 (*2*), 5 (*3*), 6 (*4*), 7 (*5*), 8 (*6*), 10 (*7*), 12 (*8*), 13 (*9*)

hydrated iron carboxylate the nucleation initiates a thermal generation of free radicals from the metal carboxylate [138]:

$$\text{M-COOR} \xrightarrow{320\ ^\circ\text{C}} \text{M} + \text{RCOO}$$
$$\text{M-COOR} \xrightarrow{320\ ^\circ\text{C}} \text{M}_x\text{O}_y + \text{RCO}$$

The free radicals can recombine or form volatile products such as CO, CO_2, H_2O, ethers etc. Generally, the free radicals attack another M–COOR molecules and initiate the chain reactions. Carboxylic acids play a dual role in the synthesis of nanocrystalline iron oxide. Firstly, the carboxyl group is a ligand to bind iron ions into the precursor complex. Secondly, free oleic acid is a stabilizing agent covered the particles and yielding the fine dispersions in organic solvents. Nickel nanoparticles (with diameter of 7 nm and product purity of 74.3 %) were obtained by the thermal decomposition of nickel acetate in the presence of hexadecylamine surfactant or by the controlled evaporation of solution of nickel-oleylamine complex [139, 140]. In some cases, it has been possible to identify the intermediates such as nanocrystals of Ni_3C [141].

Carboxylic acids derivatives of the types of potassiumferrouscarboxylate with general formula $K_3[Fe(L)_6] \cdot xH_2O$ (L is formate, acetate, propionate, butyrate) underwent thermolysis in air up to 1,173 K. After dehydration an anhydrous complex exothermally decomposed with the formation of different intermediates, in particular, of potassium carbonate. Further decomposition (above 973 K) led to $KFeO_2$. This product can be also obtained at 873 K if oxalyldihydrazide $C_2H_6N_4O_2(aq)$ as a complexing agent is added to the batch. Then a reaction mixture was ignited in a muffle [142]. Reduction of Fe(III) to Fe(II) being endothermal absorbs some heat that facilitates the formation of ferrites at more high temperatures. Such behavior is similar to thermolysis of alkaline metal ferrioxalates. Both methods allow one to obtain the stoichiometric pure ferrites at lower temperatures and shorter reaction time. Besides, the additional grinding of the initial materials is not required that is necessary in synthesis of ceramics and may cause defects of the lattice and, in turn, effect permanent magnetic properties. Thermolysis of citrate of alkaline-earth metals and tris(oxalate/maleate/malonate ferrates (III) was investigated quite well. Different reasons cause the significant attention to thermolysis of strontium and barium citrate ferrates [143, 144]. It was shown by Mossbauer spectroscopy that isothermal and non-isothermal decomposition of $M_3[Fe(C_6H_5O_7)_2]_2.4H_2O$ (M=Sr, Ba) includes three main stages [145]: dehydration of the complex, its thermal decomposition yielded α-Fe_2O_3 and metastable acetonedicarboxylate intermediate, and finally, oxidative decomposition of the intermediate and formation of the $Ba_3[Fe(C_6H_5O_7)_2]_2$ ferrite. Analysis of isomeric schift and quadrupole splitting of Mossbauer spectra allowed proving the composition, structure and quantities of the products at 673 and 923 K ($BaFe_2O_4$ and $Ba_3Fe_2O_{7-x}$).

The most detailed mechanism of thermolysis of metal carboxylates of saturated carboxylic acids has been elucidated in a number works [146–149]. Thermal transformations of Fe(II), Fe(III), Ni(II), Cu(II), and Pb(II) formates of the formula $M(HCOO)n$ and the oxalates $Fe_2(C_2O_4)_3{\cdot}5H_2O$ and $FeC_2O_4{\cdot}0.5H_2O$ are temperature-separated successive processes consisting of dehydration and subsequent decomposition of the anhydrous carboxylate. The kinetic parameters of decomposition, the pattern of dependence of the decomposition rate on the degree of salt conversion and the corresponding approximating equations are presented in Table 6.6. The kinetics of isothermal decomposition in an SGA is affected by the temperature of the experiment (T_{exp}), the rate of the preceding dehydration and the m_0/V ratio (m_0 is the sample weight, V is the volume of the reaction vessel).

Thermolysis of iron and copper carboxylates follows two consecutive macrostages to give intermediately metal carboxylates with a low oxidation state of the metal; this is caused apparently by rather low redox potentials of the Fe(III)/Fe(II) ($E^o = +0.771$ V) and Cu(II)/Cu(I) ($E^o = +0.158$ V) pairs [51].

The composition of the volatile products obtained in a high-vacuum pyrolysis of copper formate in the cavity of a mass spectrometer [148] under conditions that rule out secondary reactions indicates that the transformation is a complex multichannel process. In the early stages (up to the degree of conversion $\chi_\Sigma \sim 0.5$, χ_Σ is the fraction of the total intensity of the principal mass peaks at the end of the decomposition, Σ implies summation over all components), CO_2, CO, H_2O and HCO_2H are formed simultaneously, and $\chi CO_2 > \chi H_2O \approx \chi CO > \chi HCO_2H$; subsequently the evolution of H_2O and CO virtually stops, the rate of their accumulation being low and given by $W_{min}(\chi_\Sigma)$.

Thermolysis of Co(II) formate conditionally can be divided into two stages as well: dehydration and a subsequent decomposition. The first stage has been investigated rather in detail [150]. Thermal decomposition of the anhydrous product is accompanied by evolution of gaseous products H_2, CO and CO_2; water vapors formed yet during dehydration, as well as CH_3OH, CH_3COOH, products of thermolysis of formic acid [151–153].

Table 6.6 The kinetic parameters of thermal decomposition of metal salts of saturated carboxylic acids [149]

Compound (T_{exp}, K)	Pattern of the curve $W(\eta)$	Approximating equation $W(\eta)$	k_i, ξ_0 $(m_0/V) \times 10^3$,/g cm^{-1}	$k(T) = k_0 \exp[-E_a/(RT)]$ k_0/s^{-1}	 E_a/ kJ mol^{-1}
$Ni(COO)_2 \cdot 2H_2O$ (473–439 K)	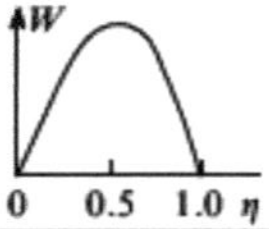	$W(\eta) = k_1(1-\eta) - k_2\eta(1-\eta)$ $(m_0/V) \uparrow -k_1 \downarrow, k_2 \downarrow; W(0) = k_1$	k_1 k_2 1.63	$1.5 \cdot 10^{20}$ $2.7 \cdot 10^{18}$	244.0 202.5
$Fe(COO)_2 \cdot 2H_2O$ (513–553 K)	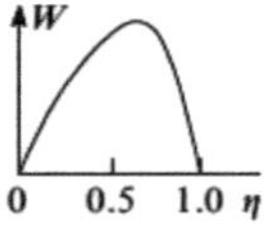	$W(\eta) = k(1-\eta)(\eta+\xi_0)^2$ $(m_0/V) \uparrow -\xi_0 \uparrow; k = \text{const}$ $W(0) = k\xi_0{}^2$	k ξ_0 $W(0)$ 2.0	$0.33 \cdot 10^2$ $5.6 \cdot 10^8$ $1.04 \cdot 10^{19}$	38.0 101.0 240.0
$Fe(COO)_3$ (513–553 K)	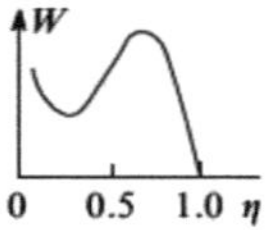	$\eta = 0 \div 0,2; W(\eta) = k_1(\eta_{1\infty} - \eta)$; $\eta > 0.2; W(\eta) = k(1-\omega)^2(\omega+\xi_0)$, where $\omega = [\eta - \eta_{1\infty}(1 - e^{-k_1 t})]/(1-\eta_{1\infty})$; $(m_0/V) \uparrow -k_1 \uparrow, k_2 \uparrow$; $\eta_{1\infty} = 0,38 = \text{const}; \xi_0 = \text{const}$; $W(0)_1 = k_1\eta_{1\infty}; W(0)_2 = k\xi_0{}^2$	k_1 k_2 ξ_0 $W(0)_1$ $W(0)_2$ 2.7	$7.1 \cdot 10^{17}$ $7.8 \cdot 10^{16}$ 0.63 $2.7 \cdot 10^{17}$ $4.9 \cdot 10^{16}$	221.0 210.5 8.4 221.0 228.0
$Cu(COO)_2$ (393–453 K)	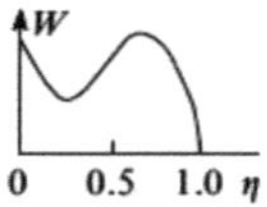	$\eta = 0 \div 0.3; W(\eta) = k_1(\eta_{1\infty} - \eta)$; $\eta > 0,3; W(\eta) = k(1-\omega)(\omega+\xi_0)^2$, where $\omega = [\eta - \eta_{1\infty}(1 - e^{-k_1 t})]/(1-\eta_{1\infty})$; $(m_0/V) \uparrow -k_1 \uparrow, k_2 \uparrow$; $\eta_{1\infty} = 0.42 = \text{const}; \xi_0 = \text{const}$; $W(0)_1 = k_1\eta_{1\infty}; W(0)_2 = k\xi_0$	k_1 k_2 ξ_0 $W(0)_1$ $W(0)_2$ 1.3	$1.3 \cdot 10^{11}$ $3.5 \cdot 10^2$ $1.2 \cdot 10^3$ $5.5 \cdot 10^{10}$ $4.2 \cdot 10^5$	137.5 50.5 35.5 137.5 72.0

(continued)

Table 6.6 (continued)

Compound (T_{exp}, K)	Pattern of the curve $W(\eta)$	Approximating equation $W(\eta)$	k_i, ξ_0 / $(m_0/V)\times 10^3$,/g cm^{-1}	$k(T)=k_0\exp[-E_a/(RT)]$: k_0/s^{-1}	E_a/kJ mol^{-1}
$Pb(COO)_2$ (478–503 K)	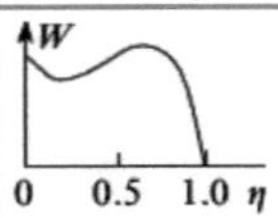	$\eta = 0 \div \eta_{W min}(0.16)$; $W(\eta) = k_1(\eta_{1\infty} - \eta)$; $\eta = \eta_{W min} \div \eta_{W max}(0.5 - 0.7)$; $W = W_{min} + k_2(\eta - \eta_{W min})$; $(m_0/V)\uparrow - W(0)\uparrow, k_2\uparrow, W_{min}\uparrow$, $W_{max}\uparrow, h_{W min}\downarrow, \eta_{W max}\uparrow$; $W(0) = k_1\eta_{1\infty}$	k_1 $\eta_{1\infty}$ k_2 $W(0)$ W_{min} 3.83	$4.7\cdot10^6$ $2.5\cdot10^4$ $4.2\cdot10^{10}$ $1.2\cdot10^{11}$ $8.7\cdot10^{11}$	74.0 39.0 104.5 113.0 117.0
Cocrystallite $Fe(COO)_2$ and $Ba(COO)_2$ [Fe]:[Ba] = 12 (543–583 K)	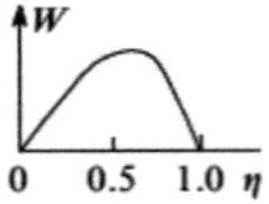	$\eta = 0 \div 0.3$; $W(\eta) = k(1-\eta)(\eta+\xi_0)^2$; $W(0) = k\xi_0{}^2$	k ξ_0 $W(0)$ 2.7	$0.37\cdot10^3$ $2.8\cdot10^4$ $1.5\cdot10^{13}$	50.5 67.0 184.0
$Fe_2(C_2O_4)_3\cdot5H_2O = 2FeC_2O_4\cdot0.5H_2O + 2CO_2 + 4H_2O$ (398–423 K)	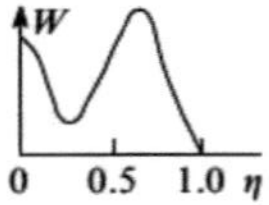	$\eta = 0 \div \eta_{W min}(< 0.2)$; $W(\eta) = k_1(1-\eta)$; $\eta = \eta_{W min} \div \eta_{W max}$; $W = W_{min} + k_2(\eta - \eta_{W min})$; $(m_0/V)\uparrow - W(\eta) \approx$ const;	k_1 k_2 W_{min} W_{max} $\eta_{W min}$ 3.83	$1.8\cdot10^4$ $1.0\cdot10^{16}$ $1.7\cdot10^{11}$ $8.3\cdot10^{15}$ $1.0\cdot10^7$	53.5 154.5 123.5 157.5 54.5
$FeC_2O_4\cdot0.5H_2O$ (573–608 K)	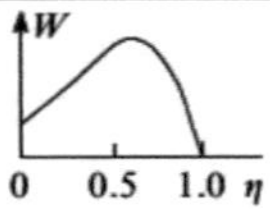	Before W_{max}: $W(\eta) = W(0) + k_1\eta; k_1\eta$; after W_{max}: $W(\eta) = k_2(1-\eta)$; $(m_0/V)\uparrow - W(0)\uparrow, k_2\uparrow, W_{max}\uparrow$, $k_1 \approx$ const	k_1 k_2 $W(0)$ W_{max} 1.93	$1.3\cdot10^{24}$ $9.6\cdot10^{17}$ $4.7\cdot10^6$ $3.1\cdot10^{18}$	318.0 251.0 127.5 253.0

Note: $W = d\eta/dt$ is the rate of conversion; $W(0)$ is the initial rate of conversion; $\eta = \alpha_t/\alpha_\infty$ is the degree of conversion; α_t, α_∞ is the number of moles of the evolved gaseous products per mole of the initial compound at time t and at the end of transformation, respectively; V is the volume of the reaction vessel; ↑, ↓ are increase and decrease of a parameter, respectively; ξ_0 is the constant depending on m_0/V and T_{exp}

The topography of the thermal decomposition of anhydrous metal formates has a number of specific features: the crystals of both the initial compounds and the solid decomposition products have a block structure with retention of their habitus. The blocks are rather porous and are 'fastened' to a crystal by the polymer with the hypothetical composition $-[-CH(R)-O-]_s-[-CH(COOH)-O-]_r-$, where $r \ll s$. Probably, the polymer formation under the SGA conditions is related to the catalytic thickening (polymerization) of the gaseous decomposition products $-CO_2, H_2, CO, H_2O, HCO_2H-$ under the action of forming catalytically active transition metal clusters (Scheme 6.2).

Scheme 6.2 Formation of polymer products during the thermal decomposition of metal formates

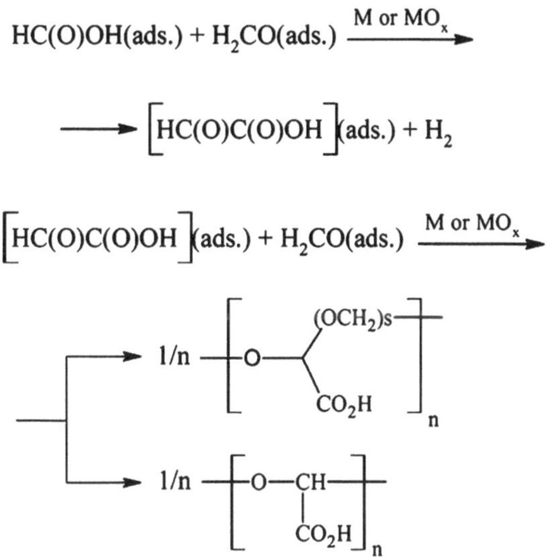

A similar pattern of formation of the polymeric product was observed previously in the thermolysis of $Fe(HCO_2)_2$ [154]. According to electron microscopy (EM) examination, decomposition of metal carboxylates derived from saturated carboxylic acids affords zerovalent metal and/or metal oxide particles of less than 100 nm size with a narrow size distribution (Table 6.7) and nearly spherical shape.

Transition metal formate samples were found to self-stabilize the disperse products. During the transformation, a less electron-dense polymeric shell playing a passivating role is formed on the surface of the electron-dense product particles (Fig. 6.11). The rate of formation of the polymeric product competes with the nanoparticle growth rate. The shell thickness is 3–5 nm and the size of the electron-dense particles is 1.5–2.5 nm. Note that no such shell is produced on the decomposition products of Pb(II) formate or Fe(III) oxalate.

A self-regulated formation of spherical particles of the 'electron-dense nucleus (metal or metal oxide) – less electron-dense shell (organic polymer)' type occurs also during the thermal decomposition of bismuth monocarboxylates with different lengths of the alkyl chain (oxo-formate, −caprylate, −laurate and -stearate) [156]. In a vacuum, inert medium or a high-boiling (423–523 K) organic solvent

Table 6.7 Average particle size and thickness of the polymeric shell for decomposition products of metal carboxylates [149, 155]

Compound	T_{ex}, K	$d^f{}_s$, nm[a]	d_{EM}, nm[b]	ΔL_{shell}, nm[c]	Product[d]
$Fe(COO)_2 \cdot 2H_2O$	543	27.0	20.0	3–5	Fe_3O_4
$Fe(COO)_3$	543	~25.0	–	–	Fe_3O_4
$Ni(COO)_2 \cdot 2H_2O$	483	~30.0	50.0	~4.0	Ni
$Cu(COO)_2$	413	~30.0	~30.0	–	Cu
$Cu(COO)_2 \cdot$MEA (see[e])	398	–	69–75	–	Cu
$Cu(N_2H_3COO)_2 \cdot 2H_2O$ (see[f])	393	–	200–300	–	Cu
$Fe_2(C_2O_4)_3 \cdot 5H_2O$	583	35.0	30.0	–	Fe_3O_4
$Pb(COO)_2$	493	~150.0	~3,000.0	–	Pb
$Pb(COO)_2 \cdot$MEA	525	–	~3,000.0 (~46.0)[g]	–	Pb

[a]Particle diameter calculated from the specific surface area data for the solid products at the end of decomposition
[b]Average particle diameter calculated from EM data
[c]Polymer shell thickness
[d]Powder X-ray diffraction and electron diffraction data
[e]MEA is monoethanolamine
[f]Hydrazinocarboxylate
[g]After ultrasonic treatment

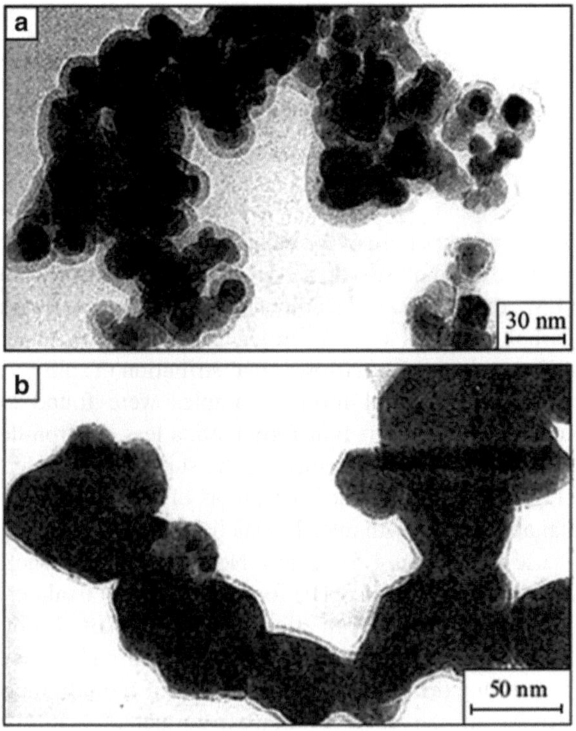

Fig. 6.11 Polymeric shell on the particles formed upon decomposition of Fe $(HCO_2)_2 \cdot 2H_2O$ (543 K) (**a**) and $Ni(HCO_2)_2 \cdot 2H_2O$ (483 K) (**b**)

Fig. 6.12 Electron microscopic image of the spherical particulars of bismuth metal formed upon reduction of bismuth stearate in the presence of high-boiling alcohol (http://www.sbras.nsc.ru/win/sbras/rep/rep2003/tom1/nim.html)

(e.g., benzyl alcohol), powdered nanocrystalline bismuth is formed, while in air bismuth oxide as various polymorphs or as a mixture with the metal is produced. Initially, nanosized (1–2 nm) bismuth particles in ordered layered structures with interlayer spacing of ~5 nm are formed and subsequently they are converted to polymer-stabilized crystalline structures, which are enlarged to 50 nm (Fig. 6.12). The polymer probably arises due to thickening of the destruction products of the carboxyl fragment.

Homogenous thin films of ZnO with a controlled morphology of wurtzite deposited in a silicone support were obtained from solutions of zinc acetate or nitrate using the ultrasound treatment [157]. The thermal decomposition of zinc acetate on a silicone substrate is carried out at relatively low temperature (473–523 K) with the formation of nanoparticles, nanowires, and nanowalls as mono- or polycrystals [158].Of interest is obtaining of ZnO by thermal decomposition at 473 K of coordination polymers of $[Zn(4\text{-}bpdb)(NO_2)_2]n$ types (where bpdb is 1,4-bis(4-pyridil)-2,3-diazo-1,3-butadiene) [159] resulted of high-crystalline nanomaterials.

To obtain homogeneous hexagonal nanocrystalline ZnO, decomposition of Zn-oleate complex were investigated by the thermogravimetric analysis [160]. The decomposition starts at 523 K and ends at 763 K. Size of nanocrystallites depends time (from 1 to 10 h) and temperature of thermolysis; nanoparticles can be redispersed in non-polar organic solvents.

Flame pyrolysis of copper formate solution in the excess of hydrogen produced copper nanoparticles covered with carbon on the following scheme [161]:

$$\begin{aligned}
&Cu(HCO_2)_2.xH_2O\ (aq) \rightarrow Cu(HCO_2)_2(I) + xH_2O\ (g)\\
&Cu(HCO_2)_2(I) \rightarrow CuO\ (cr) + H_2O\ (g) + 2CO(g)\\
&CuO(cr) + H_2(g) \leftrightarrow Cu(cr) + H_2O(g)\\
&CO(g) + H_2(g) \leftrightarrow C(g) + H_2O(g)\\
&Cu(cr) + C(g) \rightarrow Cu/C\ (50\ nm)
\end{aligned}$$

Thermolysis of oleate complexes solvating in 1-octadecene is widely applied (see, for example [162–165]). Significant efforts have been attempted to study different carboxylate salts of silver [166–170] including perfluorine carboxylates [171], oxalate [172], etc. [173]. As a rule, relatively low temperatures (393–523 K), a long-term reaction time (8–24 h), an inert atmosphere are needed to obtain metal nanoparticles. Thus, the high yield of nanoparticles is characteristic for decomposition of silver carboxylates in the presence of hydrogen. Carboxylate groups are deprotonated followed with reduction of silver ions and formation of spherical nanoparticles of 4–6 nm in diameter [174].

The conditions to control size and structure of silver nanoparticles arised from the different types of precursors [175] including coordination polymers were determined [176].

Thermolysis of metal salts of dicarboxylic acids, in particular, oxalates has been investigated for a long time (see [177–181]).

Pre-history (temperature and duration of deposition, aging, morphology) affects essentially the structure of a pyrolyzed product as in case of iron oxalate. Thermal decomposition in air produces hematite (α-Fe_2O_3), deficiency of oxygen at 773–973 leads to the formation of Fe_3O_4 nanoparticles of 35–55 nm of size [182–186]:

$$2FeC_2O_4 \cdot 2H_2O + 5/12O_2 \rightarrow 1/2Fe_2O_3 + 1/3Fe_3O_4 + 2CO + 2CO_2 + 4H_2O$$

An oxidative interaction of ethylene glycol with $Cu(NO_3)_2{\cdot}3H_2O$ with a nitric acid (pH & 1.5) additive yields polynuclear oxalate copper complexes [118, 119]. Their thermolysis at 563 K affords monovalent copper oxide (cuprite, Cu_2O), while that at 1,273 K gives CuO. Polymeric structures of a similar type can also arise in case of 1,2- and 1,3-propandiol and other metal nitrates. Meanwhile, ethylene glycol reacts with manganese oxalate $MnC_2O_4{\cdot}xH_2O$ ($x = 2$ or 3) with displacement of a water molecule giving rise to a new solvation complex, $MnC_2O_4(HOCH_2CH_2OH)$ [187]. Thermolysis of this crystal solvate in air produces a quazi-one-dimensional Mn_3O_4, or Mn_2O_3 nanowhiskers, or MnO particles (in the inert atmosphere). The metastable MnO is obtained by thermolysis of manganese acetate in an oleic acid mixture with trioctylamine [188]. Decomposition of this salt in oleylamine at 473 K furnishes a 2D structure, 10.2×6.8 nm nanoplates. The formation of MnO nanospheres from the $Mn_2(CO)_{10}$ complex with oleylamine upon the addition of trioctylphosphine was also reported [189]. The Mn_3O_4 nanoparticles with the core-shell structure [190] were obtained by thermolysis of $Mn(acac)_2$ (acac is acetylacetonate) in oleylamine [191] or by oxidation of MnO with trimethylamine-N-oxide [192–194]. Note that the MnO and Mn_3O_4 nanoparticles reveal ferromagnetic properties at low temperatures and the coercive force is 8.9 kOe at 5 K; this is essentially more than that for the bulk Mn_3O_4. Nanoparticles of MnO of 5–10 nm in size possess weak ferromagnetism at low temperatures but the antiferromagnetic phase transition of bulk MnO occurs at about 120 K. Superparamagnetic particles of iron oxide were obtained by thermal decomposition of iron acetylacetonate (at 538 K) in the presence of phenyl ether, oleylamine, and 1,2 hexadecandiol [195] or by the same way from anhydrous iron oleate (at 573 K) [196, 197].

Thermolysis of the complex $Eu[C_6H_3(CO_2)_3\text{-}1,3,5]\cdot 6H_2O$ obtained from an aqueous solution of europium nitrate and benzene-1,3,5-tricarboxylic acid affords 10–20 nm-thick one-dimensional Eu_2O_3 nanorods with a 50–100 nm width and a length ranging from hundreds nanometres to several micrometres [198–200]. At temperature below 873 K, the TGA curves show two main weight loss stages, one at 383 K (elimination of water molecules) and one at 742 K (elimination of the organic ligand); as this takes place, the one-dimensional crystalline thread of the nanomaterial being formed (Eu_2O_3) is retained. The oxide Eu_2O_3 was also obtained as nanoparticles of different shape: rods [201], nanowires [202], hollow spheres [203] and nanotubes [204]. The interest in these structures is caused by their unusual physical and chemical properties [205]. For example, europium-doped yttrium oxide (Eu^{3+} in the Y_2O_3 host lattice) is a red inorganic luminophore efficient for the use in plasma displays as hollow spheres and inefficient when has the 'core-shell' morphology.

Thermal decomposition of europium acetate at 923 K for 3 days in a steel reactor in the self-generated atmosphere affords single-phase crystalline hexagonal nanoplates of europium oxycarbonate $Eu_2O_2CO_3$ [206]. Lanthanide oxycarbonates $Gd_2O_2CO_3$, $Sm_2O_2CO_3$, $Y_2O_2CO_3$, $La_2O_2CO_3$ obtained by this way have attracted considerable attention because of their versatile applications [207, 208]. Under similar conditions, zinc acetate is converted to nanopencil-shaped luminescent ZnO [209].

Scattered data on the thermal decomposition of metal acetates in polymer matrices were published. For example, metallopolymer composites are formed in the high-temperature (1,273 K) thermolysis of cobalt acetate in the presence of polystyrene (PS), poly(acrylic acid) (PAA) or poly(methyl vinyl ketone) (PMVK) [210]. Carbonization of the organic part affords a material containing metallic cobalt clusters.

Worthy of note is the synthesis of monodisperse oxide nanoparticles from submicrone oxide powders based on controlled dissolution – recrystallization processes. Thus on slow dissolution of a haematite powder in an oleic acid – octadec-1-ene mixture followed by decomposition of the resulting salt, Fe_3O_4 and $\gamma\text{-}Fe_2O_3$ nanoparticles are formed in 3:2 ratio with an average diameter of 12.1 nm [211]. Colloid nanocrystals of transition metal oxides of well-defined sizes, shape, morphology, and crystallinity, and having ability to self-assembling are widely used in various fields, especially in catalysis, biomedicine, as sensors and so on [212, 213].

6.3.3 Thermolysis of Metal Complexes

One of the most facile and convenient methods for the preparation of monodisperse metal nanoparticles is thermal decomposition of their complexes, for example, acetylacetonates, carbonyls, and so on, in high-boiling solvents. In the presence of surfactants, oleic acid or oleylamine (in some cases, metal oleates dissolved in octadecene are used), thermolysis of these compounds results in the formation of evenly shaped monodisperse particles. Long-chain amines, thioles, carboxylic acids, phosphine oxides are efficient stabilizing agents both of nobel metal nanoparticles in different solvents (see, for example [214–218]) and nickel, cobalt, iron [219, 220], and silver [174] nanoparticles and so on. They prevent aggregation of nanoparticles formed during thermal decomposition. Usually, the process is carried out at 573 K under N_2 atmosphere. Long-chain acids or amines can serve as alternatives to surfactants. For example, stabilization of cobalt nanoparticles of diameter 25–35 nm was observed during the thermal transformation of cobalt (II) bis(salycylidenate) in the presence of oleylamine [221]. Oleylamine complex of nickel acetate was decomposed at 488 K yielding monodispersive sphere nanoparticles of nickel or its oxide [222]. Similar syntheses are numerous. Worthy of note is the formation of nanocrystalline nickel (18.1 nm) at the controlled thermolysis of ethylene diamine (en) complex $[Ni(en)_2(H_2O)_2](NO_3)$ [223]. Under static conditions an autogenic decomposition occurs at 473 K, the kinetics of its oxidation follows the Johnson–Mehl–Avrami mechanism, the activation energy is 135.1 kJ/mol. The product is stable and not oxidized up to 623 K. The kinetics of deamination is retraced for a solid phase thermolysis of tris (ethylenediamine)nickel(II) sulfate [224].

Thermal decomposition of the coordination three-dimensional polymer $[Pb(phen)(l\text{-}N_3)(l\text{-}NO_3)]_n$ containing an azide-anion ligand gave PbO nanoparticles [225]. The controlled thermolysis of $Fe_3(CH_3COO)_6(OH)_2(CH_3COO)$ in a closed reactor in an inert atmosphere at 973 K affords ferromagnetic magnetite nanoparticles [226]. Thermal decomposition of iron(III) oxalate tetrahydrate in a dynamic inert atmosphere is accompanied by the formation of small amounts of superparamagnetic iron(III) oxide. The key intermediate formed at 483 K is FeC_2O_4 [227]. The yield products are FeO (wustite), α-Fe, magnetite. Thermolysis of the starting complex in an oxidative atmosphere comprises dehydration, reduction to Fe(II) and the subsequent decarboxylation to hematite.

Components of the reaction medium can have a considerable influence on the specific growth mechanism and the nanoparticle shape. Thermolysis of iron(III) oleate in the presence of tetraoctylammonium bromide yields octahedral iron oxide nanocrystals, the transformation of the quaternary ammonium salt into trioctylammonium bromide being the rate-determining step [228, 229].

The different mechanisms of thermolysis of such precursore are postulated on the assumption of the composition and structure of the yield products though many of them are inconsistent. First of all, it concerns bipyridine complexes Zn(II) and Cd(II) [230, 231] as well as Ni(II) [232] and Fe(II) [233] (and references there in)].

Thermolysis of the bipyridine nickel complex proceeds at 723 K for 24 h in an argon stream resulting of face-centered cubic phase of Ni nanoparticles of 3.5–5.0 nm in size. Decomposition mechanism of tris(bipyridine)iron(II) dichloride in a argon stream (probably, as a nickel analog too) comprises of two stage of the ligand removing with the formation of coordinative unsaturated iron centers. This is facilated a H-depletion in the pyridine cycle and subsequent attaching the chlorine ion and reduction of the iron ion to Fe(0) nanoparticles of 50–72 nm in size (Scheme 6.3).

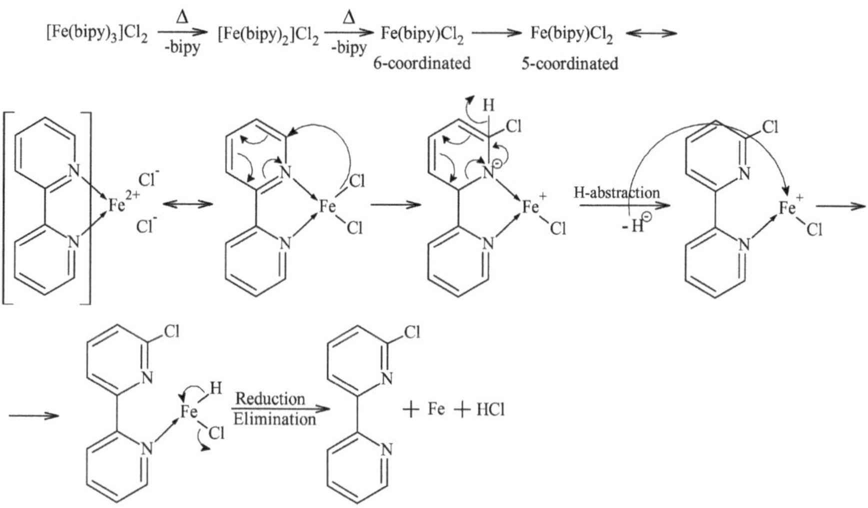

Scheme 6.3 Mechanism of thermal decomposition of tris(bipyridine)iron(II) dichloride to Fe (0) in an inert atmosphere

Most likely, thermolysis of bis-chelate azomethine 2-hydroxy- and 2-N-tosylaminobenzaldehide complexes of Cu(II), Co(II), and Ni(II) in a condensed phase at 643 and 873 K are also multi-stage [234]. A general route of thermolysis of Schiff base type chelates can be represented by Scheme 6.4.

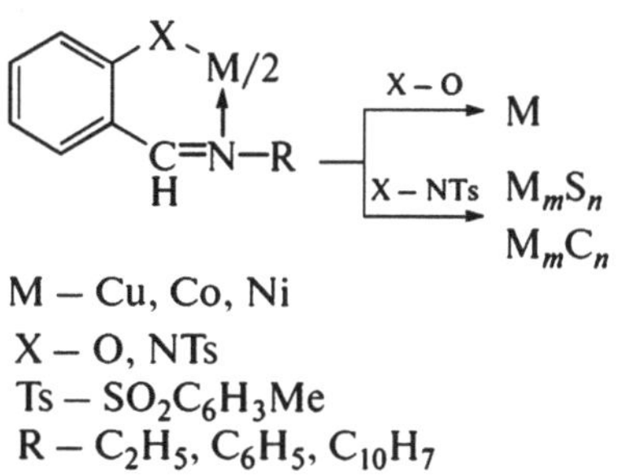

Scheme 6.4 Thermolysis of chelate complexes

For all these complexes it has been ascertained the formation of self-organizing nanocomposites comprising of crystal sulphide cobalt clusters (Co_9S_8) (thermolysis at 593 K) and Co_6S_5 (thermolysis at 873 K); Ni_3S_2 and NiC, or nanocrystalline Cu with the structure of core-shell. The Cu particles of 2–3 nm or even bigger combined into aggregates make the core; the shell is the destruction products of ligand.

Metal nanocrystalls, metal oxides, and metal sulphides were obtained by thermal decomposition of corresponding metal-surfactants [235].

It is interesting that silver nanoparticles with an average size of 60 nm and spherical morphology can be synthesized by thermolysis of 2-pyridine carboxylate complex $Ag(PPh_3)_2$(2-pyCOO) in the presence of oleic acid. Meanwhile, without the surfactant only large particles of the bulk silver of a granulated shape are formed [236].

A comparative simple route for synthesis of nanosized Co_3O_4 being one of the important p-type semiconductor with high gas-sensor, catalytic and electrochemical properties [237–239] is suggested. This is a solid-phase thermolysis of the organometallic precursor – bis[(salycilaldehydate)$_2$cobalt(II) at 773 K for 5 h in air [240, 241]. Thermolysis of N,N-(bis(salycilidene)ethylene-1,2-diamine)nickel in the absence of any template or surfactant was proven to be useful for synthesis of uniform NiO particles at temperature 700 K [242].

A simple large-scale method to synthesize high-purity Zn nanowires using a biomass as carbon source in a carbothermal process has been reported [243]. The process in which ZnO and *Eucalyptus sp.* tar pitch were used as source materials was carried out in inert atmosphere, without vacuum or catalyst, at temperatures (1,073–1,173 K) lower than those required in the carbothermal reduction of ZnO with graphite. Note that reoxidation of the Zn nanowires obtained in air at 573–673 K affords the nanostructured hexagonal ZnO with a high purity and crystallinity [244].

Thermolysis of bis-aqua-tris-salycilaldehyde zirconium(IV) nitrate in oleynimine is useful for obtaining single-phase ZrO_2 ceramics with a cubic shape [245]. The addition of triphenyl phosphine to the reaction mixture allowed one to control the size of particles. Under certain conditions zirconyl nanoparticles of the quasi-spherical shape and about 30 nm of size are formed.

Analysis of thermal transformations of metal complexes has been performed in recent work [28]. Below in Sect. 6.4 we consider the thermolysis of such metal complexes in polymer matrices as well as polymerization conjugated with a thermolysis process of acrylamide complexes is discussed (Sect. 6.5).

6.4 The Conjugate Thermolysis – Nanoparticle and Polymer Matrix Formation Process

Currently, various methods for stabilizing nanosized particle are available [246]; some of them have already been mentioned. The self-regulated stabilization of highly dispersed metal particles either as metal-containing precursors or as monomers, which are then subjected to polymerization and thermolysis, seems especially promising.

6.4.1 *Thermal Transformations of Salts of Unsaturated Carboxylic Acids*

Thermolysis of metal-containing monomers (salts of unsaturated mono- and dicarboxylic acids) is poorly studied but vigorously developing line of research and the best embodiment of the idea of nanoparticle stabilization in situ. The monomers include alkali, alkaline earth and transition metal acrylates ($MAcr_n$), maleinates ($MMal_n$), itaconates ($MItac_n$) and fumarates ($MFum_n$) [247–249], their polysalts [250, 251]. This series can be extended for salts of acetylenecarboxylic acids [252]. Such compounds capable to polymerization have been named metallomonomers.

Most of the results of investigations of thermal transformations of metal-containing monomers are qualitative [253–260].

For example, the initial decomposition temperatures of acrylates increase in the sequence of metals: Fe < Cr < Ni < Co < Mn; however, these temperatures are lower than the decomposition temperature of polyacrylic acid. Probably, this is caused by the catalytic influence of metal ions or nanoparticles on the destruction of the polyacrylate fragments. Preliminary γ-irradiation (200–600 kGy) of Ni (II), Co (II) and Cu(II) polyacrylates and increase in the irradiation dose lead to a decrease in the initial decomposition temperatures [261]. The stage mechanism of thermolysis has been proposed for complex acrylates of the composition of M(phen)$(Acr)_2(H_2O)_y$, where phen is phenanthroline, M = Mn(II) (y = 0), Ni(II) (y = 2), Cu (II) (y = 1), Zn(II) (y = 2) (TG, DTG, air, 293–1,273 K, rate heating 10 deg./min) [262]. For these acrylates, except for Zn- complex, all of the macrostages were exothermic. The solid products are Mn_2O_3, NiO, CuO and ZnO, respectively.

Studies of the thermolysis of metal salts of unsaturated carboxylic acids both in the thermal analysis (TA) and SGA modes revealed the general pattern of their transformations. They tend to decompose via a sequence of three macro stages occurring at different temperature: [155] dehydration (desolvation) of the initial monomers (403–473 K); solid-state homo- or copolymerization of the dehydrated monomer (monomers) (473–573 K); decarboxylation of the polymer thus formed to give a metal-containing phase and oxygen-free polymer matrix (at temperatures of >523 K, or for copper carboxylates >453 K) accompanied by vigorous gas evolution.

The most informative quantitative data on the kinetics and the physicochemistry of thermal transformations of such precursors were derived from a SGA process in a non-isothermal closed reactor [263–270]. The studied compounds included $Cu_2(CH_2{=}CHCOO)_4$ ($CuAcr_2$), $Co(CH_2{=}CHCOO)_2{\cdot}H_2O$ ($CoAcr_2$), $Ni(CH_2{=}CHCOO)_2{\cdot}H_2O$ ($NiAcr_2$), $Fe_3O(OH)(CH_2{=}CHCOO)_6{\cdot}3H_2O$ ($FeAcr_3$), cocrystallites $[Fe_3O(OH)(CH_2{=}CHCOO)_6]{\cdot}[Co(CH_2{=}CHCOO)_2]_{2,4}$ ($FeCoAcr_2$) and $[Fe_3O(OH)(CH_2{=}CHCOO)_6]{\cdot}[Co(CH_2{=}CHCOO)_2]_{1,5}{\cdot}3H_2O$ (Fe_2CoAcr_2)' $Co(OCOCH{=}CHCOO){\cdot}2H_2O$ ($CoMal_2$), $[Fe_3O(OH)(OCOCH{=}CHCOO)_6]{\cdot}3H_2O$ ($FeMal_3$).

The time dependence of gas evolution $\eta(t)$ for these compounds is described satisfactorily by the approximating relation

Table 6.8 Kinetic parameters of thermolysis of transition metal salts of unsaturated carboxylic acids

Compound	Texp, K	Parameter	$\eta_{1f}, \Delta\alpha_{\Sigma f} = A \exp[-\Delta H/(RT)]$		k	$k = k_0 \exp[-Ea/(RT)]$		References
			A	ΔH, kJ/mol		k_0, s^{-1}	E_a, kJ/mol	
$CuAcr_2$	463–513	η_{1f}	$1.8 \cdot 10^4$	48.1	k_1	$9.5 \cdot 10^{11}$	154.7	[263]
		$\Delta\alpha_{\Sigma f}$	3.6	12.5	k_2	$9.2 \cdot 10^{11}$	163.0	
$CoAcr_2$	623–663	η_{1f}	1.0	0	k_1	$3.0 \cdot 10^{14}$	238.3	[264]
		$\Delta\alpha_{\Sigma f}$	1.55	0	k_2	0	0	
$FeAcr_3$	473–573	η_{1f}	1.0	0	k_1	$4.2 \cdot 10^{21}$	246.6	[266]
		$\Delta\alpha_{\Sigma f}$	$1.6 \cdot 10^2$	25,5	k_2	0	0	
	573–643	η_{1f}	1.0	0	k_1	$1.3 \cdot 10^6$	127.5	
		$\Delta\alpha_{\Sigma f}$	$1.7 \cdot 10^2$	26,3	k_2	0	0	
$NiAcr_2$	573–653	η_{1f}	2.6	1.1	k_1	$1.7 \cdot 10^{17}$	242.4	[265]
		$\Delta\alpha_{\Sigma f}$	$1.4 \cdot 10^{11}$	125.4 (<613 K)				
		$\Delta\alpha_{\Sigma f}$	1.2	10.5 (>613 K)	k_2	$7.5 \cdot 10^8$	156.8	
FeCoAcr	613–663	$\eta_{1f} = 0.65(613\ K) \div 0.45(663\ K)$			k_1	$2.3 \cdot 10^{12}$	206.9	[267, 268]
		$\Delta\alpha_{\Sigma f}$	$5.25 \cdot 10^2$	7.5	k_2	$6.0 \cdot 10^8$	137.9	
Fe_2CoAcr	613–663	$\eta_{1f} = 0.50(613\ K) \div 0.3(663\ K)$			k_1	$2.6 \cdot 10^{12}$	204.8	[267, 268]
		$\Delta\alpha_{\Sigma f}$	$1.9 \cdot 10^2$	6.0	k_2	$6.6 \cdot 10^5$	125.4	
$CoMal_2$	613–643	η_{1f}	1.0	0	k_1	$1.6 \cdot 10^6$	125.4	[269]
		$\Delta\alpha_{\Sigma f}$	$1.3 \cdot 10^2$	23.4	k_2	0	0	
$FeMal_3$	573–643	η_{1f}	$0.59 \cdot 10^2$	23.4	k_1	$3.3 \cdot 10^7$	133.8	[270]
		$\Delta\alpha_{\Sigma f} = 4.78(673\ K) \div 7.4$ (643 K)			k_2	$1.0 \cdot 10^7$	110.8	

Note. $\eta = \Delta\alpha_{\Sigma t}/\Delta\alpha_{\Sigma f}$ is the degree of conversion where $\Delta\alpha_{\Sigma,t} = \alpha_{\Sigma,t} - \alpha_{\Sigma,0}$; $\Delta\alpha_{\Sigma f} = \alpha_{\Sigma,f} - \alpha_{\Sigma,0}$; $\alpha_{\Sigma,f}$, $\alpha_{\Sigma,t}$, $\alpha_{\Sigma,0}$are the final, current and initial total numbers of the gaseous decomposition products evolved per mole of the starting compound at room temperature, respectively ($\alpha_{\Sigma,0}$ is related to dehydration and polymerization processes); $\eta_{1f} = \eta(t)$, $k_2 t \to 0$, $k_1 t \to 0$.

$$\eta(t) = \eta_{1f}[1 - \exp(-k_1\tau)] + (1 - \eta_{1f})[1 - \exp(-k_2\tau)], \qquad (6.9)$$

where k_1, k_2 are the effective rate constants for the transformation ($W = d\eta/dt$) at the corresponding stage, $\eta_{1f} = \eta(t)$ for$k_2\tau \to 0$ and $k_1\tau \to \infty$, $\tau = t - t_0$, t_0 are the heating time of the sample. Note that the ratio m_0/V has almost no influence on the transformation rate, which is indicative of thermolysis in the condensed phase.

Table 6.8 summarizes the kinetic characteristics of the thermal transformations of the studied compounds. Attention is attracted by the fact that thermolysis of $FeAcr_3$ comprises two temperature-separated gas evolution regions with different kinetic parameters: low-temperature (<573 K) and high-temperature (>573 K) regions. A possible explanation of this phenomenon is the occurrence of two parallel gas evolution processes. In terms of decreasing gas evolution ability of the acrylates $MAcr_n$, metals can be arranged in the series: Cu ≥ Fe > Co > Ni. The thermal transformation of $MAcr_n$ and $MMal_n$ in the gas phase produces [254, 271] CO_2, CO, H_2, H_2O vapour, acrylic (from $MAcr_n$ thermolysis) or maleic (from $MMal_n$) acid vapour, C_2H_4 (for $CuAcr_2$), CH_4 (for $CoAcr_2$ and $NiAcr_2$). The solid products of transformation are particles of the corresponding metal and/or metal oxide embedded in the decarboxylated polymer (in the case of co-crystallized $FeCoAcr_2$ and Fe_2CoAcr_2, cobalt ferrite, $CoFe_2O_4$, is formed).

Analysis of the chemical transformation pathways shows that under the assumption of energetic non-equivalence of the M_7O bonds in unsaturated metal carboxylates, the primary decomposition step is the formation of radicals: acrylic ($CH_2{=}CHCOO^{\cdot}$) and maleic ($^{\cdot}OOCCH{=}CHCOO^{\cdot}$) radicals for acrylates and maleates, respectively. The radicals initiate polymerization, which is followed by decarboxylation of the metallopolymer. As an example, Scheme 6.5 presents the mechanism of thermal decomposition of acrylate.

Taking account of Scheme 6.5, the material balance equations and quantitative data on the yields of gaseous and volatile products of thermolysis, it was found that the composition of the solid phase changes during the transformation. It can be represented by the following balance relation: for acrylates $MO_r(CH_2CHCOO)_{p-k}(CH_2CH)_k(CHCHCOO)_{q-l}(CHCH)_l$ and for maleates $MO_r(=CHCOO)_{2p-k}(=C-)_k(=CCOO)2q-l(=C-)_l$ where before decarboxylation, $r=k=l=0$ ($r \neq 0$ for $FeAcr_3$), p and q are the numbers of internal and terminal (hydrogen-depleted) carboxylate groups, respectively.

An important problem, although rarely discussed, is the morphology and topography of the solid products of carboxylate thermolysis the knowledge of which would extend the understanding of the physicochemistry of this process. The initial $MAcr_n$ and $MMal_n$ samples are optically transparent crystal-like porous particles without crystallinity at distances comparable with the transmitted light wavelength. They have relatively great specific surface area ($S^0{}_{sp} = 9–30\ m^2/g$), which does not change significantly upon thermolysis. However, in some cases (for $CuAcr_2$, $CoAcr_2$ and somewhat for $NiAcr_2$), large aggregates are dispersed. As a result, the average particle size decreases and S_{sp} increases 2–3-fold and then decreases again due to the sintering of particles [263, 264].

Even at early stages (during the sample warming-up), the particles lose transparency and their surface becomes sugar-like, probably as a result of desolvation processes, which may attest to a significant contribution of a bulk homogeneous reaction. Thus, thermolysis of these compounds is a heterogeneous-homogeneous process [265, 266, 268].

The average size of decarboxylated metallopolymer particles (S_{sp} measurement data) is estimated as ~20–30 nm for $CuAcr_2$ and ~30–50 nm for $CoAcr_2$. Electron microscopy and electron diffraction studies [265, 269] of the sample microstructure have shown (Fig. 6.13a) a rather narrow particle size distribution, a 4–9 nm mean diameter of particles (Fig. 6.13b) and a 8–10 nm mean distance between particles in the matrix. The even particle distribution in the matrix and the narrow particle size distribution may be indicative of a high degree of pseudo-homogeneity of decarboxylation and the formation of a new phase. Note that the mean size of the

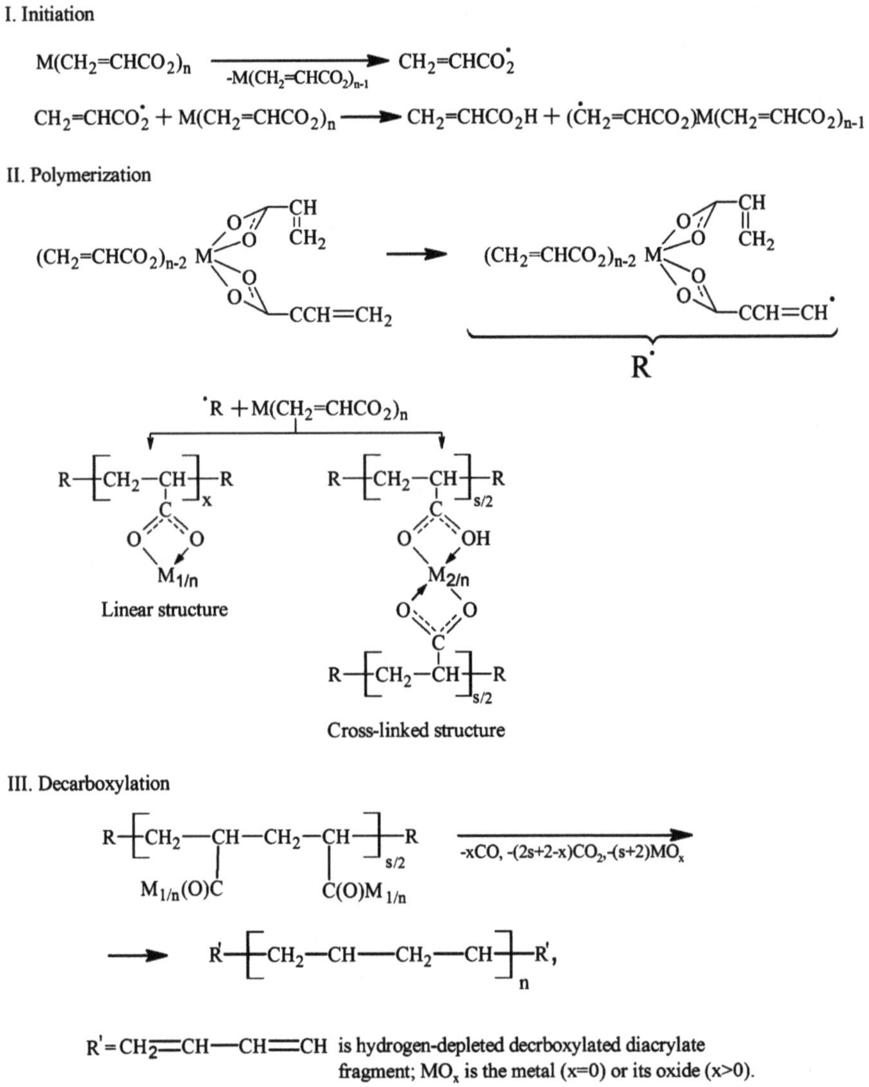

Scheme 6.5 The main routes of thermal transformations of metal acrylates

thermolysis products is much greater for the salts of saturated carboxylic acids than for the unsaturated acid salts (see Fig. 6.13b).

Of interest are lanthanide acrylates due to their specific optical properties [272–275]. The coordination vacancy of transition metals of acrylates can be occupied not only by water molecules but also by other ligands such as amines. In this

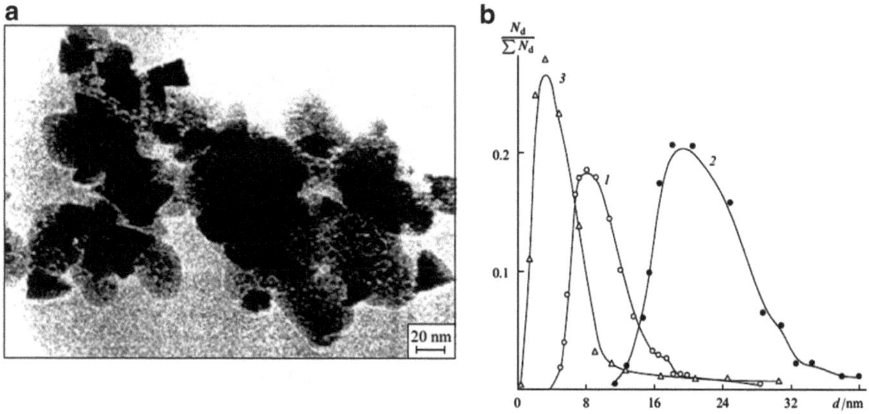

Fig. 6.13 Electron microscopic image of the thermolysis product of $Co(O_2CCH{=}CHCO_2)\cdot 2H_2O$ at 623 K (**a**) and size distribution of metal-containing particles (**b**) [269] (*1*) $Fe_3O(OH)(CH_2{=}CHCOO)_6\cdot 3H_2O$, (*2*) Fe $(HCOO)_2\cdot 2H_2O$, (*3*) Co (OCOCH=CHCOO)2H$_2$O

connection of interest is the work [276] in which the thermal stability of mixed acrylate and ethylenediamine metal complexes is investigated

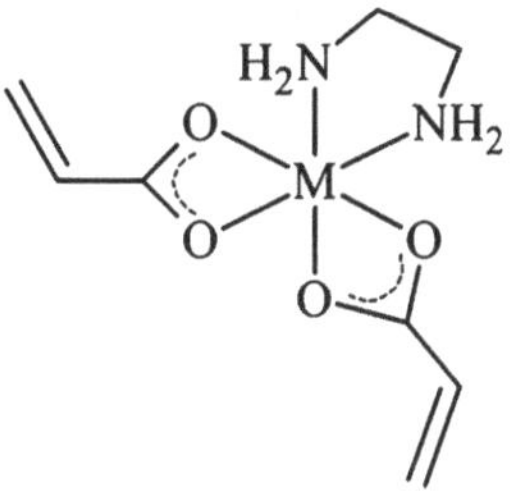

[M(en)(CH_2=CHCOO)2] · *n*H2O ((**1**) M = Ni, *n* = 2; (**2**) M = Cu, *n* = 0; (**3**) M = Zn, *n* = 2; en = ethylenediamine.

After dehydration and elimination of ethylenediamine a metal oxide yield product is formed except metallic copper arising from the copper complex. The complexes of acrylates with tetradentate tripodal ligands are known [277–279]. Such complexes reveal a superoxidedismutase-like activity, particularly, those with aromatic amines (2,2′- or 4,4′-dipyridil, 1,10-phenanthroline) which model different biological objects. However, in these works the stage of polymerization, the peculiarity of thermolysis of such complexes and, consequently, the properties of the resulting nanoparticles is ignored.

Synthetical approaches and structure of pre-thermolized carboxylates of unsaturated dicarboxylic acids are different from those of monocarboxylic acids. Firstly, already on the stage of their synthesis (as shown for metal maleates, fumarates, itaconates, acetylenedicarboxylates [252, 280]) coordination polymers are formed.

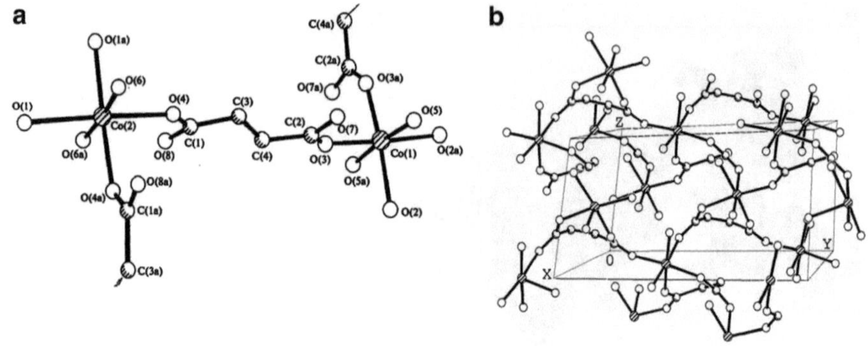

Fig. 6.14 Molecular structure of coordination polymers of Co(II) fumarate pentahydrate (**a**) and maleate (**b**)

Thus, according to the X-ray analysis unsaturated metal dicarboxylates have both monomeric (Co(II) and Fe(II) hydromaleates, $M(C_4H_3O_4)_2{\cdot}4H_2O$) and chain- (Co (II) fumarate, $CoC_4H_2O_4{\cdot}5H_2O$) or three-dimensional (Co(II) maleate, $CoC_4H_2O_4{\cdot}3H_2O$) polymeric structure (Fig. 6.14) multiple bonds of which are not coordinated with metal.

However the mechanism of their thermal transformations is similar to that of unsaturated monocarboxylic acid salts [269, 271]. One can assume that with the rise of the level of heat vibrations in the lattice of monomer the rupture of the weakest M–O bonds is the most probable. As the result the biradicals $^{\cdot}OOCCH{=}CHCOO^{\cdot}$ are formed from maleates, for instance. The formed radicals react with the metal-containing maleate fragments to give corresponding acids and H-depleted radical $R^{\bullet}$ of maleic groups by the following scheme:

$$^{\cdot}OOCCH = CHCOO^{\cdot} + RH \rightarrow HOOCCH = CHCOOH + R^{I} \tag{6.10}$$

Where $RH = (CHCOO)_2M^{n+}{}_{2/n}$, $R^{I} = (^{\cdot}CCOO)_2M^{n+}{}_{1/n}$.

The formed R^{I} initiates the polymerization to produce the linear or networked polymers.

$$R^{I} + s(OOCCH = CHCOO)M^{n+}{}_{2/n} \rightarrow R^{I} - \left[-M_{1/n}OOCCH - CHCOOM_{1/n}-\right]_s - R^{I} \tag{6.11}$$

With temperature the metal-containing fragments of formed polymers decompose to produce a metal (or its oxide) and CO_2.

$$\begin{aligned} &R^{I} - \left[-\left(M_{1/n}OOC\right)CH - CH\left(COOM_{1/n}\right)-\right]_s - R^{I} \rightarrow \\ &\quad CH \equiv C - [-CH - CH-]_s - C \equiv CH + 2(s+2)CO_2 + (s+2)M. \end{aligned} \tag{6.12}$$

The polymers resulted in the decarboxylation reaction can be additionally thermo-polymerized to form the net structure with conjugated multiple bonds.

In general, the composition of solid phase products of thermolysis can be represented as a sum of the C–H–O-fragment fractions:

$$MO_z(=CHCOO)_{2p-x}(=CCOO)_{2q-y}(=CH-)_x(\equiv C-)_H, \tag{6.13}$$

where $x = y = z = 0$ ($z \neq 0$ in the case of Fe(III)-maleate), p and q are the amounts of intra-chain and H-depleted terminal groups ($p + q = 1$), respectively.

Thermolysis of Cu(II) and Zn(II) [281] maleates and fumarates was studied by non-isothermal methods (TGA, DTG, DTA) in the temperature range 273–873 K (heating rate 5 deg./min). Dehydration of the complexes demonstrated that the thermal stability of copper and zinc fumarates was higher than the stability of the corresponding maleates, the dehydrated copper fumarates and maleates were equivalent in stability and zinc fumarate was more stable than zinc maleate.

Comparative analysis of thermolysis of geometry isomers (*cis*- and *trans*-) was carried out for Zn(II) and Cu(II) [282] and Mn(II), Co(II), and Ni(II) [281] maleates and fumarates, respectively, to study the influence of the ligand structure on the thermal stability and mechanism of thermolysis of the corresponding complexes. In the case of the nickel salts dehydration starts at 403 K and ends at 498 K. The second stage comprises the quickly oxidative decomposition (mass loss 69 % at 663 K) up to NiO. This indicates that the yield product arises directly from nickelfumarate. The following scheme has been proposed for this reaction (Scheme 6.6).

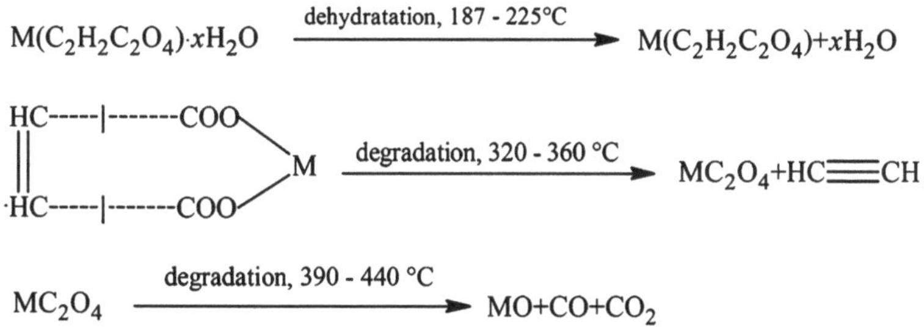

Scheme 6.6 A scheme of thermolysis of metal maleates and fumarates in air

Probably, this mechanism should be defined more precisely making allowance for a polymer structure of the salts being formed during the decomposition. Secondly, the weakest carboxylic group bond should be also considered because the decomposition of a dehydrated complex starts from its rupture.

In a study of thermolysis of ferrimaleates $M_3[Fe(Mal)_3]_2 \cdot xH_2O$ (M = Mn, Co, Ni, Cu) in a static air atmosphere at temperatures of up to 873 K, general characteristics of the decomposition process were established [258]. This is a multistage transformation starting with dehydration of ferrimaleates to ferrites; after dehydration, the Fe(III) precursor is converted to an intermediate Fe(II) derivative, $M_3[Fe(Mal)_2]_2$; after that, the monomer destruction yields iron(III) oxide and the metal

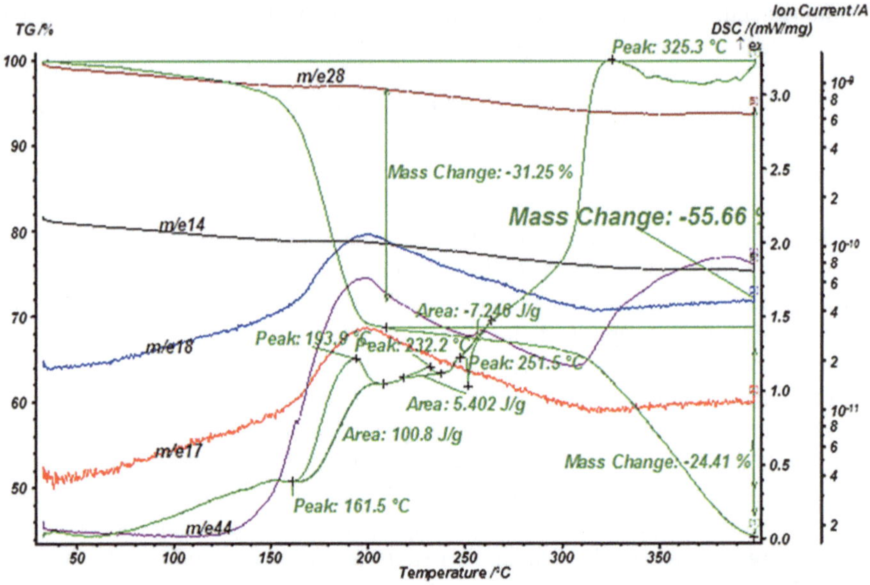

Fig. 6.15 TGA, DSC and mass-specta of Co(II) acetylenedicarboxylate (heating rate 5 deg./min, argon)

oxide (MO). The formation of the ferrite MFe_2O_4 can be described by the following reactions:

$$M_3\left[Fe(Mal)_3\right]_2 xH_2O \xrightarrow{433-548\,K} M_3\left[Fe(Mal)_2\right]_2 \xrightarrow{573-673\,K} Fe_2O_3$$
$$+ MO \xrightarrow{623-773\,K} MFe_2O_4$$

Synthesis of multimetallic ceramics via polymer-assisted transformations of unsaturated carboxylates will be analyzed in Sect. 6.6.

The major gaseous products of thermal decomposition of Co(II) acetylenedicarboxylate (CoADC) in the temperature range 413–1,223 K are CO_2, C_2H_2, C_6H_6 which can be attributed to a decarboxylation process occurred already under polymerization of the monomer salt (Fig. 6.15) [252].

In the range of temperatures 413–453 K the rate of gas evolution at thermolysis of CoADC is satisfactorily approximated by the first-order equation $W = d\eta/dt = k(1-\eta)$ where $\eta = (\alpha_t - \alpha_0)/(\alpha_\infty - \alpha_0)$ is the degree of conversion; α_t, α_0, α_∞ are the current, initial and final total numbers of the gaseous decomposition products evolved per mole of the starting CoADC, $k = 1.6 \cdot 10^{15} \exp[-35{,}800/(RT)]$, s^{-1}. It was found that the total decarboxylation of metal-containing fragments of the CoADC completes over 673 K and the total gas evolution ends at >1,073 K. In this case, a significant mass loss of the sample (up to 70 % at 1,223 K) was observed.

Thus, thermolysis of the salts of unsaturated carboxylic acids and the properties of the products attest to simultaneous "one-pot" occurrence of two processes: nanoparticle synthesis and their stabilization by the formed decarboxylated matrix of a controlled thickness.

6.4.2 Co-condensation of Metal and Monomer Vapours

It appears promising to combine thermolysis with low- temperature methods of nanoparticle insertion into polymerization-active monomers, for example, for preparing metal nanoparticles in poly(p-xylylene) films. This method implies co-condensation of vapours of the metal (and/or semiconductor) and the monomer, p-xylylene, which is prepared by pyrolysis of cyclic di-p-xylylene ([2, 2] paracyclophane), on a liquid nitrogen-cooled substrate [283–285]. The co-condensed product is subjected to low-temperature solid-state polymerization; this prevents the aggregation of metal particles at the stage of nanocomposite formation; the metal content in nanocomposite films reaches 50 vol.%. The block diagram for implementation of this method for nanocomposite preparation is shown in Fig. 6.16. The cyclophane molecules pass through the pyrolysis zone (~870 K) being converted to an active intermediate, which is deposited on a cooled substrate together with metal and/or semiconductor atoms. Then poly(p-xylylene) or its derivative is formed upon thermal polymerization, while nanoparticles or 1–20 nm clusters (depending on the chemical structure of the precursor and polymerization conditions) are formed in the polymer matrix. The particles having rather narrow size distribution are mainly located in amorphous areas of the polymer [286–290].

This method was used to stabilize the Ag, Zn, Cd, Pb[291], Cr [292], Mn [293] and other nanoparticles in poly(p-xylylene) films. The nanoparticle stabilization can be attained by both the formation of a metal cluster in the polymer matrix and the formation of *d*-metal π-complexes with the quinonoid form of the p-xylylene monomer. The complex formation between the polymer and *d*-metals brings about modification of the polymer properties, and the material becomes homogeneous rather than consisting of two phases. A similar reaction with p-xylylene can be proposed for other metals of the chromium, manganese and iron groups [293]. However, the targeted investigation of the thermal stability of these supramolecular assemblies is only at an early stage.

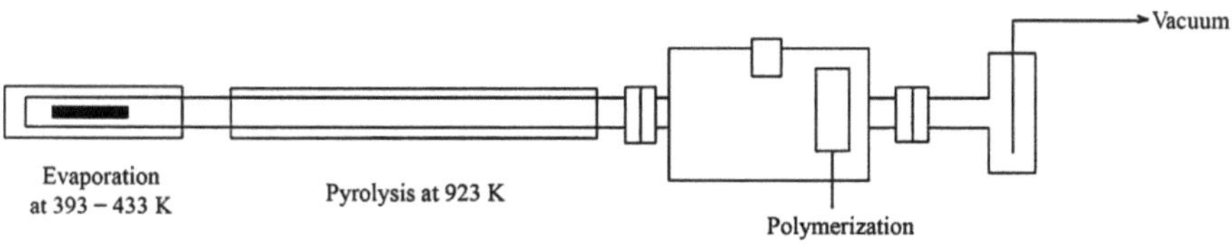

Fig. 6.16 Setup for the production of nanocomposite films [287]

6.4.3 *The Methods of Formation of Metal – Carbon Nanocomposites*

Metal - carbon nanocomposites can be prepared by pyrolysis of poly(vinyl chloride) (PVC), poly(vinyl alcohol) (PVA), PAN and cellulose hydrate fibres under carbonization conditions in the presence of metal-containing precursors. A typical procedure for the preparation of metal–carbon nanocomposites consists of several stages. The thorough mixing of the carbonaceous material with an aqueous solution of a metal salt (most often, nitrate) is followed by ultrasonic homogenization, filtration and drying in air; the process is completed by thermolysis at 870 K and pyrolysis at higher temperatures (in some cases, up to 2,070–2,270 K depending on the material composition) [294]. The chemical stability of these materials depends considerably on the nature of the initial polymer, conditions of thermolysis and the nature of additives (see, for example, Ref. [295, 296]).

In recent years, vigorous studies of carbon (graphite) nanostructured membranes have been carried out, dealing with both optimization of the selective layer and the mechanisms of molecular transport in them. The preparation of composite membranes with enhanced mass exchange (selectivity, permeability) and service (mechanical strength and thermal and chemical stabilities) characteristics was reported [297, 298]. The membranes are formed upon carbonization of phenol formaldehyde resins on heating to 973 K on ceramic supports. Tubular or flat ceramics (with pore size of 40 nm to 5 mm) based on α-Al_2O_3, TiO_2, ZrO_2, in particular those with additional metallic layers (Cr, Ti, Mo) are used as the supports. In the case of chromium additives, a graphite-resembling membrane structure is formed.

A thermally stable carbon nanocomposite with inclusion of metallic copper (nanoparticle size of 10–80 nm) is produced upon thermolysis of PAN and $CuCl_2$ in the presence of nitric acid [299]. This method is also suitable for preparing bimetallic nanocomposites. In particular, successive layering of the corresponding precursors (in 1:1 ratio) on PAN and thermolysis at 870 K results in the formation of the Au–Co/C metal–carbon nanocomposite [300]. Of some interest is the interaction of metal salts with a conjugated system of polymers. For example, $Co(acac)_2$ (acac is acetylacetonate) is predominantly coordinated with a conjugated system and slightly bounded with a terminal nitrile group of the conjugated chain. At the same time, $CoCl_2$ interacts with a conjugated system more strong playing the role of a dopant [301].

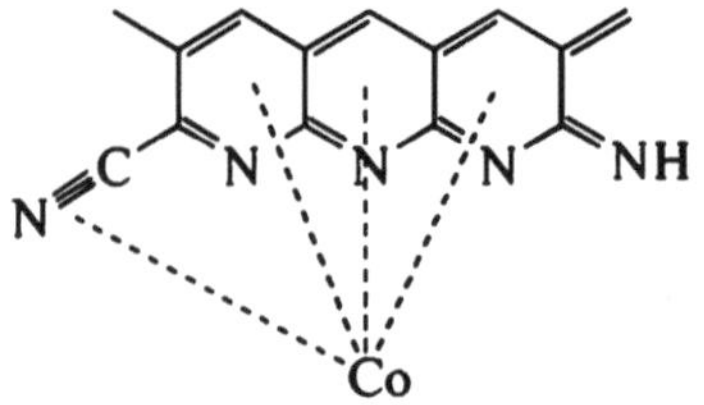

PAN/$Co(C_5H_7O_2)_2$

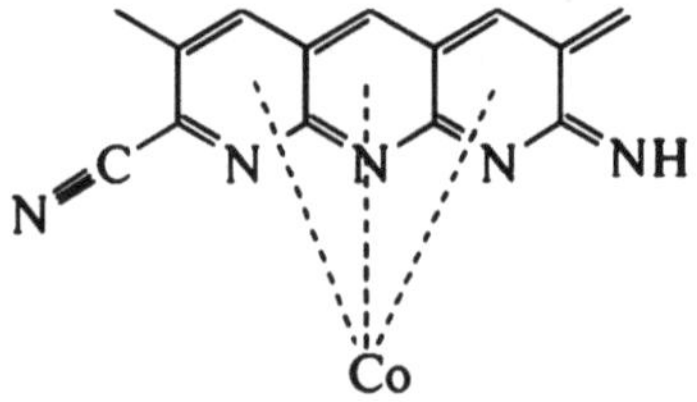

PAN/$CoCl_2$

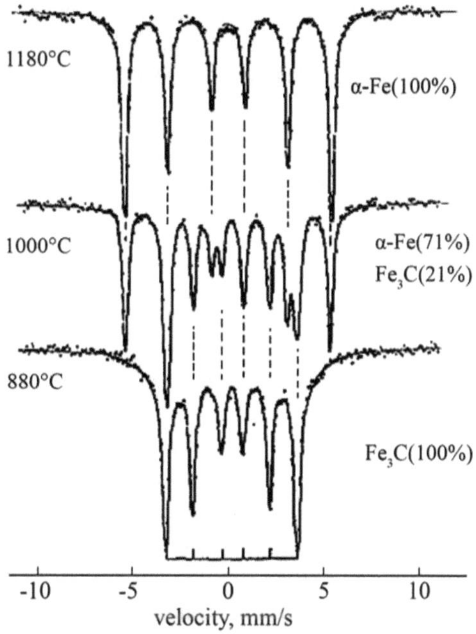

Fig. 6.17 Mossbauer spectra of the products of thermolysis of hexadienyl-[tri(carbonyl)iron] acrylonitrile copolymer (molar ratio 9.2:1) obtained at different temperatures

Comparatively on the early stages of the study of thermolysis attention was drawn the peculiarities of macromolecular complexes obtained by polymer-analogous transformations (see, for example, [302]). These macrocomplexes contain chemically bound metal compounds in the side chain. For example, the thermal treatment of hexadienyl-[tri(carbonyl)iron]acrylonitrile copolymer (the ratio of monomer units in polymer 9.2:1) (**1**) in air at 1373 K gives the uniform fine powder of nanocrystallites (80–120 nm) stabilized by a carbon matrix. Their phase composition (iron carbide or α-Fe) depends on the temperature of thermolysis (Fig. 6.17).

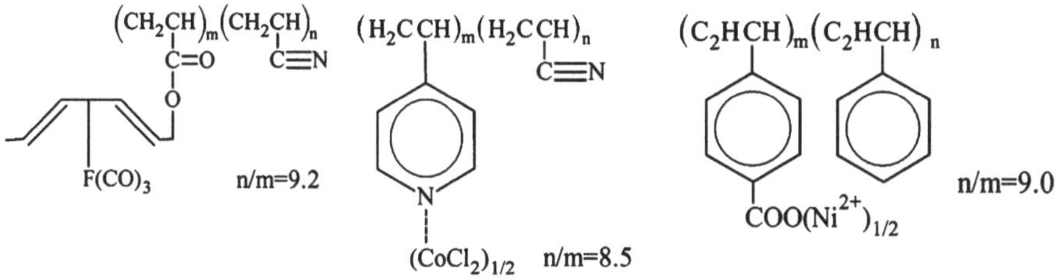

1 **2** **3**

Meanwhile, thermolysis of acrylonitrile and 4-vinylpyridine copolymers (molar ratio 8.5:1) with $CoCl_2$ or nickel-bis(styrylcarboxylate) at 1,273 K for 24 h affords ultrasmall β-cubic Ni (52 nm) or β- and α-Co (18 nm) nanoparticles.

The insertion of metal-containing precursors into glass–carbon matrices appears to be a promising approach. Thus thin films of platinum-doped glass carbon have been obtained by immobilization (at 338 K) of mononuclear complexes of ethylenebis(triphenylphosphine)platinum(0) on the poly(phenylenediacetylene) oligomer and subsequent pyrolysis (up to 873 K) (Scheme 6.7).

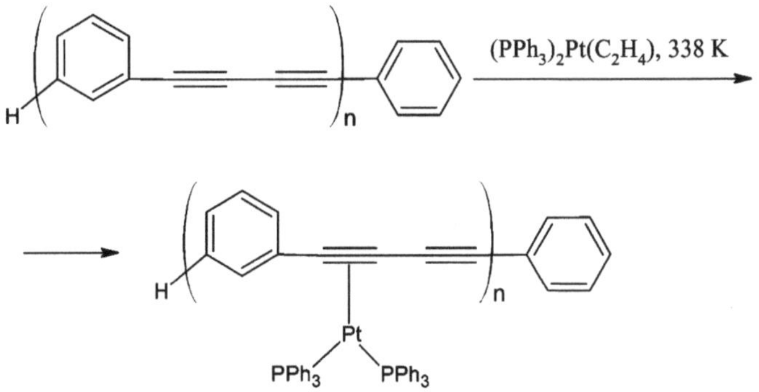

Scheme 6.7 A scheme of synthesis of ethylenebis(triphenylphosphine)platinum(0)/poly (phenylenediacetylene) oligomer

Platinum forms spherical particles with a mean diameter of 1.6 nm and a narrow size distribution [303]. It is significant that thermolysis of complexes not bound to the oligomer results in considerable enlargement (to 600 nm) of Pt particles.

Methods of this type are used rather often to obtain composite materials. For example, ceramics described as $M_xSi_yC_z$ (M = Fe, Co) was produced by high-temperature pyrolysis (1623 K) of diacetylene oligomers containing transition metal clusters [304]. Co-pyrolysis of ferrocene and thiophene (1,273–1,423 K) affords carbon nanotubes with Fe particles localized only inside the tube [305]; the metal clusters arising upon thermolysis of $Fe(CO)_5$ catalyze the formation of carbon nanoparticles [306]. Metal–carbon composites are also formed upon laser pyrolysis of a mixture of $Fe(CO)_5$ and C_2H_4, joint carbonization (1,173 K) of ferrocene and maleic anhydride or tetrachloroethylene [307, 308], controlled thermolysis of organocobalt compounds [309] and so on. Pyrolysis of poly (phenylcarbyne) $(PhC)_n$ mixed with $HAuCl_4$ at 873 K under reducing atmosphere ($Ar+10\%H_2$) affords amorphous carbon nanocomposite films containing gold nanoparticles [310]. The key stage of these processes is the nanoparticle formation and mixing with the polymer. The presence of 1 % of metal carbonyl catalyzes the extensive destruction of mesogenic pitches (specific matrices with polycondensed aromatic structures, which can function as p-ligands stabilizing metal clusters) [311, 312]. An increase in the iron content to 5 % in ethylene polymerization and thermolysis at 520 K give rise to Fe–LDPE composites (LDPE is low-density polyethylene) with a considerable portion of structures resembling layered graphite compounds [313]. As the content of nanoparticles in the polymer increases, the degree of polymer amorphization increases, which is clearly demonstrated by Table 6.9.

Table 6.9 X-Ray characteristics of Fe composite materials [313]

Polymer/iron components	RCS, nm	Content of the crystalline phase, %
Polycarbonate	8.0	65
+0.5 % Fe	9.0	60
+1.0 % Fe	9.0	60
+5.0 % Fe	10.0	50
+10 % Fe	10.0	40
Low-density polyethylene PELD	21.5	30
+1.0 % Fe	21.5	30
+5.0 % Fe	26.5	20
+10 % Fe	25.0	15

Note. RCS is the region of coherent scattering

A simple method for the preparation of ferromagnetic 1D nanofibres with a core–shell structure is based on the reaction of $Fe(CO)_5$ with elemental sulfur or tellurium at 970 K in a closed reactor under self-generated atmosphere (RAPET process) [314]. This produces FeS/C and FeSe/C nanoparticles several micrometres long with a diameter of 80–100 nm coated by a carbon shell. Presumably, under these conditions, the catalytic effect of iron is manifested, giving rise to oriented carbon nanotubes, which stabilize chalcogenide nanoparticles. The methods based on controlled carbonization of polymers followed by immobilization of metal nanoparticles (for example, to produce high-temperature filters and gas separating membranes [297]) are being currently successfully developed.

Worthy note a comparative simple method of preparing nano- and microparticles of Zn/C, Cd/C, Al/C stable in air by decomposition corresponding compounds of $Zn(C_2H_5)_2$, $Cd(CH_3)_2$, $Al(C_2H_5)_3$ at 773–1,073 K in a self-generating atmosphere [315]. Morphology of these materials is a core-shell. Such approach appears to be perspective for obtaining the similar composites based on other metals (for example, from tetraalkyltin); thermolysis of molybdenum alkoxide affords MoO_2/C [316]. From such precursors under sertain conditions (high temperature, reduction atmosphere H_2/C_xH_y) it is possible to prepare metal carbides in one-stage route. These are SiC (at 1,073 K) as nanorods, nanotubes of WC (from $W(CO)_6$ at 1,173 K), micro- and nanospheres of Mo_2C (from $Mo(CO)_6$ at 1,073 K) or even composites of SiC/C [317]. Thermolysis of powder cellulose (PC) in the presence of hydrolyzing tetraethylsilane gave rise to the nanocomposite PC/SiO_2 [318].

The above-mentioned thermolysis of solid/liquid mixtures in the case of metal/polytetrafloroethylene (M/PTFE) offers a number of the advantages such as one-stage self-supporting process with a high rate and selectivity, no additional consuming of heating, the relatively simplicity of apparatus. Structure of carbon particles containing metals and Si nanofibers formed can be adjusted by a heating regime [58].

The aerosol thermolysis in solution is an easy method to produce ceramic electrolytes of oxide fuel elements (for example, on basis of yittrium (8–10 mol. %)- or gadolinium (10–20 mol.%) stabilized zirconia) [319].

The above-analyzed data is only a brief illustration of the potential of thermolysis in the construction of metal/carbon nanocomposites.

6.5 Thermolysis of Metal-Containing Monomers in the Frontal Mode

A promising trend in the self-controlled synthesis of metallopolymer nanocomposites is frontal polymerization (FP) of solid metal-containing monomers [320–324]. Frontal polymerization is a layer-by-layer self-maintained chemical transformation where the localized zone of intense reaction (polymerization front) travels over the bulk of the polymerizing compound. As compared with the usual bulk polymerization, FP has a number of important advantages: lower energy expenditure, high rates and low reaction times, greater degrees of conversion and the possibility of the solid-state process.

The FP mode is implemented as follows. For a short period of time (~10 s), the system is subjected to thermal perturbation (ignition) by heating a small portion of the reacting volume with an electric heating device or a stationary external heat-transfer agent set to a specified temperature. Further polymerization is maintained by system inner resources proceeding in a narrow temperature range close to adiabatic heating of the reaction mixture. The appearance of a melt zone (1st order phase transition) and a colour change can be observed visually; the migration of the colour change boundary is used to monitor the reaction rate. The dynamic pattern of transformation of this polymerization wave is presented in Fig. 6.18.

The FP phenomena in solid monomers are little described in the literature [325, 326]. The first and still the only one reported case of purely thermal initiation of FP in the condensed phase is decomposition of acrylamide (AAm) complexes of transition metal nitrates $[M(CH_2{=}CHCONH_2)_4](NO_3)_n \cdot xH_2O$, where M = Cr (III), Mn(II), Fe(III), Co(II), Ni(II), Cu(II), Zn(II), Pd(II) $n = 2, 3$; $x = 0, 2$. The isothermal gas evolution rate (W) in a static non-isothermal reactor (SGA mode) as a function of the degree of conversion (η) is approximated satisfactorily (up to $\eta \leq 0{,}9$) by a first-order autocatalysis equation $W = d\eta/dt = k(1 - \eta)(\eta + \xi_0)$ where the constants $k = 4.2{\cdot}10^7 \exp[-24{,}000/(RT)]\ s^{-1}$, $\xi_0 = 1.9{\cdot}10^{-2}$.

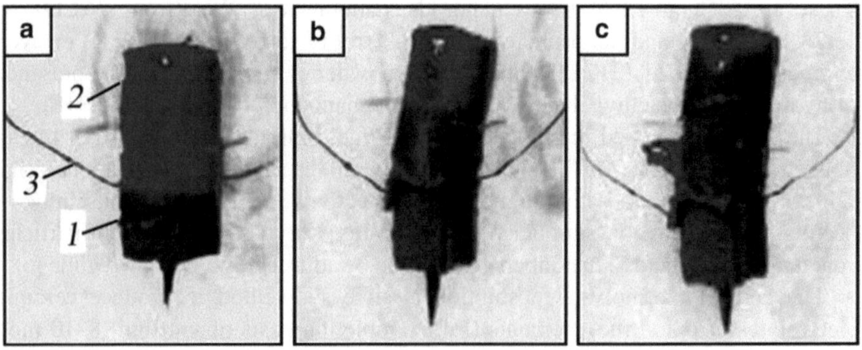

Fig. 6.18 Dynamics of the polymerization front propagation for Co(II) acrylamide complex [322]. (**a**) after 20 s, (**b**) after 45 s, (**c**) after 55 s. (*1*) monomer, (*2*) polymer, (*3*) thermocouple

The transformation includes three stages:

1. dehydration of the monomer (333 ± 358 K)

$$\left[Co(CH_2 = CHCONH_2)_4(H_2O)_2\right](NO_3)_{2\,(s)} \rightarrow \\ \rightarrow \left[Co(CH_2 = CHCONH_2)_4\right](NO_3)_{2\,(s.)} + 2H_2O)_{(liq)} - 19\ \mathrm{kJ/mol},$$

2. polymerization of the dehydrated complex (373–413 K) accompanied by the reaction of H_2O vapour with the NO_3^- anion and the NH_2 group of the AAm ligand yielding NH_3 and HNO_3 vapours.

$$n\left[Co(CH_2 = CHC(O)NH_2)_4\right](NO_3)_2 \xrightarrow{-H_2O} \\ \longrightarrow \left[Co(CH_2 = CHC(O)NH_2)_4\right]_m + NH_3 + NHO_3 + \\ + \leq 83\ \mathrm{kJ\,mol^{-1}};$$

3. thermooxidative destruction of the resulting polymer (5,453 K) through interaction with HNO_3 vapour and with reactive decomposition products thereof.

The last two stages are exothermic, which is a key cause of initiation of polymerization and development of the FP mode. The X-ray diffraction pattern of the product formed upon thermolysis of the cobalt complex at 673 K has peaks corresponding to the characteristic lines of the β-cobalt face-centred cubic lattice. The microstructure of the cobalt-containing product (Fig. 6.19) is composed of spherical nanoparticles with a mean size of 5–9 nm evenly distributed in the polymer matrix [327, 328].

Polymer-assisted synthesis of metallopolymer nanocomposites allows one to control precisely the size of nanoparticles as well their composition and structure. Thus, unlike the products of thermolysis of metal carboxylates, nanocomposites based on acrylamide complexes contain highly dispersed metal phase or, at higher temperatures of thermolysis, metal carbide phase. Such difference may be caused by the character of chemical processes occurring at thermolysis of the starting

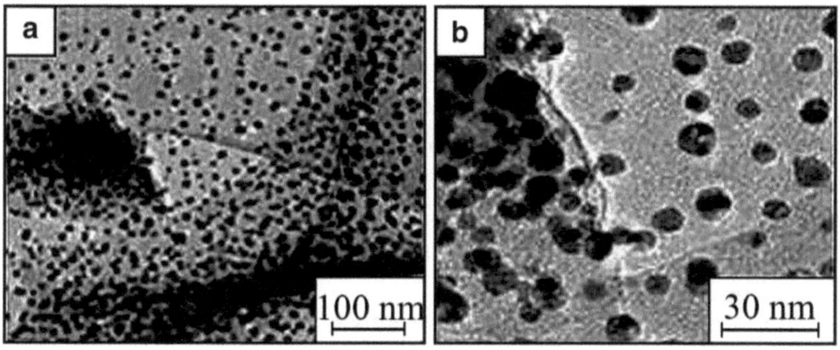

Fig. 6.19 TEM images (on different scales) of spherical cobalt nanoparticles in the polymer matrix [327]

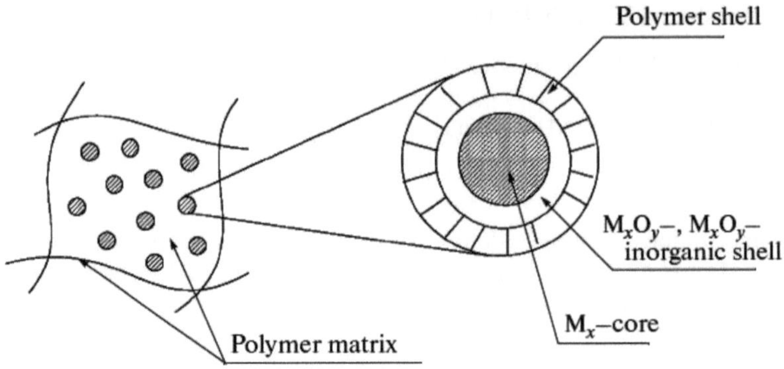

Fig. 6.20 A schematic view of the structure of nanocomposites obtained by frontal polymerization [329]

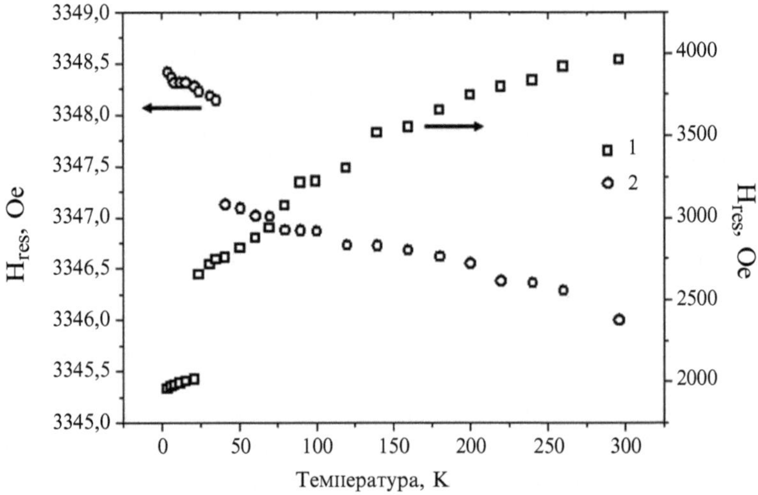

Fig. 6.21 Temperature dependences of resonant field H_{res} for (*1*) cobalt nanoparticles and (*2*) polymer shell

monomers and intermediates formed [329, 330]. The nanoparticles in such systems have a specific structure of a core-shell which comprises metal-containing core and the surface layer from a polymer (Fig. 6.20).

The correlation of the temperature dependences for ferromagnetic resonance (FMR) line widths and the resonant field H_{res} values of the polymer shell and a cobalt nanoparticle, which it surrounds, is indicative of an interaction between them (Fig. 6.21) [327, 329].

Variation of the thermolysis conditions allows adjusting the sizes and composition of nanoparticles as well as properties of nanocomposites formed. For example, it is of interest to track the magnetic properties of the products of thermal transformations of CoAAm obtained at temperatures of 673, 773, 873, and 1,073 K. The

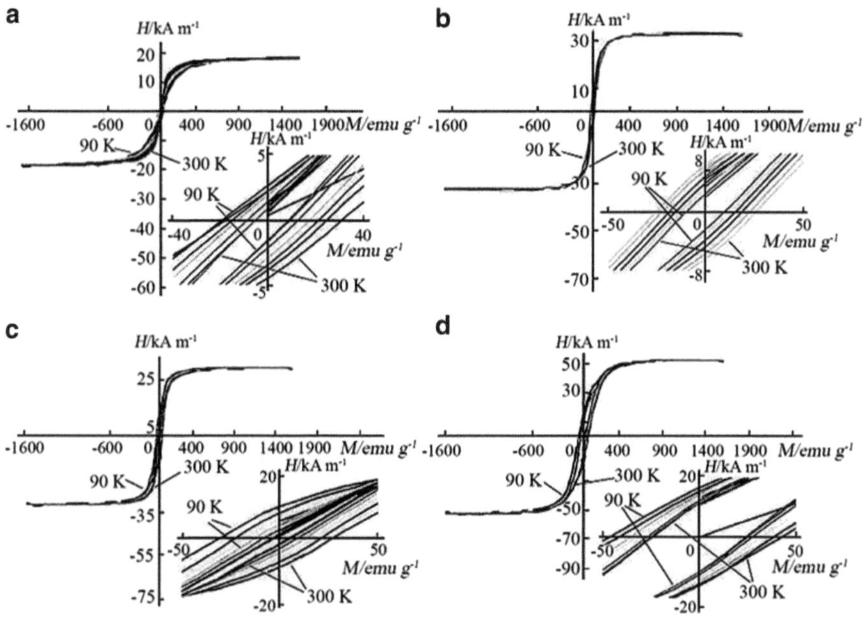

Fig. 6.22 Magnetization *vs.* magnetic field in the temperature 90–300 K for the CoAAm thermolysis products formed at 673 (**a**), 773 (**b**), 873 (**c**), and 1,073 K (**d**)

hysteresis loops measured in the temperature range of 90–300 K attest to the ferromagnetic behavior of the thermolysis products (Fig. 6.22) [330]. However, the pattern of the curves, especially in the case of thermolysis product formed at 673 K, indicate the probable presence of a disordered interfacial layer with a skewed spin structure on the surface of magnetic particles. At the same time, the nanocomposites on the basis of Ni(II) complex obtained under the same conditions reveal superparamagnetic properties with the blocking temperature $T_b \approx 40$ K (Fig. 6.23) [331]. An increase in the thermolysis temperature results in an increase in the particle size of Ni-containing nanocomposites from 6–7 nm (673 K) to 8–14 nm (773 K) and 12–17 nm (873 K). The sharp increase in *H*c for the product obtained at 873 K indicates that at this temperature, the material switches from superparamagnet to ferromagnet.

Note that this approach for synthesis of nanocomposite materials immediately during FP of metal-containing monomers is an effective method for obtaining polymer-immobilized catalysts including hybrid types [332, 333].

In parallel with FP experiments, theoretical views on the modelling of this process have also been developed. Currently, various model FP mechanisms are discussed. The development of the polymerization front in a relatively porous monomer is ideologically similar to the formation of the combustion front [334]. It is not surprising that the theory of thermal mechanism of propagation of the polymerization front (as well as the combustion front) became predominant, although alternative (diffusion) mechanisms of front development [335] and a

Fig. 6.23 Magnetization *vs.* temperature in the FC and ZFC modes in a 8 kA m^{-1} field (0.008 T) for the products of thermolysis of the acrylamide complex of Ni(II) nitrate formed at temperatures of 673 (**a**) and 773 K(**b**)

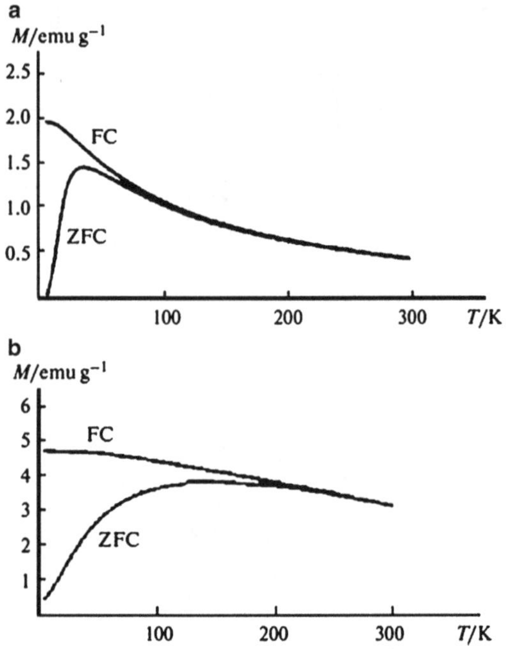

mathematical model of FP [336] were proposed. The latter concentrates on the investigation of heat transfer complicated by phase transitions in the travelling narrow layer of the system (one-dimensional conjugate problem for two semispaces with a heat evolution source at the moving conjugate boundary). An analytical functional dependence of the temperature at the wave front on the velocity of the phase transition boundary that was in agreement with experimental data was obtained.

6.6 Thermolysis of Precursors in the Polymer Matrices

There are two choices for the insertion of metal-containing precursors into a polymer matrix: *ex situ* and *in situ* [246]. The *ex situ* (in a second moment) process comprises the addition of micro- or nanoparticles of the inorganic precursor to the finished polymer and mixing. According to the *in situ* method, the initial nanoparticles are generated directly in the polymer matrix upon decomposition of the proper macromolecular metal complex or the products of its transformation.

As a rule, a polymer matrix affects the rate of decomposition of a thermolizing compound. However, at the low concentrations of polymer when the viscosity of a system is not high the kinetics of the precursor decomposition in a polymer solution and in a free polymer solution is the same. Such behavior was observed at the

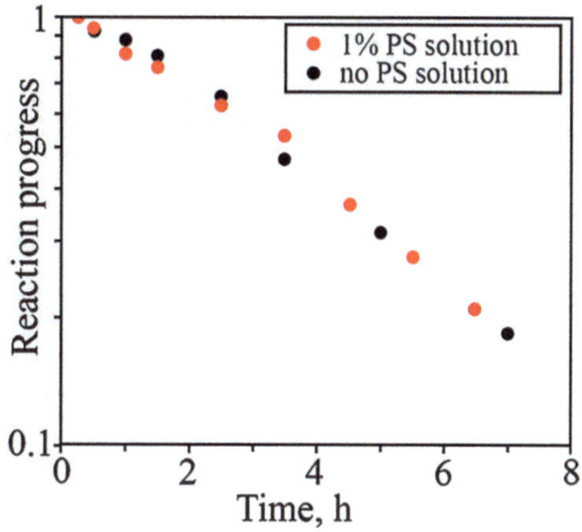

Fig. 6.24 The kinetics of decomposition of $Cr(CO)_6$ in a toluene solution in the presence and without of polystyrene (PS) [337]

decomposition of $Cr(CO)_6$ at 363 K in a polystyrene solution in toluene [337] (Fig. 6.24).

Of interest from the practical standpoint is the use of thermolysis of polymers serving as templates for the manufacture of porous inorganic membranes [22–24]. The latter are usually produced by the sol - gel method [10]. After burning-out of the organic part, 1.5–10 nm channels are formed in the SiO_2, TiO_2, SnO_2, ZrO_2, Nb_2O_5, etc. membranes. These membranes can be used in nanocatalysis or solar cells or as nano-sized reactors [338]. The use of natural or synthetic polymers as the template agents has been reported, in particular, agarose gel (1.5–2.0 mass %) [339], partially cross-linked polyacrylamide gel prepared by a reported procedure [340] or gelatin with cell diameter of 10–100 nm [341]. After the formation of inorganic Ti network, the polymer is burned out at 773–873 K; at the same stage, the membrane is sintered to form a porous monolithic structure. The TiO2 (rutile) phase with a mean particle size of 23.6 nm is formed, whereas high-temperature (1,223 K) annealing is accompanied by particle enlargement to reach a size of 50.4 nm. The general pattern of the process is shown in Scheme 6.8.

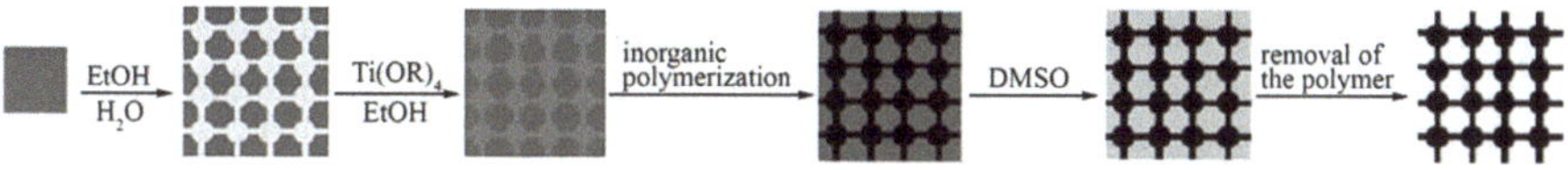

Scheme 6.8 Pattern of formation of an inorganic Ti network from a polymer gel [339]

Yet another example of this approach is the formation of nanocrystals of hydroxyapatite $Ca_{10}(PO_4)_6(OH)_2$, the main bone and tooth structural material (see Chap. 7). Aqueous solutions of $Ca(OH)_2$ and $Ca(H_2PO_4)_2$ are mixed in the presence of a sonicated colloid solution of bovine serum albumin (Mw = 66 kDa) [342]. This gives rise to amorphous nanoparticles with a fibrous network structure with a broad size distribution. Apparently, sonication turns the protein globular structure into a linear structure, and the arising $HO^{\cdot}$ radicals cross-link the chain through oxidation

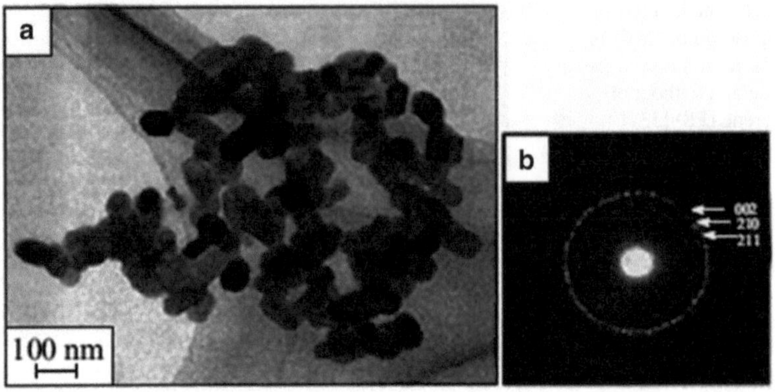

Fig. 6.25 Electron microscopic image of hydroxyapatite nanocrystals with bull serum albumin (**a**) and electron diffraction pattern (**b**) [342]

of the SH groups, while the negatively charged COO^- groups stabilize the calcium phosphate particles. The calcination of this precursor results in destruction of the protein to give nanoparticles the size of which depends on the albumin concentration. At a concentration of 5 g/L, nearly spherical 25 ± 100 nm particles and short rod-like crystals with size ranging from 40 nm × 75 nm to 70 nm × 150 nm are formed (Fig. 6.25).

During pyrolysis, polymers can also act as reactants. For example, the formation of metal carbides with the polymer matrix serving as the source of carbon has been repeatedly noted. In addition, H_2, CO and other gaseous products of polymer thermolysis serve as reducing agents for reaction mixture components.

6.6.1 Effect of Metal-Containing Components on the Destruction of the Polymer Matrix

The thermal destruction of polymers is associated with the development of the chain process that is initiated by the formation of carbon-centred radicals (R.) upon chemical bond cleavage in the macromolecule [48, 49, 343, 344]. The products of thermodestruction include low-molecular-mass volatile compounds of complex composition, in particular monomers, and a non-volatile residue, which turns into a carbonized (coke-like) material at high temperatures. During the thermooxidative destruction of polymers, hydroperoxides are formed as the primary products; they decompose to give $RO_2^{\cdot}$ radicals, which induce autocatalytic destruction of the polymer. One of the ways of stabilization of the polymer matrix that decomposes by a radical mechanism is to provide kinetic chain termination upon the reaction of an inhibitor with the $R^{\cdot}$ or $RO_2^{\cdot}$ radicals, which are thus converted to low-activity or inert products.

Table 6.10 Kinetic parameter (pre-exponential factor in the Arrhenius equation and activation energy) and decomposition temperature of metal-containing poly(methyl methacrylates) (Mol. weight $(2.7 \div 5.0) \cdot 10^5$) [345]

Metal	A/s^{-1}	Ea/kJ mol^{-1}	T_d/K
Au	$2.6 \cdot 10^4$	99.3	573
Bi	$2.3 \cdot 10^2$	78.9	598
In	$1.4 \cdot 10^2$	74.2	548
Cu	$2.4 \cdot 10^2$	70.6	513
Sb	25.0	63.9	548
Pd	15.0	61.5	523
Sn	14.0	61.0	523
Ga	4.3	51.5	473

Polymers can be stabilized by introducing acceptor agents. They either remove agents that initiate chains (oxygen, active impurities, etc.) from the polymer or deactivate these agents. An effective way of influencing the thermal and thermooxidative destruction is to introduce into the polymer matrix highly dispersed metal-containing precursors capable of both inhibiting and catalyzing the polymer decomposition. For example, introduction of pre-dispersed metal particles (up to 2 % of Au, Cu, Pd, Ga, In, Sn, Sb, Bi) into PMMA (molecular mass of 270–500 kDa) during polymerization has a pronounced effect on the kinetic parameters of destruction [345]. Probably, the metallopolymer is formed as shown below

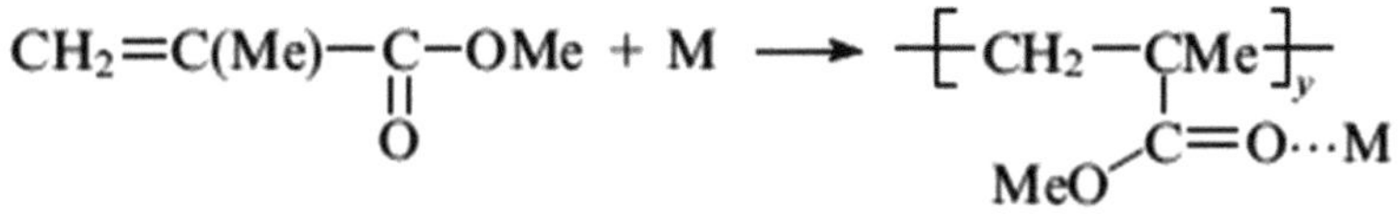

Pyrolysis of the metallopolymer is a zero-order reaction. The best results (as regards the activation energy and the temperature of the onset of decomposition) were found for the Au–PMMA system, which is, like metallic gold, is fairly stable against oxidation. In the case of Ga (readily oxidizable metal), the lowest thermal stability parameters were found (Table 6.10).

The introduction of a small amount (0.05–1.00 %) of iron atoms into LDPE provides a considerable increase in the thermal stability of the composite as compared with the pure polymer, and the decomposition temperature of the nano-α-Fe_2O_3–PS system obtained from an aqueous colloid solution of FeOOH upon stirring for many hours followed by evaporation is ~97° higher than that for pure PS [346–348].

6.6.2 *Metal Salt – Polymer Matrix System*

Thermolysis of the polymer composites with metal halides MX_n (M = Cr, Mn, Zn; X = Cl, Br) inserted in the matrix, which is characterized by a complex pattern of transformations, has been reported [349]. Apart from the task of preparing highly dispersed metal-containing products, investigations of this process are of interest for the protection of polymeric materials from the effect of flame and for the search for environmentally safe fire retardants (as alternative to toxic additives [350–352]).

Decomposition of the polymer matrix in the MX_n – PMMA composites ($MX_n = CrCl_3$, $MnCl_2$, $ZnBr_2$) occurs, first of all, as destruction. This takes places at temperatures below the destruction temperature of pure PMMA (~647–690 K): at ~430 (for $ZnBr_2$) [353], ~490 (for $MnCl_2$) [353], ~520 K (for $CrCl_3$) [354]. The replacement of $ZnBr_2$ by $ZnCl_2$ increases the destruction temperature to a value close to that of pure PMMA [354].

Heating of chromium(III) chloride as a composite with PMMA to 520–970 K leads to a ~62 % weight loss of the composite [355] caused by the removal of the volatile monomer, carbon oxides, HCl, traces of methane and unidentified low-molecular-mass organic products. Solid anhydride and chromium oxides, together with unsaturated oligomers remain in the condensed phase. Presumably [353–357] destruction of the $CrCl_3$ – PMMA system follows a radical mechanism the main stages of which (Scheme 6.9) include pre-coordination of chromium to the carbonyl group of PMMA followed by elimination of the chlorine atom; reaction of the chlorine atom with a PMMA hydrogen atom; migration of the ester methyl group to the PMMA backbone to give a chromium carboxylate salt, which decomposes to the anhydride; polymer stabilization by chromium carboxylate or by cross-linking of polymer radicals generated in the previous stages.

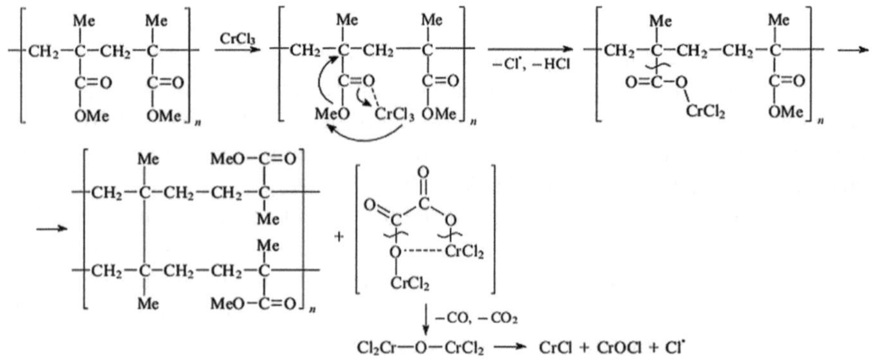

Scheme 6.9 Mechanism of the thermal destruction in the system of PMMA–$CrCl_3$

The mechanism of thermal destruction of the $MnCl_2$ – PMMA composite at 370–870 K has been discussed [357]. It is assumed that the process is initiated by the methoxyl and methyl radicals arising upon cleavage of ester bonds in PMMA; however, manganese chloride accelerates the initial depolymerization. Manganese oxide, which is one of the final solid products of thermolysis, results from the reaction of the carboxyl radical with $MnCl_2$ and the subsequent conversion to the manganese intermediate connected to the PMMA chain (Scheme 6.10).

Such processes are intensively investigated.

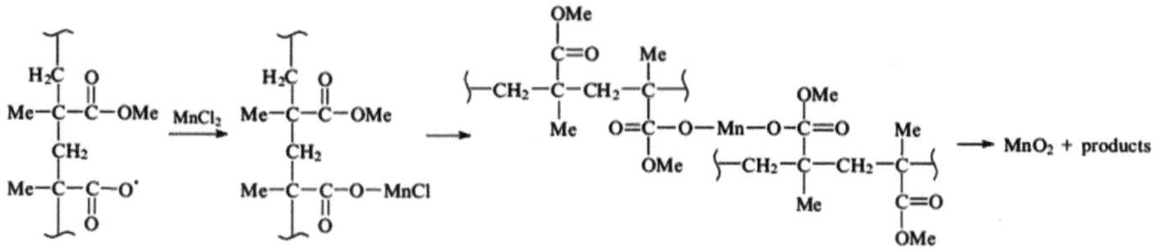

Scheme 6.10 Mechanism of the thermal destruction in the system of PMMA–$MnCl_2$

6.6.3 Thermolysis of Complex Compounds in Polymer Matrices

In recent years, considerable attention has been attracted by composites based on metal ions coordinated to the polydentate organic ligands of the framework (polymeric or inorganic) owing to their broad range of their practical applications. These materials are used as sensors [358, 359], in non-linear optics [360], for separation and molecular recognition [361], for gas storage [362, 363], in catalysis [364], in information and energy storage sources, in biomedicine, etc.

There are two basically different ways of performing thermolysis of coordination compounds in a polymer matrix, namely, the introduction of metal complexes into a traditional matrix and direct formation of a macromolecular metal complex, usually having a chelate structure.

The polymer matrices containing groups able to form complexes (amino, amido, hydroxyamino and hydroxyl groups) react with MX_n [365] (M = Cu(II), Ni(II), Co (II), Zn(II), Mn(II), VO(II), UO(II), Zr(IV), Ti(IV)) to give the corresponding adducts (Scheme 6.11). The thermal stability of metal complexes decreases in the sequence of metals Ni > Zn > Mn > Co > Cu > Ti > UO > Zr > VO, and polymers

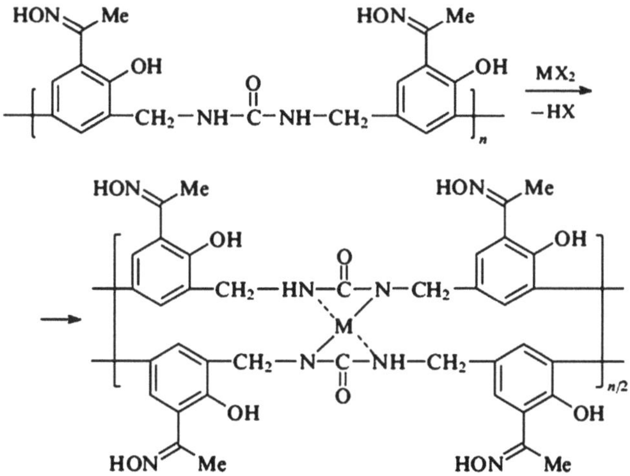

Scheme 6.11 A scheme of the formation of polymer metal chelates

that do not coordinate metals are most stable. The same sequence holds for copolymers of diaminoalkanes with the Schiff base, 5,50-methylenebis (3-bromosalicylaldehyde) [366, 367].

The thermal transformation pathways of the macromolecular metal chelates have been analyzed [368–377]. Thermolysis of copper macrochelates with polyacrylamidoxime fibres comprises three temperature-separated stages: 398–603 K (weight loss 14.3 %), 603–715 K (10.8 %) and 715–965 K (10.8 %) [378]. The major weight loss is involved in the transformation of the starting polymer. The thermal behaviour (in the temperature range of 473–1,073 K) of the macromolecular Cu(II) and Co(II) chelates with the copolymer of N-phenylmaleimide with acrylic acid was studied; the ratio of comonomers was varied and various methods were used for metal binding [379]. An increase in the number of N-phenylmaleimide units in the copolymer chain increases in some cases the thermal stability and the glass transition temperature (T_g) of the macrocomplex; however, this change is not so pronounced as might have been expected (in the optimal cases, from 623 to 646 K, and for some copolymer compositions, even a decrease to 488 K is observed).

The high-temperature pyrolytic carbonization processes of the precursors containing bulky organic ligands proceed in an unusual way. Under inert atmosphere, the ruthenium complexes $[RuL_3]X_2$ (L is bipyridine (bipy), phenanthroline; X=OH, Cl) were converted to a ruthenium – carbon composite with a ruthenium content of 20–32 mass % [380]. This is a result of multistage destruction of the organic ligand. On heating to 873 K, vigorous decomposition of the complex with evolution of its fragments, pyridinium and bipyridinium ions, is completed. This is followed by the destruction of the condensed material thus formed to evolve light hydrocarbon species. The Ru-containing composites obtained at 873–973 K have a relatively great specific surface area (424–477 m^2/g), while ruthenium clusters are almost amorphous to X-rays. The highly dispersed ($d \approx 1.5-2.0$ nm) ruthenium particles in the Ru/C composite are similar in size and are shaped like planar hexagons, which are associated. During pyrolysis, enveloping layers of turbostratic graphite-like carbon are formed around the hexagons as a result of organic ligand carbonization. The parallel-oriented turbostratic carbon layers with an interplanar spacing of ~0.34 nm are tightly pressed to ruthenium particles, thus preventing their sintering under the drastic conditions of pyrolysis (973 K). Thus, obtained composite can be represented as a set of spheres (coils) each having a ruthenium nanoparticle (core) located at its centre and turbostratic carbon layers (shell) lining the outer surface.

Metal acetylacetonate – polymer matrix systems have found wide use. During thermolysis, $M(acac)_n$ can play a dual role. On the one hand, decomposition of M $(acac)_n$ produces metal-containing nanoproducts, and on the other hand, the complex initiates depolymerization and destruction of the polymer matrix. In this respect, detailed thermovolumetric and thermogravimetric studies of the thermal decomposition of the $Co(acac)_3$ – PMMA and $Mn(acac)_3$ – PMMA systems over a broad range of $M(acac)_3$: PMMA ratios (from 1:10 to 1:400) are illustrative [349,

381, 382] Decomposition of the Co composite includes four stages corresponding to maximum temperatures of 405, 460, 540 and 610–630 K (Scheme 6.12).

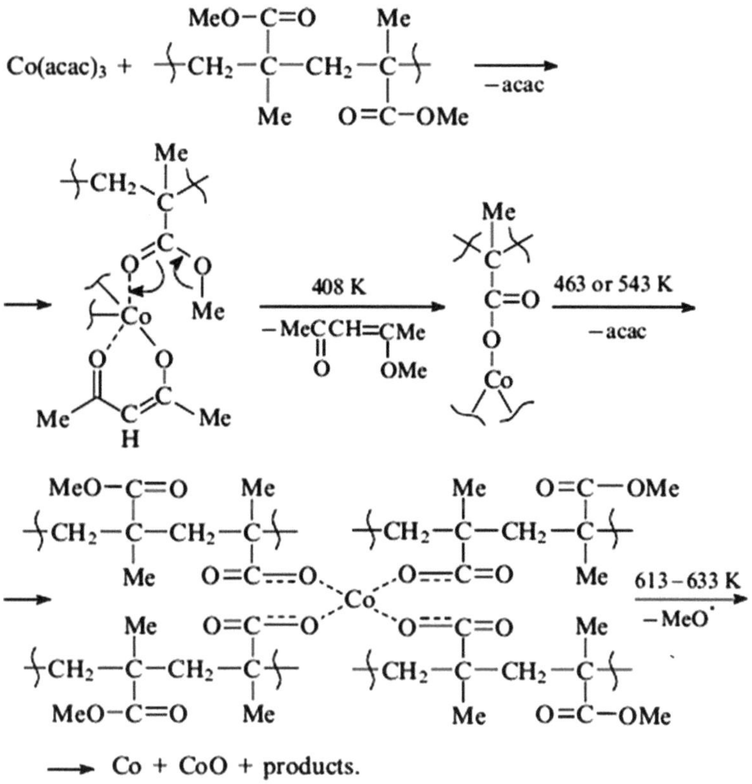

Scheme 6.12 Mechanism of the thermal destruction in the system of PMMA–Co(AcAc)$_3$

The first stage is associated with gradual reduction of Co(acac)$_3$ to give acetylacetone and Co(acac)$_2$, which is coordinated by the unsaturated terminal groups of PMMA. Decomposition of this complex is accompanied by evolution of the monomer, which is the main decomposition product. The second and third stages continue accumulation of the monomer through the reduction of Co(III) to Co(II) and subsequent depolymerization of the remaining terminal groups. Simultaneously, Co(acac)$_2$ can be coordinated by ester groups, thus accelerating their destruction to give methoxy radicals (MeO˙). The latter, in their turn, attack the ester groups thus converting them to anhydride fragments. The fourth stage is accompanied by the most pronounced gas evolution giving mainly non-condensed gases.

A similar situation is observed upon thermal decomposition of Mn(acac)$_3$ in PMMA: decomposition starts at 395 K (i.e., at a lower temperature than for pure PMMA); however, the major process takes place at higher temperatures. This is

Table 6.11 Thermal stability of PMMA and its complexes with $Mn(acac)_3$ [382]

Polymer	The number of monomer units per $Mn(acac)_3$ molecule	T_d/K	T_{max}/K
High-molecular-mass PMMA	See[a]	483	623
	50	403	638
	10	398	673
Low-molecular-mass PMMA	See[a]	493	633
	50	403	638
	10	403	673

[a]Neat polymer

confirmed by comparison of the temperatures of the onset of decomposition (T_d) and the major decomposition (T_{max}) presented in Table 6.11.

For the formation of rhodium nanocrystals with different shapes (multipods, cubes, horns, cuboctahedra), thermolysis of $Rh(acac)_3$ in the presence of polyols as reducing agents is of interest [383]. Yet another convenient method for the production of these polymer-protected monodisperse rhodium nanoparticles (size 5 ± 15 nm) is one-stage reduction of the rhodium complex Rh(acac)3 in the presence of butane-1,4-diol and polyvinylpyrrolidone (Mw = 55 kDa) at temperatures of 240 ± 300 K [384]. This gives triangular, pentagonal and hexagonal particles, which can form Langmuir–Blodgett films on the silicon surface.

The polymer metal chelates of oligo-2-[(4-bromphenylimino)methyl]phenol (OBPIMP) with Co(II), Ni(II) and Cu(II) acetates [33] have been already mentioned when we discussed the kinetic peculiarities of thermolysis:

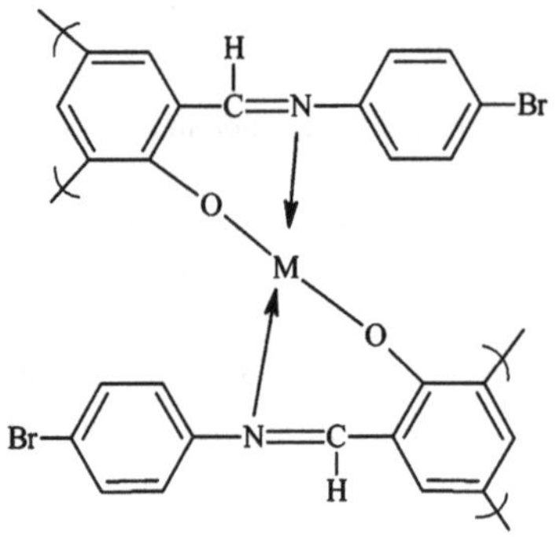

$$M = Co(II), Ni(II), Cu(II).$$

Thermodynamical and kinetic parameters of decomposition of these polymer metal chelates calculated with the Coats-Redfern equation are given in Table 6.12.

Table 6.12 Thermodynamical and kinetic parameters of thermal decomposition of the OBPIMP-chelates

Compound	Stage[a]	The reaction order, *n*	E_a, kJ/mol	ln A, s^{-1}	ΔS^*, kJ/mol	ΔH^*, kJ/mol	ΔG^*, kJ/mol
BPIMP	I (458–602 K)	0.5	76.25	13.95	−134.6	71.28	151.6
OBPIMP	I (453–581 K)	0.9	96.31	19.53	−87.52	91.76	139.6
OBPIMP-Co	I (475–590 K)	0.2	39.84	5.81	−201.6	35.31	145.0
	II (626–997 K)	0.9	31.34	0.693	−248.4	23.75	250.3
OBPIMP-Cu	I (502–626 K)	0.5	100.2	18.16	−99.75	95.29	155.2
	II (626–997 K)	1.0	65.36	6.33	−201.3	57.96	235.5
OBPIMP-Ni	I (530–656 K)	0.8	62.04	6.48	−197.3	56.76	181.9
	II (656–899 K)	0.4	48.81	6.82	−195.7	42.65	187.5

[a]I and II are the temperature ranges of the starting and 50 % mass loss

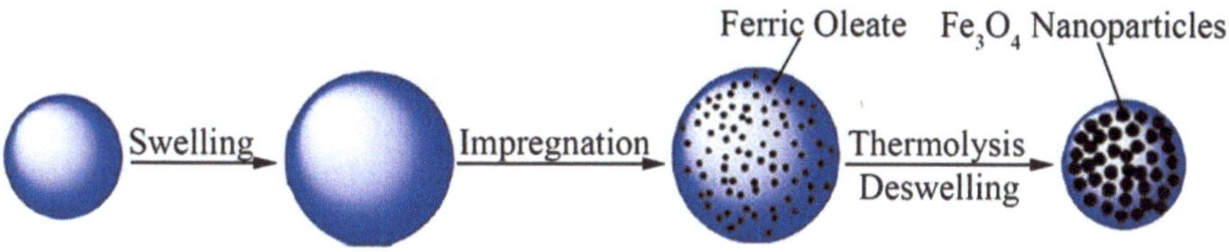

Fig. 6.26 Schematic diagram of the preparation of monodisperse magnetic polymer microspheres by swelling and thermolysis technique

Worthy of note is a recently elaborated method for the preparation of nanocomposites microspheres by combining swelling and thermolysis technique. The monodisperse polystyrene microspheres were first prepared by dispersion polymerization and swelled in chloroform. Then, ferric oleate was dispersed in chloroform as a precursor and impregnated into the swollen polymer microspheres. Subsequently, the iron oxide nanoparticles were formed within the polymer matrix by thermal decomposition of ferric oleate (at 583 K, for 0.5 h, nitrogen atmosphere) (Fig. 6.26) [385].

Iron oxide nanoparticles localized within polymer microspheres reveal unique superparamagnetic properties and can be used in various fields of biotechnology including purification of proteins, viruses, nucleic acids, in biosensors and enzyme immobilization, etc. [386–391] (see Chap. 7).

It should be noted that the polymer matrix, in turn, may influence thermal transformations of metal complexes. However, information concerning this matter is limited. Some peculiarities of thermal decomposition of $(NH_4)_6Mo_7O_{24}.4H_2O$, copper nitrate trihydrate, and polyoxymetallates in thin films (from 50 to 600 μm) of PVA can be assigned [392].

6.6.4 *Thermolysis of Organometallic Compound – Polymer Matrix Systems*

As noted above, thermal decomposition of single organometallic compounds affords metallic nanoparticles or (in oxidizing atmosphere) metal oxides. The products of thermolysis tend to be highly reactive especially during nucleation and at an early stage of growth of nanoclusters in the condensed phase. The most convenient method for the preparation of diverse metal oxide nanocrystals with well characterized size, morphology, phase composition and structure is the simple and economical non-hydrolytic bottom up process. This process includes thermolysis of alkoxometallates and their derivatives in high-boiling organic solvents (see, for example, Ref. [393]). This especially concerns the preparation of manganese oxides, which are important products for catalysis, electronics and production of magnetic materials.

Among rather simple methods, note the synthesis of vanadium oxide nanoparticles stabilized by carbon shells [394]. Thermolysis of oxovanadium triethoxide $VO(OEt)_3$ in a closed reactor and in self-generated atmosphere at 973 K affords carbon-coated nanoparticles of rhombohedral V_2O_3 with a core - shell structure and a diameter of ~55 nm. The subsequent thermal treatment of the product in air at 723 K for 2 h furnishes orthorhombic V_2O_5 ([5] [395]), (particle size ~250 nm), which is also coated. These materials have photoluminescent properties [396].

Spherical ruthenium nanoparticles are formed upon thermolysis in air of the organometallic copolymer $\{[NP(O_2C_{12}H_8)]_x[NP(OC_5H_4N)CpRu(PPh_2)_2]_{1-x}\}_n$ (Cp is cyclopentadienyl) [397].

On exposure of the cluster $[Ru_6C(CO)_{15}Ph_2PC_2PPh_2]_n$ ($n \approx 1{,}000$) to accelerated electrons, a nanochain composed of ruthenium atoms was obtained [75]. Pyrolysis of the cluster $Os_2(\mu\text{-}I)_2(CO)_6$ gives rise to rod-shaped metal particles [398]. The organometallic polymer $[Ru(CO)_4]_\infty$ with a planar structure and zigzag-like conformation is a potential precursor for the synthesis of various nanomaterials, in particular, for the formation of chain nanoparticles, upon removal of the carbonyl ligand [399]. Thermolysis of $[Ru(CO)_4]_\infty$ on a SiO_2 surface at 443 K affords ~2 mm-long and ~30 nm-thick nanofibres. Upon thermal

---

[5] In the cited work it is not paid attention to the fact that components of the reactor wall may reveal some catalytic properties. As a rule, such reactors are lined by Teflon. Among the known vanadium oxides (V_2O_5, V_2O_3, V_3O_5, VO_2 and VO), most stable is the oxide with the higher oxidation state of vanadium (V2O5), an oxidant with amphoteric properties. A comparative simple procedure was elaborated for obtaining VO_2 by thermolysis at $\leq$683 K of the precursor arised from the interaction of $VOCl_2$ with NH_4HCO_3 in solution [395].

decomposition of the cluster $Ru_3(CO)_{12}$ in the nanopores of Al_2O_3 membranes (100 nm size), nanofibres of diameter 10 nm are formed. When the temperature increases to 553 K, the structure of such items becomes irregular as like to the $Os_3(CO)_{10}(\mu\text{-}H)(\mu\text{-}OH)$ thermolysis product.

Of particular interest is the formation of refractory metal carbides upon the pyrolysis of various organometallic precursors. For example, the melting point of hafnium carbide is 4,201 $\pm$ 20 K and that of tantalum carbide is 4,256 K. Metal carbides were formed [400] upon the pyrolytic (523–623 K) elimination of the cyclopentadienyl rings from the complexes Cp_2HfR_2, where R = Cl, NEt_2, Bu_n (Ref. [401],), and upon high-vacuum thermal decomposition (523 K, ~1 Pa) of tetra (neopentyl)titanium(IV) [$Ti(neo\text{-}C_5H_{11})_4$] vapour [402]. When decomposition is performed at 873 and 1273 K, the particle size of the formed nanocomposites is in the range of 10–50 nm [403]. They have a characteristic structure comprising a dense core (metal carbide and oxide) surrounded by less dense polymeric shell.

For preparing thin-layer high-temperature coatings by centrifugation on polished silicon substrates, ceramic matrix composite films containing highly dispersed tantalum carbide were formed [404]. A solution of $Ta(OC_5H_{11}\text{-}n)_5$ with phenol formaldehyde resin additive was used in an amount specified for providing a definite C:Ta ratio. The process was carried out at moderate temperatures (1,073–1,473 K) and reduced pressure ($1 \cdot 10^{-3}$–$1 \cdot 10^{-4}$ atm). The films thus obtained represented tantalum carbide and consisted of spherical particles with a mean diameter of ~50 nm [405].

Similarly, other thermally unstable precursors – metal π-allyl complexes, carbonyls, nitrosyls, etc. decompose in the presence of polymers. Thus bis(arene) chromium chemically bonded to PAA is thermolyzed at 433 K to the oxidized species $[Ar_2Cr]^+OH$, while the polymer remains unchanged. Above 573 K evolution of cyano compounds and ammonium takes place [406] and a magnetic material is formed, while at 1,273 K product graphitization occurs to give a typical diamagnetic product.

Thermolysis of true organometallic polymers has been scarcely studied. Among the few examples, note decomposition of poly(styryl)lithium in the temperature range of 343–388 K and poly(isopropenyl)lithium (353–393 K) obtained by living anionic polymerization (initiated by $LiBu_n$) [407]. The polymers have a molecular mass of 15–30 kDa and a low polydispersity index (Mw/Mn $\approx$ 1.03). A variety of reactions take place during thermolysis of poly(styryl)lithium, for example, elimination of LiH to give branched-chain intermediates (Scheme 6.13); these reactions follow first-order kinetics.

Apparently, a similar picture is observed as carbon-coated metallic iron particles are formed from iron phthalocyanine [404]. Nanoparticles of these metals were also obtained by vacuum thermolysis of other precursors: mesityl derivatives of copper $Cu(C_6H_2Me_3\text{-}2,4,6)_5$, silver $Ag(C_6H_2Me_3\text{-}2,4,6)_4$ and gold $Au(C_6H_2Me_3\text{-}2,4,6)_5$ [408], metallocene – aromatic – acetylene complexes [409]

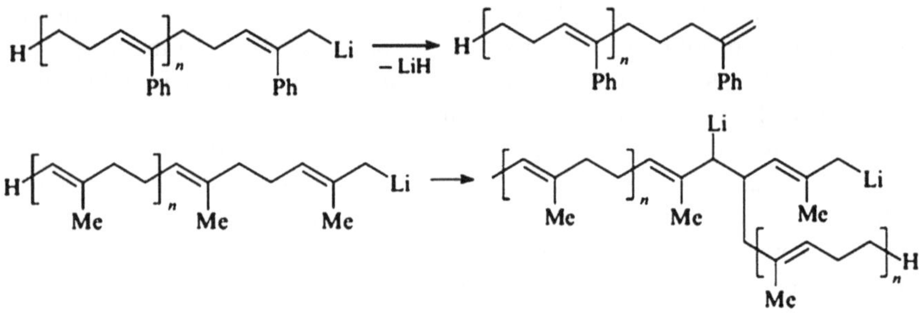

Scheme 6.13 Thermal transformations of polystyryllithium

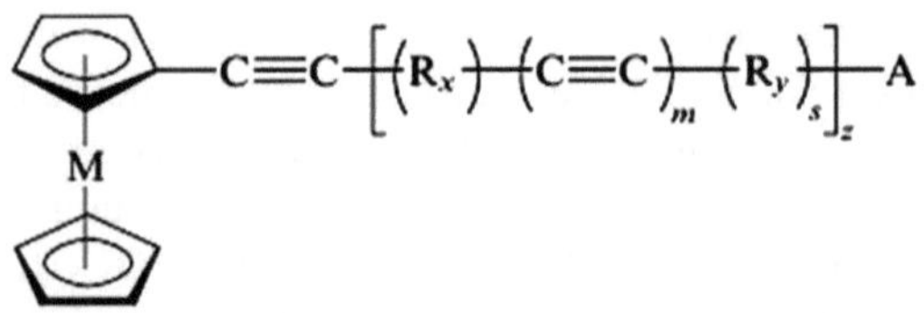

M = Fe, Sn, Co, Pd, Ni; R_x and R_y are phenylene or substituted phenylene; A = H, PhC:C, FcC ≡ C (Fc is ferrocenyl); *m*, *s*, *z* > 0.

and metal acetylides [410].

Thus, thermolysis of organometallic compound – polymer matrix systems is a multistage process. As a rule, it proceeds along two key interrelated pathways: direct decomposition of the precursor and initiation of diverse transformations of the polymer chain (migration of double bonds, depolymerization, cross-linking, destruction, etc.). The arising nanoclusters may catalyze the carbonization and graphitization processes, i.e., later stages of decomposition of organic polymers. Their mechanisms are most often not entirely clear but they are being actively studied.

Only one fact is obvious: the degree of dispersion of nanoparticles upon thermal transformations is increased as the content of polar groups in the polymer increases [411].

6.6.5 *Metal Thiolates and Sulfides in the Polymer Matrix*

Metal chalcogenides are widely used precursors for the preparation of semiconductor quantum dots (isolated nanoobjects), which show size-dependent optical properties [412]. Thermolysis of the metal thiolate precursors $M(C_nH_{2n+1}S)_x$ M (n = 3, 5, 12, 16, 18) was studied in detail. These compounds are usually obtained by the following route:

$$RSH + EtOH \rightarrow RS^{-}EtOH^{+},$$
$$nRS^{-} + M^{n+} \rightleftharpoons M(SR)_n$$

The metal–sulfur bond is fairly stable, the bond energy being in the range of 200–400 kJ/mol. The products that are formed are insoluble in common organic solvents as they are inorganic polymers.

Thermolysis of zinc, cadmium and nickel dodecanethiolates at 573 K with coordinating trioctylphosphine oxide as the solvent or without a solvent afforded ZnS and CdS nanocrystals with nanoparticle size of 1.5–3.0 nm coated by solvent molecules [413, 414] and NiS nanoparticles shaped as plates and rods [415]. Particles with similar size are formed in the thermolysis of $AgSC_{12}H_{25}-n$ [416] while in the case of $Bi(SC_{12}H_{25}-n)_3$, either layered nanostructures [417] or spheres and hexagonal plates are formed [418] as well as Ag(I), Cu(I) and Pb (II) alkanoates [419]. Thermolysis of copper dodecanethiolate in an inert atmosphere in the absence of solvent at 470–490 K results in the formation of CuS nanoparticles of diameter ~3 nm and 12.7 nm-thick faceted nanodiscs of diameter 27.5 nm formed as a result of crystallization processes [420].

Thermolysis of Pd(II) cluster with a long-chain mercaptan $[Pd(SC_{12}H_{25})_2]_6$ has been carried out in diphenylether (boiling temperature 532 K) in an argon atmosphere [421].

Palladium hexadecanethiolate $Pd(SC_{16}H_{33}\text{-}n)_2$ was decomposed in air at 520 K on the silicon surface to give 50–60 nm-thick nanofilms (the coating was applied by centrifugation or by spin coating) compatible with biomolecules [422] (Fig. 6.27).

Thermolysis of sulfur-containing derivatives of other noble metals, for example Pt, was carried out under analogous conditions [423]. Organogold derivatives $[RN\text{-}(CH_3)_3][Au(SC_{12}H_{25})_2]$ ($R = C_8H_{17}$, $C_{12}H_{25}$, and $C_{14}H_{29}$) were thermolized to spherical gold nanoparticles [424]. Similar reactions have been realized also in polymer matrices. Thus, gold(I) dodecanethiolate, a precursor of gold nanoparticles in a polystyrene matrix, was obtained in ethanol solution by the following route [425, 426]:

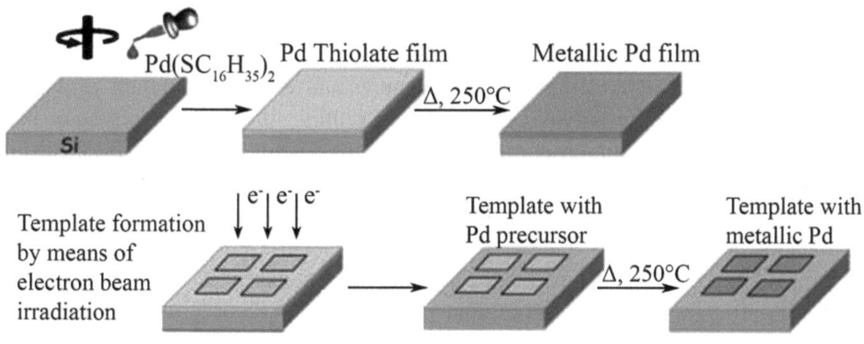

Fig. 6.27 Spin-coating palladium hexadecanethiolate on the silicon surface followed by the Pd-film formation

$$HAuCl_4 + 3(n - C_{12}H_{25}HS) \rightarrow n - C_{12}H_{25}SAu + n\text{-}C_{12}H_{25}S - SC_{12}H_{25}\text{-}n + 4HCl$$

Two moles of thiol were consumed for the reduction of Au(III) to Au(I) and the formation of disulfide as a by-product.

Note that surface-bonded gold nanoparticles are also formed upon controlled thermolysis (348 K) of the surface-grafted complex obtained by adsorption of $Me_2Au(acac)$ on partially dehydrated (on heating to 673 K) SiO_2 surface [427] while under milder conditions (at temperatures below 333 K), AuCl complexes with octadecylamine and oleylamine decompose [428]:

$$2[AuCI(RNH_2)] \rightarrow 2\ Au^0 + R_2NH + Cl_2 + NH_3,$$
$$R = n\text{-}C_{18}H_{37}, n\text{-}C_8H_{17}CH = CH(CH_2)_8.$$

The resulting zerovalent gold particles nucleate and grow to a size of 12 nm with low (8 %) polydispersity, the particles being stabilized by the amine.

The addition of a solution of $AuSC_{12}H_{25}\text{-}n$ in acetone to the PS matrix (molecular mass of 230 kDa) and intense sonication followed by casting on a glass surface produces a thin (0.1–0.3 nm) transparent film [426]. These films containing 5–10 % of dodecanethiolate and placed between aluminium foil sheets were subjected to short-term (40–105 s) thermolysis at 573 K. This gave gold nanoparticles according to the reaction:

$$2AuSC_{12}H_{25} - n \rightarrow 2Au^\circ + n - C_{12}H_{25}S - SC_{12}H_{25} - n$$

However, it cannot be ruled out that initially polynuclear complexes of the type –$(\text{–}Au^1SR\text{–})_n$ – are formed and then they are converted to nanocrystals. Finally, this gives polydisperse gold nanoparticles with a size of 1.8 nm (~150 gold atoms) evenly distributed in the PS matrix.

In the thermolysis of lead thiolate in PS, redox reaction is considered as the first stage and the formation of sulfide is the second stage [429]:

$$Pb(SR)_2 \rightarrow Pb + RS - SR \rightarrow PbS + R - S - R$$

The crucial role in stabilization of such structures and in preventing their agglomeration is assigned to the polymer matrix. The resulting film nanocomposite of lead sulfide shows luminescent properties, it is stable for several months and is promising for the use in optoelectronics. Unfortunately, the study cited [429] like many other publications, does not consider the transformations of the matrix itself. Although the remainder of the ligand is often detected in PS, it is believed [413] that there is no chemical bond between the polymer chain and the nanoparticles.

Cadmium, zinc and lead alkanethiolates introduced in PS or topaz (thermoplastic copolymer of cycloalkene with ethylene and norbornene) matrix decompose in two stages as shown below

$$M(SR)_2 + \text{polymer} \xrightarrow{\text{PhMe}} M(SR)_2/\text{polymer},$$
$$M(SR)_2/\text{polymer} \xrightarrow{570K} MS/\text{polymer} + SR_2,$$
$$M = \text{Cd, Zn, Pb; R = Alk; polymer is PS or topaz,}$$

The mean size of the formed CdS nanoparticles is 2.4 nm and that of the ZnS nanoparticles is 1.9 nm. The photoluminescent properties of the particles depend on their size and shape, the maximum luminescence decay time being observed for the products obtained at thermolysis temperature of 553 K. The structure and phase composition of CdS nanocrystallites can be controlled by the thiourea/salt ratio and by thermolysis temperature (323–773 K). The transition of the cubic phase to the hexagonal one occurred at 473–573 K, whereas a pure hexagonal phase of CdS appeared at thermolysis over 873 K. These factors including the time of sintering affect dispersity, crystallinity, and average sizes of CdS nanoparticles. It is interesting that a photocatalytic activity in the synthesis of hydrogen is optimal for a mixture of cubic and hexagonal CdS apart from its pure phases [430].

Decomposition of $Sb(n\text{-}C_{12}H_{25}S)_3$ in polystyrene at 623 K affords Sb and Sb_2S_3 nanoclusters (15–30 nm) evenly distributed in the amorphous polymeric phase [431].

As noted above [70] thermolysis of metal dithiolates to give sulfide semiconductors in thin films was performed successfully under the action of laser radiation or electron beam. The laser beam concentrating a 2 kW power was defocused over a small area (of diameter 4 mm), the pulsation time was 10–50 ms. The thin beam temperature is calculated proceeding from the polymer and beam parameters and can be controlled.

Thermolysis of the $(n\text{-}C_{12}H_{25}S)Pd/PS$ thin films occurs in 5 min at 443 K [423] and in air it can lead to the formation of metal oxides in the polymer matrix. For preventing this process, trioctylphosphine oxide is used as the stabilizing antioxidant (for example, for 2–4 nm CdS, ZnS, CdSe particles inserted *ex situ* into polyconjugated copolymers [432]). In fact, thermolysis of the $Fe(n\text{-}C_{12}H_{25}S)_2$ – PS system at 473 K for 2 min is accompanied by the formation of fully transparent light-brown film (practically complete light transmission above 550 nm) in which nanoparticles with a mean size of 10 nm x 50 nm are evenly distributed [433].

Thermal decomposition of metal alkyldithiocarbamates in the presence of alkylamines can serve as a general method for the synthesis of metal sulfide nanoparticles [434]. For example, Ag_2S nanocrystallites were prepared by direct thermolysis of the corresponding precursor in air at 473 K [435]. The recent quite comprehensive review devoted to the thermal synthesis of silver nanoparticles in polymer matrices should be mentioned [436].

Scattered data on the synthesis of heterometallic sulfides can be found in the literature. Thermolysis of a solution of copper and indium oleates in dodecanethiol in the presence of oleylamine gives Cu_2S and In_2S_3, while $CuInS_2$ is not produced in this way due to different decomposition temperatures of the components [437].

6.6.6 *Polymer-Mediated Thermal Synthesis of Mono- and Multimetallic Alloys and Ceramics*

Of interest for obtaining bimetallic nanoparticles are the complex salts containing two metals, for example, of general formula $[M_1A_n]_x[M_2B_m]_y$ where M_1 and M_2 are the central metal atoms of the complex cations, A and B are the ligands. The composition of the solid solutions obtained can be easily adjusted by the stoichiometry ratio of the precursors. Bimetallic clusters are often served as alternative in the synthesis of nanoheterostructures with a defined composition. Other well-characterized complex salts are metal oxalates, citrates, tartrates which are high-crystalline and are readily decomposed at moderate temperatures.

Usage of suitable polymers allows applying the convenient techniques such as a film coating, spinning, photolithography, etc. before the end stages of thermolysis. This enables a hard control of the structure and size uniformity of particles, their density and so on. Using the metal salts and their oxides as precursors allows obtaining amorphous powders on the sertain stages of thermolysis followed by their transformations into crystalline products.

Magnetic alloys FeCo are soft materials with high saturation magnetization (up to 2.45 T), low magnetostriction, relatively low coercive force, and high Curie temperature and energy of magnetic anisotropy (higher than the value predicted for FePt). These materials have a broad spectrum of practical applications. They are obtained using various approaches including RF-discharge plasma, chemical vapour deposition (CVD) [438], thermolysis of cobalt or iron carbonyls in the presence of surfactants [439–441]. The ferromagnetic behaviour of such optically transparent magnetic plastics corresponds to a multidomain magnetite structure; therefore, they could find use in magneto-optical technology (see, for example, Ref. [442]).

Monodispersed magnetic heterostructures MFe_3O_4 (M = Ag, Au, Pt, Pd) are well studied [443]. They were obtained by thermal decomposition of iron(III) oleate in the presence of oleylamine with a following deposition of Fe_3O_4 formed on the surface of noble metal nanoparticles. Nanoparticles $AuFe_3O_4$ having dumbbell-like shape reveal diagnostic and therapeutic features, and catalytic properties [444, 445]. The simplest route of their synthesis is thermolysis of iron pentacarbonyl in the presence of pre-formed gold nanoparticles in 1-octadecene followed by oxidation of iron in air at room temperature. An alternative way is a thermal decomposition of the mixture of iron oleate with a metal-oleyimine complex in the presence of 1,2-hexadecandiol at 583 K. Heterostructures $AgFe_3O_4$ are formed from the silver nanoparticles on the surface of an amorphous Fe_xO_y [446]. By controlled co-thermolysis of palladium and iron acetylacetonates, in the presence of oleylamine and oleic acid $PdFe_3O_4$ were obtained [447–451]. Reduction thermolysis of molecular complexes $Pd(OOCMe)_4M(OH_2)$ (M = Ni(II), Co(II), Zn(II) or $Pd(OOCMe)_4Ag_2(HOOCMe)$ and $PdM(\mu\text{-}OOCR)_4L$ (M = Co(II), Ni (II), Mn(II), Zn(II); R = Me, Bu; L = H_2O, MeCN) affords bimetallic nanoparticles PdM [452, 453].

Under heating double complex salt $[Pd(NH_3)_4][AuCl_4]_2$ in a helium atmosphere from the room temperature up to 623 K the reduction of Au(III) and Pd(II) occurred yielding heterometallic nanoparticles. The process in hydrogen was accompanied by the competitive formation and aggregation of Au and Pd individual nanoparticles up to 603 and 623 K, respectively [454].

The complexes of mixed types $[Zn(NH_3)_4][PtCl_6]$ and $[Cd(NH_3)_4][PtCl_6]$ were used as precursors for synthesis of heterometallic particles PtZn and PtCd [455].

A recent study [456] proposes a non-hydrolytic method for the synthesis of nanocrystalline magnetite (Fe_3O_4) in the presence of rigid matrices, namely, linear ω-functionalized polystyrenes (Mw from 5 to 39.4 kDa). Usually, ω-sulfopolystyrene (PS-SO_3H), ω-thiopolystyrene and ω-carboxypolystyrene (PS-CO_2H) obtained by living-chain anionic polymerization are used. Thermolysis of iron(III) acetylacetonate is carried out in hexadecane-1,2-diol at 532–537 K. ω-Carboxypolystyrene stabilizes iron nanoparticles more efficiently than PS-SO_3H; the size of magnetite nanoparticles is 3–10 nm and decreases with increase in the content of the polymer. Ceramic materials incorporating α-Fe2O3 nanocrystals were also obtained by thermolysis of hyperbranched poly(ferrocenylenesilane) [457]. The synthesis and self-organization of the polymer-coated ferromagnetic cobalt, nickel and iron nanoparticles have been analyzed in considerable detail [458, 459]. For example, the corresponding composite is obtained by thermolysis of $Co_2(CO)_8$ in dichlorobenzene in the presence of polymeric surfactants with terminal functional groups (Mw = 5–10 kDa) at temperatures of 433–453 K [460, 461]. The process occurs in two stages: at higher temperature, the metal carbonyl decomposes, while at lower temperature (below 433 K), nanoparticles grow. The mean particle size (17–21 nm) depends on the nature of the surfactant and thermolysis conditions.

Pyrolysis of cobalt carbonyl silanes comprising cobalt carbonyls or cobalt clusters gives rise to soft materials with ferromagnetic properties [462]. Thin ceramic polymetallic films, for example, alloys of magnetic CoFe nanoparticles on a SiC/C film are obtained by thermolysis of high-metallized poly (ferrocenylsilanes) with grafted cobalt clusters at 773 K [463].

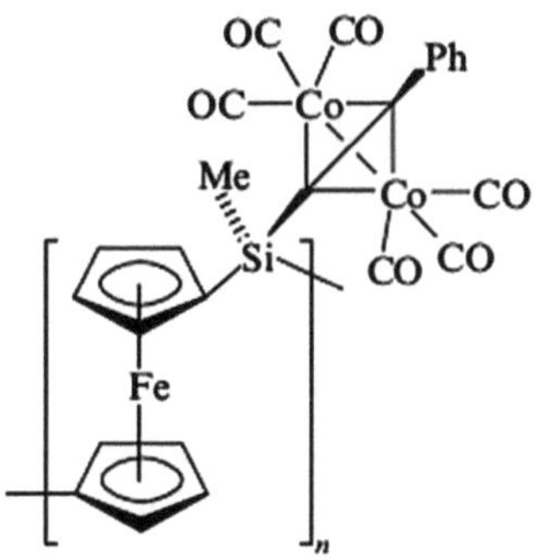

Under conditions of air oxidation at 873 K, the thickness of the ceramic film decreases from 200 to 40 nm and then remains almost invariable up to 1,173 K; as this takes place, fine superparamagnetic particles are formed on the ceramic film surface.

The formation of α-CoFe and Co(Fe)O phases was detected experimentally. In view of their magnetic properties, the thin films obtained at 773 and 873 K could find use in spintronic instruments as insulating magnetic layers.

The most convenient routes to monodisperse magnetite nanocrystals include co-precipitation in aqueous solutions of Fe(III) and Fe(II) ions, thermolysis of alkaline solutions of Fe(III) chelates in the presence of hydrazine, sonochemical decomposition of hydrolyzed Fe(II) salts followed by thermolysis [195, 464, 465].

The most widely used method for the synthesis of monodisperse inorganic ferrites described as MFe_2O_4 (M = Co, Ni, Mn, Fe, etc.) is thermal decomposition of mixed organic compounds of the corresponding metal and Fe(III), such as acetylacetonates, carbonyls, carboxylates and so on. The process is carried out in the solid phase or in high-boiling solvents with addition of surfactants (oleic acid, oleylamine and so on) [163, 466].

Thus MFe_2O_4 nanocrystals are formed upon the thermolysis of mixed oleate complexes dissolved in octadec-1-ene under N_2 at 573 K [466]. The particle size depends on the nature of the ferrite and, according to transmission electron microscopy data, it is 9, 11 and 7 nm for M = Co, Ni and Mn and 24 nm for M = Fe. Prenucleation of $CoFe_2O_4$ occurs at 523 K but without growth of nanocrystallites because the concentration of monomer is lower than the critical nucleation concentration. For reaction temperatures between 523 K and 593 K, the size and shape evolution of the nanocrystals are determined by the nucleation and growth dynamics. For temperatures in the range of 573–593 K which is above the thermolysis temperature of the mixed Co(II)-Fe(III)-oleate complex, the monomer concentration increases rapidly resulting in homogeneous nucleation. Atomic clusters of $CoFe_2O_4$ with size <2 nm are initially formed at 587 K that then grow rapidly when the temperature is raised to 593 K in less than a minute. The shape of the $CoFe_2O_4$ nanocrystals can be controlled by the aging time at 593 K, evolving from initial spherical, to spherical-to-cubic, cubic, corner-grown cubic, or starlike shapes (Fig. 6.28). Thus, by varying reaction conditions, such as the precursor concentration and the heating rate, it is possible to obtain shape-controlled monodisperse $CoFe_2O_4$ nanocrystals with high yield [466].

The magnetic saturation of $CoFe_2O_4$, $NiFe_2O_4$, $FeFe_2O_4$ (69.7, 34.2 and 58.6 emu/g, respectively) are close to theoretical values (71.2, 47.5 and 96.2 emu/g), whereas for $MnFe_2O_4$ this value is much lower (23.9 instead of 120.8 emu/g). This circumstance may be related to the decrease in the particle size during the measurements or to the formation of antiferromagnetic layer on their surface.

Thus, essential conditions for the formation of ferrites by thermolysis include: the possibility of molecular mixing of the components, their thermodynamic compatibility and similar decomposition temperatures. The failure of the formation of $CuFe_2O_4$ ferrite is attributable to the fact that already at 523 K, CuO is formed. The

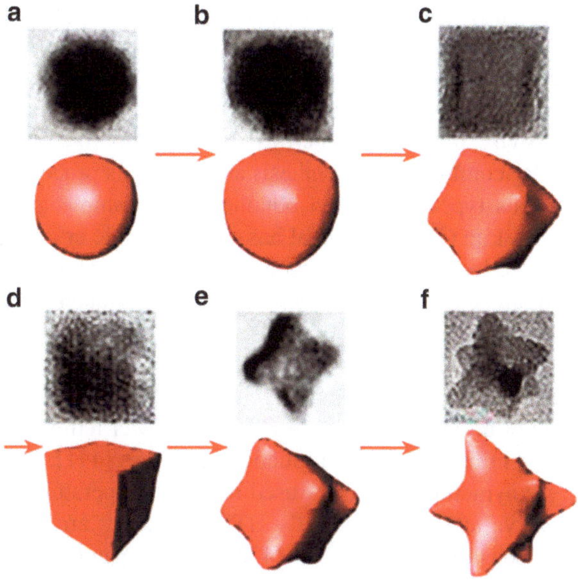

Fig. 6.28 The shape evolution of $CoFe_2O_4$ nanocrystalls. (**a**) c spherical, (**b**, **c**) spherical-to-cubic, (**d**) cubic, (**e**) corner-grown cubic and (**f**) star-like

polymer-mediated synthesis followed by thermolysis is free from this drawback. This method was used to prepare $BaSnO_3$, $BaTiO_3$, $SrTiO_3$, $NdAlO_3$, $SrBi_2Ta_2O_9$ and other nanoparticles [467].

We developed [155] a method for the synthesis of nanoparticles consisting of combined polymerization of a metal-containing monomer with the thermolysis (CMPT). High-quality $BaTiO_3$ ceramics with particle size from 10 nm to 1.5 μm, which depended on the temperature (ranging from 873 to 1,623 K) and the type of atmosphere (inert or oxidizing one) was obtained using this method from organometallic precursors [468, 469]. The traditional way for the preparation of these ceramics, for example $PbTiO_3$, is solid-state mixing of $PbCO_3$ and TiO_2 in mills and subsequent annealing at temperatures above 873 K (ex situ synthesis). However, in the latter case, the process is accompanied by the formation of toxic PbO as a side phase. The CMPT production of the lead–titanium ceramics can be described by the following sequence of reactions (Scheme 6.14) [470]:

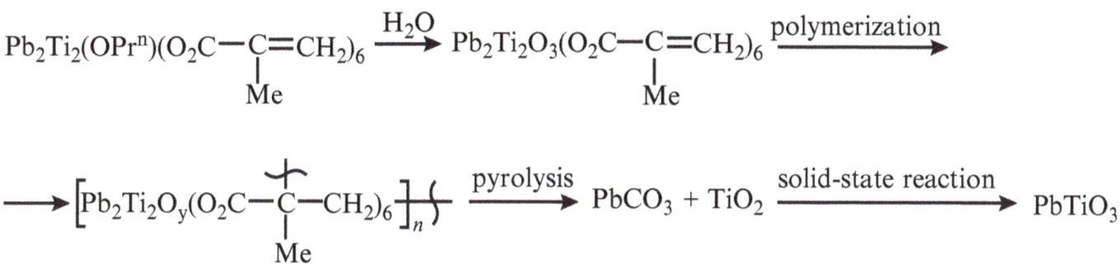

Scheme 6.14 Preparation of $PbTiO_3$ ceramics via the CMPT method

In the preparation stage of the monomer precursors (batch mixture), it is possible to add paramagnetic ions such as Mn(II), Gd(III) and Cr(III) as acetates or acetylacetonates. These additives are of interest not only for EPR measurements but also for modification of the material, which may thus acquire electrical conductivity, dielectric properties or other properties.

By combining the synthesis and pyrolysis processes, heterometallic ceramics of various types have been obtained, for example, perovskites ABO_3. The natural perovskite $CaTiO_3$ mineral has a pseudocubic crystal lattice in which large cations (A) are located at cell vertices, small cations reside at the cell centres, and oxygen ions are at the midpoints of the faces. These materials are widely used in electronics owing to their specific ferro-, piezo- and pyroelectric properties; for example, $BaTiO_3$ is used in the capacitor industry [411]. Perovskite type ferrites $M_3Fe_2O_{7x}$ (M = Sr, Ba) with particle size of 50–55 nm are formed upon thermal decomposition of molecular citrate precursors at lower temperature (873 K) than that used in conventional ceramic production processes [471].

Polymetallic ceramics used as high-temperature superconductors (HTSC) also deserve attention. Currently HTSC ceramics are widely used in microlitography, UV-sensitive sensors, etc. [472]. The conventional approach to their production comprises several stages: mixing of compounds of the corresponding metals (oxides, carbonates, oxalates or nitrates), grinding of the mixture, multistage annealing of the batch mixture, including high-temperature annealing in an oxygen atmosphere. Each stage has considerable limitations as regards reproducibility, which is related not only to the quality of grinding and mixing of the initial solid components, but also to complex physicochemical and mechanochemical transformations that take place during the material preparation. This produces both inhomogeneities at the micro level and different phases, including non-conducting ones. As a result, low-quality HTSC ceramic with smeared superconducting junction is produced. The knowledge of thermoanalytical characteristics of decomposition of the precursors of HTSC ceramics would be helpful to optimize the conditions of pyrolysis and understand the mechanism of thermal decomposition.

Different types of polymers are used for preparing HTSC materials: PMAA [473–475], PAN [476, 477], polyimides [478, 479]. poly(N,N-dicarboxymethyl) allylamine [480, 481], copolymers of acrylic acid [482, 483] etc.

Several options of the polymer-mediated synthesis followed by thermolysis to produce HTSC ceramics have been proposed: preliminary synthesis of the corresponding metal complexes via polymer analogous transformations; polymerization of traditional monomers in the presence of aqueous solutions of Y(III), Ba (II) and Cu(II) nitrates, copolymerization of the corresponding metal-containing monomers and so forth [484]. Each of the approaches has its own advantages and shortcomings. For example, the formation of the copper complex of PAA decreases the thermal stability of the polymer. The process is multistage, including copper-catalyzed rupture of the polymer chain and depolymerization, the formation of macroradicals and low-molecular-mass organic products and anhydride structures, which are decarboxylated [476, 482]. No significant differences were found between thermolysis of macrocomplexes at temperature below 613 K in argon or in air, complete decomposition taking place at 852 K.

Thermolysis of a mixture of yttrium, barium and copper nitrates (atomic ratio 1:2:3) is known [483] to include five key stages: (1) dehydration (in the temperature range of 298–459 K with an exothermic peak at 400 K); (2) dehydration with copper denitration (in the temperature range of 459–517 K with a temperature maximum at 548 K), (3) completion of copper denitration and yttrium denitration (517–585 K), (4) final yttrium denitration (585–773 K with a maximum at 648 K) and (5) barium melting and denitration (above 773 K with an endothermic peak at 848 K). The denitration of the material reaches a maximum at 913 K, and the residual weight at 1,273 K is 41.6 %, which corresponds to the composition $YBa_2Cu_3O_{6.5}$ (Fig. 6.29a).

Thermal decomposition of a PAA mixture with components of the HTSC ceramics also occurs in 5 main stages including: (1) dehydration and formation of the macrocomplex (298–419 K, the glass transition temperature of PAA is 401 K); (2) evolution of water, nitrogen oxides and CO_2 (temperature range of 419–472 K); (3) intensive evolution of water, CO_2, CO, monomeric acrylic acid (AA) and nitrogen oxides (472–603 K, DTG peaks at 535 and 552 K (Fig. 6.29b); (4) intensive destruction of PAA (603–663 K, DTG peak at 663 K). The last, fifth stage starts at 899 K and is accompanied by an exothermic peak with a maximum at 1,118 K corresponding to CO and HNO_3 evolution, decomposition of the most stable barium nitrate and $BaCO_3$ formed. The relatively large fraction of the residue (22.3 %) attests to incomplete oxidation of PAA in argon, whereas thermolysis in air inhibits the formation of barium carbonate and eliminates the fifth stage of pyrolysis. Thus, for optimization of the production process of high-quality HTSC ceramics, the initial stages of thermolysis (temperature below 773 K, heating rate 5 deg/min) should be carried out in an inert atmosphere, and the final stage (773–1,223 K, heating rate 10 deg/min) should be carried out in air. This prevents the formation of undesirable phases like $BaCO_3$, Y_2O_3, $BaCuO_2$, $Y_2Cu_2O_5$, etc.

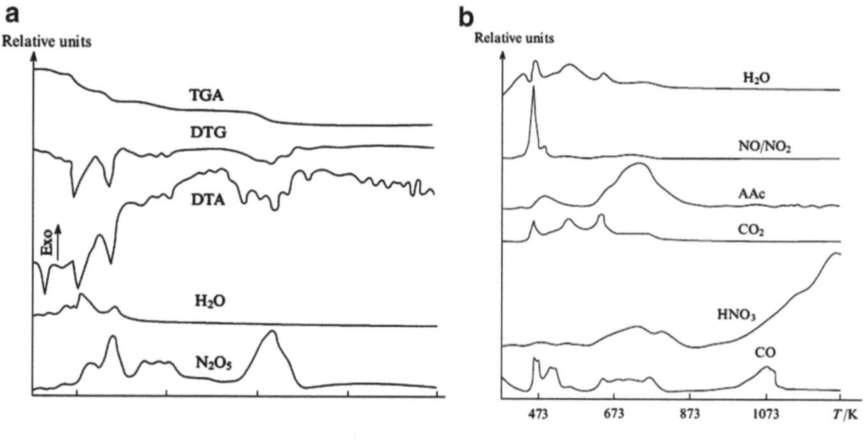

Fig. 6.29 Thermal analysis data for a mixture of metal nitrates (Y:Ba:Cu = 1:2:3) (**a**) and mass spectra of their composition with polyacrylic acid (PAA:(Y, Ba, Cu) = 2:1) recorded during the analysis (**b**) [483]. The thermal analysis was performed under argon, heating rate 5/deg min

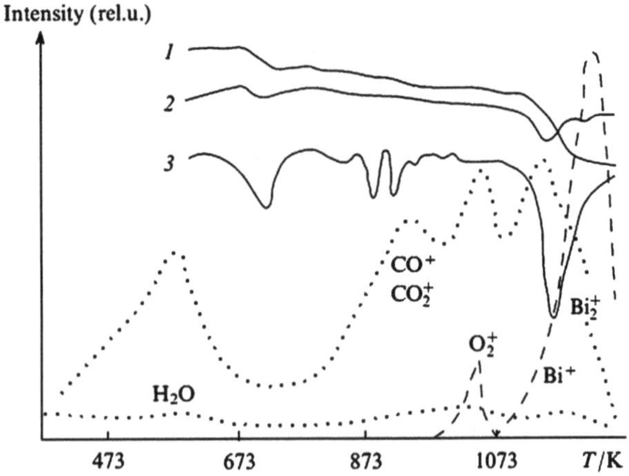

Fig. 6.30 Thermal (*continuous line*) and mass spectral (*dashed line*) analysis data for samples 2223: (*1*) TGA curves, (*2*) DTA curves, (*3*) DTGA [484]. Heating rate 4 deg/min, the CO^++CO_2^+ curves are enlarged 50-fold

The polymer-mediated synthesis has been utilized [483] to prepare various ceramics. For example, decomposition of yttrium metallopolymer at 633–793 K affords $YBa_2Cu_3O_{7-x}$ ceramics (123-ceramics), while in the case of bismuth metallopolymer, 2223 ceramics with the formula $Bi_2Sr_2Ca_{n-1}Cu_nO_{2n+4-\delta}$ ($n = 1$–3) is produced, the highest weight loss rates being achieved at 693–1,143 K. The mass spectrometric analysis of the gases evolved from the bismuth 2223 ceramics on vacuum heat treatment indicates (Fig. 6.30) that at 973–1,073 K, small amounts of oxygen and bismuth are formed, and are stabilized as O_2^+, Bi^+, Bi_2^+ ions and traces of H_2O, CO^+, CO_2^+, etc. In terms of key characteristics (superconductivity at 87 K), 100 %-magnetic shielding, electrical conductivity at room temperature of 800–1,000 $\Omega^{-1}\cdot cm^{-1}$, bulk density of 4.7 $g\cdot cm^{-3}$, critical current up to 240 $A\cdot cm^{-3}$, etc.), the yttrium ceramics obtained by the polymer-mediated method and the single-phase bismuth superconducting cuprate specimens synthesized by the combined polymerization – pyrolysis route can be assigned to the best HTSC ceramics formed under thorough oxygen annealing.

One of the promising methods of the HTSC synthesis, i.e. the YBCO-ceramics, is a thermal decomposition of novolac resins (*m*-cresol-formaldehyde) containing of corresponding metal nitrates [485–487]. The optimal thermolysis in an inert atmosphere starts at 1,023 K, however the formation of $BaCO_3$ continues in O_2 atmosphere yielding the orthorhombic YBCO-ceramics with a high superconductivity at 91 K. This polymer-mediated approach appears to be perspective for synthesis of HTSC-ceramics and requires further studies.

6.6.7 Thermolysis of Metal Carbonyls in Polymer Matrices

This is the earliest and popular method for preparing nanocomposites. Some typical examples of metal carbonyl decomposition have been noted repeatedly in this Chapter. Along with cobalt acetate, formate, acetylacetonate, the most widely used precursor is cobalt carbonyl $Co_2(CO)_8$.

The general scheme of metal carbonyl decomposition including in the presence of a polymer matrix being a stabilizing agent of the metal or its oxide nanoparticles formed can be represented as follows [337]:

$$nM(CO)_b \underset{k_{-1}}{\overset{T,\ k_1}{\rightleftarrows}} nM(CO)_{m< b-1} \underset{k_{-2}}{\overset{k_2}{\rightleftarrows}} (n/x)M_xO_y$$

where k_1, k_{-1}, k_2, k_{-2} are appropriated constants.

Among other metal carbonyls $Cr(CO)_6$, and $Fe(CO)_5$ are often used applying two methods of their decomposition in polymers: in solution and in their melt.

The most general approach comprises the dispersion of pre-synthesized nanoparticles in polymer matrices, i.e., method *ex-situ*. In the processes *in situ* more 'active' polymers which have the functional groups for complexation with a precursor are preferable.

The nucleophilic fragments of macroligands (particularly in styrene-N-vinylpyrrolidone copolymer) bring about a polymer-catalyzed decomposition of the metal carbonyl by scheme:

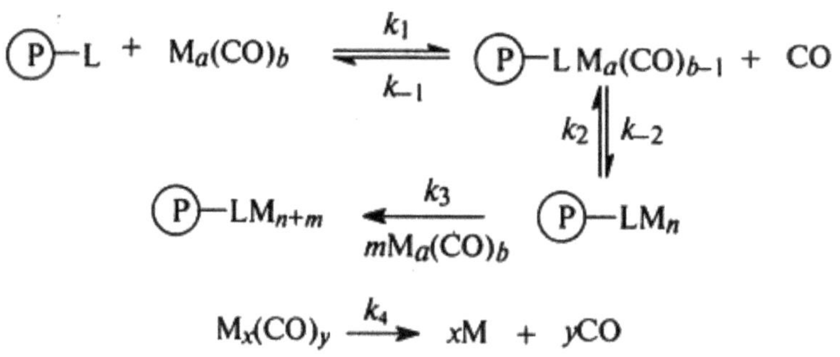

The main condition for a proper matching of a functionalized polymer is connected with the greater reaction rate on the polymer surface to reduce the decomposition in the solution to a minimum, i.e. to provide:

$$k_1 + k_2 - k_{-1} - k_{-2} > k_4$$
$$k_3 > k_4$$

If this condition is fulfilled, the main process takes place, i.e. particle growth and the formation of nanoparticles of 1–10 nm in size.

Coincidentally with the sizes of the forming nanoparticles the probability of termination of particle growth increases due to their noncovalent surface interactions with the macromolecule. The stronger the interaction the smaller the particle size; polymers having more polar groups promote the finest particles growth.

Of interest is an approach based on the decomposition of metal carbonyls in block-copolymers. Thermolysis of metal carbonyls in block-copolymers in the synthesis, for example, Cr_2O_3, Fe_2O_3, and Co_2O_3 nanoparticles in polystyrene-*b*-poly(methylmethacrylate) diblock copolymer (PS-PMMA) as a model system is studied in detail [488–492]. Results showed that the rates for reactions performed in a diblock copolymer solution are much faster than the rates of the same reactions performed in a homopolymer solution. The arrangement of the diblock copolymers in solution into spherical core-shell domains created self-assembled "nanoreactors" with PS acting as the surrounding shell and the internal PMMA domain (core) contained high precursor concentration, resulting in faster kinetics. Note that the size and shape of nanoparticles formed are similar to those generated at the thermolysis in pure PMMA solution. This may indicate that the particle morphology is determined not by the kinetic of their formation but by their interactions with their surrounding media and the presence of reactive stabilizing molecules, such that, in this case, PMMA segments, either as homopolymers or as part of a block-copolymer.

The formation of metallopolymers in a polymer melt with an addition of a high-boiling solvent [493, 494] is based on decomposition of the melt at the highest possible temperatures much exceeding the decomposition temperature of the precursor for the fastest possible and complete removal of the eliminated ligand from the reaction medium. Its main difference from 'wet' methods is that the short-range order of the structure of the initial polymer, PE, polypropylene (PP), PTFE, etc. is preserved in the melt, as opposed to solution, because in the case of the solid-state thermolysis, the polymer is destroyed to a lesser extent. The voids present in the polymer (due to density fluctuations) become accessible for the accumulation of the products. First of all, these are concentrated in the loosest disordered interspherulite areas of the organic matrix, in the space between the lamellas and at the centres of spherulites. Due to the decrease in the free polymer volume and the possible cross-linking, the segmental motion of the amorphous phase is hampered.

Cobalt-based nanoparticles (metallic cobalt, cobalt oxides, and cobalt alloys, such as CoFe) are very interesting varieties of magnetic nanoparticles. Cobalt-containing nanomaterials have the highest values of important magnetic characteristics (coercive force and saturation magnetization) among the magnetic materials [495]. In view of their relatively low cost and high efficiency, the most popular processes for manufacturing cobalt-containing nanoparticles involve the thermolysis of organocobalt compounds in organic solvents (octadecene, benzyl ether, tetraline, and heptadecane) with surfactants (fatty acids, amines, and organophosphorus compounds) [496].

X-ray powder diffraction is an effective method for studying for determining nanoparticle sizes and their phase composition in the Co nanocomposites. Examination of the X-ray diffraction pattern in Fig. 6.31 showed one metallic phase (very weak) and two oxide phases. This implies that CoO <111>, <200> and Co_3O_4 <220> phases are the major components of the sample and the metallic cobalt percentage is low, the average size of the particles is 9.0 ± 0.8 nm [461].

It cannot be ruled out that strong interaction of the polymer with the nanoparticle inserted into intermolecular voids of the matrix may lead to destruction of the crystalline portion and its transition to the amorphous state. In this material, no metal phase is detected, and it can be considered as a single-phase metallopolymer.

Using this method, the PE-based metallopolymer composites were obtained with iron clusters (nanoparticle size 9–12 and 20–22 nm) [497] with $Fe_{0.85}Mn_{0.15}$ (size ~2.5–3.0 and 20 nm) [494, 497], with ZnO and CdS clusters (mean size 4–10 nm)

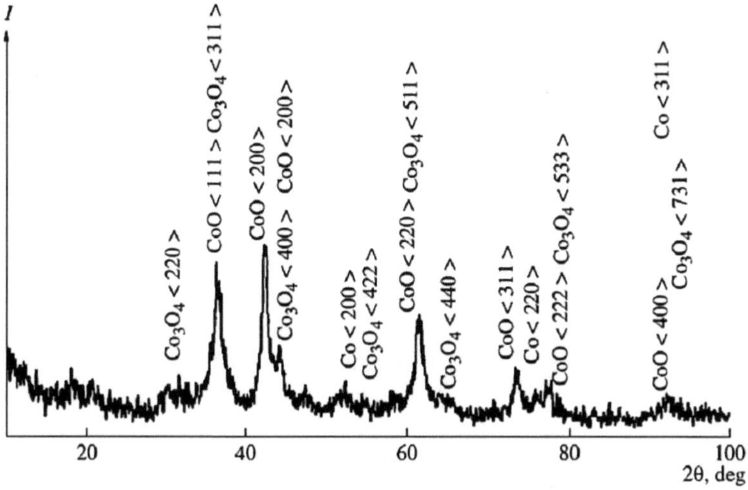

Fig. 6.31 X-ray diffraction pattern for cobalt-containing nanoparticles. CoO, Co_3O_4, and Co peaks are indexed

[493, 498]. The formed iron-containing nanoparticles exist at room temperature in the superparamagnetic state and are multiphase structures: they consist of α-Fe, Fe_3C, Fe_3O_4, Fe_2O_3 [494]. According to the existing views, these particles are constructed as a core (α-Fe) coated by a metal-containing shell (Fe_3C, Fe_3O_4, Fe_2O_3).

Thermal decomposition of $Fe(CO)_5$ in organic solvents is accompanied by the formation of intermediates – clusters of the different composition such as $Fe_2(CO)_9$, $Fe_3(CO)_{12}$ and other; therefore, sometimes triirondodecacarbonyl $Fe_3(CO)_{12}$ as precursor is used instead of $Fe(CO)_5$ [499]. $Fe(CO)_5$ decomposes into highly active atom clusters and CO in a closed reactor at 972 K in an inert atmosphere in the presence of S or Se powders. CO undergoes a Boudard disproportionation reaction $2CO \rightarrow CO_2 + C$. Under these conditions FeS and FeSe nanoparticles covered by a carbon shell are formed [314]. The layered hexagonal structure of FeS is known to be an excellent solid lubricant.

Thermal decomposition of $Fe_x(CO)_y$ in imidazolium salts with long-chain N-alkyl substituents – an ionic liquid as solvent affords iron carbide nanoparticles with the size from 2 to 15 nm [499].

Thus, thermolysis of metal carbonyls in polymer matrices is a versatile and controlled approach to obtain nanocomposites with a complex of useful properties.

6.6.8 Comparison of Different Methods of Solid-State Thermolysis of Metallopolymers

Currently, there are three key methods for the preparation of metallopolymers [365] for polymer-mediated synthesis of nanoparticles: reactions of metal compounds with linear functionalized polymers (so called polymer analogous transformations); polycondensation of the corresponding precursors to metallopolymers; polymerization and copolymerization of metal-containing monomers. Thermolysis of metallopolymers formed according to each of the methods listed above occurs in different ways.

This fact will be demonstrated in relation to hafnium-containing polymers obtained by polymer analogous transformations involving PAA (compound **1**), by condensation methods, viz., the reaction of hafnium compounds with diols (**2**–**6**) or condensation of cyclopentadienyl hafnium complexes with p-diethynylbenzene (**7**), and by polymerization of hafnium-containing monomers (**8**–**12**) [500–502]. The isothermal thermolysis of Hf(IV) – polymer systems (**1**–**12**) was studied at three characteristic temperatures: 1) 643 K (heating for 60–300 min) – the mean decomposition temperature where secondary transformations (pyrolysis of the decomposition products) are insignificant; 2) 873 K (150–200 min) – the mean temperature of the early stages of pyrolysis of the organic transformation products; 3) 1,273 K (~150 min) – the temperature of intense pyrolysis of organic products. The general pattern of controlled thermolysis of these polymers can be represented as follows:

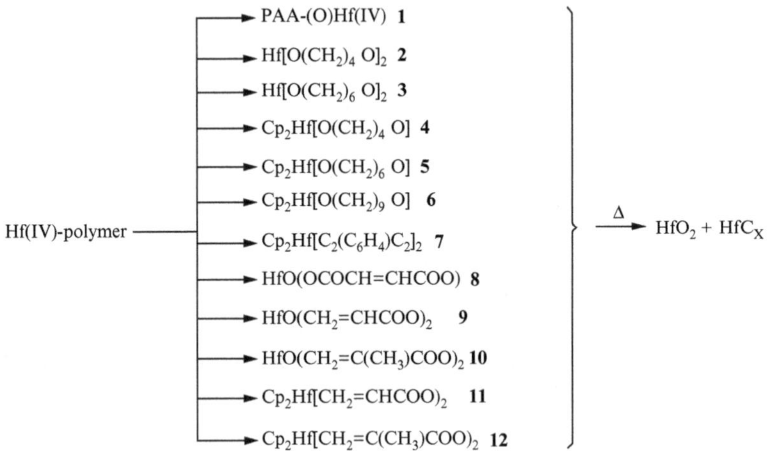

The formed composites tend to have different H:O atomic ratios covering a rather broad range and forming the series

$$14\ (\text{for } \mathbf{6}) > 11,0(\mathbf{5}) > 6,0(\mathbf{3}) > 5(\mathbf{12}) > 4,0(\mathbf{2},\mathbf{11}) > 1,2(\mathbf{9}) > 0,4(\mathbf{8}).$$

The typical regular features of the thermolysis of metallopolymers **1**–**12** are as follows. The gas evolution kinetics are similar for all of the samples (Fig. 6.32): first, the reaction rate decreases monotonically with an increase in the time of thermolysis at a constant temperature; a temperature rise for similar transformation times induces non-linear increase in the gas evolution and weight loss rates. The major gaseous product in the thermolysis of Hf carboxylate polymers **8**, **9**, **11** at 643 K is CO_2, while for polymer **11**, this is cyclopentadiene vapour. An increase in

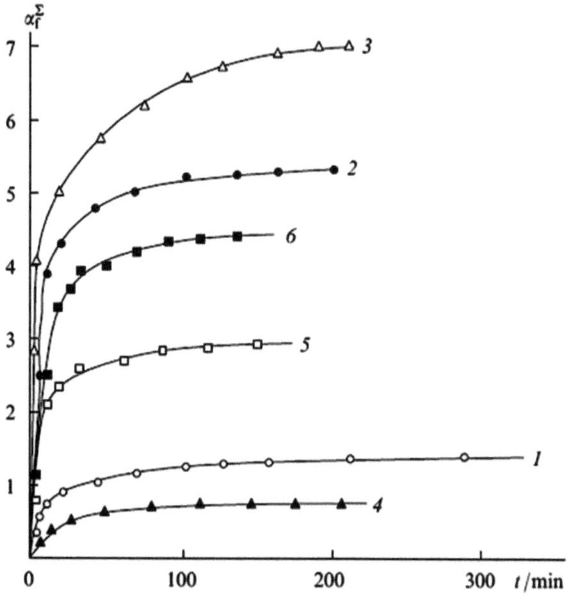

Fig. 6.32 Gas evolution kinetics during thermolysis of metallopolymers **12**: (*1*) 643, (*2*) 873, (*3*) 1,273 K and **2**: (*4*) 643, (*5*) 873, (*6*) 1,273 K [500]

the temperature of thermolysis (to 873–1,273 K) brings about considerable evolution of H_2, and in the case of Hf diol polymers **2–6**, the corresponding diols and a broad range of their oxygen-free fragments are formed even at 643 K.

The weight loss by the samples at the end of transformation is most often below the value expected for their decomposition to HfO_2 or HfC. Out of this series, only for metallopolymer **11**, the weight loss at 873 and 1,273 K is close to that observed in the decomposition of HfO fumarate to HfC: apart from the HfO_2 crystalline phases, the HfC phase is formed but no metallic hafnium (Table 6.13). The content of hafnium carbide in the composite and the degree of crystallinity of products depend on the ligand environment (the type of the bifunctional ligand used to prepare the polymer), the polymer composition and the conditions of thermolysis. At relatively low temperatures (≤643 to ~873 K), the transformation products are amorphous to X-rays; and temperature rise promotes crystallization. In view of the thermodynamic analysis of the Hf–C–O–H system, it appears unexpected [503, 504] that a well crystallized HfC phase was observed only in the thermal decomposition of polymer **8** formed from the complex $Cp_2Hf(CH_2{=}CMeCO_2)_2$ with the minimum H:O atomic ratio (0.4). Possibly, in this case, the conditions of thermolysis are far from thermodynamic equilibrium.

For a thermodynamically equilibrium $HfC_xO_yH_z$ system, the $HfO_2(s) \rightarrow HfC$ (s) transition occurs in the temperature region of 1,973–2,023 K. However, under real conditions, this transformation starts with participation of the energy-saturated highly reactive nano-sized particles formed, the reacting system being far from equilibrium. This shifts the $HfO_2(s) \rightarrow HfC(s)$ transition to lower temperatures, 873–1,273 K (see Table 6.13), for almost all types of metallopolymers.

Depending on the nature of the ligand and the conditions of thermolysis, several options of nano-sized particle conservation are available. They are related either to

Table 6.13 Effect of the thermolysis temperature on the phase composition of the solid thermolysis products of Hf-containing polymers [500]

Polymer	Temperature of thermolysis, K		
	643	873	1,273
2	Amorphous	$m\text{-}HfO_2 + HfC^*$	$m\text{-}HfO_2 + HfC^*$
3	Amorphous	$m\text{-}HfO_2 + t\text{-}HfO_2$	$m\text{-}HfO_2 + t\text{-}HfO_2^* + HfC^*$
5	Amorphous	$m\text{-}HfO_2 + t\text{-}HfO_2^* + HfC^*$	$m\text{-}HfO_2 + HfC^*$
6	Amorphous	$m\text{-}HfO_2 + t\text{-}HfO_2^* + HfC^*$	$m\text{-}HfO_2 + t\text{-}HfO_2^* + HfC^*$
8	Amorphous + $t\text{-}HfO_2 + HfC^*$	Amorphous + HfC + $t\text{-}HfO_2^*$	$m\text{-}HfO_2 + HfC^*$
9	Amorphous	$m\text{-}HfO_2 + t\text{-}HfO_2 + HfC^*$	$m\text{-}HfO_2 + t\text{-}HfO_2 + HfC^*$
11	Amorphous	Amorphous	$m\text{-}HfO_2 + t\text{-}HfO_2 + HfC^*$
12	Amorphous	Amorphous	$m\text{-}HfO_2 + t\text{-}HfO_2 + HfC^*$

Note. The asterisk marks highly disordered phases, letters m and t stand for monoclinic and tetragonal systems, respectively

destruction and subsequent carbonization of the ligand or the mixture of the precursor with the polymer at high temperature or to transformations of the polymer catalyzed by metal clusters arising during the decomposition. Generally, all transformations of this type are self-regulated processes [505].

The research into thermal transformations of precursors in the solid phase is vigorously developing. On the one hand, the transformations most often resemble the pyrolysis in high-temperature liquids [411] and it can be considered as a pseudohomogeneous process. On the other hand, they bear specific features of solid-state reactions, which is manifested as influence of the topochemistry and crystal lattice defects on the transformation rate.

6.7 Computer Modelling of the Kinetics of Nanoparticle Formation During Solid-State Thermolysis

Despite the complexity of multistage physicochemical transformations that accompany the solid-state thermolysis of metallopolymers [506] three key macroscopic stages can be distinguished:

1. thermal decomposition of the metal-containing fragment

$$ML_r \rightarrow M^* + rL; \tag{6.14}$$

2. clusterization and nanoparticle growth

$$\mathrm{M} \xrightarrow{\mathrm{M}} \mathrm{M}_2 \xrightarrow{\mathrm{M}} \mathrm{M}_3 \ldots \xrightarrow{\mathrm{M}} [\mathrm{M}_{n-1}] \xrightarrow{\mathrm{M}} [\mathrm{M}_n]; \tag{6.15}$$

3. polymerization of the demetallized ligand

$$r\mathrm{L} \rightarrow \mathrm{P}, \tag{6.16}$$

where M is the metal, M* is a metal atom or metal compound, L is ligand, M_n is metal-containing oligo(poly)mer, P is the product of polymerization or ligand destruction.

The published and experimental data mainly focus on the characteristics of the final thermolysis products (nanoparticle size distribution and spatial distribution in the matrix, physical and chemical properties of the obtained matrices, etc.). However, no studies considering the variation of the physicochemical parameters during the thermal transformation are available.

The most productive approach to the theoretical investigation of the thermolysis kinetics is computer modelling. In recent years, various solutions to problems of this type have been developed [507–509]; this allows one to follow the cluster formation dynamics in the matrix during the solid-state thermolysis, the temporal growth of particles and the topochemical behaviour of particles in the polymeric medium.

The kinetics of nucleation and particle growth are studied within the framework of the diffusion-controlled aggregation model using combined sweep and Monte–Carlo methods. As the basis of kinetic models, polymeric media of various structures, namely, isotropic (globular) and anisotropic (layer and fibrillar) structures, are considered. The aggregated particles come from the products of thermal decomposition of the metal-containing groups of the polymer chain: metal atoms or metal oxide particles. The medium in which the transformations and the motion and growth of cluster-forming particles take place is represented as a three-dimensional lattice consisting of L^3 (50 nm × 50 nm × 50 nm) cubic cells – reactive groups (1.25×10^5 reaction sites). The cell dimension (face $\alpha = 10^{-9} - 10^{-8}$ m) varies being determined by the size of the characteristic fragment of the regular polymer chain comprising one decaying metal-containing group. The determining algorithms of the model are as follows.

1. Decomposition of the metal-containing groups to give monoatomic metal particles (or oxide particles) at the rate

$$W_J = kC = C_0 k_0 \exp\left(-\frac{E_{a,J}}{RT}\right), \tag{6.17}$$

where k is the constant, C and C_0 are the current and initial numbers of the polymer reaction sites ($C_0 = 1.25 \times 10^5$ monoatomic sites), $E_{a,J}$, k_0 and T are the variable parameters (activation energy, pre-exponential factor and the temperature of the experiment, respectively). A definite (variable) probability that the decay of reaction sites is catalyzed by diffusing particles is taken into account.

2. Solid-state activated diffusion of monoatomic metal (or metal oxide) particles giving polyatomic cluster particles, the rate of diffusion being

$$W_{D,N} = D_N C^* = C^* D_1 \times N^{-1/3}, \tag{6.18}$$

where $C^* = C/a$ is the current number of the reaction sites per cell with dimension a, N is the size of N-atomic cluster, $D_1 = D_0 \exp[-E_{a,D}/(RT)]$ is the diffusion coefficient for a monoatomic particle, $E_{a,D}$ and D_0 are the variable parameters (activation energy and entropy factor, respectively). As noted above, it is assumed that a single cell cannot accommodate two (or more) separate particles, i.e., they immediately coalesce to a unified cluster.

3. Considering dissociation of clusters by the reaction

$$[M_n] \rightarrow [M_{n-1}] + M \tag{6.19}$$

to give monoatomic particles at a rate

$$W_N = k_N C_N, \tag{6.20}$$

where

$$k_N = k_{0N}\, N \exp\left(-\frac{E_{a,N}}{RT}\right) \tag{6.21}$$

depends on the cluster dimension N and decreases as the cluster dimension increases, C_N is the current concentration of the cluster particles with dimension N, $E_{a,N}$ and k_{0N} are the activation energy of the Nth cluster and the preexponential factor (equal to ~10^{13} s^{-1}), respectively. It is assumed that monoatomic particles can migrate to neighbouring cells.

4. Visualization of the formation of cluster particles. At the end of the experiment, the results are presented as the dependence of $\ln(C + 1)/\ln(C_{max} + 1)$ and $J(N)$ on t, i.e., the time dependence of the cluster size distribution ((Fig. 6.33) [506, 509] where C_{max} is the greatest number of clusters at time instant t; $J(N)$ is the cluster size distribution where J is the number of N-dimensional clusters. As a result, the accumulation kinetics of a specified Nth cluster is obtained.

The time of the numerical experiment is $t_\Sigma = Zt_p$, $t_p = -k^{-1}\ln(C/C_0)$ is the time of decay of 95.5 % of the metal-containing groups of the polymer, Z is the relative (variable) time of cluster formation ($Z \geq 1$). The values $C_0 = 5 \times 10^{26}$ particles per m^3 (for a metallopolymer with the density 1.5 g m^{-3} a chain fragment of which contains one reactive group in the cell with $a = 10^{-9}$m), $E_{a,J} =$ 167 kJ mol^{-1},

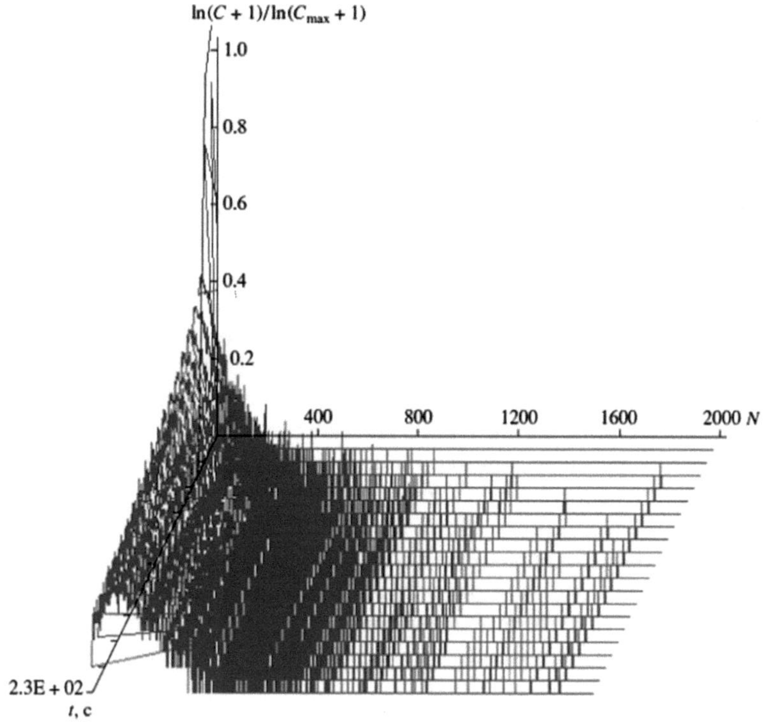

Fig. 6.33 Evolution of cluster size distribution during the computer experiment (*gl*-structure): $T = 653$ K, $Z = 5$, $W_{D,\ N} = 1 \times 10^{-18}$ particle m^2 s^{-1}, $W_J = 2.2 \times 10^{25}$ particle m^{-3} s^{-1}, $W_2 = 1 \times 10^3$ particle m^{-3} s^{-1}, and $W\infty = 4 \times 10^{-14}$ particle m^{-3} s^{-1}

$k_0 = 10^{13}\text{s}^{-1}, E_{aN} = E_{a,\ \infty} - 2(E_{a,\ \infty} - E_{a,2})/N, E_{a,\ \infty} = 420$ kJ mol^{-1} (the activation energy for the decay of the 'massive' cluster, $E_{a,\ \infty} = E_{sub}$ is the sublimation energy of the metal or its compounds) and $E_{a,2} = 170$ kJ mol^{-1} (dissociation energy of the two-particle cluster) are used as the basic kinetic parameters.

The results of the computer modelling of nanoparticle formation during the solid-state thermal decomposition of metal-containing monomers and polymers with different structural organization point to some general trends. For isotropic polymer media, an increase in the time of cluster formation leads to evolution of the particle size distribution spectrum: as Z increases, the transformation parameters being the same, the distribution maximum $J_{max}(N)$ shifts to greater N values, from $N \approx 80$ at $Z = 2$ to $N \approx 160$ at $Z = 10$, while the J_{max} value simultaneously diminishes. The formation of single clusters is a consecutive-parallel process. For example, over the experimental time $t_\Sigma = 228$ s ($Z = 5$), the yield of clusters with $N \approx 2$ and 65 passes through a maximum, while for $N \approx 125$ and 185, the number of clusters increases to reach a steady state, and smaller clusters ($1 \leq N \leq 4$ and $35 \leq N \leq 39$) do not disappear, their number reaching a steady value, apparently, as a result of dissociation of larger clusters.

An increase in the computing time results, first, in a shift of the distribution spectrum $J(N)$ to greater N values, the mean value for the N-dimensional cluster reaching ~1,500 as soon as for $Z = 100$, which may correspond to a spherical cluster with a size from 4.1 to 4.2 nm. Second, the distribution spectrum $J(N)$ becomes simpler: the polymodal distribution typical of small Z values evolves already at $Z = 100$ to a bimodal distribution with maxima at $N \approx 540 - 550$ and $N \approx 1{,}540$. Presumably, further increase in the computing time ($Z \gg 100$) may produce a unimodal particle size distribution with a much greater mean size. The considered results of modelling of the kinetics of formation of metallic clusters upon thermolysis of the metal-containing fragments of the polymeric media with different structural organization demonstrate [506, 509] the growth of nanoclusters, which may reach rather large sizes. These models are efficient and can be extended to other objects.

Of all the physical factors that can affect the behaviour of a chemical substance, temperature is most popular. Thermolysis of metallopolymers and their precursors provides almost infinite possibilities for the manufacture of various types of nanocomposites and it is the simplest and most widely used method for introducing up to 90 mass% of colloid particles of metals into a polymeric composite. It is possible, first, to subject a precursor to the thermolysis and, second, to blend it with the polymer matrix prepared beforehand (*ex situ* process). Alternatively, the researcher can perform joint thermolysis of a polymer blend with a metal compound present in a specified ratio (*in situ* process). Kinetically, the polymer-mediated synthesis of nanoparticles is a conjugate two-stage process related to multichannel transformations of the metal-containing precursor. It induces the evolution of the high-molecular-mass component: depolymerization, cross-linking, destruction, many reactions being catalyzed by the formed metal clusters [28]. The free energy of thermolysis is higher than the free energy absorbed upon the formation of metallopolymers; this is employed to perform other processes in conjugation with the thermolysis. As applied to solid-state reactions, transformations taking place in structural inhomgeneity sites (defects, dislocations) may induce other processes. The conjugate processes represent an approximation to self-assembly processes, i.e., processes with feedback [505].

The self-assembly of 'core–shell' type nanoparticles during thermolysis is one of the unique phenomena inherent in the rapidly developing supramolecular chemistry [510, 511]. Supermolecules (nanometre-sized molecules) are oligomolecular particles arising upon the intermolecular association of several components that form intricate structure of definite architectures according to the molecular recognition principle. The organization of such supermolecules is based on interaction types other than covalent bonds – hydrogen bonds, electrostatic and Van der Waals forces.

This chapter is restricted only to description of the formation of nanoparticles in organic matrices during thermolysis. Quite extended subject matter related to the thermal synthesis of inorganic matrices and the insertion of metal-containing clusters therein and production of metalloceramics remained beyond the scope of our consideration and is still waiting for being surveyed and analyzed.

References

1. A.D. Pomogailo, I.E. Uflyand, *Makromolekulyarnye Metallokhelaty* (Macromolecular Metal Chelates) (Khimiya, Moscow, 1991)
2. Y. Koltypin, N. Perkas, A. Gedanken, J. Mater. Chem. **14**, 2975 (2004)
3. X. Li, S.J. Tian, Y. Ping, D.H. Kim, W. Knoll, Langmuir **21**, 9393 (2005)
4. V. Prashant, J. Phys. Chem. B **106**, 7729 (2002)
5. X.F. Duan, C.M. Lieber, Adv. Mater. **12**, 298 (2000)
6. A.M. Morales, C.M. Lieber, Science **270**, 208 (1998)
7. Z.Q. Wang, X.D. Liu, J.F. Gong, H.B. Huang, S.L. Gu, S.G. Yang, Cryst. Growth. Des. **8**, 3911 (2008)
8. V.V. Kozlov, G.P. Karpacheva, V.S. Petrov, E.V. Lozovskaya, Vysokomol. Soedin. Ser. A **43**, 1 (2001)
9. G.P. Karpacheva, K.A. Bagdasarova, G.N. Bondarenko, L.M. Zemtsov, D.G. Muratov, N.S. Perov, Vysokomol Soedin. Ser. A **51**, 2037 (2009)
10. A.D. Pomogailo, Kolloid. Zh. **67**, 726 (2005)
11. J.H. Warner, Adv. Mater. **20**, 784 (2008)
12. H. Zhang, D.Y. Wang, B. Yang, M. Helmuth, J. Am. Chem. Soc. **128**, 10171 (2006)
13. V. Rotello, *Nanoparticles: Building Blocks for Nanotechnology (Nanostructure Science and Technology)* (Springer, Berlin, 2003)
14. A. Heilmann, *Polymer Film with Embedded Metal Nanoparticles* (Springer, Berlin, 2002)
15. P.M. Ajayan, L.S. Schadler, P.V. Braun (eds.), *Nanocomposite Science and Technology* (Wiley-VCH, Weinheim, 2003)
16. A.J. Gu, G.Z. Liang, J. Appl. Polym. Sci. **89**, 3954 (2003)
17. S.M. Pourmortazavi, I. Kohsari, M.B. Teimouri, S.S. Hajimirsadeghi, Mater. Lett. **61**, 4670 (2007)
18. N. Grassi, *Chemistry of High Polymer Degradation Processes* (Butterworths, London, 1956)
19. E. Themistou, A. Kanari, C.S. Patrickios, J. Polym. Sci.: Part A: Polym. Chem. **45**, 5811 (2007)
20. S.H. Kim, F. Nederberg, L. Zhang, C.G. Wade, R.M. Waymouth, J.L. Hedric, Nano Lett. **8**, 294 (2008)
21. J.D. Menczel, R.B. Prime (eds.), *Thermal Analysis of Polymers. Fundamentals and Applications* (Wiley, London, 2009)
22. P. Greil, Adv. Eng. Mater. **2**, 339 (2000)
23. S. Bernard, M. Weinmann, P. Gerstel, P. Viele, F. Aldinger, J. Mater. Chem. **15**, 289 (2005)
24. S. Duperrier, C. Gervais, S. Bernard, D. Cornu, F. Babboneau, C. Balan, P. Miele, Macromolecules **40**, 1018 (2007)
25. Y. Prosanov, F.F. Matvienko, Fizika tverdogo tela (Physics Sol. State) **53**, 824 (2011)
26. V.N. Kuznetsova, M.G. Malkin, J. Appl. Spectr. **67**, 762 (2000)
27. A.D. Pomogailo, V.S. Savost'yanov, *Synthesis and Polymerization Metal-Containing Monomers* (CRC Press, Boca Raton, 1994)
28. A.D. Pomogailo, A.S. Rozenberg, G.I. Dzhardimalieva, Ros. Khim. Zh. (Mendeleev Chem. J.) **53**, 140 (2009)
29. A.S. Shteinberg, *Fast Reactions in Energetic Materials: High-Temperature Decomposition of Rocket Propellants and Explosives* (Fizmatlit, Moscow, 2006)
30. D.A. Frank-Kamenetskii, *Diffuziya i Teploperedacha v Khimicheskoi Kinetike* (Diffusion and Heat Transfer in Chemical Kinetics) (Nauka, Moscow, 1987)
31. T. Cordero, J.M. Rodriguez-Maroto, J. Rodriguez-Miraso, J.J. Rodriguez, Thermochim. Acta **164**, 135 (1990)
32. A.Q. Gu, Z.L. Yu, Y.B. Li, J. Appl. Polym. Sci. **110**, 61 (2008); **114**, 911 (2009)
33. I. Kaya, A. Solguntekin, J. Appl. Polym. Sci. **113**, 1994 (2009)
34. F. Yao, Q. Wu, D. Zhou, J. Appl. Polym. Sci. **114**, 834 (2009)
35. J. Malek, Thermochim. Acta **355**, 239 (2000)

36. H. Pehlivan, F. Özmihci, T. Tihminlioglu, D. Balcöse, J. Appl. Polym. Sci. **90**, 3069 (2003)
37. H. Ismail, Z. Mohamad, A.A. Bakar, Polym. Plastics. Technol. Eng. **42**, 81 (2003)
38. H.S. Kim, H.S. Yang, H.J. Kim, H.J. Park, J. Therm. Anal. Calor. **76**, 395 (2004)
39. S.C. Turmanova, S.D. Genieva, A.S. Dimitrova, L.T. Vlaev, Express Polym. Lett. **2**, 133 (2008)
40. P.K. Roy, P. Surekha, C. Rajagopal, V. Choudhary, Express Polym. Lett. **1**, 208 (2007)
41. P.K. Roy, P. Surekha, R. Raman, V. Choudhary, J. Appl. Polym. Sci. **99**, 236 (2006)
42. O.F. Ozturk, D.A. Koze, A.N. Ay, B. Zumreoglu–Karan, J. Appl. Polym. Sci. **98**, 490 (2005)
43. F. Dogan, O.F. Ozturk, M. Yurekli, A.N. Ay, D.A. Koze, J. Appl. Polym. Sci. **106**, 1129 (2007)
44. R. Tannenbaum, S. King, J. Lecy, M. Tirrell, L. Potts, Langmuir **20**, 4507 (2004)
45. F. Yao, Q. Wu, D. Zhou, J. Appl. Polym. Sci. **114**, 834 (2009)
46. J. Opfermann, J. Therm. Anal. Calor. **60**, 641 (2000)
47. J. Opfermann, E. Kaisersberger, H.J. Flammersheim, Thermochim. Acta **391**, 119 (2002)
48. G.E. Zaikov, *Degradation and Stabilization of Polymers* (Nova Science Publication, New York, 1999)
49. G.E. Zaikov, Soros. Obrazov. Zh. (Soros Educat. J.) **12**, 48 (2000)
50. S.S. Batsanov, *Eksperimental'nye Osnovy Strukturnoi Khimii (Spravochnoe Posobie)* [Experimental Foundations of Structural Chemistry (The Handbook)]. (Izd. Standartov, Moscow, 1986)
51. R.A Lidin, L.L Andreeva, V.A Molochko, *Spravochnik po Neorganicheskoi Khimii. Konstanty Neorganicheskikh Veshchestv* (The Handbook on Inorganic Chemistry. The Constants of Inorganic Substances). (Khimiya, Moscow, 1987)
52. N. Kubota, C. Serizawa, Propellants Explos. Pyrotech. **12**, 145 (1987)
53. T. Kuwahara, S. Matsuo, N. Shinozaki, Propellants. Explos. Pyrotech. **22**, 198 (1997)
54. A. Gocmez, A.G. Yilmaz, F. Pekel, Propellant. Explos. Pyrotech **24**, 65 (1999)
55. E.-C. Koch, A. Dochnahl, Propellant. Explos. Pyrotech. **25**, 37 (2000)
56. A. Gocmez, A.G. Yilmaz, F. Pekel, Propellants Explos. Pyrotech. **24**, 65 (1999)
57. S. Cudziło, A.W. Trzciłski, Polym. J. Appl. Chem. **26**, 25 (2001)
58. A. Huczko, H. Lange, G. Chojecki, S. Cudziło, Y.Q. Zhu, H.W. Kroto, D.R.M. Walton, J. Phys. Chem. B **107**, 2519 (2003)
59. V.V. Boldyrev, *Metody Izucheniya Kinetiki Termicheskogo Razlozheniya Tverdykh Veshchestv* (The Methods of Investigation of Kinetics of Thermal Decomposition of Solids) (Tomsk University, Tomsk, 1958)
60. V.A. Logvinenko, F. Paulik, I. Paulik, *Kvaziravnovesnaya Termogravimetriya v Sovremennoi Neorganicheskoi Khimii* (Quasi-equilibrium Thermogravimetry in Modern Inorganic Chemistry) (Nauka, Novosibirsk, 1989)
61. Z.A. Vnutskikh, A.A. Fedorov, Y.S. Chekryshkin, Z.R. Ismagilov, M.A. Kerzhentsev, Khim. Inter. Ustoich. Razv **9**, 621 (2001)
62. B. Delmon, *Introduction a la Cinetique Heterogene* (Technip, Paris, 1969)
63. J. Paulik, F. Paulik, Thermochim. Acta **100**, 23 (1986)
64. P.J. Haines, *Thermal Methods of Analysis* (Blackie Academic and Professional, London, 1995)
65. S.V. Pol, V.S. Pol, A. Gedanken, Chem. Eur. J. **10**, 4467 (2004)
66. J.Q. Sun, X.P. Shen, L.J. Guo, K.M. Chen, Q. Liu, Phys. E. Low Dimens. Syst. Nanostr. **41**, 1527 (2009)
67. J.Q. Sun, X.P. Shen, K.M. Chen, Q. Liu, W. Liu, Solid State Commun. **147**, 501 (2008)
68. M. Bowtell, Adhes. Age. **40**, 62 (1997)
69. G.P. Shveikin, I.V. Nikolaenko, Theor. Found. Chem. Eng. **43**, 553 (2009)
70. F. Antolini, A. Ghezelbash, C. Esposito, E. Trave, L. Tapfer, B.A. Korgel, Mater. Lett. **60**, 1095 (2006)
71. S. Cudzilo, M. Bystrzejewski, H. Lange, A. Huczko, Carbon **43**, 1778 (2005)
72. D. Hulicova, K. Hosoi, S. Kuroda, A. Oya, Carbon **43**, 1246 (2005)

73. J. Delaunay, T. Hayashi, M. Tomita, S. Hirono, S. Umemura, Appl. Phys. Lett. **71**, 3427 (1997)
74. M.K. Corbierre, J. Beerens, R.B. Lennox, Chem. Mater. **17**, 5774 (2005)
75. B.F.G. Johnson, K.M. Sanderson, D.S. Shephard, D. Ozkaya, W.Z. Zhou, H. Ahmed, M.D.R. Thomas, L. Gladden, M. Mantle, Chem. Commun. 1317 (2000)
76. N.A. Tikhonov, I.V. Arkhangelsky, S.S. Belyaev, A.T. Matveev, Thermochim. Acta **486**, 66 (2009)
77. S.S. Belyaev, I.V. Arkhangelsky, I.V. Makarenko, Thermochim. Acta **507–508**, 9 (2010)
78. Y. Prosanov, F.F. Matvienko, Fizika tverdogo tela (Physics Sol. State) **52**(10), 2056 (2010)
79. S. Joden, J. Chem. Chem. Eng. **3**, 7 (2009)
80. L.M. Zemtsov, G.P. Karpacheva, M.N. Efimov, D.G. Muratov, K.A. Bagdasarova, Vysokomol. Soedin. Ser. A **48**, 977 (2006)
81. L.V. Kozhitov, A.V. Kostikova, V.V. Kozlov, V.I. Khursa, Izv. VUZ'ov. Materialy electronnoi tekhniki. (3), 48 (2011)
82. L.V. Kozhitov, A.V. Kostikova, V.V.Kozlov, M.Ph. Bulatov, J. Nanoelectron. Optoelectron. (7), 419 (2012)
83. Y.C. Kang, S.B. Park, I.W. Lenggoro, K. Okuyama, J. Mater. Res. **14**, 2611 (1999)
84. I.W. Lenggoro, T. Hata, F. Iskandar, J. Mater. Res. **15**, 733 (2000)
85. H.K. Kammler, L. Mädler, S.E. Pratsinis, Chem. Eng. Technol. **24**, 583 (2001)
86. E.K. Athanassiou, R.N. Grass, W.J. Stark, Nanotechnology **17**, 1668 (2006)
87. M.T. Swihart, Curr. Opin. Colloid Interf. Sci. **8**, 127 (2003)
88. J. Lacson, *Inorganic Zinc Chemicals. Chemical Economics Handbook* (electronic release) (SRI International, Menlo Park, 2000)
89. C. Pacholski, A. Kornowski, H. Weller, Angew. Chem. Int. Ed. **41**, 1188 (2002)
90. L. Schmidt-Mende, J.L. MacManus-Driscoll, Mater. Today **10**, 40 (2007)
91. S.J. Peatron, D.P. Norton, Prog. Mater. Sci. **50**, 293 (2005)
92. K. Hara, T. Horiguchi, T. Kinoshita, Adv. Mater. **15**, 838 (2003)
93. L.E. Creene, W. Law, J. Goldberger, Angew. Chem. Int. Ed. **42**, 3031 (2003)
94. T.-J. Hsueh, C.-L. Hsu, Sensors Actuators B. Chem. **131**, 572 (2008)
95. A. Khan, M.E. Kordesch, Physica E **33**, 88 (2006)
96. M.D. Liedekerke, *Pigment, Inorganic, 2.3 Zinc Oxide (Zincwhite). Ullmann's Encyclopedia of Industrial Chemistry*, 6th edn. (Wiley-VCH Verlag GmbH, Weinheim, 2001)
97. T. Tani, L. Madler, S.E. Pratsinis, J. Nanopart. Res. **4**, 337 (2002)
98. R. Riedel, A. Gurlo, E. Ionescu, Chem. Unserer Zeit. **44**, 208 (2010)
99. N. Reuge, B. Caussat, N. Joffin, J. Dexpert-ghys, M. Verelst, H. Dexpert, AIChE **54**, 394 (2008)
100. D.S. Jung, S.B. Park, Y.C. Kang, Korean J. Chem. Eng. **27**, 1621 (2010)
101. J.-C. Lin, J.W. Gentry, J. Aerosol Sci. **31**(Supl.1), 797 (2000)
102. N. Reuge, B. Caussat, Comput Chem. Eng. **31**, 1088 (2007)
103. J.R. Sohn, Y.C. Kang, H.D. Park, Jpn. J. Appl. Phys. **41**, 3006 (2002)
104. H.S. Roh, Y.C. Kang, H.D. Park, S.B. Park, Appl. Phys. A. **76**, 241 (2003)
105. I.W. Lenggoro, Y. Itoh, K. Okuyama, T.O. Kim, J. Mater. Res. **19**, 3534 (2004)
106. W.-N. Wang, A. Purwanto, K. Okuyama, in *Handbook of Atomization and Sprays*, ed. by N. Ashgriz (Springer Sci. Business Media, LLC, New York, 2011), p. 861
107. Y.C. Kang, Y.S. Chung, S.B. Park, J. Am. Ceram. Soc. **82**, 2056 (1999)
108. K. Okuyama, I.W. Lenggoro, Chem. Eng. Sci. **58**, 537 (2003)
109. W.-N. Wang, I.W. Lenggoro, Y. Terashi, T.O. Kim, K. Okuyama, Mat. Sci. Eng. B. **123**, 194 (2005)
110. Y.C. Kang, S.B. Park, Mater. Res. Bull **35**, 1143 (2000)
111. W.-N. Wang, I.W. Lenggoro, Y. Terashi, Y.C. Wang, K. Okuyama, J. Mater. Res. **20**, 2873 (2005)
112. I.W. Lenggoro, Y. Itoh, N. Iida, K. Okuyama, Mater. Res. Bull **38**, 1819 (2003)
113. N.A. Luechinger, E.K. Athanassion, W.J. Stark, Nanotechnology **19**, 445201 (2008)

114. G.B. Manelis, G.M. Nazin, Y.I. Rubtsov, V.A. Strunin, *Termicheskoe Razlozhenie i Gorenie Vzryvchatykh Veshchestv i Porokhov* (Nauka, Moscow, 1996)
115. A. Pivkina, P. Ulyanova, Y. Frolov, S. Zavyalov, J. Schoonman, Propellant. Explor. Pyrotech. **29**, 39 (2004)
116. G. Singh, D.K. Pandey, Propellant. Explor. Pyrotech. **28**, 231 (2003)
117. G. Sing, S.P. Felix, Combust. Flame **132**, 422 (2003)
118. M. Stefanescu, V. Sasca, M. Birzescu, J. Therm. Anal. Calor. **72**, 515 (2003)
119. M. Birzescu, M. Niculescu, R. Dumitru, P. Budrugeac, E. Segal, J. Therm. Anal. Calor. **94**, 297 (2008)
120. C.A. Strydom, C.P.J. Vuuren, J. Therm. Anal. **32**, 157 (1987)
121. M. Han, N.-E. Shi, W.-L. Zhang, B.-J. Li, J.-H. Sun, K.-J. Chen, J.-M. Zhu, X. Wang, Z. Xu, Chem. Eur. J. **14**, 1615 (2008)
122. M.A. El-Sayed, Acc. Chem. Res. **34**, 257 (2001)
123. Z.L. Wang, J.H. Song, Science **312**, 242 (2006)
124. H.P. Zho, Y.W. Zhang, H.X. Mai, X. Sun, Q. Liu, W.G. Song, C.H. Yan, Chem. Eur. J. **14**, 3380 (2008)
125. N. Joffin, J. Lumin. **113**, 249 (2005)
126. N. Joffin, B. Gaillier, J. Dexpert-Ghys, M. Verelst, G. Baret, A. Garsia, P. Guillot, G. Gali, R. Mauricot, S. Schamm, J. Phys. D **38**, 3261 (2005)
127. S.N. Rishikeshi, S.S. Joshi, J. Therm. Anal. Calor. **109**, 1473 (2012)
128. Z.W. Pan, Z.R. Dai, Z.L. Wang, Science **291**, 1047 (2001)
129. Z.X. Deng, C. Wang, X.M. Sun, Y.D. Li, Inorg. Chem. **41**, 869 (2002)
130. Z. Chen, L. Gao, Mater. Res. Bull **42**, 1657 (2007)
131. C.A. O'Connell, D.J. Dollimore, Thermochim. Acta **357/358**, 79 (2000)
132. G. Reichmuth, E. Dubler, Thermochim. Acta **85**, 485 (1985)
133. R. Sasikala, S.K. Kulshreshtha, J. Therm. Anal. Calor. **78**, 723 (2004)
134. A.-S. Malik, M.J. Duncan, P.G. Bruce, J. Mater. Chem. **13**, 2123 (2003)
135. R.L. Frost, W. Martens, M.O. Adebajo, J. Therm. Anal. Calor. **81**, 351 (2005)
136. A. Valor, E. Reguera, E. Torres- Garcia, S. Mendoza, F. Sanchez-Sinencio, Thermochim. Acta **389**(133) (2002)
137. A. Valor, E. Reguera, F. Sanchez-Sinencio, Powder Diffraction **17**, 13 (2002)
138. W.Y. Yu, J.C. Falkner, C.T. Yavuz, V.L. Colvin, Chem. Comm. **20**, 2306 (2004)
139. H. Wang, X. Jiao, D. Chen, J. Phys. Chem. C **112**, 18793 (2008)
140. J. Park, E. Kang, S.U. Son, H.M. Park, M.K. Lee, J. Kim, K.-W. Kim, H.-J. Noh, J.-H. Park, C.J. Bae, J.-G. Park, T. Hyeon, Adv. Mater. **17**, 429 (2005)
141. Y. Goto, K. Taniguchi, T. Omata, S. Otsuka-Yao-Matsuo, Chem. Mater. **20**, 4156 (2008)
142. B.S. Randhawa, H.S. Dosanjh, N. Kumar, J. Therm. Anal. Calor. **95**, 75 (2009)
143. B.S. Randhawa, K.J. Sweety, J. Radioanal. Nucl. Chem. **247**, 513 (2001)
144. B.S. Randhawa, M. Kaur, J. Radioanal. Nucl. Chem. **261**, 569 (2004)
145. B.S. Randhawa, M. Kaur, in Proceed. the 29th Int. Conf. on the Applications of the Mössbauer Effect (ICAME 2007) 14–19 October 2007, ed. by N.S. Gajbhiye, S. K. Date (Springer, Dordrecht, 2008), p. 1329
146. A.S. Rozenberg, E.I. Aleksandrova, Izv. Akad. Nauk. Ser. Khim. 72 (1996)
147. A.S. Rozenberg, N.V. Chukanov, Izv. Akad. Nauk, Ser. Khim. 350 (1996)
148. A.S. Rozenberg, V.R. Stepanov, Izv. Akad. Nauk, Ser. Khim. 1406 (1996)
149. A.S. Rozenberg, *Doctoral Thesis in Chemical Sciences* (Institute of Problems of Chemical Physics, Russian Academy of Sciences, Chernogolovka, 1997)
150. S.V. Davidovich, A.A. Vecher, E.A. Gusev, Thermochim. Acta **89**, 383 (1985)
151. I.V. Fedorova, V.A. Shurov, A.A. Fedorov, M.S. Gaisinovich, Zh. Prikl. Khim. **65**, 736 (1992)
152. I.V. Arkhangel'skii, L.N. Komissarova, A. Gorski, A. Kras'nicka, J. Therm. Anal. **32**, 1234 (1987)
153. A. Gorski, A. Kras'nicka, J. Therm. Anal. **32**, 1345 (1987)

154. K. Muraishi, T. Takano, K. Nagase, N. Tanaka, J. Inorg. Nucl. Chem. **43**, 2293 (1981)
155. A.S. Rozenberg, G.I. Dzhardimalieva, A.D. Pomogailo, Dokl. Akad. Nauk **356**, 66 (1997)
156. B.B. Bokhonov, Y.M. Yukhin, Zh. Neorg. Khim. **52**, 993 (2007)
157. V.V. Kireev, L.N. Demyanets, L.E. Li, V.V. Artemov, Neorg. Mater. **46**, 193 (2010)
158. R.-C. Wang, C.-C. Tsai, Appl. Phys. A **94**, 241 (2009)
159. S. Aghabeygi, F. Bigdeli, A. Morsali, J. Inorg. Organomet. Polym. **22**, 526 (2012)
160. C.S. Li, Y.N. Li, Y.L. Wu, B.S. Ong, R.O. Loutfy, Sci. China. Ser E-Tech. Sci. **51**, 2075 (2008)
161. W.J. Scharmach, R.D. Bucher, V. Papavassiliou, P. Pacouloute, M.T. Swihart, Aerosol. Sci. Technol. **44**, 1083 (2010)
162. E. Tirosh, G. Shemer, G. Markovich, Chem. Mater. **18**, 465 (2006)
163. N. Bao, L. Shen, Y. Wang, P. Padhan, A. Gupta, J. Am. Chem. Soc. **129**, 12374 (2007)
164. J.H. Park, N.M. Hwang, T. Hyeon, Nat. Mater. **3**, 891 (2004)
165. S. Sun, H. Zeng, D.B. Robinson, S. Raoux, P.M. Rice, S.X. Wang, G. Li, J. Am. Chem. Soc. **126**, 273 (2004)
166. K.J. Lee, B.H. Jun, J. Choi, Y.I. Lee, J. Joung, Y.S. Oh, Nanotechnology **18**, 335601 (2007)
167. N. Yang, K. Aoki, J. Phys. Chem. **109**, 23911 (2005)
168. M. Yamamoto, Y. Kashiwagi, M. Nakamoto, Langmuir **22**, 8581 (2006)
169. Y. Kashiwagi, M. Yamamoto, M. Nakamoto, J. Colloid Interface Sci. **300**, 16933 (2006)
170. M. Chen, Y.G. Feng, X. Wang, T.C. Li, J.Y. Zhang, D.J. Qian, Langmuir **23**, 5296 (2007)
171. S.J. Lee, S.W. Han, K. Kim, Chem. Commun. **5**, 442 (2002)
172. S. Navaladian, B. Viswanathan, R.P. Viswanath, T.K. Varadarajan, Nanoscale Res. Lett. **2**, 44 (2007)
173. L. D'Urso, V. Nicolosi, G. Compagnini, O. Puglisi, Appl. Surf. Sci. **226**(c), 131 (2004)
174. P. Uznanski, E. Bryszewska, J. Mater. Sci. **45**, 1547 (2010)
175. B. Wiley, Y.G. Sun, Y. Xia, Y. Acc, Chem. Res. **40**, 1067 (2007)
176. A. Mehrani, A. Morsali, J. Inorg. Organomet. Polym. Mater. **21**, 476 (2011)
177. D. Dollimore, D.L. Griffiths, J. Thermal Anal. **2**, 229 (1970)
178. B. Malecka, E.D. Ciesla, A. Malechi, J. Therm. Anal. Calor. **68**, 819 (2002)
179. O. Carp, L. Patron, G. Marinescu, G. Pascu, P. Budrugeau, M. Brezeanu, J. Therm. Anal. Calor. **72**, 263 (2003)
180. A. Angermann, J. Töpfer, J. Mater. Sci. **43**, 5123 (2008)
181. V.V. Boldyrev, Thermochim. Acta. **388**, 63 (2002)
182. R.L. Frost, M.L. Weier, J. Therm. Anal. Calor. **75**, 277 (2004)
183. W. Zhou, K. Tang, S. Zeng, Y. Qi, Nanotechnology **19**, 065602 (2008)
184. R. Zboril, I. Medrik, J. Pechousek, C. Gregor, J. Am. Chem. Soc. **129**, 10929 (2007)
185. M. Hermanek, M.A. Mohamed, A.K. Galwey, S.A. Halawy, Thermochim. Acta **29**, 57 (2005)
186. M. Hermanek, R. Zboril, M. Mashlan, L. Machala, O. Schneeweiss, J. Mater. Chem. **16**, 1273 (2006)
187. O.I. Gyrdasova, V.N. Krasilnikov, G.V. Bazuev, Z. Neorg, Khim. **54**, 1097 (2009)
188. M. Yin, S. O'Brien, J. Am. Chem. Soc. **125**, 10180 (2003)
189. J. Park, E. Kang, C.J. Bae, J.G. Park, H.J. Noh, J.Y. Kim, J.H. Park, H.M. Park, T. Hyeon, J. Phys.Chem. B **108**, 13594 (2004)
190. F. Jiao, A. Harrison, P.G. Bruce, Angew. Chem. Int. Ed. **46**, 3946 (2007)
191. F. Jiao, A. Harrison, P.G. Bruce, Angew. Chem. Int. Ed. **46**, 3946 (2007)
192. Y.-P. Du, Y.-P. Du, Y.-W. Zhang, L.-D. Sun, C.-H. Yan, J. Phys. Chem. C **113**, 6521 (2009)
193. A.H. Lu, E.L. Salabas, F. Schüth, Angew. Chem. Int. Ed. **46**, 1222 (2007)
194. Y.-P. Du, Y.-W. Zhang, L.-D. Sun, C.-H. Yan, J. Phys. Chem. C **113**, 6521 (2009)
195. S. Sun, H. Zeng, J. Am. Chem. Soc. **124**, 8204 (2002)
196. J. Park, K. An, Y. Hwang, J.G. Park, H.J. Noh, J.Y. Kim, J.H. Park, N.M. Hwang, T. Hyeon, Nat. Mater. **3**, 891 (2004)
197. M.M. Lin, D.K. Kim, J. Nanopart. Res. **14**, 688 (2012)

198. K. Liu, H. You, G. Jia, Y. Zheng, Y. Song, M. Yang, Y. Huang, H. Zhang, Crystal Growth. Design **9**, 3519 (2009)
199. B.L. Chen, Y. Yang, F. Zapata, G.N. Lin, G.D. Qian, E.B. Lobkovsky, Adv. Mater. **19**, 1693 (2007)
200. J.R. William, M.L. Kathryn, H.Y. An, W.L. Lin, W.B. Lin, J. Am. Chem. Soc. **128**, 9024 (2006)
201. V.G. Pol, O. Palchik, A. Gedanken, I. Felner, J. Phys. Chem. B **106**, 9737 (2002)
202. K.-L. Wong, G.-L. Law, M.B. Murphy, P.A. Tanner, W.-T. Wong, P.K. Lam, L.M. Hon-Wah, Inorg. Chem. **47**, 5190 (2008)
203. L. Zhang, J. Luo, M. Wu, H. Jiu, Q.W. Chen, Mater. Lett. **61**, 4452 (2007)
204. H. Yang, D. Zhang, L. Shi, J. Fang, Acta Mater. **56**, 955 (2008)
205. J. Goldberg, R. Fan, P.D. Yang, Acc. Chem. Res. **39**, 239 (2006)
206. V.G. Pol, M. Popa, S. Acharya, K. Ariga, P. Thiyagarajan, Inorg. Chem. **48**, 5569 (2009)
207. N. Imanaka, T. Masui, Y. Mayama, K. Koyabu, J. Solid State Chem **178**, 3601 (2005)
208. K. Koyabu, T. Masui, S. Tamura, N. Imanaka, J. Alloys Comp. **408**, 867 (2006)
209. V.G. Pol, J.M. Calderon-Moreno, P. Thiyagarajan, Langmuir **24**, 13640 (2008)
210. M.C. Alves, G. Tourillon, J. Phys. Chem. **100**, 7566 (1996)
211. C.-J. Cheng, C.-C. Lin, R.-K. Chiang, C.-R. Lin, I.S. Lyubutin, E.A. Alkaev, H.-Y. Lai, Crystal Growth Des. **8**, 877 (2008)
212. M. Epifani, J. Arbiol, R. Díaz, M.J. Perálvarez, P. Siciliano, J.R. Morante, Chem. Mater. **17**, 6468 (2005)
213. E.V. Shevchenko, D.V. Talapin, N.A. Kotov, S. O'Brien, C.B. Murray, Nature (London) **439**, 55 (2006)
214. H. Hiramatsu, F.E. Osterloh, Chem. Mater. **16**, 2509 (2004)
215. L.-M. Lacroix, S. Lachaize, A. Falqui, M. Respaud, B. Chaudret, J. Am. Chem. Soc. **131**, 549 (2009)
216. Y. Wang, H. Yang, Chem. Commun. **2545** (2006)
217. P. Uznanski, C. Amiens, B. Chaudret, E. Bryszewska, Polym. J. Chem. **80**, 1845 (2006)
218. I. Quiros, M. Yamada, K. Kubo, J. Mizutani, M. Kurihara, H. Nishihara, Langmuir **18**, 1413 (2002)
219. N. Cordente, M. Respaud, F. Senocq, M.-J. Casanove, C. Amiens, B. Chaudret, Nano Lett. **1**, 565 (2001)
220. N. Wu, L. Fu, M. Su, M. Aslam, K.C. Wong, V.P. Dravid, Nano Lett. **4**, 383 (2004)
221. M. Salavati-Niasari, F. Davar, M. Mazaheri, M. Shaterian, J. Magn. Magn. Mater. **320**, 575 (2008)
222. J. Park, E. Kang, S.U. Son, H.M. Park, M.K. Lee, J. Kim, K.W. Kim, H.-J. Noh, J.-H. Park, C.J. Bae, J.-G. Park, T. Hyeon, Adv. Mater. **17**, 425 (2005)
223. P.R. Ponminiessary, A. Vasudevan, M. Sebastian, U.A. Chennampilly, K.M.Y. Karukapadath, J. Therm. Anal. Calor. **100**, 733 (2010)
224. K.S. Rejitha, S. Mathew, J. Therm. Anal. Calor. **93**, 213 (2008)
225. Y. Hanifehpour, B. Mirtamizdoust, A.R. Farzam, S.W. Joo, J. Inorg. Organomet. Polym. **22**, 957 (2012)
226. V.G. Pol, L.L. Daemen, S. Vogel, G. Chertkov, Ind. Eng. Chem. Res. **49**, 920 (2010)
227. P. Hermankova, M. Hermanek, R. Zboril, Eur. J. Inorg. Chem. **1110** (2010)
228. A. Shavel, B. Rodrigues-Gonzalez, J. Pacifico, M. Spasova, M. Farle, L.M. Liz-Marzan, Chem. Mater. **21**, 1326 (2009)
229. Y. He, Y. Sahoo, S. Wang, H. Luo, P.N. Prasad, M.T. Swihart, J. Nanoparticle Res. **8**, 335 (2006)
230. D. Czakis-Sulikowska, J. Radwańnska-Doczekalska, M. Markiewicz, M. Pietrzak, J. Therm. Anal. Calor. **93**, 789 (2008)
231. D. Kumar, I.P.S. Kapoor, G. Singh, N. Geol, U.P. Singh, J. Therm. Anal. Calor. **107**, 325 (2011)
232. N. Parveen, R. Nazir, M. Mazhar, J. Therm. Anal. Calor. **111**, 93 (2013)

233. R. Nazir, M. Mazhar, T. Wakeel, M.J. Akhtar, M. Siddique, M. Nadeem, N.A. Khan, M.R. Shah, J. Therm. Anal. Calor. **110**, 707 (2012)
234. A.D. Pomogailo, A.S. Burlov, N.D. Golubeva, L.A. Petrova, S.A. Mishenko, S.I. Pomogailo, G.I. Dzhardimalieva, A.D. Garnovskii, Neorg. Mater. **47**, 969 (2011)
235. T. Hyeon, Y. Chung, J. Park, S.S. Lee, Y.-W. Kim, B.H. Park, J. Phys. Chem. B **106**, 6831 (2002)
236. M.M. Amini, O. Sadeghi, S.W. Ng, J. Inorg. Organomet. Polym. **23**, 826 (2013)
237. X. Wang, X.Y. Chen, L.S. Gao, H.G. Zheng, Z. Zhang, Y.T. Qian, J. Phys. Chem. B **108**, 16401 (2004)
238. J.Z.L. Zhang, H.R. Geng, L.S. Zheng, B. Du, J. Alloys Compd. **392**, 317 (2005)
239. M. Wang, C.M. Liu, H.Z. Zhang, C.P. Chen, L. Guo, Appl. Phys. Lett. **85**, 2080 (2004)
240. A. Khansari, M. Salavati-Niasari, A.K. Babaheydari, J. Clust. Sci. **23**, 557 (2012)
241. F. Mohandes, F. Davar, M. Salavati-Niasari, J. Magn. Magn. Mater. **322**, 872 (2010)
242. A. Khansari, M. Enhessari, M. Salavati-Niasari, J. Clust. Sci. **24**, 289 (2013)
243. J.V.D.S. Araujo, R.V. Ferreira, M.I. Yoshida, V.M.D. Pasa, Solid State Sci. **11**, 1673 (2009)
244. A.V. Maciel, W.N. Mussel, V.M.D. Pasa, Mat. Sci. Appl. **1**, 279 (2010)
245. M. Salavati-Niasari, M. Dadkhah, M.R. Nourani, A.A. Fazl, J. Clust. Sci. **23**, 1011 (2012)
246. A.D. Pomogailo, Ros. Khim. Zh. (Mendeleev Chem. J.) **46**, 64 (2002)
247. A.D. Pomogailo, V.S. Savost'yanov, *Synthesis and Polymerization Metal-Containing Monomers* (CRC Press, Boca Raton, 1994)
248. Y.Y. Wang, Q. Shi, Q.Z. Shi, Y.C. Gao, X. Hou, Polyhedron **19**, 891 (2000)
249. Y.Y. Wang, Q. Shi, Q.Z. Shi, Acta Chim. Sinica **58**, 675 (2000)
250. G.I Dzhardimalieva, A.D Pomogailo, Usp. Khim. **77**, 270 (2008); Russ. Chem. Rev. **77**, 259 (2008)
251. A.D. Pomogailo, G.I. Dzhardimalieva, *Monomernye I Polimernye Karboksilaty Metallov* (Monomeric and Polymeric Metal Carboxylates) (Fizmatlit, Moscow, 2009)
252. V.A. Shershnev, G.I. Dzhardimalieva, D.P. Kiryukhin, V.A. Zhorin, A.D. Pomogailo, Izv. RAN, Ser. Khim. 1649 (2013)
253. J. Skupiska, H. Wilezura, H. Boniuk, J. Therm. Anal. **31**, 1017 (1986)
254. A.S. Rozenberg, G.I. Dzhardimalieva, A.D. Pomogailo, Polym. Adv. Technol. **9**, 527 (1998)
255. Z. Wojtczak, A. Gronowski, J. Therm. Anal. **36**, 2357 (1990)
256. P.A. Vasil'ev, A.L. Ivanov, A.N. Glebov, Zh. Obshch. Khim. **68**, 535 (1998)
257. J.M. Filipović, L. Katsicas, I.G. Popović, S.L. Velickovic, T.A. Djakov, D.M. Petrovic-Djakov, J. Therm. Anal. **49**, 335 (1997)
258. B.S. Randhawa, K.J. Sweety, M. Kaur, J.M. Greneche, J. Therm. Anal. Calor. **75**, 101 (2004)
259. I.C. McNeill, A. Alston, Angew. Makromol. Chem. **261/262**, 157 (1998)
260. M. Badea, R. Olar, D. Marinescu, E. Segal, A. Rotaru, J. Therm. Anal. Calor. **88**, 317 (2007)
261. M. Zulfiqar, R. Hussain, S. Zulfiqar, D. Mohammad, I.C. McNeill, Polymer Degrad. Stability **51**, 167 (1996)
262. M. Badea, R. Jlar, D. Marinescu, G. Vasil, J. Therm. Anal. Calor. **80**, 683 (2005)
263. E.I. Aleksandrova, G.I. Dzhardimalieva, A.S. Rozenberg, A.D. Pomogailo, Izv. Akad. Nauk. Ser. Khim. 303 (1993)
264. E.I. Aleksandrova, G.I. Dzhardimalieva, A.S. Rozenberg, A.D. Pomogailo, Izv. Akad. Nauk. Ser. Khim. 308 (1993)
265. A.S. Rozenberg, G.I. Dzhardimalieva, N.V. Chukanov, A.D. Pomogailo, Kolloid. Zh. **67**, 57 (2005)
266. A.S. Rozenberg, E.I. Aleksandrova, G.I. Dzhardimalieva, A.N. Titkov, A.D. Pomogailo, Izv. Akad. Nauk, Ser. Khim. 1743 (1993)
267. A.S. Rozenberg, E.I. Aleksandrova, G.I. Dzhardimalieva, N.V. Kir'yakov, P.E. Chizhov, V.I. Petinov, A.D. Pomogailo, Izv. Akad. Nauk, Ser. Khim. 885 (1995)
268. A.D. Pomogailo, A.S. Rozenberg, G.I. Dzhardimalieva, Metal-containing polymers as precursors for the production of ferromagnetic and superconductive materials, in *Metal-*

Containing Polymers Materials, ed. by C.U. Pittman, C.E. Carraher Jr., M. Zeldin, B. Culberston (Plenum Press, New York, 1996), p. 313
269. A.S. Rozenberg, E.I. Aleksandrova, I.P. Ivleva, G.I. Dzhardimalieva, A.V. Raevskii, O.I. Kolesova, I.E. Uflyand, A.D. Pomogailo, Izv. Akad. Nauk, Ser. Khim. 265 (1998)
270. A.T. Shuvaev, A.S. Rozenberg, G.I. Dzhardimalieva, N.P. Ivleva, V.G. Vlasenko, T.I. Nedoseikina, T.A. Lyubeznova, I.E. Uflyand, A.D. Pomogailo, Izv. Akad. Nauk, Ser. Khim. 1505 (1998)
271. G.I. Dzhardimalieva, *Doctoral Thesis in Chemical Sciences.* (Institute of Problems of Chemical Physics, Russian Academy of Sciences, Chernogolovka, Moscow Region, 2009)
272. M.J. Yang, Q.D. Ling, M. Hiller, X.Z. Fun, L.H. Wang, W.G. Zhang, J. Polym. Sci. Part A: Polym. Chem. **38**, 34055 (2000)
273. V.T. Panyushkin, A.A. Mastakov, N.N. Bukov, A.A. Nikolaenko, M.E. Sokolov, J. Struct. Chem. **45**, 167 (2004)
274. F. Ramos-Lara, A.C. Lira, M.O. Ramirez, M. Flores, R. Arroyo, V. Caldino, J. Phys. Condens. Mater. **18**, 7951 (2006)
275. D.V. Kolechko, F.A. Kolokolov, A.I. Oflidi, A.A. Pikula, V.T. Panyushkin, I.E Mikhailov, G.A. Dushenko, Dokl. Chem. Part 2 441, 374 (2011)
276. M. Badea, R. Olar, D. Marinesu, G. Vasile, J. Therm. Anal. Calor. **92**(1), 205 (2008)
277. H.L. Wu, W. Ying, L. Pen, Y.C. Gao, K.B. Yu, Synth. React. Inorg. Met.-Org. Nano-Met. Chem. **34**, 1019 (2005)
278. H.L. Wu, Y.C. Gao, K.B. Yu, Trans. Met. Chem. **29**, 175 (2004)
279. H.L. Wu, Y.C. Gao, J. Coord. Chem. **59**, 137 (2006)
280. N.P. Porollo, Z.G. Aliev, G.I. Dzhardimalieva, N.N. Ivleva, I.E. Uflyand, A.D. Pomogailo, N.S. Ovanesyan, Izv. AN, Ser. Khim. 375 (1997)
281. B.S. Randhawa, M. Kaur, J. Thermal. Anal. Calor. **89**, 251 (2007)
282. P.S. Bassi, B.S. Randhawa, C.M. Khajuria, S. Kaur, J. Thermal Anal. **32**, 569 (1987)
283. S.N. Chvalun, Priroda (7), 22 (2000)
284. V.A. Sergeev, A.Y. Olenin, U.F. Titova, A.S. Kogan, A.Y. Vasil'ev, Zh. Fiz. Khim. **66**, 1921 (1992)
285. M.T. Reetz, W. Helbig, J. Am. Chem. Soc. **116**, 7401 (1994)
286. G.N. Gerasimov, V.A. Sochilin, S.N. Chvalun, L.V. Volkova, I.Y. Kardash, Macromol. Chem. Phys. **197**, 1387 (1996)
287. H. Hopf, G.N. Gerasimov, S.N. Chvalun, V.I. Rozenberg, E.L. Popova, E.V. Nikolaeva, S.A. Zavjalov, L.I. Trakhtenberg, Adv. Mater. **3**, 197 (1997)
288. E.V. Nikolaeva, S.A. Ozerin, A.E. Grigoriev, E.I. Grigoriev, S.N. Chvalun, G.N. Gerasimov, L.I. Trakhtenberg, Mat. Sci. Eng. C-Bio **8–9**, 217 (1999)
289. P.V. Morozov, E.I. Grigor'ev, S.A. Zav'yalov, V.G. Klimenko, S.N. Chvalun, Polym. Sci. Ser. A **54**, 330 (2012)
290. I.A. Boginskaya, A.V. Gusev, K.A. Mailyan, S.N. Ozerin, A.V. Pebalk, I.A. Ryzhikov, M.V. Sedova, S.N. Chvalun, J. Comm. Tech. Electr. **56**, 66 (2011)
291. V.V. Zagorskii, A.E. Nasonova, M.A. Petrukhina, G.B. Sergeev, Vestn. Mosk. Univ., Ser. 2, Khim. **36**, 159 (1995)
292. V.A. Sergeev, L.I. Vdovina, Y.V. Smetannikov, A.Y. Vasil'kov, V.N. Gurishev, Vysokomol Soedin Ser. B **34**, 50 (1992)
293. V.V. Zagorskii, S.V. Ivashko, M.A. Petrukhina, G.B. Sergeev, Vestn. Mosk. Univ., Ser. 2, Khim. **39**, 276 (1998)
294. P. Harris, S. Tsang, Chem. Phys. Lett. **293**, 53 (1998)
295. P. Serp, V. Corrias, P. Kaick Appl, Catal. A **253**, 337 (2003)
296. E.A. Mel'gunova, Y.M. Balabina, A.N. Shmakov, M.S. Mel'gunov, Zh. Fiz. Khim. **77**, 510 (2003)
297. O.K. Alekseeva, A.A. Kotenko, M.M. Chelin, Membrany **36**, 3 (2007)

298. O.K. Alekseeva, A.A. Kotenko, E.V. Nefedova, Yu.S. Nechaev, M.M. Chelin, in *Nanochastitsy v Kondensirovannykh Sredakh* (Nanoparticles in Condensed Media) (Belarus State University, Minsk, 2008), p. 154
299. V.M. Novotortsev, V.V. Kozlov, Y.M. Korolev, G.P. Karpacheva, L.V. Kozhitov, Zh. Neorg. Khim. **53**, 1097 (2008)
300. H. Nabika, M. Mizuhata, A. Kajiinami, S. Deki, K. Akamatsu, J. Electroanal. Chem. **559**, 99 (2003)
301. G.P. Karpacheva, K.A. Bagdasarova, G.N. Bondarenko, L.H. Zemtsov, D.G. Muratova, N.S. Perov, Vysokomol. Soedin. **51**, 2037 (2009)
302. H. Yasuda, S. Miyanaga, A. Nakamura, H. Sakai, J. Inorg. Organometal. Polym. **1**, 135 (1991)
303. N.L. Pocard, D.C. Pocard, D.C. Alsmeyer, R.L. McCreery, T.X. Neenan, M.R. Callstrom, J. Am. Chem. Soc. **114**, 769 (1992)
304. A.A. Shutilov, *Candidate Thesis in Chemical Sciences.* (Institute of Catalysis, Siberian Branch of the Russian Academy of Sciences, Novosibirsk, 2008)
305. S.R. Liu, Carbon **43**, 1550 (2005)
306. S. Liu, X. Tang, Y. Mastai, I. Felner, A. Gedanken, J. Mater. Chem. **10**, 2502 (2000)
307. L. Xu, W. Zhang, Q. Yang, Y. Ding, W. Yu, Y. Qian, Carbon **43**, 1090 (2005)
308. E.P. Sajitha, V. Prasad, S.V. Subramanyan, S. Eto, K. Takai, T. Enoki, Carbon **42**, 2815 (2004)
309. L.J. Zhi, Y.S. Hu, H.B. El, X. Wang, I. Lieberwirth, U. Kolb, J. Maier, K. Mullen, Adv. Mater. **20**, 1727 (2008)
310. X. Yan, T. Xu, G. Chen, X. Wang, H. Liu, S. Yang, Appl. Phys. A **81**, 197 (2005)
311. R.A. Arents, Y.V. Maksimov, I.P. Suzdalev, Y.D. Amerik, Hyperfine Interact. **56**, 167 (1999)
312. Y.B. Amerik, Y.M. Korolev, V.N. Rogovoi, Neftekhimiya **36**, 304 (1996)
313. Y.M. Korolev, A.L. Bykova, Y.B. Amerik, Vysokomol. Soedin., Ser. B **39**, 1856 (1997)
314. S.V. Pol, V.G. Pol, A. Gedanken, J. Phys. Chem. C **111**, 16781 (2007)
315. A. Gedanken, E. Luvchik, Eur. J. Inorg. Chem. **2471** (2008)
316. S.V. Pol, V.G. Pol, A. Frydman, G. Churilov, A. Gedanken, J. Phys. Chem. B **109**, 9495 (2005)
317. V.G. Pol, S.V. Pol, A. Gedanken, Eur. J. Inorg. Chem. **709** (2009)
318. A.B. Shishmakov, Y.V. Mikushina, A.S. Seleznev, O.V. Koryakova, M.S. Valova, L.A. Petrov, Russ. J. Appl. Chem. **81**, 2180 (2008)
319. E. Papastergiades, S. Argyropoulos, N. Rigakis, N.E. Kiratzis, Ionics **15**, 545 (2009)
320. V.V. Barelko, A.D. Pomogailo, G.I. Dzhardimalieva, S.I. Evstratova, A.S. Rozenberg, I.E. Uflyand, Chaos **9**, 342 (1999)
321. A.S. Rozenberg, A.V. Raevskii, E.I. Aleksandrova, O.I. Kolesova, G.I. Dzhardimalieva, A.D. Pomogailo, Izv. Akad. Nauk, Ser. Khim. 862 (2001)
322. G.I. Dzhardimalieva, A.D. Pomogailo, V.A. Volpert, J. Inorg. Organometal. Polym. **12**, 1 (2002)
323. E. Sowka, M. Leonowicz, B. Andrzejewski, A.D. Pomogailo, G.I. Dzhardimalieva, J. Alloys Comp. **423**, 123 (2006)
324. A.D. Pomogailo, G.I. Dzhardimalieva, Vysokomol. Soedin. Ser. A **46**, 437 (2004)
325. J.A. Pojman, I.P. Nagy, C. Salter, J. Amer. Chem. Soc. **115**, 11044 (1993)
326. V.S. Savost'yanov, A.D. Pomogailo, B.S. Selenova, D.A. Kritskaya, A.N. Ponomarev, Izv. Akad. Nauk SSSR, Ser. Khim. 768 (1990)
327. R.B. Morgunov, A.I. Dmitriev, G.I. Dzhardimalieva, A.D. Pomogailo, A.S. Rozenberg, Y. Tanimoto, M. Leonowicz, E. Sowka, Fiz. Tv. Tela **49**, 1436 (2007)
328. E. Sowka, M. Leonowicz, J. Kazmierczak, A. Slawska-Waniewska, A.D. Pomogailo, G.I. Dzhardimalieva, Physica B **384**, 282 (2006)
329. G.I. Dzhardimalieva, A.D. Pomogailo, N.D. Golubeva, S.I. Pomogailo, G.F. Novikov, O.S. Roshchupkina, A.S. Rozenberg, M. Leonowicz, Kolloid. Zh. **73**, 457 (2011)

330. A.D. Pomogailo, G.I. Dzhardimalieva, A.S. Rozenberg, V.A. Shershnev, M. Leonowicz, Russ. Chem. Bull. **60**, 1476 (2011)
331. M. Izydorzak, A. Skumiel, M. Leonowicz, M. Kaczmarek-Klinowska, A.D. Pomogailo, G.I. Dzhardimalieva, Int. J. Thermophys. **33**, 627 (2012)
332. N.D. Golubeva, B.K. Dyusenalin, B.S. Selenova, S.I. Pomogailo, A.K. Zharmagambetova, G.I. Dzhardimalieva, A.D. Pomogailo, Kinetika I Kataliz **52**, 250 (2011)
333. A.D. Pomogailo, G.I. Dzhardimalieva. J. Catal., 2013, 12 pages, http://dx.doi.org/10.1155/2013/276210
334. A.G. Merzhanov, V.N. Sanin, V.G. Yukhvid, Dokl. Akad. Nauk **371**, 38 (2000)
335. V.V. Ivanov, E.V. Stegno, V.P. Mel'nikov, L.M. Pushchaeva, J. Polym. Sci., Ser. A **40**, 1017 (2002)
336. L.P. Kholpanov, S.E. Zakiev, A.D. Pomogailo, Dokl. Akad. Nauk **395**, 211 (2004)
337. O. Gazit, R. Khalfin, Y. Cohen, R. Tannenbaum, J. Phys. Chem. C **113**, 576 (2009)
338. S.H. Joo, C.S. Jhoi, I. Oh, K.J. Wak, Z. Liu, O. Terasaki, R. Ryon, Nature (London) **412**, 5835 (2001)
339. X. Fan, H. Fei, D.H. Demaree, D.P. Brennan, J.M. John, S.R. Oliver, Langmuir **25**, 5835 (2009)
340. T.L. Hsiung, H.P. Wang, H.P. Lin, J. Phys. Chem. Solids **69**, 393 (2008)
341. Z. Liu, X. Sun, S. Xu, J. Lian, Z. Xiu, Q. Li, D. Huo, J.-G. Li, J. Phys. Chem. C **112**, 2353 (2008)
342. Y. Han, S. Li, X. Wang, Cryst. Res. Technol. **44**, 336 (2009)
343. N.M. Emmanuel, A.L. Buchachenko, *Khimicheskaya Fizika Stareniya i Stabilizatsii Polimerov* (Chemical Physics of Ageing and Stabilization of Polymers) (Nauka, Moscow, 1982)
344. G.P. Gladyshev, O.A. Vasnetsova, N.I. Mashukov, Ros. Khim. Zh. **35**(5), 576 (1990)
345. G.-T. Cardenas, C.-C. Retamal, Termochim. Acta **176**, 233 (1991)
346. V. Djoković, J.M. Nedeljković, Macromol. Rapid Commun. **21**, 994 (2000)
347. J. Kuljanin, M. Marinović-Cincović, S. Zec, M.I. Comor, J.M. Nedeljkovic, J. Mater. Sci. Lett. **22**, 235 (2003)
348. M. Marinović-Cincović, Č.Z.V. Aponjić, V. Djoković, S.K. Milonjic, J.M. Nedeljkovic, Polym. Degrad. Stabil. **91**, 313 (2006)
349. L.V. Ruban, G.E. Zaikov, Usp. Khim. **63**, 373 (1994) [Russ. Chem. Rev. **63** 357 (1994)]
350. N.A. Khalturinskii, Al.Al. Berlin, T.V. Popova, Usp. Khim. **53**, 326 (1984) [Russ. Chem. Rev. **53**, 197 (1984)]
351. Al.Al. Berlin, Soros. Obrazov. Zh. (9), 57 (1996)
352. R.M. Aseeva, G.E.T. Zaikov, Makromol. Chem. Makromol. Symp. **74**, 335 (1993)
353. I.C. McNeil, R.C. McGuiness, Polym. Degrad. Stabil. **9**, 2 (1984)
354. N.A. Kopylova, Y.D. Semchikov, S.D. Zaitsev, Z. Prikl, Khim. **81**, 678 (2008)
355. R.S. Beer, C.A. Wikie, M.N.L. Mittleman, J. Appl. Polym. Sci. **46**, 1095 (1992)
356. K. Kondo, M. Matsumoto, K. Okamoto, J. Chem. Eng. Jpn. **32**, 217 (1999)
357. L.E. Manring, Macromolecules **24**, 3304 (1991)
358. J.B. Beck, S.J. Rowan, J. Am. Chem. Soc. **125**, 13922 (2003)
359. Q.G. Meng, P. Boutinaud, H.J. Zhang, R. Mahiou, J. Lumin. **124**, 15 (2007)
360. S. Selvakumar, R.S.M. Kumar, K. Rajarajan, J.A.A. Pragasam, S.A. Rajasekar, K. Thamizharasan, P. Sagayaraj, Cryst. Growth. Des. **6**, 2607 (2006)
361. M.E. Kosal, J.N. Chou, S.R. Wilson, K.S. Suslik, Nat. Mater. **1**, 118 (2002)
362. K.L. Mulfort, J.T. Hupp, J. Am. Chem. Soc. **129**, 9604 (2007)
363. T.K. Maji, R. Matsuda, S. Kitagawa, Nat. Mater. **6**, 142 (2007)
364. B. Xing, M.F. Choi, B. Xu, Chem. Eur. J **8**, 5028 (2002)
365. D. Wohrle, A.D. Pomogailo, *Metal Complexes and Metals in Macromolecules* (Wiley-VCH, Weinheim, 2003)
366. M.N. Patel, V.J. Patel, Indian J. Chem., A **28**, 428 (1989)
367. M.N. Patel, D.H. Sutaria, G.J. Patel, Synth. React. Inorg. Metal-Org. Chem. **24**, 401 (1994)

368. R. Coskun, M. Yigitoglu, M. Sacak, J. Appl. Polym. Sci. **75**, 766 (2000)
369. D. Bilba, L. Bejan, L. Tofan, Croatica Chemica Acta **71**, 155 (1998)
370. B.W. Zhang, K. Fischer, D. Bieniek, A. Kettrup, React. Polym. **24**, 49 (1994)
371. R.X. Liu, B.W. Zhang, H.X. Tang, J. Appl. Polym. Sci. **70**, 7 (1998)
372. N. Kabay, H. Egawa, Sep. Sci. Technol. **29**, 135 (1994)
373. R. Lei, X. Jie, X. Jun, Z. Ruijun, J. Appl. Polym. Sci. **53**, 325 (1994)
374. E.H. Rifi, M.J.F. Leroy, J.P. Brunette, C. Schloesserbecker, Solvent. Extr. Ion Exch. **12**, 1103 (1994)
375. H. Kubota, Y. Shigehisa, J. Appl. Polym. Sci. **56**, 147 (1995)
376. N. Pekel, N. Sahiner, O. Güven, J. Appl. Polym. Sci. **81**, 2324 (2001)
377. I.H. Park, J.M. Suh, Angew. Makromol. Chem. **239**, 121 (1996)
378. G. Moroi, D. Bilba, N. Bilba, C. Ciobanu, Polym. Degrad. Stabil. **91**, 535 (2006)
379. O.G. Marambio, G. Del, C. Pizzaro, M. Jeria-Orell, M. Huerta, C. Olea-Azar, W.D. Habiche, J. Polym. Sci.:Part A. Polym. Chem. **43**, 4933 (2005)
380. N.B. Shitova, P.G. Tsyrulnikov, D.A. Shlyapin, P.S. Barbashova, D.I. Kochubei, V.I. Zaikovskii, J. Struct. Chem. **50**, 268 (2009)
381. I.C. McNeill, J.J. Liggat, Polym. Degrad. Stabil. **29**(93), 370 (1990)
382. I.C. McNeill, J.J. Liggat, Polym. Degrad. Stabil. **37**, 25 (1992)
383. S.M. Humphrey, M.E. Grass, S.E. Habas, L. Niesz, G.A. Somorjai, T.D. Tilley, Nano Lett. **7**, 785 (2007)
384. Y. Zhang, M.E. Grass, S.E. Habas, F. Tao, T. Zhang, P. Yang, G.A. Somorjai, J. Phys. Chem. C **111**, 12243 (2007)
385. C. Yang, Q. Shao, J. He, B. Jiang, Langmuir **26**, 5179 (2010)
386. R. De Palma, G. Reekmans, C. Liu, R. Wirix-Speetjens, W. Laureyn, O. Nilsson, L. Lagae, Anal. Chem. **79**, 8669 (2007)
387. M. Kuhara, H. Takeyama, T. Tanaka, T. Matsunaga, Anal. Chem. **76**, 6207 (2004)
388. S.L. Lu, G.X. Cheng, H.C. Zhang, X.S. Pang, J. Appl. Polym. Sci. **99**, 3241 (2006)
389. H. Lei, W. Wang, L. Chen, X. Li, B. Yi, L. Deng, Enzyme Microb. Technol. **35**, 15 (2004)
390. H. Qian, Z. Lin, H. Xu, M. Chen, Biotechnol. Prog. **25**, 376 (2009)
391. S. Yang, K. Lien, K. Huang, H. Lei, G. Lee, Biosens. Bioelectron. **24**, 861 (2008)
392. A.A. Ostroushko, M.Y. Sennikov, Russ. J. Inorg. Chem. **50**, 933 (2005)
393. E.V. Schevchenko, D.V. Talapin, N.A. Kotov, S. O'Brien, C.B. Murray, Nature (London) **439**, 55 (2006)
394. V.G. Pol, S.V. Pol, J.M. Calderon-Moreno, A. Gedanken, J. Phys. Chem. C **113**, 10500 (2009)
395. C. Zheng, J. Zhang, G. Luo, J. Ye, M. Wu, J. Mat. Sci. **35**, 3425 (2000)
396. Y. Wang, G. Cao, Chem. Mater. **18**, 2787 (2006)
397. C. Diaz, M.L. Valenzuela, Macromolecules **39**, 103 (2006)
398. C.X. Li, Z.Y. Zhong, W.K. Leong, Langmuir **24**, 10427 (2008)
399. C. Li, W.K. Long, Langmuir **24**, 12040 (2008)
400. P. Hooker, D.J. Tan, K.J. Clabunde, S. Suib, Chem. Mater. **3**, 947 (1991)
401. J. Cheon, L.H. Dubois, G.S. Girolami, J. Am, Chem. Soc. **119**, 6814 (1997)
402. A.V. Grafov, E.A. Mazurenko, O.V. Mel'nik, V.Y. Kofman, Ukr. Khim. Zh. **59**, 1235 (1993)
403. A.D. Pomogailo, in *Macromolecules. Design and Applications*, ed. by A.S. Abd-El-Aziz, C.E. Carraher, C.U. Pittman, M. Zeldin (Springer, New York, 2008), p. 241
404. C. Klinke, K. Kern, Nanotechnology **18**, 215601 (2007)
405. E.P. Simonenko, V.G. Sevast'yanov, N.A. Ignatov, Yu.S. Ezhov, N.T. Kuznetsov, in *Tezisy Dokladov 2-i Vserossiiskoi Konferentsii po Nanomaterialam 'NANO 2007'* (Abstracts of Reports of the Second All-Russian Conferences on Nanomaterials 'NANO 2007', Novosibirsk, 2007) p. 79
406. G.A. Domrachev, W.E. Douglas, B. Henner, L.G. Klapshina, V.V. Semenov, A.A. Sorokin, Polym. Adv. Technol. **10**, 215 (1999)
407. G.T. Viola, M. Bortolotti, A. Zazzetta, J. Polym. Sci.:Part A. Polym.Chem. **34**, 13 (1996)

408. S.D. Bunge, T.J. Boyle. US 20056929675 (2005)
409. T.M. Keller, J. Perrin, S.B. Qadri, US 20056884861 (2005)
410. J. Nishijo, C. Okabe, O. Oishi, N. Nishi, Carbon **44**, 2943 (2006)
411. A.D. Pomogailo, A.S. Rozenberg, I.E. Uflyand, *Nanochastitsy Metallov v Polimerakh* (Nanoparticles of Metals in Polymers) (Khimiya, Moscow, 2000)
412. V. Biju, T. Itoh, A. Sujith, M. Ishikawa, Anal. Bioanal. Chem. **391**, 2469 (2008)
413. L. Mirenghi, F. Antolini, L. Tapfer, Surf. Interface Anal. **38**, 162 (2006)
414. F. Antolini, T. Di Luccio, A.M. Laera, L. Mirenghi, E. Piscopiello, M. Re, Phys. Stat. Sol. **244**, 2768 (2007)
415. K. Sung, S.H. Lee, T.M. Chung, C.G. Kim, J. Nanosci. Nanotechnol. **8**, 4873 (2008)
416. Y.B. Chen, L. Chen, L.M. Wu, Inorg. Chem. **44**, 9817 (2005)
417. J. Chen, L.M. Wu, L. Chen, Inorg. Chem. **46**, 586 (2007)
418. Y. Wang, J. Chen, L. Chen, Y.B. Chen, L.M. Wu, Cryst. Growth Des. **10**, 1578 (2010)
419. M. Yamamoto, M. Nakamoto, J. Mater. Chem. **13**, 206 (2003); N. Sandhyarani, T. Predeep, J. Mater. Chem. **11**, 1294 (2001)
420. Y.B. Chen, L. Chen, L.M. Wu, Chem. Eur. J. **14**, 11069 (2008)
421. Z. Yang, A.B. Smetana, C.M. Sorensen, K.J. Klabunde, Inorg. Chem. **46**, 2427 (2007)
422. T. Bhuvana, G.U. Kulkarni, Small **4**, 670 (2008)
423. G. Carotenuto, L. Nicolais, P. Perlo Polymer Eng. Sci. 1016 (2006)
424. M. Nakamoto, Y. Kashiwagi, M. Yamamoto, Inorg. Chim. Acta **358**, 4229 (2005)
425. G. Carotenuto, B. Martorana, P.B. Perlo, L. Nicolais, J. Mater. Chem. **13**, 2927 (2003)
426. A.S. Susha, M. Ringler, A. Ohlinger, M. Paderi, N. LiPira, G. Carotenuto, A.L. Rogach, J. Feldman Chem. Mater. **20**, 6169 (2008)
427. M. Hisamoto, R.C. Nelson, M.-Y. Lee, J. Eckert, S.L. Scott, J. Phys. Chem. C **113**, 8794 (2009)
428. X. Lu, H.-Y. Tuan, B.A. Korgel, Y. Xia, Chem. Eur. J. **14**, 1584 (2006)
429. E. Holder, N. Tessler, A.L. Rogach, J. Mater. Chem. **18**, 1064 (2008)
430. N. Bao, L. Shen, T. Takata, K. Domen, A. Gupta, K. Yanagisawa, C.A. Grimes, J. Phys. Chem. C **111**, 17527 (2007)
431. F. Capezzuto, G. Carotenuto, F. Antolini, E. Burresi, M. Palomba, P. Perlo Exp, Polym. Lett. **3**, 219 (2009)
432. A. Petrella, M. Tamborra, M.L. Curri, M. Striccolli, P.D. Cozzoli, A. Adostano, J. Phys. Chem. B **109**, 1554 (2005)
433. G. Carotenuto, G. Pepe, D. Davino, B. Martorana, P. Perlo, D. Acierno, L. Nicolais, Microw. Opt. Technol. Lett. **48**, 2505 (2006)
434. Y.K. Jung, J.I. Kim, J.K. Lee, J. Am. Chem. Soc. **132**, 178 (2010)
435. T.X. Wang, H. Xiao, Y.C. Zhang, Mater. Lett. **62**, 3736 (2008)
436. D. Pullini, G. Carotenuto, M. Palomba, A. Mosca, A. Horsewell, L. Nicolais, J. Mater. Sci. **46**, 7905 (2011)
437. S. Chio, E.G. Kim, T. Hyeon, J. Am. Chem. Soc. **128**, 2520 (2006)
438. W.S. Seo, J.H. Lee, X. Sun, Y. Suzuki, D. Mann, Z. Liu, M. Terashima, P. Yang, M.V. Mcconnell, D.G. Nishimura, H. Dai, Nat. Mater. **5**, 971 (2006)
439. H. Bonnemann, R.A. Brand, W. Brijoux, H.-W. Hofstadt, M. Frerichs, V. Kempter, W. Maus-Friedrichs, N. Matoussevitch, K.S. Nagabhushana, F. Voigts, V. Caps, Appl. Organomet. Chem. **19**, 790 (2005)
440. S. Behrens, H. Bonnemann, N. Matoussevitch, A. Gorschinski, E. Dinjus, W. Habicht, J. Bolle, S. Zinoveva, N. Palina, J. Hormes, H. Modrow, S. Bahr, V. Kempter, J. Phys. Condens. Mater. **18**, 2543 (2006)
441. B.Y. Kim, I.B. Shim, Z.O. Araci, S.S. Saavedra, O.L.A. Monti, N.R. Armstrong, R. Sahoo, D.N. Srivastava, J. Pyun, J. Am. Chem. Soc. **132**, 3234 (2010)
442. Y.A. Barnakov, B.L. Scott, V. Golub, L. Kelly, V. Reddy, K.L. Stokes, J. Phys. Chem. Solids **65**, 1005 (2004)
443. F.-H. Lin, W. Chen, Y.-H. Liao, R. Doong, Y. Dong, Nano Res. **4**, 1223 (2011)

444. H. Yu, M. Chen, P.M. Rice, S.X. Wang, R.L. White, S.H. Sun, Nano Lett. **5**, 379 (2005)
445. S.H. Choi, B.H. Na, Y.I. Park, K. An, S.G. Kwon, Y. Jang, M. Park, J. Moon, J.S. Son, I.C. Song, W.K. Moon, T. Hyeon, J. Am. Chem. Soc. **130**, 15573 (2008)
446. S. Peng, C.H. Lei, Y. Ren, R.E. Cook, Y.G. Sun, Angew. Chem. Int. Ed. **50**, 3158 (2011)
447. H.W. Gu, Z.M. Yang, J.H. Gao, C.K. Chang, B. Xu, J. Am. Chem. Soc. **127**, 34 (2005)
448. Y. Jang, J. Chung, S. Kim, S.W. Jun, B.H. Kim, D.W. Lee, B.M. Kim, T. Hyeon, Phys. Chem. Chem. Phys. **13**, 2512 (2011)
449. C. Wang, C.J. Xu, H. Zeng, S.H. Sun, Adv. Mater. **21**, 3045 (2009)
450. J. Jiang, H.W. Gu, H.L. Shao, E. Devlin, G.C. Papaefthymiou, J.Y. Ying, Adv. Mater. **20**, 4403 (2008)
451. V. Mazumder, S.H. Sun, J. Am. Chem. Soc. **131**, 4588 (2009)
452. B.G. Ershova, A.V. Anan'eva, E.V. Abkhalimova, D.I. Kochubei, V.V. Kriventsov, L.M. Plyasova, I.Y. Molina, N.Y. Kozitsyna, S.E. Nefedov, M.N. Vargaftik, I.I. Moiseev, Nanotechnol. Russ **6**, 323 (2011)
453. N.S. Akhmadullina, N.V. Cherkashina, N.Y. Kozitsyna, I.P. Stolarov, E.V. Perova, A.E. Gekhman, S.E. Nefedov, M.N. Vargaftik, I.I. Moiseev, Inorg. Chim. Acta **362**, 1943 (2009)
454. T. Nedoseykina, P. Plyusnin, Y. Shubin, S. Korenev, J. Therm. Anal. Calor. **102**, 703 (2010)
455. A.V. Zadesenets, A.B. Venediktov, Y.V. Shubin, Russ. J. Inorg. Chem. **52**, 500 (2007)
456. L. Jiang, J. Kim, J. Appl. Polym. Sci. **101**, 186 (2006)
457. Q. Sun, K. Xu, H. Peng, R. Zheng, M. Haussler, B.Z. Tang, Macromolecules **36**, 2309 (2003)
458. G.Y. Yurkov, S.P. Gubin, E.A. Ovchenkov, in *Magnetic Nanoparticles*, ed. by S.P. Gubin (Wiley-VCH, Weinheim, 2009), p. 87
459. P.Y. Keng, I. Shim, B.D. Korth, J.F. Douglas, J. Pyun, ACS Nano **1**, 279 (2007)
460. G.Y. Yurkov, A.S. Fionov, Y.A. Koksharov, V.V. Kolesov, S.P. Gubin, Inorg. Mater. **43**, 834 (2007)
461. M.A. Zaporozhets, D.A. Baranov, N.A. Kataeva, I.I. Khodos, V.I. Nikolaichik, A.S. Avilov, S.P. Gubin, Russ. J. Inorg. Chem. **54**, 517 (2009)
462. M. Haussler, R. Zheng, J.W.Y. Lam, H. Tong, H. Dong, B.Z. Tang, J. Phys. Chem. B **108**, 10645 (2004)
463. K. Liu, B. Clendenning, L. Friebe, W.Y. Chan, X. Zhu, M.R. Freemann, G.C. Yang, C.M. Yip, D. Grozea, Z.-H. Lu, I. Manners, Chem. Mater. **218**, 2591 (2006)
464. G.T. Fried, G.S. Markovich, Adv. Mater. **13**, 1158 (2001)
465. R. Vijayakumar, Y. Koltypin, I. Felner, A. Gedanken, Mater. Sci. Eng. A **286**, 101 (2000)
466. N. Bao, L. Shen, W. An, P. Padhan, C.H. Turner, A. Gupta, Chem. Mater. **21**, 3458 (2009)
467. R.W. Schwartz, Chem. Mater. **9**, 2325 (1997)
468. E. Erdem, R. Bottcher, H.-C. Semmelhack, H.-J. Glasel, E. Hartmann, D. Hirsch, J. Hormes, H. Rumpf, J. Mater. Sci. **34**, 2319 (1999)
469. H. Rumpf, J. Hormes, H. Modrow, H.-J. Glasel, E. Hartmann, E. Erdem, R. Bottcher, J. Phys, Chem. B **105**, 3415 (2001)
470. H.-J. Glasel, E. Hartmann, D. Hirsch, J. Mater. Sci. **38**, 3211 (2003)
471. B.S. Randhawa, M. Kaur, Hyperfine Interact. **188**, 1329 (2008)
472. L. von Lampe, D. Schultze, F. Zygalsky, M.S. Sylverstein, Polym. Degrad. Stab. **81**, 57 (2003)
473. L. von Lampe, A. Schmalstieg, S. Götze, J.P. Muller, F. Zygalsky, H.J. Lorkowski, M. Matalla, J. Mater. Sci. Lett. **16**, 16 (1997)
474. L. von Lampe, A. Bruckner, S. Götze, Die Angew. Makromol. Chem. **251**, 157 (1997)
475. L. von Lampe, D. Schultze, F. Zygalsky, Polym. Degrad. Stabil. **73**, 87 (2001)
476. M.S. Silverstein, Y. Najary, G.S. Grader, G.E. Shter, J. Polym. Sci. Part B: Polym. Phys. **42**, 1023 (2001)
477. M.S. Silverstein, Y. Najary, Y. Lumelsky, L. von Lampe, G.S. Grader, G.E. Shter, Polymer **45**, 937 (2004)
478. S. Maeda, Y. Tsurusaki, Y. Tachiyama, K. Naka, A. Ohki, T. Ohgushi, T. Takeshita, J. Polym. Sci. Part A: Polym. Chem. **32**, 1729 (1994)

479. J.C.W. Chien, B.M. Gong, X. Mu, Y.S. Yang, J. Polym, Sci.: Part A: Polym. Chem. **28**, 199 (1990)
480. K. Naka, K. Hagihara, Y. Tanaka, Y. Tachiyama, A. Ohki, S. Maeda, J. Mater. Sci. **31**, 6389 (1996)
481. K. Naka, Y. Tachiyama, A. Ohki, S. Maeda, J. Polym. Sci.: Part A: Polym. Chem. **34**, 1003 (1996)
482. S. Dubinsky, G.S. Grader, G.E. Shter, M.S. Silverstein, Polym. Degrad. Stabil. **86**, 171 (2004)
483. S. Dubinsky, Y. Lumelsky, G.S. Grader, G.E. Shter, M.S. Silverstein, J. Polym. Sci.: Part B: Polym. Phys. **43**, 1168 (2005)
484. A.D. Pomogailo, V.S. Savost'yanov, G.I. Dzhardimalieva, A.V. Dubovitskii, A.N. Ponomarev, Izv. Akad. Nauk, Ser. Khim. 1096 (1995)
485. Y. Lumelsky, M.S. Silverstein, J. Mater. Sci. **41**, 8202 (2006)
486. I. von Lampe, D. Schultze, F. Zygalsky, M.S. Silverstein, Polym. Degrad. Stabil. **81**, 57 (2003)
487. I. von Lampe, I.S. Götze, F. Zygalsky, J. Low Temp. Phys. **105**, 1289 (1996)
488. E.H. Tadd, J. Bradley, R. Tannenbaum, Langmuir **18**, 2378 (2002)
489. S. King, K. Hyunh, R. Tannenbaum, J. Phys. Chem. B **107**, 12097 (2003)
490. R. Tannenbaum, M. Zubris, E.P. Goldberg, S. Reich, N. Dan, Macromolecules **38**, 4254 (2005)
491. R.R. Netz, D. Andelman, Phys. Rep. **380**, 1 (2003); N. Dan, Langmuir **16**, 4045 (2000)
492. E. Tadd, A. Zeno, M. Zubris, N. Dan, R. Tannenbaum, Macromolecules **36**, 6497 (2003)
493. E.G. Nebukina, A.A. Arshakuni, S.P. Gubin, Russ. J. Inorg. Chem. **54**, 1685 (2009)
494. S.P. Gubin, G.Y. Yurkov, I.D. Kosobudsky, Int. J. Mater. Product Technol. **23**, 2 (2005)
495. H.T. Yang, C.M. Shen, Y.K. Su, T.Z. Yang, H.J. Gao, Y.G. Wang, Appl. Phys. Lett. **82**, 4729 (2003)
496. T. Hyeon, Chem. Commun. **927** (2003)
497. Y.A. Koshkarov, G.Y. Yurkov, D.A. Baranov, A.P. Malakho, S.N. Polyakov, S.P. Gubin, Fiz. Tv. Tela **48**, 885 (2006)
498. M.V. Radchenko, G.V. Lashkarev, V.I. Sichkovskyi, A.A. Arshakuni, S.P. Gubin, V.O. Yuhymchuk, V. Domukhovski, T. Story, Y.P. Piryatinskii, G.Y. Yurkov, Inorg. Mater. **45**, 468 (2009)
499. D.L. Huber, Small **1**, 482 (2005)
500. A.D. Pomogailo, A.S. Rozenberg, G.I. Dzhardimalieva, A.M. Bochkin, S.I. Pomogailo, N.D. Golubeva, V.M. Grishchenko, Neorg. Mater. **42**, 164 (2006)
501. A.D. Pomogailo, in *Macromolecules. Design and Applications*, ed. by A.S. Abd-El-Aziz, C.E. Carraher, C.U. Pittman, M. Zeldin (Springer, New York, 2008), p. 241
502. A.D. Pomogailo, G.I. Dzhardimalieva, A.S. Rozenberg, V.N. Kestelman, J. Thermoplastic Comp. Mater. **20**, 151 (2007)
503. K. Hanko, G. Vass, I. Szepes, J. Organomet. Chem. **492**, 235 (1995)
504. V.A. Umilin, V.K. Vanchagova, in *Primenenie Metalloorganicheskikh Soedinenii dlya Polucheniya Neorganicheskikh Pokrytii i Materialov* (Applications of Organometallic Compounds for Production of Inorganic Coatings and Materials), ed. by G.A Razuvaev. (Nauka, Moscow, 1986), p. 58
505. Yu.D. Tretyakov, Usp. Khim. **72**, 731 (2003) [Russ. Chem. Rev. **72**, 651 (2003)]
506. A.S. Rozenberg, A.A. Rozenberg, A.V. Lankin, G.I. Dzhardimalieva, A.D. Pomogailo, Dokl. Akad. Nauk **393**, 361 (2003)
507. N.D. Zhuravlev, V.I. Raldugin, A.P. Tikhonov, Kolloid. Zh. **61**, 322 (1999)
508. A.F. Shestakov, V.N. Solov'ev, V.V. Zagorskii, G.B. Sergeev, Zh. Fiz. Khim. **68**, 155 (1994)
509. A.S. Rozenberg, A.A. Rozenberg, G.I. Dzhardimalieva, A.D. Pomogailo, Kolloid. Zh. **67**, 70 (2005)
510. J.-M. Lehn, *Supramolecular Chemistry Concepts and Perspectives* (VCH, Weinheim, 1995)
511. P.M. Zorkii, I.E. Lubnina, Vestn. Mosk. Univ., Ser. 2, Khim. **40**, 300 (1999)

Chapter 7
Bionanocomposites Assembled by “From Bottom to Top” Method

Particles, which take part in different biological processes, have a special place in nanometer composites. Rich and variable biochemistry of proteins displays key importance of a nanometer phenomenon for a working mechanism of living organisms [1, 2], and this provides a wide range of possibilities for development of new types of hybrid nanomaterials with constructible structures, functions and shapes [3–5].

Integration of nanoparticles and biomolecules, each having unique properties, on the one hand, and being in the same nanometer scale (enzymes, antigens, and antibodies of typical sizes 2–20 nm like nanoparticles), on the other hand, as of two structurally compatible classes of materials, gives resultant new hybrid nanomaterials with synergetic properties and functions (Fig. 7.1).

Interaction of nanoparticles with biopolymers (proteins, nucleic acids, polysaccharides) plays very important role in enzyme catalysis, biosorption, biohydrometallurgy, geobiotechnology, etc.). There is a great interest to nanomaterials, which can be used in biomedical and pharmaceutical applications due to their biocompatibility and biodegradation. At the same time bionanocomposites should meet some demands to plasticity of a matrix, they should have improved barrier, antimicrobial properties, ability to controlled release of bioactive substances such as antimicrobial agents, antioxidants, drugs, calcium compounds in biologically available form and their mixtures, etc. Biopolymers, such as natural and synthetic proteins, obtained by chemical methods or by genetic modification of microorganisms or plants, nucleic acids (including synthetic ones), biodegraded complex polyethers, such as polylactic acid, and its derivatives, oligo-hydroxyalkanoate, most often, polyhydroxybutyrate, their copolymers, biomedical materials, such as hydroxyapatites, are used. There is an interesting group of materials for this purpose including synthetic and natural (vegetable and animal) polysaccharides, such as cellulose and its derivatives, alginates, dextranes, acacia gum, chitosan, and any of its natural and synthetic derivatives, especially chitosan acetate, and proteins obtained from raw animal materials, as well as proteins from maize (especially zein) or soya, gluten derivatives, gelatin, casein, etc. Relatively widely used in

A.D. Pomogailo, G.I. Dzhardimalieva, *Nanostructured Materials Preparation via Condensation Ways*, DOI 10.1007/978-90-481-2567-8_7

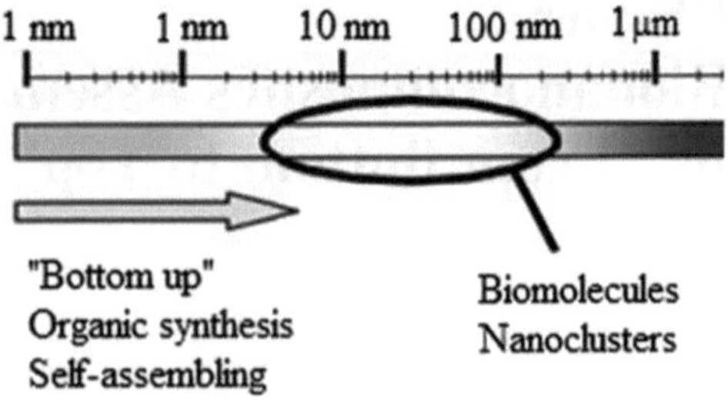

Fig. 7.1 Integration nanometer scale of nanoparticles, biomolecules and bioobjects

biomedical applications are intercalated bionanocomposites based on organically modified materials with layered structures (layered phyllosilicates, montmorillonite, etc.), including their usage for release of active ingredients, such as volatile odoriferous substances and their components (for example, lynalol, polar ether oil with anti-microbial properties).

Biocomposites also include products of concentration and biomineralization with participation of natural, also inorganic, polymers. Perfection of bioprocesses, principles of their realization and self-regulation of biosystems are still not only stunning, but permanently impel researchers to their modeling under laboratory conditions, to development of bio-imitating concepts, to designing of artificial analogues. Following, perception and copying methods of Nature, a desire to unravel its mystery bring to a new research direction, biomimetics, whose purpose is transportation of its laws into inanimate nature.

Almost all approaches and problems of nanomaterials science considered in this book more or less concern development of biocomposites. Their applications are especially important in medicine: it is a basis for progress in diagnostics and therapy on cellular and genetic levels. While modifying surfaces of drug carriers with biocompatible polymers, it is necessary to optimize functions of polymeric component, which can play a role of a binding material for therapeutic or diagnostic drugs, to provide some characteristics of drugs (solubility, bio-availability, prolongation of their action due to slow desorption of drugs from polymer matrix, their shelf life, etc.), etc.

At last, we shall note that quite many biological materials are known, such as tissues [6], fungi [7], bacteria [8], viruses [9–11], and biomolecules [12, 13], which by their spatial configuration are very close to organic-inorganic nanostructures obtained by functioning of organic fibers with nanoparticles [14].

First of all, we shall consider variants, when there is a bio-object in a system, which can initiate reduction of metal ions and thus form nanoparticles to produce nanobiocomposites.

7.1 Bioreducing Agents in the Synthesis of Nanocomposites

During the recent years biosynthesis methods have become widely used, because they are relatively simple and efficient ways of synthesis of nanoparticles in reduction processes as compared to physical or chemical methods. Fungous or mold microorganisms [15–18], bacteria [19], plant extracts [20, 21] are used as biological objects. In the base of many applications of microorganisms including bioleaching, bioremediation, bacterial corrosion, in particular, in biosynthesis of nanoparticles, there is their ability under extreme external conditions to induce specific protective mechanism of stress suppression, for example, toxicity of metal ions or metals due to a change in redox-state of ions or intercellular deposition of a metal [22].

Among the abovementioned approaches there is biochemical synthesis based on usage of biologically active components, which is in the group of the most efficient and environmentally appropriate methods.

7.1.1 Vegetable Biomasses and Extracts as Reduction Agents of Metal Ions

Production of nanoparticles with the use of various plants, plant-mediated synthesis, is a widely used technique of synthesis of nanobiocomposites. There are especially numerous examples of this type of composites based on gold and silver (Table 7.1) [23], extracellularly produced in leaf extracts of *Cinnamomum camphora* [24],

Table 7.1 Plants used in biogenic synthesis of silver and gold nanoparticles [23 and references there]

Plants	Silver and (or) gold nanoparticles	Shape (morphology)	Particle size
Medicago sativa	Au	fcc twinned, crystal and icosahedral	4–10 nm
Medicago sativa	Au	fcc tetrahedral, hexagonal platelets, icosahedral multiple twinned, decahedral multiple twined and irregular shaped.	15–200 nm
Chilopsis linearis	Au	–	1.1 nm
Pelargonium graveolens	Au	Spherical, rods, flat, sheets and triangular	21–70 nm
Cymbopogon flexuosus	Au	Triangular, hexagonal	–
Cymbopogon flexuosus	Au	Triangular	–
Avena sativa	Au	fcc tetrahedral, decahedral, hexagonal, icosahedralmultiple twinned, irregular shaped, rod shaped	–
Cicer arietinum	Au	Triangular	–
Tamarindus indica	Au	Triangular	–

(continued)

Table 7.1 (continued)

Plants	Silver and (or) gold nanoparticles	Shape (morphology)	Particle size
Triticum aestivum	Au	fcc tetrahedral, hexagonal platelets, irregular shaped, rod shaped, decahedral multiple twined, icosahedralmultiple twined	10–30 nm
Sesbania	Au	Spherical	6–20 nm
Medicago sativa	Ag	Spherical	2–20 nm
Quercetin	Ag	–	Radius 1–1.5 mm
Tetrapanax	Ag	–	< 100 nm
Capsicum annuum	Ag	–	–
Pelargonium graveolens	Ag	–	–
Aloe vera	Au, Ag	Triangular, spherical	–
Brassica juncea	Au, Ag, Cu	–	–
Emblica officinalis	Au, Ag	–	10–20 nm, 15–25 nm
Azadirachta indica	Au, Ag and Ag core –Au shell	Polydisperse, flat, plate-like, spherical, peculiar core–shell structure	5–35 nm, 50–100 nm
Cinnamommum camphora	Au, Ag	Triangular, spherical	55–80 nm

gooseberry *Emblica officinalis* [25], *Aloe vera* [21], poon *Dalbergia sissoo* [26], geraniume *Pelargonium graveolens* [27], black tea [28], etc. Biosynthesis can be conducted directly in biomass, i.e. intracellularly. This way was used to produce gold nanoparticles in alfalfa *Medicago sativa* [20] biomass, desert willow *Chilopsis linearis* [29], *Sesbania* grains [30], etc. Moreover, formation of nanoparticles can proceed in vivo. Alfalfa roots can absorb Ag(0) from agar medium and transport them into saplings, where nucleation and growth take place [31]. Similarly, the desert willow *(Chilopsis linearis)* absorbs Au (160 mg in Au/L in agar) and synthesizes gold nanoparticles in roots, stems and leafs with average size 8, 35, and 18 Å, respectively [29]. Probably, following the same general scheme: reduction and absorption of atomic Au(0) by roots of plants, transportation to some parts of a plant, growth and coalescence of nanoparticles, *Brassica juncea* biomass contained nanoparticles of Au, Ag, and Cu alloys sized from 5 to 50 nm after 14 days of growth in soil enriched with gold chloride, silver nitrate, and copper chloride [30].

It is supposed that [26], combined components of biological extracts show synergetic reduction effect during formation of nanoparticles from metal ions. Proteins, polyphenol, carbohydrates can be involved in reduction processes in these systems.

Chemical structures of the most important representatives of these substances extracted from the poon leaves *Dalbergia sissoo* are shown in the Scheme 7.1.

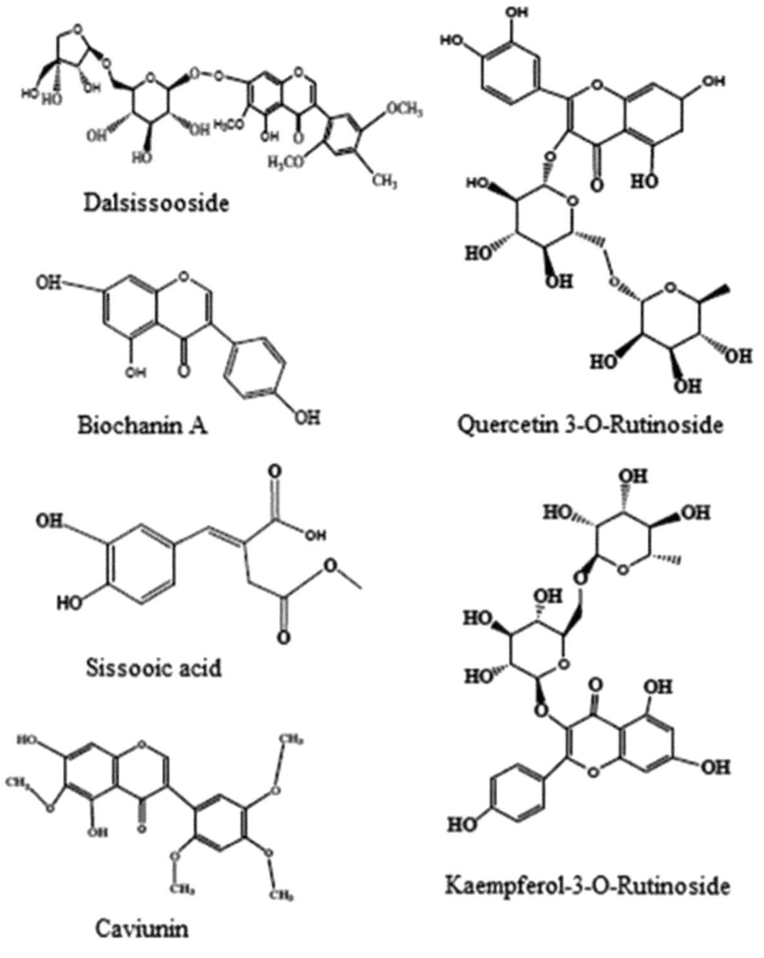

Scheme 7.1 Chemical structures of important constituents in green leaves of *Dalbergia sissoo*

It has been shown that naphthoquinones [32] and anthraquinones [33] from *F. oxysporum* can be efficient carriers of electrons during reduction of metals. Quercetin (3,5,7,3′,4′-pentahydroxyflavone) (see Scheme 7.1) is one of the components of the plant extracts which was obtained in pure state and for which reducing properties with respect to $AgNO_3$ or $Cu(NO_3)_2$ in AOT-n-alkane system were confirmed experimentally [34, 35]. It is assumed that citronellol and geraniol molecules, which are the basic components of terpenoids contained in high concentrations in the extract of *Pelargonium graveolens* [27, 36] leaves, are simultaneously stabilizers and reducing agents. Chemosorption of proteins from

P. Graveolens on the surface of Ag nanoparticles is confirmed by widening of peaks of amide line I (1,640 cm^{-1}) and of ether group C=O (1,748 cm^{-1}) in FTIR spectra of synthesized biocomposite [36]. Usually, phyto proteins similarly to many synthetic polymers (see the Sect. 2.2.4) show dual function in formation of nanobiocomposites. On the one hand, they have bioreducing ability due to amino acid residuals, such as L-tryptophan, L-tyrosine, L-arginine, L-lysine, L-asparagine acid, etc, whose reducing properties, in particular, are demonstrated with respect to gold ions [37, 38]. On the other hand, presence of various functional groups in content of proteins (hydroxyl group, carboxyl group, etc.) advances their chemosorption on the surface of nanoparticles, an important role in this case has molecular mass of a macromolecule. For example, highly molecular fraction ($M > 5$ kDa) of water soluble proteins of soy reduces $NaAuCl_4$ in water medium and stabilizes the formed nanoparticles, while low molecular proteins with molecular mass <5 kDa, though initiate formation of nanoparticles, are unable to prevent their aggregation [39], in this aspect, their behavior is similar to that of synthetic polymers, considered above.

Another group of plant components, biologically active in synthesis of nanoparticles are polyphenols (catechin gallate, epicatechin gallate, epi-gallocatechin, gallocatechin gallate, etc.). For example, water extracts of leaves and seeds of *Syzygium cumini* contain 21 and 36 mg g^{-1} of polyphenols, respectively. It is shown that sizes of formed nanoparticles depend on concentration of these components [40]. And an increase in concentration of phenol compounds in the sorghum *Sorghum spp* extract (2,010, 2,375 and 2,520 mg/L GAE[1] at the temperatures of extraction 25, 50, and 80 °C) caused respective increase in intensity of the absorption band of the obtained Ag particles at 390 nm, which directly confirmed that particularly phenol compounds were the main reducing agents in this system [41]. According to the data from Table 7.1, the produced nanoparticles are characterized by great variety of shapes. The most widely appeared is triangular morphology, including truncated (prismatic) anisotropic structures. These nanotriangles can serve building blocks for electrically conducting thin films [42]. Bioreduction of $HAuCl_4$ by extract of tamarind leaves brings to formation of highly anisotropic flat triangular structures, which are interesting for optoelectronics, photonics, and sensor devices [43]. It should be noted that precision control over the shape of formed nanoparticles and their size under conditions of biogenic synthesis is often difficult, because the mechanisms of reduction and absorption of protective agents are not quite clear. However, like in chemical reduction reactions, morphology and size of formed nanoparticles can be varied efficiently by a change in composition of biocomponents. For example, as amount of lemongrass *Cymbopogon flexuosus* extract added to $HAuCl_4$ solution increased, average size of triangular and hexagonal particles decreased, while a fraction of spherical particles with respect to them increased [44]. Average size of gold nanoparticles

[1] GAE – equivalent of gallic acid used as a standard phenol compound in spectrophotometric analysis of total concentration of phenols.

synthesized in presence of native biomass of hop was 17.3 Å, and when etherified bioreagent was used, 9.2 Å particles were produced, while hydrolyzed biomass produced nanoparticles of ~25 Å [45]. For silver nanoparticles made by treatment of silver ions by capsicum extract *Capsicum annuum* L, nanoparticle size was a time function of the reaction [46]. During 5 h spherical particles with diameter 10 ± 2 nm formed, increase in duration of the process from 9 to 13 h caused formation of nanoparticles with sizes 25 ± 3 nm and 40 ± 5 nm, respectively. It should be noted that optimization of synthesis conditions in presence of tetrapanax (rice-paper plant) biomass allowed production of silver nanocomposites with 1.8 wt % concentration and antimicrobial activity (minimal concentration of inhibition) of biogenic composite was 14.1 mg 14.1 mg (Ag) L^{-1} and 28.1 mg (Ag) L^{-1} for *Escherichia coli* and *Candida albicans*, respectively [47], which was comparable with action of colloid nanosilver.

On the whole, rates of biosynthesis are lower than in reactions of chemical reduction, and time of reaction may reach several hours, however, in some cases kinetic parameters can be compatible [48]. It is interesting to note that reduction of Au(III) ions in presence of vegetable surfactants from coconut and castor oils goes during several minutes without adding of special reducing or stabilizing agents [49]. At the same time if the chemical surfactant butyl ammonium bromide is used, Au nanostructures are formed only the next day.

The advantage of using of plant raw materials or their extracts with respect not only to chemical or physical methods of synthesis of nanocomposites, but in comparison with other biogenic processes are simplicity and minimization of a number of technological cycles, almost entire implementation of "green chemistry" at all stages of synthesis, including formation of non-toxic biocompatible materials applicable for biomedical purposes. Soft condition of synthesis, aqua media, quite high product yield make plant-mediated processes attractive for scaling, i.e. for producing a great amount of nanobiocomposites as, for example, is shown in the case of reduction of silver ions by extract of solid biomass of oil-palm (*Elaeis guineensis*), though the produced nanoparticles have wide size distribution from 5 to 50 nm [50]. Volume production can be favored by the fact that many species of raw plant materials are used for industrial purposes. A good example is *Sorghum spp*, which are used in alcohol production and other industrial products. It was recently found that *Sorghum* extracts have efficient reducing properties with respect to silver and iron ions and can provide substantial stability of colloids [41].

7.1.2 Microbiological Synthesis of Metal Nanoparticles

Microbiological synthesis of metal nanoparticles, though requires a special stage of cellular cultivation, oppositely to the abovementioned biogenic plant-mediated processes, is characterized by selectivity and high level of molecular control over metabolic processes, providing reproducible synthesis of nanoparticles of certain

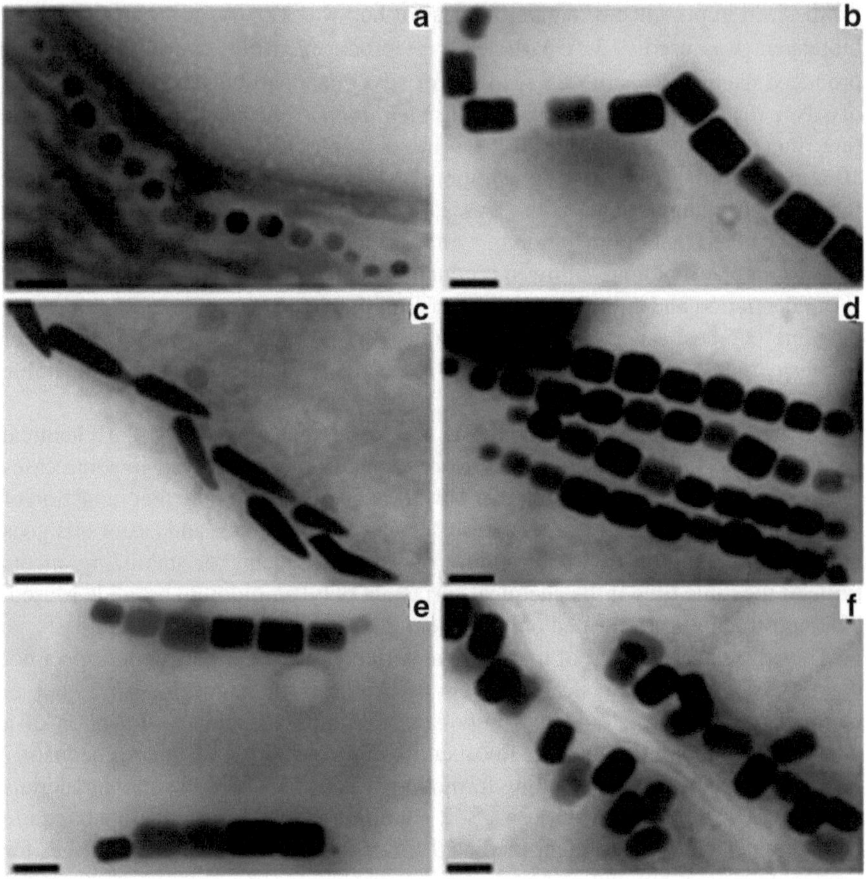

Fig. 7.2 Electron micrographs of crystal morphologies and intracellular organization of magnetosomes found in various magnetotactic bacteria. Shapes of magnetic crystals include cubo-octahedral (**a**), elongated hexagonal prismatic (**b**, **d**, **e**, **f**) and bullet-shaped morphologies (**c**). The particles are arranged in one (**a**, **b**, **c**), two (**e**) or multiple chains (**d**), or irregularly (**f**) (bar equivalent to 100 nm) [51]

sizes and structure. Evolution of microorganisms create the necessary prerequisites for development of their ability to produce spatially organized nanomaterials, among which the focus is on chemical lithography for reproducing of energy, usage of highly dispersed particles for special functions, detoxing functions for surviving in a toxic medium [51]. For example, sulfate-reducing bacteria reduce sulfates, thiosulfates, sulfites, and other sulfur containing compounds in oxidized to sulfides form. Some species of microorganisms synthesize inorganic materials at nanometer scale and integer them in functional components. Thus, magnetotactic bacteria contain intercellular chains of magnetite nanocrystals, magnetosomes (Fig. 7.2), via which they orient in geomagnetic field of Earth [52].

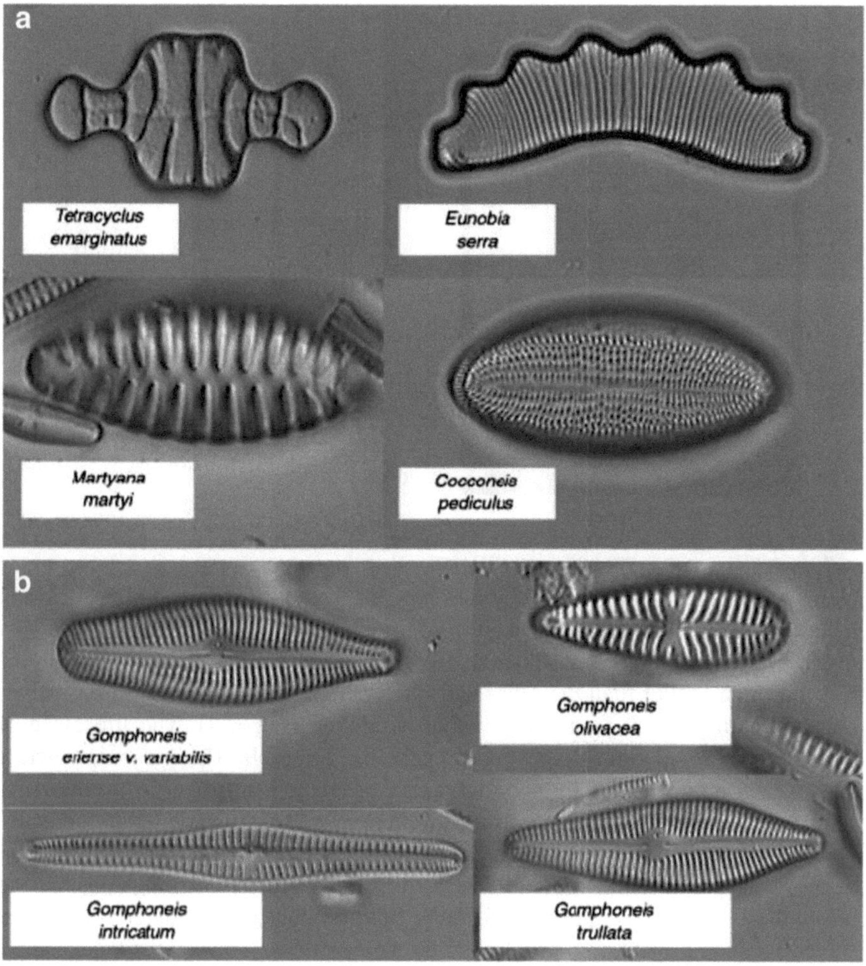

Fig. 7.3 Genera (**a**) and species (**b**) of diatoms

Depending on a type of bacteria, size of magnetosomes can vary from 35 to 120 nm. Biological mechanism of formation of magnetosomes controls accumulation of iron and biomineralization of magnetic crystals with a typical size and morphology inside membrane vesicles consisting of protein-containing lipid biolayers. Their combination with bioactive materials such as nucleic acids, enzymes, antibodies, etc. make it possible to fabricate materials with a possibility of magnetic manipulation with their biocomponents.

A probability of ordered disposition of nanoblocks of mineral crystals in intercellular or extra-cellular matrix of living organisms is clearly demonstrated by different kinds of diatoms (Fig. 7.3) [53].

Diatoms belong to a large group called the heteroconts or multi-flagella algae of *Bacillariophyta* class. They are composed of a cell and range in size from 2 μm to

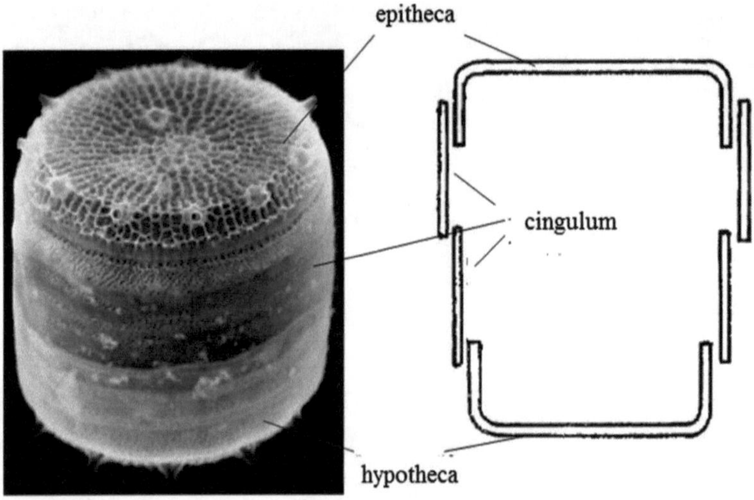

Fig. 7.4 Silica components of cell walls of *Thalassiosira tumida*

2 mm, some cells can unite into colonies [54–56]. Diatom cells are contained within a unique silica cell wall (a frustule) comprising two separate valves typically overlap one over the other like the two halves of a petri dish. One half, the hypotheca is smaller than the other half, the epitheca [57, 58]. Each of the valves contains additional ring structures called cingulum (Fig. 7.4). The life cycle of diatoms consists of three stages to compose the cell wall. The biogenic silica is synthesized intracellulary by polymerization of silicic acid taking from the environment. This material is then extruded to the cell exterior and added to the wall.

Highly specific structures such as enzymes or proteins included in bacterial membranes provide the specific interactions with reactive components during biosynthesis of nanocomposites and favor high product yield. Bacterial cells can serve as reducing agent or matrix carrier for nanoparticles. Thus, extract of bacterial culture *Rhodopseudomonas capsulata*, containing 65 % of protein, 20 % of soluble polysaccharides, and 7 % of lipids show bioreducing activity in the reaction Au (III) → Au(0) [59, 60]. At this, change in concentration of $HAuCl_4$ in the range from $2.5 \cdot 10^{-4}$ to $4.0 \cdot 10^{-4}$ M can be used to control shape of nanoparticles from spherical (10–20 nm) to nanowires (50–60 nm), respectively [60]. On the whole, the main factors of regulation of size and shape of nanoparticles in biosynthetic methods are similar to the considered above for chemical reduction: concentration of metal ions and proteins in extract, pH of a medium, time of reaction, etc. Certainly, it is necessary to take into account that additionally to specific absorption of proteins also non-specific interactions take place, which bring, for example, to isotropic growth of nanoparticles. And oppositely, by varying the ratio [$HAuCl_4$]/[extract] in the fungous strain *Rhizopus oryzae*, various shapes of gold nanoparticles were obtained (triangle, hexagonal, nanorods) [61]. Yield of the

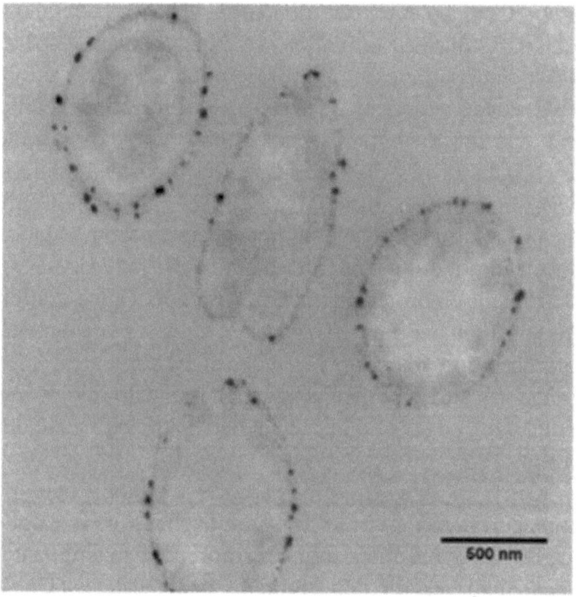

Fig. 7.5 Nanoparticles of Pd(0) in and on the outer cell parts of *Shewanella oneidensis* [64]

latter, for example, was 70–80 %. The amount of the metal adsorbed may come to 25 % of the dry substance of microbial cells [62].

It is interesting that in the case of bacterium *Shewanella oneidensis* at low concentrations of precursor there was exclusively intercellular formation of nanoparticles, and at high, predominantly deposition on walls of *S. Oneidensis* cells was observed, nanoparticles were rather coarse to 100 nm and more [63]. It is reported [64] also about deposition of Pd(0) in periplasmic space of *S. Oneidensis* (Fig. 7.5).

It should be noted that while biosorption of metal ions by microorganisms is characterized by quite high rates, bioreduction processes go far slower, and in some cases to increase their efficiency it is necessary to use donors of electrons, for example, H_2, formate, lactate, pyruvate, etc. [63, 64]. Addition of external donor of electrons was also required for synthesis of Pd nanoparticles with usage of *Desulfovibrio desulfuricans* [65] bacterium. Rather high reduction rates of Au(III) by *Pyrobaculum islandicum* in presence of H_2 is associated with hydrogenase [66]. An interesting approach is when fermentation of *Clostridium pasterianum* under anaerobic conditions causes generation of H_2, which then participates in reduction of Pd(II) to Pd(0) and the following deposits on walls of cytoplasm of a bacterium [67]. Moreover, thus produced biohydrogen can additionally serve as donor of hydrogen in catalytic activity of Pd(0). Thus, efficient catalyst can be obtained at one step.

Usually, the processes of reduction by bacteria are conducted under anaerobic conditions, and, on the whole, they are more efficient and fast than aerobic processes [68]. Though examples of the latter are also known, they are realized and important from practical point of view, for example bioreduction of $HAuCl_4$ by cyanobacteria

[69, 70] or by *Bacillus subtilis* [71]. It has been shown that intermediate product obtained during formation of intercellular Au(0) is Au(I) sulfide complex [69]. Nanoparticles of Pd(0) are obtained under aerobic and anaerobic conditions [64] under action of *S. Oneidensis*. Polyethyleneimine (PEI) or amino-enriched cationic polyelectrolytes behave both as binding molecules and reduction agents for synthesis of Au-bacterial nanocomposites similarly that when Au-composites were obtained in different ways [72, 73]. During reduction of the $HAuCl_4$ using PEI in solution the bacterial fibers become from colorless to purple that indicates the formation of Au nanoparticles on the fibers [74].

Protein environment facilities producing the complicated products both on the composition and form. Thus, when ionic silver enters in the organism it is bonded by affinic biopolymers and is reduced by biological substrates up to high dispersed metallic state [75]. Special attention is paid to soluble metal-polymer nanocomposites containing silver or other metal nanoparticles as potential antibacterial and antivirus agents.

Often in the synthesis of complex bionanocomposites individual biological molecules are preferred to whole bacterial cells. Biotemplates in addition to wide variety of shape and sizes show reproducible template structures, which make them ideal matrices for deposition of nanoparticles. DNA [76, 77] and viruses [78] biomolecules are used for production of metal nanorods, nanowires. Among proteins bacterial S-layers [79] and flagella[2] [80, 81] are successfully used. In particular, S-layers included in content of cellular wall of most bacteria are regular protein subunits, and are most widely spread structures in prokaryotic organisms. They consist of numerous copies of individual polypeptides, which spontaneously form highly organized nanoporous superlattices of different symmetry (Fig. 7.6).

Bacillus sphaericus were used to produce 405-nm gold nanoparticles spaced at 13 nm at the surface of S-layer with square periodicity [82]. Hexagonally packed S-layers of *D. Radiourans* [83] allow fabrication of ordered structures of micron sizes from regularly arranged gold nanoparticles, so that they completely follow a biotemplate structure. It is interesting to note that from the equal mixture of citrate-stabilized nanoparticles of 5 and 10 nm sizes only 5-nm colloid is bound by S-layer, while during absorption of 20-nm particles long-ordered structure is broken. Depending of shape and electrostatic properties of a biotemplate, nanobicomposite materials of various spatial arrangements are formed [53]. It is important that for modification of topographic or chemical properties of the S-layer just construction of one gene is required.

[2] Flagella – is a surface structure present in many prokaryotic and eukaryotic cells and serving for their motion in liquid medium or on the surface of solid media. Bacterial flagella thickness is 10–20 nm and length 3–5 μm. Basal body in a cellular wall drives exterior semi-rigid protein helix fiber via the body rotation, thus generating hydrodynamic force driving the cell directionally. Basal body of a flagella is a miniature electromotor, due to which bacterial cell is able to develop very high speed, 100 μm/s, i.e. more than 50 lengths of the cell body per second. The driving force takes energy from ionic gradient on the inner membrane of the bacterial cell – transmembrane potential of hydrogen or sodium ions.

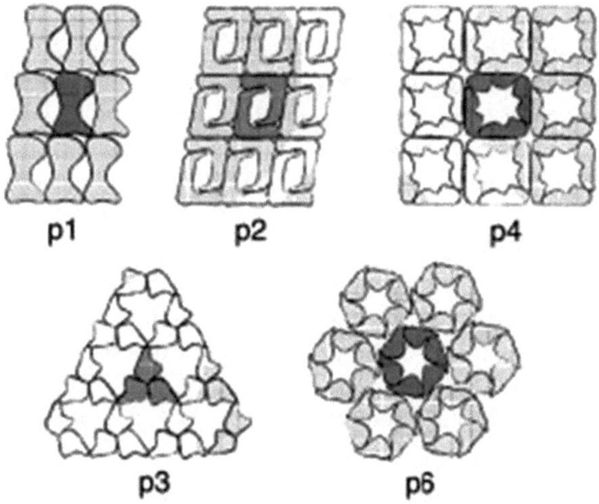

Fig. 7.6 Different types of bacterial surface layers (S layers) with a diagonal (p1, p2), square (p4), or hexagonal (p3, p6) symmetry [53]

Method of molecular engineering of biological molecules can be applied to fabricate templates having a great affinity to a certain metal ions, for example, to Au(III) [84, 85]. Introduction of a histidine fragment into flagellate protein caused a controlled deposition of Au(0) nanoparticles of 5 nm in diameter due to formation of Au(I) complex with structured amino and imidazole groups. Native flagella *Desulfovibrio desulfuricans* shows high affinity to Pd(II) complexes due to nitrogen-containing ligands, and protein fibers are completely coated by Pd (0) nanoparticles, whereas attempts to synthesize Au(0) under the same conditions failed [81]. Genetic manipulation with peptide sequence of RP437CysFliC coli allows production of RP437CysFliC line with addition of up to 12 added cysteine residuals per flagella monomer. Presence of cysteine thiol groups causes immobilizing of Au(0) nanoparticles on the surface of bacterial flagella fibers. Flagella-bound nanoparticles are in the range 20–50 nm, most suitable for catalytic applications of gold nanocomposites.

In this aspect it is also important that formation of Au-S complexes, as is known, increases catalytic activity of these systems [86].

One of the important properties of bacteria is their ability to reduce or oxidize trace elements, including toxic metals and radionuclides, which can be efficiently applied for in situ bioremediation of metal-polluted soils and water and for extraction of precious metals from dilute spent solutions [87]. To the early works in this field belong bioutilization of Pd(II) from solutions in form of Pd(0) [88] with usage of sulfate-reducing bacteria. *D. Desulfuricans* demonstrates bioreducing activity with respect to ions of the considered metals in the medium of liquid wastes and in sewage in spent car catalysts [65, 89]. Processes of reduction are implemented in electrobioreactors, containing bacterial cells immobilized on external surface of Pd-Ag electrode. Hydrogen is formed electrochemically and its transport goes through membrane of the electrode. The similar scheme was used for bioselective utilization of Au(0), Pd(0), and Cu(II): firstly Au(III) was reduced by native biomass *D. Desulfuricans,* then Pd was extracted by pre- palladinized mass (see below), and at last, Cu(II) was deposited in hydroxide and sulfates compounds using gas emission from bacterial culture [90].

Living cells show high ability to bioreduction. It follows from this that some enzymes remain active at low pH (2–3). At the same time, bioextraction of Pd from wastes (scrap) of electronic devices was ineffective, because they, as a rule, contain a great amount of Cu^{2+} ions (25 wt% or more), which inhibit hydrogenase (see below) [90]. But if bacterial cells, which were subjected to pre-palladinizing are used, Pd(0) seeds can be catalysts for further chemical reduction of Pd(II) from Cu^{2+} containing solutions. In this case extraction is not enzymatic, but autocatalytic growth of Pd(0) clusters proceeds on cells. The same is true for *E. coli* [91], however, bacterial cultures *C. necator* and *Cupriavidus metalliduran* reduced Pd without pre- palladinizing [92].

Biomineralization can be caused by dissimilating reduction of metals. For example, for oxygenless respiration Fe(III) as electron acceptor exudates Fe (II) which, in turn, inspires formation of mineral phases such as magnetite, vivianite or siderite [93]. Other products of microbiological activity can include manganese oxide, silicates, phosphates, etc. Direct transformation of impurities in heavy metals and radionuclides in nanometer mineral phases are enzymatic reduction of U (VI) to U(IV) [94, 95], and also Au(III), Ag(I), Tc(VII), Cr(VI), Se(VI)/(IV) and Pd(II) ions [96].

7.1.3 Fermentative Synthesis in the Formation of Nanoparticles

The main problems in usage of biological systems are caused by difficulties in control over fineness of particles and their morphological characteristics, difficulties with obtaining biological material, etc. Therefore, approaches of fermentative methods of synthesis of metal nanoparticles are developed. For example, water solution of $AuCl_4$ was subjected to reduction with purified sulfite reductase ferment, extracted from *Fusarium oxysporum* fungous [97].

Formation and stabilizing of nanoparticles was performed in situ in presence of phytochelatin peptide.[3] Similarly nitroreductase was extracted from *Fusarium oxysporum*, which catalyzed silver nitrate reduction to silver nanoparticles [98]. Active center of this ferment, as in the case of the abovementioned sulfite

[3] The general structure of phytochelatin is (γ-Glu-Cys)n-Gly, n = 2–11. In Greek "phyto" means it presents in plants and "chelatin" is its ability to form chelate complexes with many metals including Cd^{2+}, Pb^{2+}, Zn^{2+}, Sb^{3+}, Ag^{+}, Ni^{2+}, Hg^{2+}, $HAsO_4^{2-}$, Cu^{2+}, Sn^{2+}, SeO_3^{2-}, Au^{+}, Bi^{3+}, Te^{4+}, W^{6+} ions.

reductase is NADPH.[4] As is assumed, for catalytic formation of nanoparticles reduction of α-NADPH до α-NADP$^+$ is needed to Au(II) in intermediate processes and then of metal gold [97].

Hydrogenase activity of *D. desulfuricans* and *Escherichia coli* is found in biosynthesis of Pd(0) [88, 99] and Au(0) [68] nanoparticles. Though reduction of Au(III) is partly suppressed by Cu(II) ions, which are known as inhibitors of periplasmic, not cytoplasmic hydrogenases, finally a mechanism of biochemical ways of formation and growth of Au(0) nanoparticles stays unknown. Role of various hydrogenases in reduction of Pd(II) is studied in detail also in the case of *Desulfovibrio fructosivorans* [100]. It is assumed that enzymes serve sources of electrons for the reduction process and nucleation and growth centers of particles (see Table 7.2 [16, 27, 101–111]).

7.2 Sol-Gel Process as a Way of Production of Template-Synthesized Bionanocomposites

Bioglass® glass ceramics was first used in biopractice in early 1970s [112]. For the first time encapsulation of active ferments in sol-gel matrix was fulfilled in 1990 by mixing of biomolecules with sol-gel precursors [113], and just in several years many different hybrid bioceramic materials of this type were designed. Presently incorporation of bioactive substances into ceramic gel is widely used technique for

[4] Nicotinamide adenine dinucleotide phosphate is frequently naturally occurred co-ferment of some dehydrogenases, ferments, which catalyze oxidizing-reducing reactions in living cells. NADP accepts hydrogen and electrons of oxidized compound and serves a carrier of them.

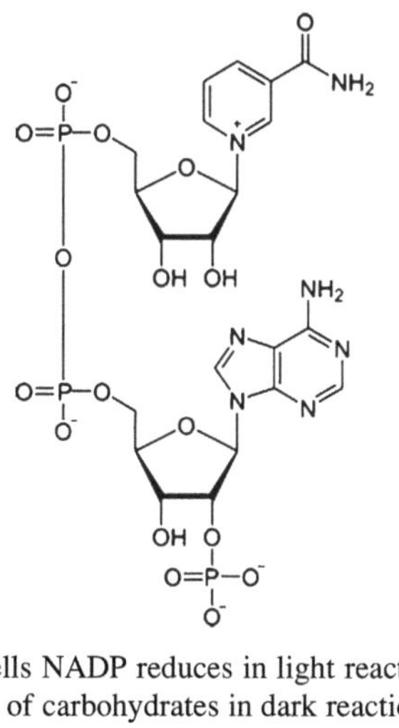

In chloroplasts of vegetable cells NADP reduces in light reactions of photosynthesis and then supplies with hydrogen synthesis of carbohydrates in dark reactions.

Table 7.2 Biosynthesis of nanocomposites

Nanoparticles	Organism	Size (nm) and shape of nanoparticles	Reaction conditions	References
Bacteria				
Ag(0)	*Pseudomonas stutzeri*	35–46, >200	$AgNO_3$	[101]
Ag(0)	*Lactobacillus sp.*	20–50, >100, 15 and 500	$AgNO_3$	[102]
Au(0)	*Rhodopseudomonas capsulata*	10–20	$HAuCl_4$, extracellular reduction	[60]
Au(0)	*Escherichia coli, Desulfovibrio desulfuricans*	20–50	2 mM $HAuCl_4$, periplasmic space, cell surface, extracellular reduction	[68]
Au(0)	*Shewanella oneidensis*	<10	$HAuCl_4$ $3H_2O$, Intracellular reduction	[63]
Au(0)	*Bacillus subtilis*	5–25	AuCl	[106]
Au (0)	*Shewanella algae*	10–20	Au(III), anaerobic conditions	[107]
Au(0)	*Lactobacillus* strains (buttermilk)	25–50, >100	$HAuCl_4$	[102]
Au(0)	*Thermomonospora, actinomycetes*	7–12	Extracellular reduction	[16]
Au(0)	*Rodococcus actinomycetes*		Intracellular reduction	[110]
Yeast				
Ag(0)	*MKY3 strain*	2–3	Ag^+, Extracellular reduction	[103]
Fungus				
Ag(0)	*Verticillium* sp.	2–20	Extracellular reduction on the cell wall	[104]
Ag(0)	*Fusarium oxysporum*	2–50	Extracellular reduction	[105]
Au(0)	*Colletotrichum* sp.	8–20	$HAuCl_4$	[27]
Au(0)	*Verticillium* sp.	2–20, 25	Au(III), intracellularly	[108]
Au(0)	*Fusarium oxysporum*	20–40	Extracellular reduction	[109]
Au(0)	*Rhizopus oryzae*	50–70, triangular, hexagonal, pentagonal, spherical, nanowires, nanorods	$HAuCl_4$ $3H_20$, Extracellular reduction	[61]
Au(0)	*Fusarium oxysporum*	7–20	$HAuCl_4$, Sulfite reductase *in vitro*	[97]

(continued)

Table 7.2 (continued)

Nanoparticles	Organism	Size (nm) and shape of nanoparticles	Reaction conditions	References
Au-Ag сплав	*Fusarium oxysporum*	8–14	Extracellular reduction	[111]
Plants				
Ag(0)	*Pelargonium graveolens* (экстракт листьев)	16–40	10^{-3} M aqua solution, $AgNO_3$	[36]
Ag(0)	*Dalbergia sissoo*	5–55, spherical	$AgNO_3$	[26]
Au(0)	*Dalbergia sissoo*	50–80, spherical, triangular, hexagonal	$HAuCl_4$ $3H_2O$	[26]

formation of bionanocomposites. Bioceramics is ideal material, because it has high rigidity, mechanical strength for fracture and impact strength. Inorganic matrices include silicon, titanium, zirconium oxides, TiO_2–cellulose composites, etc. These composites can be obtained also by dry method, like xerogels and ground powders. Based on these materials biosensors and ferment electrodes can be developed, encapsulating agents for drug delivery, adsorbents for pharmacy and cosmetic industry, photocatalysts for air and water cleaning, etc. Bioinorganic and hybrid nanostructures combine optical, electronic, and mechanical properties of inorganic materials and low cost of natural biomaterials, and are already used potentially not only in biomedical applications, but also as optical [114–118], magnetic [119, 120], catalytic and other materials [121].

Organic polymers are widely used materials for transplantation of soft tissues, though they do not have biological activity, they are most often bio-tolerant. Moderate temperatures and soft conditions of hydrolysis and condensation-polymerization of monomer alkoxides of metals and metalloids allow trapping of protein molecules without their denaturation. High stability of thus incorporated biomolecules, inertia and high S_{sp} of the matrix, its porosity facilitate heterogenization procedure, without need of covalent bonding of the matrix, and make attractive sol-gel variants for immobilizing proteins, including whole cells [122].

One more widely used among considered approaches in biomedical applications is coating of a surface of metal implants (more often Ti_6Al_4V) alloys with organic polymer, which has a great importance for integration into a bone and binding with it. Finding a mechanism of biocompatibility and effect of dynamics of physiological processes has an important place in structuring of the implant/host interface. Though titanium alloys show perfect corrosion resistant properties, metal ions can be released in physiological medium, and this can have an adverse effect on organism, especially in the case of vanadium ions.

Here were confine to analysis of the main approaches and listing of the obtained materials.

7.2.1 Biomedical Nanocomposites

Bionanocomposites consisting of ceramics and resolved polymeric implant are promising for successful regeneration of bone tissue [123]. One of typical examples is nanocomposite material based on glass ceramics and nanofiber degraded polymer of poly(lactic acid) (PLA) [124, 125].

This method can be used for encapsulation into mesostructured microspheres and cells [126]. Examples of post-synthetic modifying of ceramics (after aerosol drying and surfactant removal) are rare. This way is mainly used for development of multifunctioning vector for diagnostics and therapy, producing of incorporated drugs such as ibuprofen [127], triclosan, doxorubicin [128], alendronate, zoledronate [123], etc. and biocomponents (phospholipidsliposomes, transmembrane proteins) [129, 130].

Proteins, such as copper-zinc superoxide dismutase, myoglobin, hemoglobin, and bacterio-rhodopsin are encapsulated in porous silica gel matrix prepared by sol-gel synthesis, which strongly holds these molecules without loss of their fermentative activity and without change in their spectral properties [131]. The matrices provide transport of small molecules to reaction center and transport of reaction products, strongly hold protein molecules in pores. The same way was applied for heterogenization of glucose oxidase and peroxidase used as active solid phase elements in glucose sensor.

The same way is applied to bind antibodies for potential use in medicine, immune chromatography, immune sensing, etc. For example, sol-gel captured immunoglobulins preserve their ability to bind external antigens (2,4-dinitrophenylhydrazine) from solution. Atrazine binding properties of a sol-gel matrix doped with 10 % PEO and including monoclonal anti-atrazine antibodies are studied in detail [132]. This matrix "recognized" in solution and bound widely used atrazine herbicides. It is important that there was neither leaching of antibodies, nor non-specific physical sorption of atrazine by ceramic matrix. At that, there was no decrease in activity, at least, during 2 months, while activity in a solution decreased under these conditions to 40 %. Antibodies encapsulated in these particles can be used as sensors for recognizing of specific antigens [131]. There are examples [133] of the first successful attempts to capture catalyst antibodies by sol-gel matrix and their usage: antibodies 14D9 contained in these matrices catalyze different reactions including hydrolysis of cyclic acetals, ketals, epoxides, etc. Peroxidase trapped by silica particles shows high stability as compared to non-immobilized ferments at change in temperature and pH. Encapsulation of ferments in silica gel nanoparticles allows compensation of ferment deficiency in living systems, and their usage in medicine without hazard of allergy or proteolytic reaction with almost negligible leaching. It should be stressed that there are some more advantages of these materials: high thermal and pH stability, prevention of leaching of captured proteins, fermentative reaction, which is suitable for spectral control (both in pores and in matrix), convenient storage, a possibility of re-usage, etc. Moreover, these systems provide control over morphology and sizes of particles.

Lipidic bilayered vesicles with interior water cells are widely used as models of biomembranes in supramolecular chemistry. They are often used as nanocapsules for drugs delivery or for transfection of genetic structures of nucleic acids, ferments, they are candidates for designing of supramolecular devices.

Sol-gel technique is realized for immobilising of ferments acting as bioreactors [134], for this chemically active end groups of ferments and active bonds of ceramic dopants, for example, -Sn-Cl are used. The process of immobilizing and synthesis of these materials can be presented by the following Scheme 7.2:

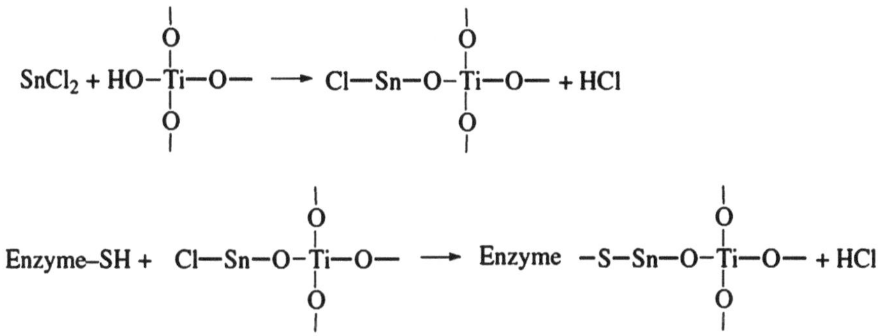

Scheme 7.2 Covalent binding of enzyme molecule with the TiO_2 surface

By this mechanism, for example, alkoholdehydrogenase is immobilised inside nanotubes of template synthesized TiO_2, which (cofactor NAD^+, phosphate buffer, pH8) is active in ethanol oxidization. Since that nanotubes are open from both ends, this configuration makes it possible to use them as a flow nanoreactor. There are numerous similar examples including covalent binding of antibodies for functioning of sol-gel films (Fig. 7.7) [135–141].

At the same time this causes formation of particles with uncontrolled sizes, because distribution of molecules captured by ceramic matrix is far from uniform, and kinetics of the catalyzed reactions does not obey to Michaelis-Menten laws [142].

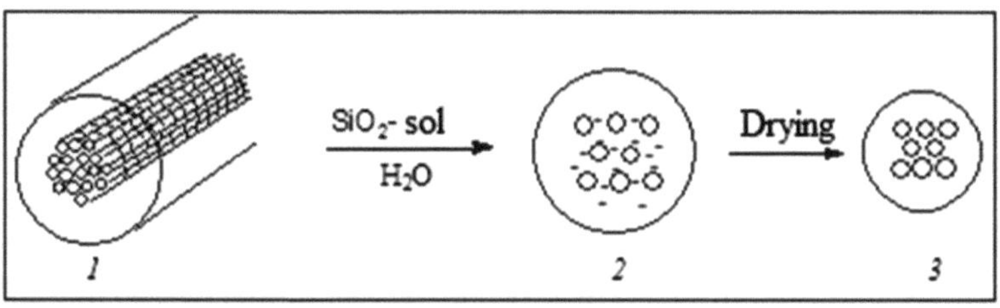

Fig. 7.7 A scheme of an organized macroporous SiO_2 structure, formed by the bacterial templates: bacterial filaments with multicellular fiber structure) (*1*); a mineralization of the inter lament space (2) and a macroporous replica formation by drying at 873 K (*3*)

Usage of polysaccharides as templates in sol-gel processes allows control over organized hybrid nanocomposites [143–145]. Formation of 3D fiber net [143] of a composite is due to hydrogen bonds formed between hydroxyl of macromolecules and products of TEOS hydrolysis. These materials are used in many fields as protective coatings, food package or structural composites as is shown in some examples in the Table 7.3. Biopolymers from marine mollusks can be fixed on silica gel obtained from different sources (silicic acid, sodium silicate, silicon alkoxide) forming hybrid materials. Carbohydrate polysaccharides (arabinogalactane, cellulose derivatives, etc.) containing active hydro- and carboxyl groups are capable to bind magnetite nanoparticles. Magnetitodextranes, dextranferrites, carboxymethyldextraneferrit obtained are used as biocompatible magnetic carriers, immune-magnetic sorbents, preparations for immune-magnetic separations of antigens. Moreover, ferroarabinogalactane possesses membrane-acting properties and is immune-response modulating agent [146, 147]. Although carbohydrates reveal weaker stabilizing effect than proteins but they have quite a number advantages. Thus, they do not denaturate at high temperatures and pH as well as in organic solvents.

Nanocomposite materials based on TiO_2 and poly-ε- caprolactone (PCL), containing 6, 12 and 24 wt% TiO_2, obtained in sol-gel process [148] have bioactive properties. Polymer is incorporated in the net due to hydrogen bonds between carboxyl groups and functional groups of inorganic matrix (see also Sect. 6.3). Kinetics of ampicillin desorption shows that the studied material releases high doses of antibiotic during first hours, and then slower release of the drug provides supplying dose to the end of experiment. When adding poly-N,N- dimethylaminoethyl methacrylate, soluble structures and insoluble composites form. In the insoluble ones there are about 70–75 % links of polymer amine absorbed on the surface of sol particles, while for the soluble ones the main fraction of the links is in loops [149, 150]. Thus, a net of polymer amine is formed bound to silica particles. This is confirmed by the examples of oligomer amines [151], synthetic oligopropylamines similar to those found in diatoms.

As in the case of traditional nanocomposites, hybrid thin films obtained by aerosol technique [152] by integration of (bio)organic component in microspheres at one stage form organic-inorganic materials of two types: class I (weak interaction between components) and class II are materials with strong interaction between components.

There are interesting biocomposites based on mixed oxides, for example SiO_2 -ZrO_2 or SiO_2-CaO-P_2O_5 in one matrix [153]. Bioactive materials based on calcium phosphates (hydroxyapatite and tri-calcium phosphate, see Sect. 7.3) and glass/ceramic glasses are widely used in dentistry and orthopaedics. The first of them, silica gel glasses considered as promising substitutes for bones and regenerated tissue matrix, because they have high compatibility with tissue (rigid and soft), osteoconductivity (they are carriers for regenerating material, which in-grows into implant), osteo-induction (ability to stimulate regeneration of bone tissue).

Among different types of bioactive glasses, sol-gel derivatives have been developed relatively recently [154, 155]. Comparison of sol-gel glasses with their

Table 7.3 Hybrid biocomposites based on silicon containing materials and polysaccharides for different applications [259, 379–385]

Molecular precursor or building block	Polysaccharide, cell, microorganism	Properties	Application	References
TEOS	Vinyl-modified guar gum	Absorbed properties, mechanical stability	Decontamination	[379]
Sodium silicate	Alginate	Nanometric scale, nontoxic for cells	Drug delivery	[259]
TMOS	Chitosan	Nontoxic for cells, proliferation	Repair of bone	[259]
Sodium silicate	Collagen	Bioactivity, nontoxic for cells	Tissue engineering	[259]
3-glycidoxypropyltrimethyloxysilane	Gelatin	Hydrophobic, biocompatibility, transparency	Antiwetting coatings	[259]
Sodium silicate	*Horseradish peroxidase* and glucose-6-phosphate dehydrogenase	High catalytic activity, Michaelis-Menten kinetics, nonleaching of enzyme	Materials for biosensors, immobilization ability, fermentative reactor	[380]
Silicone alkoxides, organoalkoxysilanes	Bovine Serum Albumin	High stability, optical isomer separation of D- and L-tryptophan	Monolithic columns for chromatography	[381]
Silicone alkoxides, organoalkoxysilanes	*Horseradish peroxidase*	Templates obtained by sol-gel method for solid phase lithography, stability, reusing	Optical waveguide of biosensors	[382]
MCM-41 (Mesopore silicate)	Cytochrome C	Nontoxic for cells, light internalization with living human cells	Transmembrane protein delivery	[259]
Organoalkoxysilanes	Microphytic alga *Chlorella vulgaris*	Long-term stability, heavy metal ion's affinity	Amperometric sensors	[259]
Silicone alkoxides	Langerhans insula	High long-term stability, immune isolation of transplant tissue with a minimal resection and fibrosis	Bioartificial organs	[383]
Sodium silicate and colloid silica	Cyanobacteria,	Long-term stability, photoactivity protection	Photobioreactor	[384]
Zeolite	Xylanolytic bacteria	Stable accumulation, easy-to-use	Biogas production	[385]

Fig. 7.8 Morphology of bioactive collagen-glassfibrillar nanocomposite [162]

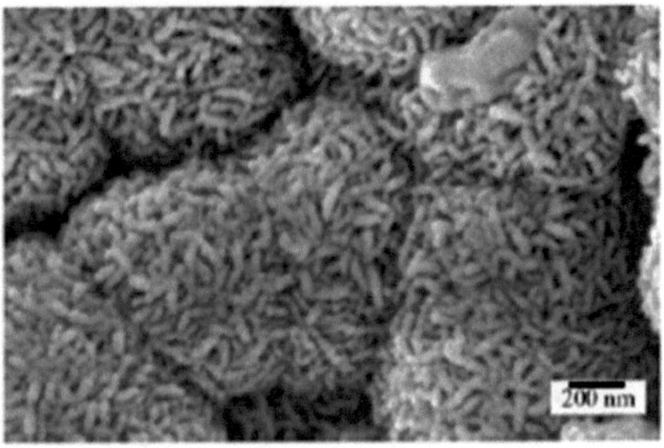

metal-derivatives analogues in a wide range of concentrations (up to high content of SiO_2) makes it possible to optimize solubility and bioactivity. Sol-gel glasses can be used as powders, coatings, porous foams, and they can show good ability to formation of bone tissue in vivo [156–159]. In this view sol-gel approach has some technological advantages [160]. This method followed by electrospinning was used to prepare bioactive glasses shaped as nanofibers [161]. Diameter of these fibers is extremely small (10–100 nm) as compared with the fibers produced by spinning form melt (usually 10–100 μm). For rehabilitation medicine a new composite biomaterial is developed [162], which consists of nanofibers and reconstructed collagen matrices (Fig. 7.8). The process is carried out as follows. Sol-gel derivative of bioactive composite ($58SiO_238CaO4P_2O_5$) was transformed by electrospinning into nanometer fibers with average fiber diameter 320 nm, which were consequently treated with hydrazine and collagen, the main organic component of bone matrix, and as a result a stitched nanocomposite was produced shaped as a thin membrane.

Bioactivity of nanocomposite *in vitro* was estimated by incubation period in SBF medium[5] [163]. There are a lot of similar examples (see, for example, [164–173]).

For biomedical application also hybrid material is used based on poly (2-hydroxyethylmethacrylate) (pHEMA) and silica gel, which is obtained by sol-gel synthesis and shows some bioactivity, for example, in proliferation [174–177]. Amount of inorganic precursor (TEOS) is taken so that in a mixture with organic monomer fraction of silica gel would be 30 wt%. There is strong interaction between phases, which inhibits interphase separation: thermal stability of pHEMA improves, silica gel content has effect on decomposition temperature and has no effect on T_g of pHEMA; swelling decreases as silica gel content decreases (10–30 wt%), while increase in the latter content causes increase in *in vitro* bioactivity, which advances formation of apatite on a surface of modified

[5] SBF is artificial surrounding tissues liquid which contains ions (pH = 7.4, 142 mM Na^+, 5 mM K^+, 1.5 mM Mg^{2+}, 2.5 mM Ca^{2+}, 125 mM Cl^-, 27 mM HCO_3^-, 1 mM HPO_4^{2-}, 0.5 mM SO_4^{2-}) in concentrations close to human plasma (in a typical experiment 50 mg of microporous composite was placed in biological activity at 37 °C up to 7 without renewing) [163].

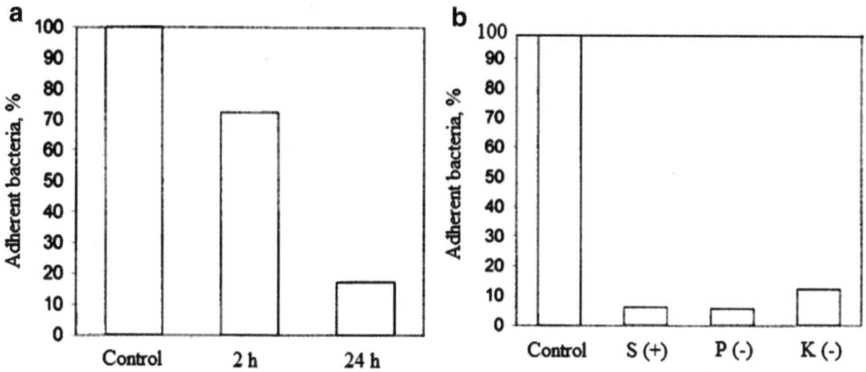

Fig. 7.9 Survival rate of gram positive (**a**, **b**) and gram negative (**b**) adherent bacteria on SiO_2 nanosol coatings with embedded colloidal silver and chlorohexidine after 2 (**a**) and 24 h (**a**, **b**) *S Staphylococcus*, *P Pseudomonas aeruginosa*, *K Klebsiella pneumonia* [171]

hydro gel soaked with biological liquid SBF. As it was noted, bioactive nanocomposites can be used for frames structuring in engineering of bone tissue [170, 172, 174, 175, 178–180].

The similar example is Ampicillin-β-lactam antibiotic active against gram-positive and gram-negative bacteria, including the form of bound at the stage of sol-gel synthesis, which is widely used for treatment of infections [181, 182] (Fig. 7.9).

The main part of application of hybrid mesoporous particles relates to multifunctional vectors. Thus, aerosol particles are important as biomedical carriers [130] in pharmacy, preparation of ointments, etc.

7.2.2 *Biocomposites as Means for Drug Transfer*

Reconstruction of damaged tissue, artificial substrate for cell growth, locally acting system for drug transfer are the most widely spread application fields of bionanocomposites. Nanoparticles used for multifunctional vectors of drugs (MVD) should have hydrophilic surface. As a rule, their size should not exceed 60 nm. Loading of drugs is a simple thermodynamic absorption process: precise amount of mesoporous microspheres is added to a drug solution and stirred from 5 to 20 min to 24–28 h. Amount of bound solution depends on a carrier structure and on force of interphase interaction between a matrix and drug molecules. In order to increase circulation time of these particles in blood flow, they are coated by hydrophilic elements, often PEGs. This, on one side, prevents their coagulation, and on the other hand, this eliminates or minimizes absorption of proteins on them, and makes a particle "invisible" for cells of immune system. For example, local transport of antimicrobial drugs in periodontal pocket has advantages caused by delivery of more drugs to a target with minimal damage for the organ [183]. Supplying of desirable constant drug concentration, a decreases in systematic level of the drug, and decrease in potential harm are also reached by MVD means. Oxides used for MVD should form strong ion-covalent interaction between oxide matrix and phosphonated form of a drug or they should form biocompatible apatite-like phase.

There is increasing interest to mesoporous silica gel materials (such as MCM-41 and SBA-15) for their usage as MVD and controlled release of a drug. Their advantages are non-toxic originality, suitable diameters of pores and high specific surfaces, abundant with Si-OH bonds. Moreover, these systems show prolonged action though their capacity to drugs is relatively low, and they typically have irregular volume morphology, which is not ideal for MVD. In this connection there is an interest in hollow silica gel spheres formed on interface boundary in oil/water emulsion. To form particles with stable lamellar mesostructure, close to spherical, silica can be integrated in interlayer areas of multilamellar vesicle. Linear polysiloxane colloids such as templates formed *in situ* during emulsion polymerization are coated with cross-linked polysilsesquioxane shell at gel formation of trimethoxymethylsilane, hollow particles form by extraction of soluble core by a solvent. Colors as functional molecules were encapsulated by a silicon net and hollow particles. Hybrid particles with core-shell structure were obtained by coating of surfaces of monodisperse polystyrene beads by homogeneous coating made of silica.

In other variant, loading of a drug is done by mesoporous microspheres. This way provides release of a bound drug up to 50–70 %. Two-stage release of a drug is determined by absorption of the drug by two mechanisms (centers): molecules of the drug are absorbed on exterior surface of microspheres or in pores, the first are released fast, the second ones more slowly. Mesoporous silica gel material (powder or thin films) under biologic conditions can be degraded during several hours, destruction rate depends on content, porosity, calcination temperature and causes loss in activity of a drug [184, 185].

New type of organic-inorganic hybrids includes organic part of biological origin bound to inorganic nano-object [186–191]. In biomedical applications [192–196] intercalated systems are widely used (Sect. 6.3). Simplicity of synthesis, versatility, biodegrading and biocompatibility of layered dihydroxides (LDHs) are especially interesting for production of bionanocomposites [197, 198]. For example, combination of alginate and maize protein (zein) provides new matrices for MVD [199], which can directly encapsulate a drug, for example, ibuprofen (IBU). These MVDs were tested for controlled extraction of ibuprofen used as a model drug under conditions of transport through gastrointestinal tract. Velocity of transfer of encapsulated drug decreases with increase in zein content, probably, due to hydrophobic character of this polymer. Ibuprofen is intercalated in [Mg_2Al]Cl by ion-exchange reaction. The procedure of obtaining of alginate-zein drugs is shown schematically in Fig. 7.10.

Controlled kinetics of ibuprofen extraction from alginate-zein biocomposite beads can be useful for different kinds of therapy. Simplicity of zein production as a film and a possibility of usage of small-sized LDHs provide broadening of this approach for immobilizing of different drugs.

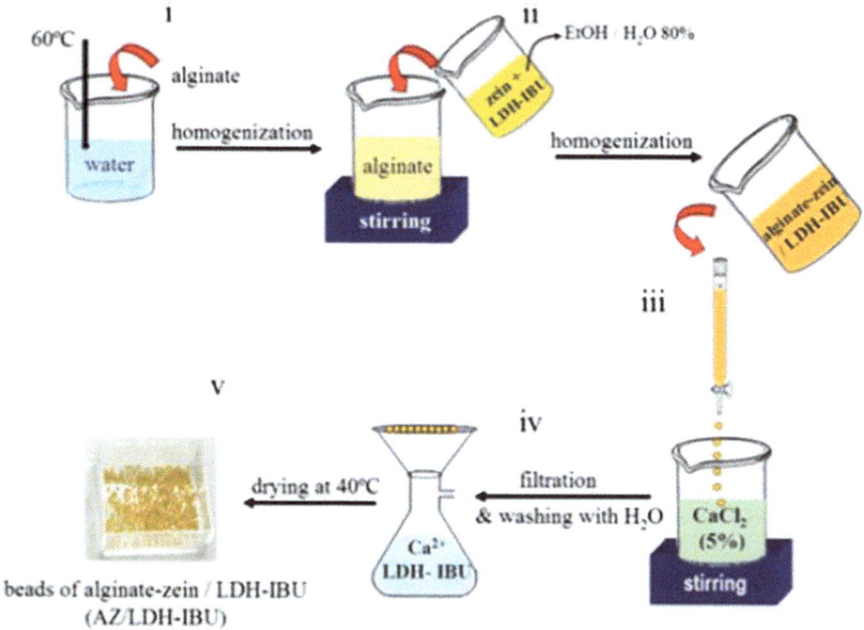

Fig. 7.10 Scheme of the general procedure employed for the preparation of the alginate–zein beads entrapping the LDH-IBU hybrid [199]

From the chemical point of view, all multifunctional vectors based on silicon dioxide (pure SiO_2, organically modified SiO_2, SiO_2-ZrO_2, SiO_2-CaO-P_2O_5 mixed oxides) are often synthesized in presence of template agents [200–209]. The SiO_2 – ZrO_2 mesoporous microparticles containing ZrO_2 between 0 and 20 mol % were far more stable in relevant biological conditions, than pure silica gel. Moreover, interesting properties were found in presence of zirconyl caused by complex formation of drugs with zirconium and by hydrophilic/hydrophobic character of the drugs (using for example hydrophobic zoledronate, containing imidazole ring or hydrophilic alendronate with propyl chain and terminal amino-group).

As regards loading of a drug, there is a common tendency: independently of a type of absorbed drug its absorption is increased as content of zirconium oxide in silica gel matrix increases. However, in composites with the highest zirconium oxide concentration (20 mol %) there are lower textural characteristics (surface area, volume and size of pores). Loading of a drug is more efficient (by 3.5 times) in the case of hydrophilic forms (alendronate), than for hydrophobic drugs (zolendronate). This result can be explained by hydrophilic neighboring of micro-particle surface provided by Si-OH and Zr-OH groups. On the other hand, hydro-philic/hydrophobic character of drugs can impact the process of its release. At the same time the process of alendronate release is its fast release followed by slower more controlled process, profile of alendronate release is more sigmoid and can be factorized into three phases: delay phase, explosion phase and saturation phase.

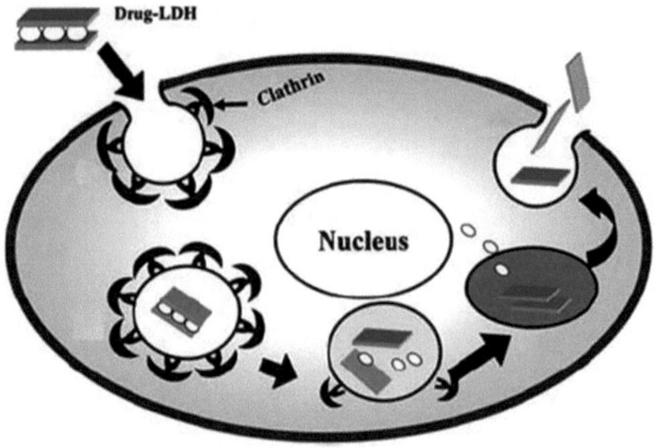

Fig. 7.11 The proposed cellular uptake mechanism of drug-LDH nanohybrids: internalization of the nanohybrid via clathrin-mediated endocytosis, transport and release of the drug in the lysosome and LDH externalization via exocytosis

This profile is caused by hydrophobic property of zolendronate, which impedes diffusion of water physiological medium into pores causing the delay phase. Immediately after release of a part of zolendronate hydrophilic/hydrophobic balance of a surface is distorted thus providing diffusion of a solvent into pores, and normal character of a drug release is accompanied by the explosion phase and saturation. It is interesting to note that amount of released drug is far lower (especially in the case of zolendronate) in the cases when drugs are strongly bound to the surface (mixed silicon and zirconium oxides), than in the case of a simple absorption (pure silica gel). Another way of control over drug release is by aggregation of liposomes with mesoporous microparticles (Fig. 7.11).

This biomimetic approach bringing to cell-like structure or "protcells" is one of elegant ways of development of contemporary multiplatform for biocompatible drug delivery. It is based on ability of membranes in a cell to control metabolism and, in particular, hinder diffusion of ions and charged hydrophilic molecules. These studies are based on electrostatic interactions under conditions of pH neutral medium between microparticles based on silicon oxide, which can be anionic in the case of non-modified silica gel or cationic in the case of silica gel modified with amines.

7.2.3 Biomineralization and Bioconcentration

Sol-gel processes is a widely spread way of formation of bionanocomposite materials by the methods similar to polymer sol-gel synthesis [210]. The main synthetic approaches developed for production of hybrid nanocomposites and discussed in detail above, are the common methods for sol-gel mineralization,

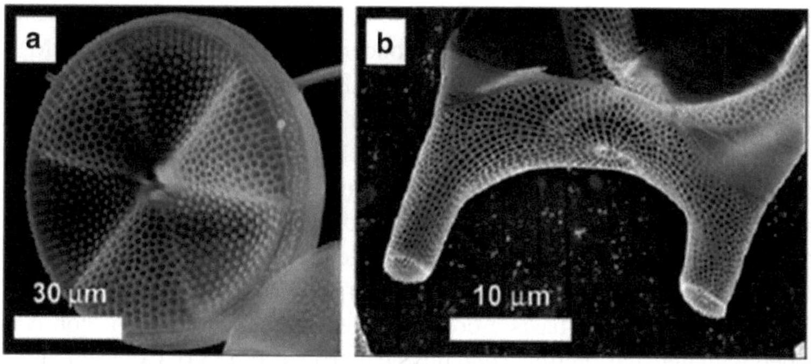

Fig. 7.12 Diatom valve view for *Actinoptychus spec.* (**a**) and *Eucampia zodiacus* (**b**) [226]

including mechanism of switching of biologically active macromolecules at the stage of formation of ceramics, glasses, and other inorganic composites (see, for example [211–214]). Many of these methods are taken from living nature and are used for realization of different bioprocesses, biosorption, biomineralization, during formation of natural composite materials with hybrid structure, for example, bone, boron-silicate glasses and diatoms, etc. [215]. Almost all metals take part in biomineralization, however, the best studied are the processes with participation of Si, Mg, Ca, Sr, Ba, and also Mn, Fe ions [216–220].

Laws of biomineralization processes are best of all studied for condensation of silicic acid in presence of water soluble polymers [149, 221–223]. Condensation of monomer silicic acid proceeds via intermediate formations of organic-silicon nanoparticles in presence of poly(allylamine) [224], poly-(1-vinylimidazol) [225]. In nature these processes happen under impact of biosiliconized organisms, diatoms and sponges, which produce more than 20 % of photo-synthetic oxygen. Diatoms are one-cell organisms with silicon exoframe, they can accumulate a great amount of silicon and use it as silica for structuring important elements of their organisms (Fig. 7.12) [226]. However, the main stages of structuring of silicon claps: capturing of silicic acid from environment, its storage in cytoplasm and transport of silica deposit to vesicle, formation of micro- and nanostructured video-specific claps are not understood. It is known that important role in biosynthesis of the claps play polymer ampholytes, silaffins, which are proteins with polyamine (3–20 nitrogen atoms) links and phosphate groups [227, 228].

Polyamines have been found in diatoms in free form, they were also extracted from silicon sponges, which stimulated studies of synthesis of new silica and composite materials and biomimetic modeling of processes in living nature.

Depending on initial configuration of polylysine (PLys) (α-spiral or β-sheet), silica nets are formed with different pore sizes consisting of aggregated spherical nanoparticles with diameter 200 nm. In the poly(allylaminehydrochloride)/phosphate – TMOS system silica also forms as a result of extraction of polymer phase in water solution [229–231].

In-cell concentration of soluble forms of silica in diatoms is from 19 to 340 mM, which exceeds significantly equilibrium concentration in solution (2–3 mM). Silica is bound with organic substances, most probably, in form of spherical particles in special silicon transport vesicles (STV), their size is about 30–40 nm [232, 233, 229]. However, it is not excluded that silicic acid presents in cytoplasm in form of soluble complexes of oligo-silicates with polymers, though it is unknown how high can be concentrations of silica in cytoplasm to do no harm to cells.

An organic matrix (template) controls nucleation processes, growth and formation of inorganic materials with perfect morphology during *bioconcentration – biomineralization*; on this basis a complex hierarchical structure of composites is developed with unusual chemical and physical properties [234]. Consequently, study of biomineralization includes solution (or, at first approach, at least understanding) of two problems: how strictly organized inorganic materials form (morphogenesis) and how these processes can be reproduced in biomimetic systems (morphosynthesis or mineralization *in situ*). There are five main stages of morphogenesis of diatoms [235–238]: formation of silica spheres of 30–50 nm in diameter (in STV); delivery of silica in STV to silicaleme and release of silica to SDV; beginning of silica deposition [239]; pore formation [240].

Understanding of morphogenesis of diatoms brought to discovery of new proteins, silaffins, which can take part in biomineralization process in SDV. Most probably, morphogenesis of porous silica structures with silaffines and polyamines has a mechanism similar to synthesis of mesoporous structures with usage of surface-active substances and block-copolymers [237, 241, 242]. Therefore, organic-silicon nanoparticles are a synthetic model of vesicles responsible for storage and transport of silica precursors in cytoplasm of diatoms, variations of composition of a stabilizing polymer and of conditions for further condensation provide a possibility for production of silica close to biogenic by composition.

Molecular recognition and molecular tectonics are very important aspects of biomineralization, however, genetic grounds of evolution of biomineralogic picture remain unknown, as is an answer to the principal question: how morphogenetic compatibility is implemented on the *biology-inorganic chemistry* interface. Most often this form of biominerals is predeterminated by spatial structure, as a result of some conformation or location of a cell. Rubber-like ormosil with dispersed Ca^{2+} ions can be obtained in sol-gel process, this composite can replace soft tissues. This relates to SiO_2-PMMA composites with incorporated Ca^{2+} ions, which show biological activity [243]. It is shown that silanol groups formed in ormosils are dominating as a factor, which controls biological activity, while the effect of dissolved Ca^{2+} ions is secondary. Moreover, in bioaggregation processes can be also included [244] such relatively complex particles as $BaTiO_3$, $SrTiO_3$, $NaNbO_3$, perovskites with ABO_3-type, synthetic analogues of which were considered in - Chapter 5. It is assumed [245] that fine mono-disperse precursors of high temperature superconducting ceramics, etc. can be obtained with the help of bacteria.

General problems of sol-gel synthesis applied to formation of organized matter include four approaches [246]: formation of self-aggregated organic matrices (transcription syntheses); cooperative assembling of ensembles, template and building blocks (synergetic syntheses); morphosynthesis, in which organized non-linear chemical neighboring, reaction fields (static, reconstructing, transiting), and their combination (integrated synthesis) are used for generation of models. This strategy (reaction ensemble → replication → metamorphism) is similar to general scheme of mineralization. It can be illustrated on examples of template-directed syntheses of ordered mesoforms and organic clays, micro-framework structures, also with usage of bacterial templates. Especially clear this appears in reproduction of hierarchic macro-structured ordered silicates, which can be shown on example of *Bacillus subtilis* multicell fibers as scaled organic templates [246].

7.3 Intercalation Processes in Development of "Green" Nanobiocomposites

Principle of action of many biosensors and transportation means is based on surface recognition of biosystems. Nanoparticles present a variety of places for selective binding with biomolecules. By development of this surface nanoparticles can be organized for surface recognition of biomolecules and cellular structures. The recent review [247] reports on progress in the field, which deals with interaction between nanoparticle and bimacromolecule (on the example of interaction with proteins, DNA) (Fig. 7.13).

Efficient and selective interactions with biomacromolecules depend on area of receptor contacts and dynamically organized structures with high binding

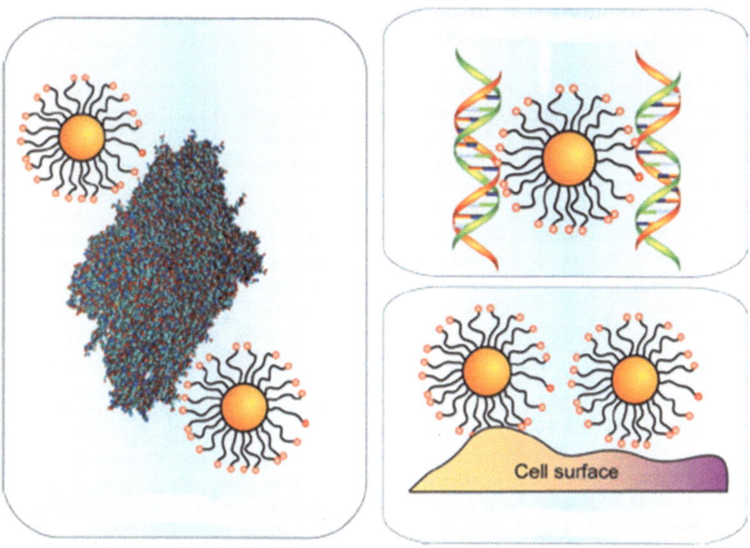

Fig. 7.13 Schematic presentation of nanoparticle interactions with proteins, DNA, and cells [247]

ability [248]. However, recognition of biomolecular structures is very difficult because of their large dimensions and complexity of the surface topology [249].

Among potential nanofillers, MMT nanoclays are widely used in producing of bionanocomposites due to their availability and deeply studied intercalation chemistry (see Chapter 5) [250]. A vast variety of modern hybrid and biohybrid materials based on clays, including those that contain living entities or their fragments, are produced. Bionanocomposites show not only improved structural characteristics, but also act as useful functional materials for ecological and biomedical purposes and other application areas. Thus, bio-objects and their components easily broaden interlayer distance of, for example, MMT (0.98 nm) to 1.10 nm (alcohols), polyethylene glycol ($M_w = 1{,}000$) to 1.11 nm, cellulose acetate butyrate to 1.13 nm, starch to 1.21 nm, glucose to 1.25 nm, etc.

In saponite L-DOPA zwitterions are vertical in interlayer distance as a monolayer of partially superimposed formations [251]. Besides, clays are ecologic and their antimicrobial properties are widely accepted in pharmaceutics, cosmetics, and food industry [252, 253]. Layered silicates from smectite family [254–258] or fiber clays, such as sepiolites are used as inorganic reinforcing components in new materials like native bone and tissues [259]. The ways of formation of biohybrid materials based on layered silicates are similar to synthetic polymers and can be presented on the example of hemoglobin intercalation in organically modified magadiite (Fig. 7.14). Preliminary intercalation of ammonium tetra-butyl widens interlayer gallery of clay (to 2.56 nm), thus facilitating access of a bulk protein molecule [140].

The most widespread types of intercalated polysaccharides, cells, microorganisms, proteins, and ferments are shown in Table 7.4.

Magnesium silicate modified by aminopropyl with intercalated DNA is exfoliated in water solutions due to protonation of amino groups with formation of

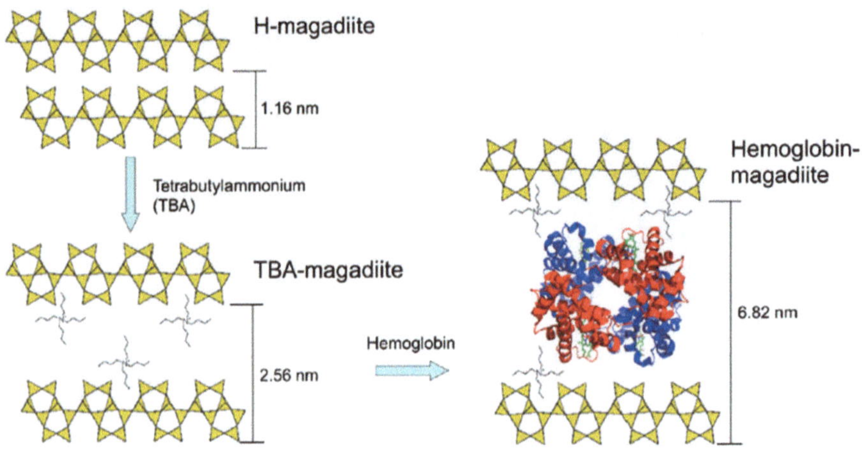

Fig. 7.14 Haemoglobin intercalated in organically modified magadiite [140]

Table 7.4 Intercallated biohybrids (polysaccharides, cells, microorganisms) for various applications [386–394]

Precursors	Polysaccharides, proteins, cells, microorganisms	Properties	Application	References
MMT, cloisite, caolin, hectorite	Starch, its derivatives	Barrier properties (for gases, water vapors), mechanical stability	Food packing	[386]
MMT	Chitosan	Adsorption and anion exchange properties	Anionic decontamination	[387]
Sepiolite	Chitosan	Anion exchange properties, mechanical stability	Potentiometric sensors	[388]
Wollastonite	Silk fibroin	Bioactivity, nontoxic for cells, enhanced mechanical properties, barrier properties (for gases, water vapors)	Tissue engineering	[389]
Cloisite-Na	Gelatin	Barrier properties (for gases, water vapors)	Food packing	[390]
Laponite	Polyphenol oxidase	Sensitivity towards citrus flavonoid, prolonged catalytic activity	Amperometric biosensors	[391]
Sepiolite	Lipasa	High stability, recycling	Fermentative reactor for biodiesel fuel	[392]
Bentonite	Algae *Ulva* sp.	High biomass loading, easy regeneration, reuse	Biosorbent for uranium (VI) recovery from water	[393]
Sepiolite	Influenza virus	Protection against antigen activity, increase the immune response	Intranasal or intramuscular vaccines	[394]

ordered meso-lamellar (a) and individual DNA molecules coated by super-thin disperse nanosheets formed due to exfoliation of the silicate by the scheme (Fig. 7.15) [260]:

However, intercalation of bio-structures is associated with some specific features. Thus, topography of light sensitive protein of bacteriorhodopsin in a purple membrane lied in 2D crystal and forming a layered composite during intercalation polymerization of methacrylate is analyzed *in situ* (Fig. 7.16) [261]. At that a polymer film can be subjected to additional cross-linking. This approach can be used for production of optically transparent biocomposites, because molecules of the purple membrane captured into the polymerized system are characterized by stability of structure and chemical composition and preserve their photochromic functionality.

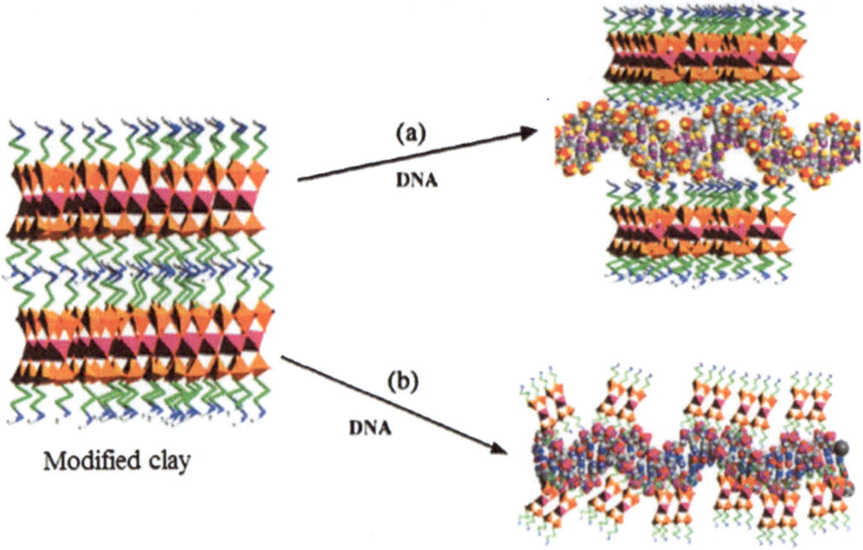

Fig. 7.15 Reassembly of aminopropyl-modified magnesium silicate organoclay layers by association with DNA leading to an ordered mesolamellar nanocomposite (**a**) or to an ultrathin organoclay covering on individual DNA molecules with dispersed nanosheets formed during exfoliation (**b**) [260]

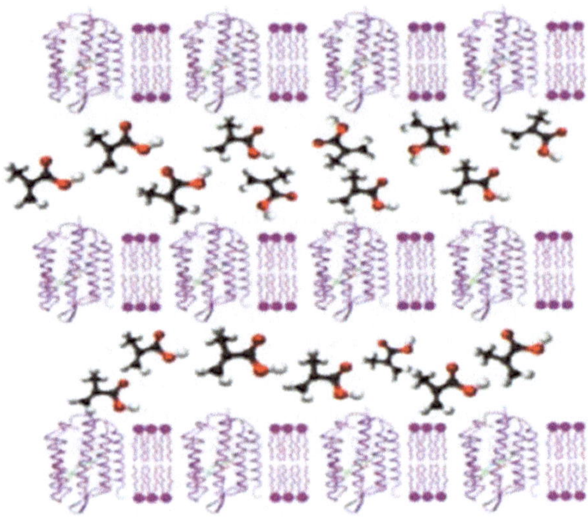

Fig. 7.16 A thin film consisting of an ordered lamellar stack of purple membrane sheets containing transmembrane bacteriorhodopsin protein molecules and intercalated monomer [261]

7.3.1 Biodegrading "Green" Bionanocomposites

One of the interesting research directions is production of biodegrading nanocomposites for preparing a frame for a tissue construction, therapy of periodontal bone defects and ridge augmentation, drug delivery (Fig. 7.17) [262].

Commercial starch, chitosan, cellulose, polylactide, poly(hydroxybutyrate) are used as biopolymers for thermoplastic matrices. The most often used biodegraded polymer for biomedical applications is poly(lactic) acid (PLA), because it can be obtained under controlled conditions, consequently, with predictable properties, such as elasticity modulus, ultimate strength, and rate of biodegradation [192, 195, 263–269]. Besides, PLA has high biocompatibility *in vivo* and ability to bone tissue regeneration (osteoconductivity) [270–272]. Stereoisomers and most interesting transformations of lactic acid including polymerization are shown in the scheme (Fig. 7.18).

Though PLA has good mechanical and physical properties, including biocompatibility and biodegradation, it does not have some fundamental characteristics of biomaterials. For example, in tissue engineering PLA not only should stimulate and support tissue growth, but in the process the growth and degradation rates of the tissue should be correlated. To some degree these properties can be controlled by OMMT introduction (Fig. 7.19) [262].

Nanobiocomposites of the PLA containing 3–5 % of organically modified MMT (OMMT) are obtained by mixing in a melt in two different mixers, mini-twin-rotor extruder, and in a mixer with inner dispensers bringing to different degree of dispersion, characterizing rheology of the melt. The obtained PLA/OMMT samples had percolation nets in the melt [273].

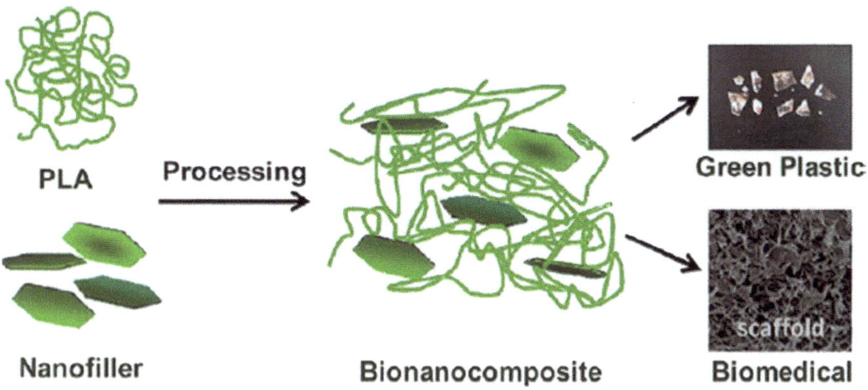

Fig. 7.17 A general scheme of synthesis of bionanocomposites based on biodegradable polymers [262]

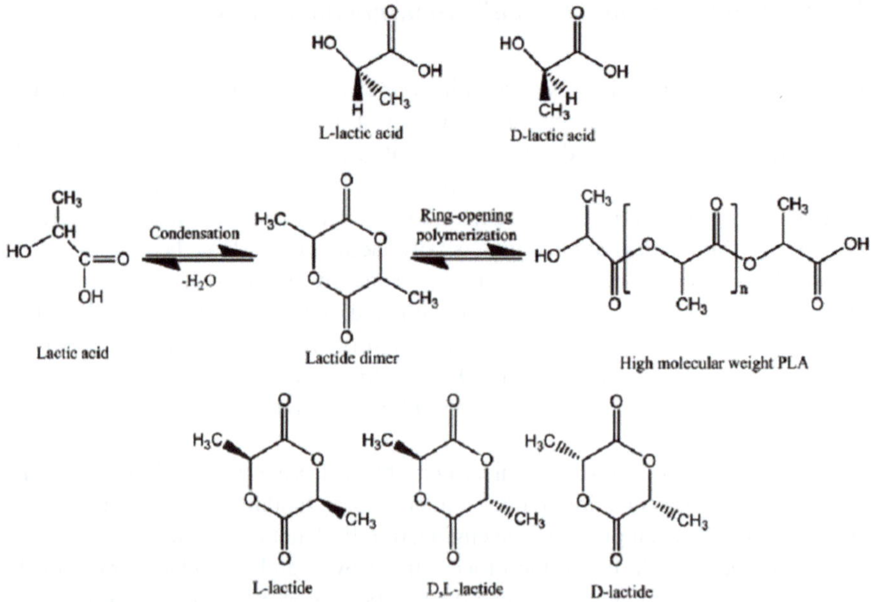

Fig. 7.18 Stereoisomers of lactic acid, and the schemes of lactic dimer, high-molecular-weight PLA and diastereoisometric forms of lactides

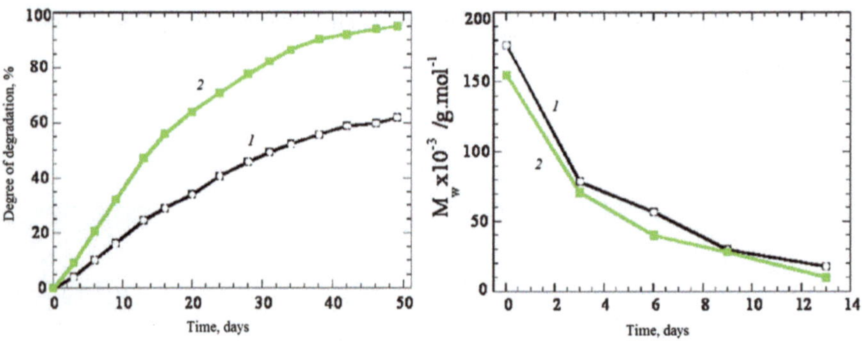

Fig. 7.19 Biodegradation ability of neat PLA (*1*) and biocomposite modified with 4 % OMMT (*2*) [262]

The main mechanism of PLA degradation includes alcoholysis/acidolysis inside, and molecular ether bond of polylactide ether groups in a chain [274]:

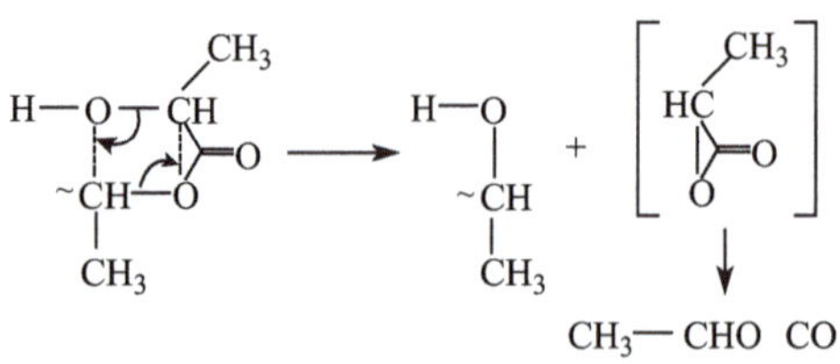

Traces of metals, which can present in a polymer (for example, like impurities or remnants of catalysts), such as Sn, Zn, Al, and Fe catalyze reactions of in/intermolecular alcoholysis/acidolysis and depolymerization at high temperatures: the more selective is a catalyst, the less efficient is its depolymerization effect at high temperatures. Though this concerns, mostly PLA, the similar behavior can be expected for other biopolymers [275–277]. Decomposition of plastified PLA filled with Na-MMT in nitrogen shows that an increase in a filler content causes a significant decrease of thermal stability: temperature of the maximal decomposition rate (according to TGA data) decreases continuously from 370 to 325 °C with increase in MMT concentration from 1 to 10 wt% [278, 279]. Biocomposites based on PLA also have flame-retardant properties [280, 281]. Modifying impact of organically modified clays on different physical and biological properties appears also in other "green" nanobiocomposites obtained from renewable sources, for example, polyethers: polyhydroxyalkanoates (PHAs), poly(3-hydroxobutyrate), copolymer poly(3-hydroxobutyrate-co-70 % 4-hydroxobutyrate [P(3HB-co-70 % 4HB)], etc. [253]. These materials are future candidates to biodegrading composites, they attract keen attention for medical applications, because they have properties which are not typical for known synthetic polymers, such as biocompatibility and biodegradation [282]. Mostly, lactides were used for studying of flame-protective properties of plastics from renewable sources: clay/polyether bionanocomposites [283, 284].

For qualitative measurements of antimicrobial activity of the nanocomposites gram-positive (*Staphylococcus aureus*) and gram-negative (*E. coli*) bacteria were used and antimicrobial properties of different PLA composites were studied using the disc-diffusion method by diameter of composite-inhibited zone against these bacteria. The poly(3-hydroxybutyrate) copolymer (P(3HB-*co*-4HB)) has shown a discernible microbial-inhibiting zone related to incorporated nano-clays. However, other poly(hydroxyalkanoate) composites did not show any activity even being incorporated in clays, maybe due to their special morphology. Most probably, antimicrobial properties of biocomposites reinforced by modified clay can be caused by antimicrobial activity of their substitutes, quaternary ammonium groups. These groups carrying alkyl chains can destroy bacterial cell membranes suppressing their metabolic activity and causing lysis of a bacterial cell with time [252]. Increase in clay concentration appears in stronger antimicrobial activity of the composites. According to [285], higher specific surface can be reached by increase in clay concentration on which more bacteria can be adsorbed. Stability of gram-negative bacteria against the composites can be explained by special features of the bacterial cellular structures. Biocomposites based on P(3HB-*co*-4HB) will be applied in medicine and pharmaceutics due to their biodegradation and biocompatibility. Thus, it seems promising usage of P(3HB-co-4HB) nanocomposites in view of *in vitro* decomposition for biomedical applications.

Laminar clays, such as kaolin contain 2D lamellas bound via intercalation (or intermediary) layer. Lamellas should, first of all, exfoliate or chemically separate; they work as nanofillers and should disperse in a polymer matrix. Organic clays based on smectites and less on sepiolites were intensely studied for production of pesticide composites, which diminish loss of bioactivity caused by volatility or photodegradation of insecticides and herbicides bound with organic clays. Another important application is removal of pollutions, in particular, those having organophilic properties.

New tends in syntheses based on organic clays are related to usage of non-toxic modifiers of biological origin, for example, lecithin or various biopolymers instead of traditional quaternary ammonium surfactants. This approach is applied not only for environmental protection, but also for biomedical purposes. Usage of these nanostructural hybrids can be practically important for development of frames for tissue engineering or as adjuvants for vaccines. The future abilities of natural two-layer aluminum silicate halloysite (with clay nanotubes) mean a possibility of filling halloysite nanotubes with active ingredients, which will provide their usage in cosmetics, scent disguise, agriculture, medicine and other areas.

As is known, synthetic apatites have high affinity to host tissues and biological activity, they bring to increase in compatibility of materials due to their chemical and structural similarity to the natural apatite crystals. Nanometer apatite crystals in mineral basis of bone tissue provide higher metabolic activity than synthetic apatites. Synthesis of apatite crystals is well described with usage of vast variety of methods, including solid phase chemical reactions, for example, mechanical-chemical from the mixture $Ca(OH)_2$-P_2O_5 and CaO-$Ca(OH)_2$-P_2O_5 CaO and $CaHPO_4$ [286–290].

We shall consider a composite consisting of gelatin-poly-D, L-lactide-hydroxy-apatite nanofiber made *in situ* by hydrothermal mineralization [291, 292]. Nanofibers of hydroxyapatite (HA) are uniformly distributed in gel in a polymer matrix. There are ion interactions between Ca^{2+} HA ions and negatively charged gel functional groups in the nanocomposite. Besides, HA plays important role of a bridge binding the polymer gel and poly-, D,L-lactide in the nanocomposite, due to which dense net forms. When the polymer is removed from the composite, pure HA crystals remain. The Fig. 7.20 shows three-phase model linking two polymer chains with inorganic

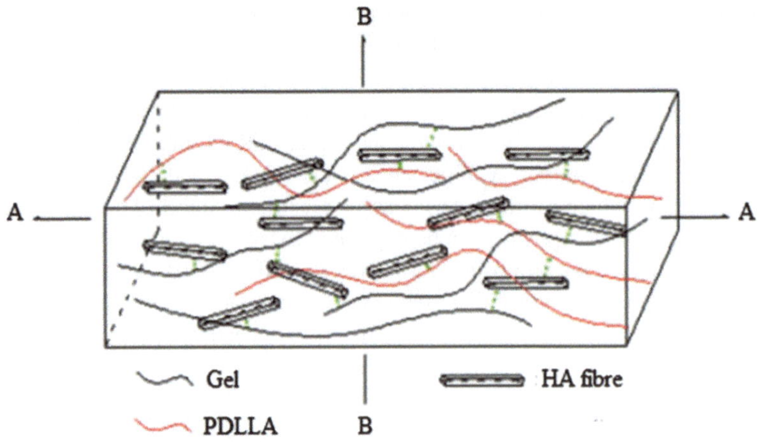

Fig. 7.20 The schematic model of interactional bonding between Ca^{2+} in HA nanofibers and negative charged functional groups in gel and PDLLA molecular chains [292]

nanofiber based on experimental results and explaining improvement of mechanical properties of the composite [293, 294].

Colloid apatite nanospheres of 2–5 nm in diameter are obtained in presence of PAA polyelectrolyte used as structurally directing agent for synthesis of calcium-deficient apatite. PAA-superpressant is a substance supressing nucleation and growth of apatite crystals during *in situ* wet chemical synthesis, widely used for production of micro and nanocrystalline apatite [295, 296] and biometric formation of its particles [297, 298].

We shall also highlight glass filler with bioactive composition with average size 320 nm generated by electrospinning process. Nanofibers with different concentrations (to 35 %) are uniformly distributed in PLA solution. The following drying and thermal pressing were used to produce dense nanocomposites, which induce fast formation of artificial hydroxycarbonate apatite layer from physiological solution. As nanofiber concentration increased from 5 to 25 %, bioactivity *in vitro* improved under artificial conditions (osteoblast response was estimated in terms of cell growth, differentiation (fitness) and mineralization).

The purpose of the future investigations in this direction is development of nanostructured, multifunctional and bio-induced hybrid materials including usage of new synthesis methods, giving of predetermined physical and mechanical properties in combination with calculated and other theoretical studies.

7.3.2 Nanocomposites Based on Polysaccharides

Polysaccharides are polymer hydrocarbon structures consisting of repeated links bound by glycoside bond, which present one of the most widely spread group of natural polymers (cellulose, starch, dextran, and chitin). Their usage in development of nanostructured hybrid materials has recently increased, because natural polysaccharides can be promising and available substitutes for non-degrading polymers due to their biodegradation, biocompatibility and availability. Hybrid nanocomposites based on polysaccharides are used for immobilizing of ferments without loss of their activity and increase in stability [233, 299, 300].

Nanocomposite based on natural montmorillonite (MMT – Cloisite) and pullulan (exopolysaccharide of microbial origin, Mn @200,000) is approbated as a suitable candidate for replacement of synthetic polymers and oxygen-barrier coatings and for production of highly wetted surfaces of new coatings [301]. In particular, oxygen barrier and wettability of US sounded pullulan composite, as well as of bionanocomposite coatings based on it, depends on volume concentration of a filler and is compatible with theoretically predicted (Table 7.5).

Cellulose consisting of repeated links of *d*-glucose structural blocks is a highly functioning, linear, rigid-linked homopolymer characterized by its hydrophility, chirality, biodegradation, wide possibilities of chemical modification, and formation of various semi-crystalline fiber morphology. According to the data of review [302], nanometer cellulose fiber materials (i.e. microfiber and bacterial cellulose) are

Table 7.5 Volume fraction of filler and oxygen permeability coefficients of coated PET and bionanocomposite coatings [301]

	$P'O_2$ [$mL \cdot \mu m \cdot m^{-2} \cdot (24\ h)^{-1} \cdot atm^{-1}$]	
Filler volume fraction	Total	Coating
0.017	883.24	142.32
0.046	659.76	83.25
0.073	476.06	50.98
0.098	332.15	31.79
0.123	228.01	20.27
0.145	163.66	13.93
0.167	139.08	11.65
0.188	154.29	13.06
0.207	209.28	18.37

promising for production of bionanocomposites, because cellulose is abundant, has high strength and hardness, light weight and is characterized by biodegradation. High-strength nanocomposites are produced [303] using sheets of bacterial cellulose, impregnated by phenol resin and compressed at 100 MPa. Amiloid fibers are one of the widespread important self-assembled molecular nanobiomaterials (nanowires, layered materials, gels, etc.) obtained by the bottom-up strategy [304].

Many companies produce and use biocomposites beginning from wood powder to vegetable fibers (such as *Ananase rectifolius*, *Cocos nucifera* L., *Agave* sp. and *Corchorus* sp.), for example, as components of seats, doors, panels, or in interior parts or in car boots: in Mercedes-BenzA-Class car there are 27 components consisting of vegetable fibers [305]. Information about synthesis, structure and properties of these modified natural materials is very expansive (see, for example, [306–308]). We shall just note that nanocellulose, nanostructural materials (such as halloysite clay nanotubes, modified betonites and montmorillonites) are examples of commercialized organic-inorganic hybrid bionanomaterials [309, 310]. Nanocomposites based on cellulose acetate, commercial organic clay (Cloisite30B), triethyl-acetate with different contents of antimicrobial agents (thymol and cinnamaldehyde) are cast from solution. Antimicrobial activity depends on content of ether oils in a nanocomposite, significant impact on antimicrobial activity also has organic clay [311]. Significant plasticizing effect was observed on thymol and cinnamaldehyde in cellulose nanocomposite. At last, active nanocomposites have shown a significant antimicrobial activity with respect to L. innocua, which was higher for the nanocomposite containing cinnamaldehyde, than in that containing thymol. Besides, the nanocomposite containing thymol showed higher antimicrobial activity than acetate cellulose films without nanofiller. Presence of a surfactant, organic clay, can contribute to increase in antimicrobial activity of cellulose acetate, propionate and butyrate. Cellulose acetate is most interesting for its biodegradation in combination with high optical transparency and high hardness, and with ability to obtain cast films [312].

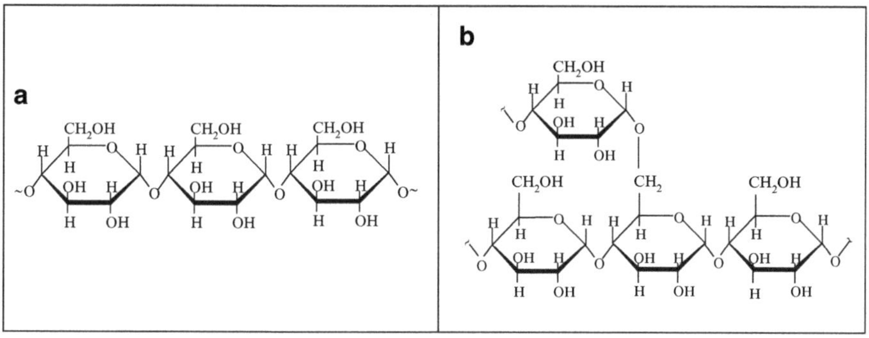

Fig. 7.21 Chemical structure of (**a**) amylose, (**b**) amylopectin

Nanofibers of bacterial cellulose (BC) are used as strong biotemplates for simple one-staged production of new nanocomposites: Au—bacterial cellulose fiber. The BC nanofibers are uniformly coated with Au nanoparticles from water suspension with usage of poly(ethylenimine) as reducing and binding by hydrogen bonds agent [313]. The possible mechanism of formation of Au-BC nanocomposites with different thickness of Au shell is proposed. The Au-BC nanocomposites are promising carriers for immobilizing ferments and for production of biosensors.

Starch grains are partially crystalline and consist mainly of two polysaccharides, glucopyranose homopolymers: (a) amylose and (b) amylopectin (Fig. 7.21). Processes of starch conversions are: acid hydrolysis, oxidization, dextrinization or pyro-conversion, and fermentative hydrolysis.

Into interplanar distance of MMT only linear polymer, amilose, can penetrate, oppositely to huge globules of branched amylopectin. At that increment of tearing load and edge wetting angle is higher for exfoliated than for intercalated nanocomposite, which is caused with higher degree of interaction of biomacromolecules with MMT particles in it (Fig. 7.22) [314–317].

Nanocomposites based on starch plasticized with glycerin are obtained during intense stirring using MMT natural smectite clay, kaolinite, hectorite or modified (by quaternary amines of fat acids of hydrated tall oil) hectorite. In all cases clay additions increase Young modulus and shear modulus. MMT and non-modified hectorite provide far higher increase in these parameters, than do kaolinite and treated hectorite, whose particles have lower aspect ratio and higher specific surface than MMT. Impact of a type of mineral on mechanical properties of a composite shows that highly hydrophylic starch molecules cannot interact with clay, and its particles cannot disperse properly [318–321].

Young modulus increases linearly with increase in intergallery space (Fig. 7.23).

Great gallery space allows large starch molecules to diffuse between layers and thus promotes more interphase interactions, which causes more intense strengthening effect [318].

Dynamic rheological properties of starch mixtures depend on their types (wheat, potato, and wax starch grains), and on origin of clays modified by different Cloisite -C: CNa^{+}, C30B, C10A and C15A used during gelatining of starch

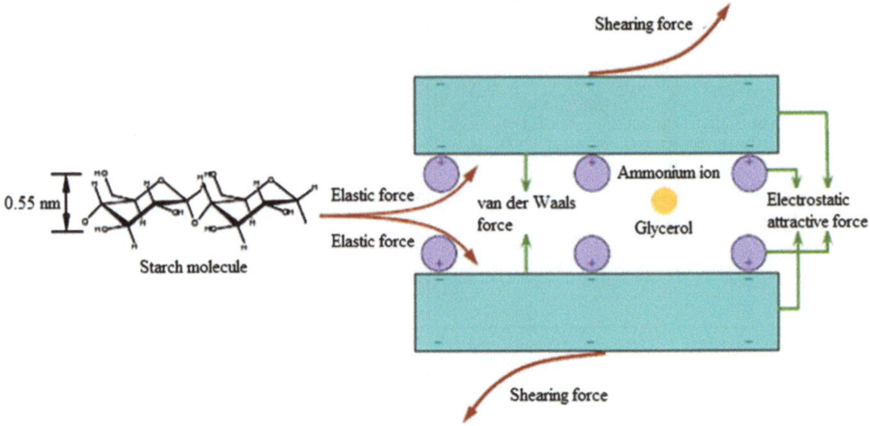

Fig. 7.22 Scheme of intercalation in the starch/clay nanocomposite [319]

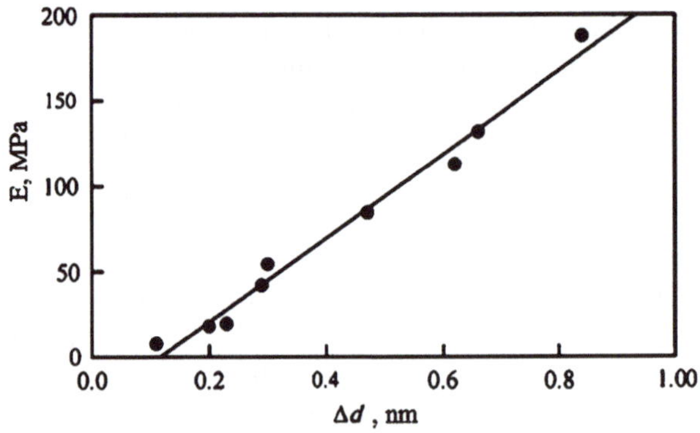

Fig. 7.23 Correlation between Young’s modulus and gallery spacing (Δd) of nanocomposites with 5 wt% of various clays [318]

Table 7.6 Mechanical properties of different starches and clays [323]

Sample	Young modulus, MPa	Stress at peak, MPa	Strain at break, %
Wheat starch	28.3 ± 1.8	2.24 ± 0.04	31.7 ± 1.5
Wheat starch + 3 wt% MMT-Na	35.6 ± 0.6	2.32 ± 0.08	27.3 ± 0.6
Wheat starch + 6 wt% MMT-Na	39.2 ± 1.4	1.90 ± 0.06	21.0 ± 0.8
Wheat starch + 3 wt% SEP-Na-sepiolite	45.3 ± 1.8	2.91 ± 0.06	36.5 ± 2.1
Wheat starch + 6 wt% Na-sepiolite	67.3 ± 2.3	2.99 ± 0.04	31.0 ± 1.0

[322]. The composite obtained by combination of wheat starch and CNa + (heated to 95 °C) has shown the highest modulus [323] (Table 7.6, Fig. 7.24). This behavior is caused by two factors: interaction between clay and amilose destroys during gelatining, and gel-like material forms. However, position of diffraction maxima does not change, which points to the fact that the clay plates remains in a pack

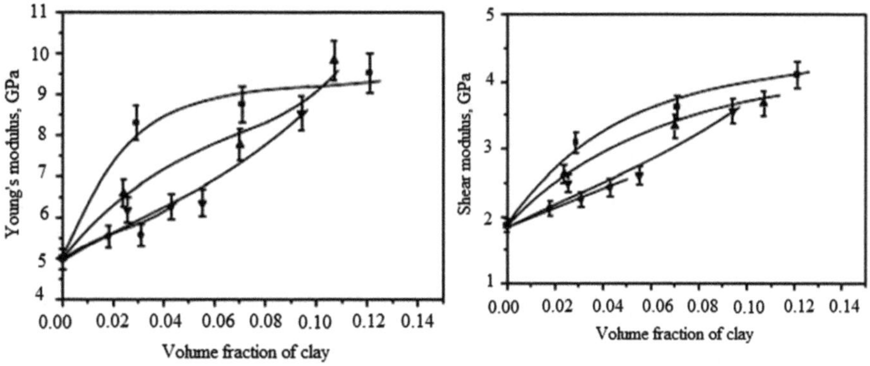

Fig. 7.24 Dependence of physical-mechanical properties of starch/clay composite on the fraction of clay (■ – starch/MMT; ▲ – starch/nonmodified hectorite; • – starch/modified hectorite; ▼ – starch/caolinite) [324]

Fig. 7.25 Diffractograms of: (*1*) MMT, (*2*) starch/MMT nanocomposites [326]

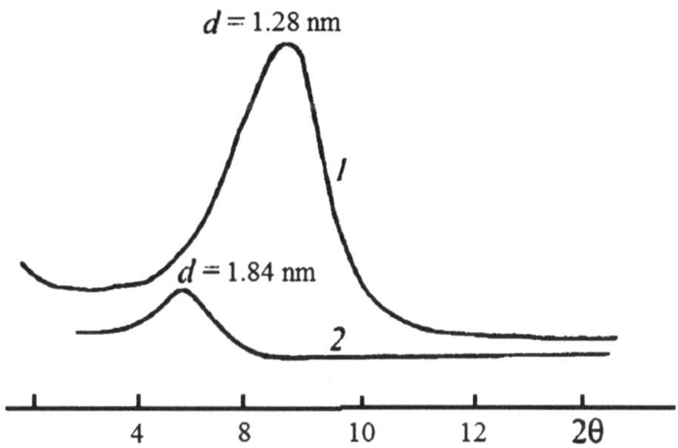

configuration, and interaction with starch molecules is only at the surface level [324].

In turn, the starch/clay mixtures are prepared in different ways. However, the most important procedures (casting, intense stirring, extrusion) relate to two different mechanisms of clay and starch mixing [317], as in synthetic polymers. Most often for this is used an extruder, since it provides a combination of shear and tensile flows. However, this instrumentation in combination with calendaring or blow molding can bring additional tension during extrusion. By this way clay nanoparticles can also be oriented in a certain direction [325].

Under conditions of intense mechanical impact in rotor-pulsation device nanocomposites are obtained based on maize starch and Na-montmorillonite (biopolymer:glycerine:MMT = 69:27.5:3.5) [326]. Interlayer distance is expanded from 1.28 to 1.84 nm (Fig. 7.25). This provides conditions for decomposition of the crystal structure of the clay mineral and formation of single plates (paste dispersion) and its distribution in biopolymer.

Aspect ratio of clay is also important factor for formation of mechanical properties. In order to reach percolation level, higher aspect ratio is needed, lower critical lengths and lower concentration of nanoparticles. Moreover, the higher degree of interlamellar exfoliation (*d*), the higher is elasticity modulus. A degree of intercalation/exfoliation depends on chemical modification of clay and its compatibility with starch molecules.

Chitosan (Ch) is multifunctional polymer based on polysaccharides including active hydroxyl and highly reactive amino groups. It is considered as optimal starting material for absorption [327–329]. Scheme of chitosan integration between layers of oxidized graphite can be presented as (Scheme 7.3):

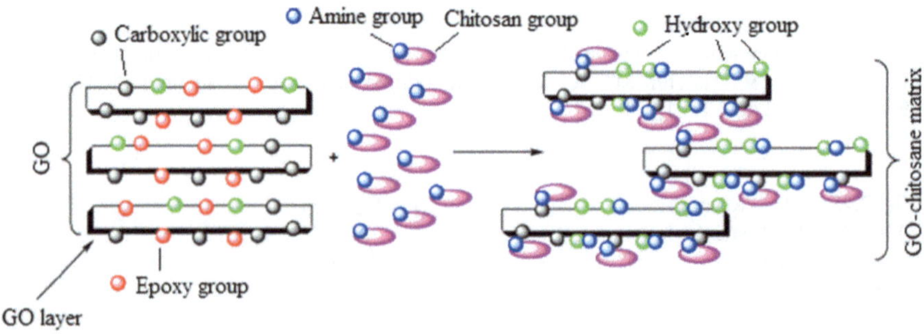

Scheme 7.3 Intercalation of chitosan into interlayer space of graphite oxide [329]

We do not consider here a range of nanocomposites based on other polysaccharides, including those of dextran range (reserve polysaccharides of yeast and bacteria formed by remains of glucose are used as substitutes for blood plasma), etc.

7.3.3 Biomedical Application of Hybrid Nanocomposites

The considered nanocomposites are important in medicine (drugs), genetic engineering (DNA, RNA), biotechnologies, (proteins, individual cells), and in food industry [330–333].

Lipidic bi-layer vesicles are well known materials, which are intensely used as supramolecular ensembles for structuring molecular devices, widely known are liposomes coated with ceramics, so called "cerasomes" [334]. Cerasomes are novel organic-inorganic hybrids consisting of liposomal membrane with ceramic surface. Cerasomes were obtained from organoalkoxysilane pro-amphiphiles (**1** and **2** in Fig. 7.26) under conditions of sol-gel reaction [335]. Diameter of nanoparticles is 70—300 nm and 20—100 nm for **1** and **2**, respectively (see above Fig. 7.26). The results show that neither cerasome-plasmid complex responsible for transfection (~70 nm), nor cerasomes without DNA (60–70 nm) are toxic. Different additional functionalities (magnetic, luminescent, sensor) can be assigned to a capsule shell by integration of nano-Fe_3O_4, q-CdS, etc. [336, 337].

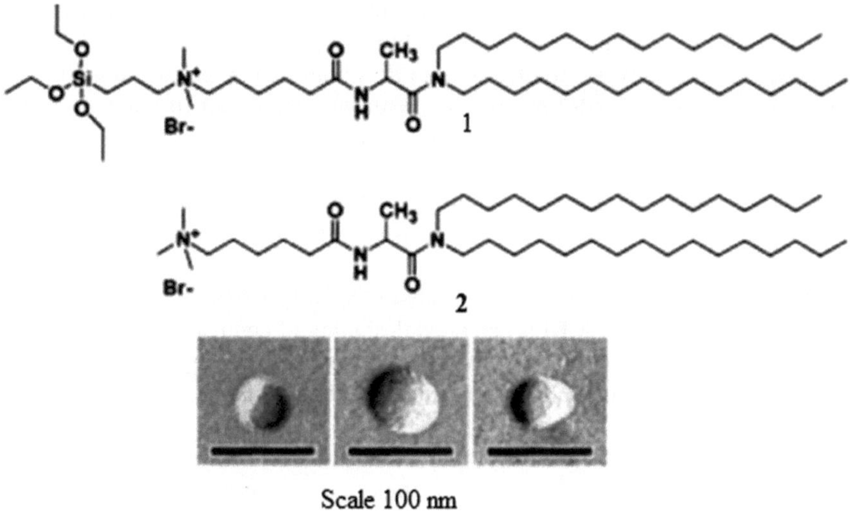

Fig. 7.26 Structure of lipids **1** and **2** and TEM microphotographs of liposomes formed with lipid **1** in water at 12.5 μM

Magnetic resonance tomography, hyperthermia, controlled drug delivery [129, 338], regeneration of bone tissue, prosthodontic treatment are accessed with usage of additional specific chemical agents: Fe_2O_3 magnetite for tomography and hyperthermia, functioning of organic silicates (binding agents) for drug delivery, for biomodifiers. One of promising applications of bionanocomposites is clinical Nuclear Magnetic Resonance Imaging (MRI). Antineoplastic chemotherapeutical drugs (chemical, biological, genetic, radiological) cause a metabolic imbalance of nucleic acids, impede biosynthesis processes and DNA functions, suppress haemopoiesis, digestion, are cardio- neuro- and nephro-toxic. The problem of development of the target drug delivery systems to a tumor cells (see Sect. 7.2.2), including magnetic carriers, is very urgent. Fixation and deposition of drugs in a tumor zone allows a significant decrease in therapeutic dose of a medicine and reduction to minimum of toxic-allergic reactions of an organism [339, 340] as compared to non-target delivery. To the methods of passive immune therapy in oncology belongs also usage of monoclonal antibodies, whose efficiency is comparable with chemotherapy at lower level of toxicity. A special place has development of magnetically controlled drugs of chemotherapeutical, diagnostic and hyperthermal activity. Two mechanisms are known by which a cell becomes multidrugresistant: increase in drugeffluxpumps through a cell membrane and increase in anti-apoptosis ways. Usage of nanotechnologies for TDDS development allowed researchers to overcome limitations of antineoplastic drugs due to increase in solubility of drugs and decrease in toxicity for healthy cells. Encapsulation of drugs in nanoparticles makes it possible to avoid drugefflux thus increasing intercellular concentration of a drug. SiRNA (small interfering RNA participating in suppression of genetic expression at the stage of transcription and translation) can disturb work of signal cellular paths blocking genetic expression and inhibiting anti-apoptosis ways with respect to tumor therapy. Usage of nano-carriers for delivery of siRNA simplifies both kidney purification and degradation of protected siRNA chains, increasing their half-lifetime in blood. Co-delivery of drugs and siRNA together in one system can be more efficient in overcoming of cancer cell resistance, than is treatment of cancer cells by separate systems carrying either siRNA or drugs. Progress in nanometer systems of co-delivery in overcoming of cellular resistance to many drugs is analyzed in a latest review [341].

Clinical implementation of tomography in the early 1980s revolutionized diagnostics. Development of MRI accelerated designing of contrast agents. Characteristics of contrast agents and methods of their detection widely vary [342]. Thus, dendrimers are involved in biomedical studies as diverse platform for production of nanomaterials with required properties [343]. Hybrid superparamagnetic nanocomposites are used in tomography for localization and diagnostics of brain and cardiac infarction, liver damages or tumors, where nanoparticles have a tendency to accumulation in high concentrations due to a difference in tissue compositions and/or endocytosis and channel processes [344]. For MRI usually superparamagnetic contrast agents are used, which consist of nanoparticles having a maghemite/magnetite core, encapsulated into polysaccharides of dextran family with trademark *superparamagnetic iron oxides* (SPIOs), available at market. Colloid nanocomposites are also known as ultrafine superparamagnetic iron oxides (USPIO) due to extremely small hydrodynamic sizes (20–30 nm) coated with dextran, trademark Sinerem®. Significant efforts are made for development of new synthetic ways of development of contrasting agents with improved properties. For example, biocompatible hybrid magnetic dispersions are obtained from Fe nanoparticles (12 nm) by continuous laser pyrolysis of $Fe(CO)_5$ vapor [345]. A new one-pot way for production of USPIO contrasting agents by thermal decomposition of iron (III) precursors has been developed [346, 347] involving magnetic nanoparticles (10 nm) with covalent coating of polyethylene glycol modified by terminal monocarboxylic groups. Hydrodynamic size of this particle is 20 nm. MRI experiments performed on rats have shown that these particles have good biocompatibility and can potentially be used as contrasting agents. Also hybrid magnetic nanoparticles were tested for MRI [348], which were obtained by thermal decomposition of iron (III) acetylacetonate in hot organic solution and then modified by 2,3-dimercaptosuccinic acid (particle sizes vary from 4 to 12 nm). Moreover, this material with conjugated ligand of a shell has shown a perfect selectivity in cancer diagnostics with tumor selective antibodies Herceptin. Magnetic particles encapsulated in liposomes (magnetoliposomes[6]) [349] are used as MRI contrasting agents ([350] and references therein). We shall consider

[6] Advantage of magnetic liposomes as compared to USPIOs is in that various biomedical functions can be provided by conjugation with biological ligands [349].

some reviews on composites based on magnetic nanoparticles,[7] [351–358] concerning their synthesis, properties, functioning and applications, including biology [359–362]. As an example we shall describe synthesis of magnetic chitosan (Chm). Pure chitosan, 2 g, was dissolved in 400 ml of acetic acid solution. Than 0.75 g of magnetic particles were added of the chitosan solution and the mixture for subjected to ultrasonic treatment for 30 min. Then glutaraldehyde was added (as linking agent) to the mixture solution for chitosan linking, in order to prevent high degree of chitosan swelling in water solution. Adsorption of synthetic color by adsorbents is considered as a simple and economic method of its elimination from water and water deposits.

Magnetic Hyperthermia is a form of local hyperthermia with a purpose to heat very local part of a body. More than 50 years ago local magnetic hyperthermia with usage of fine magnetic particles was described [363]. Being exposed to alternate magnetic field these particles can act as a local source of heat in some place of a human body, a target.

A response of magnetite nanoparticles dispersed is compared in two different media: water solutions and 2 % agarose gel modeling bone-like thermal properties [364, 365]. A significant decrease in thermal effect was observed in the case of agarose gel, because Brownian motion in it is impeded. Thermal effect of Fe_2O_3 -SiO_2 and non-capsulated Fe_2O_3, immersed in agarose is similar for both cases, which points to preserving of magnetic properties of magnemite nanoparticles at encapsulation into mesoporous microspheres.

In the recent years a hope appears to realizing of efficient method of medical treatment of cancer. Heating of magnetic oxide particles with low electric conductivity in external alternate magnetic field proceeds due to loss of paired spins in a particle during re-magnetization, or due to friction losses if particles rotate in a medium with a certain viscosity. Induction heating of magnetic oxide particles (by eddy current) is insignificant due to low electric conductivity [366]. Acceptability of magnetic nanoparticles coated with dextran for hyperthermia of mouth cavity [367] is tested in combination with genetic therapy and hyperthermia with usage of cation liposomes, i.e. liposomes filled with magnetic nanoparticles [368]. Hybrid nanocomposites with bimodal antineoplastic functionality are prepared [369] on the basis of iron and porphyrin oxides nanoparticles and are active in combined medical treatment by photodynamic therapy and hyperthermia.

A significant role in structuring of ensembles based on biomodified nanoparticles play specific bimolecular interactions [370–373]. Thus, DNA can be used for assembling of gold nanoparticles in dimers, trimers or higher

[7] In the last decade technique of preparation of magnetic microspheres has been developed and optimized, including *in situ* formation of core-shell structures, different types of emulsion polymerization, linking, etc. Most often low-dimensional magnetic particles are coated during suspension polymerization. However, these particles have a wide distribution by shape and sizes of magnetic fractions. Commercial magnetic microspheres are obtained by deposition of magnetic nanoparticles into porous polymeric latex and insulating them by a polymer layer. Though this method is laborious and time consuming, the obtained nanocomposites have high homogeneity and high saturation magnetization, and they meet biotechnological demands [351–358].

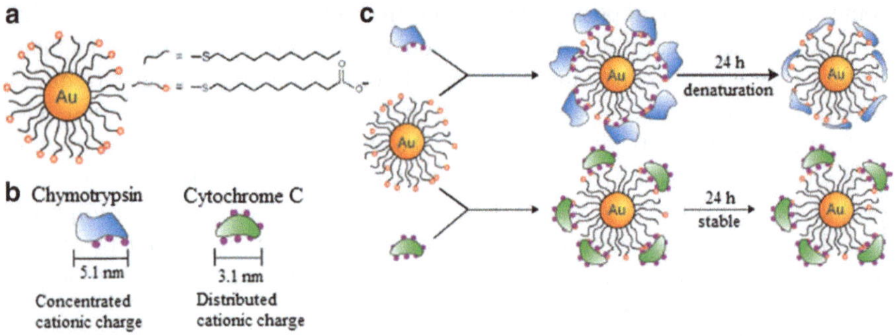

Fig. 7.27 (**a**) Mixed monolayer gold nanoparticles featuring a hydrophobic interior with carboxylate end groups. (**b**) Schematic depiction of protein electrostatic surfaces (**c**) Protein–particle assembly of gold nanoparticles with chymotrypsin and cytochrome C [374]

aggregates. Similarly, gold nanoparticles with chemosorbed antibodies [372, 373] or disulfide biotin analogues [389] can be linked by introduction of bivalent antigens and streptavidin, respectively, for formation of aggregated structures of nanoparticles.

Ensemble of nanoparticles provides access to spatial-dependent modulation of optical, electronic and magnetic properties of the ensembles for biological applications. For example, nanoparticle-protein ensemble can be used for control of interparticle space [374] and morphology by choice of respective protein size, its shape and charge. Thus, two types of proteins, cytochrome C and chymotrypsin (ChT) used for structuring composites (Fig. 7.27) show different behaviors on the surface of carboxylate functioned Au nanoparticles. ChT is developed on surface of a nanoparticle and works as linear polymer, while CC preserves its native conformation on the surface of the nanoparticle, this behavior remains in a solid state (Fig. 7.27c).

Brief analysis of the problem explains interest to action of molecular mechanisms in bionanocomposites expressed not only by biologists, but also chemists, specialists in synthesis of novel materials. A possibility of using organic-inorganic nanoparticles for synthesis of new composite biomaterials of different morphologies has been studied in detail. Stabilization of the obtained permolecular structures is realized due to multiple hydrogen, and often donor-acceptor bonds with participation of surface groups and donor atoms of organic polymers. Encapsulated ceramic nanoparticles can be used as biosensor devices and for development of vaccines, etc. The most important sides of this problem concern biomineralization of mixed-valent poly-cored structures and clusters in biology (especially iron-oxo, molybdenum-oxo, and manganese-oxo structures), the way of small molecules activation with their participation, biosorption, environment control, biomedicine [375]. Special attention of researchers is focused on biosensors with optical, electrochemical and magnetic detection systems. Magnetic properties are used in nuclear magnetic resonance and hyperthermia. Different kinds of nanocapsules are applied for drug delivery. In our opinion, usage of hybrid nanoparticles and

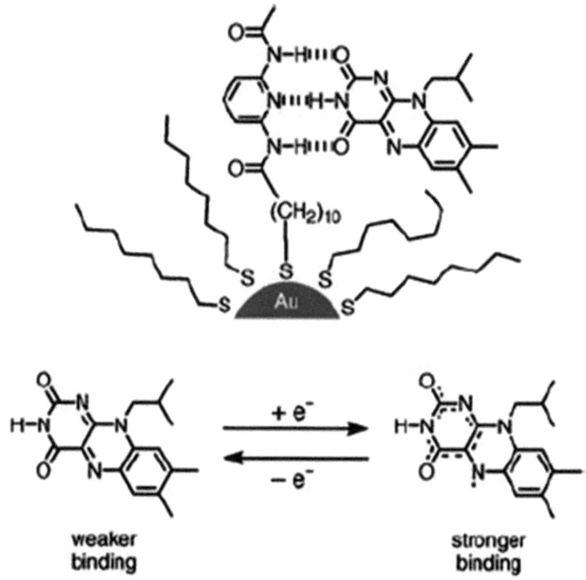

Fig. 7.28 The electrochemically controlled recognition of flavin by a pyridinediamide-functionalized nanoparticle [378]

nanocomposites in biocatalytic processes, one of mainstreams of development of bionanocomposites.

Graphene is the newest uni-atomic 2D graphite carbon system, ascending star in materials science and physics of condensed matters [376]. It was noted on the example of chitosan immobilizing in interlayer distance of low-layered graphene oxide, that this material is promising also for biological application.

Combination of metal nanoparticles with biomolecules can provide electronic or optical transformation of a biological phenomenon for development of new biosensors [377].

Due to ability of biomaterials to complementary recognition, functioning of nanoparticles with biomolecules can bring to specific interaction nanoparticle—biomolecule, i.e. to self-assembling and complex architecture. Variation of chemical properties of a biomolecule can be used for control over interaction of a modified nanoparticle with environment. Thus, a break of a bind between a bioorganic molecule and functioned surface of a gold nanoparticle can be electrochemically controlled, because during electric reduction of a derivative flavine more stable hydrogen binds are created (Fig. 7.28) [378].

Development of integrated materials based on biomolecules and inorganic nanoobjects and incorporation of these systems in functional devices is a basis of nanobiotechnology (Fig. 7.29) [22].

Thus, substantial progress has been made in this direction, however much remains to be learned. There is no general methodology for construction of bionanocomposites hierarchically organized in terms of structure and functions and true understanding of mechanism of their operating, etc.

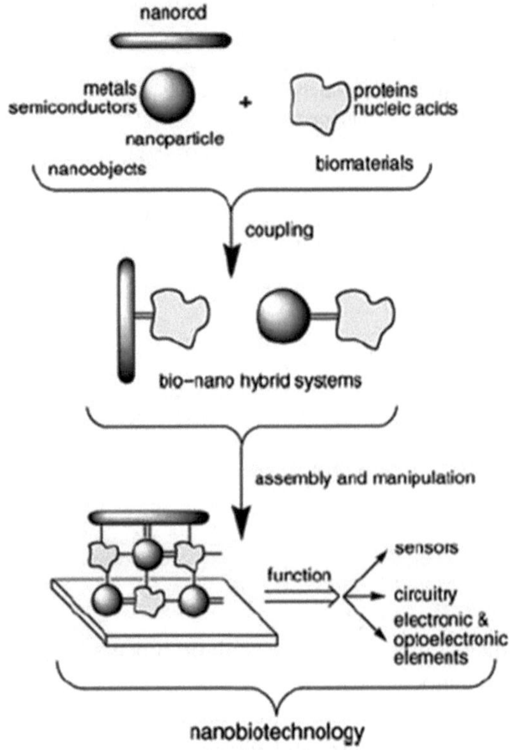

Fig. 7.29 Generation of biomolecule–nanoparticle conjugates to yield functional devices [22]

Future research will be devoted to the development of novel nano-structured, bio-inspired hybrid materials with predetermined physical and bio-mimetic properties, including the use of novel methods of synthesis, the improvement of mechanical behavior, accompanied by the application of simulation and other theoretical studies.

References

1. S. Mann, Angew. Chem. Int. Ed. **47**, 5306 (2008)
2. S. Mann, Nat. Mater. **8**, 781 (2009)
3. A.J. Patil, S. Mann, J. Mater. Chem. **18**, 4605 (2008)
4. A.M. Collins, N.J.V. Skaer, T. Cheysens, D. Knight, C. Bertram, H.I. Roach, R.O.C. Oreffo, S. Von-Aulock, T. Baris, J. Skinner, S. Mann, Adv. Mater. **21**, 75 (2009)
5. T.P.J. Knowles, T.W. Oppenheim, A.K. Buell, D.Y. Chirgadze, M.E. Welland, Nat. Nanotechnol. **5**, 204 (2010)
6. A. Singh, S. Hede, M. Sastry, Small **3**, 466 (2007)
7. A. Sugunan, P. Melin, J. Schnürer, J.G. Hilborn, J. Dutta, Adv. Mater. **19**, 77 (2007)
8. M. Gensheimer, M. Becker, A. Brandis-Heep, J.H. Wendorff, R.K. Thauer, A. Greiner, Adv. Mater. **19**, 2480 (2007)

9. S.K. Dixit, N.L. Goicochea, M.-C. Daniel, A. Murali, L. Bronstein, M. De, B. Stein, V.M. Rotello, C.C. Kao, B. Dragnea, Nano Lett. **6**, 1993 (2006)
10. K.M. Bromley, A.J. Patil, A.W. Perriman, G. Stubbs, S. Mann, J. Mater. Chem. **18**, 4796 (2008)
11. T. Li, B. Ye, Z. Niu, P. Thompson, S. Seifert, B. Lee, Q. Wang, Chem. Mater. **21**, 1046 (2009)
12. E.R. Hitzky, K. Ariga, M.Y. Lvov, *Bio-inorganic Hybrid Nanomaterials: Strategies, Syntheses, Characterization and Application* (Wiley, Weinheim, 2007)
13. M. Vallet-Regi, D. Arcos, *Biomimetic Nanoceramics in Clinical Use: From Materials to Applications* (RSC Nanoscience & Nanotechnology, Cambridge, 2008)
14. H. Dong, D. Wang, G. Sun, J.P. Hinestroza, Chem. Mater. **20**, 6627 (2008)
15. P. Mukherjee, A. Ahmad, D. Mandal, S. Senapati, S.R. Sainkar, M.I. Khan, R. Ramani, R. Pasricha, P.V. Ajayakumar, M. Alam, M. Sastry, Angew. Chem. Int. Ed. **40**, 3585 (2001)
16. A. Ahmad, S. Senapati, M.I. Khan, R. Kumar, M. Sastry, Langmuir **19**(8), 3550 (2003)
17. A. Ahmad, P. Mukherjee, D. Mandal, S. Senapati, M.I. Khan, R. Kumar, M. Sastry, J. Am. Chem. Soc. **124**, 12108 (2002)
18. P. Mukherjee, S. Senapati, D. Mandal, A. Ahmad, M.I. Khan, R. Kumar, M. Sastry, Chem. Bio. Chem. **3**, 461 (2002)
19. M. Labrenz, G.K. Druschel, T. Thomsen-Ebert, B. Gilbert, S.A. Welch, K.M. Kemner, G.A. Logan, R.E. Summons, G.D. Stasio, P.L. Bond, B. Lai, S.D. Kelly, J.F. Banfield, Science **290**, 1744 (2000)
20. J.L. Gardea-Torresdey, J.G. Parsons, E. Gomez, J.P. Videa, H.E. Troinai, P. Santiago, M.J. Yacaman, Nano Lett. **2**, 397 (2002)
21. S.P. Chandran, M. Chaudhary, R. Pasricha, A. Ahmad, M. Sastry, Biotechnol. Prog. **22**, 577 (2006)
22. E. Katz, I. Willner, Angew. Chem. Int. Ed. **43**, 6042 (2004)
23. V. Kumar, S.K. Yadav, J. Chem. Technol. Biotechnol. **84**, 151 (2009)
24. J. Huang, Q. Li, D. Sun, Y. Lu, X. Yang, H. Wang, Y. Wang, W. Shao, N. He, J. Hong, C. Chen, Nanotechnology **18**, 105104 (2007)
25. B. Ankamwar, C. Damle, A. Ahmad, M. Sastry, J. Nanosci. Nanotechnol. **5**, 1665 (2005)
26. C. Singh, R.K. Baboota, P.K. Naik, H. Singh, Adv. Mater. Lett. **3**, 279 (2012)
27. S.S. Shankar, A. Ahmad, R. Pasrichaa, M. Sastry, J. Mater. Chem. **13**, 1822 (2003)
28. N.A. Begum, S. Mondal, S. Basu, R.A. Laskar, D. Mandal, Colloids Surf. B **71**, 113 (2009)
29. E. Rodriguez, J.G. Parsons, J.R. Peralta-Videa, G. Cruz-Jimenez, J. Romero-Gonzalez, B.E. Sanchez-Salcido, Int. J. Phytoremed. **9**, 133 (2007)
30. N.C. Sharma, S.V. Sahi, S. Nath, J.G. Parsons, J.L. Gardea-Torresdey, T. Pal, Environ. Sci. Technol. **41**, 5137 (2007)
31. J.L. Gardea-Torresdey, E. Gomez, J.R. Peralta-Videa, J.G. Parsons, H. Troiani, M. Jose-Yacaman, Langmuir **19**, 1357 (2003)
32. A.G. Medentsev, V.K. Alimenko, Phytochemistry **47**, 935 (1998)
33. R.A. Baker, J.H. Tatum, J. Ferment. Bioeng. **85**, 359 (1998)
34. E.M. Egorova, A.A. Revina, Colloids Surf. A Physicochem. Eng. Asp. **168**, 87 (2000)
35. E.M. Egorova, A.A. Revina, T.N. Rostovshchikova, O.I. Kiseleva, Vestn. Mosk. Univ. Ser. Khim. **42**, 332 (2001)
36. S.S. Shankar, A. Ahmad, M. Sastry, Biotechnol. Prog. **19**, 1627 (2003)
37. X.L. Zhu, Q.L. Yang, J.Y. Huang, I. Suzuki, G.X. Li, J. Nanosci. Nanotechnol. **8**, 353 (2008)
38. S.K. Bhargava, J.M. Booth, S. Agrawal, P. Coloe, G. Kar, Langmuir **21**, 5949 (2005)
39. R. Shukla, S.K. Nune, N. Chanda, K. Katti, S. Mekapothula, R.R. Kulkarni, W.V. Welshons, R. Kannan, K.V. Katti, Small **4**, 1425 (2008)
40. V. Kumar, S.C. Yadav, S.K. Yadav, J. Chem. Technol. Biotechnol. **85**, 1301 (2010)
41. E.C. Njagi, H. Huang, L. Stafford, H. Genuino, H.M. Galindo, J.B. Collins, G.E. Hoag, S.L. Suib, Langmuir **27**(1), 264 (2011)
42. A. Singh, M. Chaudhary, M. Sastry, Nanotechnology **17**, 2399 (2006)

43. B. Ankamwar, M. Chaudhary, M. Sastry, Synth. React. Inorg. Metal-Org. Nano-Metal Chem. **35**, 19 (2005)
44. S.S. Shankar, A. Rai, A. Ahmad, M. Sastry, Chem. Mater. **17**, 566 (2005)
45. M.L. Lópeza, J.G. Parsonsb, J.R. Peralta Videab, T.L. Gardea-Torresdey, Microchem. J. **81**, 50 (2005)
46. S. Li, Y. Shen, A. Xie, X. Yu, L. Qiu, L. Zhang, Green Chem. **9**, 852 (2007)
47. F. Zeng, C. Hou, S. Wu, X. Liu, Z. Tong, S. Yu, Nanotechnology **18**, 1 (2007)
48. K. Ghule, A.V. Ghule, J.Y. Liu, Y.C. Ling, J. Nanosci. Nanotechnol. **6**, 3746 (2006)
49. M.N. Nadagouda, G. Hoag, J. Collins, R.S. Varma, Crystal Growth Des. **9**, 4979 (2009)
50. P. Velmurugan, J. Shim, S. Kamala-Kannan, K.-J. Lee, B.-T. Oh, V. Balachandar, B.-T. Oh, Biotechnol. Prog. **27**, 273 (2011)
51. N. Krumov, I. Perner-Nochta, S. Oder, V. Gotcheva, A. Angelov, C. Posten, Chem. Eng. Technol. **32**, 1026 (2009)
52. D. Schüler, R.B. Frankel, Appl. Microbiol. Biotechnol. **52**, 464 (1999)
53. C.M. Niemeyer, Angew. Chem. Int. Ed. **40**, 4128 (2001)
54. C. Ven Den Hoek, D.G. Mann, H.M. Johns, in *Algae: An Introduction to Phycology*, ed. by W.H. van de Poll (Cambridge University Press, Cambridge, 1997)
55. D. Werner, Silicate metabolism, in *The Biology of Diatoms. Botanical Monograph*, ed. by D. Werner, vol. 13 (University of California, Berkeley, 1977), p. 110
56. R.E. Lee, in *Heterokontophyta, Bacillariophyceae*, ed. by R.E. Lee (Cambridge University Press, Cambridge, 1999), p. 415
57. N. Kröger, K.H. Sandhage, MRS Bull. **35**, 122 (2010)
58. M.A. Grachev, V.V. Annenkov, Y.V. Likhoshway, Bio Essays **30**, 328 (2008)
59. S.Y. He, Z.R. Guo, Y. Zhang, S. Zhang, J. Wang, N. Gu, Mater. Lett. **61**, 3984 (2007)
60. S. He, Y. Zhang, Z. Guo, N. Gu, Biotechnol. Prog. **24**, 476 (2008)
61. S.K. Das, A.R. Das, A.K. Guha, Small **6**, 1012 (2010)
62. T. Klaus, R. Joerger, E. Olsson, S. Granqist, Proc. Natl. Acad. Sci. U. S. A. **96**, 13611 (1999)
63. S. De Corte, T. Hennebel, S. Verschuere, C. Cuvelier, W. Verstraete, N. Boon, J. Chem. Technol. Biotechnol. **86**, 547 (2011)
64. W. De Windt, P. Aelterman, W. Verstraete, Environ. Microbiol. **7**, 314 (2005)
65. P. Yong, N.A. Rowson, J.P.G. Farr, I.R. Harris, L.R. Macaskie, Biotechnol. Bioeng. **80**, 369 (2002)
66. K. Kashefi, J.M. Tor, K.P. Nevin, D.R. Lovley, Appl. Environ. Microbiol. **67**, 3275 (2001)
67. D. Chidambaram, T. Hennebel, S. Taghavi, J. Mast, N. Boon, W. Verstraete, Environ. Sci. Technol. **44**, 7635 (2010)
68. K. Deplanche, L.E. Macaskie, Biotechnol. Bioeng. **99**, 1055 (2008)
69. M.F. Lengke, B. Ravel, M.E. Fleet, G. Wanger, R.A. Gordon, G. Southam, Environ. Sci. Technol. **40**, 6304 (2006)
70. R. Brayner, H. Barberousse, M. Hernadi, C. Djedjat, C. Yepremian, T. Coradin et al., J. Nanosci. Nanotechnol. **7**, 2696 (2007)
71. T.J. Beveridge, R.G.E. Murray, J. Bacteriol. **141**, 876 (1980)
72. C. Tian, B. Mao, E. Wang, Z. Kang, Y. Song, C. Wang, S. Li, J. Phys. Chem. C **111**, 3651 (2007)
73. X. Hu, T. Wang, X. Qu, S. Dong, J. Phys. Chem. B **110**, 853 (2006)
74. R. Cui, C. Liu, J. Shen, D. Gao, J.-J. Zhu, H.-Y. Chen, Adv. Funct. Mater. **18**, 2197 (2008)
75. F. Gallyas, Hystochemistry **64**, 87 (1979)
76. J. Richter, R. Seidel, R. Kirsch, M. Mertig, W. Pompe, J. Plaschke, H.K. Schackert, Adv. Mater. **12**, 507 (2000)
77. J. Richter, Phys. E. **16**, 157 (2003)
78. W. Shenton, T. Douglas, M. Young, G. Stubbs, S. Mann, Adv. Mater. **11**, 253 (1999)
79. R. Wahl, M. Mertig, J. Raff, S. Selenska-Pobell, W. Pompe, Adv. Mater. **13**, 736 (2001)
80. A.L. Metlina, Uspekhi biolog. Khim. **41**, 229 (2001)

81. K. Deplanche, R.D. Woods, I.P. Mikheenko, R.E. Sockett, L.E. Macaskie, Biotechnol. Bioeng. **101**, 873 (2008)
82. S. Dieluweit, D. Pum, U.B. Sleytr, Supramol. Sci. **5**, 15 (1998)
83. S.R. Hall, W. Shenton, H. Engelhardt, S. Mann, Chem. Phys. Chem. **2**, 184 (2001)
84. R. Djalali, Y.F. Chen, H. Matsui, J. Am. Chem. Soc. **124**, 13660 (2002)
85. M.T. Kumara, B.C. Tripp, S. Muralidharan, Chem. Mater. **19**, 2056 (2007)
86. J.E. Bailie, H.A. Abdullah, J.A. Anderson, C.H. Rochester, N.V. Richardson, N. Hodge, J. Zhang, A. Burrows, C.J. Kiely, G.J. Hutchings, Phys. Chem. Chem. Phys. **3**, 4113 (2001)
87. J.R. Lloyd, R.T. Anderson, L.E. Macaskie, Bioremediation of metals and radionuclides, in *Bioremediation*, ed. by R. Atlas, J. Philp (ASM Press, Washington, DC, 2005), pp. 293–317
88. J.R. Lloyd, P. Yong, L.E. Macaskie, Appl. Environ. Microbiol. **64**, 4607 (1998)
89. P. Yong, N.A. Rowson, J.P.G. Farr, I.R. Harris, L.E. Macaskie, Environ. Sci. Tech. **24**, 289 (2003)
90. N.J. Creamer, V.S. Baxter-Plant, J. Henderson, M. Potter, L.E. Macaskie, Biotechnol. Lett. **28**, 1475 (2006)
91. A.N. Mabbett, D. Sanyahumbi, P. Yong, L.E. Macaskie, Environ. Sci. Technol. **40**, 1015 (2006)
92. D. Gauthier, L.S. Sobjerg, K.M. Jensen, A.T. Lindhardt, M. Bunge, K. Finster, R.L. Meyer, T. Skrydstrup, Chem. Sus. Chem. **3**, 1036 (2010)
93. J.R. Lloyd, C.I. Pearce, V.S. Coker, R.A.D. Pattrick, G. Van Der Laan, R. Cutting, D.J. Vaughan, M. Paterson-Beedle, I.P. Mikheenko, P. Yong, L.E. Macaskie, Geobiology **6**, 285 (2008)
94. Y. Suzuki, S.D. Kelly, K.M. Kemner, J.F. Banfield, Nature **419**, 134 (2002)
95. J.C. Renshaw, L.J.C. Butchins, F.R. Livens, I. May, J.M. Charnock, J.R. Lloyd, Environ. Sci. Tech. **39**, 5657 (2005)
96. J.R. Lloyd, FEMS Microbiol. Rev. **27**, 411 (2003)
97. S.A. Kumar, M.K. Abyaneh, S.W. Gosavi, S.K. Kulkarni, A. Ahmad, M.I. Khan, Appl. Biochem. **47**, 191 (2007)
98. N. Durán, P.D. Marcato, O.L. Alves, G.I.H. De Souza, E. Esposito, J. Nanobiotechnol. **3**, 8 (2005)
99. I. Mikheenko, Nanoscale palladium recovery, PhD thesis, University of Birmingham, UK, 2004
100. I.P. Mikheenko, M. Rousset, S. Dementin, L.E. Macaskie, Appl. Environ. Microbiol. **74**, 6144 (2008)
101. J.M. Slocik, M.R. Knecht, D.W. Wright, Encycl. Nanosci. Nanotechnol. **1**, 293 (2004)
102. B. Nair, T. Pradeep, Crystal Growth Des. **2**(4), 293 (2002)
103. M. Kowshik, S. Ashtaputre, S. Kharrazi, W. Vogel, J. Urban, S.K. Kulkarni, K.M. Paknikar, Nanotechnology **14**, 95 (2003)
104. P. Mukherjee, A. Ahmad, D. Mandal, S. Senapati, S.R. Sainkar, M.I. Khan, R. Parishcha, P.V. Ajaykumar, M. Alam, R. Kumar, M. Sastry, Nano Lett. **1**, 515 (2001)
105. A. Ahmad, P. Mukherjee, S. Senapati, D. Mandal, M.I. Khan, R. Kumar, M. Sastry, Colloids Surf. B Biointerfaces **28**, 313 (2003)
106. D. Fortin, T. Beveridge, in *Biomineralization*, ed. by E. Bäuerlein (Wiley-VCH, Weinheim, 2000)
107. Y. Konishi, T. Nomura, T. Tskukiyama, N. Saitoh, Trans. Mater. Res. Soc. Jpn. **29**, 2341 (2004)
108. P. Mukherjee, A. Ahmad, D. Mandal, S. Senapati, S.R. Sainkar, M.I. Khan, R. Ramani, R. Parischa, P.V. Ajayakumar, M. Alam, M. Sastry, R. Kumar, Angew. Chem. Int. Ed. Engl. **40**, 3585 (2001)
109. P. Mukherjee, S. Senapati, D. Mandal, A. Ahmad, M.I. Khan, R. Kumar, M. Sastry, Chembiochem **3**, 461 (2002)
110. A. Ahmad, S. Senapati, M.I. Khan, R. Kumar, R. Ramani, V. Srinivas, M. Sastry, Nanotechnology **14**(7), 824 (2003)

111. S. Senapati, A. Ahmad, M.I. Khan, M. Sastry, R. Kumar, Small **1**(5), 517 (2005)
112. L.L. Hench, R.J. Splinter, W.C. Allen, T.K. Greenlee, J. Biomed. Mater. Res. Symp. **2**, 117 (1972)
113. S. Braun, S. Rappoport, R. Zusman et al., Mater. Lett. **10**, 1 (1990)
114. M. Nogi, H. Yano, Adv. Mater. **20**, 1849 (2008)
115. J. Xie, Y. Zheng, J.Y. Ying, J. Am. Chem. Soc. **131**, 888 (2009)
116. V. Berry, A. Gole, S. Kundu, C.J. Murphy, R.F. Saraf, J. Am. Chem. Soc. **127**, 17600 (2005)
117. V. Berry, S. Rangaswamy, R.F. Saraf, Nano Lett. **4**, 939 (2004)
118. V. Berry, R.F. Saraf, Angew. Chem. Int. Ed. **44**, 6668 (2005)
119. S.-K. Lee, D.S. Yun, A.M. Belcher, Biomacromolecules **7**, 14 (2006)
120. B. Samanta, H. Yan, N.O. Fischer, J. Shi, D.J. Jerryc, V.M. Rotello, J. Mater. Chem. **18**, 1204 (2008)
121. K.T. Nam, D.-W. Kim, P.J. Yoo, C.-Y. Chiang, N. Meethong, P.T. Hammond, Y.-M. Chiang, A.M. Belcher, Science **312**, 885 (2006)
122. A. Vazquez, V.P. Cyras, V.A. Alvarez, Environmental silicate nano-biocomposites, in *Green Energy and Technology*, ed. by J.I. Moran, L. Avérous, E. Pollet (Springer, London, 2012), p. 287
123. C. Boissiere, D. Grosso, A. Chaumonnot, L. Nicole, C. Sanchez, Adv. Mater. **23**, 599 (2011)
124. H.-W. Kim, H.-H. Lee, G.-S. Chun, J. Biomed. Mater. Res. **85A**, 651 (2008)
125. X. Zheng, S. Zhou, Y. Xiao, X. Yu, B. Feng, J. Biomed, Mater. Res. B. Appl. Biomater. **91**, 181 (2009)
126. E.C. Carnes, D.M. Lopez, N.P. Donegan, A. Cheung, H. Gresham, G.S. Timmins, C.J. Brinker, Nat. Chem. Biol. **6**, 41 (2010)
127. E. Ruiz-Hernandez, A. Lopez-Noriega, D. Arcos, M. Vallet-Regi, Solid State Sci. **10**, 421 (2008)
128. J.W. Liu, X.M. Jiang, C. Ashley, C.J. Brinker, J. Am. Chem. Soc. **131**, 7567 (2009)
129. J.W. Liu, A. Stace-Naughton, X.M. Jiang, C.J. Brinker, J. Am. Chem. Soc. **131**, 1354 (2009)
130. T. Buranda, J. Huang, G.V. Ramarao, L.K. Ista, R.S.R.S. Larson, T.L. Ward, L.A. Sklar, G.P. Lopez, Langmuir **19**, 1654 (2003)
131. J.I. Zink, J.S. Valentine, B. Dunn, New J. Chem. **18**, 1109 (1994)
132. A. Bronshtein, N. Aharonson, D. Avnir, A. Turiansky, M. Alstein, Chem. Mater. **9**, 2632 (1997)
133. Y. Lvov, H. Möhwald (eds.), *Protein Architecture: Interfacing Molecular Assemblies and Immobilization Biotechnology* (Marcel Dekker, New York, 2000)
134. J. Woodward (ed.), *Immobilized Cells and Enzymes – A Practical Approach* (IRL Press, Washington, DC, 1985)
135. M.F. Desimone, C. Herlary, G. Mosser, M.-M. Giraud-Guille, J. Livage, T. Coradin, J. Mater. Chem. **20**, 666 (2010)
136. D. Eglin, S. Maalheem, J. Livage, T. Coradin, J. Mater. Sci. Mater. Med. **17**, 161 (2006)
137. S. Smitha, P. Shajesh, P. Mukundan, K.G.K. Warrier, J. Sol-Gel Sci. Technol. **42**, 157 (2007)
138. S. Smitha, P. Shajesh, P. Mukundan, T.D.R. Nair, K.G.K. Warrier, Colloids Surf. B. **55**, 38 (2007)
139. J. Allouche, M. Boissiére, C. Herlary, J. Livage, T. Coradin, J. Mater. Chem. **16**, 3120 (2006)
140. S. Peng, Q. Gao, Q. Wang, J. Shi, Chem. Mater. **16**, 2675 (2004)
141. Z.Y. Wang, Y. Zhao, L. Ren, L.H. Jin, L.P. Sun, P. Yin, Y.F. Zhang, Q.Q. Zhang, Nanotechnology **19**, 445103 (2008)
142. T.K. Jain, I. Roy, T.K. De, A. Maitra, J. Am. Chem. Soc. **120**, 11092 (1998)
143. Y.A. Shchipunov, J. Colloid Interface Sci. **268**, 68 (2003)
144. Y.A. Shchipunov, T.Y. Karpenko, Langmuir **20**, 3882 (2004)
145. Y.A. Shchipunov, A. Kojima, T. Imae, J. Colloid Interface Sci. **285**, 574 (2005)
146. N.A. Brusentsov, T.N. Brusentsova, Khim. Farm. Zh. **35**, 10 (2001)
147. S.A. Medvedeva, G.P. Aleksandrova, L.A. Grishchenko, N.A. Tyukavkina, Zh. Obshch. Khim. **72C**, 1569 (2002)

148. M. Catauro, M.G. Raucci, D. De Marco, L. Ambrosio, J. Biomed. Mater. Res. **77A**, 340 (2006)
149. R.I. Kalyuzhnaya, K.K. Khulchaev, V.A. Kasaikin, A.B. Zezin, V.A. Kabanov, Vysokomol. Soedin. A **36**, 257 (1994)
150. L.N. Ermakova, T.A. Aleksandrova, P.V. Nuss, A.M. Vasserman, V.A. Kasaikin, A.B. Zezin, V.A. Kabanov, Vysokomol. Soedin. A **27**, 1391 (1985)
151. V.V. Annenkov, S.V. Patwardhan, D. Belton, E.N. Danilovtseva, C.C. Perry, Chem. Commun. **1521** (2006)
152. L. Nicole, C. Boissière, D. Grosso, A. Quach, C. Sanchez, J. Mater. Chem. **15**, 3598 (2005)
153. M. Colilla, M. Manzano, I. Izquierdo-Barba, M. Vallet-Regi, C. Boissiere, C. Sanchez, Chem. Mater. **22**, 1821 (2010)
154. J.P. Zhong, D.C. Greenspan, J. Biomed. Mater. Res. **53**, 694 (2000)
155. M.A. De Diego, N.J. Coleman, L.L. Hench, J. Mater. Sci. Mater. Med. **15**, 803 (2004)
156. E.M. Santos, S. Radin, P. Ducheyne, Biomaterials **20**, 1695 (1999)
157. A.J. Salinas, A.I. Martin, M. Vallet-Regır, J. Biomed. Mater. Res. **61**, 524 (2002)
158. M. Hamadouche, A. Meunier, D.C. Greenspan, C. Blanchat, J.P. Zhong, G.P. La Torre, L. Sedel, J. Biomed. Mater. Res. **52**, 422 (2000)
159. P. Sepulveda, J.R. Jones, L.L. Hench, J. Biomed. Mater. Res. **59**, 340 (2002)
160. C.J. Brinker, G.W. Scherer, *Sol-gel Science: The Physics and Chemistry of Sol-gel Processing* (Academic, San Diego, 1990)
161. H.W. Kim, H.E. Kim, J.C. Knowles, Adv. Funct. Mater. **16**, 1529 (2006)
162. H.W. Kim, J.H. Song, H.E. Kim, J. Biomed. Mater. Res. **79A**, 698 (2006)
163. S.A. Catledge, M.D. Fries, Y.K. Vohra, W.R. Lacefield, J.E. Lemons, S. Woodard, R. Venugopalan, J. Nanosci. Nanotechnol. **2**, 293 (2002)
164. F.T. Cheng, P. Shi, H.C. Man, Scripta. Mater. **51**, 1041 (2004)
165. J.-X. Liu, D.-Z. Yang, F. Shi, Y.-J. Cai, Thin Solid Films **429**, 225 (2003)
166. H. Boettcher, J. Prakt. Chem. **342**, 427 (2000)
167. J. Musil, Surf. Coat. Techn. **125**, 322 (2000)
168. J.D. Mackenzie, E.P. Bescher, J. Sol-gel Sci. Techn. **19**, 23 (2000)
169. I. Brasack, H. Boettcher, U. Hempel, J. Sol-gel Sci. Techn. **19**, 479 (2000)
170. M.M. Pereira, J.R. Jones, R.L. Orefice, L.L. Hench, J. Mater. Sci. Mater. Med. **16**, 1045 (2005)
171. H. Boéttcher, Mat.-wiss. u. Werkstofftech **32**, 759 (2001)
172. C. Ohtsuki, T. Miyazaki, M. Tanihara, Mater. Sci. Eng. C: Biomim. Supramol. Syst. **22**, 27 (2002)
173. P. Innocenzi, M. Esposto, A. Maddalena, J. Sol-gel Sci. Technol. **20**, 293 (2001)
174. A. Costantini, G. Luciani, G. Annunziata, B. Silvestri, F. Branda, J. Mater. Sci. Mater. Med. **17**, 319 (2006)
175. R.O.R. Costa, M.M. Pereira, F.S. Lameiras, W.L. Vasconcelos, J. Mater. Sci. Mater. Med. **16**, 927 (2005)
176. C. Schiraldi, A. D'Agostino, A. Oliva, F. Flamma, A. De Rosa, A. Apicella, R. Aversa, M. De Rosa, Biomaterials **25**, 3645 (2004)
177. S.L. Huang, W.K. Chin, W.P. Yang, Polymer **46**, 1865 (2005)
178. A. Costantini, G. Luciani, B. Silvestri, F. Tescione, F. Branda, J. Biomed, Mater. Res. B Appl. Biomater. **86**, 98 (2008)
179. M.T. Reetz, Adv. Mater. **9**, 943 (1997)
180. B. Silvestri, G. Luciani, A. Costantini, F. Tescione, F. Branda, A. Pezzella, J. Biomed, Mater. Res. B. Appl. Biomater. **89B**, 369 (2009)
181. M. Catauro, M.G. Raucci, F. de Gaetano, A. Marotta, J. Mater. Sci. **38**, 3097 (2003)
182. M. Catauro, M.G. Raucci, F. de Gaetano, A. Buri, A. Marotta, L. Ambrosio, J. Mater. Sci. Med. **15**, 991 (2004)
183. T.E. Rams, J. Slots, Periodontol **10**, 139 (1996)

184. J.D. Bass, D. Grosso, C. Boissière, E. Belamie, T. Coradin, C. Sanchez, Chem. Mater. **19**, 4349 (2007)
185. C. Charnay, S. Begu, C. Tourne-Peteilh, L. Nicole, D.A. Lerner, J.M. Devoisselle, Eur. J. Pharm. Biopharm. **57**, 533 (2004)
186. E. Ruiz-Hitzky, M. Darder, P. Aranda, in *Annual Review of Nanoresearch*, ed. by G. Cao, Q. Zhang, C.J. Brinker, vol. 3 (World Scientific Publishing, Singapore, 2010), p. 149
187. S. Ha, J.A. Gardella Jr., Chem. Rev. **105**, 4205 (2005)
188. P. Colombo, R. Bettini, P. Santi, N.A. Peppas, J. Control Release. **39**, 231 (1996)
189. E. Dujardin, S. Mann, Adv. Mater. **14**, 775 (2002)
190. E. Ruiz-Hitzky, M. Darder, P. Aranda, in *Bio-Inorganic Hybrid Materials: Strategies, Syntheses, Characterization and Applications*, ed. by E. Ruiz-Hitzky, K. Ariga, Y. Lvov (Wiley-VCH, Weinheim, 2008), p. 1
191. E. Ruiz-Hitzky, P. Aranda, M. Darder, *The Kirk-Othmer Encyclopedia of Chemical Technology* (Wiley, New York, 2008), p. 8
192. E. Ruiz-Hitzky, M. Darder, P. Aranda, J. Mater. Chem. **15**, 3650 (2005)
193. M. Darder, P. Aranda, A.I. Ruiz, F.M. Fernandes, E. Ruiz-Hitzky, Mater. Sci. Technol. **24**, 1100 (2008)
194. C. Tourné-Péteilh, D.A. Lerner, C. Charnay, L. Nicole, S. Bégu, J.-M. Devoisselle, Chem. Phys. Chem. **4**, 281 (2003)
195. M. Darder, P. Aranda, E. Ruiz-Hitzky, Adv. Mater. **19**, 1309 (2007)
196. E. Ruiz-Hitzky, M. Darder, P. Aranda, K. Ariga, Adv. Mater. **22**, 323 (2010)
197. P. Gomez-Romero, C. Sanchez, *Functional Hybrid Materials* (Wiley-VCH, Weinheim, 2004)
198. E. Ruiz-Hitzky, P. Aranda, M. Darder, in *Bottom-Up Nanofabrication: Supramolecules, Self-Assemblies, and Organized Films*, ed. by K. Ariga, H.S. Nalwa, vol. 3 (American Scientific Publishers, Stevenson Ranch, 2009), pp. 39–76
199. A.C.S. Alcantara, P. Aranda, M. Darder, E. Ruiz-Hitzky, J. Mater. Chem. **20**, 9495 (2010)
200. D.G. Shchukin, T. Shutava, E. Shchukina, G.B. Sukhorukov, Y.M. Lvov, Chem. Mater. **16**, 3446 (2004)
201. P. Horcajada, A. Rámila, J. Pérez-Pariente, M. Vallet-Regí, Microporous Mesoporous Mater. **68**, 105 (2004)
202. A. Rámila, B. Muñoz, J. Pérez-Pariente, M. Vallet-Regí, J. Sol-Gel Sci. Technol. **26**, 1199 (2003)
203. A.L. Doadrio, E.M.B. Sousa, J.C. Doadrio, J. Pérez Pariente, I. Izquierdo-Barba, M. Vallet-Regí, J. Control. Release **97**, 125 (2004)
204. K. Katagiri, F. Caruso, Macromolecules **37**, 9947 (2004)
205. B. Muñoz, A. Rámila, J. Pérez-Pariente, I. Díaz, M. Vallet-Regí, Chem. Mater. **15**, 500 (2003)
206. M. Vallet-Regí, A. Rámila, R.P. Del Real, J. Pérez-Pariente, Chem. Mater. **13**, 308 (2001)
207. K.A. Fisher, K.D. Huddersman, M.J. Taylor, Chem. A Eur. J. **9**, 5873 (2003)
208. C.-Y. Lai, B.G. Trewyn, D.M. Jeftinija, K. Jeftinija, S. Xu, S. Jeftinija, V.S.-Y. Lin, J. Am. Chem. Soc. **125**, 4451 (2003)
209. H. Hata, S. Saeki, T. Kimura, Y. Sugahara, K. Kuroda, Chem. Mater. **11**, 1110 (1999)
210. A.D. Pomogailo, Colloid J. **67**, 658 (2005)
211. M. Sarikaya, C. Tamerler, A.K.-Y. Jen, K. Schulten, F. Baneyx, Nat. Mater. **2**, 577 (2003)
212. P. Calvert, P. Rieke, Chem. Mater. **8**, 1715 (1996)
213. E. Bauerlein (ed.), *The Biomineralisation of Nano- and Micro-Structures* (Wiley-VCH, Weinheim, 2000)
214. C.A. Mirkin, T.A. Taton, Nature **405**, 626 (2000)
215. A.E. Ingalls, K. Whitehead, M.C. Bridoux, Geochim. Cosmochim. Acta **74**, 104 (2010)
216. L.L. Hench, J. Am. Ceram. Soc. **81**, 1705 (1998)
217. T.P. Hoepfner, T.D. Case, Ceram. Int. **29**, 699 (2003)
218. G. Goller, H. Demirkian, F.N. Oktar, E. Demirkesen, Ceram. Int. **29**, 721 (2003)

219. D.J. Belton, O. Deschaume, S.V. Patwardh, C.C. Perry, J. Phys. Chem. B **114**, 9947 (2010)
220. C.F. Conrad, G.A. Icopini, H. Yasuhara, J.Z. Bandstra, S.L. Brantley, P.J. Heaney, Geochim. Cosmochim. Acta **71**, 531 (2007)
221. C. Gröger, K. Lutz, E. Brunner, Cell Biochem. Biophys. **50**, 23 (2008)
222. S.V. Patwardhan, Chem. Commun. **47**, 7567 (2011)
223. J.M. O'Reilly, B.K. Coltrain, Organic/inorganic composite materials, in *Polymeric Materials Encyclopedia*, ed. by J.C. Salamone (CRC Press, London, 1996), pp. 4772–4781
224. M. Sumper, Angew. Chem. Int. Ed. **43**, 2251 (2004)
225. V.V. Annenkov, E.N. Danilovtseva, Y.V. Likhoshway, S.V. Patwardhan, C.C. Perry, J. Mater. Chem. **18**, 553 (2008)
226. V.A. Palshin, Synthesis and properties of organosilicone nanoparticles, PhD thesis, IPCP RAS, Chernogolovka, 2012
227. N. Kröger, R. Deutzmann, M. Sumper, Science **286**, 1129 (1999)
228. N. Kröger, S. Lorenz, E. Brunner, M. Sumper, Science **298**, 584 (2002)
229. E. Brunner, L. Lutz, M. Sumper, Phys. Chem. Chem. Phys. **6**, 854 (2004)
230. S.V. Patwardhan, N. Mukherjee, S.J. Clarson, J. Inorg. Organomet. Polym. **11**, 193 (2001)
231. D. Belton, G. Paine, S.V. Patwardhan, C.C. Perry, J. Mater. Chem. **14**, 2231 (2004)
232. S.V. Patwardhan, S.J. Clarson, C.C. Perry, Chem. Commun. **1113** (2005)
233. Y.A. Shchipunov, Y.V. Burtseva, T.Y. Karpenko, N.M. Shevchenko, T.N. Zvyagintseva, J. Mol. Catal. B: Enzym. **40**, 16 (2006)
234. M. Starikaya, I.A. Aksay (eds.), *Biomimetics* (AIP Press, Woodburg, 1995)
235. R. Gordon, Fed. Proc. **40**, 827 (1981)
236. J. Parkinson, R. Gordon, Trends Biotechnol. **17**, 190 (1999)
237. R. Gordon, D. Losic, M.A. Tiffany, S.S. Nagy, F.A.S. Sterrenburg, Trends Biotechnol. **27**, 116 (2009)
238. R. Gordon, R.W. Drum, Int. Rev. Cyt. **150**, 243 (1994)
239. S. Hazelaar, H.J. van der Strate, W.W.C. Gieskes, E.G. Vrieling, J. Phycol. **41**, 354 (2005)
240. L. Lenoci, P.J. Camp, Langmuir **24**, 217 (2008)
241. S. Wenzl, R. Hett, P. Richthammer, M. Sumper, Angew. Chem. Int. Ed. **47**, 1729 (2008)
242. M. Sumper, S. Lorenz, E. Brunner, Angew. Chem. Int. Ed. **42**, 5192 (2003)
243. K. Tsuru, C. Ohtsuki, A. Osaka, T. Iwamoto, J.D. Mackenzie, J. Mater. Sci.: Mater. Med. **8**, 157 (1997)
244. H.A. Pohl, in *Coherent Excitation in Biological Systems*, ed. by H. Frolich, F. Kremer (Springer, Heidelberg, 1983)
245. S. Mann, J. Chem. Soc. Dalton Trans. **1**, 3953 (1993)
246. S. Mann, S.L. Burkett, S.A. Davis, C.E. Fowler, N.H. Mendelson, S.D. Sims, D. Walsh, N.T. Whilton, Chem. Mater. **9**, 2300 (1997)
247. K. Saha, A. Bajaj, B. Duncan, V.M. Rotello, Small **7**, 1903 (2011)
248. O. Keskin, A. Gursoy, B. Ma, R. Nussinov, Chem. Rev. **108**, 1225 (2008)
249. F. Spyrakis, A. BidonChanal, X. Barril, F.J. Luque, Curr. Top. Med. Chem. **11**, 192 (2011)
250. G. Gorrasi, M. Tortora, V. Vittoria, G. Galli, E. Chiellini, J. Polym. Sci. Polym. Phys. **40**, 1118 (2002)
251. M. Jaber, M. Bouchoucha, L. Delmotte, C. Methivier, J.-F. Lambert, J. Phys. Chem. C **115**, 19216 (2011)
252. J.W. Rhim, S.I. Hong, H.M. Park, P.K.W. Ng, J. Agr. Food. Chem. **54**, 5814 (2006)
253. R. Hema, A.A. Amirul, P.N. Ng, Polym. Bull. **70**, 755 (2013)
254. N.K. Mal, M. Fujiwara, Y. Tanaka, Nature **421**, 350 (2003)
255. N. Jungmann, M. Schmidt, M. Maskos, J. Weis, J. Ebenhoch, Macromolecules **35**, 6851 (2002)
256. Y. Lu, J. McLellan, Y. Xia, Langmuir **20**, 3464 (2004)
257. N. Jungmann, M. Schmidt, J. Ebenhoch, J. Weis, M. Maskos, Ang. Chem. Int. Ed. **42**, 1713 (2003)

258. O. Emmerich, N. Hugenberg, M. Schmidt, S.S. Sheiko, F. Baumann, B. Deubzer, J. Weis, J. Ebenhoch, Adv. Mater. **11**, 1299 (1999)
259. E. Ruiz-Hitzky, P. Aranda, M. Darder, M. Ogawa, Chem. Soc. Rev. **40**, 801 (2011)
260. A.J. Patil, M. Li, E. Dujardin, S. Mann, Nano Lett. **7**, 2660 (2007)
261. A.M. Collins, N.H.M. Kaus, F. Speranza, W.H. Briscoe, D. Rhinow, N. Hampp, S. Mann, J. Mater. Chem. **20**, 9037 (2010)
262. S.S. Ray, Acc. Chem. Res. **45**, 1710 (2012)
263. S. Singh, S. Sinha Ray, J. Nanosci. Nanotechnol. **7**, 2596 (2007)
264. S. Inkinen, M. Hakkarainen, A.-C. Albertsson, A. Sodergad, Biomacromolecules **12**, 523 (2011)
265. J. Ahmed, S.K. Varshney, Int. J. Food Prop. **14**, 37 (2010)
266. R.M. Rasal, A.V. Janorkar, D.E. Hirt, Prog. Polym. Sci. **35**, 338 (2010)
267. S. Joshi, J. Ind. Ecol. **12**, 474 (2008)
268. S. Sinha Ray, K. Yamada, M. Okamoto, A. Fujimoto, A. Ogami, K. Ueda, Polymer **44**, 6633 (2003)
269. S. Sinha Ray, M. Bousmina, Prog. Mater. Sci. **50**, 962 (2005)
270. R.E. Drumright, P.R. Gruber, D.E. Henton, Adv. Mater. **12**, 1841 (2000)
271. L. Avérous, E. Pollet (eds.), *Environmental Silicate Nano-Biocomposites, Green Energy and Technology* (Springer, London, 2012)
272. R.A. Hule, D.J. Pochan, MRS Bull. **32**, 354 (2007)
273. N.V. Pogodina, C. Cerclé, L. Avérous, R. Thomann, M. Bouquey, R. Muller, Rheol. Acta. **47**, 543 (2008)
274. Y.J. Fan, H. Nishida, S. Hoshihara, Y. Shirai, Y. Tokiwa, T. Endo, Polym. Degrad. Stab. **79**, 547 (2003)
275. G. Chen, J. Yoon, J. Polym. Sci. Polym. Phys. **43**, 478 (2005)
276. Q. Zhou, M. Xanthos, Polym. Degrad. Stab. **94**, 327 (2009)
277. K. Okamoto, K. Toshima, S. Matsumura, Macromol. Biosci. **5**, 813 (2005)
278. M.A. Paul, M. Alexandre, P. Degee, C. Henrist, A. Rulmont, P. Dubois, Polymer **44**, 443 (2003)
279. S. Marras, I. Zuburtikudis, C. Panayiotou, Eur. Polym. J. **43**, 2191 (2007)
280. K. Fukushima, M. Murariu, G. Camino, P. Dubois, Polym. Degrad. Stab. **95**(6), 1063 (2010)
281. S. Bourbigot, G. Fontaine, Polym. Chem. **1**(9), 1413 (2010)
282. K. Sudesh, H. Abe, Y. Doi, Prog. Polym. Sci. **25**, 1503 (2000)
283. S.Y. Lee, Biotechnol. Bioeng. **49**, 1 (1996)
284. K. Sudesh, T. Iwata, Clean **36**, 433 (2008)
285. X. Wang, Y. Du, J. Yang, X. Wang, X. Shi, Y. Hu, Polymer **47**, 6738 (2006)
286. W. Kim, Q. Zhang, F. Saito, J. Mater. Sci. **35**, 5401 (2000)
287. K.C.B. Yeong, J. Wang, S.C. Ng, Biomaterials **22**, 2705 (2001)
288. S.C. Liou, S.Y. Chen, H.Y. Lee, J.S. Bow, Biomaterials **25**, 189 (2004)
289. I. Yamaguchi, K. Tokuchi, H. Fukuzaki, Y. Koyama, K. Takakuda, H. Monma, J. Tanaka, J. Biomed. Mater. Res. **50**, 20 (2001)
290. S.C. Liou, S.Y. Chen, D.M. Liu, Biomaterials **24**, 3981 (2003)
291. H.-W. Kim, J.C. Knowles, H.-E. Kim, J. Biomed. Mater. Res. B Appl. Biomater. **74**, 686 (2005)
292. X. Zheng, S. Zhou, Y. Xiao, X. Yu, B. Feng, J. Biomed. Mater. Res. B Appl. Biomater. **91**, 181 (2009)
293. S. Nayar, A. Sinha, Colloids Surf. B. **35**, 29 (2004)
294. A. Bigi, B. Bracci, S. Panzavolta, Biomaterials **25**, 2893 (2004)
295. S.-M. Lee, E. Pippel, U. Gösele, C. Dresbach, Y. Qin, C.V. Chandran, T. Bräuniger, M.K.G. Hause, Science **324**, 488 (2009)
296. N.C. Bigall, M. Reitzig, W. Naumann, P. Simon, K.-H. van Pée, A. Eychmüller, Angew. Chem. Int. Ed. **47**, 7876 (2008)
297. S.-C. Liou, S.-Y. Chen, D.-M. Liu, J. Biomed. Mater. Res. B Appl. Biomater. **73**, 117 (2005)

298. M. Jollands, K. Gupta, J. Appl. Polym. Sci. **118**, 1489 (2010)
299. G. Crini, Prog. Polym. Sci. **30**, 38 (2005)
300. S.K. Mallapragada, B. Narasimhan (eds.), *Handbook of Biodegradable Polymeric Materials and their Applications* (American Scientific Publishers, Ames, 2006)
301. L. Introzzi, T.O.J. Blomfeldt, S. Trabattoni, S. Tavazzi, N. Santo, A. Schiraldi, L. Piergiovanni, S. Farris, Langmuir **28**, 11206 (2012)
302. I. Siró, D. Plackett, Cellulose **17**, 459 (2010)
303. A.N. Nakagaito, S. Iwamoto, H. Yano, Appl. Phys. A Mater. Sci. Process. **80**, 93 (2005)
304. I. Cherny, E. Gazit, Angew. Chem. Int. Ed. **47**, 4062 (2008)
305. www.mercedesbenz.pt/content/portugal/mpc/mpc_portugal_website/ptng/home_mpc/passengercars/home/passenger_cars_world/environment_portugal/environments/value_chain/natural_fibre.html. Accessed March 2010
306. A. Zimmer, M.J. Andrade, F.A.L. Sánchez, A.S. Takimi, Use of natural and modified natural nanostructured materials, in *Nanostructured Materials for Engineering Applications*, ed. by C.P. Bergmann, M.J. Andrade (Springer, Heidelberg, 2011), p. 157
307. D. Klemm, B. Heublein, H.P. Fink, A. Bohn, Angew. Chem. Int. Ed. **44**, 3358 (2005)
308. A.N. Nakagaito, H. Yano, Appl. Phys. A. **80**, 155 (2005)
309. M.C. Floody, B.K.G. Theng, P. Reyes, M.L. Mora, Clay Miner. **44**, 161 (2009)
310. D. Sánchez-García, A. López-Rubio, J.M. Lagaron, Trends Food Sci. Technol. **21**, 528 (2010)
311. F. Rodriguez, H.M. Sepulveda, J. Bruna, A. Guarda, M.J. Galotto, Packag. Technol. Sci. **26**, 149 (2013)
312. M.M. Meier, L.A. Kanis, J.C.D. Lima, A.T.N. Pires, V. Soldi, Polym. Adv. Technol. **15**, 593 (2004)
313. T. Zhang, W. Wang, D. Zhang, X. Zhang, Y. Ma, Y. Zhou, L. Qi, Adv. Funct. Mater. **20**, 1152 (2010)
314. Y.-L. Chung, H.-M. Lai, Carbohydr. Polym. **80**, 525 (2010)
315. A. Vazquez, V.A. Alvarez, Biodegradable nanocomposites based on starch, PCL and their blends, in *Nanocomposites: Preparation, Properties and Performance*, ed. by L. Mancini, C. Espósito (Nova Publisher, New York, 2009), pp. 133–164
316. M. Avella, J.J. De Vlieger, M.E. Errico, S. Fischer, P. Vacca, M.G. Volpe, Food Chem. **93**, 467 (2005)
317. V.P. Cyras, L.B. Manfredi, M.-T. Ton-That, A. Vázquez, Carbohydr. Polym. **73**, 55 (2008)
318. K. Majdzadeh-Ardakani, A.H. Navarchian, F. Sadeghi, Carbohydr. Polym. **79**, 547 (2010)
319. Q.-X. Zhang, Z.-Z. Yu, X.-L. Xie, K. Naito, Y. Kagawa, Polymer **48**(24), 7193 (2007)
320. A.K. Mohanty, M. Misra, L.T. Drzal (eds.), *Natural Fibers, Biopolymers, and Biocomposites* (CRC Press/Taylor & Francis Group, Boca Raton, 2006), p. 896
321. A.K. Sugih, *Synthesis and Properties of Starch based Biomaterials* (University of Groningen, Groningen, 2008), p. 155
322. B.-S. Chiou, E. Yee, G.M. Glenn, W.J. Orts, Carbohydr. Polym. **59**, 467 (2005)
323. F. Chivrac, E. Pollet, M. Schmutz, L. Avérous, Carbohydr. Polym. **80**, 145 (2010)
324. B. Chen, J.R.G. Evans, Carbohydr. Polym. **61**, 455 (2005)
325. L.N. Luduena, J.M. Kenny, A. Vazquez, V.A. Alvarez, Mater. Sci. Eng. A-Struct. Mater. Prop. Microstruct. Proc. **529**, 215 (2011)
326. N.E. Kochkina, V.A. Padokhin, Russ. J. Appl. Chem. **84**, 1451 (2011)
327. L. Chen, T. Wang, J. Tong, Trends Anal. Chem. **30**, 1095 (2011)
328. H. Sun, L. Cao, L. Lu, Nano Res. **4**, 550 (2011)
329. N.A. Travlou, G.Z. Kyzas, N.K. Lazaridis, E.A. Deliyanni, Langmuir **29**, 1657 (2013)
330. M.C. Gutierrez, M.L. Ferrer, P. Tartaj, F. Monte, in *Hybrid Nanocomposites for Nanotechnology*, ed. by L. Merhari (Springer, New York, 2009), p. 707
331. C.M. Niemeyer, Angew. Chem. Int. Ed. **40**, 128 (2001)
332. C. Sanchez, P. Gomez-Romero, *Functional Hybrid Materials* (Wiley, Weinheim, 2004)
333. G.M. Whitesides, Nat. Biotechnol. **21**, 1161 (2004)

334. K. Katagiri, R. Hamasaki, K. Ariga, J.-I. Kikuchi, J. Am. Chem. Soc. **124**, 7892 (2002)
335. K. Matsui, S. Sando, T. Sera, Y. Aoyama, Y. Sasaki, T. Komatsu, T. Terashima, J.-I. Kikuchi, J. Am. Chem. Soc. **128**, 3114 (2006)
336. D.G. Shchukin, I.L. Radtchenko, G.B. Sukhorukov, Mater. Lett. **57**, 1743 (2003)
337. N. Gaponik, I.L. Radtchenko, G.B. Sukhorukov, H. Weller, A.L. Rogach, Adv. Mater. **14**, 879 (2002)
338. E. Ruiz-Hernandez, A. Lopez-Noriega, D. Arcos, I. Izquierdo-Barba, O. Terasaki, M. Vallet-Regi, Chem. Mater. **19**, 3455 (2007)
339. K. Kostarelos, Adv. Colloid Int. Sci. **106**, 147 (2003)
340. D.A. Lavan, T. McGuire, R. Langer, Nat. Biotechnol. **21**, 1184 (2003)
341. M. Creixell, N.A. Peppas, Nano Today **7**, 367 (2012)
342. M.G. Shapiro, T. Atanasijevic, H. Faas, G.G. Westmeyer, A. Jasanoff, Magn. Reson. Imaging **24**, 449 (2006)
343. S. Svenson, D.A. Tomalia, Adv. Drug Deliv. Rev. **57**, 2106 (2005)
344. P. Tartaj, M.P. Morales, T. Gonzalez-Carreño, S. Veintemillas-Verdaguer, C.J. Serna, Synthesis, properties and biomedical applications of magnetic nanoparticles, Chapter 5, in *Handbook of Magnetic Materials*, ed. by K.H.J. Buschow, vol. 16 (Elsevier, Amsterdam, 2006), p. 403
345. O. Bomati-Miguel, M.P. Morales, P. Tartaj, J. Ruiz-Cabello, P. Bonville, M. Santos, X.Q. Zhao, Biomaterials **26**, 5695 (2005)
346. Z. Li, L. Wei, M. Gao, H. Lei, Adv. Mater. **17**, 1001 (2005)
347. W. Li, Z. Huang, J.A. MacKay, S. Grube, F.C. Szoka Jr., J. Gene Med. **7**, 67 (2005)
348. Y. Jun, Y. Hou, J.S. Choi, J.H. Lee, H.T. Song, S. Kim, S. Yoon, K.S. Kim, J.S. Shin, J.S. Su, J. Cheon, J. Am. Chem. Soc. **127**, 5732 (2005)
349. R.I. Mahato, Adv. Drug Deliv. Rev. **57**, 699 (2005)
350. S. Mornet, S. Vasseur, F. Grasset, E. Duguet, J. Mater. Chem. **14**, 2161 (2004)
351. X. Xu, S. Majetich, S. Asher, J. Am. Chem. Soc. **124**, 13864 (2002)
352. S. Lu, J. Forcada, J. Polym. Sci. A Polym. Chem. **44**, 4187 (2006)
353. F. Caruso, M. Spasova, A. Sucha, M. Giersig, R.A. Caruso, Chem. Mater. **13**, 109 (2001)
354. D. Horak, J. Polym. Sci. A Polym. Chem. **39**, 3707 (2001)
355. X. Xu, G. Friedman, K. Humfeld, S. Majetich, S. Asher, Chem. Mater. **14**, 1249 (2002)
356. C. Yang, Y. Guan, J. Xing, H. Liu, Langmuir **24**, 9006 (2008)
357. C. Yang, Y. Guan, J. Xing, H. Liu, J. Polym. Sci. A Polym. Chem. **46**, 203 (2008)
358. H. Lin, Y. Watanabe, M. Kimura, K. Hanabusa, H. Shirai, J. Appl. Polym. Sci. **87**, 1239 (2003)
359. A.H. Lu, E.L. Salabas, F. Schuth, Angew. Chem. Int. Ed. **46**, 1222 (2007)
360. J.P. Jolivet, C. Chaneac, E. Tronc, Chem. Commun. **481** (2004)
361. N.A. Frey, S. Peng, K. Cheng, S.H. Sun, Chem. Soc. Rev. **38**, 2532 (2009)
362. S. Laurent, D. Forge, M. Port, A. Roch, C. Robic, L.V. Elst, R.N. Muller, Chem. Rev. **108**, 2064 (2008)
363. R.K. Gilchrist, R. Medal, W.D. Shorey, R.C. Hanselman, J.C. Parrott, C.B. Taylor, Ann. Surg. **146**, 596 (1957)
364. S. Biyikli, M.F. Modest, R. Tarr, J. Biomed. Mater. Res. **20**, 1335 (1986)
365. K.L. Reed, T.D. Brown, M.G. Conzemius, J. Biomech. **36**, 1317 (2003)
366. I. Hilger, R. Hergt, W.A. Kaiser, IEE Proc. Nanobiotechnol. **152**, 33 (2005)
367. S. Wada, K. Tazawa, I. Furuta, H. Nagae, Oral Dis. **9**, 218 (2003)
368. A. Ito, M. Shinkai, H. Honda, T. Kobayashi, Cancer Gene Ther. **8**, 649 (2001)
369. H. Gu, K. Xu, Z. Yang, C.K. Chang, B. Xu, Chem. Commun. **4270** (2005)
370. C.A. Mirkin, R.L. Letsinger, R.C. Mucic, J.J. Storhoff, A DNA-based method for rationally assembling nanoparticles into macroscopic materials. Nature **382**, 607 (1996)
371. A.P. Alivisatos, K.P. Johnsson, X. Peng, T.E. Wilson, C.J. Loweth, M.P. Bruchez Jr., P.G. Schultz, Nature **382**, 609 (1996)
372. W. Shenton, S.A. Davis, S. Mann, Adv. Mater. **11**, 449 (1999)

373. S. Connolly, D. Fitzmaurice, Adv. Mater. **11**, 1202 (1999)
374. S. Srivastava, A. Verma, B.L. Frankamp, V.M. Rotello, Adv. Mater. **17**, 617 (2005)
375. E. Ruiz-Hitzky, P. Aranda, M. Dardera, G. Rytwo, J. Mater. Chem. **20**, 9306 (2010)
376. L. Ai, C. Zhang, Z. Chen, J. Hazard. Mater. **192**, 1515 (2011)
377. P. Alivisatos, Nat. Biotechnol. **22**, 47 (2004)
378. A.K. Boal, V.M. Rotello, J. Am. Chem. Soc. **12**, 4914 (1999)
379. V. Singh, S. Pandey, S.K. Singh, R. Sanghi, J. Sol-Gel Sci. Technol. **47**, 58 (2008)
380. R.B. Bathia, C.J. Brinker, Chem. Mater. **12**, 2434 (2000)
381. M. Kato, K. Sakai-Kato, N. Matsumoto, T. Toyóoka, Anal. Chem. **74**, 1915 (2002)
382. A. Llobera, V.J. Cadarso, M. Darder, C. Domırnguez, C. Fernarndez-Sarnchez, Lab Chip **8**, 1185 (2008)
383. E.J.A. Pope, K. Braun, C.M. Peterson, J. Sol-Gel Sci. Technol. **8**, 635 (1997)
384. J.C. Rooke, A. Leronard, B.-L. Su, J. Mater. Chem. **18**, 1333 (2008)
385. S. Weiss, M. Tauber, W. Somitsch, R. Meincke, H. Muller, G. Berg, G.M. Guebitz, Water Res. **44**, 1970 (2010)
386. A. Sorrentino, G. Gorrasi, V. Vittoria, Trends Food Sci. Technol. **18**, 84 (2007)
387. J.H. An, S. Dultz, Appl. Clay Sci. **36**, 256 (2007)
388. M. Darder, M. López-Blanco, P. Aranda, A.J. Aznar, J. Bravo, E. Ruiz-Hitzky, Chem. Mater. **18**, 1602 (2006)
389. H.L. Zhu, J.Y. Shen, X.X. Feng, H.P. Zhang, Y.H. Guo, J.Y. Chen, Mater. Sci. Eng. C **30**, 132 (2010)
390. H.J. Bae, H.J. Park, S.I. Hong, Y.J. Byun, D.O. Darby, R.M. Kimmel, W.S. Whiteside, LWT Food Sci. Technol. **42**, 1179 (2009)
391. C. Mousty, S. Cosnier, M. Sanchez-Paniagua Lopez, E. Lopez-Cabarcos, B. Lopez-Ruiz, Electroanalysis **19**, 253 (2007)
392. V. Caballero, F.M. Bautista, J.M. Campelo, D. Luna, J.M. Marinas, A.A. Romero, J.M. Hidalgo, R. Luque, A. Macario, G. Giordano, Process Biochem. **44**, 334 (2009)
393. R. Donat, S. Aytas, J. Radioanal. Nucl. Chem. **265**, 107 (2005)
394. E. Ruiz-Hitzky, M. Darder, P. Aranda, M.A. Martırn del Burgo, G. del Real, Adv. Mater. **21**, 4167 (2009)

Conclusions and Future Prospects

The performed survey has shown that metal-polymer hybrid nanocomposites is an intensely developed field of chemistry and materials science, which possesses its objects and investigation methods. Information in this field is accumulated continuously collecting the facts, which seemed odd before and completing logical structuring of new ways of production of nanocomposite materials with high marketing potential.

The considered vast factual material makes a reader sure that chemistry is a great deal of art at all stages of development and preparation of a material: processes easily "realized" on paper encounter a lot of submarine ridges, which should be taken into account and overcome on the way of development of a nanometer material.

How we see development of this direction and some of its branches?

A special attention will be drawn to metrological following of new nanomaterials: development of fine methods of monitoring, analysis, increase in sensitivity, precision of methods, increase in their express performance, reliability of new instrumentation comprising contemporary electronic basis of components.

Intellectual materials of new generation will be developed, which are sensitive to exterior impacts, adapted to remote control over their physical-chemical properties (electric, magnetic, and acoustic fields, temperature, illumination, pressure, laser and radioactive irradiation). Magnetic elastomers, metamaterials, magnetic nanostructural materials, which are used as permanent magnets, magnetic conductors, holding or bracketing elements, which can repeat roughness and curvature of ferromagnetic surfaces they are contacting with, should be designed in future. The composites with magnetically induced effect of plasticity (shape memory effect), which displace the effect of magnetic field on dielectric permeability and Young modulus of magnetic elastomers, will be developed.

The promising way will be further development of nanomaterials for stealth and hydro-acoustic technologies for design of electronic security means, first of all, for lowering detection in wide frequency range of electromagnetic radiation, for ultrasonic flow meters, electromagnetic and hydro-acoustic metamaterials, in spite

A.D. Pomogailo, G.I. Dzhardimalieva, *Nanostructured Materials Preparation via Condensation Ways*, DOI 10.1007/978-90-481-2567-8 of the fact that a versatile specification of radio-absorbing materials is being developed and even produced. Most probably, such characteristics as maximum absolute value of reflection coefficient will improve within a working range of wavelengths, as well as thickness and weight of a unit area absorber, and parameters of acoustic waves in piezoelectric structures. In order to widen a working range of these devices, multilayered structures will be used, structures containing diffraction lattices, frequency selective surfaces, active loaded dipoles, and honeycombed and cellular structures. Principally new classes of nanocomposite materials will arrive, which will allow achievement of substantial success in development of thin, light wideband devices, improvement of acoustic and electro-physical properties and the effect of external electric fields on them. Fundamental ideas of nanoparticles assembling into planar ensembles will be realized, including assembling of chemically inert nanomaterials, which involves condensation of Langmuir and SAS mixed monolayers, substrate-induced assemblage of different nanoparticles in gels (soft assemblage). In this case "one-step" technique provides high compatibility with content and size of the nanoparticles and regulated packing dense of ensemble of nanoparticles on extended surfaces.

Great advantages should be expected in integrated metal-polymer biocomposites, materials produced by sol–gel method, intercalation technologies, reduction technique, and thermolysis, which have already been mentioned in conclusions of other chapters. One of the most important properties of metal nanoparticles and their composites is their ability to enhance local optical electromagnetic near-fields and concentrate electromagnetic irradiation in subvalent zones. Overcoming of fundamental diffraction limit in optics opens broad prospects in nano-optoelectronics for development of ultra-small lasers and optoelectronic devices, and in bio-optoelectronics of sensor molecules and photo modified biological cells.

Nevertheless, till now, as a rule, it is impossible to recognize correlations between content, structural features, and properties of metal polymer nanocomposites, which in many respects restrains scientifically grounded approach to structuring of these materials and prediction of their promising properties.

Index

A
Acids
  acetic, 93, 120, 147, 443
  adamantic, 29
  Bronsted, 237
  citric, 25, 26, 36, 120, 182, 183
  dehydroascorbic, 27
  itaconic, 26, 227, 303
  Keggin, 185
  L-ascorbic, 27
  L-asparagine, 394
  methacrylic (MAA), 55, 149, 152, 153, 161, 162, 175
  Oleyic, 29
  polyacrylic (PAA), 39, 41, 45, 51, 60, 61, 76, 321, 325, 353, 362, 363, 368, 425
  polyamic (PAmA), 168, 169, 172
  polylactic (PLA), 9, 246, 389, 406, 421–423, 425
  sebacic, 64
  silicic, 143, 166, 167, 215, 398, 408, 415, 416
Agar-agar, 4
Agent
  antimicrobial, 389, 426
  bioreducing, 390–403
  chain transfer, 154
  cross-linking, 54
  immune-response modulating, 408
  organic gelling, 175
  reducing, 3, 5, 8, 15, 16, 22–28, 32–36, 42–44, 47, 56, 62–71, 74, 77, 79, 80, 94–98, 102, 107, 108, 114, 122, 186, 344, 350, 390–403, 427
  stabilizing, 3, 22, 24, 29, 30, 40, 42, 44, 62, 76, 117, 125, 146, 164, 313, 322, 365, 395
  sulfiding, 119
Agglomeration, 4, 13, 14, 20, 24, 46, 58, 123, 188, 253, 257
Aggregates
  fractal, 191, 193
  number, 49
AIBN. *See* Azobis(isobutyronitrile) (AIBN)
Alcoholysis, 153, 422, 423
Algae, 319, 397, 404
  Bacillariophyta, 397
Alkoxides
  metal, 5, 10, 141, 143, 149, 163
  silicone, 409
  tantalum, 148
  titanium, 145, 147, 151–153, 175, 194, 218
  vanadium, 153, 238
  zirconium, 149
Alkoxometallates, 352
Alloy, 7, 10, 32, 72–74, 96, 117, 120, 121, 130, 131, 304, 358–364, 366, 392, 405
Alumosilicates, 209
Aminoboranes, 27–28, 74
Atom transfer radical polymerization (ATRP), 58, 60, 155
Azobis(isobutyronitrile) (AIBN), 154, 227, 251

B
Bacteria
  actinomycetes, 404
  Bacillus sphaericus, 400
  Bacillus subtilis, 400, 404, 417

A.D. Pomogailo, G.I. Dzhardimalieva, *Nanostructured Materials Preparation via Condensation Ways*, DOI 10.1007/978-90-481-2567-8

Bacteria (*cont.*)
- *Clostridium pasterianum,* 399
- *C. necator,* 402
- *Cupriavidus metalliduran,* 402
- *Desulfovibrio desulfuricans (D. Desulfuricans),* 399, 401–404
- gram-negative, 411, 423
- gram-positive, 411, 423
- magnetotactic, 396
- *Pyrobaculum islandicum,* 399
- *Rhodopseudomonas capsulata,* 398, 404
- *Shewanella algae,* 404
- *Shewanella oneidensis (S. Oneidensis),* 399, 400, 404

Benzoyl peroxide (BPO), 161
Biological activity, 10, 11, 179, 405, 410, 416, 424
Biomimetics, 144, 390, 416
Biomineralization, 10, 141, 390, 397, 402, 414–417, 434
Bionanocomposites, 10, 11, 133, 389–436
Biopolymer, 9, 142, 389, 400, 408, 421, 423, 424, 429
Bioreduction, 10, 394, 399, 402
Biosensors, 5, 10, 120, 133, 142, 305, 351, 405, 409, 417, 419, 427, 434, 435
Biosorption, 9, 10, 389, 399, 415, 434
Biosynthesis, 390–392, 395, 398, 404, 415, 431
Block-copolymer, 5, 47–54, 59, 73, 159, 160, 174, 194, 226–229, 247, 365, 366, 416
Block-scheme, 150, 178, 180, 181, 184
Bonds
- hydrogen, 212
- ion-dipole, 212
- London attraction forces, 4, 45
- Van-der-Waals, 212

Borohydride, 25, 28, 34, 56, 74, 79
Bourdon pressure gauge, 301
Brownian motion, 47, 433

C

Cellulose
- bacterial, 425, 427
- carboxymethyl, 63, 77, 94

Ceramics
- $A^{2+}B^{4+}O_{3-}$, 184, 185
- cerasomes, 430
- heteroelemental, 5, 177–189, 195
- heterometallic, 142, 183, 362
- high temperature superconducting (HTSC), 185, 362–364, 416
- stabilized yttrium-zirconium (YSZ), 183
- YBCO, 364

Chalcogenides, 1, 6, 95, 118, 143, 144, 217, 220, 271–276, 337, 354
Chemosorption, 47, 393, 394
Chitosan, 47, 76, 119, 166, 389, 409, 419, 421, 430, 433, 435
Clay
- bentonite, 207, 212, 215, 254, 257, 419
- cloisite
  - 30B, 209, 210, 254, 261, 266, 426, 427
  - C10A, 427
  - C15A, 427
  - Na, 419, 427
- halloysite, 424, 426
- kaolin, 207, 231, 233, 423, 427
- laponite, 207, 211, 258, 419
- montmorillonite (MMT), 5, 10, 11, 127, 205–207, 209–212, 214–216, 218, 221–223, 227, 231–233, 237, 242, 244, 248–252, 258–261, 266, 268, 273, 418, 419, 421, 423, 425–427, 429
- saponite, 207, 208, 211, 218, 418
- wollastonite, 419

Cloisite, 209, 210, 254, 261, 266, 419, 425–427
Clusters
- Ag_2+, 17, 115
- Ag_3, 22
- $Ag_8$2+, 17, 18
- Au_{25}(glutathione)$_{18}$, 17
- oxometallate, 151
- silicon-based oxo, 152
- SiO_2, 166

Coatings
- hybrid, 169
- polyamide-silica gel, 169
- smart, 60

Coefficient
- diffusion, 130, 145, 258, 261, 372
- effective, 261
- heat transfer, 297

Coil, 50, 274, 348
Combustion, 253, 294, 299, 300, 308, 341
Complexes
- alkoxy zirconium, 173
- catecholate, 143
- metal, 7, 32, 41–43, 47, 100, 102, 109, 114, 117, 149, 238, 269, 273, 289, 292, 322–324, 329, 333, 338, 342, 347, 351, 353, 358, 362
- phenanthroline, 182
- polymer, 9, 74, 110, 121, 124, 149, 162, 333, 335, 339, 347–352

Composites
- organic-inorganic, 166, 174
- poly(hydroxyalkanoate), 423

polymer-inorganic, 175, 259
TiO_2–cellulose, 405
Computer modelling, 7, 370–374
θ-Conditions, 50
Constants, 25, 28, 33, 79, 92, 95, 107, 113, 119, 142, 145, 146, 165, 167, 179, 193, 213, 214, 217, 241, 242, 290, 291, 293, 297–300, 308, 316, 326, 338, 365, 368, 371, 412
kinetic, 146
Curie temperature, 358
Cyclization, 303, 304
Cytochrome C, 409, 434

D
Decomposition, 7, 23, 24, 27, 80, 94, 96, 100, 103, 106, 108, 109, 144, 169, 185, 193, 220, 251, 253, 265, 267, 271, 291, 292, 297–300, 308–310, 312–315, 317–327, 331, 332, 337–339, 342, 343, 345, 346, 348–354, 357, 358, 360, 362–367, 369–371, 373, 410, 423, 429, 432
Degree
deacetylation, 76
polymerization, 47–50, 166
quaternization, 166
Dendrimers
branching number, 67
dendron, 44
generation, 42
Deoxyribonucleic acid (DNA), 11, 400, 417–420, 430, 431, 433
Dextranes, 9, 389, 425, 430, 432, 433
Diameters, 4, 13, 37, 38, 45, 46, 50, 62, 64, 76, 94, 95, 99, 111, 118, 119, 121, 124, 148, 155, 164, 185, 193, 219, 248, 258, 268, 271, 273, 274, 290, 302, 305, 313, 318, 320–322, 327, 336, 337, 343, 352, 353, 355, 357, 395, 401, 410, 412, 415, 416, 423, 425, 430
hydrodynamic, 57
Diatoms, 397, 398, 408, 415, 416
Dielectric loss tangent, 92, 93, 95
DNA. *See* Deoxyribonucleic acid (DNA)
Drugs, 8, 10, 11, 54, 120, 256, 389, 390, 405–409, 411–414, 421, 430–432, 434
multifunctional vectors (MVD), 411, 412

E
E. coli. See Escherichia coli (E. coli)
Effect
catalytic, 100, 108, 164, 337
"focusing," 19
polyelectrolyte, 60
Electrospinning, 142, 410, 425
Equation
Arrhenius, 299, 345
Avramy-Erofeev, 303
Coats-Redfern, 350
Escherichia coli (E. coli), 395, 402–404, 423
Ethylene glycol, 26–27, 71, 92–95, 98, 99, 122, 155, 182, 183, 292, 320
Exfoliation, 5, 6, 205, 206, 211, 215, 222, 223, 225, 227, 229, 230, 245, 249, 251, 252, 256, 257, 262, 266, 267, 269, 271, 273, 274, 419, 420, 430

F
Faraday, M., 4, 23
Ferrites, 94, 95, 183, 313, 314, 327, 331, 332, 360, 362, 408
Ferromagnetic resonance (FMR), 340
Films
polytetrafluoroethylene, 173
polyurethane, 190
transparent, 175, 182, 267, 356, 357
Flagella, 397, 400, 401
Forces
attraction, 4
coulomb, 20
London attraction, 4, 46
repulsive, 48
Van-der-Waals, 3, 47, 113, 271, 374
Fourier transformation infrared spectroscopy (FTIR), 69, 145, 270, 292, 394
Fractal, 113, 190–195
dimensionality (D), 40, 191–193
Frank-Kamenetsky, D.A., 294
Frontal polymerization (FP), 338–342
Fungi
Colletotrichum sp., 404
Fusarium oxysporum, 393, 402, 404, 405
Rhizopus oryzae, 398, 404
Verticillium sp., 404

G
Glass transition temperature (Tg), 109, 154, 241, 242, 255, 257, 260, 266, 348, 363, 410
Graphene, 95, 220, 270, 303, 308, 435

Growth, 1, 3, 4, 8, 10, 13–24, 26, 34, 35, 37, 38, 40, 42, 45, 46, 54, 68, 71, 72, 79, 80, 91, 94, 106, 112, 115, 116, 122, 123, 128, 143–145, 148, 151, 166, 183, 186, 193, 230, 232, 238, 287, 294, 311, 312, 317, 322, 352, 360, 365, 371, 374, 392, 398, 402, 403, 411, 416, 421, 425

H

Heating
  hydrothermal, 95
  microwave, 92, 94–96
Heteroadagulation, 288
Heterocoagulation, 218
Heteroelements, 5, 177–189
Heterometallic sulfides, 357
Heterostructures, 7, 72, 133, 358
Histogram, 35
Hydrazine hydrate, 16, 17, 25, 31, 32, 35, 44, 76, 77
Hydrogels, 54–58, 117, 118, 256
Hydrolysis, 5, 10, 15, 25, 60, 121, 143–150, 152, 154, 164, 173, 175, 177, 179, 183, 186, 190, 192, 218, 220, 228, 267, 268, 310, 405, 406, 408, 427
Hydroxyapatite (HA), 10, 343, 344, 389, 408, 424
Hydroxylation, 143
Hyperthermia, 11, 431, 433, 434

I

Imidization, 170–172
Impact toughness, 260
Incapsulation, 226
Indium-tin oxide (ITO), 108
Interactions
  energy, 222
  Fermi-Dirac, 270
  intermolecular, 6, 58, 221
  interphase, 11, 142, 155, 256, 411, 427
  π- π, 174
Intercalation, 3, 5–7, 10, 127, 142, 163, 206–222, 224–267, 269–274, 276, 417–436
  sonochemical, 127
Interpenetration networks (IPN), 161, 175

L

Lamellae, 48
L-Arginine, 394
Laser ablation, 91, 132–133, 302
Law
  Boltzmann, 293
  Michaelis-Menten kinetics, 407
Layered double hydroxide (LDH), 208, 211, 215, 216, 226, 412–414
Lithography
  chemical, 396
  dip-pen, 142
L-tryptophan, 394, 409
L-tyrosine, 394

M

Macrokinetics, 290, 298
Magnetism, 180
Magnetite, 180, 322, 358–360, 396, 402, 408, 431–433
Magnetization, 341, 342, 358, 366, 433
Magnetosomes, 396, 397
Masterbatch, 223, 243, 251, 254
Matrix
  oxygen-free polymer, 7, 325
  polymer, 3–7, 10, 40–62, 76, 100, 117, 120, 124, 128, 130, 141, 142, 164, 175, 193, 205, 206, 212, 222, 226, 255, 256, 288, 324–337, 339, 342, 344–348, 351, 352, 354–357, 365, 374, 390, 423, 424
Maya blue dye, 2
MCM-41, 147, 151, 186, 409, 412
Mechanism
  coalescence, 79
  ion-exchange, 57, 272
  Ostwald ripening, 96
  spinodal decomposition, 193
  thermolysis, stages, 7, 314, 322, 325, 331
Membrane, 8, 10, 11, 58, 78, 141, 169, 175, 243, 268, 300, 301, 334, 337, 343, 353, 397, 398, 401, 408, 410, 414, 419, 420, 423, 430, 431
Merzhanov, A.G., 294
Metal carbides
  hafnium, 8, 181, 182, 302, 353, 368, 369
  tantalum, 8, 148, 151, 181, 182, 353
  titanium, 8, 108, 142, 147, 148, 151–154, 165, 170, 175, 176, 179, 192–194, 218, 230, 274, 302, 305, 353, 361, 405
Metal carbonyls, 336, 359, 364–367
Metal complexes, 7, 32, 41–43, 47, 100, 102, 109, 114, 117, 149, 238, 269, 273, 289, 292, 322–324, 329, 333, 338, 342, 347, 351, 353, 358, 362
  macromolecular, 7, 289, 335, 342, 347

Metal containing monomers, 7, 100, 289, 325, 338–342, 361, 368, 373
Metal-organic frameworks (MOFs), 30
Metal thiolates, 354–357
Methods
absorption spectroscopic, 18, 39, 40, 74
Avramy-Erofeev, 291
Coat–Redfern, 291
differential scanning calorimetry (DSC), 209, 289, 295, 296, 298, 332
differential thermal analysis (DTA), 209, 289, 296–298, 331, 364
differential thermogravimetry (DTG), 289, 296, 325, 331, 363
electrospray mass spectrometry, 16
electrospray photoelectron spectroscopy, 16
emission spectroscopic, 16, 119
Flynn-Wall–Ozawa (FWO), 291, 292
isothermal, 294, 300
Kissinger, 291
molecular engineering, 10, 401
Monte–Carlo, 225, 371
non-isothermal, 291, 331
Pechini, 149
pulse radiolysis, 117
thermogravimetric (TG), 289, 296–298, 331
thermomechanical analysis (TMA), 298
van Krevelen (vK), 291
volumetric, 295, 300–302
Micelles
block-copolymers, 47, 49, 51–53
critical concentration (CMC), 50
cylindrical, 48, 160
Microphase separation, 48, 54, 167
MMT. *See* Montmorillonite (MMT)
Model
"diffusion growth," 17
LaMer, 14
Monte–Carlo, 225
oriented structure, 223
Ostwald ripening, 19, 22, 38, 62, 72, 106, 144
pseudo-multicomponent, 290
pseudo-single-component, 290
pseudo-two-component first-order stage-separated, 291
"seed-mediated growth," 19, 34
thermodynamic, 14–15
Molecular dynamics, 225, 246
Molecular recognition, 268, 347, 374, 416
Molecular sieves, 147, 185, 217
Montmorillonite (MMT), 5, 10, 11, 127, 205–207, 209–212, 214–216, 218, 221, 223, 227, 232, 233, 237, 242, 244, 248–252, 256, 258–261, 266, 268, 273, 418, 419, 421, 423, 425–427, 429
Morphosynthesis, 416, 417

N

Nafion, 58, 162
Nanocomposite
hybrid organic-inorganic, 156, 206, 239
layered, 5, 129, 131, 258, 260
layered silicates, 6, 142, 206, 207, 221, 222, 225, 257, 266
polymer, definition, 2, 3, 5, 51, 65, 117–118, 127–129, 147, 158, 175, 190, 230
self-assembling, 6, 267
Nanocrystals, 15, 19, 20, 22, 23, 71, 75, 94–96, 180–182, 224, 273, 274, 313, 321, 322, 324, 343, 344, 350, 352, 355, 356, 359–361, 396
Nanoparticles
bimetallic, 121, 131, 358
copper, 34, 39, 45, 125, 308, 320
decahedrons, 391, 392
fcc tetrahedral, 391, 392
fcc twinned, 291
flat, 391, 392
gold, 4, 8, 18, 23, 34, 39, 40, 45, 47, 50–52, 59, 64, 67, 71, 74, 78, 102, 103, 107, 108, 112, 121, 124, 127, 133, 164, 185, 187, 189, 190, 267, 303, 336, 355, 356, 358, 359, 391, 392, 394, 398, 400, 401, 403, 427, 433–435
hexagonal platelets, 391, 392
irregular shaped, 391, 392
palladium, 8, 33, 37, 58, 64, 72, 79, 106, 120, 126, 399, 401
rods, 94, 321, 355, 391, 392
self-organized, 4, 267, 310
sheets, 391
silver, 32, 33, 35, 37, 39, 40, 43, 44, 56, 57, 64, 65, 67, 69, 72, 73, 99–102, 108, 109, 111, 112, 117, 118, 121, 131, 132, 177, 187, 189, 320, 322, 324, 357, 358, 391–392, 394, 395, 400, 402
spherical, 40, 46, 121, 304, 320, 339, 352, 355, 415
star-like, 45, 46, 190

Nanoparticles (*cont.*)
tetrapodlike, 34
titania, 179
triangular, 391, 392
Nanoreactors, 5, 47–54, 60, 64, 71, 164, 225, 366, 407
Nitroreductase, 402
Nuclear Magnetic Resonance Imaging (MRI), 11, 431, 432
Nucleation, 1, 3, 4, 8, 10, 14–17, 19–21, 34, 37, 42, 54, 62, 71, 72, 79, 80, 91, 94, 106, 116, 122, 123, 128, 145, 147, 180, 190, 193, 311–313, 352, 360, 371, 392, 403, 416, 425
Nylon, 5, 130, 205, 241, 243, 244

O

Oleylamine, 29, 320–322, 356–358, 360
Organically modified silicates (ORMOSIL), 148, 151, 252, 416
Organoalkoxysilanes, 409, 430
Ostwald ripening, 19, 22, 38, 62, 72, 96, 106, 144
Oxolation, 143, 145, 148
Oxopolymers, 147, 158, 173

P

PAA. *See* Polyacrylic acid (PAA)
PAMAM. *See* Polyamidoamines (PAMAM)
PAN. *See* Polyacrylonitrile (PAN)
PAn. *See* Polyaniline (PAn)
Parameter
Flory-Huggins, 222
Huggins, 48
segregation, 48
thermodynamic, 71, 221, 350, 351
PBT. *See* Polybutylene terephthalate (PBT)
PCL. *See* Poly- ε-caprolactone (PCL)
Pechini method, 149
PEG. *See* Polyethylene glycol (PEG)
PEI. *See* Polyethylene imine (PEI)
PEO. *See* Polyethylene oxide (PEO)
Percolation, 191, 206, 270, 421, 430
Perovskite, 94, 95, 178, 183, 209, 211, 215, 274, 362, 416
Phenanthroline, 182, 325, 329, 348
Photocatalysis, 177
Photolysis, 97, 98, 104, 108, 112, 114
Phytochelatin, 402
PLA. *See* Polylactic acid (PLA)
Plants
Agave sp., 426
alfalfa *Medicago sativa,* 392
Aloe vera, 40, 392
Ananase rectifolius, 426
black tea, 392
Brassica juncea, 392
Chilopsis linearis, 391, 392
Cinnamomum camphora, 391, 392
Cocos nucifera L., 426
Corchorus sp., 426
Cymbopogon flexuosus, 391, 394
Dalbergia sissoo, 392, 393, 405
Emblica officinalis, 392
Pelargonium graveolens, 391–394, 405
Sesbania grains, 392, 427
Sorghum sp., 394, 395
Syzygium cumini, 394
Tamarindus indica, 391, 392
Tetrapanax, 392, 395
Triticum aestivum, 392
Plastification, 242
PMMA. *See* Poly(methyl methacrylate) (PMMA)
Polarization, 93, 155, 181, 190
Poly(2-hydroxyethylmethacrylate) (pHEMA), 410
Poly(allylamine), 415
Poly(allylaminehydrochloride), 415
Poly(ferrocenylsilanes), 159, 160, 359
Poly(methyl methacrylate) (PMMA), 44, 77, 128, 130, 162, 164, 167, 175, 226, 231, 232, 241, 243, 248, 254, 255, 300, 345, 346, 348–350, 366
Poly(n-butyl acrylate), 162
Poly(vinylpyridine) (PVPy), 52, 73, 74
Poly(vinylpyrrolidone) (PVP), 29, 30, 42, 43, 45–47, 68, 71, 76, 77, 94, 96, 113, 125, 126, 162, 175, 243, 244, 271
Polyacrylic acid (PAA), 39, 41, 45, 51, 54, 60, 61, 76, 100, 176, 321, 325, 353, 362, 363, 368, 425
Polyacrylonitrile (PAN), 99, 100, 234, 271, 302–305, 334, 362
Polyamidoamines (PAMAM), 61, 67–69
Polyaniline (PAn), 117, 120, 127, 234, 236–240, 243, 272, 274
Polybutylene terephthalate (PBT), 260, 261
Polycondensation, 64, 148, 153, 158, 173, 231, 289, 368
Poly- caprolactone -caprolactone (PCL), 192, 210, 265, 408
Polyethylene
high density polyethylene (HDPE), 252
low density polyethylene (LDPE), 224, 229, 260, 292, 336, 345

Polyethylene glycol (PEG), 47, 61, 64, 65, 94, 121, 264, 271, 411, 432
Polyethylene imine (PEI), 99, 100, 116, 272, 400
Polyethylene oxide (PEO), 5, 47, 51, 147, 158, 162, 174, 175, 205, 209, 241, 243–245, 257–259, 271, 272, 406
Polyhedral oligosesquioxane (POSS), 155
Polyhydroxybutyrat, 389
Polylactic acid (PLA), 246, 389, 406, 421–423, 425
Polymerization
  anionic, 353, 359
  atom transfer radical (ATRP), 58, 60, 155
  dispersion, 127, 161, 245, 351
  electrochemical, 127
  emulsion, 6, 127, 226, 412, 433
  frontal (FP), 338–342
  gamma-induced, 117
  living radical, 155
  oxidation, 236–237, 272
  oxidizing-reduction, 6
  photochemical, 233
  plasma, 128, 131, 132
  postintercalation, 234
  RAFT, 55
  redox-intercalation, 233–241
  ring-opening, 233
  "ship-in-the-bottle," 205
  in situ, 5, 6, 54, 58, 100, 110, 205, 206, 225–227, 231, 240, 251, 270, 272, 419
  in situ intercalation, 225, 226, 245, 257
  solid-state, 289, 325, 333
  topochemical, 228
Polymers
  amphiphilic, 53, 64
  block-copolymers, 47–54, 59, 73, 160, 174, 365, 366, 416
  conductive, 234, 238, 239
  dendrimer, 67
  hafnium-containing, 368
  hydrogels, 54–58, 118, 256
  metal-containing, 288, 302, 325, 330, 342, 345, 346, 348, 371, 373, 374
  natural
    agar-agar, 4
    cellulose, 4, 9, 63, 175, 276, 337, 389, 421, 425–427
    gummiarabic, 4, 9
    humic substances, 5, 63
    polysaccharides, 5, 9, 63, 175, 289, 398, 408, 409, 418, 419, 425–430, 432
    starch, 4, 11, 29, 418, 419, 421, 425, 427–430
  organo-element, 61–62
  organometallic, 74, 288, 302, 324, 352–354
  star-like, 45, 46, 50, 51, 71, 190, 292, 310, 361
  surfactants, 4, 23, 41, 49–51, 79, 96, 122, 132, 222, 242
Polyoxymetalates (POM), 33, 37, 168, 352
Polypropylene (PP), 165, 222, 224, 246–250, 292, 366
Polysilsesquioxane, 170, 412
Polyvinyl acetate (PVA), 162
Polyvinyl alcohol (PVA), 4, 23, 36, 99, 116–118, 121, 149, 288
Poly-(1-vinylimidazol) (PVI), 166, 415
POM. *See* Polyoxymetalates (POM)
Potential
  Madelung, 181
  redox, 17, 28, 31–33, 36, 72–73, 103, 116, 275, 314
ζ-Potential, 166
PP. *See* Polypropylene (PP)
Prepolymers, 163, 170
Processes
  aerosol, 3, 5, 141, 146, 179, 305, 408
  autocatalytic, 15, 106
  carbonization, 303, 304, 321, 334, 336–348, 354, 370
  electrospinning, 142, 410, 425
  exfoliation, 5, 6, 205, 206, 211, 215, 222, 223, 225, 227, 229, 230, 245, 249, 251, 252, 256, 257, 262, 266, 267, 269, 271, 273, 274, 419, 420, 430
  flame pyrolysis, 91, 320
  graphitization, 353, 354
  intercalation, 3, 5–7, 10, 127, 141, 142, 163, 205–222, 224–233, 236–267, 269–274, 276, 287, 417–436, 450
  metathesis, 175
  plant-mediated, 9, 391, 395
  polyolic, 4, 26, 95, 146
  redox, 58
  "soft" template, 186
  spray, 5, 7, 141
  topotactic in-situ, 225
Pullulan, 425
PVI. *See* Poly-(1-vinylimidazol) (PVI)
PVP. *See* Poly(vinylpyrrolidone) (PVP)

PVPy. *See* Poly(vinylpyridine) (PVPy)
Pyrolysis
IR, 287, 304
linear (LP), 7, 298–300

Q
Quantum dots, 13, 120, 354
Quaternization, 166
Quercetin, 392, 393

R
Radiolysis, 16, 97, 98, 113–115, 117–121
Radionuclides, 9, 401, 402
Radius
gyration (Rg), 50–53
hydrodynamic (Rh), 29, 50, 52, 53, 75, 106, 124, 188, 304, 350
RAPET. *See* Reaction under Autogenic Pressure at Elevated Temperature (RAPET)
Reaction
Boudard disproportionation, 367
condensation, 144
coupling, 158, 174, 175, 190, 192
decarboxylation, 322, 325, 327, 330, 332
heterophase, 36
Hoffman elimination, 253
hydrolysis, 5, 10, 15, 25, 60, 121, 141, 143–150, 152, 154, 164, 173, 175, 177, 179, 183, 186, 190, 192, 218, 220, 228, 267, 268, 310, 405, 406, 408, 427
interchannel, 234
oxolation, 143, 145, 148
polyetherification, 183
polymer-analogous, 69, 176
sonochemical, 122–127, 287, 360
thermomechanical activation, 3–4
topochemical, 4, 5, 147, 228, 274–276, 371
Reaction under Autogenic Pressure at Elevated Temperature (RAPET), 301, 337
Resins
epoxy, 6, 62, 266, 267
macroporous, 57, 58
microporous, 57, 58
novolak, 169–171, 266
Rhodamine B, 182

S
SAXS. *See* Small angle X-ray scattering (SAXS)
Self-organization, 47, 48, 160, 310, 359
Semenov, N.N., 294
Silicates
aluminosilicate, 207
basal distance, 207
bentonite, 207, 212, 215, 254, 257, 419
gallery, 208, 222, 225, 246, 248, 252, 256
hectorite, 206–208, 211, 215, 218, 230, 243, 258, 419, 427, 429
interlayer distance, 209, 212, 213, 218, 226, 249, 252, 257, 258
interplanar distance, 211–215, 249, 270
montmorillonite (MMT), 5, 10, 11, 127, 205–207, 209–212, 214–216, 218, 221, 223, 226, 227, 232–234, 237, 242, 248–252, 256, 258–261, 263, 266, 268, 273, 390, 418, 419, 421, 423, 425–429
phyllosilicate, 207, 208, 215, 390
saponite, 207, 208, 211, 218, 418
smectite, 6, 206–209, 211, 418, 423, 427
Sinerem®, 432
Small angle X-ray scattering (SAXS), 106, 135, 193
Smectite, 6, 206–209, 211, 418, 423, 427
Sonochemistry, 122
Specific conductivity, 236
Specific surface, 36–38, 118, 147, 157, 164, 305, 318, 327, 348, 423, 427
Spin coating, 166, 355
Spinel, 180
SPIOs. *See* Superparamagnetic iron oxides (SPIOs)
Spiropyran, 182
SPR. *See* Surface plasmon resonance (SPR)
Spray-pyrolysis, 3, 180, 305–308, 310
Squalane, 146
Stabilization
electrostatic, 20, 21, 102, 146
self-regulated, 7, 324
steric, 4, 20, 41, 46, 155
Stefan problem, 133
Stress at break, 252
Structure
anisotropic, 4, 33, 40, 96, 216, 238, 371, 394
belt-like, 311
bicontinuous, 192
brushes, 58–60, 74, 100
coil, 274
core-shell, 7, 19, 47, 51, 72, 95, 96, 102, 117, 121, 124, 130, 131, 180, 321, 324, 337, 340, 392, 412, 433
crew-cut, 48
Dawson, 33

dendrite-like, 113
diamond-like, 287
double-cylinder-type, 53
exfoliated, 223, 225, 233, 262
inclusion, 271–276
intercalation, 247
isotropic, 371
Keggin, 33
lamellar, 219, 220, 244, 412
microsphere, 406
monolithic, 343
nanodomain, 54
one-dimensional channel, 216
periodic mesoporous, 147, 206
perovskite, 183
petal-like, 52
"sandwich," 216
spinel, 180
three-dimensional, 21, 62, 154, 164, 177, 218
turbostratic, 348
Sulfite reductase, 402–403, 405
Superconduction transition temperature (T_c), 271
Superlattice, 48, 271, 400
palygorskite, 2
Superparamagnetic iron oxides (SPIOs), 321, 322, 432
Supersaturation, 14, 15, 19, 20, 72, 91, 305
Surface plasmon resonance (SPR), 39, 40, 73, 116
Surfactants
AOT, 35, 177, 186, 393
non-ionogenic, 48, 49
Plurafac LF, 44
Pluronic, 44
Synthesis
fermentative, 402–403
flame spray-pyrolysis, 180, 305, 306
hydrothermal, 94, 95, 127, 133, 207, 215, 216, 218
microbiological, 395–402
microwave–polyolic, 95
plant-mediated, 391
plasmochemical, 8, 128–129
polymer-mediated, 7, 289, 358–364, 367, 374
polymer sol-gel, 10, 142, 161–177, 287, 414
solvothermal, 13, 127, 216, 307
sonochemical, 124–126, 287
transcription, 416

T

Teflon, 130, 131, 267, 352
Temperature
Curie, 358
glass transition (Tg), 109, 154, 241, 242, 255, 257, 260, 266, 267, 348, 363, 410
lower critical solution (LCST), 55, 56
superconduction transition (T_c), 271
Templates
micellar, 3
nanoreactors, 5, 47–54, 60, 64, 71, 164, 225, 366, 407
two-layered cylinders, 53
Tensile modulus, 252
Terpenoids, 393
Tetramethoxysilane (TMOS), 144, 157, 167, 173, 218, 409, 415
TGDMA. *See* Triethylene glycol dimethacrylate (TGDMA)
Thermolysis
isothermal, 290, 294, 300, 301, 314, 338, 368
kinetics, 295, 301
non-isothermal, 291, 294, 296, 297, 301, 314, 324, 331, 338
products, 327, 331, 335, 339, 342, 346, 352
stage, 292, 358, 363
Triethoxysilane (TEOS), 144, 154, 158, 162, 165, 166, 169, 172–174, 185, 186, 192, 193, 195, 259, 408–410
Triethylene glycol dimethacrylate (TGDMA), 155, 156, 174
Turbidity, 189
Turkevich, J., 4, 13, 23

U

Ultrasonic frequency, 122, 123
Ultrasound power, 125

V

Vesicles, 10, 48, 397, 407, 412, 415, 416, 430

W
Wavelength, 39, 97, 99, 113, 119, 132, 193, 194, 209, 273, 450
Wide-angle X-ray scattering (WAXS), 209, 210

X
Xerogel, 142, 147, 163, 181, 183, 188, 217, 238, 243, 405
X-ray analysis, 330

Y
Yield
  quantum, 103, 109
  radiation-chemical, 8, 97, 113–121
Young modulus, 427, 428

Z
Zeldovich, Ya.B., 294
Zeolite, 147, 207, 217, 293, 409

MIX
Papier aus verantwortungsvollen Quellen
Paper from responsible sources
FSC® C105338

If you have any concerns about our products,
you can contact us on
ProductSafety@springernature.com

In case Publisher is established outside the EU,
the EU authorized representative is:
Springer Nature Customer Service Center GmbH
Europaplatz 3, 69115 Heidelberg, Germany

Printed by Libri Plureos GmbH
in Hamburg, Germany